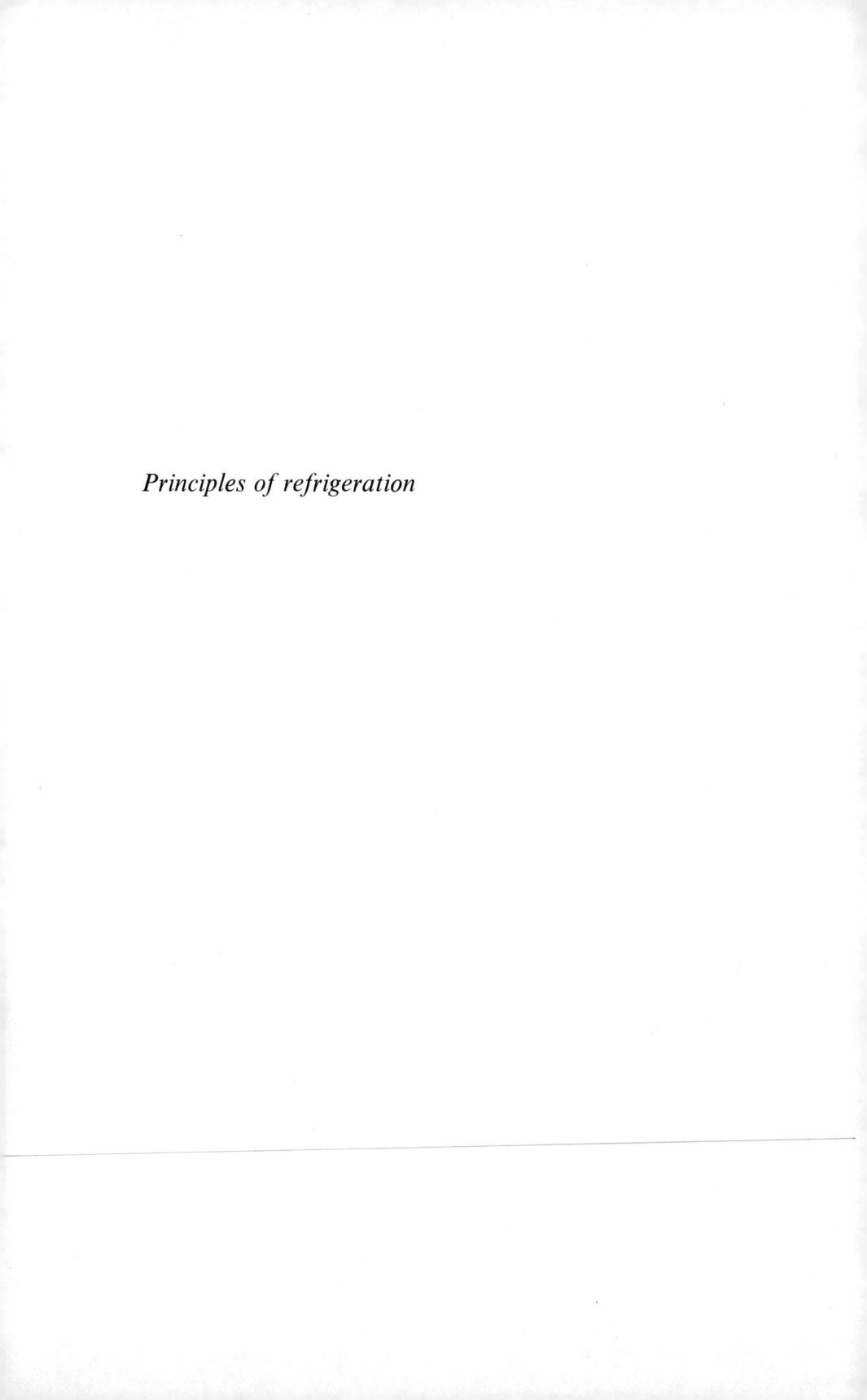

Principles of refrigeration

Principles of refrigeration

W. B. GOSNEY
*Professor of Refrigeration Engineering in the University of London,
King's College London*

CAMBRIDGE UNIVERSITY PRESS
Cambridge
London   New York   New Rochelle
Melbourne   Sydney

Published by the Press Syndicate of the University of Cambridge
The Pitt Building, Trumpington Street, Cambridge CB2 1RP
32 East 57th Street, New York, NY 10022, USA
296 Beaconsfield Parade, Middle Park, Melbourne 3206, Australia

© Cambridge University Press 1982

First published 1982

Printed in the United States of America

British Library cataloguing in publication data

Gosney, W. B.
Principles of refrigeration.
1. Refrigeration and refrigerating machinery
I. Title
621.5'6   TP492   80-42210
ISBN 0 521 23671 1

CONTENTS

Preface		ix
Notes on Units and Symbols		xi
1	**Methods of producing cold**	**1**
	1.1 Introduction	1
	1.2 Methods of refrigeration	2
	1.3 Refrigeration by evaporation	3
	1.4 The vapour compression principle	6
	1.5 The vapour absorption system	13
	1.6 Refrigeration by expansion of air	17
	1.7 The vapour jet system	19
	1.8 Thermo-electric refrigeration	20
	1.9 Secondary refrigerants	22
	1.10 Refrigerating capacity	28
	1.11 Coefficient of performance and heat ratio	29
	1.12 Refrigeration and thermodynamics	31
	1.13 Internal and external régimes	39
	1.14 The heat pump	44
	1.15 The cold multiplier	47
2	**Machinery and plant for vapour compression systems**	**50**
	2.1 Introduction	50
	2.2 Compressors	51
	2.3 Reciprocating compressors	53
	2.4 Rotary positive displacement compressors	70
	2.5 Turbo-compressors	81
	2.6 Condensing equipment	86
	2.7 Evaporators	98
	2.8 Expansion valves: evaporator feed regulation	106
	2.9 Refrigerants	121

	2.10 Refrigerant piping	128
	2.11 Auxiliary equipment	135
	2.12 Line controls	144
	2.13 Electric motors	151
	2.14 Automatic control of refrigerating capacity	157
	2.15 Safety	166
	2.16 Methods of defrosting	168
3	**The vapour compression system**	**172**
	3.1 The cycle of operations	172
	3.2 Work and heat transfers	174
	3.3 Adiabatic and reversible compression	178
	3.4 The pressure–enthalpy diagram	181
	3.5 Thermodynamic data: tables and charts	183
	3.6 The use of minimum data	192
	3.7 Comparison with the Carnot cycle	194
	3.8 The effect of evaporating and condensing temperatures on the cycle quantities	202
	3.9 The effect of refrigerant properties on the cycle quantities	206
	3.10 Subcooled liquid	213
	3.11 Superheated vapour	219
	3.12 Dependence of the volumic refrigerating effect on the suction state	228
	3.13 Dependence of the isentropic coefficient of performance on the suction state	231
	3.14 Operation near the critical point	235
	3.15 The effects of heat transfers and pressure drops between the components	240
	3.16 The effect of foreign substances in the circuit	247
	3.17 By-passing of hot gas	250
	3.18 The computation of thermodynamic properties	254
4	**Compression processes**	**267**
	4.1 Introduction	267
	4.2 The ideal reciprocating compressor	267
	4.3 The ideal compressor with clearance	276
	4.4 The real compressor: volumetric efficiency	286
	4.5 Shaft power and mean effective pressure	300
	4.6 Scale effects in reciprocating compressors	305
	4.7 Screw compressors	307
	4.8 Steady-flow compression	313
	4.9 Stagnation enthalpy, pressure and temperature	322
	4.10 Centrifugal compressors	324
	4.11 Generalised characteristics	333

Contents vii

	4.12 Effects of Reynolds and Mach numbers	342
	4.13 Peak performance characteristics	345
	4.14 Capacity reduction	348
5	**Multi-stage vapour compression systems**	**351**
	5.1 Introduction	351
	5.2 Compression in stages	352
	5.3 Desuperheating between stages	355
	5.4 Multi-stage expansion	361
	5.5 Combined liquid cooling and desuperheating	366
	5.6 The choice of intermediate pressure	369
	5.7 The balance between the stages	370
	5.8 Refrigeration at different temperatures	374
	5.9 Dual-effect compression	375
	5.10 The manufacture of solid carbon dioxide	385
	5.11 Cascade systems	391
6	**The vapour absorption system**	**401**
	6.1 Introduction	401
	6.2 Composition of mixtures	401
	6.3 Vapour pressure of solutions	403
	6.4 The cycle of operations	406
	6.5 Enthalpy of liquid and vapour	411
	6.6 Steady-flow analysis	414
	6.7 The effect of inefficiencies in the cycle	419
	6.8 Solutions of ammonia in water	420
	6.9 Composition of the vapour	423
	6.10 Bubble and dew points: distillation	424
	6.11 Pressures and temperatures in the aqua-ammonia cycle	430
	6.12 Enthalpy of aqua-ammonia	432
	6.13 The complete enthalpy–composition diagram	440
	6.14 Steady-flow analysis of the aqua-ammonia cycle	444
	6.15 The rectifier	448
	6.16 The effects of incomplete rectification	453
	6.17 Further study of distillation: Keesom's method	455
	6.18 Vapour to liquid heat exchange	462
	6.19 The Platen–Munters system	462
	6.20 Heating and cooling media for absorption systems	465
	6.21 Performance of absorption systems	467
	6.22 Other refrigerant–absorbent combinations	469
	6.23 Resorption systems	471

7	**Ideal gas cycles**	**474**
	7.1 The ideal gas state	474
	7.2 The constant pressure cycle	477
	7.3 The effect of compressor and expander inefficiency	483
	7.4 The constant pressure cycle with internal heat exchange	486
	7.5 The constant pressure cycle with sub-atmospheric pressures	494
	7.6 Aircraft cooling systems	495
	7.7 The Stirling cycle	497
	7.8 The vortex tube	502
8	**Cryogenic engineering**	**508**
	8.1 The field of low temperatures	508
	8.2 The minimum work required to make cryo-products	525
	8.3 The Joule–Thomson effect: real gases	535
	8.4 Regenerative cooling: the Linde–Hampson process	540
	8.5 The use of expansion machines: the Claude process	558
	8.6 The separation of gases	572
	8.7 Heat transfer equipment	596
	8.8 Liquefied natural gas	616
	Appendix	627
	References	635
	Index	645

PREFACE

This book is based on lectures given over a number of years to undergraduate students of mechanical engineering, together with additional material from a postgraduate course. The standard of treatment is therefore appropriate to these levels. In order to make the book useful to a wider circle of readers, however, the necessary principles are briefly resumed at each stage and enough information is given to make the worked examples intelligible. I hope that this will make the book of use not only to students but also to others entering or working in the refrigeration industry; and that substantial parts of it will be found readable by those who have only a general interest in the subject.

The choice of topics and the space allocated to them has been decided on the grounds of their importance in practice. Consequently the treatment of the vapour compression system takes up a large part of the book, whilst minor methods such as vapour-jet and thermo-electric refrigeration receive only brief mentions, despite their considerable theoretical interest. The current interest in absorption systems and the importance of cryogenic processes demand the fairly detailed treatment of these which is provided, much of it not readily available in other texts.

Nevertheless it has been impossible to treat every topic of importance in one book. As originally planned the book was to have covered the application of refrigeration methods in practice, including load estimation, insulation, heat transfer calculations and system balance. At a late stage this scheme was found to be impracticable: the book would have been much too long and some of the topics would have been of limited interest to many readers. Consequently they will form the subject matter of another book now in course of preparation.

The calculation of refrigeration problems requires a large amount of

Preface

numerical data. Since full thermodynamic tables would have occupied much space, better devoted to explanation, and since this is not intended to be a data book, I have presented only the data required for the worked examples as needed. For the treatment of absorption systems and air separation processes I have prepared diagrams in SI units, which I hope will serve to make the examples intelligible.

This is intended to be a book for learning from, not a work of reference: it is for students, not for experts. I have tried to write the book which I should myself have wished to read when I first entered the industry. Consequently some readers may find that certain points appear to be unduly laboured, and that the obvious is stated again and again. I can only reply that many years of experience in teaching the subject have taught me that it is not usually the difficult points that students fail to grasp but the very simple fundamental ones, and these I have tried to explain. The reader will find that most chapters begin at a leisurely pace, which quickens after a few pages.

The lists of references at the end of the book are not exhaustive. Apart from those which cite the first main treatment of a topic, they are supposed to suggest further sources of reading material – journals and books – and references which themselves contain a good bibliography have been favoured.

I should like to thank the companies and individuals who have helped me. Acknowledgment is made to named companies in the appropriate places. Friends in the industry who have provided information and advice are too numerous to list, and a small selection of names would be unfair since it might saddle them with the blame for errors which are entirely my own. Nevertheless I thank them all.

Lastly I should like to thank my wife for putting up with me during the years that the book has been in preparation.

December 1980

W. B. Gosney
King's College London

NOTES ON UNITS AND SYMBOLS

1. The data and worked examples presented in this book are in SI units. I have however, somewhat reluctantly, used the bar ($= 10^5 \text{N/m}^2$) as the unit of pressure instead of the basic SI unit N/m^2. The bar seems to be too strongly entrenched in the published refrigerant tables and in the markings of pressure gauges to disregard it.

2. The equations in the book are nevertheless independent of any particular choice of units, because every physical quantity is regarded as the product of a pure number, the numeric, and a unit denoted by a unit symbol. The unit symbol is treated as an algebraic entity.
   Equations express relations between the physical quantities, not between the numerics.

3. The numbers which appear on the scales of graphs and in the columns of tables are the numerics of the quantities specified in the legends. For example the volume of a fluid which is compressed according to the law $pV = \text{constant}$, when $p_1 = 1$ bar, $V_1 = 6 \text{ m}^3$, would be tabulated:

   | p/bar | 1 | 2 | 3 | 4... |
   | V/m^3 | 6 | 3 | 2 | 1.5... |

   and if this information were presented graphically the legends on the axes would be so marked. The practice follows that given in *Quantities, Units, and Symbols*, 2nd edn, The Royal Society, London, 1975, where it is explained further.

4. Symbols for thermodynamic quantities follow the usage current in British and US texts. In particular a lower case letter stands for a specific quantity, i.e. a quantity divided by mass, whilst a capital letter stands for the quantity itself. Thus:

   V volume v specific volume
   U internal energy u specific internal energy

Notes on units and symbols

H	enthalpy	h	specific enthalpy
S	entropy	s	specific entropy
C	heat capacity	c	specific heat capacity
Q	heat	q	specific heat (heat/mass)
W	work	w	specific work

The exception to this practice is the specific gas constant, denoted by R following common usage.

Molar quantities are denoted by lower case letters with a bar, except for the molar gas constant $\bar{R}$.

Time rates of flow or transfer are denoted by a dot over the letter symbol. Thus:

Q	heat (transfer)	$\dot{Q}$	heat (transfer) rate
W	work	$\dot{W}$	power
V	volume	$\dot{V}$	volume flow rate
m	mass	$\dot{m}$	mass flow rate
n	amount of substance	$\dot{n}$	flow rate of amount of substance

Subscripts to denote phases are:

f saturated liquid   g saturated vapour   i saturated solid

and:

h_{fg} denotes the specific enthalpy of evaporation, etc.

Other frequently used symbols are:
- p absolute pressure in general, or partial pressure of a gas in a mixture
- P total absolute pressure of a gas mixture
- T absolute thermodynamic temperature
- t empirical temperature on a conventional scale
- $\mathscr{V}$ velocity
- ε coefficient of performance, defined by eq. (1.11.1)
- η efficiency, defined in the text as required, with subscript
- ζ heat ratio, defined by eq. (1.11.5)

Other symbols and subscripts are defined in the text as needed.

5 Molar mass, denoted by M, has been used in preference to 'molecular weight' or 'relative molecular mass'. The distinction is that M has units (e.g. kg/kmol) and that it is applicable to mixtures as well as pure substances.

6 The data for the examples on ammonia are based on the equations given in Circular 142 of the National Bureau of Standards, 'Tables of Thermodynamic Properties of Ammonia', 1923, converted to SI units using the definition of the Btu given therein:

1 Btu(NBS 1923)/lb = (4.183/1.8) kJ/kg

A table of saturation properties of ammonia, and a pressure – enthalpy chart for the superheated vapour, are given in the Appendix. Detailed tables in SI units may be obtained from the author.

The data for examples on R12, R22 and R502 are reproduced with permission of Du Pont from *Thermodynamic Properties of...'Freon' 12 Refrigerant,....'Freon' 22 Refrigerant and ...'Freon' 502 Refrigerant*, available from Du Pont or from 'Freon' Refrigerant distributors. Pressure–enthalpy charts for these refrigerants are given in the Appendix.

1

Methods of producing cold

1.1 Introduction

The science and art of refrigeration is concerned with the cooling of bodies or fluids to temperatures lower than those available in the surroundings at a particular time and place. One method of doing this is to use ice transported from colder regions. Another method is to store ice from the winter for use in the summer, or to make ice during cold nights for use during the day. Such methods have been used for centuries in some parts of the world, particularly in the Middle East and in India. Ice houses for storing the ice were constructed with whatever insulating materials were available. The most spectacular of these older ice houses are those of Iran, recently described by Beazley (1977). Smaller ones were built in Europe and America during the eighteenth century. Those in Britain were mostly on the estates of large country houses for storing ice from the winter.

Whether organised as a trade or not, it seems probable from literature references that ice has been available, at least to those who could afford it, for several centuries. When Saladin heard of the illness of Richard I of England during the Third Crusade, he is said to have sent a caravan to mountains far away to bring ice which he then took to Richard. According to Mitford (1966), when Louis XIV was besieging Lille in 1667, the commander defending the city heard that the French had no ice and he sent some over the lines with the message that the siege was going to last many months and he would not like to think of the King going without ice. At this time the courtiers at Versailles were said to require five pounds of ice per day each in the summer, though there is no indication of its origin. Joseph Addison refers to iced desserts in the *Tatler* in March 1719; and Boileau in his *Repas Ridicule* satyrised a banquet at which there was no ice at the height of the summer.

Possibly these are isolated examples of how ice could be obtained by those with power and money. The credit for making ice available to a much wider circle of buyers is usually given to Frederic Tudor, who, about 1806, cut ice from the Hudson River and the ponds of Massachusetts and sold it. He exported it eventually to the West Indies, India, Australia and Europe. In Calcutta, Tudor's ice could undercut the price of native ice manufactured during the night by exposing vessels of water laid on an insulating layer of straw to the night sky. Considerable quantities were shipped to the southern states of the Union until this trade was interrupted by the Civil War. Britain received ice from the USA, sold by the Wenham Lake Ice Co. from their shop in the Strand, London, and despatched to various parts of the country by rail. At a later date ice was imported from Norway, some of it directly to fishing ports, for use by trawlers at sea and to send the fish to the towns.

The scale of the ice trade in North America was evidently considerable. In 1854, 156 000 tons were exported from Boston alone. The ice houses were described as vast buildings, of which there were 50 in the neighbourhood of Forest Pond in Massachusetts. Sawdust and shavings were the principal insulating materials, but because of their relatively poor insulating properties the walls had to be about 1 metre thick.

Fuller descriptions of the American ice trade are given by Cummings (1949) and Anderson (1953).

The trade in natural ice continued long after the invention of mechanical refrigeration, with the usual claims and counterclaims about the relative merits of natural and manufactured ice. A small amount of natural ice was still being imported into Britain from Norway after World War I, but the trade finally ceased about 1930.

Refrigeration as it is known today is almost entirely produced by artificial means. In the most common process a substance, called the refrigerant, is made to undergo a sequence of operations by which its temperature is lowered sufficiently for it to be able to accept heat from some bodies or fluids which are to be refrigerated. In this chapter a review of the several methods of bringing the refrigerant to the desired low temperature is given. The more important of these methods are then studied in later chapters.

1.2 Methods of refrigeration

The principal methods of refrigeration available are: (a) vapour compression; (b) vapour absorption; (c) air cycle; (d) vapour jet; and (e) thermo-electric.

The vast majority of plants of all sizes from domestic refrigerators to large industrial systems use the vapour compression principle. The other systems are used in special circumstances. For example, the absorption system is used in large chemical plants, in air-conditioning and in some domestic refrigerators. Because it needs an input of heat at a moderately high temperature to drive it, it finds applications where such heat is readily available or where mechanical power is not available.

The air cycle, in which the temperature of air is reduced by an expansion process in which work is done by the air, was used for many years as the principal method of refrigeration at sea, chiefly on account of the inherent safety of the method. It is now used for cabin cooling in aircraft.

The vapour jet system had a certain vogue at one time as a method which could be operated directly by a supply of steam at a moderate pressure, and which required no toxic gases, but it is now little used.

Thermo-electric refrigeration works on the principle of the Peltier effect; i.e. the cooling effect produced when an electric current is passed through a junction of two dissimilar metals. With the materials so far available its efficiency is rather low, but it has many uses in circumstances where efficiency does not matter much, as in very small refrigerators for specimens on microscope stages, instruments for measuring dew point, and a few others.

In both vapour compression and vapour absorption systems the refrigerating effect is produced by making a liquid boil at a suitably low temperature. The difference between the two systems lies in the means used to cause the boiling, and to restore the vapour so formed to the liquid state again. The principles common to both of these systems are discussed in the next section.

1.3 Refrigeration by evaporation

The fact that a volatile fluid such as ether chills the skin of the hand when it evaporates has long been known. In 1755 William Cullen, Professor of Chemistry in the University of Edinburgh, used this effect to make ice. He placed some water in thermal contact with ether under the receiver of a vacuum pump. The operation of the pump caused accelerated evaporation of the ether and its temperature was lowered sufficiently to freeze the water.

The principles behind Cullen's method are twofold. First: every liquid exerts a vapour pressure, i.e. has a tendency to turn into vapour. A liquid in a container empty of all gases but its own vapour comes to equilibrium with the vapour at a value of the pressure known as its saturated vapour pressure, or simply *vapour pressure* for short. The vapour pressure increases with temperature, as illustrated in Fig. 1.3.1 for some substances used as refrigerants. The vapour in the space is known as *saturated vapour*. If vapour is removed from the space, by pumping it away as in Cullen's experiment, some of the liquid must evaporate to replace it. At this point the second principle comes into operation. A liquid when it turns into vapour needs heat, called 'latent heat' by Joseph Black who was the first to measure it shortly after Cullen's experiment. The name latent was supposed to indicate that the heat taken in did not cause any rise in temperature if the pressure remained constant. In modern terms latent heat is called *enthalpy of evaporation*. When, therefore, a liquid is forced to evaporate by pumping away its vapour, the enthalpy of evaporation must first be conducted to the

Fig. 1.3.1. The saturated vapour pressure of some substances used as refrigerants. At a pressure of 1 atmosphere (= 1.013 25 bar) the temperature of liquid and vapour in equilibrium is the temperature of the normal boiling point. Ammonia with a boiling point of $-33\,°C$ is said to be more volatile than ether with a boiling point of $34\,°C$.

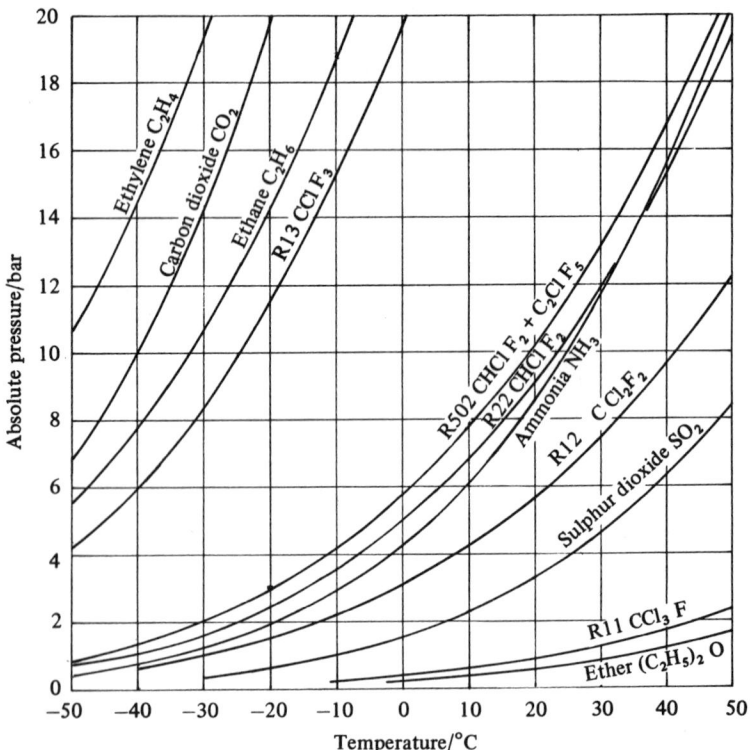

1.3 Refrigeration by evaporation

liquid. If enough heat is forthcoming from external sources, the temperature and pressure of the liquid need not change. But if the heat is not forthcoming from external sources, it must be obtained from the part of the liquid which does not evaporate, and the temperature of this part falls. A temperature difference is thus established between the liquid remaining and the surroundings, and this can be used to reduce the temperature of other bodies or fluids.

The event described above, in which a temperature difference is established between two bodies initially at the same temperature, is not a natural spontaneous process. It is against all experience for, say, two pieces of iron initially at the same temperature and isolated from their surroundings to come spontaneously to different temperatures. The impossibility of this happening spontaneously is the essence of the Second Law of Thermodynamics. In one of its forms due to Clausius the Law states that it is impossible for a continuously acting machine to take heat from a body at one temperature and give out heat to a body at a higher temperature without leaving some effect in the surroundings. In Cullen's experiment, the 'effect' required in the surroundings takes the form of a continuous supply of work to the vacuum pump and a continuous supply of ether. The supply of work can be arranged, but the supply of ether all the time would certainly be inconvenient. To make the machine work continuously some means must be found for using the ether over again by condensing it back to liquid.

The ether vapour discharged by the vacuum pump cannot immediately be condensed to the liquid form because that would require a temperature as low as that of the evaporating liquid. But by increasing the pressure of the vapour, i.e. by compressing it, the vapour can be made to condense at a higher temperature, and in fact at a temperature high enough for it to give out heat to the ambient air, or to water at about the ambient temperature. The knowledge of how to turn vapours or gases into liquids by compression followed by condensation was being gathered during the second half of the eighteenth century. J. F. Clouet and G. Monge liquefied sulphur dioxide in 1780, and ammonia was liquefied by van Marum and van Troostwijk in 1787. The idea of putting together the principles of refrigeration by evaporation and liquefaction by compression seems to have been first suggested by Oliver Evans of Philadelphia, but it is not clear whether he tried it or not. The first complete description of a refrigerating machine working in a cycle and which was actually made is given in a patent specification of 1834 by Jacob Perkins, an American inventor working in London. This machine is the prototype of all subsequent vapour compression systems.

1.4 The vapour compression principle

The claim which Perkins made in his specification of 1834, on which he was granted British Patent 6662 was 'an arrangement of apparatus or means... whereby I am enabled to use volatile fluids for the purpose of producing the cooling or freezing of fluids, and yet at the same time constantly condensing such volatile fluids, and bringing them again and again into operation without waste'.

The arrangement which Perkins proposed is shown in Fig. 1.4.1 taken from his specification, and in diagrammatic form in Fig. 1.4.2. There are four main components. The volatile fluid in the *evaporator* is made to boil by pumping away vapour with the pump or *compressor*. The fluid, or refrigerant, in the evaporator is in thermal contact with the substance which it is desired to refrigerate, e.g. water, brine or air, and takes heat from it. The vapour drawn from the evaporator by the compressor is raised to a higher

Fig. 1.4.1. The apparatus described by Jacob Perkins in his patent specification of 1834. The refrigerant (ether or other volatile fluid) boils in the evaporator B taking heat from the surrounding water in the container A. The pump C draws vapour away and compresses it to a higher pressure at which it can condense to liquid in tubes D, giving out heat to the water in the vessel E. The condensed liquid flows through the weight-loaded valve H, which maintains the difference of pressure between the condenser and the evaporator. The small pump above the valve H is used for charging the apparatus with refrigerant.

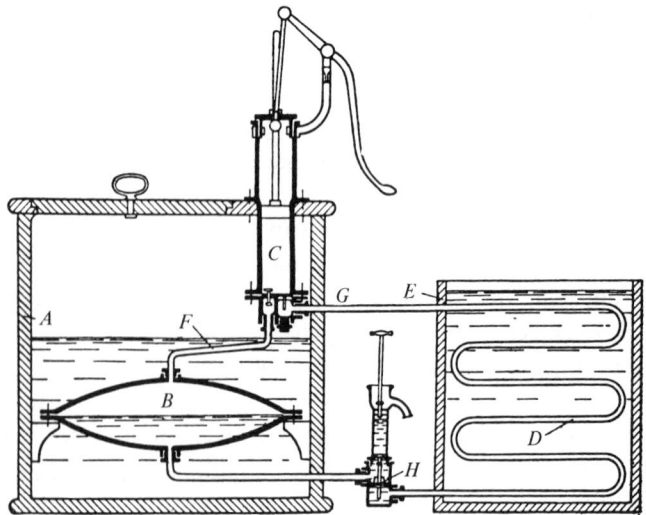

1.4 The vapour compression principle

pressure and discharged to the *condenser*, in which it gives out heat to cooling water and liquefies. The liquid is then fed back to the evaporator for re-use. To maintain the difference of pressure between the condenser and the evaporator a restriction known as the *expansion valve* is interposed between them.

According to a much later account by Sir Frederick Bramwell (1882), a machine was made to Perkins' design using as refrigerant an evil smelling liquid obtained from the distillation of India rubber, and 'in the height of the summer' it made a small quantity of ice. Bramwell, who was an apprentice at the time, has described how they wrapped the ice in a blanket and carried it to Perkins' lodgings in a cab one warm summer's evening. The machine was in all probability driven by hand, and no copies appear to have been made, nor was there any commercial exploitation whatever.

Perkins' machine seems in fact to have aroused absolutely no interest. It is not mentioned in the literature of the time and seems to have remained unknown throughout nearly fifty years until Bramwell described the work, almost casually, in a letter to the *Journal of the Royal Society of Arts*. An appendix to a paper by A. C. Kirk to the Institution of Civil Engineers in 1874, compiled by J. Bourne, contains a list of 170 references to refrigeration, but there is no mention of Perkins. Few basic inventions of modern times are so badly documented.

Fig. 1.4.2. The essential components of the vapour compression system.

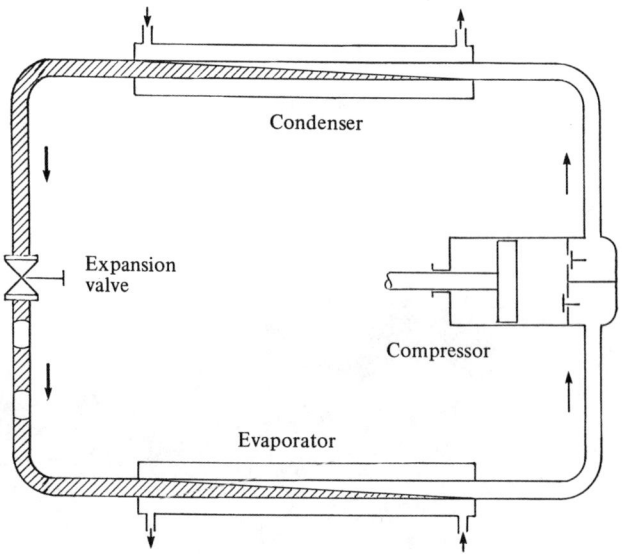

The man principally responsible for making the vapour compression principle into a workable machine was James Harrison, born at Renton, Scotland, in 1815 or 1816. He had little formal technical training except what he gained by attending chemistry classes at Anderson's College, forerunner of the present University of Strathclyde, whilst serving an apprenticeship as a printer. He then worked in London for a while before emigrating to Australia in 1837, settling first at Sydney, then at Melbourne and Geelong, where he became a journalist and publisher. It is not clear whether Harrison knew of Perkins' work or not. According to one account he noticed the chilling effect of ether while using it to wash type. Some time about 1850 he had invented a hand-operated machine which made ice. In 1856 Harrison came to London to take out British Patents 747 of 1856 and 2362 of 1857, and to have better machines made. One of these was shipped

Fig. 1.4.3. Harrison's ether machine as made by Daniel Siebe. The engine for driving the table-type of compressor is not shown. The condenser is in the vat on the right. The shell-and-tube evaporator is in the box immediately to the left of the compressor, with the brine pump driven from a pulley on the compressor shaft. The brine circulates through a trough, on the extreme left, holding the ice-moulds. (*The Engineer*, 1861, Vol. 11, p. 231.)

1.4 The vapour compression principle

to Australia and was used by Harrison to set up his ice works first at Geelong and then at Melbourne. A copy of this one was made in Sydney where the manufacture was taken up by P. N. Russell. In England the machines were made by Daniel Siebe, who installed one in the paraffin works of Young, Meldrum and Binny, of Bathgate, Scotland, to crystallise paraffin wax from shale oil. Daniel Siebe also exhibited a machine at the International Exhibition in London in 1862, similar to that shown in Fig. 1.4.3, and from this time the manufacture was on a more or less regular basis, by Siebe and others. Until the advent of ammonia and carbon dioxide, ether machines were widely used in breweries and for ice-making. They continued to be favoured in India for many years owing to their lower condensing pressure compared with other refrigerants at the high cooling water temperatures encountered, and were manufactured for this market until the present century (Oldham, 1946–7).

Another pioneer of the vapour compression system using ether was Alexander Catlin Twining (1801–4), who had a machine working in Cleveland, Ohio, in 1856, capable of making 2000 lb of ice in 20 hours. The ice was offered for sale, but seems to have been unable to compete with the natural product, and no development of Twining's method on any scale is recorded.

The use of ether, with a normal boiling point of 34.5 °C, implies a pressure in the evaporator less than atmospheric pressure, with consequent danger of air leaking inwards and forming a potentially explosive mixture. On the other hand, the pressure in the condenser was not high and strong construction was not needed. Charles Tellier introduced dimethyl ether, with a much higher vapour pressure, boiling point − 23.6 °C, in Paris in 1864, and Raoul Pictet in Geneva used sulphur dioxide, boiling point − 10 °C, in 1874. Dimethyl ether did not come into general use, but sulphur dioxide was an important refrigerant for some sixty years.

A major advance was made by Carl von Linde in Munich in the 1870s by the introduction of ammonia. This has a much higher vapour pressure than the substances mentioned above, as indicated by its lower boiling point of − 33.3 °C, and pressures of ten atmospheres and more were needed in the condenser. But Linde was able to overcome the mechanical problems, and since then ammonia has been the most important refrigerant for large plants.

Carbon dioxide, of still higher vapour pressure, was tried by Linde, and also by T. S. C. Lowe in the USA, but its use is mainly associated with the name of Franz Windhausen of Berlin, who used it in 1886. The carbon dioxide system requires very high pressures, of the order 80 atmospheres, in the condenser, so very heavy construction is needed. Windhausen's machine was taken up by J. and E. Hall Ltd. in England, and was further developed by Alexander Marcet and Everard Hesketh. Because of the safe nature of carbon dioxide, it became the principal refrigerant for use in ships for fifty years, and has only been displaced by new refrigerants since about 1955.

These newer refrigerants are the fluorinated derivatives of the hydrocarbons methane and ethane. They had been known as chemical compounds since the late nineteenth century, but their properties as refrigerants were first investigated by Thomas Midgley in the laboratories of General Motors, Detroit, USA, 1929–30. The one which first appeared promising was dichlorodifluoromethane, CCl_2F_2 with a boiling point of $-29.8\,°C$. This compound was given the trade name 'Freon 12'. Since then many other derivatives have been marketed and numbered according to a standard nomenclature which is given below.

Compounds derived from saturated hydrocarbons such as methane and ethane, having the general formula:

$$C_a H_b Cl_c F_d$$

where $b + c + d = 2a + 2$, are allocated the number

$$(a-1)(b+1)d$$

When $(a-1)$ is zero, as for derivatives of methane, it is omitted.

Example 1.4.1

$CHClF_2$ is given the number 22.

C_2ClF_5 is given the number 115.

Isomers, i.e. compounds containing the same atoms and having the same molar masses, occur within the derivatives of ethane and higher hydrocarbons. They are distinguished by the order of symmetry, which is determined by summing the relative atomic masses attached to each carbon atom. The most symmetrical is given simply the number $(a-1)(b+1)d$, the next in order of symmetry has a suffix lower case 'a', then 'b' and so on.

Example 1.4.2

$CClF_2-CClF_2$ is given the designation 114.

CCl_2F-CF_3 is given the designation 114a.

Unsaturated compounds are indicated by placing the number 1 before the number derived as above.

1.4 The vapour compression principle

Cyclic compounds are denoted by an upper case 'C' placed before the number. For example the cyclic compound octafluorcyclobutane is denoted C318.

Brominated compounds are indicated by a suffix upper case 'B' after the number, followed by the number of bromine atoms. For example, $CBrF_3$ is designated 13B1, i.e. the same as 13 with the chlorine atom replaced by a bromine atom.

The families of fluoro-chloro-compounds derived from methane and ethane are set out in Fig. 1.4.4 in terms of their normal boiling points. Compounds in commercial use are marked with their designating numbers according to the above scheme.

The fluorinated refrigerants are now manufactured by several different firms under various trade names, e.g. 'Freon', 'Arcton', and so on. As a general designation it is now agreed to use the word 'Refrigerant', sometimes abbreviated to 'R', followed by the identifying number.

The principal fluorinated refrigerants in use at present are:
Refrigerant 12, CCl_2F_2 (boiling point $-29.8\,°C$), is used in almost all

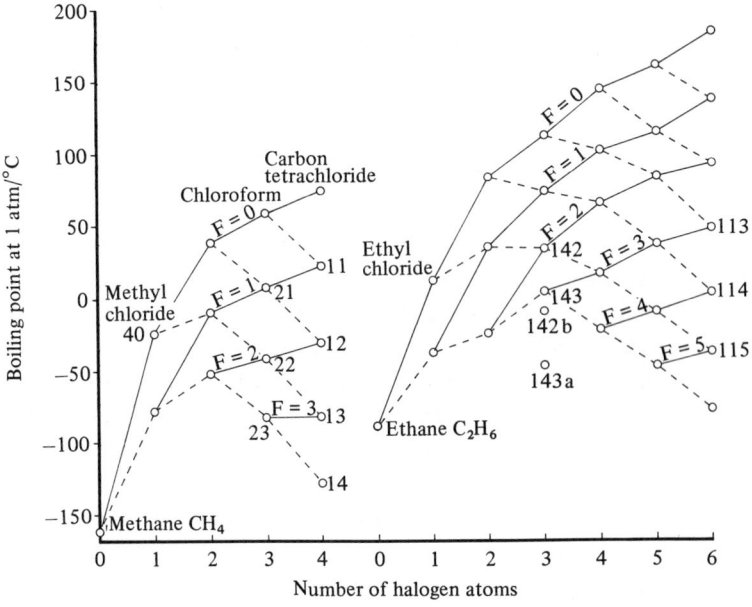

Fig. 1.4.4. Halogenated derivatives of methane and ethane. The full lines join compounds having equal numbers of fluorine atoms ($F = 1, 2$, etc.); the broken lines join compounds having equal numbers of chlorine atoms. R142b and R143a are isomers of R142 and R143.

domestic refrigerators and in many small commercial refrigerators.

Refrigerant 22, $CHClF_2$ has a higher vapour pressure than R12 (boiling point $-40.8\,°C$) and is used in commercial and industrial plant of all sizes, and in marine installations.

Refrigerant 11, CCl_3F, is a low vapour pressure (boiling point $23.8\,°C$) substance, and is used in systems employing turbo-compressors.

Refrigerant 13, $CClF_3$ (boiling point $-81.4\,°C$), and Refrigerant 14, CF_4 (boiling point $-128\,°C$), have very high vapour pressures and are used in systems for very low evaporating temperatures.

Refrigerant 502 (boiling point $-45.4\,°C$) is a mixture of Refrigerant 22 and Refrigerant 115, $CClF_2-CF_3$ (boiling point $-39.1\,°C$) in the proportions 0.488/0.512 by mass. It is a constant boiling mixture or azeotrope, meaning that it distils without change of composition. R502 has properties somewhat similar to those of R22, but it has a lower rise of temperature during compression and it can be used for producing low temperatures for food freezing without resorting to two-stage compression. It is also a very dense gas and is suitable for use in systems with hermetic compressors in which the motor windings are cooled by the refrigerant vapour.

Some of the fluorinated refrigerants such as R12 and R22 have been in extensive use now for nearly 50 years and have been thought to be very safe in all respects, apart from the fact that, like carbon dioxide and other heavy gases, they displace air upwards in confined spaces and can cause suffocation. For this reason they must always be treated with respect, particularly since most of the fluorinated refrigerants have very little smell.

Since 1974, however, it has been claimed that the release of large amounts of R11 and R12 into the atmosphere, most of which has come from aerosol spray cans and not from refrigerators, would eventually deplete the earth's ozone layer by the catalytic action of chlorine in the stratosphere. This would increase the amount of biologically active ultra-violet radiation arriving at the earth's surface and increase the incidence of skin cancer. As a result some states have banned the use of R11 and R12 in spray cans. Others are contemplating a ban or a phased reduction. It is not yet clear whether the amounts used in, and ultimately lost from, refrigerators will constitute a threat to the environment, and what if anything will be done about it. A readable account of the matter is given in '*The Ozone War*' by L. Dotto and H. Schiff, Doubleday, 1978. The scientific aspects are discussed in *Pollution Paper* No. 15, H.M.S.O., London, 1979.

The fluorinated refrigerants came into use in response to a demand for 'safe' refrigerants for use in small automatic refrigerating plants in shops, dairies, hotels and homes. The early refrigerators had been of a kind which required skilled attention more or less continuously. To make them suitable for homes they had to be automatic. Developments in this direction began about 1900 and domestic refrigerators came onto the market in the USA about 1917. These automatic machines incorporated two features: one was a means of regulating the opening of the expansion valve to match the flow rate of refrigerant demanded by the compressor, and the other was a thermostat to switch off the compressor motor when the cold space had been brought down to temperature. The refrigerants used were sulphur dioxide, a very toxic gas, and methyl chloride, a reasonably safe refrigerant but slightly flammable and also somewhat reactive with some metals and lubricating oil. Both sulphur dioxide and methyl chloride were eventually superseded by the fluorinated hydrocarbons.

Automatic operation, though first introduced in small systems, was later extended to large ones, and today it is unusual for a large plant not to have at least some automatic features.

Although automatic in day-to-day operation, compressors and systems still needed some attention from time to time, perhaps to replace refrigerant which had leaked out, or to attend to the stuffing box or gland which sealed the system at the point where the crankshaft emerges from the crankcase. In the early 1930s the so-called hermetic construction was introduced. The motor and compressor were directly coupled on the same shaft and mounted together inside the refrigerant circuit. No shaft seal was needed and leakage was virtually eliminated. This type of construction is now universal on domestic refrigerators and on most small systems: it is also commonly used for larger machines.

1.5 The vapour absorption system

An alternative method of removing vapour from the surface of a boiling liquid is to absorb it by some substance with which the vapour reacts chemically or in which it dissolves readily. For example, water vapour is rapidly absorbed by sulphuric acid, and this was the basis of the method by which Sir John Leslie made ice artificially in 1810. He placed vessels, one containing water and one containing strong sulphuric acid, under a bell jar which was then evacuated by a pump. In time a layer of ice formed on the surface of the water. The explanation is basically the same as that for Cullen's experiment. The water was made to evaporate by reducing the pressure of the vapour over it and in so doing it gained its enthalpy of evaporation from the remaining water. The temperature dropped until eventually the water froze.

The role of the air pump here should be understood because it is not absolutely essential to the process; it merely speeds it up. Even in the atmosphere, water evaporates if its saturation vapour pressure corresponding to its temperature is greater than the partial pressure of water vapour in the air. Evaporation requires enthalpy of evaporation (latent heat) and the temperature of the water remaining is lowered. This is the principle of the wet bulb thermometer, used for measuring humidity. On a dry day with the air temperature above 0 °C it is possible for the water on the wet bulb to freeze because of the evaporation. The passage of water vapour from the water surface is impeded however by the presence of the air, and by removing this the process can be speeded up. In Leslie's experiment the water vapour formed was absorbed by the acid so that the atmosphere inside the bell jar was kept dry. It is clear that if ice is to be made the acid must maintain a partial pressure of water vapour inside the jar which is less than the saturated vapour pressure of water at 0 °C.

Leslie's method became the basis of several commercial machines for making small amounts of ice for cooling wine, some of which were in use until comparatively recently. They require, however, a periodic recharge of sulphuric acid. The operation can be made continuous by withdrawing the acid from the vacuum vessel and concentrating it by boiling. The strong acid is then returned to the vacuum vessel. This type of machine was designed by Windhausen in 1878 and worked with some success (Hopkinson, 1882), but it was never very popular. It was used both for direct manufacture of ice by evaporation and also to produce chilled water. In this system water clearly acts as the refrigerant; the sulphuric acid is called the *absorbent*.

Meanwhile a more important version of the absorption system had appeared in 1859, the continuous aqua-ammonia system invented by Ferdinand Carré (1860), with ammonia as the refrigerant and water as the absorbent. It is well known that water is a strong absorbent for ammonia vapour, and if an evaporator is placed in communication with a vessel through which a stream of water is flowing, ammonia is absorbed and the pressure of the vapour is reduced. The ammonia in the evaporator is made to boil and its enthalpy of evaporation can be taken from external objects. The solution of ammonia formed in the absorber is regenerated by raising it to a higher pressure and boiling it, at which point ammonia vapour is given off and can be condensed. By rectification the ammonia vapour can be made almost pure, but it is not necessary to make pure water, so the liquid which is returned to the absorber vessel is in fact a weak solution of ammonia in water.

An early absorption system of this type is shown in Fig. 1.5.1, and the main components of the system are shown diagrammatically in Fig. 1.5.2.

1.5 The vapour absorption system

Fig. 1.5.1. Carré's continuous absorption machine. *A* is the boiler or generator with safety valve, *B* the condenser, and tank *C* contains the evaporator pipes. The vapour goes to the absorber *D* where it is absorbed by weak solution from the boiler. The strong solution so formed passes through the heat exchanger *E* to the boiler, under the action of the pump. *F* is a header tank for the cooling water.

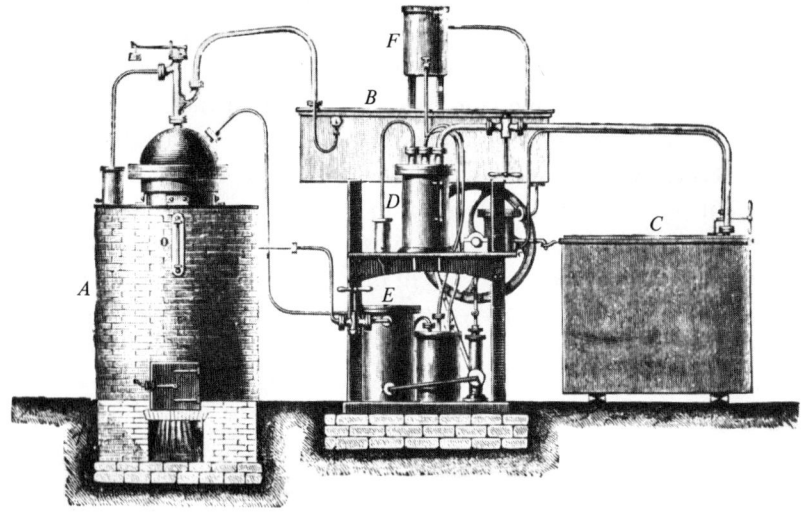

Fig. 1.5.2. The essential components of the vapour absorption system.

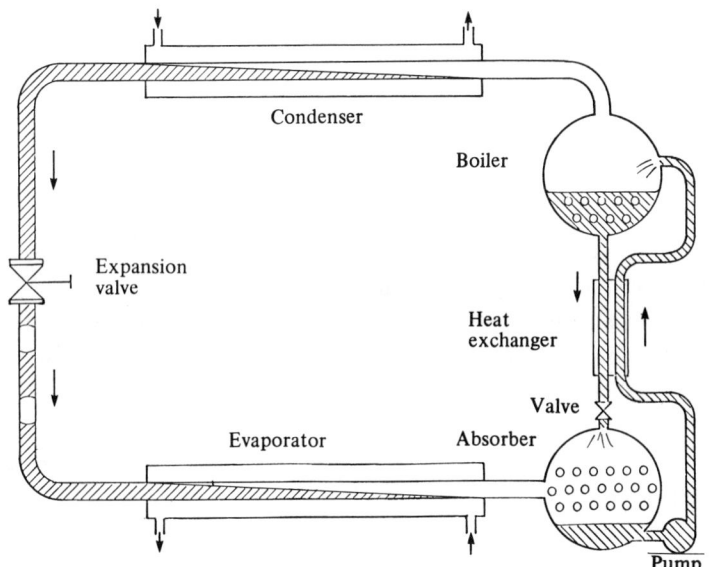

Comparing Fig. 1.5.2 with the vapour compression system, Fig. 1.4.2, it will be seen that the components condenser, expansion valve and evaporator are the same in each. The compressor, however, is replaced by the combination of *absorber, solution pump, heat exchanger, boiler* (or *generator*) and *liquid valve*. This group of components 'sucks' vapour from the evaporator, and delivers high pressure vapour to the condenser, just as the compressor does. The absorber is fed with weak solution which absorbs the ammonia vapour. The absorption of ammonia by water is a strongly exothermic process, and if no cooling were provided the temperature would rise and absorption would stop. The absorber must therefore be cooled, usually by the same water as is used to cool the condenser, and the absorber thus operates at about the same temperature as the condensing temperature. The strong solution formed in the absorber is pumped to a higher pressure by the solution pump and delivered to the boiler or generator via a heat exchanger. In the generator, the strong solution is boiled by heating it, and the vapour given off is rectified to nearly pure ammonia and delivered to the condenser. The weak solution formed in the generator is of course hot, and so the heat exchanger is interposed between the generator and absorber. The weak solution is cooled towards the absorber temperature and the strong solution is warmed towards the generator temperature. To maintain the difference in pressures between the generator and absorber, a valve is installed in the pipe just before the weak liquid enters the absorber.

The early absorption systems were heated directly by coal, but eventually steam heating of the generator was introduced. In recent times there has been a move back to direct firing by oil or natural gas.

The system invented by Carré was used for many years in basically the same form, and it played a most important part in the early development of refrigeration. As the success of the ammonia compression system became clear, the compression system became the dominant one. The absorption system was eclipsed for many years, except in domestic refrigerators, but it has now come back into use in large plants.

A special form of the aqua-ammonia system has been used in many domestic refrigerators. In this arrangement the solution pump and the expansion valves are eliminated by making the pressure uniform throughout the whole system. This is done, following an idea of Geppert in 1899, by charging the low-pressure side with a non-condensable gas which adds its pressure to that of the ammonia vapour, to make a total pressure equal to that in the generator and condenser. Geppert had tried air as the non-condensable gas, but had been unable to obtain a sufficient rate of evaporation, the conditions being more or less like those in Leslie's experiment, but without the vacuum pump. He had to use a fan to circulate the air, and this defeated the object of making a system without any moving parts. In 1922 Carl Munters and Balzar von Platen, whilst students at the Royal Institute of Technology, Stockholm, devised a system in which hydrogen was used as the non-condensable gas, its circulation between evaporator and absorber being ensured by the differences in density caused by the different concentrations of ammonia in it. In addition they used the 'bubble pump' or coffee percolator principle to raise the strong solution to the top of the boiler and provide a hydrostatic head to feed weak solution to the absorber. The system could be manufactured as a hermetically sealed one before this became possible for vapour compression refrigerators, and it had considerable success for domestic refrigerators. Developments of the sealed compression system, however, have once again eclipsed the absorption system for this purpose.

Another form of absorption system, using water as the refrigerant and a solution of lithium bromide in water as the absorbent, is a direct descendant of Windhausen's method, and works in exactly the same way, the sulphuric acid being replaced by a strong solution of lithium bromide. The system is much used to make chilled water for air conditioning systems.

1.6 Refrigeration by expansion of air

When a gas such as air at a high pressure is allowed to expand, i.e. increase its volume, in such a way that it does work on a piston, and at the same time it is kept isolated from heat, its temperature falls. This fact became generally known during the eighteenth century, but the idea of using it for cooling things did not appear until 1828 when Richard Trevithick described a process based on it. The first cold air machine was constructed by John Gorrie in Florida in 1844, and this machine is sometimes claimed as the first successful refrigerator. It is not clear what happened to Gorrie's machine, or how many were made. At least one was brought to London about 1852 and was used for making ice, but no more seems to have been heard of it. The first air machine of any size was that developed by Alexander Carnegie Kirk in 1862. It was based on a superior principle to that of Gorrie, being a reversal of the Stirling heat engine, and it was the prototype of the now very successful Philips air liquefier. After Kirk, inventors reverted to the simpler idea of Gorrie, and in 1875 Paul Giffard perfected the open type of machine (Lightfoot, 1881). The principle is shown in Fig. 1.6.1. Air from a cold room is drawn into a cylinder and

Methods of producing cold 18

compressed. During this process the temperature of the air rises with increasing pressure. The hot air is then delivered to a heat exchanger where its temperature is reduced by cooling water. The compressed air is then expanded in a cylinder against a piston doing work, and the temperature of the air falls. The cold air is discharged to the cold room where it refrigerates the stored produce and rises in temperature again.

The work done by the air in the expansion cylinder is used to provide some of the work needed to drive the compressor. The steam engine employed for this purpose on ships was usually mounted on the same base as the compression and expansion cylinders and directly coupled to them.

Notable improvements to the cold-air machine were made by T. B. Lightfoot, A. Haslam, and by Henry Bell and James Coleman of Glasgow. The vessel 'Strathleven' which brought the first cargo of frozen meat from Australia to England in 1880 was fitted with one of their machines (Coleman, 1882).

The open system described above was the principal method for marine

Fig. 1.6.1. Schematic diagram of the cold-air system. The compressor draws air direct from the cold room, and delivers it under pressure to a heat exchanger where the air is reduced in temperature by cooling water. The air is then expanded doing work and its temperature falls. The expander and compressor cylinders are coupled together, the additional power required being supplied by an engine.

The engine is not shown in the diagram, but it was usually on the same base-plate directly coupled to the compressor and expander. The compressor and expander are shown as double-acting machines, which was their common form in cold air systems.

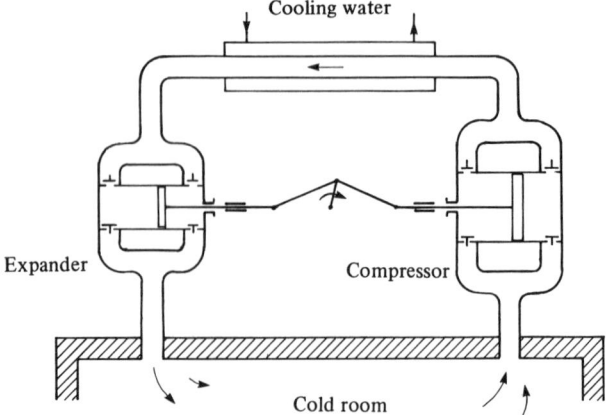

refrigeration for over twenty years, during which time it was improved in many respects. It was eventually displaced by the vapour compression machine. A closed form of cycle employing much higher pressures was developed by Frank Allen of New York under the name of the 'dense air' machine.

In recent times the cold air system has come into use again for cooling aircraft cabins. One of the disadvantages of the old systems was the bulk of the apparatus using piston compressors and expanders. Small turbo-machines can handle large volume flow rates however, and these are employed in aircraft. In a jet aircraft a supply of compressed air is available from the main engines, so a separate compressor is not needed. The air is simply expanded in a turbo-expander and the cold air is distributed to the cabins.

1.7 The vapour jet system

The principle used in the vapour jet system is that of cooling by evaporation, exactly as in the vapour compression and absorption systems. The difference lies in the method used to draw the vapour from the evaporator and compress it.

The device which is used to do this is called an *ejector* or *thermo-compressor*, invented by Sir Charles Parsons originally for pumping the air out of the condensers of steam power plants. It is shown diagrammatically in Fig. 1.7.1. A supply of high-pressure vapour, usually steam, passes

Fig. 1.7.1. Refrigeration system using a steam ejector or thermo-compressor. The suction vapour is drawn by entrainment in the jet of high velocity steam leaving the nozzle. The mixture is then decelerated in the diffuser and the pressure rises up to condensing pressure.

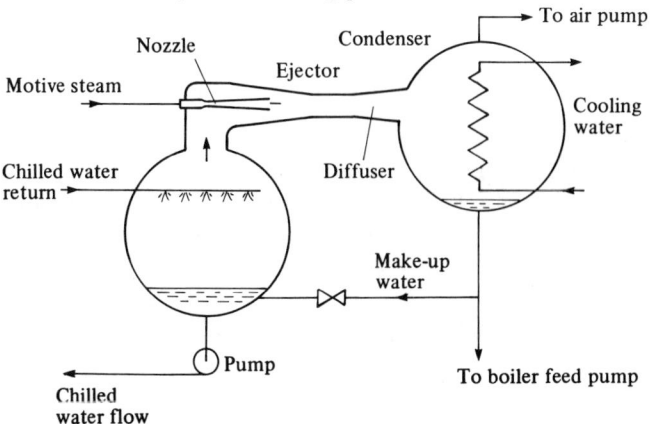

through one or more nozzles in which it acquires a high velocity whilst expanding down to evaporator pressure. This steam, known as the motive steam, shares its momentum with vapour from the evaporator so that the resulting mixture has sufficient velocity to move against the pressure gradient up to the condenser pressure in a *diffuser*, in which the mixture is decelerated. Both the motive vapour and the vapour drawn from the evaporator are condensed, and the condensate is then divided into two flows, one to feed the evaporator and the other to supply the boiler through a feed pump.

The steam jet system was developed principally by Maurice Leblanc about 1910 in Paris. Using steam, the system demands a fairly good degree of vacuum tightness on the low-pressure side, but ejectors to remove air from the condenser are still needed. More recently other vapours have been tried in ejector cycles for special purposes. The ejector has also been used as a low stage 'booster' to a reciprocating compressor.

The vapour jet system has a use in certain circumstances. Being driven by heat instead of mechanical or electrical work, it is applicable where these are not readily available. At one time it was used for making chilled water for air-conditioning systems, but has largely been displaced in this field by the absorption system using water and lithium bromide on account of the superior efficiency of the absorption system. The water and lithium bromide system, however, is not really suitable for working as an open system, and to avoid contamination of the solutions the chilled water is usually in a closed circuit. When this is not possible the vapour jet system is more suitable. For example, it is used in vacuum cooling of foodstuffs where the water in the food is the actual refrigerant and cools the food by taking its latent heat from it. Another use is for cooling the aggregates in mass concrete.

Further information on vapour jet refrigeration is given by Forster (1954) and Vahl (1966).

1.8   Thermo-electric refrigeration

Refrigeration by this method is based on the discovery of Peltier in 1834 that when an electric current passes around a circuit composed of two different metals one junction becomes cool and the other becomes warm. With pure metals the effect is comparatively small and is largely swamped by the temperature rise due to the resistance of the conductors and by the conduction of heat from the warm junction to the cool one. Nevertheless, using bismuth and antimony, Lenz is said to have made a small amount of ice by this method in 1838.

1.8 Thermo-electric refrigeration

For pure metals a low thermal conductivity is associated with a low electrical conductivity, so that if a metal is chosen with a low thermal conductivity the loss by conduction from one junction to the other is small but the loss due to resistance is large. The effectiveness depends principally on the thermo-electric power, and for pure metals this is rather small.

In recent times semiconductors have been discovered which have high thermo-electric powers, and these have made possible small working refrigerators. Semiconductors are of two types named according to whether the current is carried by electrons (n-type) or by 'holes', i.e. vacancies for electrons (p-type). They are made by doping very pure substances with minute amounts of impurities to provide the current carriers. A possible explanation for Lenz's success in making ice, which few have been able to repeat using bismuth and antimony, is that his materials had accidental impurities. A semiconductor used today for cooling purposes is bismuth telluride, Bi_2Te_3.

A cooling element is shown in Fig. 1.8.1 made from n- and p-type materials. The two blocks are formed into a circuit using copper bridges and connectors. The copper itself plays no part in the process, except as a conductor. A source of low voltage direct current is needed. Since each element needs only a fraction of a volt, many of them are connected in series to form a module, with all the warm junctions on one side and all the cool junctions on the other.

Thermo-electric cooling is at present practicable only for small refrigerators because of its low efficiency. Its further development is held up by the lack of better materials, and it is not certain that these are in fact possible, even in principle.

Further information is given by Ioffe (1957) and Goldsmid (1964).

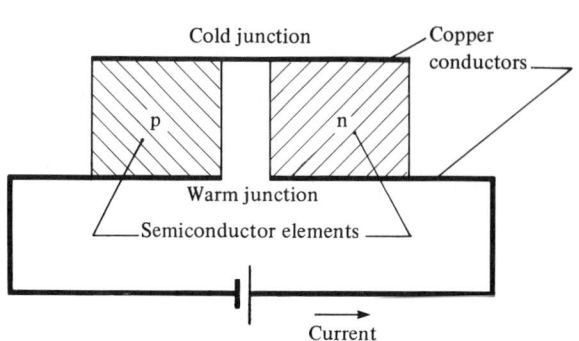

Fig. 1.8.1. A thermo-electric element consisting of p- and n-type semiconductors.

1.9 Secondary refrigerants

In the vapour compression or absorption system the cooling effect produced may be applied directly to the bodies or fluids which are to be refrigerated, such as when the evaporator carrying boiling refrigerant is placed inside a cold room. This practice is known as *direct expansion*. On the other hand, the refrigeration may be needed at places more or less distant from the evaporator, and a *secondary refrigerant* is used to distribute the refrigeration. The secondary refrigerant itself is cooled in the evaporator and is then pumped to the various places where refrigeration is wanted.

The advantages of using a secondary refrigerant are that the refrigerant in the primary system can be confined to a small area, the plant room, with reduced risk of leakage because of the smaller number of pipe joints needed; the whole refrigerating plant can be factory assembled, so that only pipework for the secondary refrigerant need be done on the site; and the problem of controlling and distributing the refrigeration is somewhat simplified.

The secondary refrigerant must obviously remain liquid at the temperature desired. For air-conditioning installations, water is a suitable liquid. Chilled water at about 6–7 °C is produced in the plant room and distributed to unit coolers or other forms of heat exchanger throughout the building. For lower temperatures, various non-freezing solutions or pure liquids are used. The principle types are as follows.

(i) Solutions of inorganic salts in water. The most widely used is calcium chloride, but magnesium and sodium chlorides are also used for certain purposes. The lowest temperature to which an aqueous solution can be cooled before total solidification takes place is the eutectic temperature (see below), which for these salts is: $CaCl_2$, -55 °C; $MgCl_2$, -17 °C; NaCl, -21 °C. But it is not possible to use the solutions at these temperatures because the surface which cools the solution must be at a still lower temperature and deposition of solid (ice + salt) would take place thereon. Calcium chloride brine can be used down to about -40 °C, provided that it is strong enough.

Brines have relatively high densities and specific heat capacities, so that the volume rate of circulation required for a given refrigerating capacity is low. They are, however, rather corrosive, especially in weak aerated solutions which are slightly acidic. A strong brine should be used and it should not be oxygenated by splashing through air. The returns to a tank should enter below the surface.

Hydrometers for measuring brine density are often calibrated in degrees

1.9 Secondary refrigerants

Twaddell. The density ρ is given by:

$$\frac{\rho}{\text{kg/m}^3} = 1000 + 5(°\text{Tw}) \tag{1.9.1}$$

For $-30\,°\text{C}$, a calcium chloride brine of density $1275\,\text{kg/m}^3$ should be used, measured at room temperature.

Though strong brines without additives have been used successfully in systems with steel pipe, a corrosion inhibitor is usually recommended. This may be $1.5–2.0\,\text{g/dm}^3$ of sodium dichromate dihydrate, $Na_2Cr_2O_7.2H_2O$. This makes the brine acid, and it must be neutralised by the addition of enough caustic soda to bring the pH value to above 7.0.

(ii) Methanol (methyl alcohol) and ethanol (ethyl alcohol) alone or in aqueous solution. Methanol freezes at $-97\,°\text{C}$ and ethanol at $-117\,°\text{C}$. The densities and specific heat capacities of the pure liquids are less than those of calcium chloride brine, but the liquids are usable to lower temperatures. Because of their flammability precautions are required, especially when the plant is not running, with the methanol or ethanol at room temperature. The aqueous solutions are rather viscous.

(iii) Aqueous solution of higher alcohols, principally ethylene and propylene glycol. These have a eutectic temperature of about $-60\,°\text{C}$ with a mass fraction of glycol in the solution of 0.6. The densities and specific heat capacities are relatively high, but the solutions are viscous. They are also rather corrosive and must be inhibited.

(iv) Pure organic liquids such as trichlorethylene (freezing point $-86\,°\text{C}$) and R11 (freezing point $-111\,°\text{C}$). These have high densities and low viscosities but their specific heat capacities are low.

(v) Anhydrous ammonia can be used down to temperatures near its freezing point of $-77.7\,°\text{C}$, but the vessels and pipework must be capable of withstanding the vapour pressure of the ammonia at ambient temperature when the plant is not in use. Liquid ammonia has a higher specific heat capacity than water, but a lower density.

Solutions of ammonia in water are also used. These have a lower vapour pressure which does not require pressure-tight construction, but they are rather viscous at low temperature.

When the temperature of a weak solution of a salt, alcohol, glycol or ammonia, in water is reduced, no change takes place until the temperature falls below $0\,°\text{C}$. At some temperature, depending on the concentration, pure water ice is deposited. This temperature, shown as a function of the mass fraction of the solute in the solution, is plotted in Fig. 1.9.1 for common salt as the curve $W-E$. It is sometimes loosely called a freezing point, but it must be remembered that the whole mass of solution does not freeze there. The curve $W-E$ is in fact the solubility curve of ice in the solution, and represents the temperature at which ice and solution are in equilibrium at the strength given. A much stronger solution, on the other hand, when reduced in temperature deposits crystals of the solute or of one of its hydrates, and equilibrium between the solute and solution is represented by the curve $E-G$, which extends to the right ultimately to the temperature of the fused salt, but with some breaks and discontinuities owing to the formation of intermediate compounds. The lines $W-E$ and $G-E$ intersect at a definite temperature and mass fraction known as the eutectic point. At this temperature, the solute, ice and solution of eutectic strength can all exist in equilibrium together.

A solution of eutectic strength freezes at a fixed temperature just like a pure liquid, and in so doing it gives out a definite latent heat or enthalpy of solidification. The eutectic ice so formed is a mixture of crystals of water ice and of the solute. For ionic salts the solid solute is usually in the form of a hydrate. For sodium chloride, for example, the compound $NaCl.2H_2O$ is deposited, and for calcium chloride it is $CaCl_2.6H_2O$.

Fig. 1.9.1. Equilibrium diagram for a solution of common salt (sodium chloride, NaCl) in water. E is the eutectic point at which a solution of mass fraction ξ_{eut}, ice and solid solute can all exist in equilibrium together. Below 0.15 °C, sodium chloride solidifies as the hydrate $NaCl.2H_2O$.

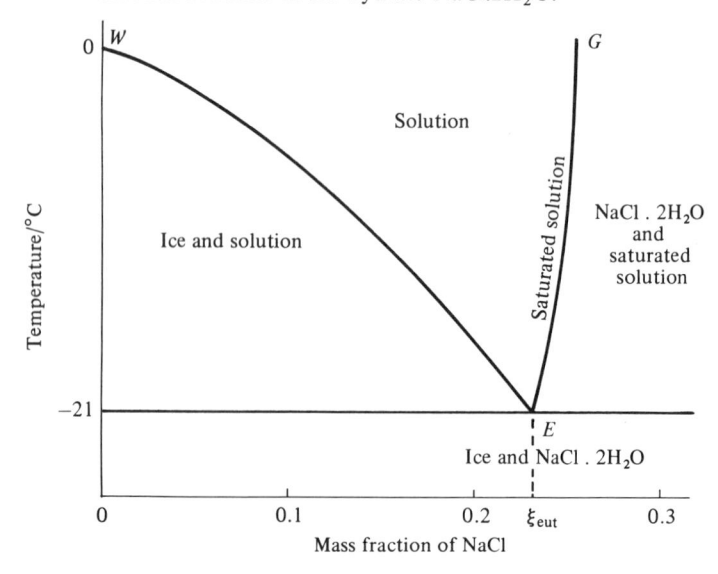

1.9 Secondary refrigerants

The enthalpy of melting of a eutectic ice is made up of the enthalpies of melting of the water ice and of the solute hydrate, plus or minus heat of solution. When the eutectic ice melts the components dissolve in each other, and if heat of solution is evolved when this happens it provides a part of the enthalpy of melting, so reducing the heat which can be absorbed from external sources, i.e. from the things to be cooled. On the other hand, if heat of solution is absorbed during the process of dissolving, as happens with some hydrates, then it adds to the heats of melting and increases the heat which can be absorbed from external sources.

Eutectic ice is used as a means of storing refrigeration when a temperature lower than that of ordinary water ice is wanted. It is often used in 'eutectic plates' to refrigerate vehicles delivering frozen foods to shops. The eutectic ice is frozen during the night whilst the vehicle is at its depot, by plugging a refrigeration unit on board the vehicle into the mains supply.

Some substances, of which ammonia is one, exhibit several eutectic points, corresponding to the formation of intermediate hydrates, as exhibited later in Fig. 6.8.1.

Eutectic behaviour is also of interest in connection with the freezing of foodstuffs containing water, and solutes such as salt or sugar.

Some eutectic points and specific enthalpies of melting are given in Table 1.9.1.

The disadvantage of using a secondary refrigerant, whether a fluid like brine or a solid like eutectic ice, is that it introduces a further difference of temperature between the ultimate things to be cooled and the primary refrigerant. As will be seen in section 1.13, this inevitably entails an increase in the power required to drive the process.

Table 1.9.1. *Eutectic mixtures (Solvent: water)*

Solute	Formula	Mass fraction of solute	Temperature °C	Specific enthalpy of melting of eutectic ice kJ/kg
Ammonia	NH_3	0.33	−100	175
		0.57	−87	310
		0.81	−92	290
Barium chloride	$BaCl_2$	0.22	−7.6	
Cane sugar	$C_{12}H_{22}O_{11}$	0.62	−14.5	
Calcium chloride	$CaCl_2$	0.32	−55	212
Sodium chloride	NaCl	0.23	−21	235
Sodium sulphate	Na_2SO_4	0.04	−1.2	335

Expendable refrigerants

For some purposes it is more convenient, instead of installing refrigerating machinery, to purchase refrigeration in the form of ice, liquid or solid carbon dioxide, or liquid nitrogen. Ice still has some uses today, for example on fishing boats, but it is not so important as formerly. The lower temperatures needed for the manufacture or transport of frozen foods can be met by carbon dioxide or nitrogen.

Solid carbon dioxide ('dry ice') has long been an article of commerce in blocks of about 15 kg which are sliced and crushed for use. Liquid carbon dioxide freezes at $-56.6\,°C$, and at this temperature its vapour pressure of 5.18 bar is above atmospheric pressure, as shown on the phase equilibrium diagram in Fig. 1.9.2. Consequently carbon dioxide can only exist at atmospheric pressure as a solid or as a vapour. The equilibrium temperature at standard pressure is $-78.5\,°C$. At this temperature it changes directly from solid to vapour, a change of phase called sublimation. The specific refrigerating effect, which is the same as the specific enthalpy or latent heat of sublimation, associated with this change is given in Table 1.9.2.

The vapour produced at $-78.5\,°C$ still has some cooling capacity for

Fig. 1.9.2. The equilibrium diagram for the phases of carbon dioxide. Liquid can exist only above the pressure 5.18 bar. At ordinary atmospheric pressure, about 1 bar, solid carbon dioxide sublimes to vapour directly.

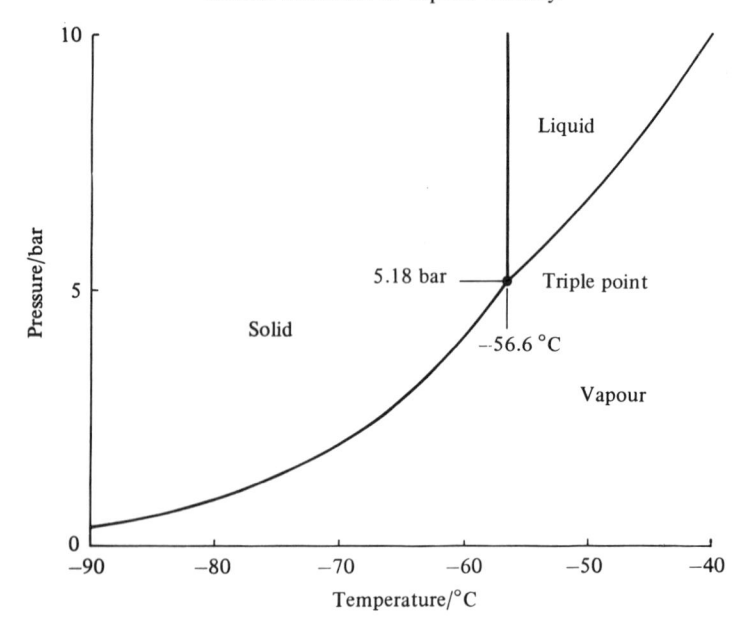

1.9 Secondary refrigerants

things above that temperature, and the additional refrigerating effect on warming up to $-20\,°C$, as well as the total, is given in the table.

Instead of buying solid carbon dioxide at $-78.5\,°C$, which must be stored in insulated containers and which suffers some loss, liquid carbon dioxide under pressure in steel cylinders may be used. This can be stored at ambient temperature without loss, but its specific refrigerating effect is not so great as that of the solid. It is convenient for occasional use on a small scale, but more commonly now liquid carbon dioxide is bought and stored at a temperature of about $-16\,°C$ and a (gauge) pressure of about 21 bar. When the liquid is let down to atmospheric pressure through a valve, a fraction of it flashes into vapour and the remaining fraction is ejected as finely divided snow. This may be sprayed directly over foodstuffs to freeze them or to maintain their frozen state.

In recent years liquid nitrogen has come into use. It is delivered and stored as liquid at a temperature of $-195.8\,°C$ at atmospheric pressure. The specific refrigerating effects are shown in the table. It is important to ensure with nitrogen that the refrigerating effect of the vapour is fully used, since it is of the same order as the enthalpy of evaporation.

Although expendable refrigerants eliminate the need for a large refrigerating plant in many cases, quite elaborate storage facilities have to be provided. For liquid carbon dioxide at $-16\,°C$, an insulated pressure vessel is needed and a small refrigerator is usually provided to reduce the loss of carbon dioxide caused by heat gains through the insulation. Liquid nitrogen does not need a pressure vessel but it has to be very well insulated.

In each case it is necessary to take delivery of a large quantity if a low price is sought, and tanks holding up to 50 tonne are common. The concentrated load may require piled foundations.

Table 1.9.2. *Specific refrigerating effects of expendable refrigerants* †

	t °C	$(\Delta h)_a$ kJ/kg	$(\Delta h)_b$ kJ/kg	$(\Delta h)_a + (\Delta h)_b$ kJ/kg
Ice	0	333		
Liquid CO_2 at $25\,°C$	-78.5	152	47	199
Liquid CO_2 at $-16\,°C$	-78.5	264	47	311
Solid CO_2	-78.5	571	47	618
Liquid nitrogen	-195.8	199	185	384

† $(\Delta h)_a$ is the specific refrigerating effect available from the change of phase.
$(\Delta h)_b$ is the specific refrigerating effect available from warming up the vapour to $-20\,°C$.

1.10 Refrigerating capacity

The measure of the job done by a refrigerating machine is the rate of heat transfer to its evaporator or cold part, from the bodies or fluids which are being refrigerated. This rate of heat transfer will be called the *refrigerating capacity*. It is commonly expressed in one of the following systems of units:

SI   W
Technical metric   kcal/h
British   Btu/h

$$1\,W = 0.859\,845\,kcal/h = 3.412\,14\,Btu/h \qquad (1.10.1)$$

Plant capacities in practice range from a few microwatts in very low temperature refrigerators to several megawatts in large plants.

Only SI units are used in this book, but mention must be made of an important historical unit known as the 'Ton of Refrigeration'. It was supposed at one time to be the rate of heat transfer required to make one short ton (2000 lb) of ice per day from water at $0\,°C$, but was subsequently exactly defined by the American Society of Refrigerating Engineers as:

$$1\text{ Ton of Refrigeration} = 12\,000\text{ Btu/h} = 200\text{ Btu/min} \qquad (1.10.2)$$

In terms of other units:

$$1\text{ Ton of Refrigeration} = 3516.85\,W = 3023.95\,kcal/h \qquad (1.10.3)$$

Other definitions of a 'Ton of Refrigeration' have been proposed from time to time and this has caused some confusion in the past.

The 'Ton' is often used in general terms to indicate the order of size of a plant, e.g. a 100 Ton plant, or 'fractional tonnage' plants (small commercial ones). It also happens that at the temperatures of evaporation and condensation used in air-conditioning, 1 Ton of refrigeration requires a power of about 1 horse-power, which gives a useful way of visualising the size of a plant. (N.B. This relation is approximate only. It must not be used for estimating power.)

It is important when specifying refrigerating capacity to define exactly where it is to be measured, i.e. across which 'control surface' the stated heat transfer rate occurs. In a direct expansion system, if the control surface encloses only the evaporator, then heat transfer to the expansion valve and suction pipe to the compressor or absorber are not included. If the valve and suction pipe are brought inside the control surface, then the refrigerating capacity gives the rate of heat transfer to the surfaces between the inlet to the expansion valve and the inlet to the compressor or absorber. This latter rate may be called the gross refrigerating capacity. In a plant with a secondary refrigerant, one must specify whether the capacity is measured

1.11 Coefficient of performance and heat ratio

on the primary circuit between specified points or on the secondary circuit; and, if a pump or fan is used to circulate brine, water or air, whether the power for this is to be included as an equivalent heat transfer rate.

Other names for what is called refrigerating capacity above are also to be met, and one of these will be used from time to time, namely the *refrigerating load*. This also means the rate of heat transfer to the evaporator or cold part of the plant, but it places the emphasis on what the plant is asked to do rather than on what it can do. It will be said, for example, that the load is increased or decreased, put on or taken off, meaning that the temperature of the things being cooled, or the rate of heat transfer from them, is increased or decreased. The refrigerating plant reacts to match the load, by an increase or a decrease of the refrigerating capacity. This use of the word 'load' is the same as its use in connection with engines. Applying load to an engine means making a demand for more power by increasing the torque, i.e. by attempting to stop the shaft turning. The engine reacts by giving more power, providing it is within its capacity to do so. At a steady state, with no acceleration, the power equals the load of course. But the difference of emphasis is clear: load is the demand which is made of the engine, power is what the engine gives.

Another term in use is *refrigeration duty*. It has at least two meanings. One is the same as refrigerating capacity, meaning the rate of heat transfer at a particular instant to the evaporator. Another meaning is the specified refrigerating capacity, i.e. what is asked for by the buyer of the plant, usually at a specified set of conditions. The conditions may be internal ones, such as evaporating and condensing temperatures, temperature of liquid at the expansion valve, and temperature of the vapour leaving the evaporator and entering the compressor. On the other hand, the conditions may be external ones: temperature of cooling water, rate of flow, temperature of cooled fluid and rate of flow, for example.

Finally, the term *refrigerating effect* is also in use, meaning either refrigerating capacity or the heat transfer to the evaporator associated with the passage of a certain amount of refrigerant. In this book the term refrigerating effect will only be used in this second sense: a heat transfer associated with the passage of a certain amount of refrigerant through the evaporator. It will most often refer to the passage of unit mass, and then will be called the specific refrigerating effect.

1.11 Coefficient of performance and heat ratio

The efficiency of a refrigerating plant is measured, like that of an engine, by the ratio of the useful effect produced to the means used to produce it. For all refrigerators the useful effect is the refrigerating capacity.

Methods of producing cold

The means used to produce this useful effect differ, however, between different types.

For the vapour compression system, and for the cold air and thermoelectric systems, the means take the form of power delivered to the machine to keep it working. The efficiency may therefore be expressed as

$$\varepsilon = \dot{Q}_e/\dot{W} \qquad (1.11.1)$$

where $\dot{Q}_e$ is the refrigerating capacity and $\dot{W}$ is the power. Subscript e is to indicate that the heat transfer rate $\dot{Q}_e$ is towards the refrigerant in the evaporator. The ratio ε is called the coefficient of performance.

To be absolutely clear, one must specify at what point the power is to be measured. For example, with an electric motor driving the compressor of a vapour compression system, either the electric power or the mechanical shaft power may be stated, giving different values of ε. In this book the method of driving the compressor will usually be left out of consideration, and $\dot{W}$ normally taken as the shaft power to the compressor.

Unlike the thermal efficiency of an engine, which cannot exceed unity, the coefficient of performance can have values much larger than one.

Coefficient of performance, since it is the ratio of two quantities, heat and work, which have the same dimensions is essentially dimensionless. This is clear enough in the SI system, since the units of heat and work are both joules. In other unit systems, coefficient of performance is often expressed in mixed units. For example, in the technical metric system the units used are frequently kcal/Wh. Since

$$1\,\text{kcal/Wh} = 1.163 \qquad (1.11.2)$$

ε can be changed from one form to the other as desired.

In British units, coefficient of performance is often expressed in Tons of Refrigeration per horsepower, or as the reciprocal, horsepower per Ton. Again

$$1\,\text{hp} = 42.407\,\text{Btu/min} \qquad (1.11.3)$$

approximately and

$$1\,\text{hp/Ton} = \frac{42.407\,\text{Btu/min}}{200\,\text{Btu/min}} = 0.21203 \qquad (1.11.4)$$

For example, given $\varepsilon = 5.0$

$$\varepsilon = 5.0 = 5.0\,\frac{\text{kcal/Wh}}{\text{kcal/Wh}} = \frac{5.0}{1.163}\,\text{kcal/Wh} = 4.299\,\text{kcal/Wh}$$

In British units

$$\varepsilon = 5.0 = 5.0 \frac{\text{Ton/hp}}{\text{Ton/hp}} = (5.0)(0.21203) \text{ Ton/hp} = 1.060 \text{ Ton/hp}$$

and

$$\frac{1}{\varepsilon} = \frac{1}{1.06} \text{ hp/Ton} = 0.943 \text{ hp/Ton}$$

For absorption and steam jet refrigerators, the input is mainly in the form of a heat transfer to the boiler, neglecting the small power supplied to the feed pump. The efficiency of such systems may therefore be expressed as a ratio:

$$\zeta = \dot{Q}_e/\dot{Q}_b \qquad (1.11.5)$$

where $\dot{Q}_b$ is the rate of heat transfer to the boiler. The ratio ζ will be called the heat ratio.

Since absorption systems are often driven by a supply of steam which is used to heat the boiler, another way of expressing efficiency is sometimes used. This is the ratio of the refrigerating capacity to the mass flow rate of the steam. Provided that one knows the heat given out by the steam in condensing, i.e. the change of enthalpy of the steam in passing through the heating tubes in the boiler, the mass flow rate of steam may be turned into a heat transfer rate to the boiling solution. The enthalpy change of the steam is simply the enthalpy of evaporation in most cases, assuming that the steam is supplied dry and saturated and that the condensate leaves without much subcooling.

Unlike the coefficient of performance of a vapour compression refrigerator, which is often greater than one, the heat ratio of a simple absorption refrigerator is always less than one. The reason is simply that it needs as much heat, and in fact more, in the boiler to drive the ammonia out of solution as it does in the evaporator to change it from liquid to vapour. In each place enthalpy of evaporation has to be supplied. A further amount, enthalpy of solution, has to be supplied in the boiler to separate the water and ammonia. A still further amount has to be supplied to bring about the rectification of the ammonia vapour to make it nearly anhydrous.

1.12 Refrigeration and thermodynamics

The operation of a refrigerator may be described by the effects which it produces in its surroundings, disregarding entirely how it works. The complete plant is then regarded simply as a 'black box' or control

Methods of producing cold

surface across whose boundaries heat and work transfers are observed. A vapour compression plant, for example, takes in heat from fluids or bodies which are to be refrigerated, it takes in work via the shaft of the compressor, and it gives out heat to the water or air in the condenser. The cold air system and the thermo-electric refrigerator perform similarly, except that in the case of the thermo-electric refrigerator the work input is electrical. If one includes a motor in the vapour compression or cold air plant, the three machines are exactly similar in the effects which they produce around them. They will be called simply work-operated refrigerators.

The absorption system and the steam jet system on the other hand need heat transfer from something at a much higher temperature than the surroundings to drive ammonia out of the solution in one case and to generate steam at pressure in the other. There is also a small work transfer required to drive the solution pump or the feed pump, but this is usually negligible. These will be called heat-operated refrigerators.

The block diagram for a work-operated refrigerator is shown in Fig. 1.12.1. Disregarding incidental heat gains and losses from the pipes, the principal transfers are: (1) acceptance of heat Q_1, say, from bodies or fluids to be refrigerated at a relatively low temperature t_1; (2) acceptance of work W, say, to drive the plant, either mechanical or electrical depending on whether the motor is supposed to be enclosed in the box or not; and (3) rejection of heat Q_2, say, to cooling water or air at a temperature t_2 higher than t_1.

For a continuously operating system of this kind, the First Law of

Fig. 1.12.1. Block diagram of a work-operated refrigerator such as a vapour compression system, showing the heat and work transfers. Heat Q_1 is taken from bodies or fluids at a low temperature t_1, work W is taken in and heat Q_2 is given out at the higher temperature t_2. For a refrigerator, t_2 is the temperature of the environment, cooling air or water, to which heat can be rejected. For a heat pump, section 1.14, t_1 may be the temperature of the environment, air or water, from which heat can be accepted.

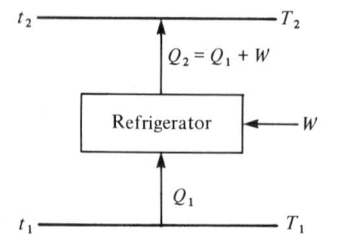

1.12 Refrigeration and thermodynamics

Thermodynamics states that if the heat and work transfers are measured over a period of time for which the contents of the control space have exactly the same state at the beginning and the end, then:

$$Q_1 + W = Q_2 \tag{1.12.1}$$

If the machine operates under steady conditions, i.e. its thermodynamic state is constant at all times, then the heat transfers may be expressed as heat rates, $\dot{Q}_1$, $\dot{Q}_2$, and the work may be expressed as power $\dot{W}$, and:

$$\dot{Q}_1 + \dot{W} = \dot{Q}_2 \tag{1.12.2}$$

The Second Law of Thermodynamics, in the form due to Clausius given above in section 1.3, states essentially that the work W cannot be zero. But it goes further than this and shows that there is a certain maximum coefficient of performance, Q_1/W, which cannot possibly be exceeded without contravening the Law. Carnot showed that the maximum possible efficiency of an engine working between two temperatures depends only on these temperatures, and by a similar argument it can be shown that the maximum possible coefficient of performance of the work-operated refrigerator depends only on the temperatures t_1 and t_2. It is independent of the manner in which it works and the material used as refrigerant. Consequently this ideal refrigerator can act as a thermometer and tell us something about the relation of one temperature to another. For practical reasons the absolute thermodynamic temperature scale, on which temperatures are denoted by upper case T, is defined by the equation:

$$Q_1/T_1 = Q_2/T_2 \tag{1.12.3}$$

where Q_1 and Q_2 are the heats taken in and rejected at absolute temperatures T_1 and T_2, respectively, by this ideal machine.

Equation (1.12.3) gives a numerical value for the ratio of two absolute temperatures but does not determine the size of the degree or the measure of a difference of temperatures. A labelled reference point is therefore needed, and this has been chosen by international agreement as the temperature of the triple point of water, i.e. the temperature at which water liquid, solid ice and vapour can exist together in equilibrium. This point is defined as:

$$T(\text{triple point of water}) = 273.16 \text{ K} \tag{1.12.4}$$

where K is the symbol for the unit of absolute thermodynamic temperature, called the kelvin.

The conventional Celsius scale of temperature is related to the absolute scale by:

$$0\,°C = 273.15\,K \tag{1.12.5}$$

i.e. the triple point of water substance has the Celsius temperature 0.01 °C exactly.

It will be noted that according to this definition of the Celsius scale the freezing point of water at standard atmospheric pressure has to be determined experimentally. It is not necessarily 0 °C exactly. Nor is the boiling point of water at standard pressure exactly 100 °C. They are, however, very close to these values and the differences do not matter in practice.

Temperature, whether measured on the kelvin scale or on the Celsius scale, is the same fundamental quantity, and is recognised as such in the SI system, along with mass, length, time and electric current. It is strictly illogical to use two different symbols for it. Nevertheless this will be done for practical convenience, and temperatures measured above a datum of 273.15 K i.e. on the Celsius scale, will be denoted by lower case t, whilst temperatures on the kelvin scale will be denoted by upper case T; just as it might be convenient possibly to measure the height of Mt. Everest from the centre of the earth in one case, and above mean sea level in another, and use different symbols for the results.

The difference between two temperatures will always have the units kelvins of course, just as the difference of two mountain heights would be in metres wherever they were measured from. Thus:

$$40\,°C - 25\,°C = 15\,K$$

Correspondingly, the unit of any physical quantity which contains temperature difference will be expressed in terms of kelvins, for example specific heat capacity has the units J/kg K, and thermal conductivity has the units W/m K.

In refrigeration at moderately low temperature, Celsius temperatures are commonly used, and this practice is followed here except where absolute temperature T is required to express the basic physical law as in the equations below.

The definition of absolute temperature immediately gives a value for the maximum possible coefficient of performance of a refrigerator operating between T_1 and T_2. For by eq. (1.12.1) the coefficient of performance ε is:

$$\varepsilon = \frac{Q_1}{W} = \frac{Q_1}{Q_2 - Q_1} \tag{1.12.6}$$

1.12 Refrigeration and thermodynamics

and in the ideal case Q_1 and Q_2 are related by eq. (1.12.3). Hence the maximum possible coefficient of performance is:

$$\varepsilon_C = \frac{T_1}{T_2 - T_1} = \frac{1}{(T_2/T_1) - 1} \tag{1.12.7}$$

The subscript C for Carnot is used here to denote the maximum possible value of ε, because Carnot proposed one hypothetical cycle of operations whereby this maximum value could be realised in principle.

Suppose that the absolute temperature T_2 is held constant and T_1 is allowed to diminish. According to eq. (1.12.7) the value of ε_C decreases as the ratio T_2/T_1 increases, ultimately as T_1 becomes very small towards zero. This limiting value of T_1 is the zero of the absolute thermodynamic scale of temperature, or absolute zero for short. The Second Law of Thermodynamics does not indicate whether it can ever be attained or not,

Fig. 1.12.2. The Carnot coefficient of performance, ε_C, plotted to a base of the lower temperature with several constant higher temperatures as the parameters on the curves.

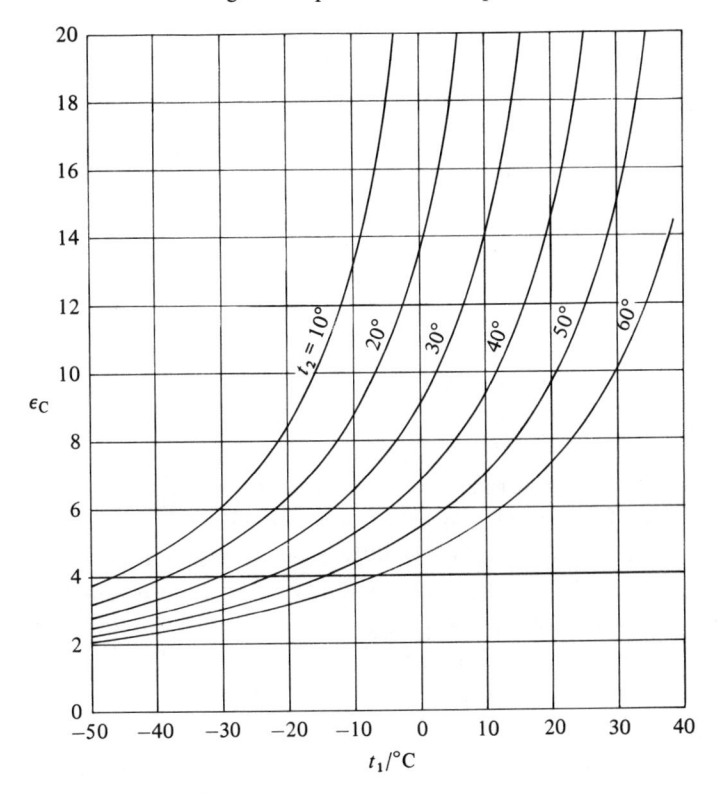

but it is clear that in the absence of a perfect insulator it is not possible to maintain a body at absolute zero because infinite work would be needed to drive a refrigerator capable of accepting the heat transfer through the insulation.

Equation (1.12.7) is of fundamental importance for understanding the work requirements of refrigerators because the coefficient of performance of a real machine, though always less than ε_c, varies with the temperatures in a similar manner. In Fig. 1.12.2, ε_C is plotted to a base of the low temperature t_1 with the higher temperature t_2 as a constant parameter on each of the curves. This method of displaying the variation of a quantity with the lower and higher temperatures will be much used in later chapters. The range of low temperatures shown goes down to those which are used in food freezing and some industrial process, and up to those used in industrial drying. The range of the higher temperature represents ambient temperatures in various climates where refrigeration might be required and also some higher ones used industrially.

The first point evident from the figure is the rapid increase in ε_C at any constant t_2 as t_1 increases towards it. As can be seen, very high coefficients are possible when the temperature difference is small. Conversely, the coefficient is quite low when the temperature difference is large, indicating that a large power is needed to give a certain refrigerating capacity under these conditions.

The second point is that at low values of t_1 the value of the higher temperature t_2 has relatively little effect on the coefficient of performance, whereas at higher values of t_1 it has a marked effect. For example, at $t_1 = -50\,°C$ the effect of changing t_2 from 40 to 50 °C is to reduce ε_c by only 10%, but at $t_1 = 25\,°C$ the same change reduces ε_c by 40%.

The temperatures t_1 and t_2, or absolute temperatures T_1 and T_2, are the temperatures of the body being refrigerated and of the medium to which heat can be rejected. The temperatures of the refrigerant itself of course have to be different from these: the evaporating temperature t_e, say, has to be lower than t_1; and the condensing temperature t_c say has to be higher than t_2. The effects of these differences are discussed in the next section. Supposing for the moment that these differences are absent, the higher the ratio T_c/T_e, the higher is the ratio of the condensing pressure to the evaporating pressure. Compression of a given volume of vapour over a high pressure ratio clearly requires more work than compression over a low pressure ratio. But the same volume of vapour has a lower density at a lower evaporating pressure and produces less refrigeration than the same volume at a higher evaporating pressure. Consequently the real coefficient of

1.12 Refrigeration and thermodynamics

performance of a vapour compression refrigerator follows approximately the same variation with evaporating temperature as the Carnot coefficient.

The Carnot coefficient of performance can be used to give a rough estimate of the actual coefficient of performance by multiplying it by some efficiencies:

$$\varepsilon = \eta_r \eta_{isen} \varepsilon_C \qquad (1.12.8)$$

where η_r represents the efficiency of the refrigerant cycle related to the Carnot coefficient calculated using T_e and T_c for T_1 and T_2, respectively, and η_{isen} is the isentropic efficiency of the compressor, defined in Chapter 3. If the efficiency of an electric motor is to be taken into account the result is to be multiplied by this efficiency.

An approximate expression for η_r for ammonia, also useful for R12 and R22, has been given by Linge (1966). Within the range of T_e covered by Fig. 1.12.2:

$$\eta_r = 1 - \frac{t_c - t_e}{265 \text{ K}} \qquad (1.12.9)$$

when the liquid at the expansion valve is saturated. When there is some subcooling to t_u, say, η_r has to be multiplied by the factor:

$$1 + \frac{t_c - t_u}{250 \text{ K}}$$

The efficiency of the compressor depends on many factors, but it is often within the range 0.5 to 0.8. Motor efficiency depends on the size of the motor and on the load: at full load it is about 0.7 for small motors and about 0.95 for large ones.

It must be emphasised that the above procedure is intended solely for the purpose of obtaining an approximate estimate of the coefficient of performance. Chapter 3 deals in detail with the determination of η_r, and Chapter 4 with the calculation of the actual power.

The ideal refrigerator, which has the maximum coefficient of performance, is specified as one which is thermodynamically reversible. This means that if it is operated in the reverse direction the effects which are produced in the surroundings in the forward direction can be entirely annulled. For example, the refrigerator of Fig. 1.12.1 would be reversible if, when it is run as a heat engine taking in heat Q_2 and rejecting heat Q_1 and delivering work W, these quantities have exactly the same magnitudes, but opposite direction, as when the machine is run as a refrigerator. In other words, a reversible refrigerator produces effects in the surroundings which can be completely undone by running the machine the other way. For a full discussion of reversibility, reference should be made to standard works on thermodynamics, e.g. Keenan (1941).

For a heat-operated refrigerator the block diagram is as shown in Fig. 1.12.3. Heat Q_1 is accepted at a low temperature t_1, heat Q_2 is given out at a warmer temperature t_2, and heat Q_3 is taken in at the still warmer temperature t_3. The first Law of Thermodynamics gives:

$$Q_1 + Q_3 = Q_2 \tag{1.12.10}$$

or for steady conditions:

$$\dot{Q}_1 + \dot{Q}_3 = \dot{Q}_2 \tag{1.12.11}$$

For the ideal refrigerator with the maximum possible heat ratio Q_1/Q_3, the Second Law of Thermodynamics gives:

$$\frac{Q_1}{T_1} + \frac{Q_3}{T_3} = \frac{Q_2}{T_2} \tag{1.12.12}$$

The maximum possible heat ratio is then obtained by using this equation to eliminate Q_2 from eq. (1.12.10):

$$\zeta_C = \frac{T_1}{T_2 - T_1} \frac{T_3 - T_2}{T_3} \tag{1.12.13}$$

Fig. 1.12.3. Block diagram of a heat-operated refrigerator such as an absorption system when the small amount of work to drive the solution pump is neglected. Heat Q_1 is taken from bodies or fluids at the low temperature t_1, heat Q_3 is taken from, for example, steam or combustion gases at the high temperature t_3, and heat Q_2 is given out at the intermediate temperature t_2. For a refrigerator, t_2 is the temperature of the environment. For a heat pump, section 1.14, t_1 may be the temperature of the environment.

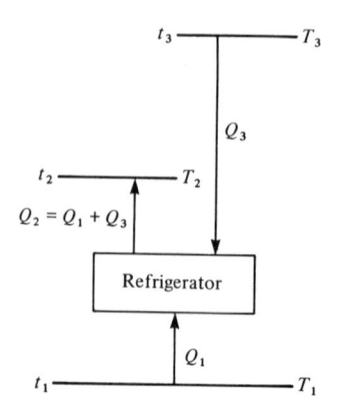

1.13 Internal and external régimes

The first factor on the right-hand side of eq. (1.12.13) is the maximum coefficient of performance of a refrigerator operating between absolute temperatures T_1 and T_2; the second factor is the maximum efficiency of an engine operating between the absolute temperatures T_3 and T_2. This is to be expected since a refrigerator which works on a heat transfer at a high temperature is evidently equivalent to a work operated refrigerator driven by a heat engine.

According to eq. (1.12.13) the maximum heat ratio can be greater than unity provided that T_3 is high enough. The heat ratio of ordinary absorption refrigerators is, however, always less than unity. This discrepancy occurs because, although a high temperature T_3 may be obtainable from steam or from combustion, the absorption refrigerator cannot take advantage of it, and the temperature of the boiling solution in practice is a good deal lower. Furthermore, the solution cannot accept heat at a constant temperature, and the considerations of the next section have to be applied.

1.13 Internal and external régimes

The Carnot coefficient of performance of eq. (1.12.7) is the maximum attainable when refrigerating bodies or fluids at the constant absolute temperature T_1 and giving out heat to (warmer) bodies or fluids at the constant absolute temperature T_2. Two of the conditions for achieving this maximum coefficient are that there should be no difference between T_1 and the temperature of the refrigerant in the system whilst taking in heat, and no difference between T_2 and the temperature of the refrigerant whilst giving out heat.

It is impossible, however, to take in or give out heat with no difference of temperature at all. By allowing a sufficiently large surface area for heat transfer, the difference can be made small but never eliminated entirely. In practice, because of economic considerations, there is always a difference in temperature between the fluids in the evaporator of the order 5–15 K, and a similar value in the condenser. In the evaporator the refrigerant must be 5–15 K colder than the fluids being refrigerated; in the condenser the refrigerant must be 5–15 K warmer than the cooling medium accepting heat from the refrigerant.

To emphasise these differences, the temperatures of the bodies or fluids external to the refrigerant circuit will be called the temperatures of the external régime, whilst those of the refrigerant itself will be called the temperatures of the internal régime. The temperatures of the external régime are often outside the control of the designer of the plant. They are determined by the user, e.g. the temperature required in a cold store, and by the site conditions, e.g. the temperature of cooling water available. The temperatures of the internal régime, however, are usually chosen by the designer, based on economic temperature differences between the refrigerant and the external régime, or on other considerations which relate to the application of the plant.

The difference in temperature between the external and internal régimes causes a loss in coefficient of performance which will be illustrated by an example.

Example 1.13.1

Suppose that the constant temperatures of the external régime are:

$$t_1 \approx -10\,°C \qquad T_1 = 263\,K$$
$$t_2 \approx 25\,°C \qquad T_2 = 298\,K$$

The maximum coefficient of performance is, by eq. (1.12.7):

$$\varepsilon_C = \frac{263}{298 - 263} = 7.51$$

If differences in temperature of 10 K exist between the internal and external régimes, the refrigerant takes in heat at the constant temperature 253 K, and the refrigerant gives out heat at the constant temperature 308 K. The coefficient of performance then cannot possibly exceed that of a reversible refrigerator operating between the temperatures 253 K and 308 K, i.e.

$$\frac{253}{308 - 253} = 4.60$$

This is a serious reduction in the coefficient of performance which is attainable, and it indicates the need to maintain temperature differences between the internal and external régimes which are not unnecessarily large.

So far it has been assumed that the temperatures of the bodies or fluids in the external régime remained constant whilst heat was taken in or given out. This however is rarely the case in practice. Cooling water rises in temperature as it passes through the condenser of a vapour compression plant. The temperature rise can be made indefinitely small by increasing the flow rate sufficiently, but this condition is not approached in practice because of the cost of the water or of pumping it. At the low temperature end of the cycle, it often happens that the things to be refrigerated must change in temperature, e.g. food which is being frozen must first be cooled to its freezing point, then frozen over a temperature range, and finally cooled further to the temperature at which it is stored.

1.13 Internal and external régimes

In these circumstances, if the refrigerant remains at a constant temperature whilst accepting heat, as for example refrigerant boiling at a constant pressure, and remains at a constant temperature whilst rejecting heat, as for example refrigerant condensing at a constant pressure, it is impossible merely by increasing the heat transfer surface to reduce indefinitely the differences in temperature between the internal and external régimes.

Example 1.13.2
Suppose, for example, that the cooling water rises in temperature from 20 °C to 30 °C in passing through the condenser. The refrigerant must condense at a (constant) temperature not less than 30 °C however much heat transfer surface is provided. This means that a substantial part of the heat of condensation is given out from refrigerant at 30 °C to water cooler than 30 °C. If food to be frozen is initially at 10 °C, and its temperature is to be reduced eventually to -20 °C, any refrigerant doing this by boiling at a constant temperature must do so at a temperature not higher than -20 °C.

For refrigerating duties of this type, when the temperature of the external régime is not constant, the Carnot cycle and the coefficient of performance eq. (1.12.7) using the extreme values of the temperatures in the external régime for T_1 and T_2 are unduly pessimistic as standards of performance. In example 1.13.2, with $t_1 \approx -20$ °C, $T_1 = 253$ K, $t_2 \approx 30$ °C, $T_2 = 303$ K, the coefficient of performance would be only:

$$\frac{253}{303 - 253} = 5.06$$

In principle one can aim at a value much better than this by making the refrigerant in the cycle follow the variations of the temperature in the external régime, as first suggested by Lorenz (1894). One way of doing this is to use a mixture of refrigerants, the boiling point of which changes with composition as the more volatile component is boiled away. Another method is to use a gas as the refrigerant.

The criterion of perfection for such machines can be deduced from the condition that no entropy is generated when all operations are carried out reversibly. The change of entropy of a body in going reversibly from a state denoted by subscript 'a' to a state 'b' is defined as:

$$S_b - S_a = \int_a^b \frac{dQ}{T} \qquad (1.13.1)$$

For the special case of a reversible process at constant temperature T:

$$S_b - S_a = \frac{Q}{T} \tag{1.13.2}$$

When the temperature varies but the heat capacity C of the body is a constant during the reversible process, $dQ = C\,dT$ and eq. (1.13.1) can be integrated to give:

$$S_b - S_a = C \ln \frac{T_b}{T_a} \tag{1.13.3}$$

These equations apply to a body of mass m, say. For unit mass similar equations apply except that Q is replaced by $q = Q/m$, and C is replaced by $c = C/m$, the specific heat capacity.

Denoting a steady mass flow rate of substance being refrigerated by $\dot{m}_1$ with a change of specific entropy from s_1' to s_1'', and giving up heat q_1 per unit mass; and denoting the steady mass flow rate of the (warmer) cooling fluid by $\dot{m}_2$ with a change of specific entropy from s_2' to s_2'', accepting heat q_2 per unit mass, the equation of entropy production for reversible operation is:

$$\dot{m}_1(s_1' - s_1'') = \dot{m}_2(s_2'' - s_2') \tag{1.13.4}$$

Defining average absolute temperatures $T_{1,\mathrm{av}}, T_{2,\mathrm{av}}$ by:

$$T_{1,\mathrm{av}} = \frac{q_1}{s_1' - s_1''} \tag{1.13.5}$$

$$T_{2,\mathrm{av}} = \frac{q_2}{s_2'' - s_2'} \tag{1.13.6}$$

eq. (1.13.4) may be written:

$$\frac{\dot{m}_1 q_1}{T_{1,\mathrm{av}}} = \frac{\dot{m}_2 q_2}{T_{2,\mathrm{av}}} \tag{1.13.7}$$

The power required is given by:

$$\dot{W} = \dot{m}_2 q_2 - \dot{m}_1 q_1 \tag{1.13.8}$$

and the coefficient of performance is therefore:

$$\varepsilon_\mathrm{L} = \frac{\dot{m}_1 q_1}{\dot{m}_2 q_2 - \dot{m}_1 q_1} = \frac{T_{1,\mathrm{av}}}{T_{2,\mathrm{av}} - T_{1,\mathrm{av}}} \tag{1.13.9}$$

The subscript 'L' indicates that the coefficient of performance is calculated according to the Lorenz concept, and ε_L will be called the Lorenz coefficient of performance.

1.13 Internal and external régimes

When a fluid has a constant specific heat capacity c over the temperature range concerned, its average absolute temperature for the purpose of eq. (1.13.9) is simply its logarithmic mean absolute temperature. For:

$$q = c(T'' - T') \qquad (1.13.10)$$

and:

$$s'' - s' = c \ln \frac{T''}{T'} \qquad (1.13.11)$$

Hence:

$$T_{av} = \frac{T'' - T'}{\ln(T''/T')} \qquad (1.13.12)$$

Example 1.13.3
Equation (1.13.9) will be used to determine the Lorenz coefficient of performance (c.o.p.) appropriate to the type of refrigerating duty mentioned above: cooling of a food from 293 K($\approx 10\,°C$), freezing it at 272 K($\approx -1\,°C$), and further cooling to 253 K($\approx -20\,°C$) using cooling water rising in temperature from 293 K($\approx 20\,°C$) to 303 K($\approx 30\,°C$) to accept heat. The specific heat capacity of the unfrozen food will be taken as 3.2 kJ/kg K, that of the frozen food as 1.7 kJ/kg K, and the enthalpy of freezing as 240 kJ/kg.

For the cooling water the average absolute temperature is simply the logarithmic mean:

$$T_{2,av} = \frac{303 - 293}{\ln(303/293)} K = \frac{10}{0.03356} K = 298.0 \text{ K}$$

For the food, the denominator of eq. (1.13.5) will be evaluated first.

$$s'_1 - s''_1 = \left(1.7 \ln \frac{272}{253} + \frac{240}{272} + 3.2 \ln \frac{293}{272}\right) kJ/kg\,K$$

$$= 1.243 \text{ kJ/kg K}$$

The heat per unit mass is:

$$q_1 = [1.7(272 - 253) + 240 + 3.2(293 - 272)] \text{ kJ/kg}$$

$$= 339.5 \text{ kJ/kg}$$

The average absolute temperature of acceptance of heat for the purpose of eq. (1.13.9) is therefore by eq. (1.13.5):

$$T_{1,av} = \frac{339.5 \text{ kJ/kg}}{1.243 \text{ kJ/kg K}} = 273.1 \text{ K}$$

The Lorenz c.o.p. is by eq. (1.13.9):

$$\varepsilon_L = \frac{273.1}{298.0 - 273.1} = 10.97$$

compared with 5.06 for the c.o.p. of a Carnot cycle operating between the extreme temperatures.

This is the absolute maximum c.o.p. which could be achieved for the above duty by an ideal reversible refrigerator in which the temperature of the refrigerant followed exactly the temperatures of the food and the cooling water so that no temperature differences at all occurred between the external and internal régimes. This is not of course possible in practice. But the Lorenz c.o.p. shows what can be aimed at. For example, an improvement on the single-stage method of refrigeration envisaged in the Carnot c.o.p. would be a two-stage process in which the pre-cooling of the food and its freezing was done by one refrigerator with a constant internal temperature for heat acceptance as close to $-1\,°C$ as possible, followed by a second stage performed by another refrigerator with a constant internal temperature as close to $-20\,°C$ as possible. Possibilities of this kind exist with multi-stage compression systems treated in Chapter 5.

The assumed relation between the heat given out by the food and the temperature is a simplification introduced for the purpose of illustrating the method. The enthalpy of freezing of foods is in fact given out over a range of temperature. The determination of $T_{1,av}$ then requires the evaluation of the entropy change by graphical integration on a plot of specific enthalpy h versus $1/T$.

1.14 The heat pump

It is clear from the considerations of section 1.12 that the heat discharged at the condenser of a work-operated refrigerator is the sum of the heat taken in at the evaporator and the work supplied to drive the machine. This suggests that the refrigerator can be used as a heating machine, as first pointed out by Lord Kelvin (Thomson, 1852). If one places the evaporator of a vapour compression system out of doors and the condenser inside a building, the heat discharged into the building is greater than the work supplied to drive the machine, the difference coming from the outside. Supposing the work to be provided in the form of electrical work, using this in electric resistance heaters would produce a heat transfer exactly equal to the electrical work, whereas using the work to drive a refrigerator gives a potential heat transfer which is much larger.

1.14 The heat pump

The multiplying effect of the heat pump is expressed by the ratio of the heat rate $\dot{Q}_c$ from the condenser to the power taken, $\dot{W}$, i.e.

$$\varepsilon_{\text{(heat pump)}} = \dot{Q}_c / \dot{W} \tag{1.14.1}$$

which is called the coefficient of performance of the heat pump. The term coefficient of performance is in common use for machines whose purpose is refrigeration. For machines whose purpose is heating it must be made clear that the term is defined by eq. (1.14.1). In this book the term coefficient of performance and the symbol ε without qualification will be used only for the ratio defined by eq. (1.11.1).

It is important to specify whether $\dot{W}$ in eq. (1.14.1) is mechanical or electrical work, particularly for small machines driven by electric motors, which are relatively inefficient compared with large ones.

An ideal heat pump is defined in the same way as an ideal refrigerator, and from eq. (1.12.3) the coefficient of performance is:

$$\varepsilon_{\text{C(heat pump)}} = \frac{T_2}{T_2 - T_1} = \frac{1}{1 - (T_1/T_2)} \tag{1.14.2}$$

where subscript C denotes Carnot as before.

An absorption refrigerator can also be used as a heat pump. In this case the heat ratio is defined as:

$$\zeta_{\text{(heat pump)}} = \dot{Q}_2 / \dot{Q}_3 \tag{1.14.3}$$

where $\dot{Q}_3$ is the rate of heat transfer to the system at the high temperature T_3. The corresponding expression for the heat ratio of an ideal heat pump is then:

$$\zeta_{\text{C(heat pump)}} = \frac{T_2}{T_2 - T_1} \frac{T_3 - T_1}{T_3} \tag{1.14.4}$$

As for the refrigerator, this expression is seen to be the product of the ideal coefficient of performance of a heat pump working between temperatures T_1 and T_2, and the efficiency of an ideal engine working between absolute temperatures T_3 and T_1.

It will be observed that the second term on the right-hand side of eq. (1.14.4) contains T_1 instead of T_2 which appeared in eq. (1.12.12). The reason for this indicates the basic difference between the refrigerator and the heat pump. The refrigerator's job is to produce temperatures lower than those in the immediate environment, whereas the heat pump's job is to produce temperatures higher than those in the neighbourhood. The temperature of the immediate environment is taken as the temperature at which one can accept or reject heat without having to pay for it, except for fans or pumps to move air or water to which heat can be rejected or accepted as required. In the expressions for the coefficient of performance of refrigerators T_2 represents this temperature of the environment, whilst T_1 is the lower temperature which the refrigerator produces. In the case of the heat pump T_1 represents the temperature of the environment whilst T_2 is the higher temperature at which the heat pump delivers heat.

A combination of a work-operated refrigerator and a heat engine, for example a steam power plant, constitutes a heat-operated heat pump, and it has to be judged with reference to the heat ratio given by eq. (1.14.4). This combination is in fact much more attractive than an absorption heat pump because as a rule it can give a better heat ratio. Furthermore, there is the possibility of utilising heat which the power plant, for unavoidable reasons, rejects above the environmental temperature.

The heat-driven heat pump described above uses heat at a relatively high temperature T_3 to produce a larger heat transfer at the lower (but still higher than that of the environment) temperature T_2. It is worth noting, however, that it is possible to use heat at a relatively low (but higher than that of the environment) temperature T_2 to produce a heat transfer at the high temperature T_3. The block diagram of this device, which might be called a 'temperature elevator', is shown in Fig. 1.14.1. An entropy analysis

Fig. 1.14.1. Block diagram of a 'temperature elevator'. Heat Q_2 is taken in at temperature t_2; heat Q_3, necessarily less than Q_2, is given out at the higher temperature t_3 and heat Q_1 is rejected at the temperature t_1 of the environment.

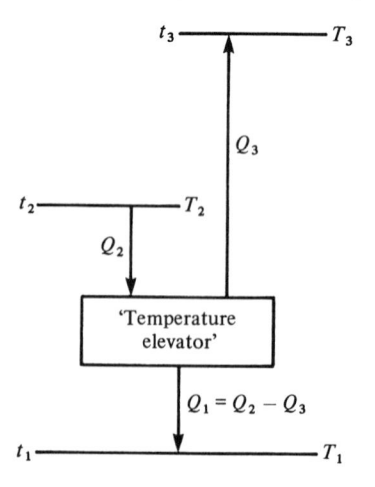

1.15 The cold multiplier

shows that the maximum value of the ratio Q_3/Q_2 is given by:

$$\left(\frac{Q_3}{Q_2}\right)_C = \frac{(1/T_1)-(1/T_2)}{(1/T_1)-(1/T_3)} = \frac{T_3}{T_3-T_1}\frac{T_2-T_1}{T_2} \quad (1.14.5)$$

where subscript C stands for Carnot as before to denote the maximum possible value according to the laws of thermodynamics. The first form of the ratio shows that Q_3/Q_2 is necessarily less than unity because:

$$1/T_2 > 1/T_3$$

Nevertheless, a device of this kind could find uses in some circumstances, for example in using low temperature heat from a flat plate solar collector and upgrading it to provide hot water, as an alternative to using a concentrating solar collector. It could be realised by employing an engine working between the temperatures T_2 and T_1 to drive a vapour-compression heat pump working between T_1 and T_3, but a system incorporating the principle of the absorption refrigerator could also be used, as described next.

When the heat pump has to take heat from a fluid falling in temperature, and to deliver heat to a fluid rising in temperature, the considerations of section 1.13 relating to the Lorenz cycles of comparison must be applied. This means that for fluids of constant specific heat capacity the temperatures used in eq. (1.14.2) and (1.14.4) are to be the logarithmic mean absolute temperatures, as defined by eq. (1.13.12).

1.15 The cold multiplier

An interesting application of the principles of thermodynamics to refrigeration is the 'cold multiplier', first envisaged by Nolcken (1930) and developed by Maiuri (1932). It was shown in section 1.13 that heat transfer across a temperature difference involves a reduction in the possible coefficient of performance, and one attempts to keep the temperature difference between the refrigerant and the things being refrigerated as small as possible. When the refrigerant is necessarily evaporating at a much lower temperature, as for example when solid carbon dioxide at about $-78\,°C$ is used to refrigerate goods at $-20\,°C$, the temperature difference is considerable and unavoidable. The large temperature difference of 58 K represents a potential which is wasted. It would be possible, however, to operate a heat engine taking in heat at ambient temperature, $20\,°C$ say, rejecting heat at $-78\,°C$ and delivering useful work. This work could be used to drive a refrigerator operating between $-20\,°C$ and ambient temperature.

The cold multiplier of Nolcken and Maiuri works like an absorption refrigerator in reverse. The block diagram is the same as that of the temperature elevator, Fig. 1.14.1. T_1 is the low temperature of the carbon dioxide, T_2 is the temperature of the goods being refrigerated and T_3 is the temperature at which heat can be rejected to cooling water or air. The ratio which is now of interest is Q_2/Q_1 and a heat and entropy analysis shows that the maximum value of this is:

$$\left(\frac{Q_2}{Q_1}\right)_C = \frac{(1/T_1)-(1/T_3)}{(1/T_2)-(1/T_3)} \tag{1.15.1}$$

Since $(1/T_1) > (1/T_2)$ the maximum value of the multiplying factor is always greater than unity.

Example 1.15.1
Using the figures suggested above:

$t_1 \approx -78\,°C \qquad T_1 = 195\,K$
$t_2 \approx -20\,°C \qquad T_2 = 253\,K$

and

$t_3 \approx 20\,°C \qquad T_3 = 293\,K,$

determine the maximum value of the multipying factor.
By eq. (1.15.1):

$$\left(\frac{Q_2}{Q_1}\right)_C = \frac{(1/195)-(1/293)}{(1/253)-(1/293)} = 3.18$$

In the system as constructed by Maiuri a factor of about half the ideal factor was obtained.

The possible applications of devices like those of Fig. 1.12.3 and Fig. 1.14.1 can be derived systematically by putting the ambient or environmental temperature, T_0 say, equal to T_1, T_2 and T_3 in turn. The maximum value of the relevant heat ratio is given by:

$$\left(\frac{Q_n}{Q_m}\right)_C = \left|\frac{(1/T_m)-(1/T_0)}{(1/T_n)-(1/T_0)}\right| \tag{1.15.2}$$

where m and n can take the values 1, 2 or 3. The following scheme can then be drawn up, showing the maximum values of the heat ratios in relation to unity.
Devices like Fig. 1.12.3.

(a) $T_1 = T_0$, $(Q_2/Q_3)_C > 1$. Example: absorption heat pump.
(b) $T_2 = T_0$, $(Q_1/Q_3)_C >$ or < 1. Example: absorption refrigerator.
(c) $T_3 = T_0$, $(Q_1/Q_2)_C < 1$.

1.15 The cold multiplier

Devices like Fig. 1.14.1.

- (d) $T_1 = T_0$, $(Q_3/Q_2)_C < 1$. Example: temperature elevator.
- (e) $T_2 = T_0$, $(Q_3/Q_1)_C >$ or < 1.
- (f) $T_3 = T_0$, $(Q_2/Q_1)_C > 1$. Example: cold multiplier.

Devices like (c) and (e) have not yet, so far as is known, been used. (c) would be a machine capable of using the moderately low temperature of ice to produce a much lower temperature: whilst (e) could use the low temperature of ice to produce a temperature higher than that of the environment. As already remarked in section 1.14, (d) offers the possibility of upgrading solar heat.

It was assumed above that any work input to the system is negligible compared with the heat transfers, and this is generally true in absorption machines. It will be seen in section 6.19 that by using a non-condensable gas in the system it can be made to operate without any liquid pumps and thus have no work input.

Although all of the above devices are possible in principle, the design of them for specific applications depends on finding suitable combinations of substances to use in an absorption system.

2

Machinery and plant for vapour compression systems

2.1 Introduction

The aim of this chapter is to provide brief descriptions of the commonest types of plant used in vapour compression systems. It is not intended as a detailed guide to the construction and maintenance of refrigerating plants, but rather as a background to the more theoretical treatment which follows in later chapters. Most of the material presented is the common stock of knowledge of the industry and consequently few references to published work are given.

Except in special circumstances components for refrigeration plants are not specially built for a particular job. They are usually standard components from a range of products, with a technical performance which is specified in catalogues. These components are manufactured by specialist firms by mass production methods where possible. This aspect of refrigeration engineering is basically production engineering with the addition of some special techniques such as leak testing.

In very large plants components may be produced to order in individual designs. This is particularly so of evaporators, which can take many different forms according to the product to be cooled. But in most plants even the evaporators tend to be standard products.

The selection of the components and their integration into a system may be done by a manufacturer of appliances, for example of domestic refrigerators or room coolers, and in these cases the manufacturer may produce some of the components himself. Apart from factory-made appliances, however, most plants are built by a contractor, sometimes to his own design, sometimes to an overall design of a consultant engaged by the client. The contracting firm orders the components, has them delivered to site and erects the plant. Finally, he commissions it, i.e. charges it with

2.2 Compressors

refrigerant, adjusts the controls, etc. and sees that everything is working properly. During this period he frequently trains the client's staff in the operation of the plant.

These two functions, manufacture of the parts and their assembly into a system, are sometimes carried out by the same firm, and indeed this was the general practice in the earlier days of the industry. Increasingly the different knowledge and skills required for manufacture and installation, including different management techniques needed for factory and site work, have tended to bring about the separation of the functions of product manufacturer and contractor.

Some of the machinery and plant described in this chapter is no longer of current manufacture, but it is thought worth while to mention it, partly because there are many old systems still in operation, and partly because it is often useful to know how things were done in the past in order to appreciate why they are done in a particular way today.

2.2 Compressors

The function of the compressor in a refrigerating system is to draw vapour from the evaporator, so causing a low pressure therein, at which the refrigerant can boil to give the desired temperature, and to raise the pressure of the vapour and deliver it to a condenser where the vapour can be condensed by the cooling water or cooling air.

Compressors can be divided into two main types:

(i) positive displacement machines and
(ii) roto-dynamic machines.

In a positive displacement compressor a volume of gas is taken into the machine and trapped inside a space which is then deformed in such a way that its volume diminishes. During this process the pressure of the gas rises, and when the pressure has risen to the required value the gas is pushed out against the pressure in the delivery pipe. A positive displacement machine is not therefore a steady-flow machine: it aspirates and delivers the gas in pulses.

A roto-dynamic machine operates by subjecting a steadily flowing stream of gas to forces which result in a continuous rise of pressure. This can be done in radial flow or axial flow. Almost all of the compressors used in refrigeration are of the radial flow type, known as centrifugal compressors, but which are usually called 'turbo-compressors' in refrigeration practice. Axial flow compressors are only used in special circumstances, principally in liquefaction plants for natural gas.

The principal types of positive displacement compressors are:

(i) reciprocating;
(ii) rotary with sliding vanes; and
(iii) rotary screw compressors.

These will be described in the following sections. Before doing so, however, the difference between a gas compressor and a liquid pump will be made clear. A true compressor subjects the gas to a change in volume whilst it is inside the machine, whereas a liquid pump cannot do this because a liquid is almost incompressible. The rise of pressure in a compressor takes place continuously while the volume is being reduced. In a liquid pump the rise in pressure takes place almost instantaneously when the liquid inside the machine is exposed to the pressure of the liquid already in the delivery pipe or, in some types of pump, as soon as the volume begins to diminish and a delivery valve opens. There is no appreciable back-flow of liquid into the machine from the delivery pipe whilst the pressure rises, because the liquid volume cannot be changed. If this principle were used for a gas, a back-flow of gas from the delivery pipe would occur whilst the pressure of the gas inside the machine was rising. This is evidently wasteful because the gas which flows back has to be compressed twice. The essential feature of a true compressor is that the pressure of the gas inside the machine is raised to the delivery pressure before this gas is exposed to that in the delivery pipe.

A liquid pump can be used as a gas compressor, but it is not very efficient. At low pressure ratios the loss may be small, however, and machines like Root's blower, which have no internal deformation of the volume, are used for such applications. On the other hand, a true compressor cannot be used as a liquid pump because it is impossible to reduce the trapped volume of liquid without huge forces being applied.

A positive displacement compressor generates a certain volume for each revolution of the driven shaft which is called the swept volume. The term comes from piston compressors in which this volume is that swept out by the pistons during the suction stroke, but it is applied to all types of machines. It will be convenient to call the product of the swept volume and the rotational speed of the driven shaft the swept volume rate.

The ratio of the actual volume of gas taken in, expressed at the inlet pressure and temperature, to the swept volume is called the volumetric efficiency of a positive displacement compressor.

The term swept volume has no meaning for a roto-dynamic compressor, nor has the term volumetric efficiency in the above sense. It is sometimes used however with a different definition to express the internal leakage of such machines.

2.3 Reciprocating compressors

The operating principle of a reciprocating compressor is shown in Fig. 2.3.1. By the outward movement of a piston in a cylinder, vapour is drawn in through a suction valve which is opened automatically by the pressure difference. The vapour flows in during the suction stroke as the piston moves towards outer dead centre, more or less filling the cylinder volume at point A with vapour at suction pressure, i.e. approximately evaporator pressure. The piston moves inwards again and the suction valve is closed by a spring, trapping the vapour. The pressure then rises as the enclosed volume diminishes. This continues until the pressure inside the cylinder reaches the pressure in the delivery pipe, i.e. approximately condenser pressure, point B, when the pressure inside begins to force the discharge valve open against a spring. Delivery of vapour to the delivery pipe continues as the piston moves on inwards to inner dead centre. At this point C delivery should be complete and the discharge valve closed by its spring as the piston moves outwards again. There is still some vapour at delivery pressure remaining inside the cylinder at inner dead centre because it is impossible to make the piston sweep through the whole volume of the

Fig. 2.3.1. Principle of operation of a reciprocating compressor. The piston is shown moving outwards, drawing in vapour through the suction valve, opened by the pressure difference.

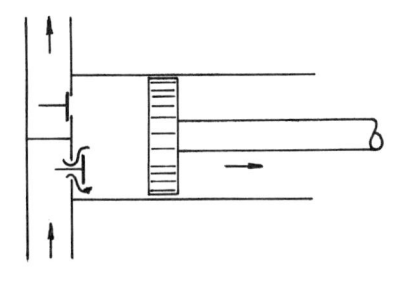

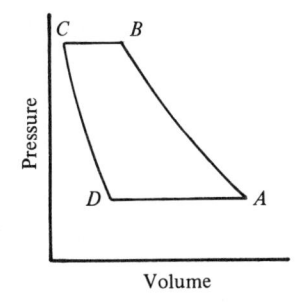

cylinder. Space is needed to accommodate the valves, and to allow for manufacturing tolerances. The clearance space, usually expressed as a fraction of the swept volume, is of the order 3–4% in modern high-speed compressors. Because of this vapour in the clearance space, the pressure inside the cylinder does not immediately fall to suction pressure as the piston begins to move outwards, but follows an expansion process in which the suction pressure is reached at some point in the stroke, D. Until the pressure falls to suction pressure the suction valve cannot open. The effect is that instead of taking in a volume of vapour equal to the swept volume $(V_A - V_C)$, a smaller volume $(V_A - V_D)$ is drawn from the suction pipe. To make this volume as large as possible, the clearance fraction must be as low as possible, and this limits the choice of the type of valve which can be used.

Furthermore, the small clearance between the piston and the cylinder head means that high-speed refrigeration compressors are susceptible to damage in the event of slugs of liquid or oil arriving accidentally in the cylinder. On all but small compressors accommodation for this has to be provided by a lifting safety head on the cylinder.

A common early form of ammonia compressor was the horizontal double-acting machine, shown in Fig. 2.3.2. It followed contemporary steam engine design in many respects, being double acting (i.e. compressing on both sides of the piston) with the refrigerant vapour confined to the cylinder by a stuffing box around the piston rod. The crosshead, connecting rod, crank and flywheel (which was also the driving wheel, not shown) were in the open. To minimise leakage through the stuffing box, a gap formed in the packing by a spacer ring, known as the lantern ring, was filled with oil under pressure. Thus only oil could leak outwards between the packing and the rod. The slight leakage of oil inwards served to lubricate the piston and cylinder walls. The valves in the early machines were often of the poppet type, shown in Fig. 2.3.2, though later machines used ring-plate valves. Cylinder diameters were large by present standards, with bores up to 500 mm, and strokes up to 1200 mm. The rotational speed was consequently limited to about 50 rev/min, for the largest machinery, corresponding to a mean piston speed of 2 m/s. The clearance in these machines was quite small, often less than 0.5% of the swept volume. No safety head was provided because with the generous valve areas and low speed they were able to pump a mixture of vapour and liquid without trouble.

The horizontal compressor, originally driven by a steam engine and later by electric motor, was manufactured until a few years ago, and many still

2.3 Reciprocating compressors

Fig. 2.3.2. Section and plan of a horizontal double-acting compressor for ammonia, *c.* 1920. 1 – suction manifold; 2 – discharge manifold; 3 – poppet type suction valves with access covers; 4 – delivery valves and covers; 5 – lantern ring; 6 – oil separator; 7 – non-return valve in delivery pipe; 8 – gauge connection; 9 – pipe to oil reservoir; 10 – crossover connection between suction and delivery pipes.

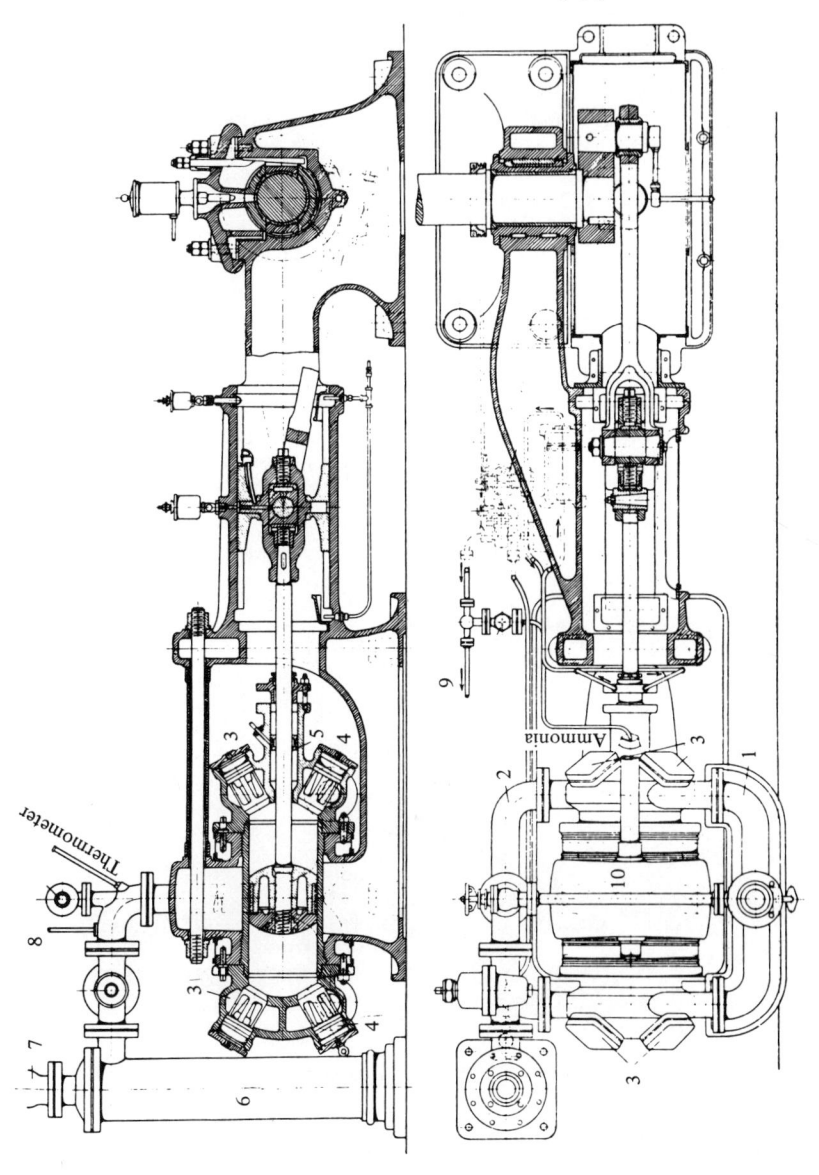

remain in service. Because it occupied a lot of floor space, and also because a convenient arrangement was to mount a steam cylinder horizontally, directly coupled to a vertical compressor cylinder, the vertical compressor was introduced. This was at first a double-acting machine, but was eventually made single-acting, i.e. with compression taking place on one side only of the piston. The crosshead was eliminated and the piston rod became a connecting rod driving directly on to the gudgeon pin from the crank pin.

The piston was lengthened to take the side thrust caused by the obliquity

Fig. 2.3.3. VSA compressor for ammonia with forced lubrication and mechanical shaft seal, c. 1950. Bore 165 mm, stroke 152.5 mm. Refrigerating capacity, at the maximum speed of 550 rev/min, 94 kW at $-15\,°C$ evaporating temperature and $30\,°C$ condensing temperature, with a shaft power of 30 kW. The suction and delivery stop valves and the crossover valves are mounted in the control valve chest.
Suction vapour passes through ports in the cylinder wall to the undersides of the pistons and through valves in the pistons into the cylinders. The hole in the cylinder wall shown above the left-hand piston is a partial duty port whose function is described in section 4.3. (By courtesy of UDEC Refrigeration Ltd.)

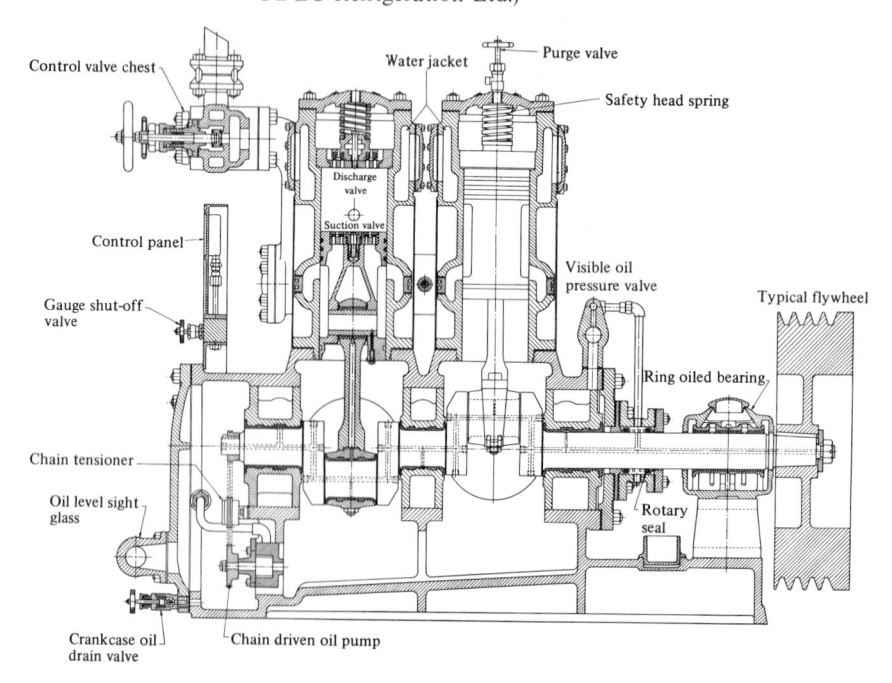

of the connecting rod, giving the so-called trunk piston. With this design it is no longer possible to seal the refrigerant vapour from the atmosphere at the connecting rod, which oscillates from side to side, and it is necessary to enclose the whole of the motion work in a pressure-tight crankcase, with a stuffing box at the place where the shaft emerges. The crankcase pressure is equalised to the pressure in the suction pipe through a connecting port. In all but the smallest machines forced lubrication is adopted. A feature of many designs is the incorporation of the suction valve in the piston crown, allowing a greater area for flow, with the discharge valve in the head. Compressors of this type, with the direction of vapour flow from piston to head, are described as 'uniflow', as contrasted with 'return flow' machines in which both suction and discharge valves are in the cylinder head. The cylinder head itself is held down by a heavy spring which allows the head to lift on occasion and discharge any liquid slugs to the delivery pipe.

In the early horizontal machines wet vapour was compressed, and the temperature at the end of compression was little if any higher than condensing temperature. By the time the vertical single-acting machine came into general use, dry compression of the vapour was the rule, with the vapour state going into the superheat region during compression, and quite high delivery temperatures. It was necessary to provide water jackets around the cylinders of ammonia compressors. A section of a typical vertical single-acting (VSA for short) compressor of uniflow design is shown in Fig. 2.3.3.

VSA compressors came gradually into general use during the first quarter of the century and they have been made ever since, with some modifications. The principal changes have been in the design of the valves, the replacement of the stuffing box by a mechanical seal, and an increase in the rotational speed with consequent reduction of cylinder size and piston stroke.

When the fluorinated hydrocarbon refrigerants were introduced in the early 1930s, the traditional design of valve used for ammonia was not adequate for the new gases. Because of the much higher densities of the vapour, the pressure drops through the valves were excessive and caused an increase in power. Newer designs of valves were adopted, generally of the ring-plate type, (Fig. 2.3.4(a)). The valve lifts were larger than for ammonia, the spring force was lightened and the mass of the moving parts was reduced. In small machines valves are usually flexing reeds, (Fig. 2.3.4(b)), either cantilevered or retained at both ends.

Prevention of leakage of the new gases was found to be more difficult than for ammonia. Because of their higher molecular weights the mass flow rate through a leak of a given size at a given pressure difference is more than for ammonia, and the powerful solvent action of the liquids tends to find out any cracks blocked with oil or dirt. Furthermore, leakage is more expensive because of the higher cost of fluorinated refrigerants. The older type of stuffing box or packed gland was found to be unsatisfactory for the new refrigerants, and mechanical seals were used instead, based on a superior principle. Wear on two concentric cylindrical surfaces mating together can only be taken up by changing the radius of one of the surfaces. In a packed gland this is done by compressing the relatively soft packing onto the shaft, but it requires frequent attention. Wear between two flat surfaces, however,

Fig. 2.3.4. (a) Section through a ring-plate discharge valve. The two valve rings are shown, in the open position, in full black. When closed they are pressed lightly onto the seats by the coil springs, three or more per ring arranged around the ring. The ports in the lower and upper plates may be circular or, more usually, kidney shaped holes. For a discharge valve the ports in the lower plate contribute to the clearance volume. For a suction valve, the ports in the other plate contribute similarly.

Larger valves have three or more rings. There are several modern variations of the design. In a common form, the rings are joined together to form a flexible plate clamped at the centre.

(b) Principle of a cantilevered reed-type suction valve. Discharge valves are made as a flexing beam with a slot at one support to allow bending, and a cover plate held down by springs to restrict the lift, as seen in Fig. 2.3.6.

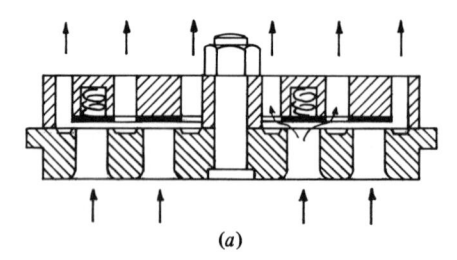

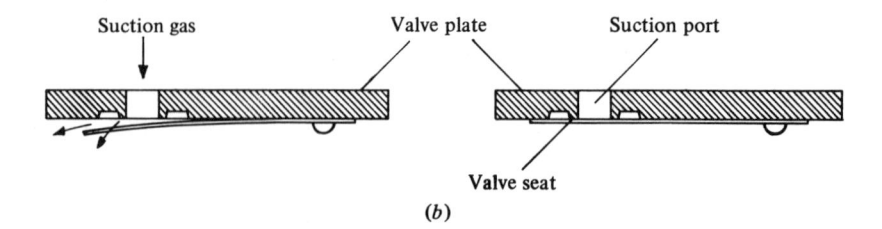

2.3 Reciprocating compressors

Fig. 2.3.5. (a) A bellows-type of shaft seal. 1 – stationary ring sealed into the housing by the flexible bellows 3, with pressure provided by the stationary spring 4. 2 – rotating seal ring, sealed onto the shaft by the synthetic rubber angle ring 5. The stationary ring is often made of bronze, whilst the rotating ring is made of a material with a hard surface such as nitrided steel.

(b) A Flexibox five-component Type-R seal. 1 – stationary ring sealed into the housing by the O–ring 3, with pressure provided by the rotating spring 4. 2 – rotating seal ring sealed to the shaft by the O–ring 5. (By courtesy of Flexibox Ltd. who also make balanced seals.)

(c) Diagram showing the nomenclature for eq. (2.3.1). The means of sealing the stationary ring into the housing are not shown.

(d) A seal incorporating a step in the shaft in which the contact pressure can be made independent of the crankcase pressure.

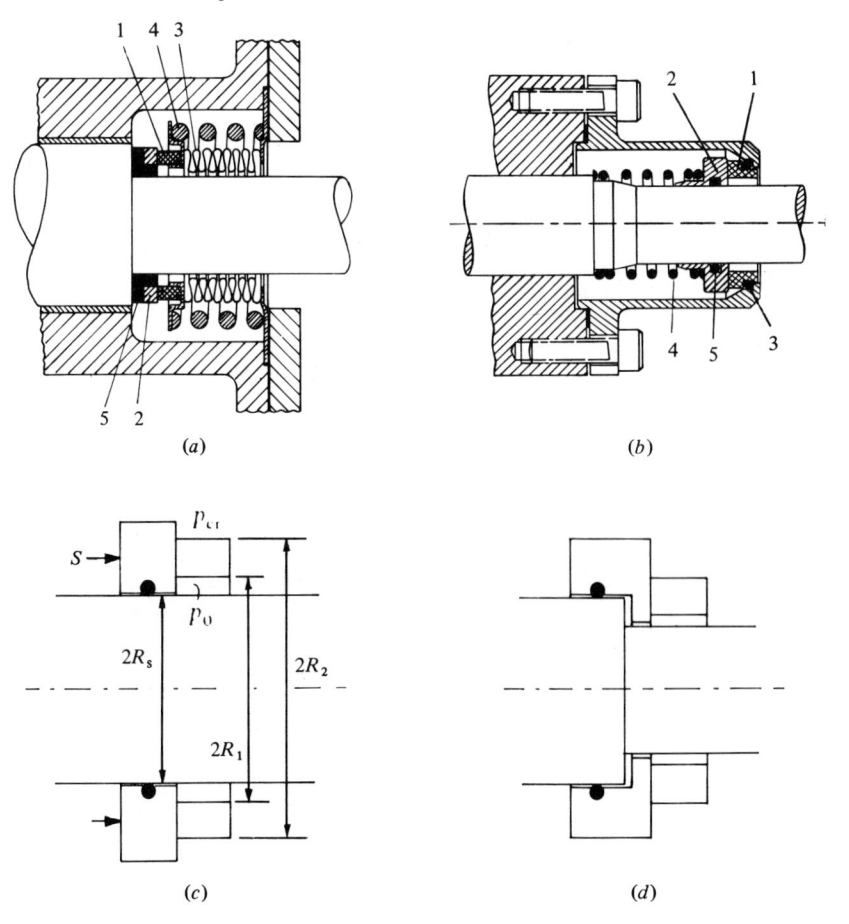

can be compensated simply by moving them together, which can be done automatically by a spring. This is the principle of the mechanical seal. By providing an ample supply of oil to the sealing faces, boundary lubrication if not full film lubrication can be obtained and the wear kept within reasonable limits. A hard material such as steel or cast iron can be used for one ring, and a fairly soft material such as phosphor bronze or carbon can be used for the other. The soft one can be made replaceable.

Two designs of seals for small machines are shown in Fig. 2.3.5.

Fig. 2.3.6. A small open-type compressor suitable for drive by V-belts on the flywheel, not shown. Bore 41 mm, stroke 38 mm, speed 850 rev/min. A – delivery valve, reed-type with plate for limiting lift, and liquid release springs; B – valve plate; C – magnet for collecting small abrasive iron particles; D – suction strainer; E – delivery passage to service valve, not shown; F – oil ring; K – suction port with reed-type suction valve under the valve plate; L – oil feed to gudgeon pin; M – oil passage through crankshaft; N – thrust washer; O – spiral oil pump; P – stationary seal ring attached to the metallic bellows. (By courtesy of Hall-Thermotank Products Ltd.)

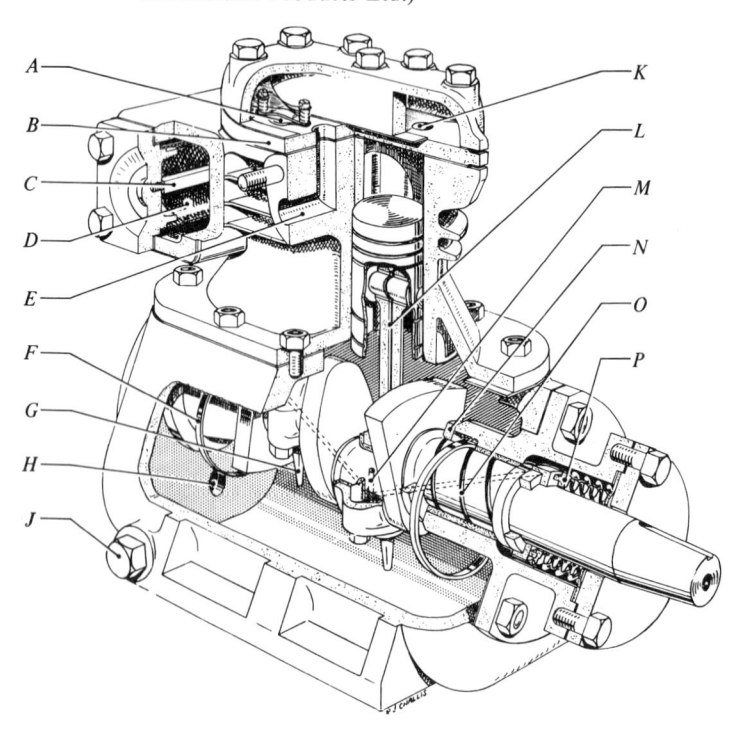

2.3 Reciprocating compressors

Figure 2.3.5(a) has a flexible metal bellows carrying the stationary ring, which seals against another ring carried by the shaft. A spring pushes the two together. In Fig. 2.3.5(b) the rotating ring is sealed by an O-ring on the shaft and the stationary ring by an O-ring in the housing. Figure 2.3.6 shows a small compressor employing a bellows seal.

When the pressure in the crankcase is higher than atmospheric pressure there is a thrust on the crankshaft tending to push it out. In some small machines this thrust is taken by the spring of the mechanical seal, leaving a net thrust pushing the crankshaft in, which is taken by a thrust bearing at the other end of the shaft, often simply a ball, which locates the shaft axially. In the design of Fig. 2.3.6, and in all large compressors, the crankshaft is located axially by thrust bearings at each end which take the thrust in one direction when the pressure inside the crankcase is above atmospheric pressure, and in the other direction when it is below.

Because the pressure in the crankcase of a refrigerant compressor is approximately the same as the evaporating pressure as a rule, it can have widely different values depending on the evaporating temperature. The mechanical seal has to be designed to accommodate these variations so that there is not too much pressure on the seal faces at one extreme, nor too little at the other extreme which would allow the faces to separate. Referring to a seal like that of Fig. 2.3.5(c), the equation expressing the equilibrium of the forces is:

$$S + p_{cr}(A_2 - A_s) = p_0(A_1 - A_s) + \tfrac{1}{2}(p_{cr} + p_0)(A_2 - A_1) + F \qquad (2.3.1)$$

where S is the force exerted by the spring, p_{cr} and p_0 are the crankcase and atmospheric pressures respectively, $A_2 = \pi R_2^2, A_1 = \pi R_1^2$ and $A_s = \pi R_s^2$ where R_2 and R_1 are the radii of the outside and inside of the seal contact area, and R_s is the radius of the shaft on which the O-ring seals. F is the force acting between the seal faces. At standstill this force comes entirely from the physical contact between them, but as the speed rises hydrodynamic forces are generated which tend to separate the faces and which gradually replace the contact force. It is assumed in eq. (2.3.1) that the pressure distribution across the seal face is linear with radius. If a logarithmic distribution is assumed, the logarithmic mean of p_{cr} and p_0 is obtained on the right-hand side instead of the arithmetic mean; in practice it makes little difference.

If the force F is to be entirely unaffected by changes in the crankcase pressure, terms containing p_{cr} on the right and left-hand sides of eq. (2.3.1) have to be equal, giving the condition:

$$A_s = \tfrac{1}{2}(A_1 + A_2) \qquad (2.3.2)$$

Clearly the construction shown in Fig. 2.3.5(c) does not meet this condition since A_1 there is larger than A_s. A step in the shaft between the O-ring and the seal face, as shown in Fig. 2.3.5(d), is needed to make the contact pressure independent of the crankcase pressure.

Equation (2.3.2) applies also to a bellows seal if instead of A_s the effective area of the bellows is used. A bellows seal can be made independent of p_{cr} on a plain shaft by correct sizing of the bellows in relation to the face of the stationary seal ring.

In practice it may not be necessary to balance a seal completely. For many small compressors the unbalanced form is satisfactory, but larger compressors for use over a wide range of crankcase pressures usually have some degree of balance incorporated.

Seals for large compressors employ a more elaborate form of construction than those shown in Fig. 2.3.5(a) and (b). A seal for a large compressor is shown in Fig. 2.3.7.

Fig. 2.3.7. Mechanical seal for the compressor shown in Fig. 2.3.8. 1 – oil return orifice plug; 2 – seal setscrew; 3 – oil passage through shaft; 4 – joint; 5 – seal retainer; 6 – multiple springs; 7 – wedge ring; 8 – rotating seal; 9 – gland seal housing; 10 – sealing faces; 11 – stationary seal; 12 – joints; 13 – gland housing cover; 14 – joint. X is the clearance dimension. (By courtesy of Hall-Thermotank Products Ltd.)

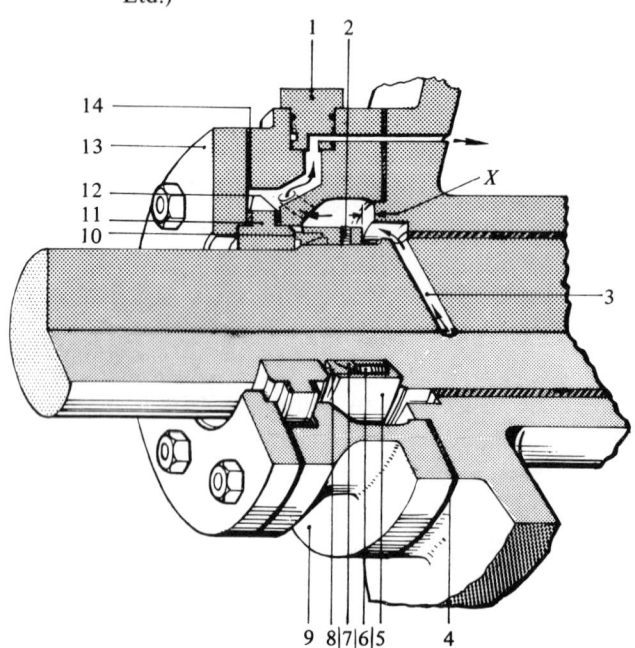

2.3 Reciprocating compressors

The VSA machine was made originally with large cylinder diameters, similar to those of the horizontal compressor, but the piston stroke was shortened because of the height of the machine. As rotational speeds were increased, the piston stroke had to be reduced still further in order to avoid excessive pressure drops through the valves. To maintain the same total

Fig. 2.3.8. A modern V-block compressor, with eight cylinders of bore 127 mm and stroke 102 mm, suitable for use on ammonia and the fluorinated refrigerants. Speed 1000 rev/min. 1 – delivery manifold; 2 – pistons of aluminium alloy; 3 – cast iron crankcase and cylinder housing; 4 – crankshaft of spheroidic graphite iron, dynamically balanced, carried by white metal lined main bearings, and axially located by steel backed white metal thrust washers; 5 – crankcase oil cooler, direct expansion; 6 – cast-iron cylinder liners; 7 – connecting rods, steel, with white metal big-end bearings and lead bronze small-end bushes; 8 – suction strainer; 9 – unloading equipment, for lifting the suction valves; 10 – oil pump and Auto Kleen strainer; 11 – safety spring; 12 – annular ring type suction valve. The shaft seal, not visible, is similar to that shown in Fig. 2.3.7 (By courtesy of Hall-Thermotank Products Ltd.)

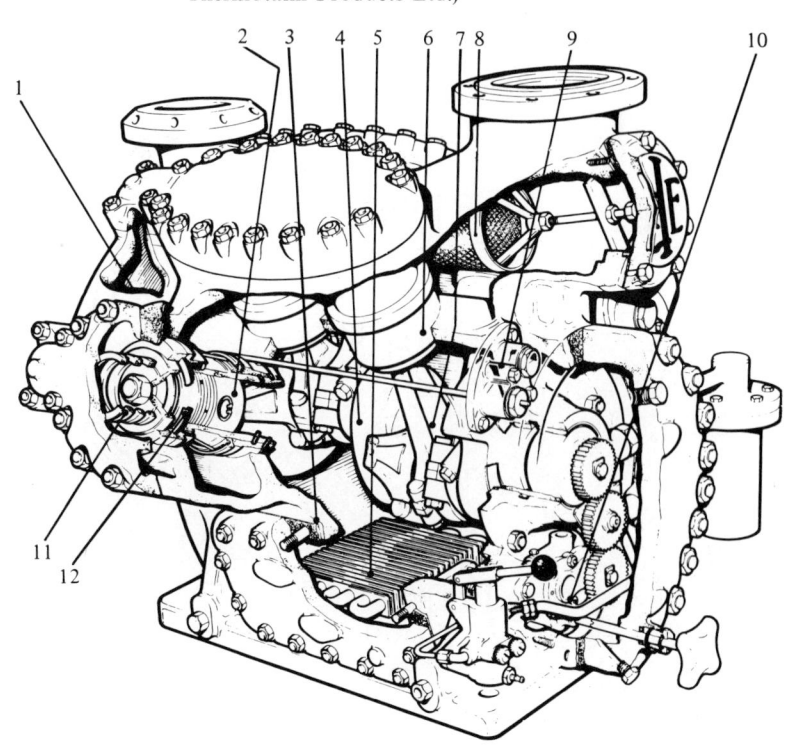

swept volume the number of cylinders was raised, leading eventually to the arrangement of these in V formation, and then in three and four bank arrangements. The present generation of reciprocating compressors are mostly of this design, a typical example being shown in Fig. 2.3.8. A common construction of suction valve is in the form of a ring of ports outside the circumference of the cylinder, with an annular valve ring held down by light springs (Fig. 2.3.9). This leaves the whole of the cylinder area to

Fig. 2.3.9. Detail around a cylinder of the compressor shown in Fig. 2.3.8. 1 – safety head spring; 2 – delivery valve securing nut; 3 – centre piece; 4 – delivery valve guard; 5 – suction valve guard; 6 – spacer ring; 7 – cylinder head joint; 8 – suction strainer; 9 – duty reduction gear operating rod; 10 – push rods; 11 – stud; 12 – piston; 13 – moving ring; 14 – compression rings; 15 – scraper ring; 16 – liner; 17 – sleeve; 18 – suction valve; 19 – suction valve spring; 20 – delivery valve; 21 – delivery valve spring; 22 – locking plate; 23 – locking screw; 24 – top cover.

The suction valve 18 is held down by a number of coil springs 19. During normal operation it is lifted by gas pressure. At this condition the push rods 10 are positioned obliquely between the suction valve and the ring 13. Movement of the rod 9 to the right brings the push rods vertical and lifts the suction valve off its seat. (By courtesy of Hall-Thermotank Products Ltd.)

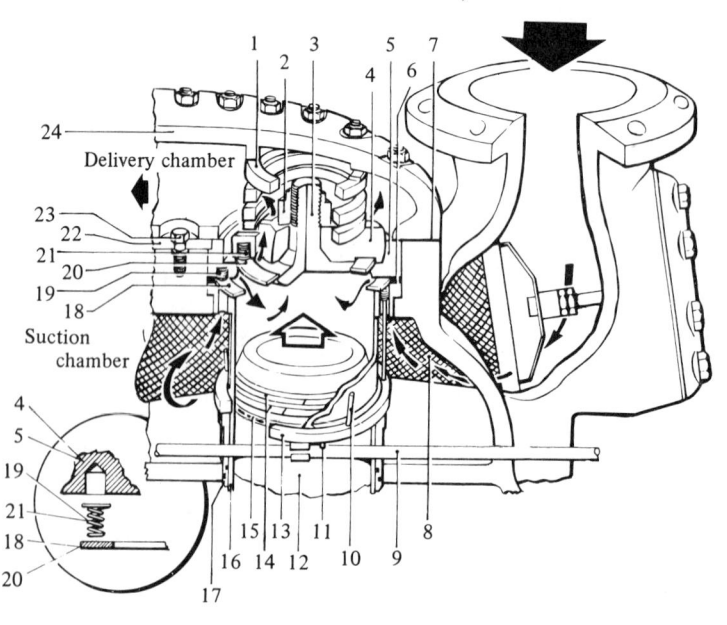

2.3 Reciprocating compressors

accommodate the discharge valve in the head. By means of push rods the suction valve rings can be held off their seats so that the compressor can be started without pressure loads on the pistons and bearings. The valve rings are allowed to relax on to the seats when the normal lubricating oil pressure has been developed by the pump. The same equipment of push rods and actuating pistons is also used to regulate the pumping capacity of the compressor by lifting individual valve rings on one or more cylinders as the demand for refrigeration falls off.

For compressors pumping the fluorinated hydrocarbons, a special problem is posed by the solubility of their vapours in lubricating oil. When a compressor is not operating, the pressure in the crankcase is relatively high and the oil is cold. Under these conditions the vapour readily dissolves in the oil. When the machine is running, the crankcase pressure falls and the oil temperature rises. The vapour comes out of solution causing foaming of the oil and possible carry-over to the cylinders via the equalisation port. This phenomenon is often audible as a knock on start-up. One remedy is to keep the oil refrigerant solution in the crankcase warm during standstill by a

Fig. 2.3.10. An internally sprung single cylinder hermetic compressor, suitable for domestic refrigerators up to about 120 dm^2 internal volume. Bore, 17 mm; stroke, 9.4 mm; swept volume 2.13 cm^3; speed 2900 rev/min on 50 Hz supply. Refrigerating capacity on R12, 60 W; electric power consumption 58 W at an evaporating temperature of $-15\,°C$, condensing temperature 55 °C, with suction vapour entering the case at 32 °C and saturated liquid before expansion. (By courtesy of Danfoss.)

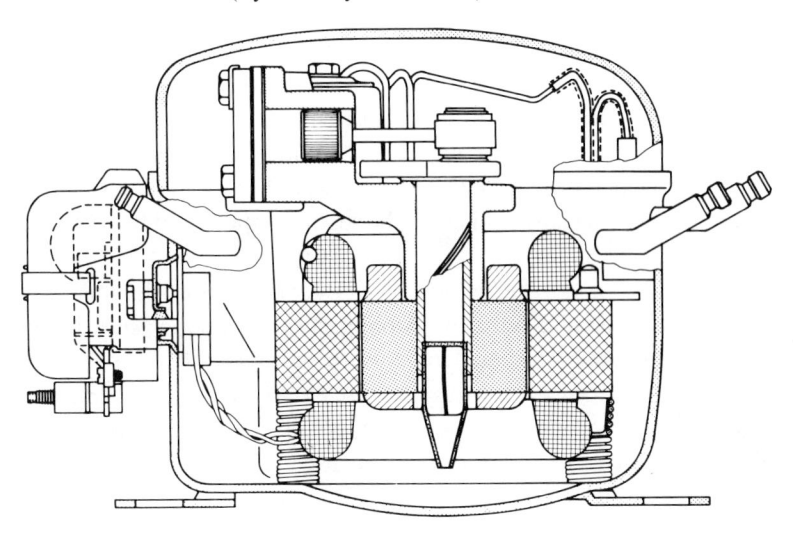

small electric heater. After the compressor has been running some time the oil becomes warm, and for ammonia and R22 it may then be necessary to provide an oil cooler in the crankcase. Another point which must be borne in mind is the effect of dissolved refrigerant on the oil, i.e. reducing the viscosity.

The problem of leakage at the shaft was largely overcome by the mechanical seal which eliminated the need for day-to-day attention. But attention at longer intervals was still necessary, and the belt driving the compressor from the motor needed tightening from time to time. For domestic refrigeration these were definite disadvantages. Some makers eliminated the belt drive by direct coupling of the motor to the compressor but retaining the shaft seal. With the introduction of the hermetic compressor the shaft seal was eliminated. The motor and compressor are directly coupled on the same shaft and the combination is installed inside a welded steel shell with the motor inside the refrigerant circuit (Fig. 2.3.10). All pipes and connections are made by welding or brazing. The conductors to the motor are taken through the shell by sealed terminals, originally with synthetic rubber bushes but now incorporating fused glass. The refrigerant and oil charges in the system last for the life of the unit. Some early hermetic systems for domestic refrigerators used sulphur dioxide as refrigerant, but R12 is now universal. At the same time the 'capillary tube' was introduced in place of a mechanical expansion valve, so that the whole system could be sealed up without any screwed connections.

One of the problems with hermetic compressors is the cooling of the motor. In earlier designs the stator was pressed into the steel shell and some opportunity was provided to lose heat by conduction to the case and then to the outside air. It was still necessary, however, to cool the motor windings and rotor, and this was done by allowing the suction vapour to pass through them. In some later designs the stator is not even in contact with the shell, the whole motor/compressor assembly being mounted on springs inside the shell in order to reduce the noise. In this event almost all the heat from the motor has to be taken by the suction vapour and by oil circulating inside.

Because the mass rate of flow of refrigerant vapour diminishes as the evaporating pressure falls, the cooling effect of this vapour diminishes. Consequently the winding temperature increases at lower evaporating temperatures. Hermetic compressors used in freezers therefore may have to find additional means of heat rejection. This is often done by cooling the oil. The vapour delivered is partially condensed in a section of the condenser and brought back into the shell to cool the oil. The condensed liquid boils

2.3 Reciprocating compressors

Fig. 2.3.11. A single cylinder hermetic compressor, bore 38.6 mm, stroke 22.2 mm, speed 2900 rev/min on 50 Hz supply suitable for freezers and commercial low-temperature applications. Refrigerating capacity 1070 W, electric power consumption 810 W, at an evaporating temperature of $-25\,°C$, condensing temperature $45\,°C$, on R502. (By courtesy of Prestcold Ltd.)

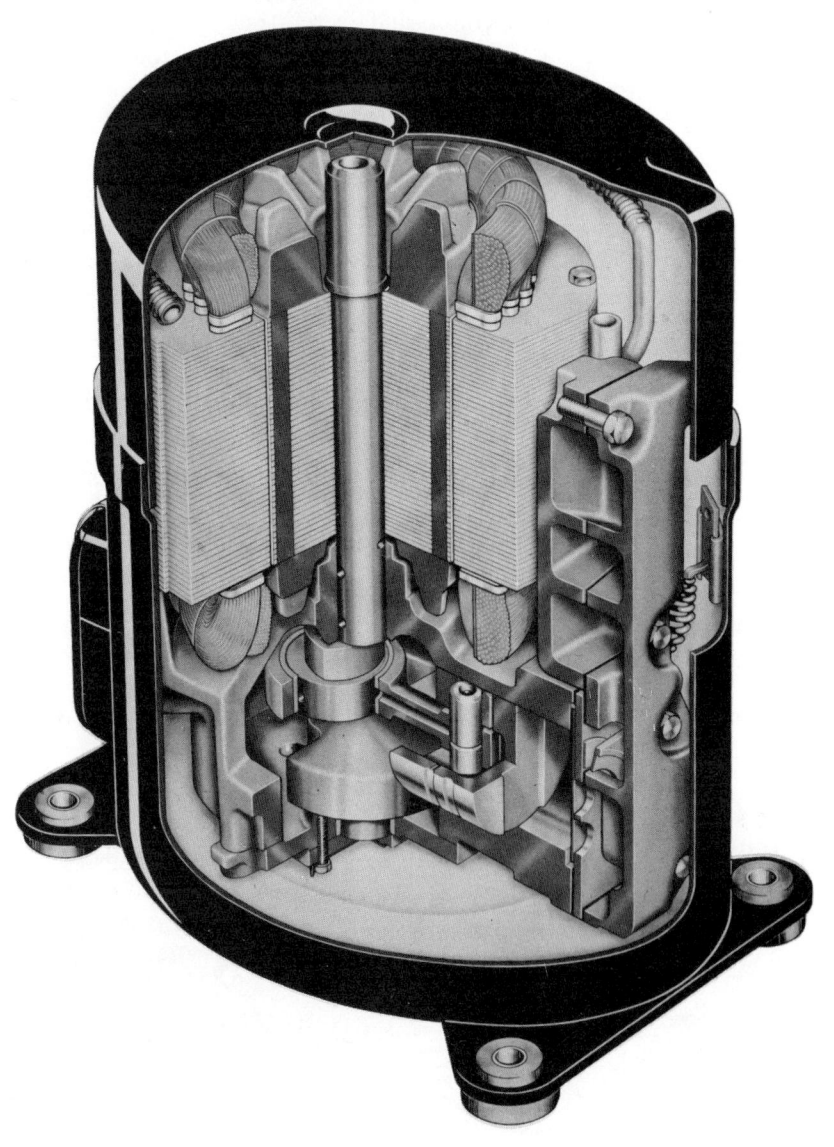

Machinery and plant for vapour compression systems 68

off inside a coil of pipe in the oil and goes back to the condenser proper where it is then totally condensed and passes on to the evaporator.

With all the available means of cooling, the temperatures attained in the windings of hermetic motors are still high, and special materials for the insulation of the wire have to be used.

The idea of hermetic construction has been extended to higher capacities than those for domestic refrigerators and freezers (Fig. 2.3.11). These larger machines are usually sold as 'hermetic' compressors, but they are often provided with suction and delivery stop valves which have screwed connections to take flared copper tube. The systems in which they are used are thus not hermetically sealed.

The disadvantage of the hermetic compressor is that it is not possible to service it without cutting open the shell. Provided that a high standard of cleanliness is maintained during assembly of the system, the compressor can last the life of the system. For systems assembled in the field it may not

Fig. 2.3.12. An accessible ('semi-hermetic') compressor, with cylinder heads and valves removable for inspection and service. The motor is cooled partly by suction vapour entering by the flanged connection at the right-hand end, and partly by direct conduction through the case to the fins. Eccentrics are used instead of cranks, a common practice in small compressors. (By courtesy of Dunham-Bush Ltd.)

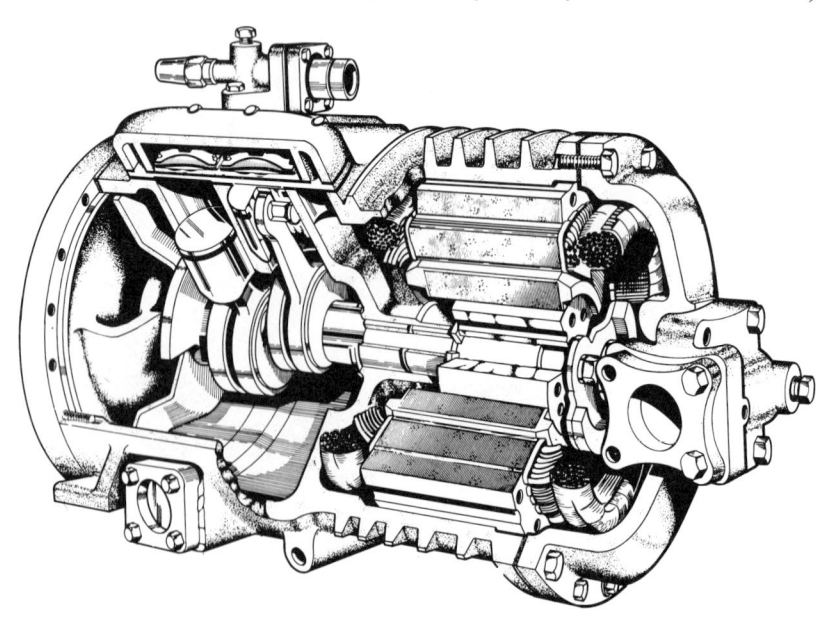

2.3 Reciprocating compressors

be possible to achieve the required standard, however, and it may subsequently be necessary to open the compressor for service. A compromise between the open and the hermetic compressor, known variously as a 'semi-hermetic', 'accessible' or 'serviceable' compressor enables this to be done. An example is shown in Fig. 2.3.12. This machine has a cylinder head, and a valve plate which is accessible from the outside by removing the head bolts. The flanged end of the case can be removed so that the rotor and shaft can be withdrawn. The stator is in direct contact with the case so that some heat can be dissipated to the surrounding air, which can be augmented by a fan or by a cooling coil carrying water wound between the fins. The 'semi-hermetic' compressor is made in larger sizes than the hermetic ones. Some manufacturers offer the same basic compressor in open-type construction, with a shaft seal, or as a semi-hermetic machine.

A disadvantage of hermetic and semi-hermetic compressors for larger systems is that in the event of a motor burning out the decomposition products of the windings and slot insulation are distributed through the system. Thorough cleaning is required before the system can be operated again.

In principle a reciprocating compressor can rotate in either direction, but in practice this may be fixed by the fact that the oil pump will deliver in one

Fig. 2.3.13. A service valve used on small compressors. A – connection to pressure gauge, B – connection to condenser or evaporator, C – port to match the hole for delivery or suction in the compressor body. (*a*) Shows the valve on the back seat as in normal operation, with the compressor open to the system. Connection A would normally be covered by a blank flare nut at this time. (*b*) Shows the valve on the front seat with the compressor isolated from the system for repair. For testing, with a pressure gauge connected to A, the valve is first back-seated and then cracked open a little. When the valve is not in use the spindle should be covered by a cap.

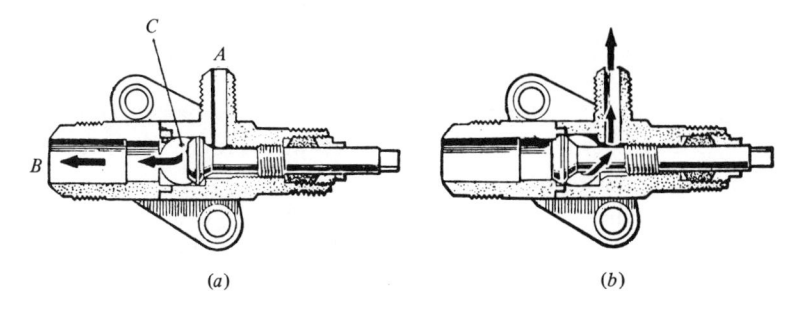

(*a*)   (*b*)

direction only. Large open-type machines therefore usually have a specified direction of rotation. With hermetic and semi-hermetic compressors driven by three-phase motors, the direction is not known. It is necessary therefore to provide an oil pump capable of delivering oil whichever way it rotates. The problem does not arise with small compressors driven by single-phase motors for which the direction of rotation is fixed by the winding connections inside the shell.

Apart from the smallest hermetic compressors, some or all of the following features may be supplied with a machine: suction and delivery stop valves; pressure gauges for suction, delivery and lubricating oil pressures; cross-over connections and valves enabling the compressor to draw from the condenser and deliver to the evaporator; suction gas strainer; oil filter, cleanable without opening the system; electrical cut-outs for abnormally high delivery pressure or oil temperature, and for low suction pressure and low oil pressure; a bursting disc or relief valve to relieve a high pressure on the delivery side to the suction pipe; and arrangements for unloading some or all of the cylinders by holding the suction valves open.

Small commercial compressors are usually provided with a special form of suction and delivery stop valve (Fig. 2.3.13) which enables a pressure gauge to be fitted temporarily for the purpose of diagnosis and service. When the spindle is screwed right out, the sealing member is on the back seat and shuts off the connection to the gauge. This is the normal running position. After the gauge is connected, the spindle is eased forward slightly so opening the gauge to the system. If it is necessary to do repair work to the compressor it can be isolated from the system by screwing the spindle right forward to seal on the front seat. In this position the delivery and/or suction pipes are sealed off.

2.4 Rotary positive displacement compressors

Rotary vane compressors

There are two principal types of rotary compressor with sliding vanes in use, illustrated in Fig. 2.4.1 (*a*) and (*b*). In the single vane type (*a*) the axis of rotation O is eccentric to the geometric axis of the rotor O', but coincides with the axis of the barrel. In the multi-vane type (*b*) the axis of rotation O coincides with the geometric axis of the rotor but is eccentric with respect to the axis of the barrel O'.

In the single vane compressor, sealing between the high- and low-pressure vapour has to be achieved at the line of contact between the vane and the rotor, between the vane and its slot, between the rotor and the barrel along

2.4 Rotary positive displacement compressors

their line of contact, and between the rotor and the end plates. Close tolerances to ensure small clearances at these places are therefore required. Wear by sliding of the rotor against the barrel can be reduced by making the outer part of the rotor a ring which rotates on the inner part of the rotor and which rolls around the inside of the barrel. This is the usual form of construction for single vane compressors.

It will be observed from Fig. 2.4.1(a) that a delivery valve is provided but that a suction valve is not necessary. The suction space is filling while compression and delivery in the left-hand space in the diagram are proceeding. The effect of the small volume of clearance gas left in the delivery port is somewhat different to its effect in a reciprocating compressor, where it retards the opening of the suction valve. In the single vane compressor the suction space is already full of gas at the end of delivery and the clearance space gas simply mixes with the gas about to be compressed. With radial delivery as shown, the clearance gas could push some of the suction gas out through the suction port as soon as the line of contact of the roller and the barrel reaches the delivery port, but with delivery through the end plate this cannot happen. Consequently the volumetric efficiency of a single vane compressor does not fall off rapidly with increasing pressure ratio as does that of a reciprocating compressor.

Fig. 2.4.1. Principle of operation of (a) a single vane rotary compressor and (b) a multi-vane rotary compressor. The single vane compressor requires a delivery valve, whereas in the multi-vane compressor the valving action is brought about by the geometry of the machine.

In (a) the ports are shown for clarity in the cylindrical wall of the barrel, but they are often located in the end plate as shown by the broken line circles.

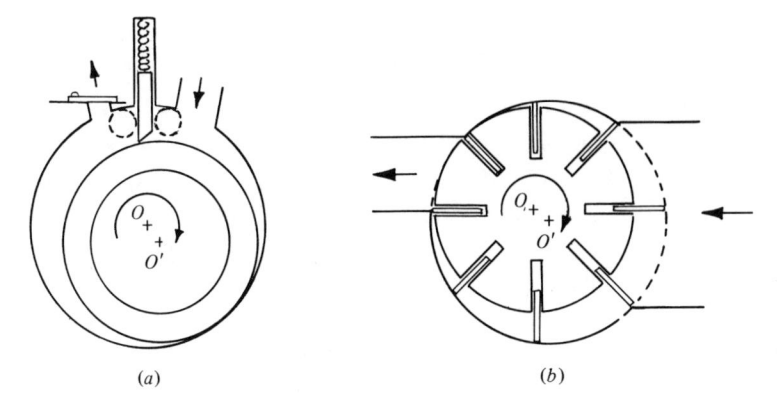

(a)   (b)

For the same reason a single vane rotary compressor performs much better than a reciprocating one as a vacuum pump.

In the multi-vane compressor the sealing between the high and low pressures occurs at the lines of contact between the vanes and the barrel, between the vanes and the slots, and at the ends. Across the lines of contact between the vanes and the barrel and across each slot, the pressure difference is of course only a part of the total pressure difference across the machine, unlike the case of the single vane compressor. There is no need for the rotor itself to touch the barrel, but the radial clearance should be made small in order to reduce the volume of high-pressure vapour carried round to the suction side which would prevent the intake of new gas and reduce the volumetric efficiency. The vanes are commonly made of a non-metallic material and during running are flung outwards, no springs being used. At high speeds the force is so great that to prevent them bearing too heavily on the barrel they are retained by rings.

The multi-vane compressor requires neither suction nor delivery valves, the inlet and outlet of gas being entirely controlled by the geometric arrangement. A result of this is that the compressed gas is opened to the delivery pipe when it has been compressed through a fixed volume ratio which is decided by the geometry of the machine. The pressure ratio corresponding to this 'built-in' volume ratio depends on the index of compression of the gas. The built-in pressure ratio may not be the same as that against which the machine is operating at a particular time, and losses occur.

Because the vanes are not forced against the barrel when the machine is at standstill, gas can flow back through the machine to the suction side from the delivery side. This equalisation of pressure around the rotor is useful in that the motor does not have to provide a large starting torque. On the other hand, it is wasteful to allow an amount of hot compressed gas to flow back to the evaporator in a refrigerating system. A non-return valve in the delivery pipe will prevent this whilst still allowing pressure equalisation around the rotor.

Rotary compressors of both types have been used in small and large versions. They were much used at one time for domestic refrigerators in sealed systems. In larger sizes their main field of application has been for the low stage or booster compressor in multi-stage systems, because of their ability to handle large volume flow rates when compared with a reciprocating compressor of the same bulk, which in turn is owed to the high rotational speeds which are possible. Some manufacturers have produced versions of the multi-vane compressor with oil injection which are suitable

2.4 Rotary positive displacement compressors

for the whole range of refrigeration duties. These machines have largely been superseded by screw compressors.

Twin-screw compressors

The twin-screw compressor was invented by Lysholm in the 1930s (Lysholm, 1943), but it was not applied to refrigeration until some twenty years later.† Since that time the design has undergone considerable development.

The principle of operation is shown in Fig. 2.4.2. Two rotors, a male and a female, mesh together inside a casing between two end plates in which suction and delivery ports are provided. The suction port is shown by a broken line at the top end of the rotors in the diagram. In the position shown the four flutes lettered A, B, C and D on the female rotor, and the three interlobe spaces a, b, c on the male rotor are open to suction and filling with gas. These flutes and spaces are closed at the delivery end by the end plate there. Rotation of the rotors in the direction shown cuts off A from suction and brings it into the position shown as occupied by E, whilst space c comes into that occupied by d. The gas in these spaces is now trapped between the end plates. Further rotation carries these volumes of gas around until the lobes and flutes mesh on the side facing the viewer and compression begins. The gas in flute F, for example, is being compressed since it cannot yet escape at the delivery end and the sealing line between the rotors is advancing downwards. Compression continues until the flutes and interlobe spaces meet a delivery port in the end plate and the compressed gas flows into the delivery pipe. For moderate pressure ratios there may also be a port in the casing allowing some radial discharge, as indicated by the broken line in Fig. 2.4.2.

Because the opening of the delivery port is determined entirely by the geometry of the machine, there is a certain built-in volume ratio, and a

---

† Although Lysholm introduced the modern form of twin-screw compressor, he attributed the invention to Krigar, patented in 1878. In fact a similar machine was described by F. Reuleaux in 1875 and attributed to Révillion about 1830 (*The Kinematics of Machinery*, translated by A. B. W. Kennedy, London, 1876. Dover reprint, New York, 1963).

Reuleaux's Chapters 9 and 10 on 'Chamber crank chains' and 'Chamber wheel trains' are worth studying even today by anyone who thinks that they have invented a new form of rotary compressor.

The term 'screw' compressor, though commonly used, is not very helpful for an understanding of the working of the machine. It is better to think of it as a pair of helical gears with special tooth profiles and a large wrap angle, i.e. the angle through which the radius vector generating the helix rotates between one end of the gear and the other.

corresponding built-in pressure ratio for a particular gas. If the pressure in the flutes and spaces has not reached the pressure in the delivery pipe when the port is uncovered, gas flows back into the flutes and spaces. On the other hand, if the pressure inside the flutes and spaces at the instant of opening exceeds that in the delivery pipe, gas rushes out suddenly in an unresisted expansion. In either event there is a loss of efficiency, which will be treated in section 4.7.

Ideally the built-in pressure ratio should match the actual pressure ratio for maximum efficiency, but this is not possible in practice because a compressor in any given installation may have to operate at different pressure ratios, depending on the evaporating and condensing tempera-

Fig. 2.4.2. Principle of operation of a twin-screw compressor. Gas enters the flutes and interlobe spaces at the top through the suction port in the end plate shown by the broken line, and is subsequently reduced in volume and expelled from the machine by the remeshing of the rotors on the side facing the viewer. The viewpoint of the rotors in this figure is indicated in Fig. 2.4.3.

With the rotation and flow as shown in this figure and in Fig. 2.4.3, the male rotor is right-handed and the female rotor left-handed, as produced by some makers. Others make the machine with opposite handedness.

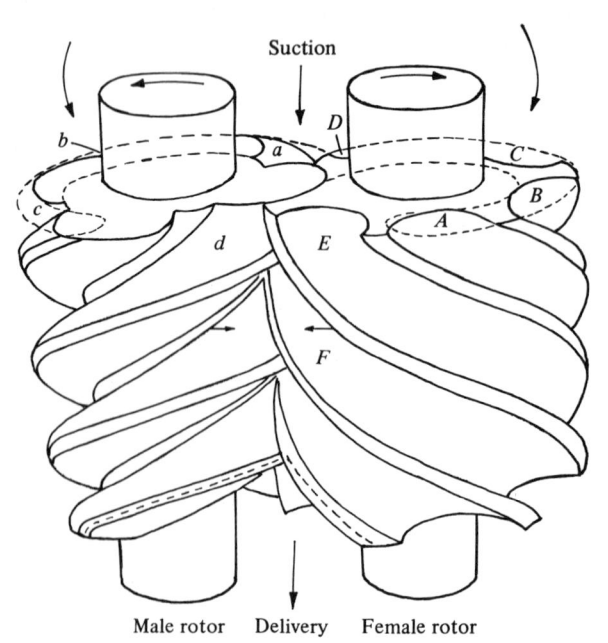

2.4 Rotary positive displacement compressors

tures. Nevertheless there is usually one pressure ratio which corresponds to average or design conditions, and the machine can be designed for this. A further compromise has to be adopted in that manufacturers of screw compressors offer only two or three built-in volume ratios as a rule, and one of these has to be chosen even if it is not quite right. For example, when three ratios are offered, the low one may be chosen for air-conditioning duties, the medium one for ice-making and the high one for food freezing.

These different built-in volume ratios can be obtained by altering the position of the radial outlet port. In Fig. 2.4.2 lowering the broken lines down the page delays the point of release and thus the built-in ratio.

Examination of Fig. 2.4.2 shows that the lower flutes and spaces facing the viewer contain compressed or partly compressed gas, whereas those on the side remote from the viewer are mostly filled with gas at suction pressure. This unbalanced pressure distribution causes loads on the bearings, the bottom ones (not shown) in Fig. 2.4.2 being more heavily loaded than the top ones. It also causes bending of the rotors. At the same

Fig. 2.4.3. Diagram showing the arrangement of the sliding valve for capacity reduction. The slot begins at 80% of the swept volume. The rotors are omitted from the right-hand view, but the direction of the helices on the female rotor is shown. The approximate extent of the axial delivery port is shown in the left-hand view by the broken line.

The rotor profiles shown in this figure are the older symmetric ones. The current asymmetric profiles are shown in Fig. 4.7.1.

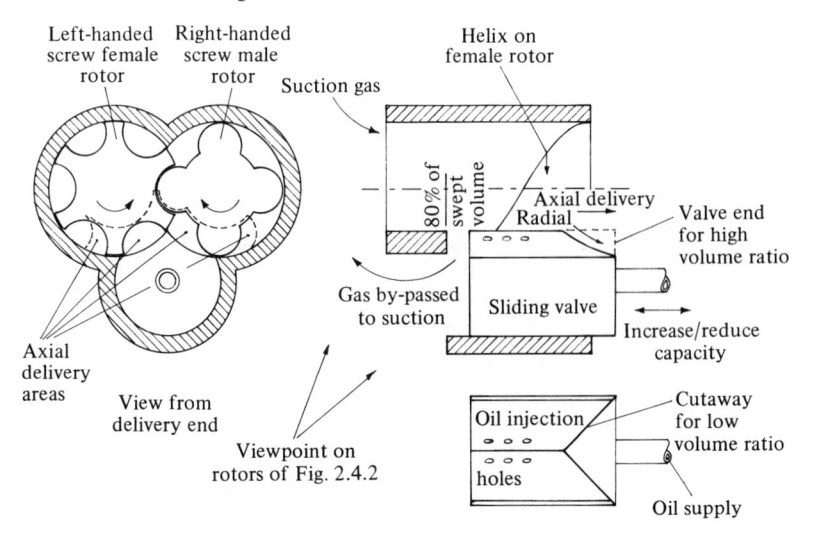

Fig. 2.4.4. The full line in the top graph shows the theoretical change of the volume trapped in the machine as the sliding valve is moved to the right in Fig. 2.4.3. Because of throttling through the slot the actual volume of gas aspirated follows the broken line.

The bottom part of the figure shows how the built-in volume ratio changes with movement of the sliding valve. For a high built-in volume ratio of 4.5 there is no radial delivery port, i.e. no V-shaped cutaway in the sliding valve, and the point of release is controlled by the axial port in the end plate, which is fixed. Hence the built-in volume ratio diminishes. The line for the built-in volume ratio 2.5 shows the change when a radial port is provided by a V in the sliding valve. Until the notch of the V reaches the delivery face, release is controlled by the sliding valve position and the built-in volume ratio increases. Thereafter release is controlled by the axial port and the built-in volume ratio diminishes.

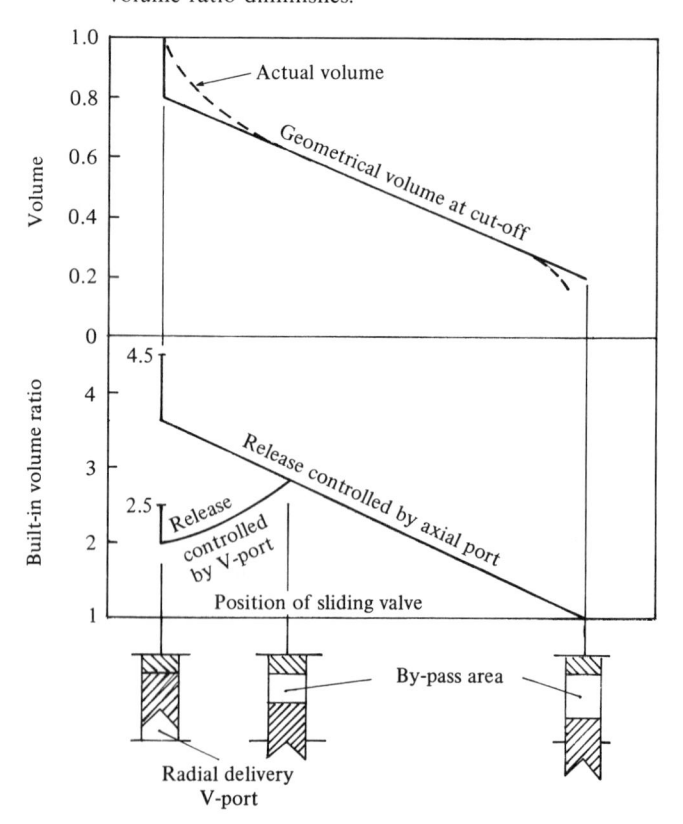

2.4 Rotary positive displacement compressors

time there is an axial load caused by the different pressures at the two ends of the rotors which has to be resisted by thrust bearings and by a piston under oil pressure.

Refrigeration compressors usually have 4 male lobes and 6 female flutes, as shown in the diagram. This choice gives approximately equal stiffness of the rotors in bending and a high swept volume per unit size of casing. For higher pressure applications in chemical processes the $6+8$ and other combinations are used. These give stiffer rotors at the sacrifice of some swept volume.

Reduction of the pumping capacity is achieved by a slot in the casing on the compression side, i.e. the side facing the viewer in Fig. 2.4.2. The slot is opened by a sliding valve which can be moved in the casing, as shown in Fig. 2.4.3. The fixed edge of the slot is usually located at 80% of the swept volume. Hence at the first movement of the sliding valve the capacity would drop instantaneously to 80% of the maximum if there were no resistance to flow through the slot, as shown by the full line in the upper graph of Fig. 2.4.4. Because of the resistance, the actual volume flow rate does not fall off immediately but follows the broken line shown. Movement of the slide to the right reduces the volume flow rate further down to a minimum of 15 or 20% of the maximum.

The sliding valve carries the cutaway which provides the radial delivery port already mentioned. This is shown by the full line in Fig. 2.4.3, whilst a sliding valve which provides only axial outlet is shown by the broken line. Assuming that the axial port in the end plate remains fixed, then different but smaller built-in volume ratios can be obtained by modifying the V in the sliding valve, the casing, rotors and end plates remaining the same. The rotors and delivery end plate are then designed for the highest built-in volume ratio desired, and smaller values obtained by fitting different sliding valves.

Movement of the sliding valve to reduce capacity, however, changes the built-in volume ratio. Considering first a valve without a radial port for a high built-in ratio, as soon as the slide is moved the trapped volume falls to 80% and the built-in volume ratio also falls to 80%. As the slide is moved to the right the trapped volume falls still further and so does the built-in volume ratio. This is shown in Fig. 2.4.4 for a built-in volume ratio of 4.5.

When a radial port is provided, for a lower built-in ratio, the initial fall to 80% still occurs at the first movement of the slide. Thereafter a rise takes place because the movement of the V in the slide delays the point of release by more than the reduction of the volume trapped at the beginning of compression. This rise continues until the radial port is closed and the axial one only is available. Thereafter the built-in ratio diminishes as the slide moves to the right.

In the first twin-screw compressors used in refrigeration, the flutes and lobes were symmetrical†, but modern machines use an asymmetric profile, shown in Fig. 4.7.1. This has the advantage of improving the sealing line between the rotors and also of giving slightly higher swept volume.

When first introduced into refrigeration in the 1950s, the twin-screw compressor was operated as a dry machine with only minimal or no lubrication of the rotors. By using accurately cut timing gears, the rotors were supposed to be prevented from touching each other. These machines had quite large internal leakage, however, and had to be run at speeds of the order 5000 to 8000 rev/min to achieve good volumetric efficiencies. The large mass flow rate through the relatively small casing made effective cooling of the gas during compression impossible. Their use was limited therefore to pressure ratios against which they could pump without excessive leakage or excessive delivery temperature.

These limitations were removed by injecting oil into the gas inside the machine just after the start of compression. In the current generation of machines for refrigeration use, the timing gears are omitted and, with oil injection, pressure ratios of up to 20 or so are possible using ammonia without exceeding a delivery temperature of 100 °C, and with satisfactory volumetric efficiency when running at rotor tip speeds of 50–60 m/s. The machines are seldom used at the pressure ratio cited, however, because the benefits from the use of compound compression, treated in Chapter 5, are not then available.

The oil is injected through holes in the sliding valve, shown diagrammatically in Fig. 2.4.3. The volume of oil is about $\frac{1}{4}$% of the swept volume, with another approximately equal amount being supplied to cool the bearings. At an evaporating temperature of $-15\,°C$, the mass flow rate of oil would be about equal to the mass flow rate of ammonia, and relatively more at lower evaporating temperatures. A large and effective oil separator is needed to prevent this oil from passing to the condenser and low-side. The oil from the separator is cooled, either by water or by direct evaporation of refrigerant.

From an operational point of view, twin-screw compressors are usually held to be reliable compared with reciprocating machines, and to run for longer intervals between major overhauls. Figures of 30 000 to 40 000 hours have been quoted. They can also be manufactured for large refrigerating capacities which are beyond the range of current high-speed reciprocating

† Lysholm's early machines had asymmetric rotors.

2.5 Turbo-compressors

vapour entering the eye. Absolute tightness is not necessary here and it is usually a labyrinth type of seal allowing a small leakage.

It is shown in Chapter 4 that the specific work absorbed and the enthalpy rise produced by the impeller depends on the square of the tip speed of the impeller. A high tip speed can be obtained by using a large impeller or a high rotational speed or both. The size of the impeller may be limited because it is disproportionate to the capacity of the machine or because of low efficiency when the mass flow rate is low. To produce the required rise in specific enthalpy and pressure, the rotational speed must be increased. With

Fig. 2.5.2. A four-wheel turbo-compressor with provision for suction at an intermediate pressure for carrying a load at an intermediate temperature or for improving the cycle efficiency as described in Chapter 5. The compressor can be driven directly by a steam turbine or via external speed increasing gearing by an electric motor. The refrigerating capacity is varied by suction throttling down to an intermediate level and by hot gas by-pass below this. Three versions of this design provide a range of refrigerating capacity on refrigerant R13B1 of 600–2000 kW at an evaporating temperature of $-20\,°C$. The minimum evaporating temperature is $-45\,°C$. The smallest machine 4N-260 has an impeller diameter of 260 mm and a speed range of 7760–9700 rev/min. The largest machine 4N-350 has an impeller diameter of 350 mm and a speed range of 5550–6940 rev/min. (By courtesy of Sulzer Bros. (UK) Ltd.)

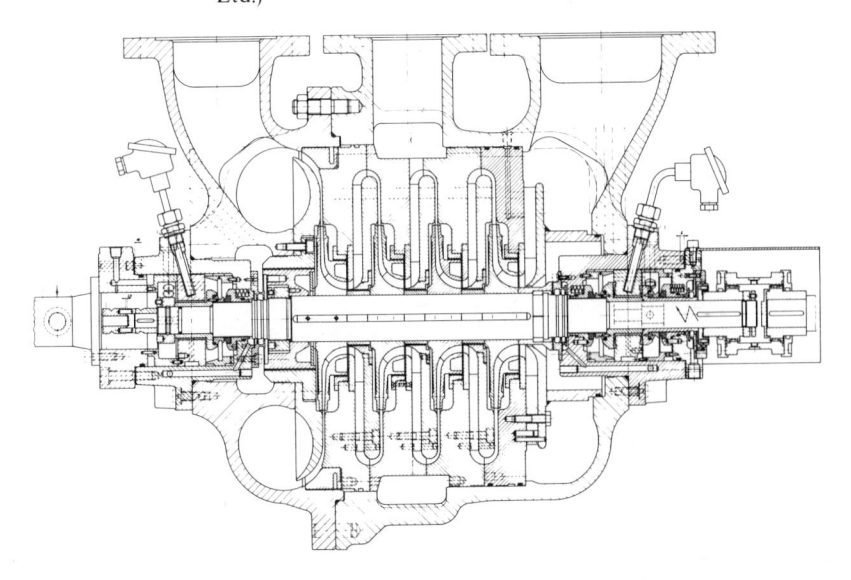

direct coupling to an electric motor, the maximum speed available is 2900 rev/min on a 50 Hz alternating current supply, or 3500 rev/min on 60 Hz, using two-pole motors. If these speeds are not high enough, a speed increasing gear must be used, or a high-frequency supply may be specially generated. When steam turbine drive is available there is no problem.

The limitation on impeller diameter imposes a lower limit on the refrigerating capacity obtainable with a given refrigerant for a certain maximum rotational speed. For air-conditioning duties this lower limit is currently of the order 250–350 kW.

Turbo-compressors can be built in very large sizes, however. A refrigerating capacity of 30 MW or more can be provided on a single shaft for air-conditioning large complexes of buildings, and for chemical plants.

If the specific work required for a particular duty would require too large an impeller or an impossible rotational speed, the compression can be done in stages using two or more impellers. This offers opportunities of improving the cycle, as explained in Chapter 5, by the use of liquid pre-cooling at the intermediate pressure. The usual arrangement of impellers in multi-stage machines is on the same shaft facing the same way (Fig. 2.5.2). This requires transfer passages to bring the vapour from the volute of one stage to the eye of the impeller of the next. One successful design of hermetic compressor with two stages has used the impellers back to back (so balancing the thrusts approximately) at opposite ends of the motor.

Reduction of the pumping capacity may be achieved by throttling the suction vapour or raising the condensing pressure to increase the pressure difference across the machine, or a combination of both. With steam turbine drive, changing the speed is a favoured method. The compressors of medium size used in air-conditioning frequently use pre-rotation vanes in the inlet duct. The principle is to give a rotation to the vapour stream in the same direction as the impeller and reduce the relative velocity between them. The volume flow rate scooped in by the impeller is thus reduced, and there is also a reduction in power. Another method is to alter the diffuser vanes.

Lubrication of the bearings and seal is invariably by force feed from a pump, driven by a separate motor, arranged to start some time before the impeller begins to rotate and to shut down after the impeller has coasted to rest. There may also be a coupled pump which takes over when the impeller has run up to speed. If there is no coupled pump, an arrangement such as a pressurised reservoir must be provided to ensure lubrication during run-down in the event of a power failure.

It is possible to distinguish two main classes of turbo-compressor on the

2.4 Rotary positive displacement compressors

compressors. They are not available for small capacities, however, in which field the reciprocating machine still holds its own.

The machine with timing gears is still manufactured for general gas compression purposes and is widely used in the chemical industry, as well as for providing oil-free air in various process industries. With water injection for cooling, gases like chlorine and mixtures containing acetylene can be compressed.

The development of twin-screw compressors for refrigeration applications has been described by Perry & Laing (1961), Laing & Perry (1964), Laing (1969), Lundvik (1967) and Lundberg (1968, 1975).

Single-screw compressors

In recent years the single-screw compressor, invented by B. Zimmern, has been introduced. It consists of a screw rotor meshing with two star wheels whose function is to seal the flutes on the rotor, as shown in Fig. 2.4.5. The same sequence of induction, compression and delivery takes place simultaneously in the top half (shown) of the rotor and the under half. Consequently the pressure forces bending the rotor and loading the bearings are balanced, unlike those in the twin-screw compressor. It is also possible to design the machine so that the axial load is small.

Reduction of capacity is provided by sliding valves which allow gas to pass back to the suction space. At the same time movement of the valves

Fig. 2.4.5. Principle of operation of the single-screw compressor. At (*a*) gas entering from the right fills the spaces formed between the rotor, the casing and the star wheels. The volume of one of these spaces is shown being reduced in (*b*). Eventually the reduced space communicates with a delivery port and the gas is delivered to the left at (*c*). The same sequence of operations proceeds on the underside of the rotor.

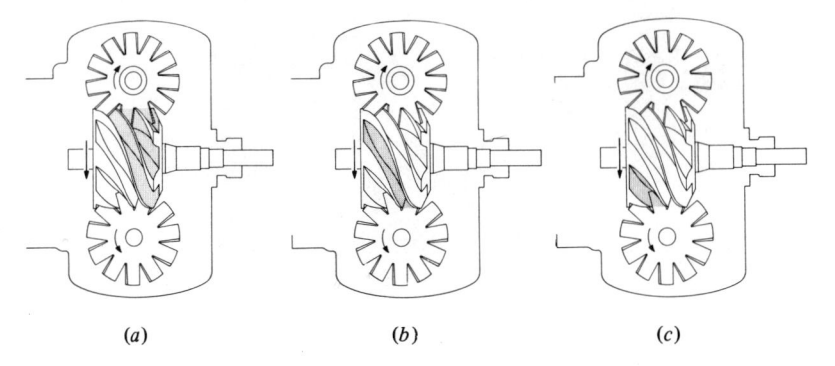

(*a*)   (*b*)   (*c*)

alters the positions of the delivery ports so that the built-in volume ratio remains approximately constant. The construction of a machine is shown in Fig. 2.4.6. The development of the single-screw machine is described by Clarke, Hodge, Hundy & Zimmern (1976) and Hundy (1978).

Fig. 2.4.6. Cutaway view of a single-screw compressor. The rotor 1, mounted in a roller bearing at the drive end and in angular contact bearings at the other end, is driven by the shaft passing through the seal at 3 and engages with the star wheels 6 whose shafts are carried by the ball bearings 13. Gas enters via the suction flange 2 and is delivered via a fixed delivery port (not visible) and the variable delivery port 8 controlled by the sliding valve 7 to the delivery space 9 and thence to the delivery flange (not shown). This can be selected for upward or downward delivery as required. Oil is supplied via connection 4 through tubes (not shown) to the drive-end bearing and shaft seal and to the injection holes 5 in the sliding valve. The sliding valves are moved by the hydraulic piston 10 which also works the position indicator 11 via a helically grooved shaft. The space 12, protected against leakage from the delivery pressure by a labyrinth seal on the end of the rotor, is vented to suction pressure to reduce the end thrust on the rotor. (By courtesy of Hall-Thermotank Products Ltd.)

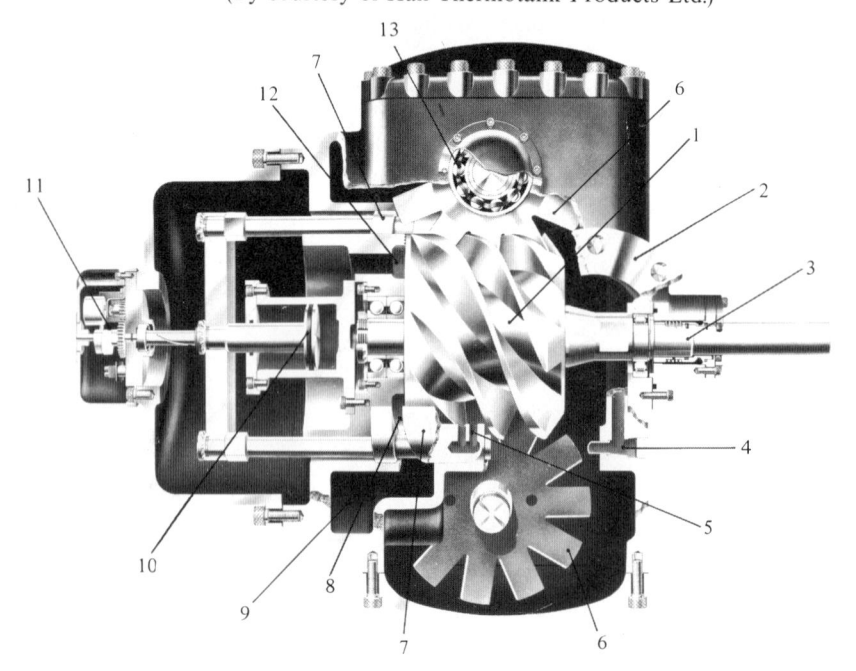

direct coupling to an electric motor, the maximum speed available is 2900 rev/min on a 50 Hz alternating current supply, or 3500 rev/min on 60 Hz, using two-pole motors. If these speeds are not high enough, a speed increasing gear must be used, or a high-frequency supply may be specially generated. When steam turbine drive is available there is no problem.

The limitation on impeller diameter imposes a lower limit on the refrigerating capacity obtainable with a given refrigerant for a certain maximum rotational speed. For air-conditioning duties this lower limit is currently of the order 250–350 kW.

Turbo-compressors can be built in very large sizes, however. A refrigerating capacity of 30 MW or more can be provided on a single shaft for air-conditioning large complexes of buildings, and for chemical plants.

If the specific work required for a particular duty would require too large an impeller or an impossible rotational speed, the compression can be done in stages using two or more impellers. This offers opportunities of improving the cycle, as explained in Chapter 5, by the use of liquid pre-cooling at the intermediate pressure. The usual arrangement of impellers in multi-stage machines is on the same shaft facing the same way (Fig. 2.5.2). This requires transfer passages to bring the vapour from the volute of one stage to the eye of the impeller of the next. One successful design of hermetic compressor with two stages has used the impellers back to back (so balancing the thrusts approximately) at opposite ends of the motor.

Reduction of the pumping capacity may be achieved by throttling the suction vapour or raising the condensing pressure to increase the pressure difference across the machine, or a combination of both. With steam turbine drive, changing the speed is a favoured method. The compressors of medium size used in air-conditioning frequently use pre-rotation vanes in the inlet duct. The principle is to give a rotation to the vapour stream in the same direction as the impeller and reduce the relative velocity between them. The volume flow rate scooped in by the impeller is thus reduced, and there is also a reduction in power. Another method is to alter the diffuser vanes.

Lubrication of the bearings and seal is invariably by force feed from a pump, driven by a separate motor, arranged to start some time before the impeller begins to rotate and to shut down after the impeller has coasted to rest. There may also be a coupled pump which takes over when the impeller has run up to speed. If there is no coupled pump, an arrangement such as a pressurised reservoir must be provided to ensure lubrication during run-down in the event of a power failure.

It is possible to distinguish two main classes of turbo-compressor on the

2.5 Turbo-compressors

vapour entering the eye. Absolute tightness is not necessary here and it is usually a labyrinth type of seal allowing a small leakage.

It is shown in Chapter 4 that the specific work absorbed and the enthalpy rise produced by the impeller depends on the square of the tip speed of the impeller. A high tip speed can be obtained by using a large impeller or a high rotational speed or both. The size of the impeller may be limited because it is disproportionate to the capacity of the machine or because of low efficiency when the mass flow rate is low. To produce the required rise in specific enthalpy and pressure, the rotational speed must be increased. With

Fig. 2.5.2. A four-wheel turbo-compressor with provision for suction at an intermediate pressure for carrying a load at an intermediate temperature or for improving the cycle efficiency as described in Chapter 5. The compressor can be driven directly by a steam turbine or via external speed increasing gearing by an electric motor. The refrigerating capacity is varied by suction throttling down to an intermediate level and by hot gas by-pass below this. Three versions of this design provide a range of refrigerating capacity on refrigerant R13B1 of 600–2000 kW at an evaporating temperature of $-20\,°C$. The minimum evaporating temperature is $-45\,°C$. The smallest machine 4N–260 has an impeller diameter of 260 mm and a speed range of 7760–9700 rev/min. The largest machine 4N–350 has an impeller diameter of 350 mm and a speed range of 5550–6940 rev/min. (By courtesy of Sulzer Bros. (UK) Ltd.)

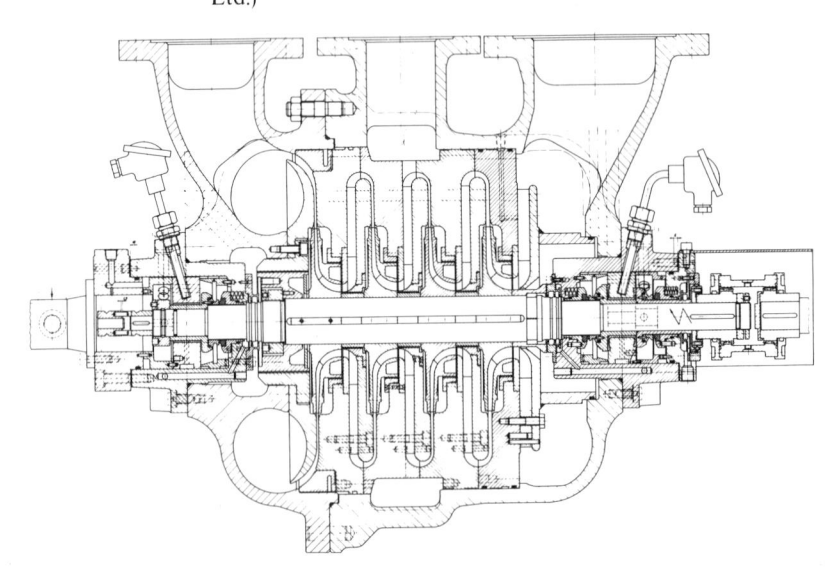

2.5 Turbo-compressors

The term 'turbo-compressor' as understood in refrigeration usually means a centrifugal compressor. The development of this from the centrifugal pump goes back to the beginning of this century and is associated with the French engineer Auguste Rateau (1863–1930). Its introduction into refrigeration is due to Willis Haviland Carrier (1876–1950) in 1921.

The principle of the turbo-compressor is shown in Fig. 2.5.1. Vapour is taken into the centre, known as the *eye*, of a rotating impeller, which is a wheel having passages, either internal or between it and a casing, leading from the eye to the periphery. In these passages the vapour undergoes a rise in pressure because of centrifugal force, and at the same time an increase in absolute velocity. Leaving the impeller, the vapour flows spirally outwards in the *diffuser*. The function of the diffuser is to decelerate the vapour and increase the pressure. The vapour from the diffuser is finally collected in a chamber known as the *volute* or *scroll* and discharged to the delivery pipe.

The partitions inside the impeller which divide the flow into more or less sector-shaped channels are usually called the blades. In refrigeration compressors these are either backward curved, used in traditional designs, or radial, used in many modern single-stage designs for air-conditioning.

Fig. 2.5.1. Principle of the centrifugal compressor. Vapour enters axially at 1 and flows through the passages 3 in the impeller 2. During the flow through the impeller the pressure rises and the absolute velocity increases. In the stationary diffuser 4 a further rise in pressure occurs with a decrease in the absolute velocity. The vapour is collected in the scroll or volute 5 and is delivered at 6.

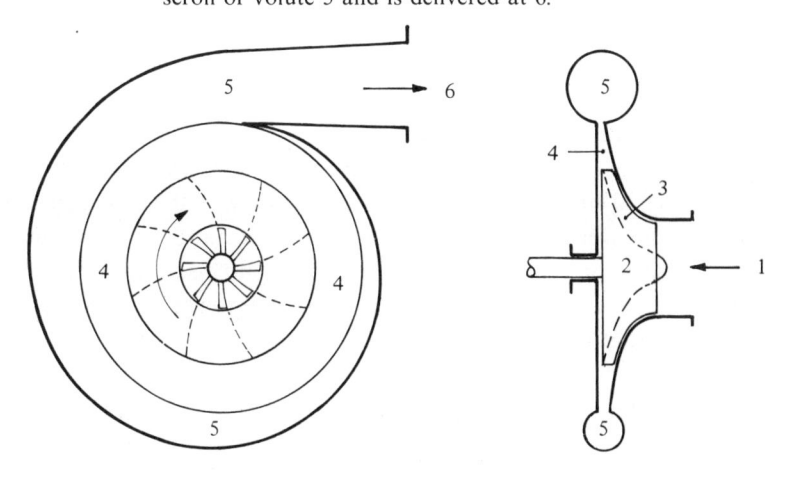

Radial bladed impellers produce a greater pressure rise for a given tip speed of the impeller, but put more duty on to the diffuser.

Because the impeller does not fit the casing exactly, the pressure at the outlet of the impeller is established in the space between the back plate and the casing, and between the front side or shroud, if present, and the casing. Across the area of the eye suction pressure acts. As a result there is an unbalanced force pushing the impeller towards the incoming vapour, which must be taken by a thrust bearing or a balancing piston.

Diffusers are of two kinds, vaneless and vaned, and sometimes a combination of the two is used. The vaneless diffuser is simply an annular space in which a free vortex is established. The moment of momentum remains constant and as the radius increases the velocity diminishes, with a consequent rise in pressure. In a vaned diffuser there are guide vanes forming passages which give the correct area for flow. The disadvantage of the vaneless diffuser is that it occupies considerable space because the radius ratio has to equal the desired velocity ratio. On the other hand, it is possible to diffuse from supersonic velocities without shock, which occurs in vaned diffusers at the leading edges of the vanes when the vapour leaving the impeller has an absolute velocity greater than sonic. Apparently, however, shock is not so harmful to efficiency as was once thought, and it is accepted as a normal thing in the vaned diffusers of some modern compressors. Turbo-compressors for air-conditioning duties are expected to operate over a wide range of capacity, about full load down to 10%, and to improve performance over this wide range diffusers are often adjustable. The width of a vaneless diffuser may be varied by moving one of the cheeks, or the angular setting of the vanes in a vaned diffuser may be adjustable.

The impeller shaft must be sealed where it emerges from the casing, to prevent leakage of refrigerant. Because the rotational speeds are normally high, the rubbing speeds are also high between the sealing surfaces, and special designs of seals are used. If the compressor is driven via speed increasing gears it is often advantageous to put the gears inside the pressure-tight casing and the shaft seal on the low-speed shaft. The seal may be eliminated entirely by hermetic construction with the driving motor and gears, if any, enclosed inside the casing. The motor is then cooled by refrigerant vapour or by evaporation of liquid refrigerant. When open construction is employed, a common practice is to use two seals, a running one and another which can be tightened up when the machine is at standstill.

In addition to the shaft seal which prevents refrigerant from escaping to the outside, a seal is needed between the shroud of the impeller and the

2.5 Turbo-compressors

market at present. The large heavily built traditional type of machine for gas liquefaction, chemical plant and large air-conditioning schemes, is usually conservatively rated with backward curved blades, multi-stage, versatile and adapted to a variety of drives. In recent years the demand for compressors for air-conditioning having capacities of the order

Fig. 2.5.3. A single wheel turbo-compressor driven at 17 500 or 20 000 rev/min via integral epicyclic gearing by a two-pole electric motor, 2950 rev/min on 50 Hz supply. 1 – impeller, 216 mm diameter (larger model 312 mm diameter); 2 – rotor boss nut; 3 – adjustable diffuser vanes for capacity regulation, operated by rotation of the ring 4 from the diffuser actuator 5; 6 – motor shaft end driving the internally geared rim 7, planet wheels 8 and the sun wheel on the impeller shaft; 9 – gear casing and 10 – oil supply. Normally used with Refrigerant 12; range of capacity on water chilling duties 600–1400 kW (larger model 1400–2600 kW). Minimum evaporating temperature $-20\,°C$. (By courtesy of Sulzer Bros. (UK) Ltd.)

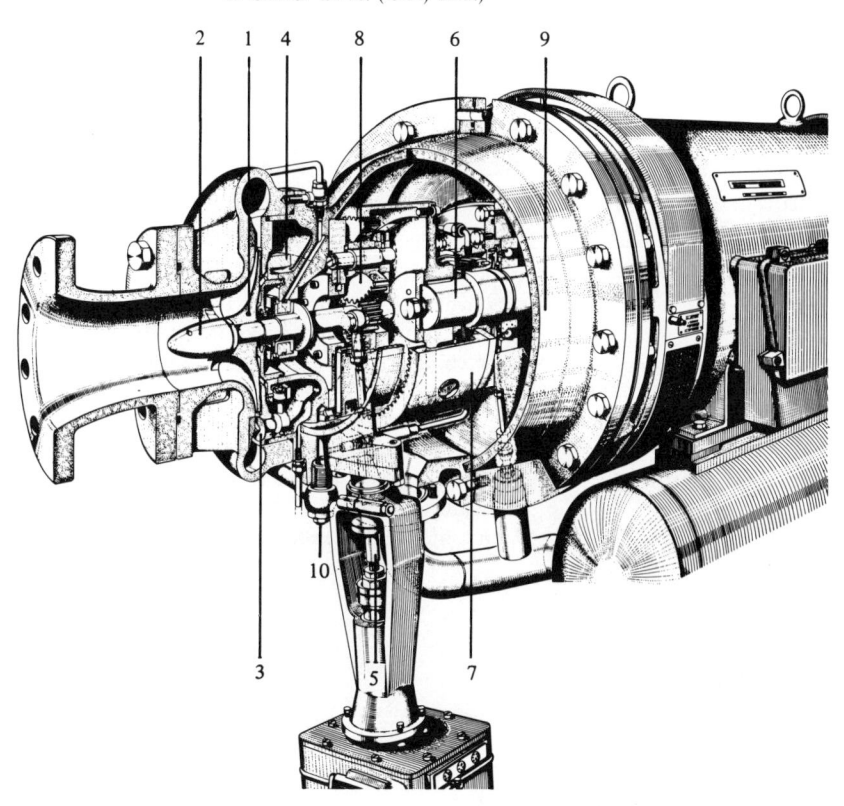

350–2000 kW has stimulated the mass production of high-speed, single-stage, radial bladed machines, offering high capacity in a small size and at relatively low cost. A machine of this type is shown in Fig. 2.5.3.

2.6 Condensing equipment

The arrangement and design of condensers depends much on the cooling medium which is available. Three principal methods are water cooling, air cooling and evaporative cooling.

The principal form of water-cooled condenser in large sizes is the horizontal shell-and-tube condenser, shown in Figs. 2.6.1 and 2.6.2. Refrigerant vapour enters the shell at the top and condenses on the outside of tubes through which water flows. The condensed refrigerant drains off the tubes and collects at the bottom of the shell whence it drains to a receiver. The water enters the tubes from the end covers, also called headers or water boxes, and makes one or more passes through the length of the condenser. The object of using several passes is to increase the velocity of the water in the tubes for the same total quantity of water, and thereby to improve the coefficient of heat transfer. For ammonia the coefficient of heat transfer on the condensing side is high and of the same order as the coefficient on the water side, but for the fluorinated refrigerants the condensing coefficient is much lower. In these circumstances it is beneficial to increase the area on the condensing side of the tubes by fins. Since the tubes have to be inserted through the holes in the tube plates during manufacture, and because it may

Fig. 2.6.1. Diagrammatic arrangement of a horizontal shell-and-tube condenser. Some tubes are omitted at the top of the shell to provide an area for the vapour to flow over the nest of tubes. In large condensers two or three inlets may be provided. The arrangement of water flow shown, with the water passing twice through the length of the condenser, is called a two-pass arrangement. By further subdivision of the end covers more passes can be provided, as in Fig. 2.6.2. The aim is to increase the velocity of the water without making the condenser too long.

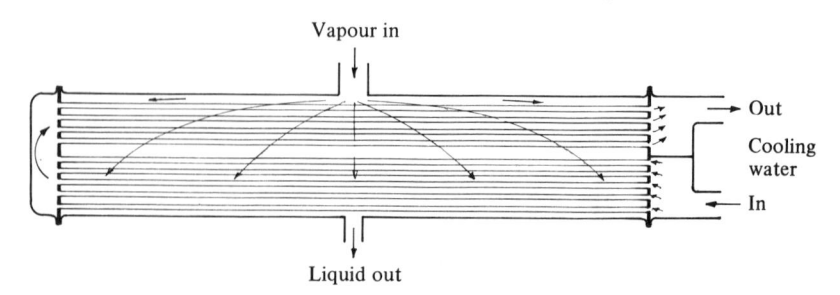

2.6 Condensing equipment

be necessary to take one out for replacement at some time, the fins have to be formed within the outside diameter of the tube, as shown in Fig. 2.6.3, and this is known as low-fin tube.

Fig. 2.6.2. Construction of shell-and-tube condensers. (*a*) Refrigerant vapour enters at 1 and refrigerant liquid leaves at 2. The cooling water enters the end with bolted cover 5 at 3 and leaves at 4. The tubes are expanded into the tube-plates 6, which are welded to the shell at each end, and they are supported in the middle by the plate 7. The pass plates 8, corrugated to fit between the closely spaced tubes, provide eight passes of water flow. 9 is the connection for the non-condensable gas purge line, 10 is a vent connection to the shell and 11 is an air purging cock on the water side.

(*b*) Alternative method of fitting the end covers using a proprietary 'Victaulic' joint for easy removal for tube cleaning. This design is intended for fluorinated refrigerants and has low-finned tubes like Fig. 2.6.3. Design (*a*) for ammonia has plain tubes. (By courtesy of Hall-Thermotank Products Ltd.)

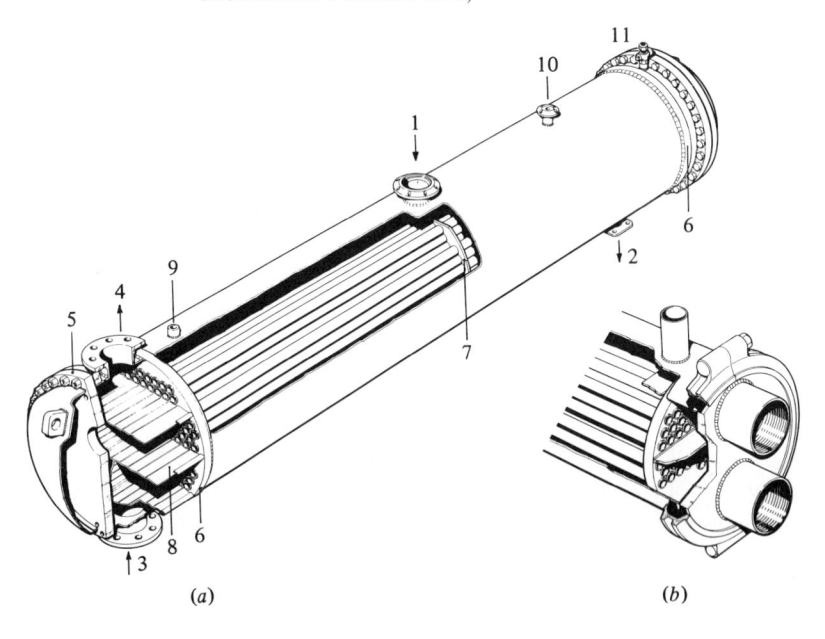

Fig. 2.6.3. Low-fin tubing to pass through the holes in the tube plates during assembly.

Condenser tubes for the fluorinated refrigerants may be of copper or brass, but for ammonia, steel is necessary. This presents a problem when sea water is the coolant, because it is rather corrosive to steel. Tubes of two layers of different metal have been used, steel on the outside for ammonia, bonded to a copper-based alloy on the inside for the sea water. Another method is to protect the steel by using a sacrificial anode of zinc or other metal with similar electrochemical properties. It may be worth noting that condenser tubes are quite sensitive to the effects of corrosion, partly because they are kept warm, and partly because they are thin to save weight and space and to improve heat transfer. Because they are not very stiff, the tubes are supported at intervals in a long condenser by tube support plates. At one time the tubes were commonly fixed into the tube plates at the ends by ferrules or other screwed fixings, but are now usually expanded, welded or brazed.

High water velocities are desirable for good heat transfer and also to inhibit the deposition of dirt on the interior walls of the tubes. Unfortunately, with soft metals like copper velocities higher than about 2 m/s tend to cause erosion with some waters. Higher values can be used with steel, depending on the acceptable pressure drop.

In all condensers attention should be paid to the drainage of liquid from the surface of the tubes. The main resistance to heat transfer occurs in the film of condensate around the tubes, and this should not be allowed to thicken. The tubes are staggered as much as possible therefore to minimise the number of tubes in a vertical tier. Except under special conditions to be described later, liquid should not stand in the bottom of the shell but should have free drainage to the liquid receiver. This can usually be assured by keeping the velocity in the drain pipe less than about 0.5 m/s. Another method is to provide a vapour balance pipe between the top of the condenser and the receiver. 'Backing-up' of liquid in the condenser often occurs in practice as a temporary condition owing to the liquid receiver becoming warmer than the condenser. It can be verified by feeling the condenser shell, the bottom of which will feel colder than the sides and top. Backing-up is not in itself harmful, but if there is not enough refrigerant in the plant to maintain the liquid seal to the pipe from the receiver to the expansion valve, vapour will enter this pipe and the evaporators will not receive their correct flow rate of refrigerant. The evaporating temperature falls and the compressor delivers less vapour to the condenser, where the pressure tends to fall further. The original condition is thus aggravated. Positive control of the condensing and receiver pressures by the methods to be described, which use a regulated amount of back-up, is the way to avoid the problem.

2.6 Condensing equipment

Small water-cooled condensers up to a refrigerating capacity of a few kilowatts are often constructed as a shell-and-coil assembly. The water flows through a coil of tube and the refrigerant condenses on the outside. The coil is mounted in a shell the bottom of which may act as the liquid receiver. Shell-and-coil condensers are mostly used with fluorinated refrigerants for which copper tube, easily formed into a coil, is suitable. They are not so easily cleaned as shell-and-tube condensers since it is not possible to push a rigid cleaning rod through them, and chemical cleaning has to be employed.

Small shell-and-coil condensers are often grouped on a baseplate with the compressor and motor, an arrangement which is called a condensing unit.

Horizontal shell-and-tube and shell-and-coil condensers are both best suited to a clean water supply. Although the interior of the tubes can be cleaned this entails putting the condenser out of service. Vertical shell-and-tube condensers were very common at one time and still have their uses for very dirty water because they can be cleaned in service. The tubes do not run full and are not under water pressure. The arrangement is shown in Fig. 2.6.4. Refrigerant condenses on the outside of the tubes, and water trickles down the inside from a trough at the top. The tubes can be cleaned

Fig. 2.6.4. Diagrammatic arrangement of a vertical shell-and-tube condenser. Water trickles over the weirs formed by the tops of the tubes and down the inside of the tube wall.

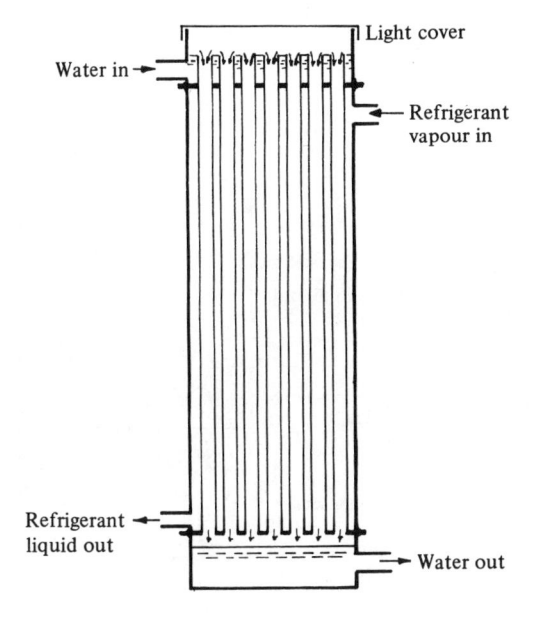

at any time with long brushes. The coefficient of heat transfer is not so high as in horizontal condensers because of the lower water velocity and because of the thicker condensate film which forms on the long vertical tubes. The tubes are usually of wider bore than those in horizontal condensers. They do not need to be of small bore because the velocity of the water inside them is determined by gravity and the height of the tubes. Hence a cheaper construction with fewer tubes, which are also more easily cleaned, can be used.

Another type of condenser useful for dirty water is the atmospheric condenser. It consists of a vertical grid of horizontal tubes with return bends, across which water trickles from a trough at the top. Refrigerant vapour is admitted to the tube or tubes in parallel at the top and drains to the bottom. In some designs there are intermediate take-off points for the condensed liquid to prevent the lower tubes in the grid from running full bore. If this happened the vapour would not have access to the cooled wall

Fig. 2.6.5. (a) Figure showing the principle of a cooling tower. Water is sprayed or trickled over the packing, also called the 'fill', of wooden slats or plastic corrugated sheets, in counter-flow with a current of air.

(b) An evaporative condenser. It differs from the cooling tower only in that the water trickles over the pipes carrying the condensing refrigerant instead of over the fill.

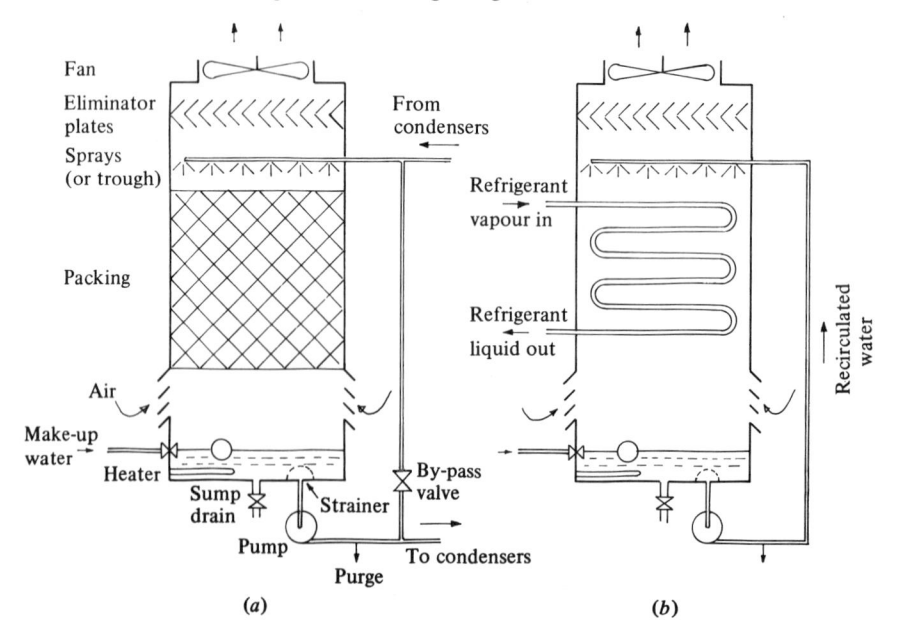

2.6 Condensing equipment

of the tube and this part of the pipe would be wasted. Though it could produce some subcooling of the liquid refrigerant, this is not so valuable as a rule as using the surface to condense refrigerant and so reduce the condensing pressure.

Water for condensers may be available from a nearby river or well, but in cities it is likely that mains water will have to be used. For a large plant the cost of running this to waste would be prohibitive, so it has to be recirculated after reducing its temperature in a cooling tower. The principle is shown in Fig. 2.6.5(a), and an actual tower is shown in Fig. 2.6.6. Water is sprayed by nozzles, or drips from a trough, over surfaces of wood or plastic in counterflow with air. In large towers for power stations the air flow is induced by natural convection, but in the towers used on roof tops for refrigeration plant the air flow is almost always assured by a fan. Both induced draft, with the fan at the top as shown, and forced draft with the fan at the bottom are used. The temperature of the water is reduced by the

Fig. 2.6.6. An all-plastic cooling tower with rotating water distributors. Plastic is often used because of the difficulty of combating corrosion in metal towers. (By courtesy of Hall-Thermotank Products Ltd.)

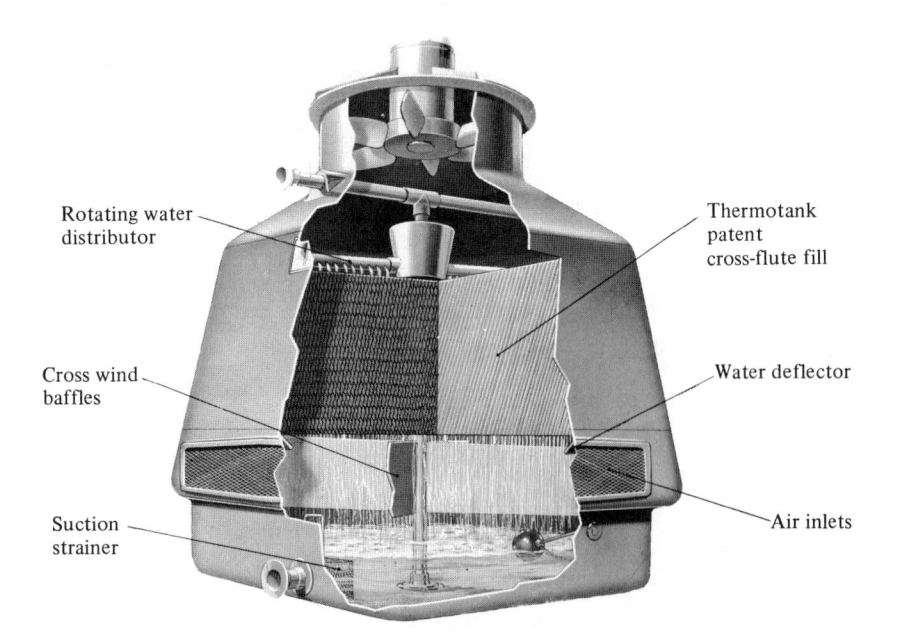

evaporation of part of it to within a few degrees of the wet bulb temperature of the ambient air.

Another way of economising on water is to use an evaporative condenser, Fig. 2.6.5(b). This is like a cooling tower except that the water is sprayed over tubes carrying the condensing refrigerant. The water is recirculated from the sump to the spray nozzles or trough, make-up water being provided by a float controlled valve. A disadvantage of the evaporative condenser is that it is necessary to bring the refrigerant pipes to the roof or wherever the condenser is installed, with long runs of pipe and more joints to leak. With a cooling tower the refrigerant circuit can be confined to a small area of the engine room. On the other hand, the evaporative condenser can achieve a condensing temperature which is closer to the wet bulb temperature of the outside air because it eliminates the necessary rise in temperature of the cooling water passing through shell and tube condensers. Furthermore, it is less susceptible to freezing in cold weather.

In both cooling towers and evaporative condensers it is essential to supply more make-up water than is actually evaporated and to purge some of the sump water to waste. This has to be done to prevent the accumulation of dissolved solids which enter with the make-up water. A simple calculation shows the amount entering. Suppose there is a cooling system circulating about 100 000 kg/h of water; approximately 1/100 of this is evaporated and make-up water at the rate of 1000 kg/h is required. For a very hard water the dissolved solids may be 400 in 10^6 (400 p.p.m.), and the solids entering with the make-up water will amount to 0.4 kg/h, or about 67 kg/week. The water vapour carries away no solids, so that these would simply accumulate in the cooling system and ultimately fur it up.

The amount of water purged depends on the concentration rise which is allowable in the system before scaling occurs. Denoting the ratio of the concentration in the water in circulation to that in the make-up water by r, a balance of the mass of dissolved solids shows that the amount of water to be purged as a fraction w of the water evaporated is given by:

$$w = \frac{1}{r-1} \tag{2.6.1}$$

and the make-up water has thus to be:

$$1 + w = \frac{r}{r-1} \tag{2.6.2}$$

times the mass of water evaporated.

2.6 Condensing equipment

A part of w may be carried away as droplets from the top of the tower or evaporative condenser (windage), but this is not enough as a rule and most of it has to be deliberately purged from the system to waste. This can be done by an automatic valve responding to a signal indicating the total solids in solution in the circulating water.

The allowable concentration ratio r depends mainly on the carbonate hardness (temporary hardness) of the make-up water. If this is nearly saturated with calcium carbonate, dissolved as bicarbonate, any rise in temperature will theoretically cause deposition of carbonate scale. Since the warmest part of the system is in the condenser tubes of a shell-and-tube condenser or outside the condenser tubes in an evaporative condenser, these are the places where scale deposition occurs. Chemical treatments are available, such as glassy polyphosphates or filming amines, which either retard the deposition of carbonate or which ensure that the scale forms in a soft layer which is flushed off the surfaces by the water flow. Even with these treatments in use, the concentration ratio r should not exceed a value of 1.5–2.0 with very hard UK waters, and eq. (2.6.1) then shows that the fraction w purged has to be between 1 and 2 times the water evaporated.

Water softening can be used to reduce the amount of purge required. The relatively insoluble calcium carbonate is replaced by soluble sodium carbonate and the concentration ratio can be allowed to rise considerably, but not of course indefinitely. Purging is still required to carry away dissolved solids at the rate at which they enter the system, but because they are carried away in a stronger solution less purge water is needed.

In addition to scale formation, corrosion, growth of algae and industrial dirt often cause trouble in cooling systems. Corrosion is usually countered by ensuring that the pH value of the water in circulation is slightly on the alkaline side, say 7.0–7.5, though in special circumstances some extra treatment may be needed. If water softening is employed it is common practice to leave a little residual hardness as a counter to corrosion, of the order $20/10^6$. Green algae are a problem mainly in the older type of tower in which water is distributed from troughs at the top which are open to the atmosphere and sunlight. Continuous or regular periodic dosing with toxic chemicals is required to keep algae in check. Dirt accumulates in the cooling system as it is washed out of the atmosphere by the sprays, and much of it ends up as a thin layer of slime in the condenser tubes. The particles are usually too fine to be stopped by the pump suction strainer. Very little can be done to prevent fouling of the tubes by dirt, and they have to be cleaned periodically.

Air-cooled condensers

In an air-cooled condenser, refrigerant condenses inside the tubes and air passes over the outside. Because the coefficient of heat transfer on the air-side is always much lower than that on the refrigerant side it is worth using fins on the air-side. Various types of construction have been used, but the most usual form today is that shown in Fig. 2.6.7, in which a grid or grids of tube carry plate fins held on by expansion of the tube. If the fins are punched from sheet metal, often aluminium, with holes in them to fit the tubes, it is clear that the fins cannot be assembled onto the grid when all the return bends are in position. Hence the grid has to be made from pre-formed U-tubes onto which the fins are assembled and the return bends subsequently brazed on. In some designs the fins are punched with elongated holes to pass over a pair of tubes, and in others they are slotted to allow assembly from the side. Such fins can be fitted to a tube grid bent into final shape, but at the expense of losing contact between the fin and the tube over half the circumference of each tube.

The flow of air through the condenser is usually provided by a propellor or aerofoil type fan. Small condensers are frequently mounted upright on the same baseplate as the compressor, drive motor and liquid receiver, and the fan is attached to the end of the motor shaft, usually by screws onto the drive pulley. Such an arrangement on a baseplate is called a condensing unit. Small condensing units may be sited inside a building near the

Fig. 2.6.7. Arrangement of an air-cooled condenser with three tube grids in parallel, fed from a header and draining to a header. The end plates, of thicker material than the fins, are provided with bolt holes for mounting the condenser. The width W is called the finned length, and the area $W \times H$ is called the face area.

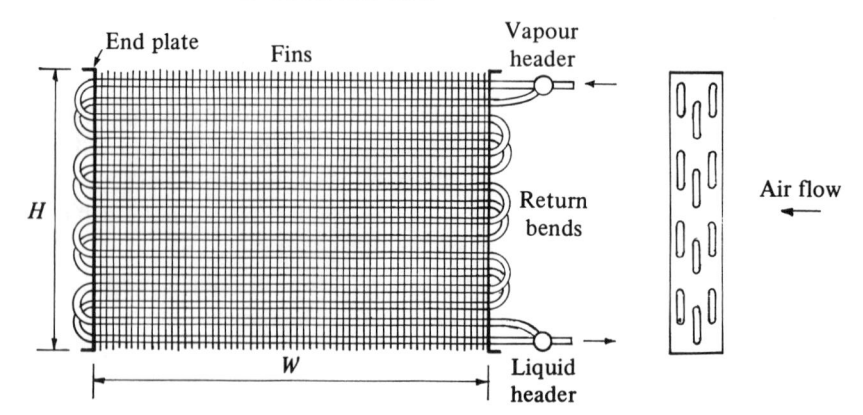

2.6 Condensing equipment

evaporators which they serve, for example on the top of a small cold room. Large condensing units, however, have to be remotely sited, for example on the roof of a building, to ensure sufficient supply of cooling air to them.

Until some years ago most air-cooled condensers were used in this way as part of a condensing unit. The increasing cost of water and the troubles associated with cooling water systems have encouraged the use of air-cooled condensers in much larger sizes, and these are mounted away from the compressors and motors which may occupy a special plant room. The condensers are often sited on the roof of the building, usually with the tube grid in the horizontal position with the air flow in a vertical direction. With this layout it is possible to hide the condensers from view at ground level by a low parapet wall.

Small hermetically sealed refrigeration systems such as those used in domestic refrigerators have no drive shaft for a fan and so have to employ a separate fan motor or to dispense with forced circulation of air. A condenser cooled by natural convection, called a 'static' condenser, is the usual practice in small domestic refrigerators. It is constructed as a rule from a grid of tube with wire fins which is fitted at the back of the cabinet, or sometimes as a grid of tube brazed to the steel sheet which forms the back of the cabinet and acts as a fin. In some designs the grid of tube is attached to the inside surface of the cabinet side panels, an arrangement called a 'skin condenser'.

Larger hermetically sealed systems used in the bigger freezers and room coolers employ a fan with its own motor.

Condenser pressure control

Although it was seen in Chapter 1 that a low condensing pressure favours a high coefficient of performance, it is sometimes necessary to maintain a high condensing pressure for practical reasons.

When the condensing temperature varies with the outside ambient air temperature, as in systems using water-cooled condensers with cooling towers, evaporative condensers and air-cooled condensers out of doors, the condensing pressure during the cold seasons of the year can be quite low. If the expansion valve is sized to pass the required flow rate of refrigerant under these conditions, i.e. when the difference in pressure across it is low, it will be much too large when the condensing pressure is high, so causing erratic operation and hunting.

The problem is most severe in installations which operate at a relatively high design evaporating temperature, i.e. in those for air-conditioning using direct expansion of the refrigerant. For example, the evaporating temperature for some computer rooms may be of the order 15°C. In a temperate climate the ambient air temperature will be lower than this during much of the year, and although the condensing temperature is higher than the ambient air temperature, the pressure difference across the valve may sometimes vanish completely.

For water-cooled condensers with cooling towers the condensing pressure can be maintained by reducing the flow rate of water through the condenser. An automatic regulating valve sensitive to the condensing pressure is used. With very little heat going to the tower, however, there is a risk of the water freezing there, and some further measures are usually needed. The fan or fans may be switched off or run at reduced speed, the air flow may be restricted by dampers, or sections of the tower may be closed down. An electric heater in the sump is provided as a last resort.

With evaporative and air-cooled condensers, reduction of the air-flow may be used to control the condensing pressure directly. Reducing the speed of the fans or switching them off is often ineffective, however, because natural convection may be strong enough to depress the condensing pressure, especially at low load. Air control dampers can be more effective but are likely to be more expensive if they are automatically operated by position motors.

An effective and relatively cheap method of maintaining condensing

Fig. 2.6.8. Two arrangements for maintaining condensing pressure at low ambient temperatures. In (a) the valve B should be close to the condenser to make the liquid back-up as quickly as possible.

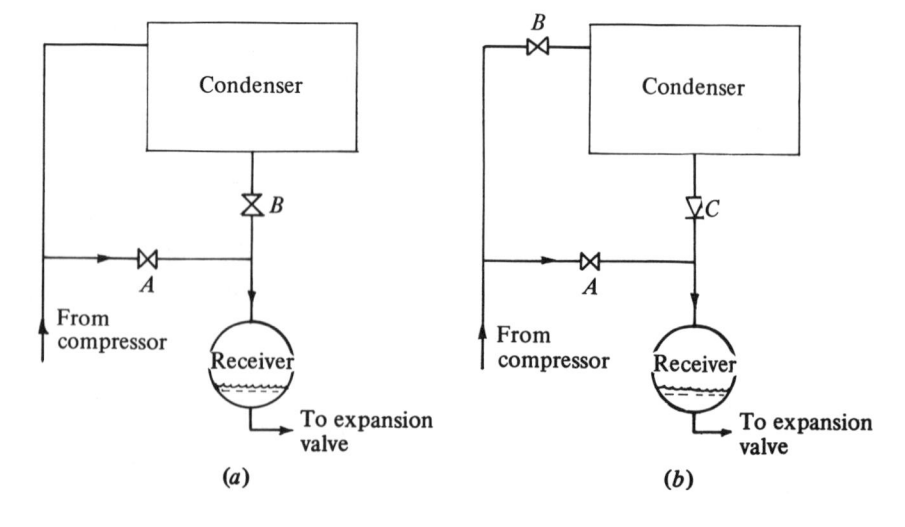

2.6 Condensing equipment

pressure is shown in Fig. 2.6.8(a). The condenser normally drains into a liquid receiver whence the liquid goes to the expansion valve. Valve A shown is an automatic valve responsive to downstream pressure, i.e. receiver pressure, arranged to close as the receiver pressure rises. Valve B is an automatic valve responsive to upstream pressure, i.e. condenser pressure, arranged to close when the condenser pressure falls. At start-up with low pressure in both condenser and receiver, A is open and B is closed. Vapour is pumped into the condenser and the receiver, and the pressure in these rises. At high ambient temperature the controlled pressure is eventually approached and A begins to shut and B to open. At controlled pressure A is closed and B is fully open, and the condensed liquid drains freely into the receiver. At low ambient temperature, however, the closed valve B retains liquid in the condenser and reduces the surface area available for heat transfer and condensation. Heat transfer from the retained liquid still takes place of course, but the effect of this is simply to subcool the liquid, not to reduce the condenser pressure. The condenser pressure therefore continues to rise as the condenser fills with liquid. The hot gas pumped into the receiver condenses there and raises the receiver temperature and pressure. As the controlled pressure is approached valve B begins to open and A to close. An equilibrium is eventually reached with the condenser partly filled and with some vapour going into the receiver directly. This method is commonly called a 'condenser back-up' method, though there is of course no actual flow of liquid from the receiver to the condenser. It should be noted that there must be enough refrigerant charge in the system to ensure that there is still liquid in the receiver when the condenser is partly filled.

In systems which operate intermittently, cycling on a thermostat, the above method does not respond fast enough to give a rapid rise of receiver pressure when the compressor is first switched on. If the evaporator works on a pump-down cycle, as described in section 2.14, with a low-pressure cut-out switching the motor off, the evaporator can be pumped down and the compressor stopped before any refrigeration has been done. In these circumstances the method shown in Fig. 2.6.8(b) can be used. Valve B is of the same type as B in Fig. 2.6.8(a) but placed before the condenser. At start-up hot gas is pumped into the receiver only, because B is closed and the non-return valve C prevents flow of gas up to the condenser. The receiver therefore acts as the condenser until the pressure is high enough to open valve B and to begin to close valve A.

Condensers cooled by town mains water are also provided with water regulating valves, though, as a rule their purpose is to prevent waste of water rather than to raise condensing pressure. In particular, when the compressor is switched off the condenser pressure falls and the automatic valve shuts off the water.

2.7 Evaporators

The function of the evaporator is to generate vapour with as high a heat transfer coefficient as possible without priming, i.e. without carrying liquid droplets over into the suction pipe to the compressors. Before describing the main types, attention will be drawn to the difference between the heat transfer coefficient to a boiling liquid and that to a dry vapour. Because of the greater thermal conductivity of the liquid the heat transfer coefficient to liquid is much greater, by a factor of ten or more, than that to a vapour. Hence the aim of evaporator design is to provide conditions in which liquid can be brought into contact with the pipe walls and to remove the vapour formed as rapidly as possible.

The three principal types of evaporator are shown diagrammatically in Fig. 2.7.1. They are called (a) dry expansion or 'once-through', (b) recirculation or 'flooded' and (c) shell-and-tube evaporators.

(a) *Dry expansion evaporators.* In such an evaporator refrigerant enters at one end of a tube as a two-phase mixture from the expansion valve, and is totally evaporated by the time it reaches the other end of the tube. The pipe may be immersed in a fluid to be cooled or in air, still or moving. Dry expansion was formerly common in large plants for direct cooling of cold rooms and for ice making by submerging the evaporator in the brine tank.

Fig. 2.7.1. The three principal types of evaporators: (a) 'once-through' or dry expansion type in which refrigerant is totally boiled off at the end; (b) recirculation type, also called 'flooded' type, in which liquid remaining at the end of the tube is returned to the inlet, with separation of vapour; (c) shell-and-tube type, the refrigerant boiling outside tubes carrying the fluid to be refrigerated.

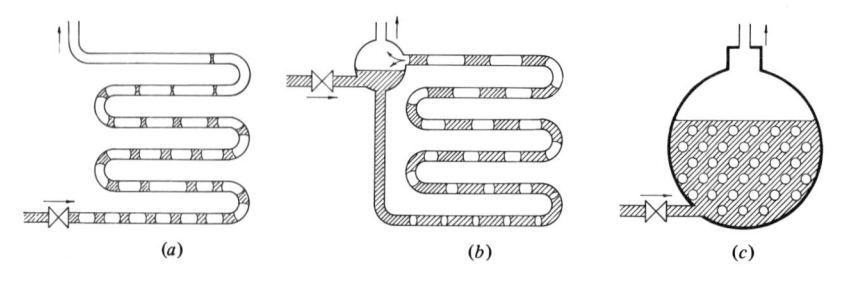

2.7 Evaporators

The expansion valve was regulated by hand to give the desired vapour state, wet or dry, at the outlet. For the early forms of compressor which could take in wet vapour without harm, the arrangement was quite suitable and good wetting of the internal surface of the coil could be obtained. When dry vapour is required at the outlet for modern types of compressor, it is not so efficient because an appreciable length of the tube has to be used to dry the vapour and ensure that no liquid droplets remain. Nevertheless the simplicity of the dry expansion evaporator has much to commend it, and evaporators of this type are very widely used in the smaller sizes.

Under normal operating conditions the volume of liquid in a dry expansion evaporator is about 15–20% of the internal volume of the tubes. Assuming that this liquid is uniformly distributed along the tubes, and occupying a segment of it, the wetted periphery would be about 30–34% of the circumference of the tube, and the effective area for heat transfer would be 30–34% of the internal surface area. The rest of the area is wasted. It is possible, however, by increasing the mass flow rate through a fixed cross-sectional area of tube to make the flow become annular and to wet a

Fig. 2.7.2. One method of feeding an evaporator for air cooling with six circuits, using a distributor. The tubes from the distributor are made of equal length to equalise resistance. The outlets of the circuits are manifolded into the vertical header, the suction being taken from the top or from the bottom.

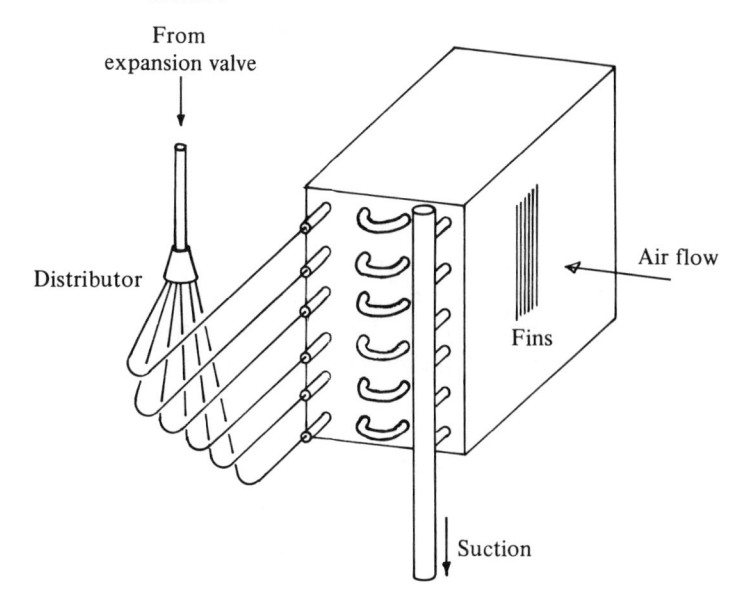

larger proportion of the surface. This of course causes a greater pressure drop which is discussed further in Section 3.15.

Dry expansion is now principally used in conjunction with the thermostatic expansion valve, described in the next section, which regulates the flow rate of liquid entering to give slightly superheated vapour at the exit. The main field of application is in small evaporators for cooling air in cold rooms. The evaporator is constructed of tubes with fins, somewhat similar in construction to the condenser shown in Fig. 2.6.4, but with a different arrangement of the tubes for the refrigerant. In a condenser the object of the flow arrangement is to remove liquid from the tube surface as much as possible; in an evaporator the object is to keep liquid in contact with the surface. Furthermore, it is more difficult to distribute refrigerant equally between parallel paths in an evaporator than in a condenser, and this often requires a more complicated interlacing of the coils. Many different arrangements are in use, but a common one is indicated in Fig. 2.7.2. The distributor apportions liquid to the several paths, usually called 'circuits', with the general direction of flow of refrigerant horizontal and opposed to that of the air stream. The refrigerant vapour is brought into thermal contact with the warmest air so that a small amount of superheat can be obtained to operate the thermostatic expansion valve.

(b) *Recirculation evaporators.* In the recirculation evaporator the refrigerant is not totally evaporated in the pipes. A mixture of liquid and vapour leaves the end of the heat transfer section. The mixture is taken into a separator vessel, known as an accumulator, suction separator or surge drum, in which disengagement of the vapour can occur. The vapour goes off to the compressor and the liquid remaining passes through the tubes again. The form of evaporator shown in Fig. 2.7.1(b) is called a gravity recirculation type because the circulation is produced by the difference in the specific weights of the liquid in the downcomer and the two-phase mixture in the riser. In practice several risers in parallel, in the forms of coils or serpentine grids, may be provided between a bottom header fed by a large downcomer and a top header which discharges into the surge drum. The rate at which refrigerant is admitted to the drum is regulated by an expansion valve controlled by a float or other device. A large gravity circulation evaporator for cooling water or a process liquid is shown in Fig. 2.7.3.

Much better wetting of the internal surface of the evaporator is obtained with the recirculation method than with dry expansion, the fraction of liquid in the pipes by volume being of the order 50%. This type of

2.7 Evaporators

evaporator is sometimes referred to as 'flooded' for that reason.

With gravity circulation it is necessary for the drum to be above the level of the coils or grids. This restriction may be removed if the liquid is fed to the coils by a pump. Pump circulation is often used when refrigeration has to be distributed to several places in a building, such as the various evaporators of a cold store, or processing stations in ice-cream or food factories. The surge drum is situated near the compressors, with the pump(s) below it,

Fig. 2.7.3. A large gravity recirculation evaporator for water chilling duties. The water flows across the vertical refrigerant tubes and is forced by the pass plates to cross them several times to increase the velocity. The float valve sends a signal to the pilot valve which controls liquid refrigerant inlet to the drum. The hand expansion valve can be used during start-up or service. The other valves are isolating stop valves. (By courtesy of UDEC Ltd.)

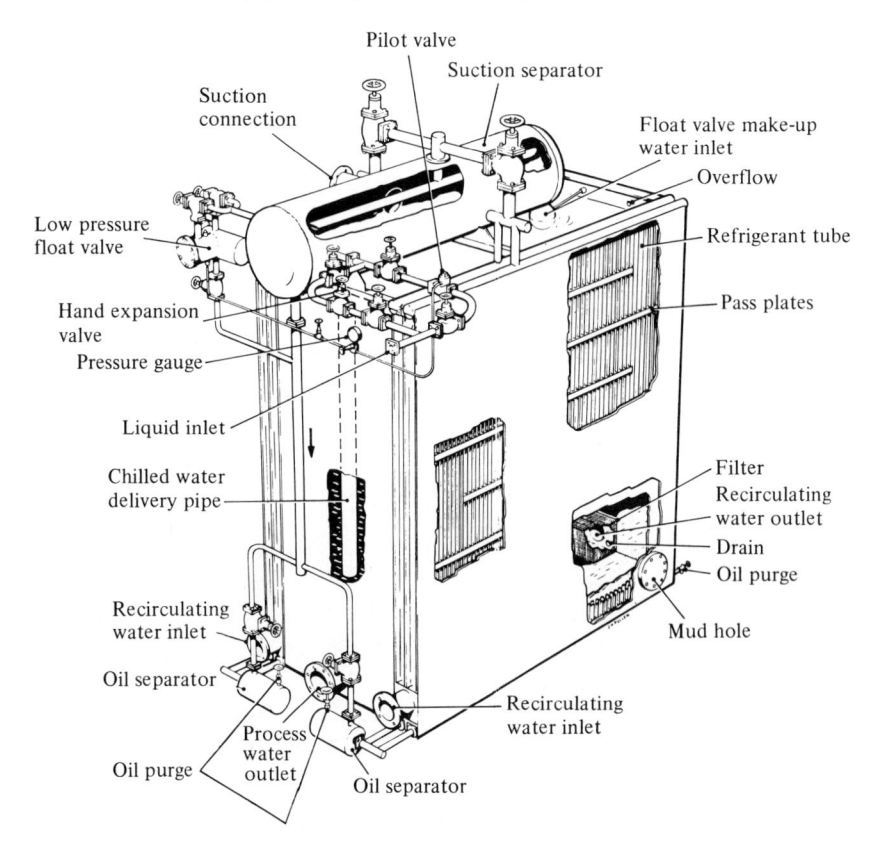

delivering liquid into a common header from which branches take off to the several evaporators, as shown in Fig. 2.7.4. The supply to each evaporator is set either by an orifice of a calculated diameter or by a hand regulating valve which is tuned to the demand on the evaporator during commission-

Fig. 2.7.4. Diagrammatic arrangement of the piping around the surge drum in a pump circulation system. Cold liquid from the surge drum is pumped to the flow header and thence to the several evaporators. The flow through each evaporator is trimmed by a hand regulating valve HRV1. A calibrated orifice may be used instead. The valve arrangements at the evaporator for automatic control and hot gas defrosting are shown in Fig. 2.16.1. The relief valve RV1 may be required, with automatic operation, to relieve the pump delivery to the drum when the evaporators are all shut off. The by-pass line enables an evaporator to be drained of liquid for service. In some systems the evaporators are fed from the top. The liquid supply to the drum is shown as passing through a level controlled expansion valve, but it might be an electronic level detector controlling a solenoid valve. The expansion valve is by-passed by the hand expansion valve HRV2 for use during service of the float valve. RV2 is the main relief valve for the low side of the system. The stop valves shown are for isolating purposes, for service or maintenance.

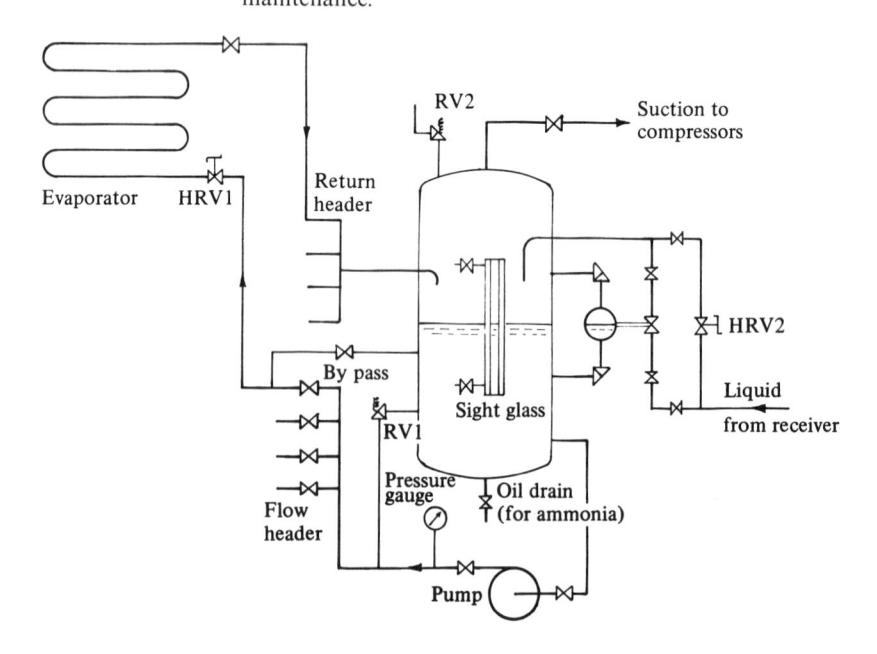

2.7 Evaporators

ing of the plant. The mass rate of circulation is several times the evaporation rate, so that good flooding of the tubes is obtained. The flow in the tubes may be upward, downward or horizontal. Upward flow requires more charge and a bigger surge drum, but gives better heat transfer rates than downward flow.

Refrigerant pumps are of positive displacement or centrifugal type, with a shaft seal or packed gland, or with hermetic construction, usually called 'canned' pumps. Since the liquid refrigerant in the surge drum is saturated, i.e. at its boiling point, it is liable to cavitate in the pump, i.e. to flash into vapour, which stops the pump working. It is necessary to have sufficient height of liquid over the pump suction flange to prevent this, the height depending on the type of pump. The value is usually given by the pump manufacturer as a 'net positive suction head', meaning the required difference between actual pressure and saturation pressure expressed as the height of a column of water. In calculating the actual positive suction head the pressure drop in the downcomer should be taken into account. The design of the suction piping is discussed by Lorentzen (1963) who has also treated the design of the pump circulation system in a later paper (1976).

A popular type of positive displacement pump is shown in Fig. 2.7.5. When a centrifugal pump is used, an anti-vortex plate is often required in

Fig. 2.7.5. The Douglas–Conroy pump. The pins $D1$, $D2$ and $D3$ are fixed to the driven plate and drive the drum G. The crescent H is stationary. The pins $D1$, $D2$, $D3$ are drilled to relieve fluid trapped in the jaws of G during engagement.

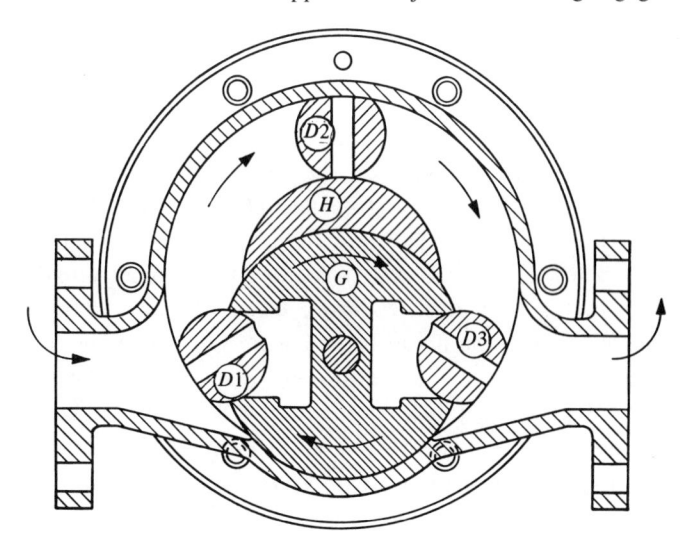

the surge drum to prevent the transmission of the suction vortex up to the liquid surface.

Surge drums for pump circulation are made both horizontal and vertical. In either event the design has to be based on the following principles. When the liquid level in the drum has the lowest value there must still be enough liquid to provide a suction head for the pump. Above this level there is a space capable of holding the total charge of the connected evaporators, and above this level there is room for disengagement of liquid droplets from the vapour to occur. The velocity allowed in the disengagement zone is of the order 0.5 m/s, and this fixes the diameter of the drum.

A special form of circulation evaporator which uses the kinetic energy of the jet of refrigerant leaving the expansion valve is occasionally used. It is generally similar to Fig. 2.7.1(b), but the two-phase mixture from the valve is injected into the bottom of the downcomer in a direction to promote circulation up the heated tubes. Large air-cooling evaporators of this type were described by Griffiths (1936), and a similar system is commonly used in the jackets of scraped surface heat exchangers, known by the proprietary name 'Votators'. They are used for chilling margarine and other viscous materials during manufacture, and for freezing ice cream.

(c) *The shell-and-tube evaporator.* Figure 2.7.1(c), is a (usually) horizontal vessel containing boiling liquid refrigerant through which pass tubes carrying the fluid to be cooled. The construction is similar to that of the shell-and-tube condenser, with two or more passes for the liquid, except that the evaporator shell may not be tubed so near to the top in order to provide a space for disengagement of liquid droplets. This is usual in small evaporators, but in large ones it may be more economical to provide one or more vapour domes, as shown diagrammatically in Fig. 2.7.1(c), for disengagement, and use more of the cross-section of the shell for tubes. Liquid refrigerant is admitted to the shell by an expansion valve controlled by a float or other level-sensing device. The point of admission of the refrigerant is sometimes below the liquid level and sometimes above. Admission below the liquid level gives more turbulence and stirring with somewhat better heat transfer rates, especially at low difference of temperature between the liquid being cooled and the refrigerant. On the other hand the admission of the large volume of vapour which forms in the expansion valve below liquid level causes violent surges if the valve does not modulate properly, i.e. tends to give all or nothing. These surges tend to cause 'hunting' of the system, i.e. more or less periodic fluctuations of suction pressure.

2.7 Evaporators

A special type of shell-and-tube evaporator has been used with fluorinated refrigerants. Instead of covering the tubes with liquid refrigerant, it is sprayed over them by a pump drawing from the bottom of the shell where there is a (low) liquid level, and to which unevaporated liquid drains back. The purpose is to overcome the 'static head penalty' which results from the fact that in a deep pool of liquid the pressure at the bottom is higher than at the top and hence the bottom liquid has to be raised to a higher temperature before boiling. It also serves to reduce the charge of liquid.

Fig. 2.7.6. Construction of a dry expansion water chiller for use with a thermostatic expansion valve on fluorinated refrigerants. Refrigerant from the expansion valve enters at 1 and leaves totally evaporated at 2. The end cover 3 is bolted to the tube plate 4 and compresses gaskets on each side of the division plate 5 which has pass plates arranged to provide four passes of the refrigerant. The water enters at 6 and leaves at 7 after making a zig-zag passage across the nest of tubes, directed by the polythene segmental baffle plates 8 held in position by the spacers 9. 10 is insulation and 11 is a pocket into the water side to take the bulb of a thermostat for a low-temperature cut-out.

For other secondary refrigerants and for operation at lower temperatures an alternative division plate can be used to provide two passes and so reduce pressure drop on the refrigerant side. (By courtesy of Hall-Thermotank Products Ltd.)

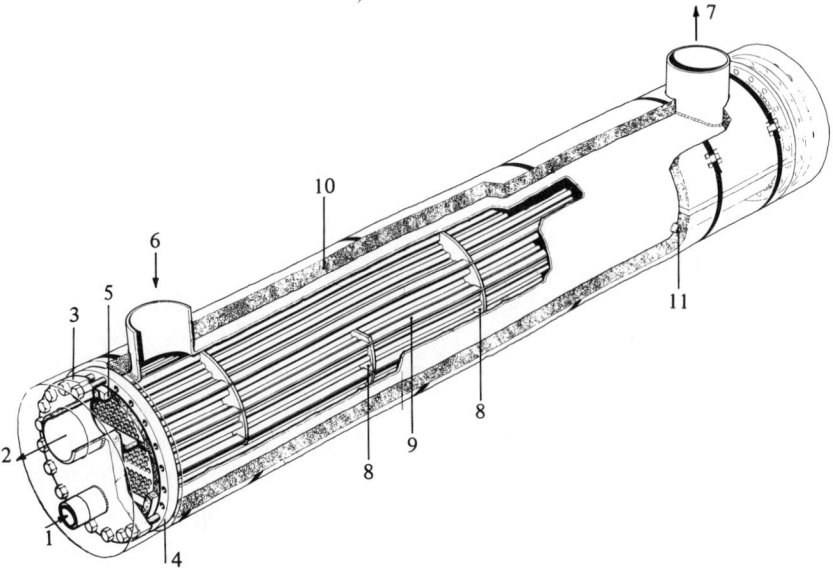

The ordinary shell-and-tube evaporator is the most widely used type for industrial liquid cooling duties, particularly with ammonia. For chilled water production in air-conditioning plants it has the disadvantage of being somewhat susceptible to damage by freezing of water inside the tubes. The formation of ice in a tube reduces the flow rate through that tube and encourages further formation. Furthermore a relatively large charge of refrigerant is required. For air-conditioning duties, and others requiring chilled water at about 4.5 °C or higher, a common design has the refrigerant boiling inside the tubes and the water in the shell. Because of the low water velocity across the section of the shell it is necessary to provide baffles which make the water traverse the tubes several times (Fig. 2.7.6). Admission of refrigerant to the tubes is regulated by a thermostatic expansion valve, so that these water chillers function like type (a) above. It will be seen in the next section that the thermostatic valve requires a superheat of about 4.5 K to operate it, and consequently it is not possible to chill water below 4.5 °C without the evaporating temperature falling below 0 °C, when freezing would occur. The gravity recirculation evaporator shown in Fig. 2.7.3, with level control, does not have this limitation and it can produce chilled water at about 1.7 °C.

Water chillers must be protected against freezing, not only during operation by a thermostat to stop the refrigeration when the temperature approaches 0 °C, but during shut down in cold weather by electric heaters under the insulation, which should be connected to the supply in such a way that they cannot be inadvertently switched off.

2.8 Expansion valves: evaporator feed regulation

The expansion valve is essentially an orifice or other restriction which maintains the pressure difference between the two sides of the plant and in which throttling expansion of the refrigerant takes place. Because the flow rate of refrigerant varies with the demand in the evaporator, however, the orifice has to be adjustable to suit the conditions. Originally this was done by hand, and hand regulating valves are still used for special purposes. They may be incorporated in plants for use in the event of the automatic controls being out of service, or for use during start-up.

A hand regulating valve has a tapered needle which can be moved in and out of an orifice by a screwed spindle. In large valves the handwheel carries a pointer which moves over a graduated circle so that the exact position of the needle can be determined and returned to when necessary. It should be noted that a hand expansion valve is not a stop valve and should not be used as such. Screwing the needle hard down into the orifice will damage it

2.8 Expansion valves: evaporator feed regulation

and the orifice edge. A stop valve for interrupting the supply of liquid completely is usually placed in advance of a hand expansion valve.

In modern refrigerating plants the flow rate of liquid into the evaporator is regulated by automatic means. The term 'expansion valve' often implies the valve itself and the means of regulating it. The fundamental requirement is to admit liquid refrigerant to the evaporator at the rate required by the load and by the pumping rate of the compressor, at the same time keeping the evaporator surface as fully wetted as possible. The method of doing this depends on the type of evaporator. Four main types of control will be described: (*a*) superheat control, (*b*) evaporator pressure control, (*c*) high side level control and (*d*) low side level control. Of these (*a*), (*b*) and (*c*) are applicable to dry expansion evaporators, whilst (*c*) and (*d*) are applicable to flooded and shell-and-tube evaporators.

Before describing these methods in detail, the hand regulation of a dry expansion evaporator will be considered. Supposing an evaporator like that of Fig. 2.7.1(*a*), with a hand regulating valve, let the valve initially be set to be just open, i.e. admitting very little refrigerant to the evaporator. This refrigerant is then easily boiled off and the compressor pumps the vapour away. Since there is little flow, however, the compressor initially can pump away the vapour formed faster than liquid is admitted. Hence the pressure in the evaporator falls to a value at which the density of the vapour at the suction makes the mass flow rate through the compressor equal to that through the valve. At the same time the evaporator has little liquid in it. At this condition there is a relatively large difference between the temperature of the fluid being cooled by the evaporator and the boiling temperature of the refrigerant. This state of working is inefficient, because the vapour has to be pumped over a larger pressure ratio to bring it up to the condensing pressure. Furthermore, the poor wetting of the evaporator surface means that the rate of heat transfer is low and not much cooling is being done. If for example brine were being cooled, the operator would notice that it was coming out not much lower in temperature than it went in.

In practice the operator would open the expansion valve to admit more refrigerant to the evaporator. Observing the pressure gauge on the suction pipe he would then notice a rise in pressure, because the compressor can handle the increased flow rate now with an increased density of vapour at the suction. At the same time the operator would see the temperature of the brine leaving become lower. Further opening of the valve would again produce a higher suction pressure and more cooling of the brine. The plant is now working more efficiently.

If the operator kept opening the valve further, however, there would come a point when he would notice that the suction pipe was beginning to frost over, indicating to him that surplus liquid was now leaving the evaporator and going to the compressor. Early compressors were capable of pumping this wet vapour but modern compressors are not. So the valve has to be shut back again. The optimum condition is when the suction pressure has the highest value, indicating good wetting of the evaporator, consistent with no liquid returning to the compressor. Ideally one would like the vapour leaving the evaporator to be just dry to make the best use of the evaporator surface.

(a) *Superheat control.* In principle the judgement exercised by the operator above could be done automatically by measuring the suction pressure, measuring the temperature of the vapour leaving the evaporator, and comparing this temperature with the saturation temperature of the refrigerant corresponding to the suction pressure. When the difference is zero there is no superheat, but the refrigerant leaving the evaporator might be wet. Hence one has to accept a few degrees of superheat to ensure that the state is just on the dry side of saturation, and to provide a signal to operate the valve.

How this is done in the thermostatic expansion valve is shown

Fig. 2.8.1. Principle of operation of a thermostatic expansion valve. The needle in the orifice is positioned by the difference in pressure, acting against the spring, between the volatile fluid in the bulb and the refrigerant in the evaporator. The flexible element is shown as a corrugated diaphragm but it is often a metal bellows.

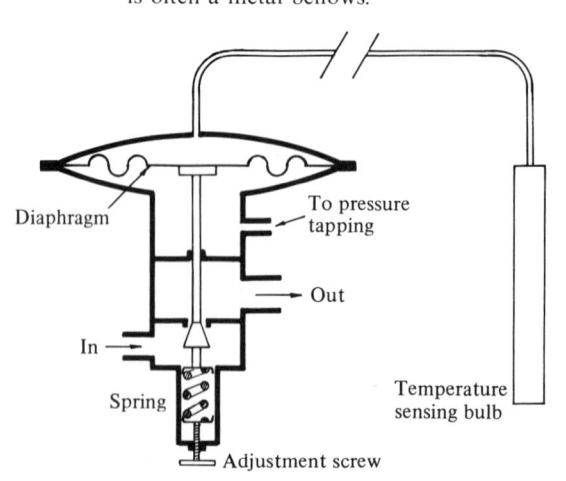

2.8 Expansion valves: evaporator feed regulation

diagrammatically in Fig. 2.8.1. A needle which varies the area for flow through an orifice is connected to a flexible diaphragm or bellows. This in turn is positioned by the difference of two forces acting against a spring. The two forces are: the pressure, acting on the topside of the diaphragm, which is produced by a volatile fluid in a bulb or 'phial' clamped to the other end of the evaporator, and the pressure in the evaporator acting on the underside of the diaphragm. The bulb assumes approximately the temperature of the vapour leaving the evaporator and produces a pressure which is a function of that temperature. The evaporator pressure is a function of the boiling temperature in the evaporator. Hence the force which positions the needle is a function of the superheat of the vapour leaving.

Supposing that the bulb is charged with the same fluid as the refrigerant in the system, the conditions are shown on a plot of vapour pressure against temperature in Fig. 2.8.2(a). The spring force F, say, divided by the area of the diaphragm, A, represents an equivalent pressure which has to balance the difference between the pressure in the bulb p_b and the pressure in the evaporator p_e, i.e.

$$F/A = p_b - p_e \tag{2.8.1}$$

Fig. 2.8.2. The operation of a thermostatic expansion valve shown on pressure–temperature graphs. (a) In a straight charged valve the fluid in the bulb is the same as the refrigerant in the system, and the same spring force F gives a lower superheat at the higher evaporating temperatures. (b) In a cross-charged valve the fluid in the bulb is not the same as the system refrigerant and it is possible to choose it so that the superheat remains approximately constant. (c) A valve with a maximum operating pressure (MOP) can be made by charging the bulb and bellows volume with a limited amount of liquid. This dries out at point D as the temperature rises, and the force applied to open the valve does not increase so rapidly thereafter. The controlled superheat therefore increases.

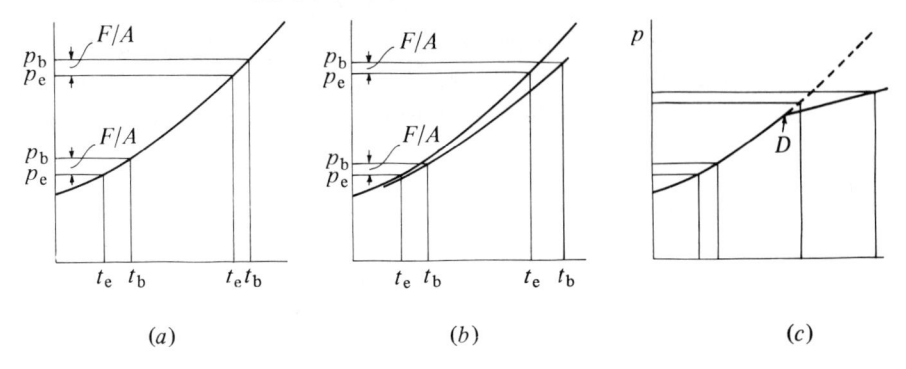

at equilibrium. It is clear from the shape of the curve that the same spring force corresponds to different values of the superheat ($t_b - t_e$) at different places on the curve. The valve would regulate the liquid supply to give a lower superheat at high evaporating temperatures, i.e. tend to flood the evaporator, and to give a higher superheat at lower evaporating temperatures, i.e. tend to starve the evaporator. This is a characteristic of a valve whose bulb is charged with the same fluid as the refrigerant, called a 'straight' charged valve.

By charging the bulb with a different fluid to that in the system, it is possible to correct the error and to achieve almost constant superheat at different evaporating temperatures, as shown in Fig. 2.8.2(b). Such a valve is called 'cross charged'. Another way of doing this is used in the 'double bellows' valve, in which the bellows which takes the bulb pressure is somewhat smaller than that which takes the evaporating pressure.

In some applications of thermostatic expansion valves it is desirable to limit the maximum evaporating pressure, i.e. make the valve shut down above a certain pressure. As shown in Chapter 4, a reciprocating compressor has a power characteristic with a maximum at a certain evaporating pressure. When a refrigerator such as an ice-cream or frozen-food cabinet is being brought down to temperature on first switching on, the power required would pass through this peak if the evaporator were kept fully supplied with liquid. It would be uneconomical to provide an electric motor capable of providing this peak power for the very few times that it is needed. A motor capable of providing the normal running power only can be used if the expansion valve is made to starve the evaporator and limit the pressure during the pull-down period when the heat load on the evaporator is high. This is done by charging the bulb of the expansion valve with a limited mass of charge, so that it is entirely vapour above the desired maximum evaporating pressure and temperature. Below this temperature liquid and vapour are present in the bulb and the normal pressure–temperature relation is obtained. This is shown in Fig. 2.8.2(c), where point D is the dry-out point of the charge. Above this temperature the controlled superheat increases considerably and the valve tends to shut. A valve of this type is called 'limit charged' or 'dry charged' to a certain maximum operating pressure or MOP. It is also possible to produce a similar effect by mechanical means incorporating a lost motion device in the valve mechanism.

Limit charged valves are unfortunately liable to a failure known as 'reversal' or 'cross ambient' failure, which can occur when the head of the valve is allowed to become cooler than the bulb. In that event the small

2.8 Expansion valves: evaporator feed regulation

diagrammatically in Fig. 2.8.1. A needle which varies the area for flow through an orifice is connected to a flexible diaphragm or bellows. This in turn is positioned by the difference of two forces acting against a spring. The two forces are: the pressure, acting on the topside of the diaphragm, which is produced by a volatile fluid in a bulb or 'phial' clamped to the other end of the evaporator, and the pressure in the evaporator acting on the underside of the diaphragm. The bulb assumes approximately the temperature of the vapour leaving the evaporator and produces a pressure which is a function of that temperature. The evaporator pressure is a function of the boiling temperature in the evaporator. Hence the force which positions the needle is a function of the superheat of the vapour leaving.

Supposing that the bulb is charged with the same fluid as the refrigerant in the system, the conditions are shown on a plot of vapour pressure against temperature in Fig. 2.8.2(a). The spring force F, say, divided by the area of the diaphragm, A, represents an equivalent pressure which has to balance the difference between the pressure in the bulb p_b and the pressure in the evaporator p_e, i.e.

$$F/A = p_b - p_e \qquad (2.8.1)$$

Fig. 2.8.2. The operation of a thermostatic expansion valve shown on pressure–temperature graphs. (a) In a straight charged valve the fluid in the bulb is the same as the refrigerant in the system, and the same spring force F gives a lower superheat at the higher evaporating temperatures. (b) In a cross-charged valve the fluid in the bulb is not the same as the system refrigerant and it is possible to choose it so that the superheat remains approximately constant. (c) A valve with a maximum operating pressure (MOP) can be made by charging the bulb and bellows volume with a limited amount of liquid. This dries out at point D as the temperature rises, and the force applied to open the valve does not increase so rapidly thereafter. The controlled superheat therefore increases.

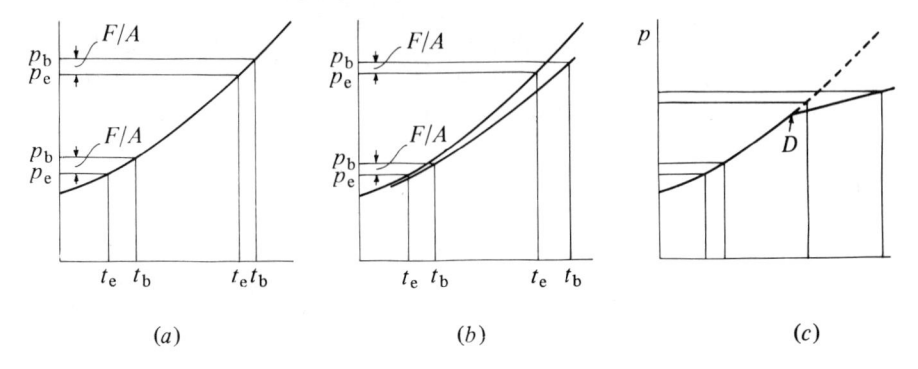

at equilibrium. It is clear from the shape of the curve that the same spring force corresponds to different values of the superheat $(t_b - t_e)$ at different places on the curve. The valve would regulate the liquid supply to give a lower superheat at high evaporating temperatures, i.e. tend to flood the evaporator, and to give a higher superheat at lower evaporating temperatures, i.e. tend to starve the evaporator. This is a characteristic of a valve whose bulb is charged with the same fluid as the refrigerant, called a 'straight' charged valve.

By charging the bulb with a different fluid to that in the system, it is possible to correct the error and to achieve almost constant superheat at different evaporating temperatures, as shown in Fig. 2.8.2(b). Such a valve is called 'cross charged'. Another way of doing this is used in the 'double bellows' valve, in which the bellows which takes the bulb pressure is somewhat smaller than that which takes the evaporating pressure.

In some applications of thermostatic expansion valves it is desirable to limit the maximum evaporating pressure, i.e. make the valve shut down above a certain pressure. As shown in Chapter 4, a reciprocating compressor has a power characteristic with a maximum at a certain evaporating pressure. When a refrigerator such as an ice-cream or frozen-food cabinet is being brought down to temperature on first switching on, the power required would pass through this peak if the evaporator were kept fully supplied with liquid. It would be uneconomical to provide an electric motor capable of providing this peak power for the very few times that it is needed. A motor capable of providing the normal running power only can be used if the expansion valve is made to starve the evaporator and limit the pressure during the pull-down period when the heat load on the evaporator is high. This is done by charging the bulb of the expansion valve with a limited mass of charge, so that it is entirely vapour above the desired maximum evaporating pressure and temperature. Below this temperature liquid and vapour are present in the bulb and the normal pressure–temperature relation is obtained. This is shown in Fig. 2.8.2(c), where point D is the dry-out point of the charge. Above this temperature the controlled superheat increases considerably and the valve tends to shut. A valve of this type is called 'limit charged' or 'dry charged' to a certain maximum operating pressure or MOP. It is also possible to produce a similar effect by mechanical means incorporating a lost motion device in the valve mechanism.

Limit charged valves are unfortunately liable to a failure known as 'reversal' or 'cross ambient' failure, which can occur when the head of the valve is allowed to become cooler than the bulb. In that event the small

2.8 Expansion valves: evaporator feed regulation

amount of liquid in the bulb can distil over to the valve head. If this happens it is the temperature of the head which determines the pressure in the bulb/tube system. The low pressure is then insufficient to open the valve. The remedy is to keep the valve head as warm as possible. This is done in some designs by circulating the warm refrigerant around the head via internal passages in the body of the valve.

'Adsorbent' charged valves contain activated charcoal with carbon dioxide gas in the bulb as the temperature sensing device. Since the gas cannot condense at the usual evaporating temperatures these valves are free from reversal.

For large evaporators of the dry expansion type, it may be found that one long tube would give too much pressure drop. A number of parallel paths or 'circuits' are then used. These can be fed by a single expansion valve fitted with a distributor, as indicated in Fig. 2.7.2. Distributors are of either centrifugal or pressure-drop types, though most combine the two methods to some extent. In the centrifugal type a vortex is created and tappings are taken from the periphery. The pressure-drop type uses quite a large part of the pressure difference between the condenser and the evaporator to create flow through a number of nozzles. The success of both arrangements depends on symmetry of the circuits. The distributor should be positioned with its axis vertical, and the feeder tubes for the circuits should be made the same length.

The type of valve shown in Fig. 2.8.1, with a tapping to take the pressure at the other end of the evaporator, is called 'externally equalised'. An externally equalised valve must be used whenever there is a significant pressure drop in the evaporator, and whenever a distributor is used. If the pressure drop is small a separate tapping is not necessary and the pressure on the downstream side of the orifice may be applied directly under the diaphragm. A bleed hole to the downstream chamber is required to sense the pressure there. A valve of this type is called 'internally equalised'.

In the discussion of superheat above it was assumed that the superheat was specified at one fixed position of the needle. A convenient reference position is when the needle is just on the seat with no reaction between needle and seat. The superheat at this point is called the 'closed superheat', and is usually of the order 4–5 K. To move the needle to its fully open position an increase of superheat of some 2 K or so is needed, giving an open superheat of 6–7 K. This range of 2 K is the throttling range or proportional band. The smaller it is the more sensitive the valve is, but the more liable to hunting or instability.

A point to note concerning the application of thermostatic expansion valves is that the temperature of the fluid being cooled must be sufficiently above the evaporating temperature to produce the required operating superheat, as already mentioned in connection with a dry expansion water chiller. Thermostatic expansion valves are not very suitable when a close

Fig. 2.8.3. Construction of an internally equalised thermostatic expansion valve. A – thermal bulb charged with a volatile fluid, generating pressure transmitted through the capillary tube B to the topside of the flexible diaphragm C; D – needle controlling flow, through the orifice and valve seat E; F – superheat spring exerting force upwards on the diaphragm; G – push rod transmitting force from diaphragm to needle; H – spring maintaining contact in the diaphragm/push rod/needle link; J – superheat adjusting screw; K – inlet strainer; L – capsule carrying the needle/orifice assembly. The arrow M shows the direction of flow which is up through the orifice and down outside the capsule to the outlet. (By courtesy of Teddington Refrigeration Controls Ltd.)

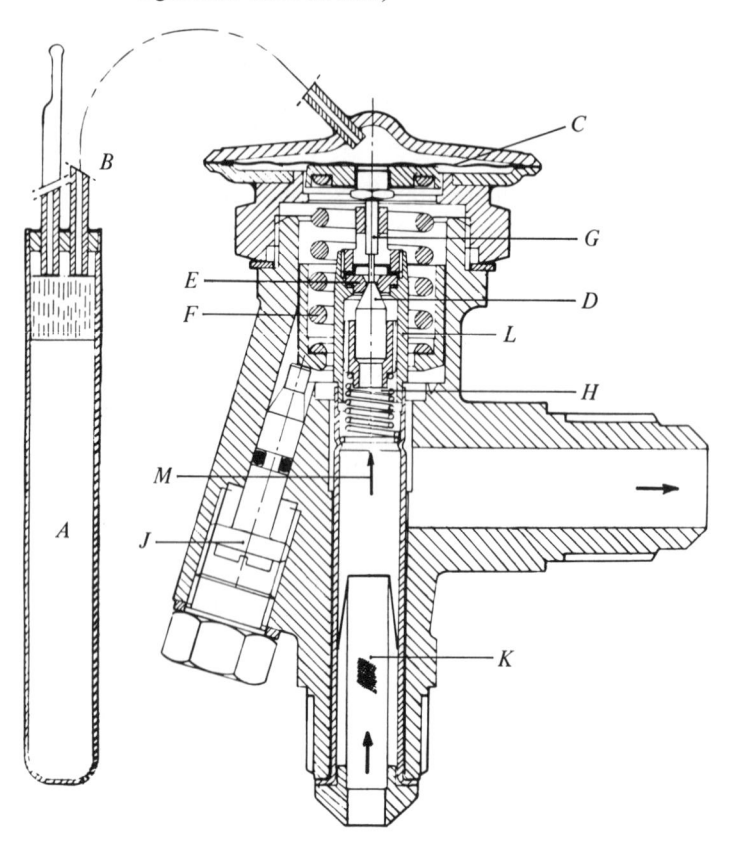

2.8 Expansion valves: evaporator feed regulation

approach between the fluid cooled and the evaporating temperature is desired. Advantage should of course be taken of counter flow to give the highest superheat.

A way around this difficulty is to use a suction to liquid line heat exchanger, as described in Chapter 3, putting the bulb of the thermostatic valve after the exchanger. This introduces extra expense if the heat exchanger would not otherwise have been contemplated, and possible instability in operation.

Thermostatic valves have been made in many different constructions, of which one is shown in Fig. 2.8.3. A bellows is also often used instead of a diaphragm. A longer stroke of the needle is possible with increased sensitivity, but an extra expense.

(b) *Evaporator pressure control.* The valve used to do this maintains an approximately constant pressure in the evaporator, opening when the evaporating pressure falls and closing when it rises. Since its introduction pre-dated that of the thermostatic expansion valve it became known as the 'automatic expansion valve', though the thermostatic valve would now better deserve that name.

The principle is shown in Fig. 2.8.4. Evaporating pressure acts on a diaphragm or bellows against a spring, opening and closing the valve. By compressing the spring the controlled pressure can be increased.

The constant pressure valve does not fulfill the basic requirement of

Fig. 2.8.4. Principle of operation of an 'automatic' expansion valve. As the evaporating pressure rises, the valve spindle is moved upwards against the spring and the valve moves towards closure.

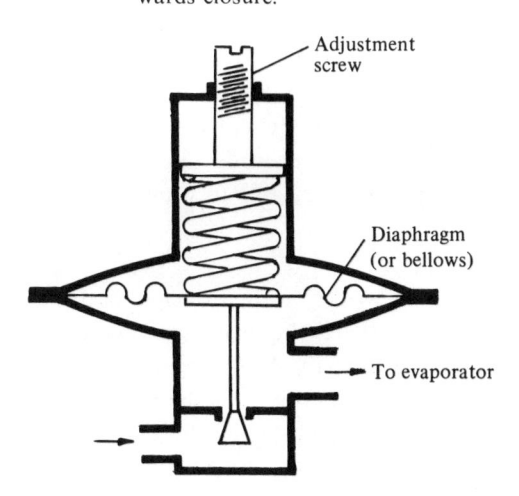

feeding liquid to the evaporator to match the load with good wetting of the evaporator surface. Suppose for example that steady conditions exist initially with the evaporator containing the correct amount of liquid to utilise the surface without carrying back droplets to the compressor. On an increase of load the rate of boiling increases and the pressure tends to rise because the compressor cannot pump away the extra vapour. The rise of pressure, however, tends to close the valve and bring the pressure back to its original value. The valve is now admitting less refrigerant to the evaporator than the compressor is pumping away, and the volume of liquid in the evaporator falls. The valve thus starves the evaporator of refrigerant at the time of increased load when it is wanted most. Putting matters another way, a constant evaporating pressure is not what is required when the temperature of the fluid being cooled can vary. As already seen, the highest evaporating pressure consistent with no liquid going to the compressor is what is wanted.

For the opposite change of condition, a decrease of load, the valve opens to admit too much liquid to the evaporator and tends to fill it, with the possibility of liquid carry-over to the compressor.

Constant pressure valves nevertheless have been used successfully in applications in which the load does not vary too widely, for example in ice-cream conservators which, once brought down to temperature under supervision of a serviceman, thereafter run at nearly constant conditions. With on–off control of the compressor, the flooding back can be prevented by clamping the thermostat bulb to the suction pipe. This is rapidly chilled when liquid reaches it and the thermostat switches the motor off. The charge in the system may be limited so that even when it is all in the evaporator there is not enough left to damage the compressor seriously.

The constant pressure valve is useful in experimental refrigerating plants when it is convenient to hold the suction pressure constant for a test, whilst adjusting the load on the evaporator to give the desired superheat.

Both the thermostatic valve and the constant pressure valve use atmospheric pressure as a reference pressure, and the superheat and evaporating pressure, respectively, are liable to wander with barometric changes.

(c) *High side level control.* This works on the principle of draining liquid from the condenser as it forms so that no liquid is stored in the system on the high side of the expansion valve. For a fixed refrigerant charge in the system, and neglecting variation of the mass of vapour in the pipework, the mass of liquid in the evaporator is determined. Naturally, no receiver or vessel

2.8 Expansion valves: evaporator feed regulation

holding a large amount of liquid can be used in the high side, though the system may incorporate a normally empty receiver into which the charge may be pumped for service; nor, unless special distribution arrangements are made, can several evaporators be used in parallel.

A high side level control consists of a small chamber provided with a float valve, as shown in Fig. 2.8.5. The float is linked by a lever to a slide-valve (or in some makes a needle valve) in such a way that a rise of level in the chamber opens the valve and allows liquid to pass to the evaporator. In small systems the throttling expansion may take place at the control orifice. In

Fig. 2.8.5. Section through a Witt high-side float controlled valve.
A – inlet from condenser; B – outlet to evaporator; C – shut-off valves with stem caps; D – connection for purging or for a gas balance line, with shut-off valve cap; E – by-pass pipe explained in text; F – float; G – drain valve with stem cap; H – slide valve controlled by float position; J – lever for opening slide valve manually. The liquid level is shown conventionally and does not correspond to the actual position of the float in the drawing which is at the bottom with the slide valve closed. (By courtesy of Bie Engineering Ltd.)

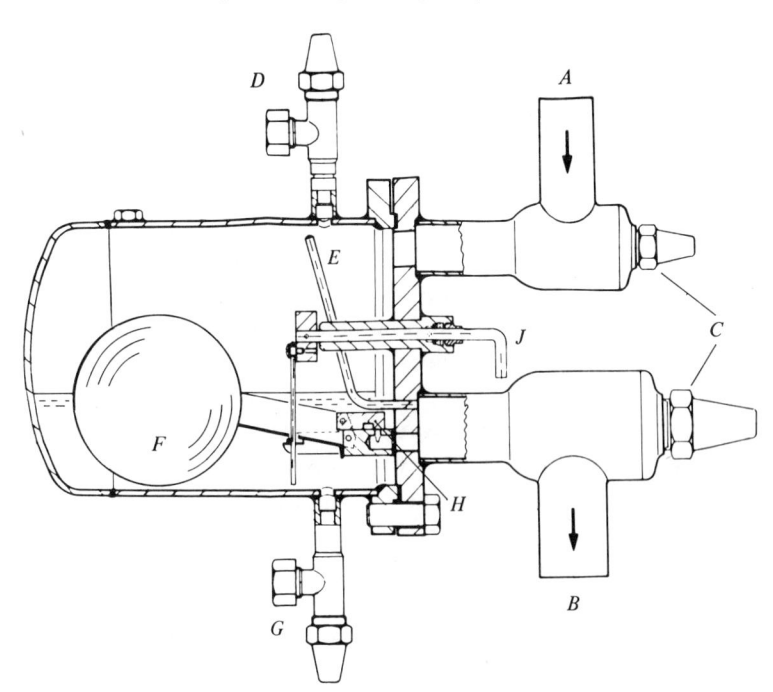

large ones, to avoid having to insulate a cold feed pipe, a fixed expansion valve may be provided immediately ahead of the evaporator.

The by-pass or vent pipe shown has two functions. It allows noncondensable gas which is carried down from the condenser to escape into the evaporator, for if this gas were to remain in the float chamber its pressure would be added to the saturation pressure and this would prevent drainage of the condenser. In addition the vent pipe makes possible the location of the float chamber above the level of the condenser if desired. By leaking a small amount of vapour through the vent pipe the pressure in the float chamber can be made slightly less than that in the condenser so that flow from the condenser can occur in any position.

It may be noted that the working of a high side float valve is basically the same as that of a steam trap allowing condensed water to drain from a heater battery or radiator without passing live steam to waste.

By holding an approximately constant mass of refrigerant in the evaporator, the high side float control keeps the evaporator surface well wetted at the conditions for which the plant is charged. Supposing that enough charge is put in to fill the evaporator when there is not much load and the liquid is just boiling quietly, a large load coming on would then produce vigorous boiling and an expansion of the total contents of the evaporator. There would be a possibility of liquid being carried over to the compressor. To avoid this a separator drum or knock-out pot would be required. On the other hand, if the plant were charged so the evaporator could readily hold all the liquid and vapour bubbles mixture at high load, the level would be somewhat lower at low load and the evaporator would not be fully wetted.

Clearly the operation of high side level control depends on having the correct charge in the system. There is no margin for leakage, but this is not a serious objection in modern plant.

High side float control has been used on all sizes of plant in both dry expansion and flooded systems.

(*d*) *Low side level control.* Already mentioned in connection with recirculation and shell-and-tube evaporators, this operates so that when the level in the evaporator falls the valve opens and admits refrigerant. The original type of level sensing device was a float, but other means are often used now. One of these is a partly immersed rod of an electric conductor which has a capacity to earth depending on the height of the liquid, acting as a dielectric, on it. The capacity is used as a signal to operate the expansion valve. Another device is a heated bulb containing a volatile fluid

2.8 Expansion valves: evaporator feed regulation

and its vapour. It is located at the desired level in the evaporator. When the bulb is surrounded by vapour, the heating element raises the temperature and pressure of the fluid in the bulb sufficiently to open the valve by acting on a diaphragm or bellows. When the bulb is covered by liquid it is cooled and the pressure inside the bulb falls, allowing the expansion valve to be closed by a spring.

When a float is used as a level sensor it may operate the valve directly through a mechanical linkage, as a pilot valve to work a larger expansion valve as described below, or to open and close limit switches which in turn close and open a solenoid valve to shut-off or admit liquid to the evaporator.

In small evaporators the float is often placed inside the evaporator and a mechanical linkage is used to work the valve. This was a common system at one time in domestic and other small refrigerators. In larger systems the float is usually placed in a separate float chamber rather than in the surge drum or evaporator. It can then be isolated for service without shutting down the evaporator, which is kept working via a hand expansion valve in parallel with the float. This arrangement is seen in Fig. 2.7.3. At the same time the separate float chamber senses a mean level in the evaporator without being subjected to the turbulence caused by the boiling there.

The level in the drum or evaporator is transmitted to the float chamber by pipes from the top and bottom, called the balance lines or equalising lines. The condition for hydrostatic equilibrium is that there must be no flow along these pipes. They should not therefore form part of any liquid supply or vapour take-off pipes. Furthermore, a wrong reading of the level will be obtained if there is any considerable evaporation going on in the liquid balance line or inside the float chamber. This may happen because the liquid is, as a rule, colder than the surrounding air. If evaporation occurs there has to be a flow of liquid from the drum and there cannot be hydrostatic equilibrium. The usual insulation of the float chamber is sufficient to prevent this in most cases, but the point is worth bearing in mind whenever one attempts to measure the level of a cold saturated liquid in a vessel by reproducing the level in an external chamber or sight glass.

Pilot operation of expansion valves

In large expansion valves the opening and closing forces are considerable and it is not practical or economical to make the signal (level, superheat, etc.) sensing device capable of generating them. Instead the signal sensor operates a small pilot valve which in turn controls the main 'slave' or servo-valve.

Two arrangements for doing this are shown diagrammatically in Fig. 2.8.6. In each a piston A provided with a bleed hole B is connected by a spindle to the valve plate or needle C which moves to alter the orifice between it and the seat D. The pilot valve admits low (evaporating) pressure in Fig. 2.8.6(a) and high (condensing) pressure in Fig. 2.8.6(b) to the upper side of the piston. A screwed spindle E is provided in each case for opening the valve manually.

In the valve of diagram (a) the pressure on the underside of the piston is the upstream (condensing) pressure. The refrigerant can flow through the bleed hole to the chamber F and thence through the pilot valve to the low pressure in the evaporator. The pressure in the chamber F depends on the extent of opening of the pilot valve. When this is closed the pressure in chamber F rises to the upstream pressure, and with the pressures across the piston balanced the spring shuts the main valve. When the pilot valve is fully open the pressure in chamber F is lower than that under the piston and the pressure difference across the piston forces the main valve open. At intermediate positions of the pilot valve the valve plate C can reach an equilibrium position partly open.

The control action which is obtained depends on what signal operates

Fig. 2.8.6. Diagrammatic arrangements of two kinds of pilot-operated valve. In (a) the upstream pressure opens the valve against the spring when the space above the piston is connected by the pilot valve to the downstream (evaporating) pressure. In (b) the highside pressure admitted over the piston by the pilot valve opens the valve against the spring.

Though intended here to illustrate the action of pilot-operated expansion valves, the same figures apply to other valves described later.

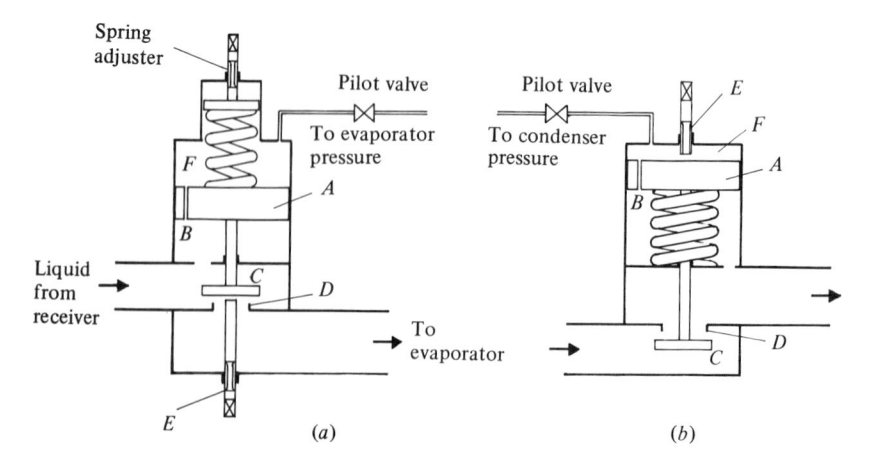

2.8 Expansion valves: evaporator feed regulation

the pilot valve. It may for example be a solenoid valve energised by a float switch in a flooded evaporator. The opening of the pilot valve when the level falls would then open the main valve to admit liquid to the evaporator. A low side float valve used as a pilot valve would produce the same result. This arrangement is shown in Fig. 2.7.3. A thermostatic expansion valve would operate the main valve to maintain approximately constant superheat.

In the valve of diagram (b) the pressure on the underside of the piston is the downstream (evaporating) pressure. Refrigerant flows through the pilot valve from the high side to chamber F and then through the bleed hole. Again the pressure in chamber F depends on the extent of opening of the pilot valve, and the valve plate C is positioned accordingly. The pilot valve with this arrangement could be a solenoid valve, a high side float valve or again a thermostatic expansion valve.

Pilot operated valves are used in other applications in refrigerating plant and will be mentioned again later.

The combination described above is called direct acting, because the main valve opens when the pilot opens. The slave valve can also be made to give reverse action, i.e. close when the pilot valve opens.

Orifice size.

The calculation of the size of orifice needed in an expansion valve to pass a given mass flow rate of refrigerant is usually based on the ordinary hydraulic formula:

$$\frac{\dot{m}}{a} = C_d \left[\frac{2(p_c - p_e)}{v_f} \right]^{1/2} \tag{2.8.2}$$

Fig. 2.8.7. Nomenclature for the derivation of the minimum area for flow between a sharp-edged orifice and a conical needle.

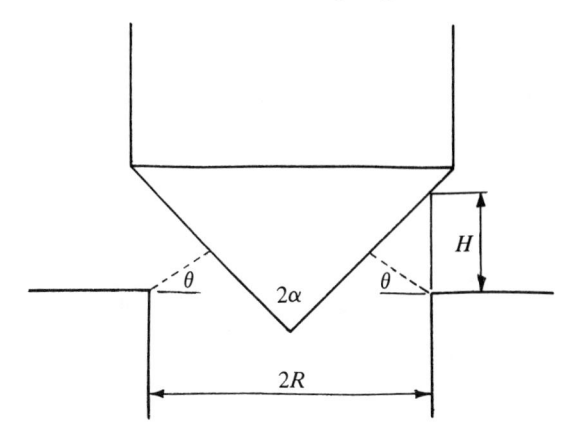

where $\dot{m}$ is the mass flow rate, a is the smallest area for flow, p_c and p_e are the condensing and evaporating pressures respectively, v_f is the specific volume of the liquid refrigerant and C_d is a discharge coefficient, of the order 0.6 for sharp-edged orifices. Even for saturated liquid this expression holds quite well. Although the condition of thermodynamic equilibrium would lead one to expect that flashing into vapour would occur in the orifice as soon as the pressure begins to fall, tests show that the liquid continues in a metastable state until it has left the orifice.

When eq. (2.8.2) is applied to a thermostatic expansion valve with a distributor, the pressure drop across this must be taken into account.

The area at the minimum cross-section of the flow depends on the position of the valve needle. Referring to Fig. 2.8.7, which shows a needle in the form of a cone with an included angle of 2α, and an orifice of radius R, the surface area a of the frustum of a cone based on the orifice and intersecting the needle cone is given by:

$$\frac{a}{\pi R^2} = \frac{H}{R} \frac{\sin \alpha}{\cos(\alpha - \theta)} \left[2 - \frac{H \sin \alpha \cos \theta}{R \cos(\alpha - \theta)} \right] \tag{2.8.3}$$

where θ is the angle at the base of the frustum and H is the lift of the needle. The area a has a minimum value when θ satisfies:

$$\frac{2R}{H \sin \alpha} = \frac{\sin \theta}{\sin(\alpha - \theta)} + \frac{2 \cos \theta}{\cos(\alpha - \theta)} \tag{2.8.4}$$

Taking different values of θ, the corresponding H/R for minimum area can be found from eq. (2.8.4) and the minimum areas from eq. (2.8.3). The curve of minimum area against lift can then be plotted. It may be noted that to satisfy eq. (2.8.4), θ is always less than α, i.e. the frustum of minimum area is flatter than the cone which is orthogonal to the needle cone.

The 'capillary tube' restrictor

Small, hermetically sealed systems often do not have an expansion valve at all. Its place is taken by a fixed restriction in the form of a length of small bore tube ('capillary tube'), about 1 mm diameter in domestic refrigerators. It acts as a fixed orifice, but providing it is sized correctly, it works as an open loop control. At a given condensing pressure there is one value of the evaporating pressure at which the tube will pass the correct amount of refrigerant to equal the flow rate which the compressor pumps at the given condensing pressure. No liquid receiver is used with a capillary tube, so that it keeps the condenser drained of liquid at this condition and the refrigerant charge is all in the evaporator. Below this evaporating pressure, for a given condenser pressure, the flow rate through the tube is greater (because the pressure difference across it is greater) whilst the pumping capacity of the compressor is less (because of the reduced density of the vapour). There is thus an imbalance between the tube and the compressor which is resolved by by-passing some vapour from the condenser to the evaporator through the tube. This is of course wasteful, because work has been expended in compressing vapour which does no useful refrigeration; but the waste can be tolerated in small systems. At evaporating pressures above that at which the tube and compressor are in balance, the tube cannot pass the mass flow rate which the compressor is sending to the condenser, and liquid accumulates in the condenser accompanied by depletion of the evaporator charge. The covering of condensing surface by liquid reduces the surface available for condensing heat transfer and forces up the condensing pressure, thus tending to increase the flow rate through the tube again. The reduction of evaporator charge reduces the wetted surface there and the heat transfer rate, so tending to bring the evaporator pressure down again and with it the mass flow rate pumped by the compressor. A balance can again be attained, but with wasteful results, because the condenser pressure is higher than it need be and the evaporating pressure is lower.

2.9 Refrigerants

Any reasonably volatile substance which is liquid at the temperature desired in the evaporator can be used as a refrigerant, but in practice the choice is limited by factors such as toxicity, cost, flammability, chemical stability, etc. which will now be considered. The thermodynamic properties of refrigerants are treated in Chapter 3.

(a) *Toxicity*. This is usually established by experiments on animals and by inference from any available data about the effects on humans. Refrigerants have been commonly graded on a scale established by the National Board of Underwriters of the USA, according to which grades 1–6 are listed, the higher the grade number the less toxic the substance.

Of refrigerants which have been in common use sulphur dioxide is the most toxic, being placed in Group 1, indicating that exposure to a concentration of $\frac{1}{2}$–1% for five minutes is sufficient to produce death or serious injury. Ammonia is much less toxic, Group 2 ($\frac{1}{2}$ hour lethal at $\frac{1}{2}$–1%) and carbon dioxide is still safer in Group 5, along with the hydrocarbons methane, ethane and propane. The fluorinated refrigerants are in Groups 4, 5 or 6. R12 is one of the safest, Group 6.

It is now recognised, however, that the rating of a gas in the atmosphere for short-term toxicity in relatively high concentrations is not a realistic guide to the possible danger to people working in it, and more attention is now paid to the Threshold Limit Values, TLV for short. The TLV–TWA signifies the Time Weighted Average concentration to which nearly all workers may be repeatedly exposed for 8 hours per day or 40 hours per week without adverse effect. The TLV–STEL indicates the Short Term Exposure Limit, i.e. the maximum concentration to which workers can be exposed for a period up to 15 minutes without intolerable irritation, chronic or irreversible tissue damage, or narcosis sufficient to increase accident proneness. Some values for refrigerants are given in Table 2.9.1 from the Health and Safety Executive Guidance Note EH 15/80. The STEL values are tentative. Whatever the threshold limit value, it is always wise not to undergo more exposure to any refrigerant than is absolutely necessary. Refrigerant from systems should always be purged into the open air, even when it would not be an obvious nuisance, like ammonia. This precaution is particularly important for R22 which is suspected of being a possible carcinogen. This has not yet, however, been confirmed (1981).

(b) *Flammability*. The hydrocarbons methane, ethane and propane are of course highly flammable. They are not used therefore in ordinary refrigeration practice, although in most other respects they are very suitable. Propane, for example, has properties rather like those of ammonia and is cheaper. In the petroleum industry where fire precautions are stringent, hydrocarbons are commonly used as refrigerants since they are cheap and readily available from the products in many cases.

Carbon dioxide and sulphur dioxide are non-flammable. Ammonia can be made to burn in an atmosphere of oxygen, and it forms weakly explosive mixtures with air when the proportions by volume lie between 16 and 25%. Because these concentrations are rarely met in practice, ammonia in the

Table 2.9.1. *Threshold limit values*

	Volume parts per 10^6	
	TWA	STEL
Ammonia	25	35
Carbon dioxide	5000	15 000
Sulphur dioxide	2	5
Methyl chloride	100	125
R11, R12, R22, R114	1000	1250

2.9 Refrigerants

past has been treated as effectively non-flammable. But from time to time explosions have occurred when ammonia has leaked from refrigerating plants, and it is recommended that naked lights should not be used in rooms where there is a risk of an ammonia leak, and that sufficient ventilation to prevent a high concentration from forming should be provided. Flame-proof motors are not normally demanded.

With regard to the halogenated hydrocarbons, those with a sufficient number of the hydrogen atoms substituted by fluorine, chlorine or bromine are non-flammable. All of the fluorinated refrigerants in common use belong to this class. When, however, a substantial proportion of hydrogen is left in the molecule, as in methyl chloride CH_3Cl, some flammability is to be expected. Methyl chloride is not easy to ignite, but there have been explosions associated with it.

(c) Smell and indication of leaks. The powerful smells of ammonia and sulphur dioxide are very good indications of leaks and provide strong incentives to mend them. In this respect a strong smell is an advantage. On the other hand, it is usually argued that a strong smell from what might be quite a small leak and which is not dangerous, would be liable to cause panic in crowded places. For this reason ammonia is seldom used in air-conditioning systems.

None of the other common refrigerants has a pronounced smell. Carbon dioxide can be tasted when in sufficient quantity, and some people can smell the fluorinated refrigerants. Strong smelling chemicals such as acrolein have occasionally been added to refrigerants to make it easier to detect leaks.

A strong smell indicates a leak, but it does not locate it. The smell of a light gas such as ammonia is rapidly diffused through the engine room and one can only tell the general area from which it comes. Various chemical and physical methods are in use for locating leaks.

The simplest and oldest test is to apply soap solution to the suspected place and to look for bubbles. This method is not applicable to ammonia, however, because of the solubility of ammonia in water, and the usual method is a burning sulphur stick which gives dense white clouds when ammonia is present. The fumes from strong hydrochloric acid may also be used.

Leaks of ammonia into cooling water, unless they are very large, cannot be detected by smell. One can smell ammonia in water at something between $50/10^6$ and $100/10^6$ (parts per million) by mass. A plant of about 100 kW refrigerating capacity taking cooling water at the rate of about

5 kg/s could thus lose 20 kg of ammonia per day before any could be detected by smell in the cooling water.

Nessler's solution, potassium mercuri-iodide in excess of alkali, is very sensitive to ammonia, however. It is made by adding a saturated solution of mercuric chloride in distilled water (about 5 to 6 g in 100 cm^3 are needed) to a solution of 12 g of potassium iodide in 50 cm^3 of distilled water until a permanent red precipitate forms. A solution of 40 g of potassium hydroxide in 100 cm^3 of distilled water is then added. A further small amount of the mercuric chloride solution is then added until a permanent turbidity results. The mixture is set aside to clarify and decanted. It can be kept in a dark bottle with a rubber stopper.

One or two drops of the solution are added to about 10 cm^3 of the water to be tested. At $25/10^6$ of ammonia and above, a distinct turbidity of a brown or yellow colour is obtained. At lower concentrations the turbidity is not so apparent but a definite colour is still noticeable. By using a comparison tube of pure water, concentrations below $1/10^6$ can be detected with ease.

With recirculated cooling water it is necessary to test the water entering and leaving the condenser and to make a visual comparison.

Nessler's solution can be used to test for ammonia in sodium chloride brine, but in calcium chloride brine a dense white precipitate is produced even when ammonia is not present. Unless a large amount of ammonia is present, the yellow colour is obscured. Sodium carbonate can be added to the sample to precipitate the calcium, and the resulting clear solution tested for ammonia. This is not so easy in practice, however, especially when the brine is already discoloured by rust. Another method is to add a little caustic soda to the brine and to warm it gently. Ammonia in the vapour given off can be detected by the pink colouration of paper moistened with a 1% solution of phenolphthalein in alcohol.

For small plants containing fluorinated refrigerants, the soap test is not sensitive enough, and other methods have been devised. One of these is the so-called halide lamp, based on the Beilstein test for organic chlorine. Air containing a halogenated refrigerant drawn into a flame impinging on copper at dull red heat produces a bright green or purple colour (Nolcken, 1948). Electronic methods based on the increase in emission of positive ions from a heated platinum electrode when organic halogens are present (White, 1950), are also available.

In plants using halogenated refrigerants there is usually some oil dispersed around the entire system, and where a leak occurs, oil leaks out as well. This can be detected by ultra-violet light which causes the oil to

2.9 Refrigerants

fluoresce. It is necessary to perform the test in the dark, so it is only suitable for factory use on small systems.

Factory assembled systems for domestic refrigerators contain such a small amount of refrigerant that the most stringent tests for leaks are necessary. Methods in use are to draw a vacuum in the system and to note any change in pressure over a number of hours, and to surround the evacuated system with a hydrogen atmosphere and search for traces of hydrogen inside with a mass-spectograph (Murdoch, 1957).

Refrigerant which leaks out of a plant has of course to be made good, and the cost of replacing the loss may be an item to be reckoned with in large plants. For a given difference of pressure and a given size of leak, the mass rate of leakage increases with the molar mass of the substance. When the fluorinated refrigerants were first introduced, the methods of sealing against ammonia and sulphur dioxide were found to be inadequate, and new devices were introduced such as rotary seals, instead of stuffing boxes, and packless valves, instead of packed gland valves. Furthermore, the powerful solvent action of the fluorinated refrigerants on grease made them much more effective leak seekers than ammonia, and leak testing standards had to be raised considerably.

(*d*) *Behaviour with materials of construction.* Some refrigerants react with common materials of construction, the principal incompatibilities being ammonia and copper, and methyl chloride and aluminium. Ammonia can be used with iron, steel and aluminium, and methyl chloride with iron, steel and copper. The fluorinated refrigerants can be used with iron, steel, copper and aluminium. Alloying elements in aluminium, particularly magnesium and manganese, occasionally give trouble, however, and should be avoided.

Reactions in refrigeration systems are aggravated by the presence of water, and every attempt is made to keep water out. In the presence of water and oil, for example, methyl chloride and R12 can form hydrochloric acid which takes some copper into solution from pipes and deposits it on cast iron surfaces, such as pistons, causing seizure. The problem has largely been solved by adequate dehydration of the system before charging.

In systems of hermetic construction, in which the motor and its windings are inside the refrigerant circuit, the refrigerant must be inert to the various non-metallic materials used for the motor insulation.

(*e*) *Chemical stability.* This means that the refrigerant should not dissociate at the temperatures encountered in the plant. The refrigerant in the system is never absolutely pure of course, being contaminated with lubricating oil and traces of water, and it is exposed to metallic impurities such as iron oxide which can catalyse decomposition. Carbon dioxide and sulphur dioxide are very stable, and so are the pure saturated hydrocarbons, methane, ethane, etc. Ammonia is usually regarded as stable, but it does seem to decompose slightly under certain circumstances, producing gases which have to be purged from the plant. The fluorinated hydrocarbons in general use are very stable, R22 and R502 being somewhat more so than R12 at high temperatures.

(*f*) *Behaviour with lubricating oil.* In most refrigerating machines oil is used not only to lubricate the bearings and running surfaces, but also to prevent refrigerant leakage by sealing gaps, e.g. between the piston rings and the cylinder bore, and between a valve plate and its seat. The oil is thus intimately mixed with the refrigerant, and a part of it is always present in the vapour delivered to the condenser, except in compressors of special oil-free construction. Besides possible chemical reactions, there are important physical relations to consider. These are concerned with the questions of miscibility of the liquid phases, and the solubility of refrigerant vapour in the oil.

Ammonia and oil are almost immiscible, meaning that in a vessel of liquid two layers can form, a top layer consisting mainly of ammonia with a small amount of dissolved oil, and a bottom layer consisting mainly of oil with a small amount of dissolved ammonia. Oil which has been delivered by the compressor, and has escaped past an oil separator, can therefore be drained off from the bottom of the vessel at intervals, with suitable precautions against the dissolved ammonia coming out of solution when the pressure on it is released.

On the other hand, R12 and mineral oil are miscible in all proportions and no separation into two layers occurs in a vessel containing liquid. For some other fluorinated refrigerants, typically R22, there may be a critical solution temperature above which miscibility is complete and below which it is only partial. When the critical solution temperature lies between the condensing and evaporating temperatures oil cannot be separated in the receiver, but the solution splits into two layers in the evaporator if sufficient oil is present. With fluorinated refrigerants the oil-rich layer is on the top.

The solubility of refrigerant vapour in oil is important in the crankcase of the compressor. Dissolved refrigerant dilutes the oil and reduces the viscosity. This must be taken into account in the design of the bearings and in the selection of the oil. The inorganic refrigerants ammonia and sulphur

2.9 Refrigerants

dioxide have a negligible effect on oil viscosity, but vapours of the hydrocarbons and halogenated hydrocarbons dissolve in mineral oil to a considerable extent under the pressure existing in the crankcase and cause oil dilution. Another important effect is the foaming of oil in the crankcase mentioned in section 2.3.

Miscibility of liquids and solubility of vapours are distinct though related phenomena which should not be confused. Thus an oil may dissolve two vapours almost equally up to a certain pressure. But if they have different miscibility limits it will be possible by increasing the pressure to form stronger solutions of one of the refrigerants than of the other.

(g) *Behaviour with water*. Water has very different miscibilities with different refrigerants, and this fact has an important bearing on the operation of plants.

With ammonia, water is miscible in all proportions, and forms solutions with very low freezing points. Water is never therefore deposited as ice inside a system. If the temperature falls below $-78\,°C$, the freezing point of pure ammonia, crystals of pure ammonia are deposited when water is present in moderate amounts (see Fig. 6.8.1).

In Refrigerant 12, on the other hand, the solubility of water is minute, only about 0.0026% by mass at $0\,°C$. Any water in the system therefore, above the small amount which can be dissolved by the refrigerant, must be present as free water. This can freeze wherever the temperature drops to $0\,°C$. Freezing would usually occur in the expansion valve or capillary tube in small systems and the ice deposited would block the orifice. It is imperative therefore that no water above what can be held in solution should be in the system, and for R12 this is a very small amount.

Components, tubes and fittings are usually dehydrated in the factory, as is the complete system when it is factory assembled. In assembly every care has to be taken to exclude water. Finally a drier containing silica gel or a synthetic zeolite ('molecular sieve') is usually inserted in the system. In R22 the solubility of water is somewhat larger, about 0.06% at $0\,°C$. This level of water in solution is undesirable not only on account of the water blocking the expansion valve, but because of the possibility of chemical reaction between refrigerant, oil and water.

(h) *Cost*. On a mass basis, the cheapest refrigerants are the pure hydrocarbons, which are unfortunately not suitable for general use on account of their flammability. Next come the inorganic substances carbon dioxide and ammonia, and the most expensive are the fluorinated compounds. The fluorinated ones in large volume production such as R12 and R11, which are used as aerosol propellants in greater quantities than as refrigerants, are almost as cheap as ammonia, but the more unusual ones are still relatively expensive, especially those with substantial proportions of fluorine or bromine in the molecule.

Though refrigerants are bought by mass, the first cost of charging a plant depends on the volume of liquid required, which is principally the volume required in the evaporator to give adequate flooding to achieve good heat transfer. On a volume basis, the fluorinated refrigerants become even more expensive than ammonia because of their high liquid densities; and the pure hydrocarbons become relatively cheaper on account of their low liquid densities.

Storage and handling of refrigerants

Apart from the hazards mentioned above associated with flammability and toxicity, all refrigerants must be carefully stored and handled if danger is to be avoided.

Refrigerant cylinders received from the supplier are charged, in the UK, with a volume of liquid which would fill the cylinder completely when the temperature reaches 45 °C, with a few per cent safety margin. Above this temperature very high pressures are developed. Hence cylinders should not be stored in warm places or where there is a definite fire risk. They must be carefully moved to avoid damage to them, and firmly secured when in use by properly designed supports or straps to prevent them falling over, with possible damage to the cylinders and injury to people. The valves of cylinders should be treated with respect and in no circumstances should they be used for lifting them by cranes or slings, etc.

In use every precaution must be taken to ensure that refrigerant goes only where it is wanted, in the system, by providing tight connections and not being careless over leaks. Liquid refrigerant impinging on the skin can cause frostbite because of the very low temperature which may be attained on sudden release of pressure. Furthermore, it must be remembered that even relatively safe refrigerants of the fluorinated type displace air and cause a suffocating atmosphere near the floor because of their high vapour density.

2.10 Refrigerant piping

Small refrigerating systems employing the fluorinated hydrocarbon refrigerants are usually piped up with copper tube of a specially soft type. It is supplied dehydrated internally and sealed at the ends, and in this

2.10 Refrigerant piping

form is known as 'refrigeration grade' tube. When a length is cut off the tube remaining should be sealed by hammering or by a plastic cap to exclude dirt and moisture, the great enemies in small systems. Because of the small orifices in expansion valves and the small bore of 'capillary' tubes, which are easily blocked by dirt or ice, it is imperative to keep dirt and moisture out of the system from the beginning.

The standard sizes of refrigeration grade copper tube in the UK and the USA, and also commonly used over much of the world, have fractional inch actual outside diameters in the series: $\frac{3}{16}, \frac{1}{4}, \frac{5}{16}, \frac{3}{8}, \frac{1}{2}, \frac{5}{8}, \frac{3}{4}$ and 1 inch. The wall thickness for tubes in the UK is usually 20 SWG (0.036 inch = 0.91 mm) for tubes up to $\frac{5}{8}$ inch and 18 SWG (0.048 inch = 1.22 mm) or 19 SWG (0.040 inch = 1.02 mm) for $\frac{3}{4}$ and 1 inch outside diameter (o.d.) tube.

Metric o.d. size tubes are also in use in the series: 6, 10, 12, 15, 18, 22 and 26 mm, with wall thickness 1 or 1.5 mm.

When the tube has to be connected to a component, for example a shut-off valve, compressor, etc., the joint is made by either (a) a flare fitting or (b) a capillary fitting.

(a) *Flare fitting.* The Society of Automotive Engineers (SAE) flare, now also an American Standard, is most widely used. A 45° flare formed on the end of the tube is compressed between a male cone on the component and a female cone in a flare nut, as shown in Fig. 2.10.1. The sealing occurs between the flare and the male cone and the tool used to produce the flare has to provide an accurate polished cone on the inside of the tube flare.

Standard flare nuts and fittings have Unified Fine (UNF) threads according to the following table.

Tube o.d (inch)	Thread (inch)	(per inch)	Spanner (inch AF)
$\frac{1}{4}$	$\frac{7}{16}$	20	$\frac{3}{4}$
$\frac{3}{8}$	$\frac{5}{8}$	18	$\frac{7}{8}$
$\frac{1}{2}$	$\frac{3}{4}$	16	1
$\frac{5}{8}$	$\frac{7}{8}$	14	$1\frac{1}{8}$

Fittings for $\frac{3}{4}$ inch o.d. tubes are available but they are not much used, being expensive and rather heavy for manual flaring. They have a $1-\frac{1}{16}$ inch diameter thread of Unified form, 14/inch.

Other standards for flare fittings are also in use. One of these is a mixed metric and inch standard for millimetre size o.d. tubes but with UNF threads on the fittings which take metric spanners, as follows.

Tube o.d. (mm)	UNF thread (inch)	Spanner (mm AF)
6	$\frac{7}{16}$	16
10	$\frac{5}{8}$	22
12	$\frac{3}{4}$	25
15	$\frac{7}{8}$	28
18	$1\frac{1}{16}$, 14/inch (non-standard)	36
26	$1\frac{3}{8}$	43

The above sizes are most widely used at present for metric series tubes, but another standard based on the metric constant pitch thread of 1.5 mm is proposed.

A wide variety of hot stamped brass fittings is available, e.g. double-male connectors for joining two tubes already flared, either equal or unequal; double-female connectors (swivel nuts) for joining two components with male flares, and so on. To close a pipe ending in a flare a plug is screwed into the flare nut on the pipe. To close a male flare on a component a flare

Fig. 2.10.1. An SAE flared connection between a copper pipe and a brass fitting. The seal is made between the inside of the flare on the tube and the conical seat on the fitting.

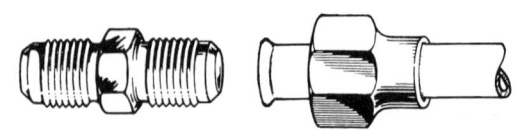

2.10 Refrigerant piping

nut is used to grip a soft copper gasket (bonnet) over the end of the flare.

Connections to compressors and vessels are made by tapping the boss to receive a screwed plug and sealing the threads with thread-sealing compound (traditionally litharge and glycerine) or nowadays with polytetrafluorethylene (PTFE) tape. The most widely used thread for this purpose is the National Pipe Thread (NPT, an American Standard, formerly called the Briggs thread). It has a 1 in 16 taper. The size is specified by the nominal bore. Connectors screwed NPT at one end and taking an SAE flare at the other are available.

(b) *Capillary fitting.* Capillary fittings have joints made by solder between the outside of the tube and the inside of the fitting. The clearance is sufficiently small to draw in the molten solder and fill the gap by capillary action, hence the name.

The fittings are supplied either with or without the solder, i.e. 'plain' or 'solder ring'. The solder ring fitting carries a ring of solder in the bore which is melted by the heat of a torch when the tube is in position. Fittings on the market are available with hard or soft solder, but the soft solder variety is not suitable for refrigeration work except at low pressures.

When a solder joint is made to a component there is a possibility of damaging the component by the heat of the torch. Some components have extended ends so that the joint is made some distance away. In others it may be possible to remove the temperature sensitive part of the component. But for many components such as thermostatic expansion valves and solenoid valves this is not possible, and the considerable heat needed in making a big joint would damage them. In such cases the component is flanged and a companion flange is soldered to the pipe with the component absent. It is bolted up subsequently. The flanges are mated with a tongue and groove joint described below. The use of flanges also facilitates the removal of a component for replacement or service when the pipes are large. With small ones it is possible to spring the pipes sufficiently to release the solder socket from the component, but this cannot be done with large ones without disturbing the rest of the pipework. With flanges it is only necessary to spring the pipes by the depth of the groove in the flange.

Steel tubes have to be used in ammonia plants. They are also commonly used in plants using other refrigerants when the pipe size is above about 1 inch. At one time wrought iron was used exclusively for ammonia piping, then at a later date seamless low-carbon steel was used, but now welded pipe to certain specifications is admissible. The material has to be selected with regard to the temperature of operation. Unlike copper, which retains its strength and resilience to low temperatures, carbon steel undergoes embrittlement at temperatures which may be as high as 0 °C depending on the grade.

Steel tubes are specified dimensionally by a nominal bore up to 14 inch (356 mm) diameter and by the outside diameter above that. The nominal bore is not the actual inside diameter but is in fact an oblique way of stating the outside diameter. At one time it meant the o.d. of a tube on which a standard pipe thread could be cut. For example, a 1 inch n.b. tube could accommodate a 1 inch British Standard pipe thread, which has a major diameter of 1.309 inch. The nominal bore is now related to a specified outside diameter which is given in the standards for different kinds of pipe.

The thickness of steel pipe is specified by an actual dimension, by a gauge number according to some standard, e.g. Standard Wire Gauge (SWG), by a Class letter (now virtually obsolete) or by a Schedule number. The Schedule number comes from American Petroleum Institute standards and denotes roughly the strength of the pipe. For example, Schedule 40 pipe, commonly used in refrigeration systems, is suitable for approximately the same pressure whatever the diameter.

Steel pipes were traditionally joined by flanges. It was found at an early date that the ordinary flat faced flanges as used for steam could not seal adequately against refrigerants because not enough pressure could be

Fig. 2.10.2. Two forms of screwed flange connection for steel tubes. (*a*) The gasket is trapped in the groove of the female flange by the tongue on the male flange. The flanges are screwed onto the pipe and sealed, originally by solder filling the recesses, later by threadsealing compound or tape, or by a welded fillet. (*b*) A metallic ring joint in which the ring is pulled against the ends of the pipes by the flanges.

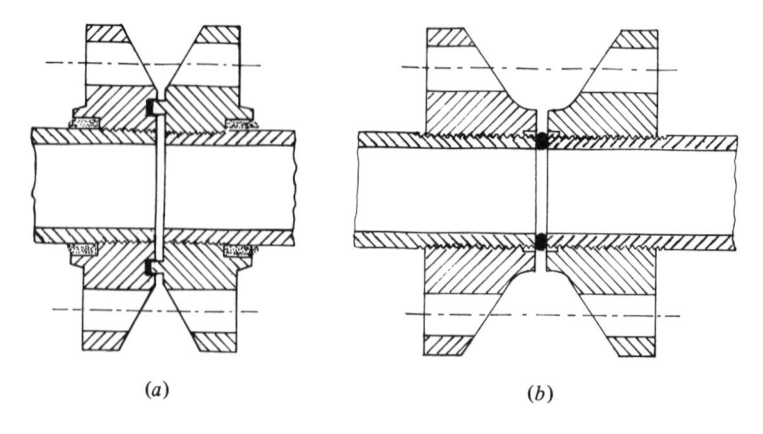

(*a*)   (*b*)

2.10 Refrigerant piping

applied by the bolts to compress the gasket. Hence a trapped gasket of narrow width in a recess, called a 'tongue and groove' joint is used. The classic de la Vergne ammonia joint is shown in Fig. 2.10.2(a). The flanges are screwed onto the pipe ends and the gasket, compressed asbestos, is trapped in the groove on the female flange by the tongue on the male flange. The recess between each flange and the pipe was filled with solder. At a later date the solder recesses were omitted and leakage along the threads was prevented by using taper threads and a thread sealing compound, or by a fillet weld between the flange and the pipe. The function of the weld in this case is not of course to provide strength but simply to seal the joint.

A later type of screwed flange joint is shown in Fig. 2.10.2(b). The tube ends are accurately squared-off and compress between them a metallic ring,

Fig. 2.10.3. Types of welded flange joints. (a) Loose flanges pull the welded ruffs carrying the tongue and groove together. (b) Socket welded flanges. (c) Common form of butt welded flange used today.

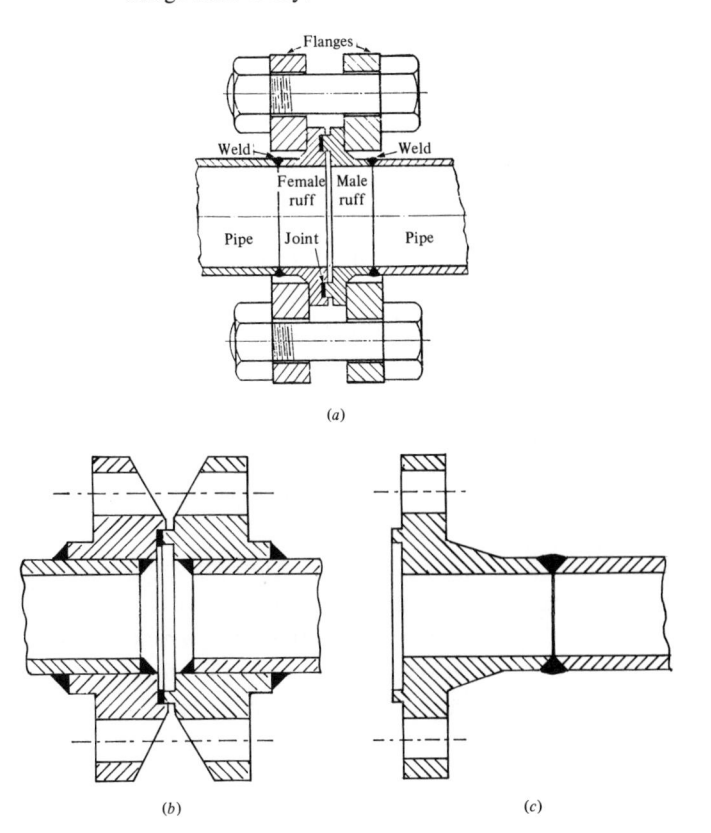

aluminium for ammonia and copper for fluorinated hydrocarbons, under the pull of the flange bolts. It will be noted that in this joint the threads do not perform any sealing function; they merely transmit the pull of the bolts to the pipe. A ring joint may also be used at a junction with a component.

For many years now pipework has been largely assembled by welding. Pipe to pipe joints are usually butt joints, though welding sockets are used in smaller sizes. Flanges still have to be used at the junctions with compressors, pumps, expansion valves, etc., and anywhere that a subsequent break has to be made for maintenance or extension. Three forms of welded flange joints are shown in Fig. 2.10.3. In (a) male and female ruffs are welded to the pipe ends after slipping on loose flanges. Rotation of the flanges to align the bolt holes is easy with this design. In (b) the flanges themselves are welded to the pipe ends. To align the bolt holes the flanges could be tacked whilst in position and then dismantled for final welding. The commonest form today is the butt welded flange shown in (c). Only the male flange is shown, as might be used for mating to a valve.

Screwed socket connections are used for small pipes, particularly for gauge lines, etc. Proprietary compression fittings are also used. These employ a trapped ring similar to that shown in Fig. 2.10.2(b) compressed by a union nut between faces, instead of by flanges. Compression fittings which rely on deformation of the pipe wall are not recommended for ammonia, though they are used to some extent with the fluorinated hydrocarbons.

A special type of steel tube ('Bundy tube') is used on some factory assembled systems for R12 and 22. It is made by rolling a sheet of steel with a brazing shim into a tube and brazing in a furnace. Another type of thin-wall seam welded steel tube is also used in factory-made systems.

Flexible tubes are used in special circumstances. Wire braided Neoprene and similar materials are used to connect the movable plates in plate freezers to the expansion valve or surge drum. Nylon or polyamide tubing is used in small beer coolers to connect the condensing units to the bottle shelves, etc.

Aluminium tube is used in domestic refrigerators. It is also often used in components such as evaporators for both ammonia and the fluorinated hydrocarbons. Evaporators of domestic refrigerators are often of aluminium. It can be flash welded to the copper tube forming the rest of the system or joined by a swaged joint with Epoxy resin.

The use of stainless steel is justified for system assembly only when special conditions, e.g. low temperature, prevail, on account of its cost and the

2.11 Auxiliary equipment

difficulty of welding it. Thin wall stainless steel tubing is used in some evaporators and condensers.

2.11 Auxiliary equipment

In addition to the essential components already described, an actual plant has a number of other components for operational purposes or for improving the efficiency. Some of these will now be described.

Liquid receiver

In most plants liquid from the condenser drains into a storage vessel or receiver. The exceptions already noted are when a high side level control or a capillary tube restrictor is used. The purpose of the receiver is to provide a reservoir of charge to accommodate changes of liquid volume in the evaporator, to hold some charge in reserve so that in the event of leakage it is not necessary to put more into the plant every day or so, and to store the refrigerant from other components of the plant when these have to be emptied for repairs or service.

The receiver is a horizontal cylindrical vessel in most plants, though sometimes vertical in very small ones. It has a connection for the main liquid line from the bottom, and one for the drain pipe from the condenser at the top. A sight glass to indicate the liquid level is usually provided.

The outlet from the receiver is controlled by the main liquid stop valve, sometimes called the king valve.

In many small plants the lower part of a water-cooled shell-and-tube condenser is used as liquid storage space.

Heat exchanger

The vapour leaving the evaporator is of course cold, but not cold enough to do any more useful refrigeration directly. For example, if a cold room is being maintained at a temperature of $-10\,°C$ by evaporation of refrigerant in pipes at $-20\,°C$, it is not possible to warm the vapour by heat transfer from the air in the room by more than 10 K. The vapour leaving, at $-10\,°C$ at the most, could still gain heat from the ambient atmosphere outside the room, but this would not be useful refrigeration. Furthermore, condensation would occur on the cold pipe going to the compressor, causing a nuisance and eventually corrosion, so this pipe would have to be insulated.

With some refrigerants, principally R12 and R502, it is beneficial to use the cold vapour to reduce the temperature of the liquid going to the expansion valve from the condenser or liquid receiver. This is done in a counterflow heat exchanger, usually finned on the vapour side because of the much lower coefficient of heat transfer on that side than on the liquid side. The heat exchanger is placed near the evaporator, so that the vapour going back to the compressor is already warm and insulation of the suction pipe is not necessary.

The heat exchanger, by reducing the temperature of liquid at the expansion valve, obviously increases the refrigerating effect of each unit mass of refrigerant flowing to the compressor. On the other hand, it increases the specific volume of the vapour and hence reduces the mass flow rate associated with a given volume flow rate at the compressor. It is shown in Chapter 3 that for R12 and R502 the net effect is an increase of refrigerating capacity from a given volume flow rate at the compressor. The net effect for ammonia is a decrease in refrigerating capacity, and consequently heat exchange is not used with this refrigerant. In any event the delivery temperature for ammonia is already high enough even without heat exchange.

With reciprocating compressors the volumetric efficiency is improved by superheat of the suction vapour, and this reinforces the increase of capacity for R12 and R502. The effect of pressure drop in the exchanger on the refrigerating capacity of the compressor must always be considered.

In hermetic and semi-hermetic compressors the suction vapour is used to cool the motor windings, and it is less effective for this purpose if it has already been warmed up in a heat exchanger. Nevertheless, motors for small compressors for domestic refrigerators are so highly rated, and run at such high winding temperatures, of the order 140 °C, that satisfactory cooling is still achieved. These high motor temperatures required a careful investigation into the properties of the materials at these temperatures in the presence of refrigerant and oil. It has been found that R12 is generally less troublesome in highly rated hermetic systems than R22.

Oil separator and still

Most positive displacement compressors discharge some lubricating oil with the vapour, and as much as possible should be separated before the condenser. The simplest separator is a vessel in which the vapour velocity is low enough for the oil droplets to settle by gravity, though in some designs centrifugal effects are used. The oil collected may be drained off through a manually operated valve, or returned automatically to the compressor by a float-controlled valve in the separator.

2.11 Auxiliary equipment

What happens to the oil which gets past the separator depends on the refrigerant. Ammonia, as already noted, is only slightly miscible with mineral oil, and an oil layer, containing some dissolved ammonia, forms at the bottom of the receiver, whence it may be drained off from time to time. (Caution! This operation must be performed with care. It sometimes happens that the valve becomes blocked with a plug of congealed oil which suddenly gives way.) Oil is not usually a problem in ammonia plants therefore, provided a suitable oil which can stand the discharge temperature without forming hard resins on the valves is used.

Liquid R11 and R12 are miscible in all proportions with mineral oil and no separation is observed either in the receiver or in the evaporator. R11 is only used however with turbo-compressors, and these pump very little oil. In R12 plants incorporating reciprocating compressors oil tends to become concentrated in the evaporator. If allowed to go on unchecked this would starve the compressor of oil eventually, and also alter the thermodynamic properties of the refrigerant in an unfavourable manner. Oil in solution lowers the vapour pressure of the refrigerant, producing the effects described in section 3.16. Furthermore, in flooded evaporators the oil in solution impairs the heat transfer coefficient; in dry expansion evaporators its effect has not yet been clearly determined, some workers finding an increase in the heat transfer coefficient, but of course only for relatively small oil concentrations.

In dry expansion evaporators an equilibrium between oil coming in and oil leaving can usually be attained by ensuring sufficient velocity of vapour in the suction pipe (about 3 m/s in horizontal and 6 m/s in vertical pipes) to entrain droplets of oil and carry them back to the compressor. The dissolved refrigerant in the oil droplets entails a small loss of refrigerating capacity if it is not evaporated with useful intake of heat. When a suction/liquid line heat exchanger is used the dissolved refrigerant is boiled off in that, cooling liquid going to the expansion valve.

In recirculation and shell-and-tube evaporators it is not so easy to entrain droplets in a controlled manner. One practice is to provide a permanent bleed of oil–refrigerant solution from the drum and return this to the compressor after boiling out the refrigerant by thermal contact with the warm liquid line. This may be done in the usual suction–liquid line heat exchanger by passing the bled solution into the suction pipe, or in a separate heat exchanger, called an oil still or rectifier.

The rate at which the oil–refrigerant solution has to be removed from the evaporator is determined by an oil mass balance on the entering and leaving streams. Expressing the composition of the mixture as the mass fraction of oil:

$$\xi_{oil} = \frac{m_{oil}}{m_R + m_{oil}} \qquad (2.11.1)$$

where m_{oil} and m_R are the masses of oil and refrigerant, respectively, in a sample, then at equilibrium:

$$\dot{m}_i \xi_{oil,i} = \dot{m}_b \xi_{oil,b} \qquad (2.11.2)$$

where $\dot{m}_i$ is the mass rate of the flow from the expansion valve and $\dot{m}_b$ is the mass rate of the bled solution. It is assumed in this case that no oil is carried back by the suction vapour. Naturally the solution should be bled off at a point where the value of $\xi_{oil,b}$ is highest, but unfortunately the required data on the behaviour of oil and refrigerant are not always available. With R22 and oil the oil-rich layer is at the top as mentioned in section 2.9 and the bleed-off point should be somewhere just under the liquid level.

Oil for use in refrigerating plants has to be dewaxed to remove solids which would be deposited in the orifice of the expansion valve and on the heat transfer surface in the evaporator. Ordinary or multigrade engine oils are not suitable, and special refrigeration grade oils are available. For charging into mass-produced small hermetic systems the oil is also deaerated and dehydrated.

Oil intended for systems employing fluorinated refrigerants has to have sufficient viscosity to ensure proper lubrication even when the oil is diluted with the low-viscosity liquid refrigerant. Consequently an ammonia oil is not usually suitable for fluorinated refrigerants.

Liquid separator

As well as the surge drum or accumulator which acts as a liquid separator in flooded and pump circulation systems, liquid separators may be provided in the suction line of any system with the aim of protecting the compressor by preventing any substantial amount of liquid from entering the suction, because of malfunction of the liquid feed controls, return of the condensate in hot gas defrost systems, and so on. The separator is simply a vessel in which the velocity of the vapour is reduced to a value low enough for droplets to settle, about 0.7 m/s or less. Eliminator plates or scrubbers may also be incorporated.

The main problem with separators is what to do with the liquid which collects in them. Since it is under suction pressure it cannot drain directly to the receiver but must be pumped there. In many systems the liquid is evaporated by a heater coil through which warm liquid going

2.11 Auxiliary equipment

to the expansion valve passes. Liquid separators are also called 'knock-out pots'.

Non-condensable gas purger

For several reasons non-condensable gases are to be found inside refrigerating plants. Modes of entry for air are by direct leakage into a

Fig. 2.11.1. A non-condensable gas purger. The evaporator coil is first fed with liquid refrigerant from the main liquid line via valve K. After a time this is closed and the coil is fed via valve L with liquid which has condensed in the vessel.

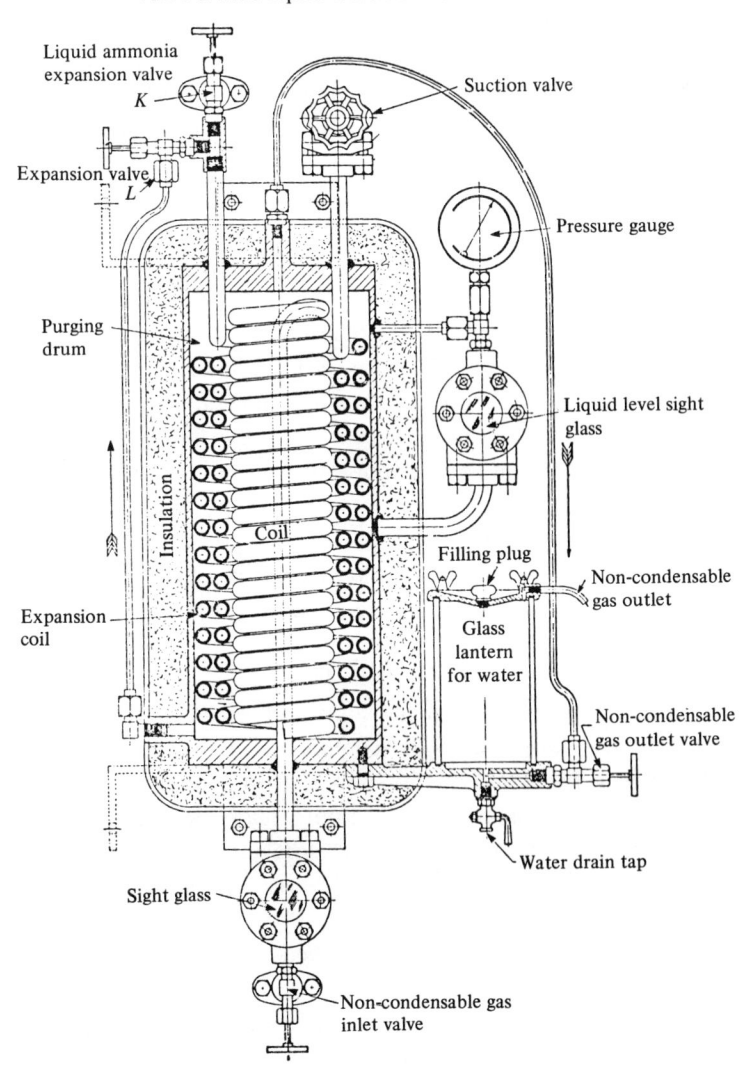

part of the system below atmospheric pressure, in solution in the lubricating oil put in from time to time, and as air left inside a pipe or component after assembly or after opening the system for service. In good practice residual air is reduced to a minimum by evacuating the system after assembly or service. Old ammonia plants sometimes seem to generate other gases inside them, possibly by dissociation of the ammonia under the influence of metallic catalysts in the system.

The effect of non-condensable gases is treated in Chapter 3. They may be removed by blowing from the condenser or receiver, but refrigerant vapour is lost at the same time. This is permissible in very small plants, provided the refrigerant is not discharged into a confined space, but not in large ones because the refrigerant may be expensive, toxic or flammable. The vapour–gas mixture is therefore purged through a vessel cooled by an evaporator coil fed with liquid refrigerant via a small expansion valve. The mixture is cooled at constant total pressure, which is near the pressure in the high side of the system, and the partial pressure of the refrigerant is reduced whilst the partial pressure of the non-condensable gas increases. The mixture can then be discharged. It is usual to discharge ammonia gas through water.

Provided that the purger vessel can be installed above the liquid receiver, the liquid which condenses in it can drain there by gravity. It may be more convenient, however, to have the purger at a lower level than the receiver. Fig. 2.11.1 shows a purger which can be installed anywhere. It uses the liquid condensed from the purged gas mixture to feed the evaporator coil in the purger vessel, and so does not have to drain to the receiver.

The pressure in the condenser of some plants, e.g. those using R11 with turbo-compressors, will be less than atmospheric pressure at certain times. A small compressor is then necessary to compress the gas–vapour mixture up to atmospheric pressure before cooling and purging to atmosphere.

Sight Glasses

These are used to indicate the level of liquid refrigerant in a vessel, of oil in the crankcase of the compressor or in an oil separator, and sometimes in the liquid line to show that the liquid passing contains no entrained vapour. Sight glasses are not usually fitted as standard to very small plants such as domestic refrigerators and small commercial systems, but a liquid line sight glass can be a very useful adjunct for servicing purposes.

Sight glasses for levels may be installed directly in the side of the vessel, but this weakens the structure and is not common except on small vessels. More usually they are connected to the vessel by balance pipes at top and bottom to convey the static pressures. When used on a liquid receiver, in

2.11 Auxiliary equipment

which the liquid refrigerant is generally warmer than ambient temperature, there is no need to insulate the balance pipes, and false readings caused by vaporisation are not likely to occur. When used on vessels containing liquid much colder than ambient temperature the liquid balance pipe should be insulated. Readings from such level gauges should always be treated with caution on account of the possibility of error caused by vaporisation and vapour bubbles in the liquid pipe.

The fittings for sight glasses must contain self-sealing ball valves to operate in the event of glass breakage, except in very small plants containing non-toxic and non-flammable refrigerants. The glass should also be surrounded by a metal grid or a substantial plastic guard.

Sight glasses for level gauging may be installed above or below the level of liquid in the vessel, by using the arrangement shown in Fig. 2.11.2. Instead of transmitting the static pressure of liquid at the bottom of the vessel by a liquid column, it is transmitted by a vapour column. This is ensured by providing a heated length of pipe which effectively prevents liquid from getting past this point. This pressure of vapour is conveyed to one limb of a U-tube manometer in any position, the other limb being connected to the balance pipe from the top of the vessel. Oil may then be used as the

Fig. 2.11.2. Arrangement for remote measurement of a liquid level by a manometer or pressure transducer. The vaporising coil prevents liquid from entering the riser of the bottom balance line. The static pressure at the level of the coil is then transferred by the vapour column to the right-hand limb of the U-tube.

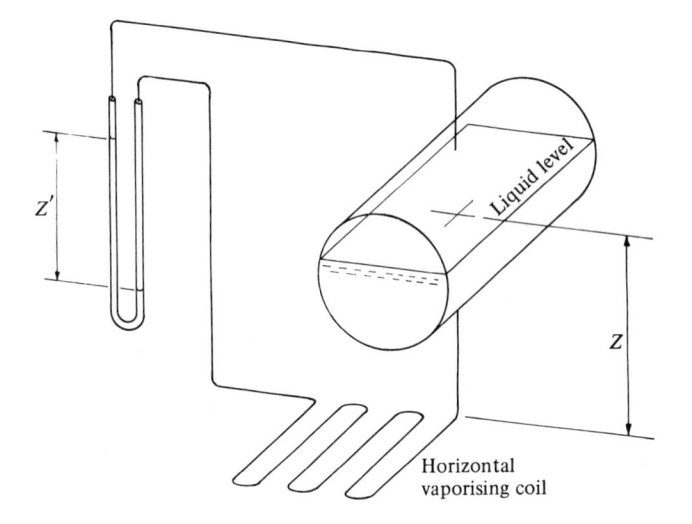

manometric fluid. It is necessary to calibrate the gauge readings to give the height of liquid refrigerant in the vessel.

Analysis of the static pressures in the circuit shows that:

$$Z(\rho_f - \rho_g) = Z'(\rho_f' - \rho_g) \tag{2.11.3}$$

where Z and Z' are the heights shown in Fig. 2.11.2, ρ_f and ρ_f' are the densities of the liquid in the drum and the manometric fluid respectively, and ρ_g is the density of the vapour. If $\rho_f = \rho_f'$, $Z = Z'$, but usually a different fluid is used in the manometer. Then provided that ρ_g is small compared to ρ_f and ρ_f':

$$Z\rho_f = Z'\rho_f' \tag{2.11.4}$$

This arrangement is particularly suitable for reading levels in vessels containing low temperature liquids, since sufficient heating of the refrigerant in the bottom balance pipe is then provided by the ambient air. It also eliminates the difficulties mentioned above regarding the uncertainty of vapour bubbles in the conventional arrangement.

Instead of the U-tube manometer a sensitive pressure transducer may be used with remote indication of the level.

A sight glass in the liquid line may be used in small plants before the expansion valve to indicate that there is sufficient charge in the plant, instead of a level gauge on the liquid receiver. In some proprietary designs the sight glass incorporates a device which indicates the moisture content of the circulating refrigerant by a change of colour. The readings of liquid line sight glasses, as indications of the sufficiency of refrigerant charge, must be interpreted with due regard to the distribution of liquid in the system. For example at low load conditions the evaporator is relatively full of liquid with few vapour bubbles, and the receiver is consequently depleted. On high load the opposite occurs, the receiver is relatively full and the sight glass shows full liquid flow. Another reason for variation of level in the receiver is provided in those systems which achieve some regulation of condensing pressure in cold weather by causing liquid to back-up into the condenser so as to reduce condensing surface.

Strainer and dehydrator.

A strainer in the liquid line from the receiver is used to protect the orifice in the expansion valve, and this is especially desirable in small plants. A dehydrator is combined with the strainer in plants using R12, R22 and R502 because the solubility of water in these refrigerants is extremely small (26, 60 and 22×10^{-6} respectively by mass at 0 °C), and minute amounts of water are sufficient to saturate the refrigerant. Any excess water remains as

2.11 Auxiliary equipment

free water going around the system, and if the evaporating temperature is less than 0 °C the water will freeze at the expansion valve orifice or the outlet of a capillary tube and block it. Water also appears to be an essential component in the reactions between refrigerant and oil which lead to chemical degradation and eventual attack on the materials of construction, expecially in sealed systems. Great care is taken during the assembly of all small systems to exclude moisture.

Dehydrators are usually of the adsorbent type using silica gel or synthetic zeolites known as molecular sieves.

Pressure gauges

Though not essential to the working of a plant, pressure gauges give the most important indications of the conditions of operation and are fitted to all plants above a certain size. They are fitted to small plants when required, via service valves like that shown in Fig. 2.3.13.

The commonest type of gauge is the ordinary Bourdon gauge with a linkage to the pointer incorporating a toothed sector and a pinion. When connected to the pulsating suction and delivery pressures the teeth on the sector can wear rapidly. Pulsations should therefore be damped out by using fine bore tubing for the gauge connection, by an orifice ('pressure snubber') or by having the gauge cock just cracked open. Gauges with helicoidal drive to the pointer are said to have better wearing properties under pulsating conditions.

Ordinary pressure gauges read gauge pressure of course, i.e. the pressure difference between the system and the atmosphere. Gauges with an absolute vacuum reference are available for special purposes, but are seldom used in normal systems. Nevertheless, gauges are often marked with a scale of saturation temperature for the refrigerant in the system, and this can only be accurate at standard atmospheric pressure unless it is specially marked for a particular altitude. Even so the normal variations of atmospheric pressure, with extremes of about 900 mbar and 1080 mbar at sea level compared with the standard value of 1013.25 mbar, can cause significant errors, especially at low suction pressures, in the inferred evaporating and condensing temperatures. Hence whenever a gauge reading is made the barometer should be read at the same time, or the atmospheric pressure taken from a weather map in the newspapers, and used to convert to absolute pressure.

Pressure gauges are constructed either for positive gauge pressure, used in the high side of a system, or as compound gauges, i.e. reading from negative (vacuum) to positive gauge pressures, for use in the low side of a system. Neither of these is very accurate in the region of zero gauge pressure. This is unfortunately a region where accuracy is often desired to ascertain the evaporating temperature, or to determine the specific volume of the vapour at suction to check the volumetric efficiency of the compressor. Consideration should therefore be given to using a mercury manometer for testing a plant using fluorinated refrigerants, but mercury is not to be used with ammonia on account of the possible formation of explosive compounds.

The calibration of pressure gauges drifts with time, and they should be re-calibrated at intervals using a dead-weight tester and/or a standard mercury manometer.

The makers usually calibrate a pressure gauge in its vertical position with the Bourdon tube full of vapour. If the tube becomes full of liquid, an error is introduced which can be significant at low pressures. This is liable to happen when a gauge is connected to a refrigerant pressure for which the saturation temperature is higher than ambient temperature. The vapour then condenses in the gauge tube, which gradually fills with liquid. Whilst this is going on the gauge does not read the connected pressure but the saturation pressure of the refrigerant at approximately ambient temperature. As a rule some air is trapped in the gauge tube and filling is incomplete. Anomalous gauge readings are often caused by the movement of vapour and its condensation inside the line connecting the gauge to the system.

2.12 Line controls

Shut-off valves

Apart from the shut-off valves provided on the compressor for isolating it from the system for service and during shut-down of the plant, most systems have a number of other valves which serve various purposes. The principal one of these valves is the stop valve in the liquid line immediately after the receiver, often called the 'king valve'. It is used to shut off the outlet from the receiver so that the refrigerant charge can be pumped out of the rest of the system and into the receiver in order to service or repair other components. The operation of transferring the charge is known as 'pumping down', and it is performed by shutting the king valve and allowing the compressor to run, so eventually pumping the charge out of the evaporator. If a plant is to be out of use for some time it is usual to pump it down in this way to reduce the possibility of leakage. It may also be done automatically prior to defrosting an evaporator.

Further stop valves may be provided in the system, depending on its size, to isolate components separately.

A typical stop valve of the older type for ammonia is shown in Fig. 2.12.1.

2.12 Line controls

The spindle is sealed against leakage by packing which can be compressed by the gland via the gland nuts. An important feature is that when the spindle is screwed out the upper side of the movable part of the valve seals against the cover so that the packing can be replaced without losing gas. The body and cover were made of malleable cast iron or of alloy cast iron, and the spindle was often chromium plated to reduce corrosion.

When valves of this type were used on systems with fluorinated refrigerants it was found to be almost impossible to make the packing tight enough to prevent leakage and yet allow the spindle to operate. Consequently the valves for these refrigerants were provided with a screwed cap sealing against a gasket which could be fitted when the hand wheel was removed. Only when the valve was to be opened or closed was the cap removed. Later designs embodied O-ring seals on the spindle and this

Fig. 2.12.1. A typical ammonia valve of the older packed type, with flanges, male at one end, female at the other. The cage which is free to rotate on the end of the spindle carries a soft inset ring which seals on the seat in the body. Rotation of the spindle does not turn the inset ring and rub it against the seat during closure. Screwing the spindle out eventually seals it on a conical seat in the cover so that the stuffing box is relieved of pressure for repacking.

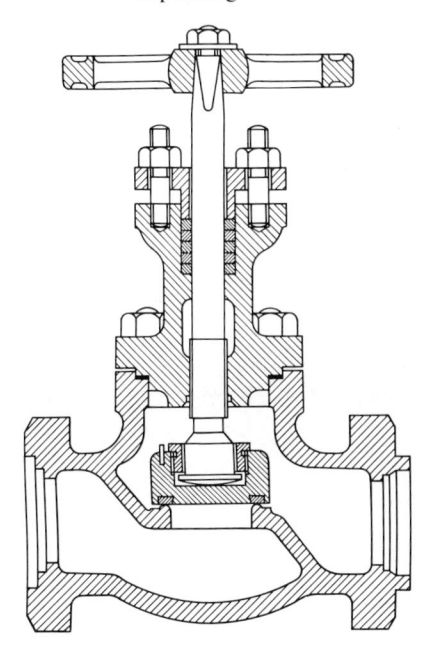

rendered the sealing cap unnecessary. A recent design of stop valve suitable for ammonia and other refrigerants is shown in Fig. 2.12.2. It is of partly welded construction, with two O-rings sealing the spindle.

Line stop valves for small systems piped with copper tube for fluorinated refrigerants are often of a special type known as 'packless' valves. The movement of the valve spindle is transmitted to the sealing member through a metal bellows or diaphragm so that there is no possibility of leakage. Since the spindle can only press on the sealing member, not pull it, a spring is needed to move the sealing member the other way. Usually the action is such that the movement of the spindle closes the valve, the spring opening it when the push from the spindle is relaxed. When there is a large pressure difference across the closed valve it should be installed so that the higher pressure helps the spring.

Modern developments in O-rings have enabled small valves with

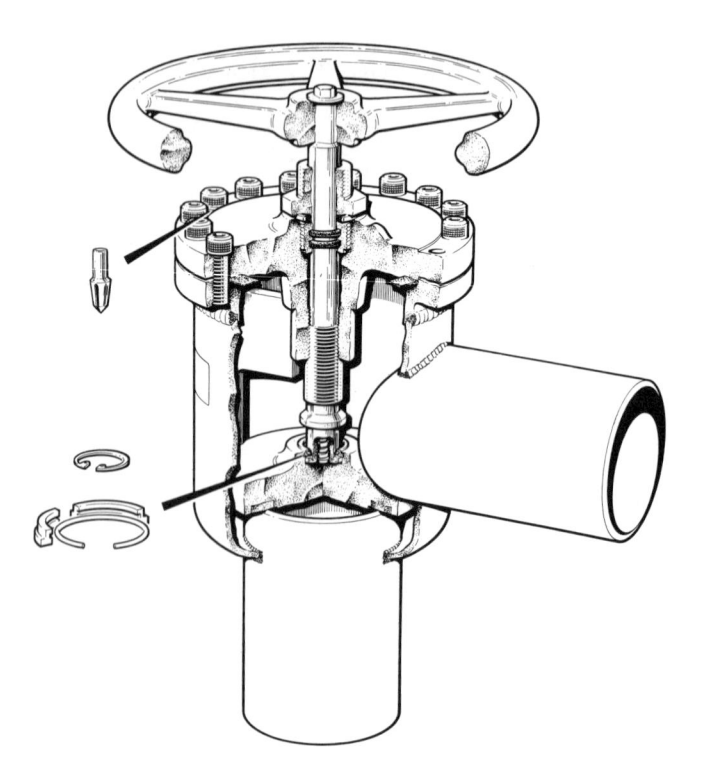

Fig. 2.12.2. A modern stop valve of partly welded construction for welding to the pipework, providing a large free area when open. (By courtesy of Hall-Thermotank Products Ltd.)

virtually no leakage to be made and these have replaced packless valves in many circumstances.

Solenoid valves (*Magnetic valves*)

These are valves which can be opened or closed in response to an electrical signal given, usually, by a thermostat. In the small sizes the sealing member of the valve is directly attached to an iron armature which can move inside a solenoid. When the solenoid is energised the armature is lifted and the valve is opened. When the circuit is broken the valve closes under gravity or with the help of a spring. To increase the impulsive force to open the valve some lost motion may be incorporated so that the armature acquires some momentum before contacting the valve stem.

Because of the relatively small pull available from the armature, larger solenoid valves are made pilot operated. The principle of operation is the same as that shown in Fig. 2.8.6(*a*) for a pilot operated expansion valve, where the valve controls the connection between the space above the piston and the downstream pressure. The pilot valve may be part of the main valve for medium sizes, or it may be a separate component, as indicated in Fig. 2.8.6, for the larger sizes.

Because of the required pressure difference across the piston to hold the valve open, a pilot operated valve must have a pressure drop across it, even at vanishingly small flow rates, unlike a directly operated valve for which the pressure drop diminishes with flow rate.

The action of a solenoid valve described above is called 'direct' or 'normally closed' action, i.e. the valve opens when the circuit is made. If a valve is required to close when the circuit is made, i.e. it is 'normally open', it can be specially constructed for this. The same result can be achieved by inserting a reversing relay in the electric circuit. When the control circuit is made, the reversing relay breaks the circuit that operates the solenoid valve.

Solenoid valves are commonly used in multi-evaporator installations, where several evaporators are operated from a common suction line with one or more compressors. A solenoid valve in the liquid line to each evaporator is opened when the thermostat associated with an evaporator demands refrigeration. Other uses are to direct hot gas to an evaporator for defrosting, as shown in Fig. 2.16.1, and for the regulation of refrigerating capacity of compressors, as in Fig. 2.14.3 and Fig. 2.14.4.

Back pressure valves (*Evaporator pressure regulating valves*) (*EPRV*)

These are installed in the suction pipe between an evaporator and compressor to throttle the flow so as to maintain a pre-set minimum upstream (evaporating) pressure. The principle of operation is shown in Fig. 2.12.3(a). The diaphragm or bellows has the same effective area as the valve disc, and consequently the forces caused by the downstream pressure cancel each other and the position of the valve stem and disc is determined by the upstream pressure acting against a spring. As the upstream pressure rises, the valve opens.

Because the valve has to pass vapour, the orifice has to be relatively large in order to avoid undue pressure drop when the valve is supposed to be wide open. Consequently, in the larger sizes they are often made as pilot operated valves working according to the principle shown in Fig. 2.8.6(b). The valve remains shut until the upstream pressure applied to the top of the piston is able to open it against the spring. A pilot operated valve like this can be made responsive to other variables if desired, for example to the

Fig. 2.12.3. (a) Operating principle of an evaporator pressure regulating valve. The effective diameter of the bellows B is the same as that of the orifice, and the forces on the valve stem from the downstream (suction) pressure cancel. The valve stem responds to upstream pressure and is closed by the spring when this falls below a value which can be set by the adjusting spindle A. (b) Principle of a crankcase pressure regulating valve. The forces from the upstream (evaporator) pressure cancel and the valve responds to downstream (suction) pressure. When this rises above a preset value the valve closes.

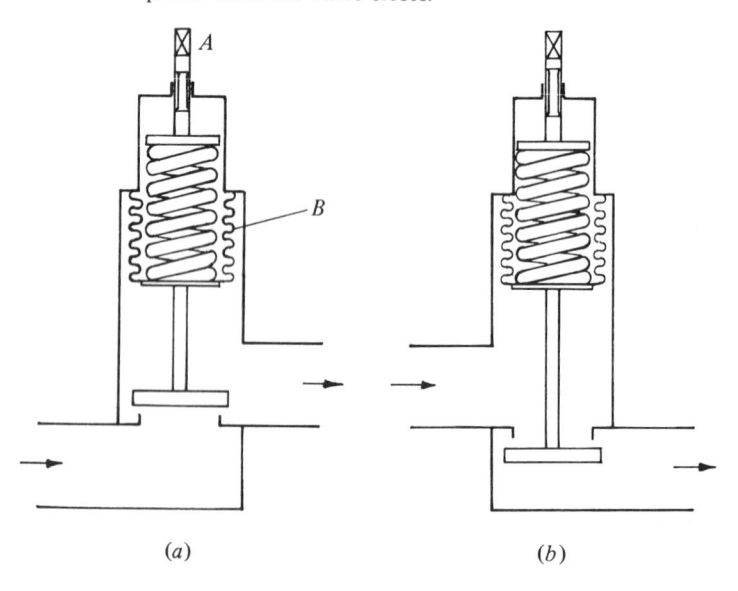

(a) (b)

2.12 Line controls

temperature of chilled water delivered by an evaporator, by providing a pilot valve which is temperature sensitive.

An example of the use of an evaporator pressure regulating valve occurs when one compressor draws vapour from two evaporators which may be at different temperatures. Suppose one evaporator at a temperature of $-20\,°C$ is holding ice-cream whilst the other is chilling bottles of soft drinks with an evaporating temperature just above zero. The pipes from the two evaporators join together before the compressor, but the vapour coming from the warmer evaporator first passes through a back pressure valve. This holds up the pressure in that evaporator to a value suitable for chilling, and throttles the vapour down to the pressure in the colder evaporator before mixing with the vapour from it. The supply of liquid to the expansion valve of each evaporator is controlled by a solenoid valve so that refrigeration for each is assured independently. The scheme is shown in Fig. 2.12.4.

Fig. 2.12.4. Arrangement for operating evaporators at different temperatures from one compressor. Solenoid valves SV1 and SV2 respond to their respective thermostats in the colder and warmer compartments, admitting liquid to each evaporator as required through the thermostatic expansion valves TEV1 and TEV2. An evaporator pressure regulating valve EPRV in the suction from the warmer evaporator holds up the evaporating pressure and temperature there. A non-return valve NRV prevents return flow and condensation of vapour in the colder evaporator when there is a high load on the warmer evaporator with a corresponding suction pressure which is too high for the colder evaporator.

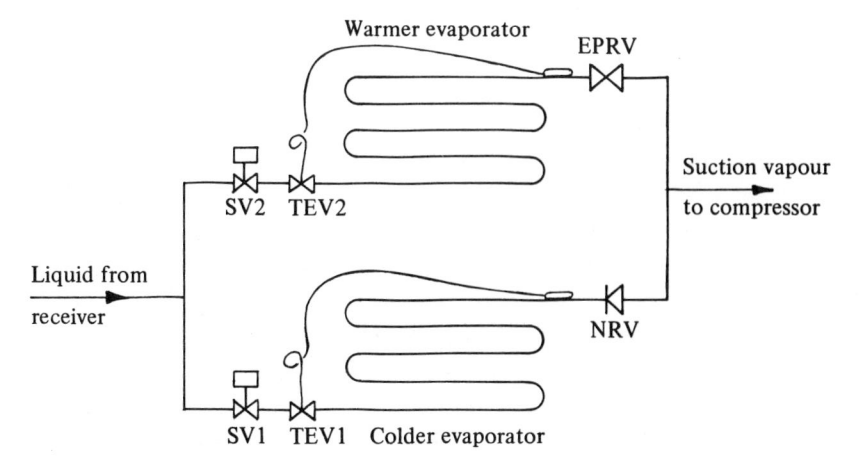

Another use is as a safety precaution in a plant for chilling water, where the evaporating temperature must not be allowed to fall below 0 °C on account of the danger of freezing.

The throttling action of a back pressure valve reduces the effective refrigeration capacity of a compressor by reducing the density of the vapour aspirated. Thus it enables a compressor to come into balance with an evaporator at a reduced load without having too low an evaporating temperature for the purpose.

Crankcase pressure regulating valves (CPRV)

These are also installed in the suction line and throttle the flow of vapour to the compressor, but they are constructed so as to close when the downstream pressure rises and open when it falls. The construction is similar to that of back pressure valves except that the forces from the upstream pressure are arranged to cancel each other, and the valve stem is moved by the downstream pressure, as shown in Fig. 2.12.3(b).

The object of a crankcase pressure regulating valve is to limit the suction pressure at the compressor so that the power taken is within the full load rating of the motor provided.

Non-return valves (check valves)

These allow flow in one direction only, through a port controlled by a ball or disc, or by a hinged flap in large valves.

One example of their use in hot-gas defrost systems is shown in Fig. 2.16.1. Another use is in the scheme already described of two evaporators operated from one compressor with an evaporator pressure regulating valve in the line from the warmer evaporator. A check valve is installed in the line from the colder evaporator to prevent the higher pressure vapour from the warmer evaporator condensing inside the colder one. Check valves may also be installed in the delivery pipes from several compressors operating in parallel to prevent vapour, at condensing pressure, from condensing in the delivery pipe of a compressor which is not running.

Hot gas injection valves (Compressor by-pass valves)

One way of reducing the refrigerating capacity of a compressor applied to an evaporator is to by-pass some of the delivered vapour back to the suction side. If the hot gas is injected straight into the suction pipe this has the effect of raising the delivery temperature considerably. By injecting the hot gas into the evaporator just downstream of the expansion valve this rise can be avoided. The total flow from the compressor then goes through the evaporator but the heat from the hot gas cancels out some of the refrigerating effect.

The injection valve may be made responsive to the variable which it is desired to control, for example evaporator pressure in order to prevent freezing in a water chiller, or directly to the temperature of something being refrigerated.

Liquid injection valves

These are used to desuperheat vapour by injecting liquid into it. Their principal use is to reduce the temperature of the vapour entering a stage in ammonia systems with two or more stages of compression. This in turn reduces the temperature of the vapour delivered by that stage, and also, as will be seen in Chapter 5, improves the refrigerating capacity and the coefficient of performance of the system.

2.13 Electric motors

The great majority of refrigerating compressors are driven by alternating current electric motors, the characteristics of which have an important bearing on the design and operation of the system. This description will be limited to the commonest forms of motors, i.e. induction motors, though synchronous motors are often used on large plants because of their better power factor.

The three-phase induction motor, which is used for compressors above a

Fig. 2.13.1. Diagrammatic curve showing the torque developed as a function of the rotational speed for a three-phase motor.

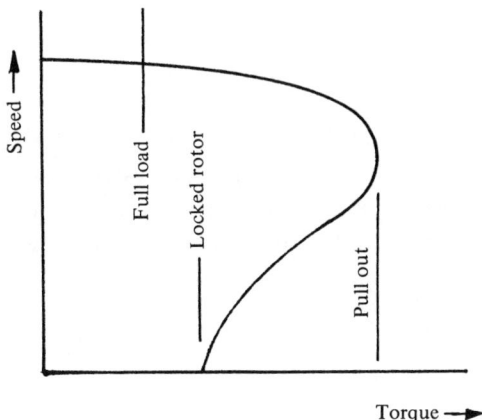

shaft power of about 1 kW, contains three stator windings displaced 120° from each other. To each winding is applied the voltage from one phase of the supply, which is displaced 120° in time from the other two phases. The resulting magnetic field has a constant magnitude but rotates at the supply frequency. The rotor, on the motor shaft, is provided with conductors which are cut by the rotating field, and a torque is developed so long as there is a difference between the rotational speeds of the field and the rotor. This difference, which amounts to about 5% at full load conditions, is known as the slip. On a frequency of 50 Hz with a slip of 5% the rotor would rotate at a speed of 47.5 rev/s = 2850 rev/min.

By providing six windings, each displaced 60° from the next ones, the field can be made to rotate at half the supply frequency, giving a motor speed of about 1425 rev/min, and similarly with nine windings a rotor speed of about 950 rev/min can be provided.

The relation between torque and speed of a three-phase motor is shown in Fig. 2.13.1. At rest a torque is developed, called the locked rotor torque. This torque must clearly be greater than the torque applied to the shaft if the motor is to start. As the speed increases, the torque increases to a maximum, known as the pull-out torque. The motor cannot normally run continuously at this torque, however, because it would overheat. The greatest torque which the motor can develop continuously depends on the manner in which it is cooled and the temperature of the air which flows over it, as well as on the insulating material used in it. At a standard air temperature, the torque which can be provided continuously is the full-load torque. Corresponding to the full load torque is a full-load power, which is the value stamped on the nameplate. The pull-out torque is about $2\frac{1}{2}$ times full-load torque for most motors.

A motor can deliver more than full-load torque and power for short periods, and this is often required when driving refrigerating compressors. The power taken by these is very variable, and it is not always economic to provide a motor big enough to take the maximum power ever encountered. Provided the overload is of short duration, the motor does not have time to overheat. Overload is most likely to occur when the system is being pulled down from ambient temperature, or when a sudden load is put on the evaporator. Under such conditions the power rises to a peak, as will be seen in Chapter 4, and it may be necessary to provide some means of limiting this power, as already explained in connection with the thermostatic expansion valve. It should be noted that the pull-out torque of the motor must be sufficient to ensure that the compressor keeps turning at all possible evaporating and condensing pressures. Other methods of limiting motor

2.13 Electric motors

power are also used. In fact any of the capacity reduction devices so far described can be used for this purpose.

Motors for hermetic or semi-hermetic compressors in which the windings are cooled by the suction vapour passing over them have a temperature characteristic which is quite different to that of open motors. For an open motor the worst condition is high evaporating pressure. For a hermetic motor the worst condition may be very low evaporating temperature when the mass flow rate of vapour over the windings is insufficient, because of the low density of the vapour, to keep them cool. Because of the variable cooling obtained at different evaporating pressures, the nameplate rating of a hermetic motor is a rather arbitrary quantity.

Single-phase induction motors are used on small compressors in places where a three-phase supply is not available. A single-phase winding, however, does not develop a rotating field but a unidirectional pulsating one, which creates no tangential force on stationary rotor conductors, and therefore no inherent locked rotor torque. If the rotor is given an impulse in either direction, however, it will run and develop torque. The nudge to start the rotor is obtained by providing a separate stator winding displaced in space from the main winding, and in which the current is also displaced in time. The combination of the two fields gives a rotating field which cuts the rotor conductors and gives a starting torque. Some means of providing the

Fig. 2.13.2. Diagrammatic torque–speed for a single-phase motor. The curve for the run winding alone indicates no starting torque. Combined with the start winding a torque at zero speed is developed. When the speed passes a value at which the run winding alone can develop enough torque, the start winding can be disconnected.

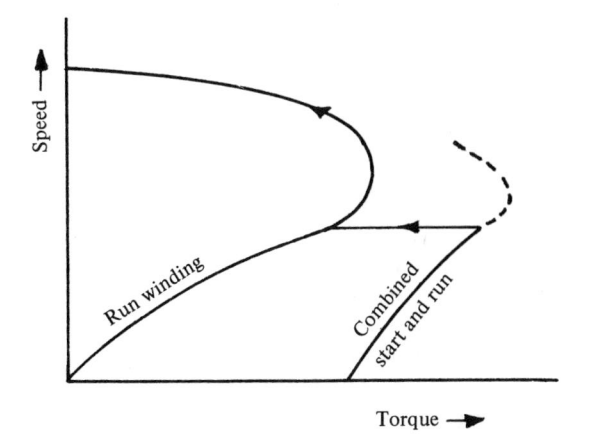

phase difference, and of cutting out the start winding when the rotor has run up to speed, are needed.

The phase difference is obtained by (i) connecting a capacitor in series with the start winding, or (ii) making the start winding of relatively high resistance. Method (i) gives a high starting torque, about four times full load torque. It is used in systems in which the compressor has to start under load, as for example when a thermostatic expansion valve is used which does not allow the pressures to equalise when the machine is not running. Method (ii) is used in systems employing capillary tube restrictors, e.g. domestic refrigerators, in which the pressures can equalise when the compressor is not running.

In some motors the capacitor and start winding are kept in circuit during normal running, so that the motor operates as a two-phase one. More usually, however, the capacitor is cut out when the motor has run up to speed. In open type motors this is done by a centrifugal switch. The torque curve is shown in Fig. 2.13.2.

In hermetic compressors it is not possible to incorporate a mechanical switch on the shaft, so an electrically operated switch outside the case has to be used. One conventional arrangement is shown in Fig. 2.13.3(a). The high

Fig. 2.13.3. (a) Conventional circuit employing a magnetically operated starting relay for starting a hermetic motor. (b) Circuit employing a PTC thermistor permanently in the start winding circuit. (By courtesy of Danfoss.)

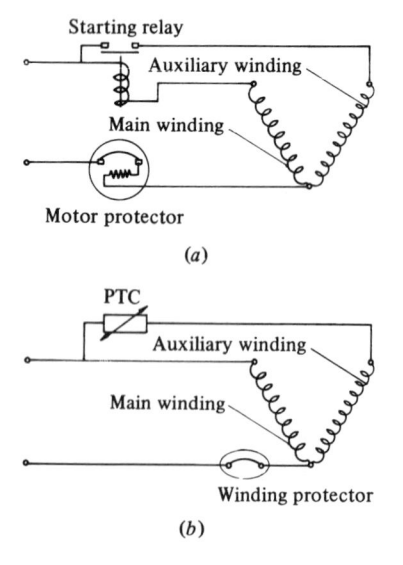

2.13 Electric motors

current through the stator when the rotor is at rest passes through a solenoid which attracts an armature and makes the start winding contacts. As the speed rises, the current falls away and the contacts open.

A new device employs a thermistor, i.e. a resistor with a marked change in resistance with temperature, in this case an increase, and hence the designation PTC is used for positive temperature coefficient. A typical curve for a PTC is shown in Fig. 2.13.4, and the circuit is shown in Fig. 2.13.3(b). The PTC is permanently in the circuit. When the motor is switched on the high current through the start winding warms up the PTC and its resistance rises, so reducing the current to a low value.

The direction of rotation of a three-phase motor can be reversed by exchanging two of the phase leads in the terminal box. In a hermetic motor one cannot see the direction of rotation and the compressor must be capable of running either way. In a single-phase motor the direction is predetermined by the way in which the start and run windings are connected inside the case.

Protection of the motor against overload is necessary to ensure that the motor does not become so hot that degradation of the insulation takes place. Originally this was done by making the overload device sensitive to current only, so that it would trip when a current capable of ultimately damaging the windings was attained. A resistance in thermal proximity to a

Fig. 2.13.4. Resistance versus temperature for a typical PTC thermistor. (By courtesy of Danfoss.)

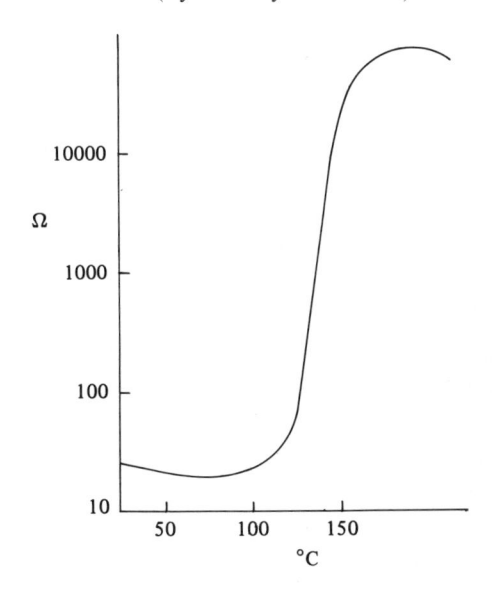

bimetallic strip was a common arrangement. More recently the overload device has been placed in the compressor windings to respond to their temperature directly. It is still necessary to retain some current sensitivity, however, to protect against the locked rotor condition, i.e. when the motor refuses to start for some reason, during which the temperature in some spots of the motor might rise so rapidly that the temperature sensitive overload device might not respond in time. Thermistors are also used as overload devices.

Motor overload protectors may be of the automatic resetting type, or arranged for manual reset by pressing a button. WARNING! Never touch a motor without isolating it from the supply, because it may have cut out on its overload or other automatic control and start up automatically at any time.

Fuses are used to protect the wiring to a motor. They do not themselves provide adequate motor protection against overload, although they give some degree of protection against short circuits. In single phase circuits the fuse must be installed in the live conductor only. In three-phase circuits a fuse is installed in each conductor. If one of these fuses goes whilst a motor is running, or a break occurs in one of the phases for any reason, the motor can continue to run as a single-phase motor and this will lead to overheating. Consequently protection against 'single-phasing' should be provided to disconnect the motor in the event of a break in one of the phases.

The frames and all exposed metal parts of motors must be effectively earthed. In favourable conditions, i.e. short distance and wet ground which give a low impedance between the machine and the earthed point of the distribution network, an earth fault will blow the fuses. When the ground impedance is large, however, it is necessary to use an earth leakage trip which is sensitive to low earth leakage currents.

The design and installation of electrical supply systems is the province of the specialist electrical engineer. This part of the work is often sub-contracted to a local firm which is familiar with the codes and conditions applying in their territory.

Small motors, both single-phase and three-phase up to about 5 kW may be started by switching them directly onto the supply. Although the current drawn from the supply during the first few instants of starting is much higher than full load current, this may not be too serious for small motors. Large motors, however, draw such a high current that it is usually necessary to provide some means of limiting this during starting. The usual method for three-phase motors is by 'star-delta' starting. The phase windings are initially connected in star and receive only $1/\sqrt{3}$ of the phase voltage. When the motor is running, the windings are reconnected in delta and receive the full voltage. This is done automatically by a 'star-delta starter'. Star-delta switching naturally gives a reduced starting torque and the compressors are therefore usually arranged to start in the unloaded condition, i.e. with the suction valves held open or, for screw compressors, the capacity reduction slides in the minimum capacity positions.

In most plants the motors are started automatically in response to the making of a circuit by a thermostat or other device. It is feasible for very small motors, for example those in domestic refrigerators, to allow the thermostat contacts to carry the full motor current, as explained in the next section, but for large motors this is not practicable. The thermostat then carries only a small control current which operates a relay, known as a 'contactor', to close the main motor circuit.

2.14 Automatic control of refrigerating capacity

A number of automatic controls which regulate the internal running conditions of a refrigeration system have already been described. In addition to these almost all modern plants have automatic controls which enable the plant to adjust itself to changes of load without continuous attention.

The simplest form of control, used in domestic and other small refrigerators, is to stop the compressor when the load is satisfied, for example when the temperature inside a cold room is low enough, or when the evaporating temperature and pressure are low enough. The period when the compressor is running is called the 'on-cycle' and the period when it is off the 'off-cycle'.

The switching of the compressor motor is done by a thermostat or by a pressostat. A thermostat consists of a vapour pressure thermometer, or sometimes a bimetallic one, which senses the temperature which it is desired to control, and which produces a movement to operate a switch. As the temperature rises to a set value, called the 'cut-in' temperature, the switch closes and starts the compressor. When the temperature falls to some lower value, the 'cut-out' temperature, the switch opens and stops the motor. The difference between the cut-in and cut-out temperatures is called the 'differential' of the thermostat, and this can often be varied by making an adjustment. The level of temperature at which the thermostat operates is called the 'range', and this can also be adjusted. For example, a thermostat might have a cut-in temperature of 5 °C and a cut-out temperature of 4 °C, a differential of 1 K. By changing the range the same instrument might be made to work at a cut-in temperature of say 8 °C with a cut-out temperature of 7 °C, keeping the same differential. Changing the differential as well, it might work at a cut-in temperature of 8.5 °C and a cut-out temperature of 6.5 °C. The range is usually adjustable by a knob on the instrument, marked with a temperature scale or by an arbitrary scale of numbers from coldest to warmest. The differential change requires an internal adjustment. In cheap thermostats for special purposes such as domestic refrigerators the possible differential adjustment may be quite small and just sufficient for setting in the factory to a stated value.

The vapour pressure thermometer works on the same principle as already described in connection with thermostatic expansion valves. It has the advantage over the bimetallic strip type in producing more force for the

Fig. 2.14.1. Diagrammatic arrangement of a thermostat or pressostat. For a thermostat the bellows expands and contracts in accordance with the pressure developed in a feeler bulb containing a volatile fluid which is situated at the point where the temperature is to be sensed. A pressostat is connected directly into a system to sense, for example, evaporating pressure. The force from the bellows is balanced by the range spring, and by screwing the adjustment up or down the temperature or pressure level at the midpoint of the differential can be raised or lowered. The snap-action spring ensures rapid opening of the contacts by becoming unstable when the bellows have contracted to the cut-out point. The differential can be altered by turning the nuts to change the differential gap.

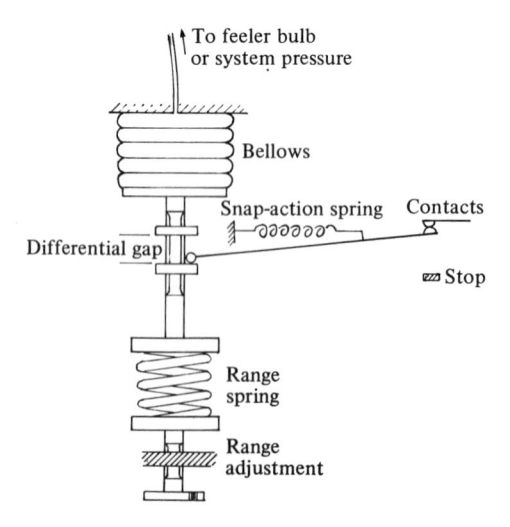

2.14 Automatic control of refrigerating capacity

operation of the switch, and is consequently widely used in thermostats which switch the motor directly. A bimetallic strip, however, can react to changes in temperature faster than a vapour pressure thermometer and is often used for more sensitive instruments working with automatic controllers. Resistance thermometers are also used.

It will be noted that a refrigerator thermostat performs the opposite function to a heater thermostat, which breaks contacts when the temperature rises and makes them when it falls.

Whichever method of sensing temperature is used, a smooth movement is generated. If there were no slackness or backlash in the mechanism connecting the temperature sensor to the switch there would be no differential: the contacts would break at the same temperature as they make. The result would be very rapid cycling of the motor on and off and overheating of the windings because of the high current during starting. To provide a differential some lost motion or backlash is incorporated in the mechanism. For example, the screwed rod shown in Fig. 2.14.1 carries two nuts which move the lever operating the switch. The differential can be changed by changing the distance between the nuts. The range is altered by increasing or reducing the force from the spring.

Without further devices the movement of the contacts at the time of make and break would be slow, and considerable arcing and damage to the contacts would occur if they carried the full motor current. To prevent this a rapid make and break device is incorporated. Two types are in common use: a toggle switch operated by an off-centre spring which is unstable in its mid-position and which moves rapidly one way or the other when it is given a slight push: and a permanent magnet which applies a sudden pull to a piece of iron mounted on the same spring as the contact when it comes within range. More expensive thermostats may use a mercury switch, which consists of a tube containing some mercury with sealed-in contacts at one end. The tube is pivoted at the middle and is unstable in a horizontal position, the mercury tending to collect either at the end with the contacts, so making the circuit, or at the other end so breaking the contacts. Even when arcing occurs for a brief time in a mercury switch it does no damage, simply vaporising mercury which eventually recondenses.

When a very close differential is required it is impossible to obtain a rapid make and break. Consequently, thermostats with fine differentials are not as a rule able to carry much current. They must be used in conjunction with a relay or contactor. The thermostat contacts carry only the signal current for the relay, of the order of 30 mA perhaps, which passes through the solenoid of the relay and brings the main contacts in. The same is true when large motors have to be switched, since it is not practicable to make the thermostat contacts capable of carrying large currents.

Instead of a thermostat a switch responding to evaporator pressure is often used in commercial refrigerators. It works in the same manner as the vapour pressure thermostat except that it senses evaporating pressure directly instead of sensing the pressure in the feeler bulb. Such a pressostat sets the swings of the evaporating pressure and hence of the evaporating temperature within definite limits. Provided that the load on the evaporator does not vary too much and that the temperature of condensation is approximately constant, satisfactory control of the temperature in a cabinet can be obtained. When used on air-cooled units such switches should have a different setting for winter and summer. The refrigerating capacity of the compressor is different at low and high condensing temperatures, and cut-out would occur at different cabinet temperatures if the pressostat were not reset.

The on and off method of control gives satisfactory results so long as the thermal lags between the evaporator and the air or product whose temperature is to be controlled are long enough to attenuate the swings of evaporator temperature. This is often the case in cold rooms with a considerable mass of stored produce inside them, when a fluctuation of 1 K or so in the air temperature is acceptable. If the thermal lags are short, as in a liquid chiller, or if a much closer regulation of temperature is desired, on and off control is not very good because the motor would have to start and stop at very frequent intervals with consequent overheating. Very small motors are allowed to start up to ten times per hour or more, but large motors only once or twice.

This difficulty can be avoided if the capacity of the compressor can be adjusted continuously or in steps to meet the demand without shutting down completely. The methods of doing this in modern piston compressors by holding the suction valves open on some cylinders have been mentioned in section 2.3, by a slide valve in screw compressors in section 2.4, and by pre-rotation or other methods for turbo-compressors in section 2.5. Some other methods will be mentioned in Chapters 3 and 4.

From the point of view of the control system, these methods can be considered as (a) stepped regulation of capacity, by putting cylinders out of action or even closing down one machine out of several, and (b) stepless regulation such as is available in a screw compressor. Two different types of thermostatic action are used in these different circumstances.

Considering first control of a system with stepped capacity regulation, suppose that in a simple case three capacities are available, full, half and

2.14 Automatic control of refrigerating capacity

zero. Clearly above a certain temperature or pressure which is to be controlled, all of the capacity must be on, and below a certain lower temperature zero capacity is required. Somewhere between there is a change-over from full to half and vice versa. The thermostat thus has to initiate some action at three different temperatures, as indicated diagrammatically in Fig. 2.14.2.

Denoting the two halves of the compressor by A and B, when the temperature is above t_3 both A and B are in. If the temperature falls to t_2 one half, say A, is taken out. If it falls further to t_1, then B is taken out. One cannot, however, switch B in again immediately the temperature rises above t_1 because this could produce the rapid cycling condition already mentioned with on–off control. B therefore stays out until temperature t_2 is reached when it is cut in. As the temperature continues to rise, A is brought in again when t_3 is reached.

This type of control is called sequential or step control. In the simple case

Fig. 2.14.2. Sequential control of two equal capacity steps, denoted by A and B.

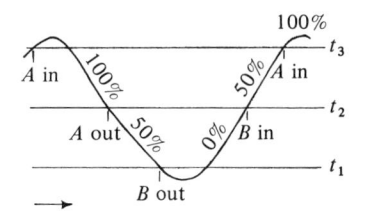

Fig. 2.14.3. Arrangement of valves for cutting out compressors, or single cylinders of a compressor in which the delivery pipes are not manifolded. Opening solenoid valve SV_B allows the pressure at the delivery of B to fall to the suction pressure, the return of high-pressure gas being prevented by the non-return valve. Similarly, opening SV_A by-passes compressor A.

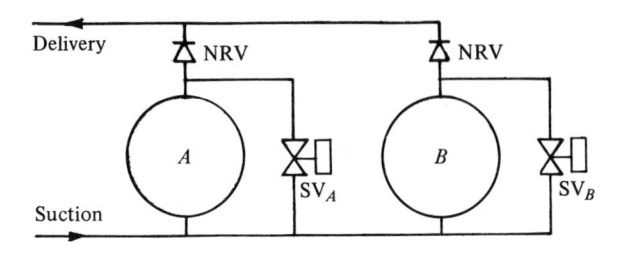

considered it could be arranged by two ordinary thermostats, one controlling A with a differential $(t_3 - t_2)$, the other controlling B with a differential $(t_2 - t_1)$, but these may be combined in the same instrument to give a two-stage thermostat. In more complicated examples a sequential controller is used. It receives an electrical signal from a temperature sensing device and initiates action to bring in capacity steps as required.

The same principles can be applied to whole compressors, shutting them down as required to match the load. This may produce too rapid cycling for the drive motors to accommodate in some circumstances, but the capacity can be reduced without stopping them by the arrangement of by-passes shown in Fig. 2.14.3. for two compressors, with two by-pass valves and two non-return valves. On opening SV_B the machine B simply pumps vapour at suction pressure around the loop, and on opening SV_A machine A does the same. Because of pressure drops in the pipes and friction in the machines they still take some power of course, but this may be better than too frequent starting and stopping. The same idea can be applied to any reciprocating compressor in which separate connections can be made to the cylinders.

In a modern installation of any size there might be several compressors, each provided with some degree of capacity regulation so that the capacity steps are quite small in terms of the total capacity, so giving the possibility of close control of temperature. In principle it is necessary to regulate the capacity of only one of several equal machines, but in practice it is common to provide capacity regulation on all so that the base load is not always carried by the same machines, and to enable any compressor to be used as standby.

Fig. 2.14.4. Arrangements for moving the capacity control slide of a screw compressor by admitting oil pressure to one side or the other of an actuator piston. (The solenoid valves are shown diagrammatically on their sides in this figure and the last. They are normally used vertically.)

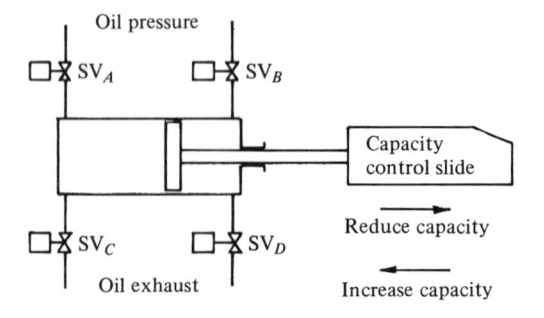

2.14 Automatic control of refrigerating capacity

Turning now to the regulation of a compressor with a stepless variation of capacity from the maximum to a certain minimum of 10 or 15%, which is typical of a screw compressor, various control systems are in use. The following method is one of the simplest and is known as floating control with a neutral zone. Referring to Fig. 2.14.4 the capacity control slide is moved to the right to reduce capacity or to the left to increase capacity by oil pressure acting on the left- or right-hand face respectively of the actuator piston shown. When the solenoid valves SV_A and SV_D are open, oil under pressure is admitted to the left-hand side of the cylinder and oil is exhausted from the right-hand side, thus moving the piston to the right. Similarly, when these valves are shut and SV_B and SV_C are open the piston moves to the left. So far no control action is involved, merely the means of moving the piston.

Floating control with a neutral zone is illustrated in Fig. 2.14.5 where a hypothetical variation of some controlled variable t, which may be the temperature of a chilled fluid or evaporating temperature, is shown. As the temperature rises to t_4, valves SV_B and SV_C are opened and the piston is

Fig. 2.14.5. Control of capacity by the method of floating control with a neutral zone. The heavy line denotes a stationary piston and hence stationary control slide. The variation of temperature shown is hypothetical and intended only to show the control sequence. In practice it is possible with suitable choice of the piston speed to reach a steady condition within the neutral zone $t_3 - t_2$. By widening the differentials of the thermostats they can be made to overlap so that t_2 is higher than t_3.

The variable t may be an actual temperature of a fluid but it is often the evaporating temperature sensed by pressure directly.

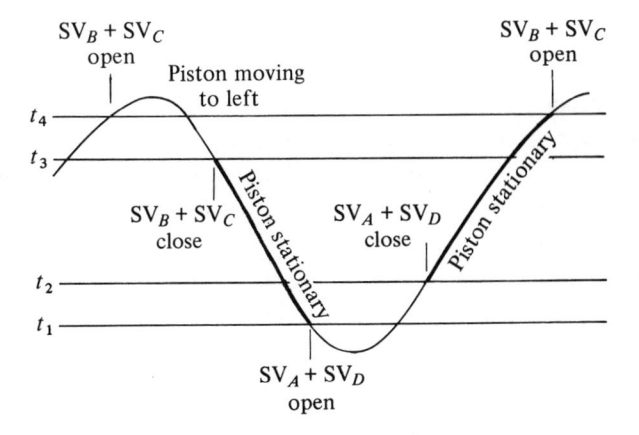

moved to the left to increase capacity. It continues to move until the temperature has fallen to t_3 when valves SV_B and SV_C are closed. Supposing that the temperature continues to fall, the piston remains stationary until temperature t_1 is reached, when valves SV_A and SV_D are opened. The piston now moves to the right and the capacity diminishes. If the temperature now rises as a result, when it reaches t_2 valves SV_A and SV_D are shut and the piston stops.

The manner in which this control system responds to changes in load depends on the speed at which the piston moves in relation to the speed at which changes of the controlled temperature can occur. If the piston moved immediately from one end of the cylinder to the other the result would be on–off control (supposing the minimum to be zero and not 10 or 15%), between the differential $(t_4 - t_1)$. But by making the speed of the piston such as to match the changes in load it is possible to achieve a steady condition with the temperature somewhere in the neutral zone $(t_3 - t_2)$ for a large part of the time. The speed of the piston movement is governed by the pressure of the oil, and by throttling the oil supply the speed can be set to an appropriate value.

In a more elaborate system the piston is moved in a series of jerks or pulses, the pulses being more rapid when the temperature is further away from the neutral zone.

The simple control action shown in Fig. 2.14.5 can be produced by thermostats or pressostats. Supposing the solenoid valves to be of the normally closed type, i.e. they open when the solenoid is energised, one of the thermostats operating between t_3 and t_4 and controlling valves SV_B and SV_C is a normal refrigerator thermostat, i.e. makes the circuit when the temperature rises to t_4 and breaks it when the temperature falls to t_3. The other thermostat is a heater thermostat, i.e. makes the circuit when the temperature falls to t_1 and breaks it when the temperature rises to t_2. These two thermostats can be combined in one instrument.

Four solenoid valves are shown in Fig. 2.14.4, but in some circumstances only two are needed. Because of the construction of the slide valve in a screw compressor the pressure of the delivery gas tends to push it towards the left, i.e. towards the maximum capacity position. At certain combinations of suction and delivery pressures, the valve SV_B can be dispensed with and valve SV_D replaced by a permanent bleed.

The simple system described is adequate in most circumstances, but when control is critical a proportional control system can be used. The position of the slide is then determined by an error signal, i.e. the difference between the actual value and the desired value of the controlled variable.

2.14 Automatic control of refrigerating capacity

When several evaporators carrying different loads operate from one compressor or from a common suction main, on–off control of liquid supply can be applied to each individual evaporator by means of a solenoid valve responsive to the temperature of the load which it is desired to control as already shown in Fig. 2.12.4. For example the several chambers of a cold store require individual control of temperature, which is obtained by interrupting the refrigerant liquid supply to an evaporator as required.

When the solenoid valve closes and interrupts liquid to an evaporator, the liquid inside the evaporator is gradually pumped away as vapour, and the evaporator is put out of action. The pressure inside remains equal to the common suction pressure, but since there is little or no liquid refrigerant in it, the rate of heat transfer is low. When all the space temperatures are satisfied all the solenoid valves are closed. The pressure in the suction main drops rapidly and the drop in pressure can be used as a signal to stop the compressor.

When no capacity regulation is provided, the evaporating temperature in the evaporators still in use gradually falls as others are put out of action, and this may be undesirable, perhaps because it could cause freezing in a water chiller or because too low an evaporating temperature in a cold room would cause low humidity and dehydration of produce. The advantage of being able to vary the capacity of the compressor is that evaporators can be cut out without depressing the evaporating temperature in the evaporators remaining in action. Clearly in such an instance the suction pressure itself would serve as the signal to bring about capacity reduction, either by cutting out cylinders of reciprocating machines or by moving the slide of a screw compressor, by cutting out machines, or by a combination of both, as already described.

In conclusion, it may be worth remarking that in the great majority of plants relatively simple means of control can work successfully because a vapour compression refrigeration system has an inherent stability with respect to changes of load. The reason for this is that the effect of an increase of load on an evaporator is to cause increased rate of boiling and a rise in the evaporating pressure. This rise in pressure causes a rise in the density of the gas and hence an automatic increase in the rate of pumping of refrigerant around the circuit and an increase in the refrigerating capacity, which is just what is required to meet the new load provided that the new evaporating temperature is acceptable. This situation is quite different from that in an engine where with a fixed fuel supply a reduction of load, i.e. a reduction in the torque opposing the turning of the shaft, leads to an increase of speed which is only ultimately limited by friction. An engine is inherently unstable with respect to changes of load, and a more elaborate control system incorporating some feedback is required to hold the speed approximately constant.

2.15   Safety

The safety of people operating, using or merely being in the neighbourhood of refrigerating plant is of the first importance. Many countries have national standards which establish minimum requirements for safety, e.g. British Standard 4434–1969: Requirements for Refrigeration Safety, and American National Standard B9.1–1971: Safety Code for Mechanical Refrigeration. In some countries the national standard is legally enforceable, i.e. it is an offence to build plant not complying with it. In others it may be enforced by insurance companies or by the fact that compliance with it gives some defence in the event of an accident. Other standards and codes also embody safety provisions for special cases, e.g. The Rules of Lloyd's Register of Shipping 1973, and The Institute of Refrigeration's code: Mechanical Integrity of Vapour Compression Refrigeration Systems for Plant and Equipment Supplied and Used in the United Kingdom, 1978. For liquefied gases at low temperatures the Cryogenics Safety Manual 1970 of the British Cryogenics Council should be consulted.

Safety comes from good design, construction and management of refrigerating plant. Accidents in the strict sense of the word, i.e. purely chance happenings such as an earthquake in a region never known to suffer from them in the past, are rarely the cause of damage or injury. In most cases the event could have been reasonably foreseen by someone, and it is the duty of the designers, erectors and operators to foresee what might occur in all likely combinations of circumstances.

A number of features are incorporated in refrigerating plants to ensure safety in the event of human or mechanical failure. These are of three main kinds: (*a*) electrical cut-outs, whose function is to interrupt the power at conditions which though not yet dangerous would eventually become so; (*b*) devices to relieve excessive pressures; and (*c*) arrangements to minimise danger and loss of life when something does go wrong. Whether a plant has all or only some of these features depends on its size and the possible consequences of failure. The smallest plants are usually exempt from the following requirements.

(*a*) *Electric cut-outs.* On compressors electric cut-outs interrupt the power when abnormal conditions begin, such as low oil level or pressure, high oil

2.15 Safety

temperature, low suction or high delivery pressure, high delivery temperatures, etc. They may be wired to call attention to the fault by visual and/or audible alarms. The compressor, being the pressure imposing element, is naturally the important component to protect, but the devices may be installed so as to give earlier warning, for example by giving warning of the stoppage of cooling water or of the cooling tower fans.

(*b*) *Pressure relief devices.* In compressors above a certain size a bursting disc or relief valve which relieves pressure from the delivery side to the suction side is often fitted, to protect against start-up with the delivery valve closed. The bursting disc only relieves pressure from one part of the system to another, but it is necessary to provide against excessive pressure in the whole system. This could occur, for example, if as the result of a fire the vessels containing liquid refrigerant were raised to a high temperature. Relief valves venting to atmosphere are therefore provided on any section of the plant which can be isolated. Relief valves are installed in pairs, connected to the plant by a dual stop valve which makes it impossible to shut off both valves at the same time, though either can be shut off and removed for test separately. The area for flow in the relief valves has to be adequate considering the size of the plant and the likely hazards, and is specified in the codes.

Another pressure relieving device is a fusible plug in a pressure vessel which melts at abnormal temperature and allows the charge to escape.

It must always be remembered that cold liquid refrigerant can generate very high pressure even in ambient temperature if it is locked up in a pipe with no vapour space. Stop valves in liquid lines should be located with this in mind, and when it is possible to valve off a section of pipe attention should be drawn to the danger.

(*c*) *Ultimate protective arrangements.* Arrangements to be provided against the event of a burst or explosion depend on the nature of the refrigerant. With non-toxic and non-flammable refrigerants it may only be necessary to empty the building. Adequate exits should be provided and marked. To prevent further damage it should be possible to shut the machinery down by switchgear outside the plant room. With ammonia there is an additional hazard of fire or explosion, and the need to enter the area to rescue people. Respirators suitable for use in ammonia must be provided outside the areas likely to be affected. Because many plant rooms are unattended now it is common practice to install automatic detectors to give audible alarm and to start exhaust fans.

2.16 Methods of defrosting

When the surface temperature of an evaporator cooling air is higher than 0 °C, the moisture deposited simply drains off, though at high air velocities some of it may be blown off as droplets. A cooler operating in this way is called a 'sweating' cooler. This is the normal condition of cooling coils in air-conditioning plants. These coils, because they do not have to accommodate frost, can be made with close spacing of about 2.5 mm between the fins.

Below 0 °C the moisture stays on the surface as frost which has two effects: (a) it covers the surface with an insulating layer, and (b) it fills the spaces between the fins and reduces the volume flow rate of air through them. The space between the fins in coolers subject to frosting is consequently made larger than for sweating coolers, e.g. 6 mm instead of 2.5 mm. In some coolers the fins at the front which meet the air first are widely spaced to take most of the frost, the fins on the rows further back being spaced more closely.

In all modern plants provision is made for automatic defrosting at the required intervals. The methods in use are:

(i) by sparging the cooler with water, possibly warm water from the condensers;

Fig. 2.16.1. Connections for hot gas defrosting with pump circulation. SV–solenoid valve, HRV–hand regulating valve, RV–relief valve, NRV–non-return valve. In normal running liquid is fed to the evaporator via HRV, which is set once and for all during commissioning, SV2, and leaves by SV3. During defrost hot gas enters by SV1 the drip tray coils, then via NRV to the top of the evaporator and leaves by RV, SV2 and SV3 being closed.

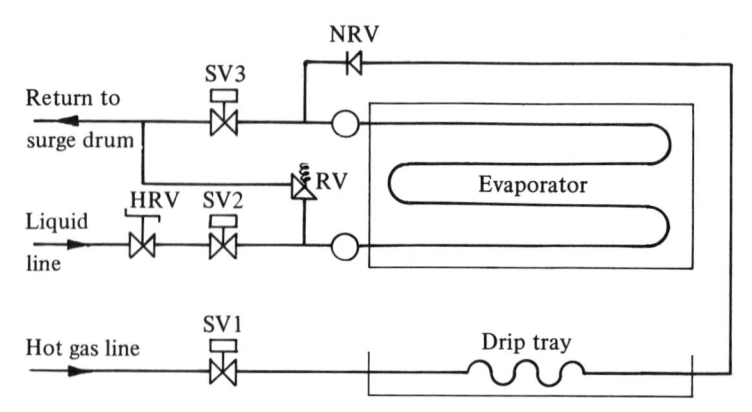

2.16   Methods of defrosting

(ii) by heating with warm brine or glycol which passes through tubes built into the coil block, or through tubes inside the refrigerant tubes;

(iii) by electric heating, the elements being in tubular form in the coil block; and

(iv) by using hot gas from the compressor, introduced inside the evaporator pipes.

Sparging with water is not so common nowadays. The piping has to be so arranged that no water lies in the pipes inside a cold room. This is done by a three-way control valve outside the room. With the hand lever in one position water is admitted to the spray pipes above the cooler inside the room. In the other position the water is shut off to the sprays, but the spray pipes are connected to a drain outside the room. Heating with warm glycol or brine is also little used now.

Electric defrost and hot gas defrost are the most usual methods today. Standard cooling coils are available with the heaters built into the finned coil assembly, and also in the drip tray underneath. These heaters take considerable power for short periods, and add to the maximum demand on the supply, which has to be paid for as a rule.

With hot gas defrosting, solenoid valves are used to re-route the vapour delivered by the compressor into the evaporator instead of into the condenser. The arrangement of these for an evaporator with pump circulation of refrigerant is shown in Fig. 2.16.1. In its simplest form hot gas defrosting can only be used when there is more than one evaporator operated by a compressor. With only one evaporator, there is no load on the compressor when it is defrosted and the delivery pressure falls so that hot vapour is not delivered.

Various ways around this difficulty are available. Some thermal storage can be incorporated in the system, a tank of water for example, which is warmed up during normal operation by doing part of the condensing duty, and which can provide a load for the compressor during defrosting. Another method, known as 'reverse cycle defrosting', converts the system into a heat pump and makes the condenser operate as the evaporator. A special reversing slide valve shown in Fig. 2.16.2 is used in the circuit.

When several evaporators are operated from one compressor they can be defrosted in turn, the others providing the load.

The fans on a cooler are switched off during defrosting, and are prevented by a delay switch from starting until the layer of water left on the cooler surface has had time to freeze again.

Defrosting is commonly initiated by a clock with electric contacts, though other methods have been tried and are used to some extent. These include flow switches to detect a change in the rate of air flow through the cooler, pressure switches sensing an increased delivery pressure at the fan outlet when it is of the blow-through type, or reduced suction pressure when it is of the suck-through type, and photoelectric cells detecting increased reflection from the cooler when it is covered with frost.

Termination of defrost can be done by the clock also, but termination by rise of temperature of the evaporator is often used.

Fig. 2.16.2. Connections for reversed cycle defrosting. With the reversing slide valve in the position shown by the full line, the compressor is drawing vapour from the condenser and delivering it to the evaporator. Normal operation occurs when the slide valve is in the position shown by the broken line.

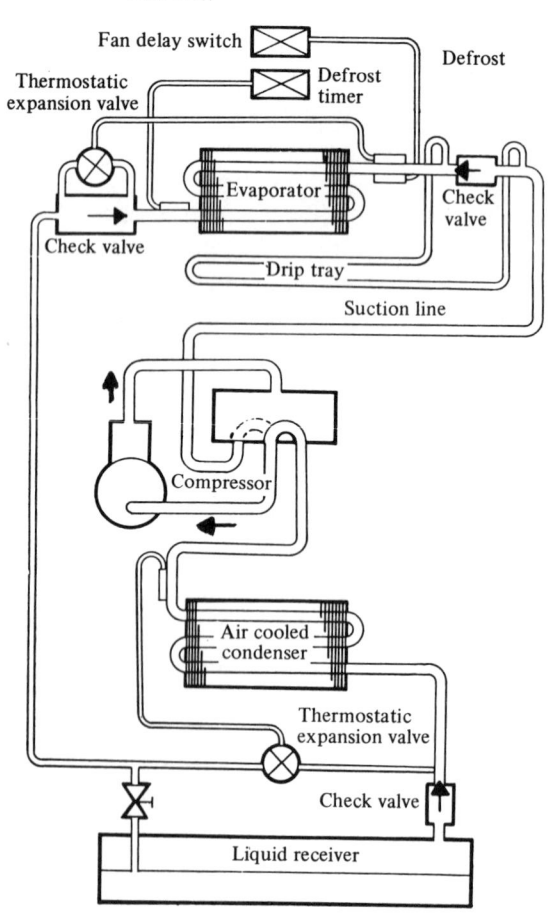

2.16 Methods of defrosting

Although defrosting on a timed cycle is commonly used, it is inefficient, especially for electric defrost. More elaborate methods are used in some advanced domestic refrigerators. For example, the clock may be arranged to run only when the door is open, recognising that most of the moisture comes in through the open door: or the number of times the door is opened may be counted and the figure used to decide when defrosting is needed.

A method of eliminating the need for periodical defrosting is to sparge the cooler continuously with brine or glycol solution. This is valuable in process air coolers which cannot be interrupted for defrosting. Because the moisture dissolves in the solution, it gradually becomes weaker and it has to be reconcentrated at intervals. In some installations the reconcentration has been done continuously by circulating the solution to the cooling towers or evaporative condensers. In food plants efficient eliminator plates are needed to prevent the brine or glycol from contaminating the produce.

3

The vapour compression system

3.1 The cycle of operations

The essential components of the vapour compression system are shown in Fig. 3.1.1. In the *evaporator* a volatile fluid, the *refrigerant*, boils at a temperature t_e, say, low enough to be useful, i.e. lower than the temperature of the bodies or fluids which are to be refrigerated. The boiling temperature of the refrigerant depends on the pressure p_e, say, in the evaporator, according to the relation expressed by the vapour-pressure versus temperature curve, shown diagrammatically in Fig. 3.1.2. The pressure in the evaporator must be reduced sufficiently therefore to give the desired temperature. This reduction of pressure is caused by the suction of the compressor, which draws the vapour away from the evaporator. The vapour induced by the compressor is then raised to a higher pressure p_c, say, at which condensation to a liquid can take place at a temperature t_c. During

Fig. 3.1.1. Essential components of the vapour compression system. t_b is the temperature of the fluid being refrigerated, t_w is the temperature of the coolant, and 1–4 correspond to positions shown in Fig. 3.1.2(b).

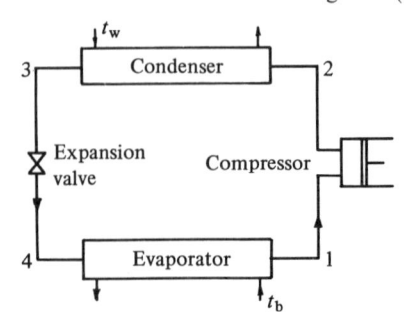

3.1 The cycle of operations

condensation at constant pressure the refrigerant gives out heat to a cooling medium which may be water or air. The temperature of condensation t_c must clearly be higher than that of the cooling medium. The liquid at pressure p_c and temperature t_c or a little less, is then admitted to the evaporator through a throttle valve or *expansion valve*, which is necessary to maintain the difference in pressure between the condenser and evaporator. During its passage through the expansion valve, from pressure p_c to pressure p_e, the temperature of the liquid falls to t_e, and some of the liquid turns into vapour. This happens because the liquid at the entry to the expansion valve is at or very near its boiling point for the pressure p_c, and as the pressure is reduced in the valve by viscous effects and by acceleration, the liquid can no longer exist in equilibrium as a liquid and must split into two phases. The vapour which forms has to absorb heat (enthalpy of evaporation), this being obtained from the liquid which is left. The refrigerant leaving the valve is therefore a two-phase mixture at the temperature t_e. The liquid fraction of this mixture boils away in the evaporator, taking in its enthalpy of evaporation as heat from the bodies or fluids being refrigerated. The vapour fraction of the mixture, commonly called the *flash gas*, plays no further part until it is compressed again, having served its purpose in reducing the temperature of the liquid fraction from t_c to t_e. The flash gas has important practical effects, however, caused by the fact that although by mass it is a small fraction of the total flow (typically about 15%), by volume it is a very large fraction (typically 98% for a refrigerant like ammonia).

Fig. 3.1.2. (a) Vapour pressure versus temperature curve, showing evaporating temperature t_e lower than that, t_b, of the fluid being refrigerated, and condensing temperature t_c higher than that, t_w, of the cooling water or air. (b) The cycle on pressure versus specific volume coordinates.

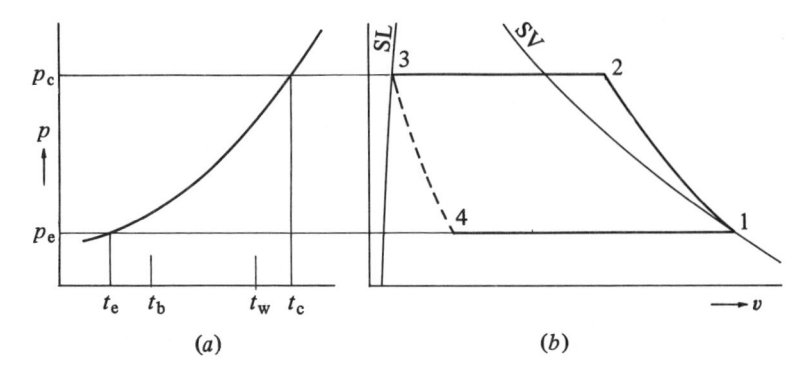

To illustrate the change in state of the refrigerant around the circuit, the specific volume v is shown on the right of Fig. 3.1.2 with the states numbered to correspond with the numbers in Fig. 3.1.1. The ordinate of the diagram is pressure, and the two curves marked SL and SV show the specific volumes of saturated liquid (i.e. liquid on the point of boiling) and saturated vapour (i.e. vapour on the point of condensing), respectively. To begin with, it is supposed that the vapour leaving the evaporator is dry and saturated, point 1, and that liquid leaving the condenser and entering the expansion valve, point 3, is also saturated.

1–2 represents compression of the vapour from pressure p_e and specific volume v_1 to the higher pressure p_c and the lower specific volume v_2. The state at 2 is shown for a refrigerant like ammonia, i.e. in the region of superheated vapour. The temperature t_2 at 2 is higher than the saturation temperature t_c corresponding to the pressure $p_2 = p_c$. Before this vapour can condense it must first be desuperheated to saturated vapour at the constant pressure p_c, during which process its temperature falls from t_2 to t_c. Condensation then takes place at constant pressure p_c to saturated liquid at 3. Process 3–4 takes place through the expansion valve, the pressure falling from p_c to $p_4 = p_e$, and the specific volume increasing from v_3 to v_4 because of the formation of some vapour. The process 3–4 is represented by a broken line on the diagram because the exact states traversed by the fluid are not known. As will be shown, the point 4 can be located, but not the intermediate states. The state at 4 can be specified by a dryness fraction or quality, x_4, defined as the fraction by mass of the total flow which is vapour. The specific volume of the mixture is the sum of the liquid volume and the vapour volume, i.e.

$$v_4 = (1 - x_4)v_{f,e} + x_4 v_{g,e} \tag{3.1.1}$$

where $v_{f,e}$ and $v_{g,e}$ are the specific volumes of saturated liquid and saturated vapour, respectively, at the evaporating temperature t_e.

The boiling of the remaining liquid in the evaporator is represented by the process 4–1 at the constant pressure p_e, in which the specific volume increases from v_4 to v_1, and the refrigerant is brought back to the state from which the description started. The refrigerant is said to undergo a *cycle of operations*, which will be called simply the vapour compression cycle.

3.2 Work and heat transfers

During the cycle work is done on the refrigerant in the compressor, and heat transfer to and from it occur at various places. The principal heat transfers occur in the evaporator and in the condenser, but there will often

3.2 Work and heat transfers

be some in the compressor and in the pipework. For the moment these extraneous heat transfers in the pipework will be neglected.

The rate of heat transfer to the refrigerant in the evaporator, denoted by $\dot{Q}_e$, is the refrigerating capacity of the plant. The rate of work transfer to the refrigerant in the compressor, denoted by $\dot{W}_c$, has to be provided and paid for to keep the plant going. The ratio of these:

$$\varepsilon = \dot{Q}_e / \dot{W}_c \qquad (3.2.1)$$

has already been defined in Chapter 1 as the coefficient of performance. Denoting the heat transfer rate from the refrigerant in the condenser by $\dot{Q}_{con}$, and the heat transfer rate from the refrigerant in the compressor by $\dot{Q}_c$, then at steady conditions by the First law of Thermodynamics:

$$\dot{Q}_e + \dot{W}_c = \dot{Q}_{con} + \dot{Q}_c \qquad (3.2.2)$$

This equation is the heat and work balance for the whole system.

Under steady conditions the First Law can be applied to each component separately. The steady-flow energy equation is:

$$\dot{Q} + \dot{W} = \dot{m}(h_2 + \tfrac{1}{2}\mathcal{V}_2^2 + gz_2) - \dot{m}(h_1 + \tfrac{1}{2}\mathcal{V}_1^2 + gz_1) \qquad (3.2.3)$$

where $\dot{Q}$ and $\dot{W}$ are the rates of heat and work transfer across a closed surface surrounding the component, known as the control surface, $\dot{m}$ is the steady mass rate of flow in and out, h is specific enthalpy, $\mathcal{V}$ velocity and z is the height above a datum level in a gravitational field of force g acting on unit mass. The subscripts 1 and 2, respectively, denote the conditions at the points where the fluid enters and leaves the control surface. The signs of $\dot{Q}$ and $\dot{W}$ are understood as positive when the directions of heat and work are towards the inside of the control surface.

For most purposes it is not necessary to include all of the terms in eq. (3.2.3). In refrigeration plant the z terms can always be omitted when considering a particular component, though they may be important in connection with the pipework in some cases (see Example 3.15.2). The terms in velocity, representing specific kinetic energy, can usually be left out, though it is always advisable to check some values to verify this.

Equation (3.2.3) will be applied to each component of the vapour compression system in turn, stating the assumptions made. A general assumption will be that there is no change of state in the pipework connecting the components. This is not of course true usually, and it is not in fact necessary if it is understood that the subscripts 1, 2, 3 and 4 refer to the states at entry and exit of the component considered. But to duplicate every exit state by another subscript for the entry state at the next component would be unnecessarily elaborate.

Compressor
Entry state 1, exit state 2.
$$\dot{W} = \dot{W}_c.$$
$$\dot{Q} = -\dot{Q}_c.$$

It is convenient to call heat *from* the compressor, $\dot{Q}_c$ positive, as it usually is. Kinetic and gravitational terms are neglected. Then:

$$\dot{W}_c = \dot{m}(h_2 - h_1) + \dot{Q}_c \tag{3.2.4}$$

Condenser
Entry state 2, exit state 3.
$$\dot{W} = 0.$$
$$\dot{Q} = -\dot{Q}_{con}.$$

It is convenient to call heat *from* the condenser positive. Kinetic and gravitational terms are neglected. Then:

$$\dot{Q}_{con} = \dot{m}(h_2 - h_3) \tag{3.2.5}$$

Expansion valve
Entry state 3, exit state 4.
$$\dot{W} = 0$$

$\dot{Q}$ is small and is neglected. There is usually in fact a heat transfer to the refrigerant, because the body of the valve becomes colder than the surroundings. But the surface area is not large, and the quotient $\dot{Q}/\dot{m}$ is found to be small in relation to other terms, particularly in relation to the change of specific enthalpy in the evaporator.

Kinetic and gravitational terms are neglected. Within the orifice of the valve appreciable kinetic energy is developed. But provided that the control surface is drawn so as to cut the flow well downstream of the orifice, where the kinetic energy has been dissipated by viscous effects, the error is small. Then:

$$0 = \dot{m}(h_4 - h_3) \tag{3.2.6}$$

or:

$$h_4 = h_3 \tag{3.2.7}$$

This equation determines the state at the outlet of the valve and enables point 4 in Fig. 3.1.2 to be located. The specific enthalpy of the two-phase mixture at 4 is:

$$h_4 = (1 - x_4)h_{f,e} + x_4 h_{g,e} \tag{3.2.8}$$

where $h_{f,e}$ and $h_{g,e}$ are the specific enthalpies of saturated liquid and saturated vapour respectively at pressure p_e. Hence solving for x_4:

$$x_4 = \frac{h_3 - h_{f,e}}{h_{g,e} - h_{f,e}} \tag{3.2.9}$$

The specific volume at 4 is then found by eq. (3.1.1).

Evaporator
Entry state 4, exit state 1.

$$\dot{W} = 0$$
$$\dot{Q} = \dot{Q}_e$$

Kinetic and gravitational terms are neglected. Then:

$$\dot{Q}_e = \dot{m}(h_1 - h_4) \tag{3.2.10}$$

By eq. (3.2.7):

$$\dot{Q}_e = \dot{m}(h_1 - h_3) \tag{3.2.11}$$

Equation (3.2.11) may be obtained directly by drawing the control surface around the evaporator and expansion valve. In this event it is not necessary to neglect the small heat transfer to the valve if it is understood as being included in $\dot{Q}_e$. This often corresponds to the real situation, for example when the expansion valve is inside the refrigerated space.

The important quantity, refrigerating capacity, $\dot{Q}_e$, is made up of two factors: the mass flow rate of refrigerant, $\dot{m}$, which depends on the size of the compressor and how fast it can pump the vapour away; and the difference of specific enthalpy $(h_1 - h_3)$, which depends on the refrigerant and on the working conditions but not on the size of the plant. The term $(h_1 - h_3)$ will be called the *specific refrigerating effect*.

The mass flow rate $\dot{m}$ is associated at any point with a volume flow rate $\dot{V}$, where:

$$\dot{V} = \dot{m}v \tag{3.2.12}$$

and v is the specific volume of the refrigerant at that point. The volume flow rate is of special interest at entry to the compressor, because it is related there to the size of the compressor, i.e. to the linear dimensions of the machine. Inserting the specific volume at entry to the compressor, v_1, into eq. (3.2.12) gives:

$$\dot{V}_1 = \dot{m} v_1 \qquad (3.2.13)$$

for the volume flow rate $\dot{V}_1$ at the compressor entry. Substituting for $\dot{m}$ from eq. (3.2.13) in eq. (3.2.11):

$$\dot{Q}_e = \dot{V}_1 \frac{h_1 - h_3}{v_1} \qquad (3.2.14)$$

The refrigerating capacity may thus alternatively be considered as made up of two factors: $\dot{V}_1$ the volume flow rate of vapour at entry to the compressor, which depends mainly on the dimensions and speed of the machine; and the term $(h_1 - h_3)/v_1$, which depends only on the refrigerant and the working conditions. The term $(h_1 - h_3)/v_1$ will be called the *volumic*† *refrigerating effect*.

3.3 Adiabatic and reversible compression

For some types of compressors the mass flow rate through them is large in relation to the surface available for heat transfer from the surroundings, and the ratio $\dot{Q}_c/\dot{m}$ may be negligibly small in relation to the change of specific enthalpy $(h_2 - h_1)$. In this case eq. (3.2.4) becomes:

$$\dot{W}_{c,ad} = \dot{m}(h_2 - h_1) \qquad (3.3.1)$$

A process in which the heat transfer is zero is called an adiabatic process, and eq. (3.3.1) gives the power required for adiabatic compression. It depends only on the mass flow rate and the initial and final states in the process. The coefficient of performance for adiabatic compression is, from eq. (3.2.1), (3.2.11) and (3.3.1):

$$\varepsilon_{ad} = \frac{h_1 - h_3}{h_2 - h_1} \qquad (3.3.2)$$

The coefficient of performance thus depends only on the states at the four principal points of the cycle.

If in addition to the condition that the process is adiabatic, it is stipulated that the process is thermodynamically reversible, then the entropy of the refrigerant remains constant during compression, i.e.

$$s_2 = s_1 \qquad (3.3.3)$$

---

† The word 'volumic' has been adopted, following a suggestion by Professor E. J. Le Fevre, to denote a quantity such as heat, work or change of enthalpy, divided by the volume of substance. It is analogous to 'specific' denoting a quantity divided by the mass of substance.

3.3 Adiabatic and reversible compression

The coefficient of performance then becomes:

$$\varepsilon_r = \frac{h_1 - h_3}{(h_2 - h_1)_s} \tag{3.3.4}$$

where the subscript s outside the bracket denotes that the difference of specific enthalpy $(h_2 - h_1)$ is evaluated for the condition of eq. (3.3.3). The subscript r denotes that the coefficient of performance is evaluated for reversible adiabatic, i.e. isentropic compression. ε_r is an important property of a refrigerant which will be studied later.

Reversible compression is an ideal which cannot be achieved in practice. The importance of the concept lies in the fact that for an adiabatic process from a given initial state to a given final pressure the specific work is least when the process is reversible. Reversibility implies that compression is performed in such a way that when it is completed the original state may be regained by an expansion process in which the same series of states is traversed, and exactly the same work is given out as in the original process. This could only happen if the force applied to a piston enclosing some fluid inside a cylinder exactly balanced the force due to the pressure of the fluid. At this condition nothing would happen however. In order to initiate motion the piston force must be increased (for compression) or relaxed (for expansion). A reversible process can therefore only be a concept of a series of equilibrium states. A real process, completed in a finite time, must depart from equilibrium.

It must be emphasised that when a constant entropy process is adopted as a standard for minimum work, it is understood that compression is adiabatic. If heat transfer from the vapour being compressed can take place, then the work can be reduced because the volume diminishes at a more rapid rate. When the heat transfer rate is sufficient, isothermal compression is approached, in which the temperature does not rise. Isothermal compression (reversible) is often adopted as a standard for air compressors because the air is initially at room temperature and it is possible in principle to cool it during compression. But the vapour entering the compressor of the cycle illustrated in Fig. 3.1.2 is cold, and there is no possibility of cooling it during the first part of compression. If anything colder were available, there would be no need for the refrigerating plant! Towards the end of compression, when the vapour has become warm, then cooling is possible in principle. But for practical reasons, including the short time that the vapour is in contact with the cylinder wall, and the low heat transfer coefficient from a dry gas, it is not possible to achieve a large heat transfer rate in a high-speed modern compressor. This question will be discussed further in section 4.4.

The power required for isentropic compression is given by eq. (3.3.1) with the condition of constant entropy:

$$\dot{W}_{c,\text{isen}} = \dot{m}(h_2 - h_1)_s \tag{3.3.5}$$

It is thus the product of the mass flow rate and the change in specific enthalpy. The quantity $(h_2 - h_1)_s$ will be called the *specific work of isentropic compression*. In terms of the volume flow rate at the compressor inlet, $\dot{V}_1$, using (3.2.13) in eq. (3.3.5) gives:

$$\dot{W}_{c,\text{isen}} = \dot{V}_1(h_2 - h_1)_s/v_1 \tag{3.3.6}$$

The term:

$$(h_2 - h_1)_s/v_1$$

will be called the *volumic work of isentropic compression*. It represents the work needed to compress unit volume of vapour isentropically from state 1 to condensing pressure, and is therefore related to the power needed to drive a compressor of given dimensions at a certain speed. The volumic isentropic work has the dimensions of pressure, for:

$$1 \text{ J/m}^3 = 1 \text{ Nm/m}^3 = 1 \text{ N/m}^2 \tag{3.3.7}$$

It will be seen in Chapter 4 that the volumic isentropic work is the same as the mean effective pressure of an ideal reciprocating compressor without clearance.

The performance of real compressors is often specified in terms of that of an isentropic compressor by an isentropic efficiency:

$$\eta_{\text{isen}} = \frac{(h_2 - h_1)_s}{(\dot{W}_c/\dot{m})} \tag{3.3.8}$$

where $(\dot{W}_c/\dot{m})$ is the specific work for the real compressor operating between the same initial state 1 and the same final pressure $p_2 = p_c$. The power taken by the real compressor is then:

$$\dot{W}_c = \frac{\dot{m}(h_2 - h_1)_s}{\eta_{\text{isen}}} = \frac{\dot{W}_{c,\text{isen}}}{\eta_{\text{isen}}} \tag{3.3.9}$$

The isentropic efficiency is not an efficiency in the usual sense of the word, meaning something out divided by something in. Isentropic efficiency compares two processes, a real one which happens, and an ideal one which does not and cannot happen. Isentropic efficiency is useful for giving a rapid and very rough estimate of power, since approximate values are known for various kinds of compressors. Values range from about 0.5 for very small compressors used in domestic refrigerators to about 0.8 for large screw

3.4 The pressure–enthalpy diagram

compressors. The value changes rather rapidly with conditions however, and isentropic efficiency is not recommended as a method for determining the required size of a driving motor.

3.4 The pressure–enthalpy diagram

The relations of the last section may conveniently be displayed on a diagram having pressure and specific enthalpy as coordinates. Such a diagram was proposed by Richard Mollier. Following a proposal of H. L. Callendar, the pressure scale is often plotted logarithmically. This has the advantage of opening up the diagram at low pressures, but it changes the shape of the lines representing the constant values of various thermodynamic properties, as well as the shape of the saturation lines. The main differences are that the saturation line in the neighbourhood of the critical point is much flatter when the pressure is plotted logarithmically, and that the lines of constant specific entropy in the superheat region become concave downwards instead of concave upwards. The diagrammatic sketches given in this book, except where otherwise mentioned, are drawn as if the pressure were to a logarithmic scale.

Figure 3.4.1 shows a skeleton diagram for a substance like R22. Below the critical point CP the saturated liquid line and saturated vapour line enclose a two-phase ('wet') region between them. To the left of the saturated liquid line lie states which have a lower temperature than saturation

Fig. 3.4.1. Diagrammatic representation of a pressure versus specific enthalpy diagram with a logarithmic pressure scale. Drawn approximately to scale for R22.

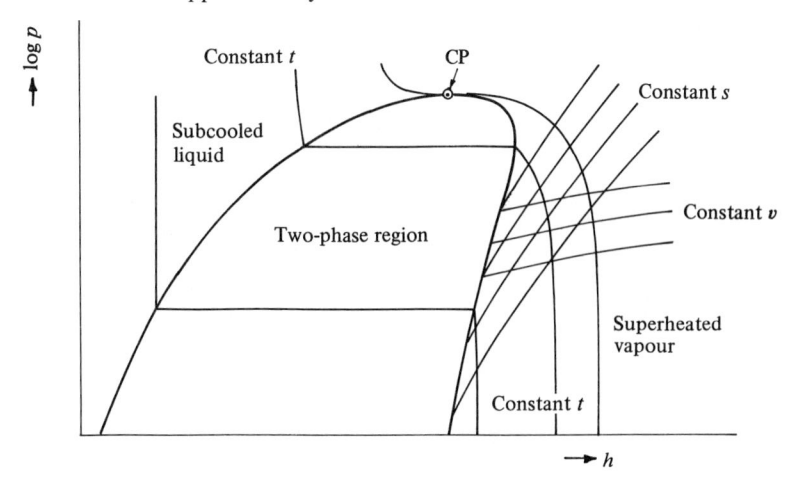

temperature at a given pressure. These are states of subcooled liquid. To the right of the saturated vapour line lie states which have a higher temperature than saturation temperature at a given pressure. These are states of superheated vapour. The area to the left of the liquid line is called the subcooled liquid region, and the area to the right of the vapour line is called the superheated vapour region. Within the two-phase region the horizontal lines of constant pressure are also lines of constant temperature. In the superheat region the lines of constant temperature leave the saturation line as indicated. As the pressure diminishes in the superheat region, the lines of constant temperature tend to become lines of constant enthalpy, i.e. vertical on the diagram, indicating that the vapour is beginning to behave like an ideal gas with its enthalpy independent of pressure. Lines of constant specific entropy are shown in the superheat region, as are lines of constant specific volume.

The vapour compression cycle of Fig. 3.1.2 is shown on diagrammatic (log) pressure–enthalpy coordinates in Fig. 3.4.2. Beginning with point 1 at entry to the compressor, saturated vapour is compressed from pressure $p_1 = p_e$ to pressure $p_2 = p_c$, leaving the compressor at point 2 in the superheat region. Reversible adiabatic compression at constant entropy would be represented by the broken line 1–2'. The point 2 for real compression may lie either to the right or left of 2', depending on the nature of the actual compression process. If the actual process is adiabatic (i.e. no heat transfer), point 2 must necessarily lie to the right of 2' because the entropy must increase in an adiabatic process and with it the enthalpy. On the other hand, if there is heat transfer from the vapour during compression, the entropy can be reduced and point 2 may then lie to the left of 2'. With certain types of compressors it is sometimes found that, according to the measured temperature at the compressor delivery, states 2 and 2' nearly

Fig. 3.4.2. The (diagrammatic) vapour compression cycle on the p–h diagram.

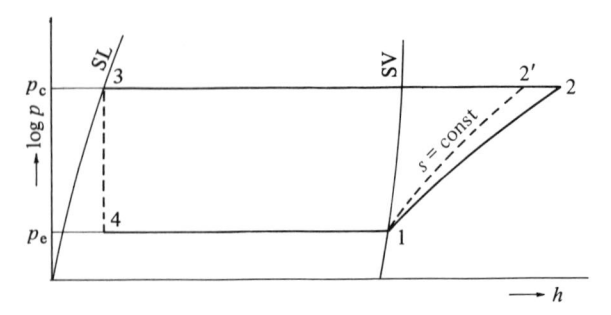

3.5 Thermodynamic data: tables and charts

coincide, indicating no change in the entropy during compression. *This does not mean that the compression process is a reversible one.* It means simply that the effect of irreversibility during compression, tending to cause an increase of entropy, has been more or less compensated by a heat transfer from the vapour, causing a decrease of entropy.

From point 2 the vapour is desuperheated at constant pressure to a state of saturated vapour and then condensed to liquid at point 3. Pressure drop in the expansion valve takes place at constant specific enthalpy according to eq. (3.2.7) to point 4. Again the process is shown as a broken line because the intermediate states of the process are not known; in fact not all parts of the refrigerant pass through the same set of states, but they all come together at the end, point 4, where the fluid has negligible velocity at entry to the evaporator.

3.5 Thermodynamic data: tables and charts

To make use of the equations derived above it is necessary to have data about the refrigerant, in the form of tables, accurately drawn charts, or equations. The use of equations is considered in section 3.18. Here the discussion will be mainly about tables and charts.

Numerical values of specific enthalpy and specific entropy have to be reckoned above some datum state at which these quantities are each assigned an arbitrary value, often zero. There is no generally agreed datum state, unfortunately, and several conventions are in use.

Tables in British units have usually been based on a datum of saturated liquid at $-40\,°C$, with the aim of eliminating negative values in the most common range of temperatures. In tables of continental origin in technical metric units the same goal has been achieved by assigning the values $h_f = 100\,\text{kcal/kg}$ and $s_f = 1.000\,\text{kcal/kg K}$ to saturated liquid at $0\,°C$. In SI units the same datum state may be used with $h_f = 200\,\text{kJ/kg}$ and $s_f = 1.000\,\text{kJ/kg K}$, as in the Du Point 'Freon' tables.

These datum states are obvious from an inspection of the tables, but there are some in use which are not self-evident. One of these is the crystalline solid at the absolute zero of temperature at which h and s are put equal to zero. This of course avoids negative values at higher temperature, but it has the disadvantage of giving relatively large values in which it is necessary to carry many significant figures to achieve the desired precision in such quantities as the specific enthalpy of evaporation.

Another convention, used mainly in cryogenic tables, is the state of the ideal gas at absolute zero. This usually makes all the values of s and h for the liquid negative, unless an arbitrary constant is added.

So long as one sticks to the same set of tables the choice of datum does not matter of course, but it becomes very important when data has to be taken from different sets of tables. Unfortunately, there are some tables which have been reproduced without stating the datum, and a certain amount of guessing is required.

The first table in a set for a refrigerant is usually the table of saturation values, i.e. values of saturation pressure, specific volumes or densities, specific enthalpies and specific entropies of saturated liquid and vapour, in terms of temperature as argument. Occasionally pressure is used as the argument.

The second table contains the data for the superheat region, giving specific volume, enthalpy and entropy at different pressures and temperatures. To save space in tabulation the superheat may be used as argument instead of temperature. The superheat is defined as the difference between the temperature t of the vapour and the temperature t_s of saturated vapour at the same pressure, i.e. $(t - t_s)$.

In order to be conveniently usable with the saturation table, the values of pressure in the superheat table should preferably correspond to those in the saturation table, i.e. if the argument in the saturation table is temperature the tabulated pressures in the superheat table should be the saturation pressures at the given temperatures.

Tables for fluorinated refrigerants are available from the manufacturers of these substances. Abridged tables for some refrigerants are given by Haywood (1972) and Mayhew & Rogers (1980) in SI units.

When using tables it is important not to round-off values prematurely. Because the enthalpy and entropy have arbitrary datum values, the number of significant figures needed to give the desired precision may be quite large and at first sight unnecessary. Clearly, enough figures should be retained to give differences of h and s precisely. In the examples, the full figures have been retained until the end, when the final answers are rounded off.

For many purposes an accurately drawn pressure–enthalpy chart is useful, especially for determining the value of the specific enthalpy at condensing pressure for the case of isentropic compression. Few charts are available on a large enough scale, however. The use of charts is not emphasised in this book because most work is performed digitally nowadays. As a sketched diagram nevertheless, the graphical display of thermodynamic properties is an invaluable aid towards understanding a problem.

The use of tables is illustrated in the following examples, based on the NBS tables for ammonia.

3.5 Thermodynamic data: tables and charts

Example 3.5.1
Using the tables for ammonia, the various quantities defined in this chapter will be determined for the cycle of Fig. 3.4.2 with isentropic compression, given the conditions:

evaporating temperature $\quad t_e = -10\,°C$

condensing temperature $\quad t_c = 30\,°C$

From the table of saturation properties, the pressures are found to be:

evaporating pressure $\quad p_e = 2.908$ bar

condensing pressure $\quad p_c = 11.66$ bar

The state at 1 is located as saturated vapour at $-10\,°C$, giving:

$$h_1 = h_g(-10\,°C) = 1431.6\text{ kJ/kg}$$
$$s_1 = s_g(-10\,°C) = 5.4717\text{ kJ/kg K}$$

At point 3, saturated liquid at $30\,°C$, the specific enthalpy is:

$$h_3 = h_f(30\,°C) = 322.8\text{ kJ/kg}$$

The specific refrigerating effect is therefore:

$$h_1 - h_3 = (1431.6 - 322.8)\text{ kJ/kg} = 1108.8\text{ kJ/kg}$$

At 1 the specific volume is:

$$v_1 = v_g(-10\,°C) = 0.4185\text{ m}^3/\text{kg}$$

and the volumic refrigerating effect is therefore:

$$\frac{h_1 - h_3}{v_1} = \frac{1108.8\text{ kJ/kg}}{0.4185\text{ m}^3/\text{kg}} = 2649\text{ kJ/m}^3$$

To determine the state at the end of reversible adiabatic compression, 2', the condition of constant entropy is used,

$$s'_2 = s_1 = 5.4717\text{ kJ/kg K}.$$

In the superheat table at a pressure of $p_2 = p_c = 11.66$ bar, two superheated states are located:

Superheat	Specific enthalpy	Specific entropy
K	kJ/kg	kJ/kg K
55	1621.8	5.4484
60	1634.5	5.4838

between which the state at 2' must lie, because the specific entropy is between 5.4484 and 5.4838. Assuming that linear interpolation is permissible, the superheat at 2' is:

$$t'_2 - t_s = 55\,\text{K} + 5\frac{(5.4717 - 5.4484)}{(5.4838 - 5.4484)}\,\text{K}$$

$$= 55\,\text{K} + 5(0.658)\,\text{K} = 58.3\,\text{K}$$

Therefore:

$$t'_2 = t_s + 58.3\,\text{K} = 30\,°\text{C} + 58.3\,\text{K} = 88.3\,°\text{C}$$

The specific enthalpy at 2' is, by linear interpolation:

$$h'_2 = 1621.8\,\text{kJ/kg} + 0.658(1634.5 - 1621.8)\,\text{kJ/kg}$$

$$= 1630.2\,\text{kJ/kg}$$

The specific work of isentropic compression is:

$$(h_2 - h_1)_s = (1630.2 - 1431.6)\,\text{kJ/kg} = 198.6\,\text{kJ/kg}$$

and the volumic work of isentropic compression is:

$$\frac{(h_2 - h_1)_s}{v_1} = \frac{198.6\,\text{kJ/kg}}{0.4185\,\text{m}^3/\text{kg}} = 474.6\,\text{kJ/m}^3$$

The coefficient of performance ε_r is:

$$\varepsilon_r = \frac{(h_1 - h_3)}{(h_2 - h_1)_s} = \frac{1108.8}{198.6} = 5.58$$

The decrease of specific enthalpy in the condenser is:

$$h'_2 - h_3 = (1630.2 - 322.8)\,\text{kJ/kg} = 1307.4\,\text{kJ/kg}$$

To determine the state at point 4 leaving the expansion valve, eq. (3.2.9) is used to find x_4, with data from the tables:

$$h_{f,e} = h_f(-10\,°\text{C}) = 135.1\,\text{kJ/kg}$$
$$h_{g,e} = h_g(-10\,°\text{C}) = 1431.6\,\text{kJ/kg} \quad (= h_1)$$
$$h_3 = h_f(30\,°\text{C}) \quad\ = 322.8\,\text{kJ/kg}$$

and:

$$x_4 = \frac{h_3 - h_{f,e}}{h_{g,e} - h_{f,e}} = \frac{322.8 - 135.1}{1431.6 - 135.1} = 0.1448$$

The specific volumes of saturated liquid and vapour are:

$$v_{f,e} = v_f(-10\,°\text{C}) = 0.001534\,\text{m}^3/\text{kg}$$
$$v_{g,e} = v_g(-10\,°\text{C}) = 0.4185\,\text{m}^3/\text{kg}$$

3.5 Thermodynamic data: tables and charts

and the specific volume of the two-phase mixture at 4 is by eq. (3.1.1):

$$x_4 = (1 - x_4)v_{f,e} + x_4 v_{g,e}$$
$$= [(1 - 0.1448)(0.001534) + 0.1448(0.4185)] \, m^3/kg$$
$$= [0.00131 + 0.06060] m^3/kg = 0.06191 \, m^3/kg$$

The liquid fraction thus occupies only:

$$0.00131/0.06191 = 0.0212 = 2.12\%$$

of the volume of the two-phase mixture.

Example 3.5.2

For the conditions of example 3.5.1 a refrigerating capacity of 500 kW is required. Determine the mass flow rate of ammonia around the circuit, the volume flow rate at entry to the compressor, the power to the compressor and the rate of heat transfer from the refrigerant in the condenser, assuming (a) reversible and adiabatic compression, and (b) an isentropic efficiency of 0.7 and adiabatic compression.

$$\dot{Q}_e = 500 \, kW$$

The mass flow through the evaporator is:

$$\dot{m} = \frac{\dot{Q}_e}{h_1 - h_3} = \frac{500 \, kW}{1108.8 \, kJ/kg} = 0.4509 \, kg/s$$

The volume flow rate at entry to the compressor is:

$$\dot{V}_1 = \dot{m} v_1 = (0.4509 \, kg/s)(0.4185 \, m^3/kg) = 0.189 \, m^3/s$$

(a) *Reversible adiabatic compression.* The power is:

$$\dot{W}_{c,isen} = \dot{m}(h'_2 - h_1) = (0.4509 \, kg/s)(198.6 \, kJ/kg)$$
$$= 89.6 \, kW$$

The rate of heat transfer from refrigerant in the condenser is:

$$\dot{Q}_{con} = \dot{m}(h'_2 - h_3) = (0.4509 \, kg/s)(1307.4 \, kJ/kg)$$
$$= 590 \, kW$$

It will be noted that this quantity is also equal to the sum of the refrigerating capacity and the power to the compressor, in accordance with eq. (3.2.2) for the case of $\dot{Q}_c = 0$.

(b) *Isentropic efficiency* 0.7. The power taken by the compressor is now:

$$\dot{W}_c = \frac{\dot{W}_{isen}}{\eta_{isen}} = \frac{89.6}{0.7} kW = 128 \, kW$$

The vapour compression system

Assuming that compression is adiabatic, though not reversible, the change of specific enthalpy across the compressor is:

$$h_2 - h_1 = \frac{(h_2 - h_1)_s}{\eta_{isen}} = \frac{198.6}{0.7} = 283.7 \text{ kJ/kg}$$

and the specific enthalpy of the ammonia leaving the compressor is:

$$h_2 = h_1 + 283.7 \text{ kJ/kg} = (1431.6 + 283.7) \text{ kJ/kg}$$
$$= 1715.3 \text{ kJ/kg}$$

The decrease in specific enthalpy in the condenser is:

$$h_2 - h_3 = (1715.3 - 322.8) \text{ kJ/kg} = 1392.5 \text{ kJ/kg}$$

and the rate of heat transfer in the condenser is:

$$\dot{Q}_{con} = \dot{m}(h_2 - h_3) = (0.4509 \text{ kg/s})(1392.5 \text{ kJ/kg})$$
$$= 628 \text{ kW}$$

Again it is noted that because compression is adiabatic, the rate of heat transfer in the condenser is the sum of the refrigerating capacity and the power to the compressor.

Example 3.5.3

For the conditions of example 3.5.1, the volume flow rate at compressor inlet is 0.1 m³/s. Determine the refrigerating capacity and the power, assuming reversible adiabatic compression.

The mass flow rate of refrigerant is:

$$\dot{m} = \frac{\dot{V}_1}{v_1} = \frac{0.1 \text{ m}^3/\text{s}}{0.4185 \text{ m}^3/\text{kg}} = 0.2389 \text{ kg/s}$$

and the refrigerating capacity is:

$$\dot{Q}_e = \dot{m}(h_1 - h_3) = (0.2389 \text{ kg/s})(1108.8 \text{ kJ/kg})$$
$$= 265 \text{ kW}$$

Alternatively, using the volumic refrigerating effect:

$$\dot{Q}_e = \dot{V}_1 \frac{h_1 - h_3}{v_1} = (0.1 \text{ m}^3/\text{s})(2649 \text{ kJ/m}^3)$$
$$= 265 \text{ kW}$$

The isentropic power is:

$$\dot{W}_{c,isen} = \dot{m}(h_2 - h_1)_s = (0.2389 \text{ kg/s})(198.6 \text{ kJ/kg})$$
$$= 47.5 \text{ kW}$$

3.5 Thermodynamic data: tables and charts

Alternatively, using the volumic isentropic work:

$$\dot{W}_{c,\text{isen}} = \dot{V}_1 \frac{(h_2 - h_1)_s}{v_1} = (0.1\,\text{m}^3/\text{s})(474.6\,\text{kJ/m}^3)$$
$$= 47.5\,\text{kW}$$

The power for compression which is not isentropic, and the heat transfer rate in the condenser are found as in the last example, using the new value of the mass flow rate.

Interpolation

With all tables it is necessary to interpolate values between those tabulated. Provided that the intervals of tabulation are small, linear interpolation is usually permissible, as was done in the examples just given. Given four values such as:

$$x_1 \quad x_2$$
$$y_1 \quad y_2$$

of any two quantities, an intermediate value y corresponding to a given x is found from:

$$\frac{x - x_1}{x_2 - x_1} = \frac{y - y_1}{y_2 - y_1} \tag{3.5.1}$$

These fractions are called the 'proportionate parts'.

Linear interpolation is precise enough when the differences between the successive tabulated quantities are nearly the same. When they are not, it is advisable to use a second-degree interpolation. Given six values:

$$x_1 \quad x_2 \quad x_3$$
$$y_1 \quad y_2 \quad y_3$$

where the x intervals are the same, i.e. $x_2 - x_1 = x_3 - x_2 = \Delta x$ say, the value of y corresponding to a given x is:

$$y = y_1 + \frac{(x - x_1)}{\Delta x}\left[(y_2 - y_1) + \frac{1}{2}\left(\frac{x - x_1}{\Delta x} - 1\right)(y_3 - 2y_2 + y_1)\right]$$
$$\tag{3.5.2}$$

As a rule x will represent temperature or pressure, with even intervals of tabulation, whilst y will represent one of the properties specific volume or density, specific enthalpy and entropy. When the first differences in y are equal, i.e. when $y_3 - y_2 = y_2 - y_1$, eq. (3.5.2) reduces to eq. (3.5.1).

Any method of interpolation is better when the quantities are linear or nearly so. Best of all is when one can discover the law relating the variables. In the absence of a definite numerical equation which would render interpolation certain, the quantities should be chosen in a rational manner. Two examples may be given.

(a) *Vapour pressure.* It is often necessary to find the vapour pressure at an arbitrary temperature. With close tabulation, linear interpolation may be used, but a better method is to assume a law of the form:

$$\ln p = A - \frac{B}{T} \qquad (3.5.3)$$

between the tabulated points. Then:

$$\ln(p/p_1) = \frac{(1/T_1) - (1/T)}{(1/T_1) - (1/T_2)} \ln(p_2/p_1) \qquad (3.5.4)$$

It is of course immaterial whether natural or ordinary logarithms are used here. Equation (3.5.4) gives a much better value than linear interpolation of pressure against temperature, and is often better than second degree interpolation.

(b) *Specific volume.* The specific volume has great practical importance because it is directly connected with the performance of a positive displacement compressor in converting a volume flow rate to a mass flow rate. In the tables it has to be interpolated at a constant pressure between two temperatures, or at a constant temperature between two pressures.

At a tabular pressure, provided that the temperature intervals are not too great, linear interpolation of specific volume is usually precise enough, though second-degree interpolation using eq. (3.5.2) is better.

At a tabular temperature linear interpolation of specific volume with pressure is not very accurate. It is better as a rule to interpolate linearly with density against pressure. The best simple method however is interpolation of the residual volume, also called the excess volume. This is defined as the difference between the true specific volume and the specific volume calculated from the ideal gas equation, i.e.

$$v_{\text{res}} = \frac{RT}{p} - v \qquad (3.5.5)$$

where R is the specific gas constant for the substance, given by $\bar{R}/M$ in which $\bar{R} = 8314\,\text{J/kmol K}$ is the molar gas constant and M is the molar mass. The residual volume changes slowly with pressure and temperature and it can be interpolated linearly.

3.5 Thermodynamic data: tables and charts

When the pressure $p_2 = p_c$ at delivery from the compressor is not the same as one of those tabulated in the superheat table, in order to locate the state 2 having the same entropy as state 1 it is necessary to interpolate two ways, between tabular pressures and between tabular temperatures. In these circumstances the following method will be found useful.

According to the thermodynamic identity for any infinitesimal process:

$$\mathrm{d}h = T\,\mathrm{d}s + v\,\mathrm{d}p \tag{3.5.6}$$

and for a process at constant entropy:

$$\mathrm{d}h = v\,\mathrm{d}p \tag{3.5.7}$$

since $\mathrm{d}s = 0$.

Suppose that the nearest tabular pressure to p_2 is slightly less than p_2, p_i say. Then the increment of specific enthalpy in continuing at constant entropy the compression from pressure p_i to pressure p_2 is:

$$(h_2 - h_i)_s = \int_i^2 v\,\mathrm{d}p \tag{3.5.8}$$

where h_i is obtained from the table at pressure p_i having the same entropy as state 1. Equation (3.5.8) shows that the correction to be made to h_i in order to obtain h_2 is the value of the integral. Provided that the difference in pressure $(p_2 - p_i)$ is not large, the integral can be evaluated approximately as:

$$v_i(p_2 - p_i)$$

where v_i is the specific volume at the end of isentropic compression to pressure p_i.

A better approximation, which is still simple to use, is:

$$p_i v_i \ln(p_2/p_i)$$

This is based on the assumption that over the small interval of pressure the difference between the isentropic curve on p–v coordinates and the curve $pv = \text{const}$ is small. As a refinement of the method one can find the value of the index of isentropic compression up to the tabulated pressure and use this to calculate the supplement given by eq. (3.5.8). The expressions for this are given in Chapter 4.

Finally, it should be remembered that any extensive tabulation of thermodynamic properties produced by hand-setting will probably contain misprints: in fact almost certainly will. In an important quantity such as density or specific volume this can have embarassing consequences. It is a good plan to check first and sometimes second differences around any figure taken from a table. Tables produced by computer printout are free from misprints as a rule, but they may have other defects such as printed data extrapolated from a region where one equation applies into another: e.g. extrapolating superheat data into the two-phase region.

As will be seen in section 3.18, thermodynamic tables should be produced from minimum data in order to avoid inconsistencies which can lead to impossible results, e.g. a value for the coefficient of performance of a cycle which is higher than the Carnot value for the same evaporating and condensing temperatures. Many of the older tables had such internal inconsistencies, and unfortunately some modern ones are not entirely free from them.

3.6 The use of minimum data

When tables of properties of the refrigerant are not available, it is still possible to estimate the coefficient of performance ε_r if the following four quantities are known:

specific enthalpy of evaporation at t_e: $h_{fg,e}$;
specific enthalpy of evaporation at t_c: $h_{fg,c}$;
specific heat capacity of liquid, assumed constant: c_f;
specific heat capacity of superheated vapour at constant pressure, assumed constant: c_p.

For convenience the specific enthalpy and specific entropy of saturated liquid are arbitrarily set to zero at evaporating temperature. Then:

$$h_1 = h_{fg,e} \tag{3.6.1}$$

$$s_1 = h_{fg,e}/T_e \tag{3.6.2}$$

where T_e is the absolute temperature of evaporation.

Proceeding up the saturated liquid line to point 3, and assuming that the liquid is incompressible:

$$h_3 = c_f(T_c - T_e) \tag{3.6.3}$$

$$s_3 = c_f \ln(T_c/T_e) \tag{3.6.4}$$

where T_c is the absolute temperature of condensation. The specific refrigerating effect is therefore:

$$h_1 - h_3 = h_{fg,e} - c_f(T_c - T_e) \tag{3.6.5}$$

The state at 2 for isentropic compression will be denoted without primes here.

3.6 The use of minimum data

The condition at point 2 is found by equating entropies at 1 and 2. The specific entropy at 2 is given, in terms of the yet unknown absolute temperature T_2 at 2 as:

$$s_2 = c_f \ln \frac{T_c}{T_e} + \frac{h_{fg,c}}{T_c} + c_p \ln \frac{T_2}{T_c} \tag{3.6.6}$$

With $s_2 = s_1$, eq. (3.6.6) can be solved for T_2. The specific enthalpy at 2 is then:

$$h_2 = c_f(T_c - T_e) + h_{fg,c} + c_p(T_2 - T_c) \tag{3.6.7}$$

The specific work of compression and the coefficient of performance are then determined from eqs. (3.3.1) and (3.3.4).

Provided that the condensing temperature is some way below the critical temperature, the specific heat capacity of the liquid is very nearly constant, and compressibility of the liquid can also be neglected. On the other hand, the specific heat capacity at constant pressure for most vapours is not constant, even at one given pressure. The accuracy of the above procedure therefore depends on whether the value of c_p used is a good average one for the range considered.

It should be noted that the enthalpy of a vapour, unlike that of an ideal gas, depends on pressure as well as temperature. It is therefore *not* correct to determine the change of specific enthalpy during compression as:

$$c_p(T_2 - T_1)$$

as can be done for an ideal gas when treating, for example, gas turbine cycles.

Example 3.6.1

The following properties of a refrigerant are known:

h_{fg} at $-5\,°C = 410\,kJ/kg$

h_{fg} at $30\,°C = 370\,kJ/kg$

$c_f = 1.6\,kJ/kg\,K$

c_p of vapour $= 1.0\,kJ/kg\,K$

Determine the coefficient of performance for a cycle with reversible adiabatic compression for:

$t_e = -5\,°C \qquad T_e = 268.15\,K$

$t_c = 30\,°C \qquad T_c = 303.15\,K$

Putting (arbitrarily) $h_f(-5\,°C) = 0$, and $s_f(-5\,°C) = 0$,

$$h_1 = 410\,\text{kJ/kg}$$
$$h_3 = (1.6\,\text{kJ/kg K})(35\,\text{K}) = 56\,\text{kJ/kg}$$
$$h_1 - h_3 = (410 - 56)\,\text{kJ/kg} = 354\,\text{kJ/kg}$$
$$s_1 = \frac{410\,\text{kJ/kg}}{268.15\,\text{K}} = 1.5290\,\text{kJ/kg K}$$
$$s_2 = (1.6\,\text{kJ/kg K})\ln\frac{303.15}{268.15} + \frac{370\,\text{kJ/kg K}}{303.15\,\text{K}}$$
$$+ (1.0\,\text{kJ/kg K})\ln\frac{T_2}{303.15\,\text{K}}$$
$$= \left[1.4168 + \ln\frac{T_2}{303.15\,\text{K}}\right]\text{kJ/kg K}$$

Since $s_1 = s_2$:

$$1.5290 = 1.4168 + \ln\frac{T_2}{303.15\,\text{K}}$$

from which:

$$T_2 = 339.1\,\text{K}$$
$$T_2 - T_c = 339.1\,\text{K} - 303.15\,\text{K} = 36\,\text{K}$$

and:

$$h_2 = [56 + 370 + 1.0(36)]\,\text{kJ/kg}$$
$$= 462\,\text{kJ/kg}$$

Then:

$$h_2 - h_1 = (462 - 410)\,\text{kJ/kg} = 52\,\text{kJ/kg}$$

and:

$$\varepsilon_r = 354/52 = 6.81$$

N.B. If $h_2 - h_1$ had been calculated *incorrectly* as $c_p(T_2 - T_1)$, the value would have been 71 kJ/kg, instead of 52 kJ/kg.

3.7 Comparison with the Carnot cycle

The coefficient of performance of an ideal reversible refrigerator taking in heat at the constant absolute temperature T_e and giving out heat

3.7 Comparison with the Carnot cycle

at the constant absolute temperature T_c is given by:

$$\varepsilon_C = \frac{T_e}{T_c - T_e} \qquad (3.7.1)$$

For the cycle considered in example 3.5.1,

$$T_e = (-10 + 273.15)\,\text{K} = 263.15\,\text{K}$$
$$T_c = (30 + 273.15)\,\text{K} = 303.15\,\text{K}$$

and:

$$\varepsilon_C = 263.15/40 = 6.58$$

The value of ε_r for the vapour compression cycle was found to be 5.58. In this section the reasons for this difference are discussed.

A Carnot cycle between T_e and T_c is shown on temperature–specific entropy coordinates in Fig. 3.7.1. The figure is drawn for a substance like ammonia which has approximately symmetrical saturated liquid and saturated vapour lines. For comparison with the vapour compression cycle, points 1 and 3 are chosen as saturated vapour and saturated liquid, respectively. The first process in the Carnot cycle is reversible adiabatic compression from state 1, i.e. at constant entropy. Compression is continued isentropically until the temperature of the vapour reaches T_c, at the point marked as 2C. It will be noted that this process does not reach condenser pressure p_c. The second process is a reversible one at constant temperature T_c during which heat transfer from the refrigerant occurs. This process requires first a reversible isothermal compression from state 2C

Fig. 3.7.1. The Carnot cycle 1–2C–3–4C on temperature versus specific entropy coordinates.

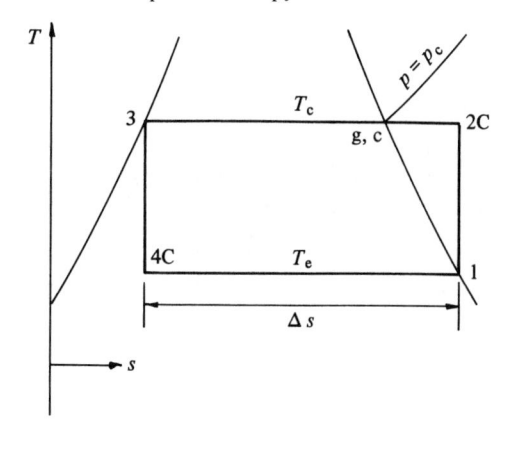

during which the pressure rises to p_c at the saturation line, marked as point g, c, followed by reversible isothermal condensation at constant pressure p_c to the saturated liquid state at 3. The next process is reversible adiabatic expansion, i.e. at constant entropy, in which work is done by the fluid, to the state marked as 4C. Finally there is a reversible isothermal evaporation at constant pressure p_e in which heat transfer to the fluid restores the original state 1.

Let the entropy width of the cycle be:

$$\Delta s = s_1 - s_{4C} = s_{2C} - s_3 \qquad (3.7.2)$$

The heat transfer in any reversible process is represented on the temperature–entropy diagram by the area, down to absolute zero, under the curve showing the path of the process. The heat transfer from the refrigerant in process 2C–3 is then:

$$T_c \Delta s$$

and the heat transfer to the refrigerant in process 4C–1 is:

$$T_e \Delta s$$

Since these are the only heat transfers in the cycle, the net work done on the refrigerant is their difference:

$$(T_c - T_e)\Delta s$$

which is represented by the area enclosed by the cycle on the T–s diagram.

Fig. 3.7.2. Hatched areas on the T–s diagram showing (a) specific refrigerating effect, (b) change of specific enthalpy in the condenser and (c) specific work of reversible adiabatic compression.

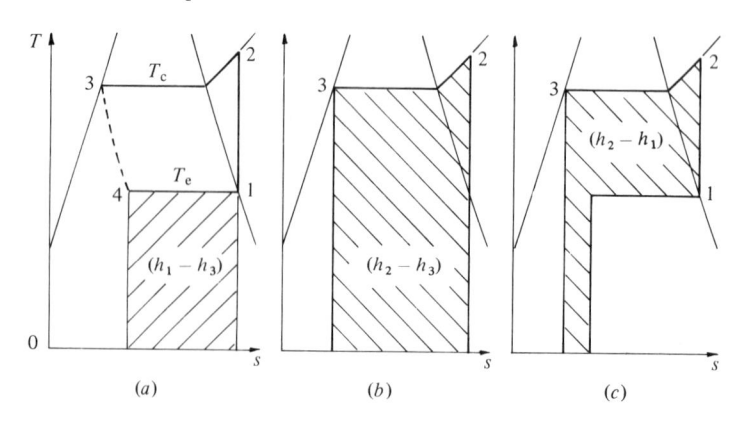

3.7 Comparison with the Carnot cycle

The coefficient of performance is then obtained, in agreement with eq. (3.7.1) as:

$$\varepsilon_C = \frac{T_e \Delta s}{(T_c - T_e)\Delta s} = \frac{T_e}{T_c - T_e}$$

This does not of course constitute a proof of eq. (3.7.1). The fact is that the absolute scale of temperature is defined in such a way that eq. (3.7.1) is, by definition, the coefficient of performance of a Carnot cycle operating between T_e and T_c.

The vapour compression cycle with reversible adiabatic compression is shown in Fig. 3.7.2(a) on T–s coordinates. The specific refrigerating effect:

$$h_1 - h_4 = h_1 - h_3$$

is represented on the diagram by the area down to absolute zero under the line for the evaporation process, i.e. under the line 4–1. This area is shown hatched in Fig. 3.7.2(a). The change of specific enthalpy in the condenser:

$$h_2 - h_3$$

is represented by the area down to absolute zero under the line 2–3. This

Fig. 3.7.3. Carnot cycle superimposed on the vapour compression cycle for the same states at 1 and 3. The hatched rectangle shows the loss of specific refrigerating effect: it is equal to the loss of specific work of expansion. The hatched 'triangle' shows the extra work caused by carrying on adiabatic compression to state 2 instead of changing over to isothermal compression at 2C.

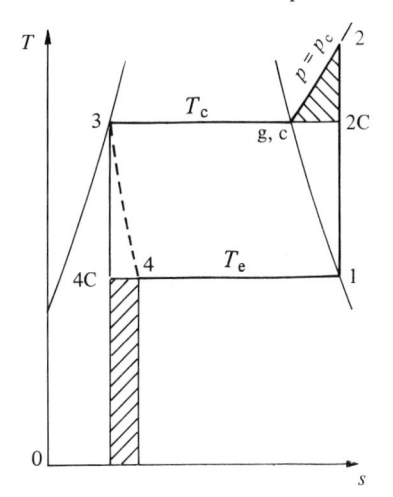

area is shown hatched in Fig. 3.7.2(b). Since these are the only heat transfers in the cycle, the specific work done on the refrigerant during compression is given by their difference:

$$(h_2 - h_3) - (h_1 - h_3)$$

and this is represented by the difference between the areas shown hatched in (b) and (a). This difference is shown hatched in Fig. 3.7.2(c).

The cycle of Fig. 3.7.2 deviates from the Carnot cycle of Fig. 3.7.1 in two ways, shown in Fig. 3.7.3 where the cycles are superimposed.

Firstly, the irreversible process of expansion through the valve 3–4 in the vapour compression cycle causes an increase of entropy:

$$s_4 - s_3 = s_4 - s_{4C}$$

whereas in the Carnot cycle the expansion takes place reversibly at constant entropy. The specific refrigerating effect is thus reduced by the amount:

$$T_e(s_4 - s_{4C}) = h_4 - h_{4C}$$

represented by the hatched rectangle under the line 4C–4. In the Carnot cycle work is obtained from the expansion process 3–4C which can be used to help drive the compressor, whereas in the vapour compression cycle no work is obtained from the expansion process. The specific work lost is equal to the work obtained from the expansion process in the Carnot cycle. From the steady-flow energy equation, this work is:

$$h_3 - h_{4C} = h_4 - h_{4C}$$

The loss of work is thus equal to the loss of specific refrigerating effect, and is again represented by the hatched area in Fig. 3.7.3.

Despite the theoretical advantages of using reversible expansion from 3–4C, it is hardly ever attempted in practice for ordinary vapour compression cycles, because the extra cost of an expansion engine or turbine is not worth even the theoretical gain; and when account is taken of inefficiency in the expander the gain becomes rather small. When the difference between T_e and T_c is large, the loss of refrigerating effect from throttling becomes important, but other methods to be treated in Chapter 5 are available to reduce the loss.

The second way in which the cycle of Fig. 3.7.2 deviates from the Carnot cycle of Fig. 3.7.1 is the continuation of adiabatic compression to pressure $p_2 = p_c$, and resulting temperature T_2, instead of changing over to isothermal compression when point 2C is reached. Comparison of the areas representing work in Fig. 3.7.1 and Fig. 3.7.2 shows that the extra work

3.7 Comparison with the Carnot cycle

because of this is the area shown hatched below the desuperheating line in Fig. 3.7.3., i.e.

$$(h_2 - h_{g,c}) - T_c(s_{2C} - s_{g,c})$$

The extra work caused by continuation of adiabatic compression to state 2 instead of changing to isothermal compression at 2C can be eliminated, in principle, in two ways. One is shown in Fig. 3.7.4(a). Instead of taking dry saturated vapour from the evaporator, the state at entry to the compressor lies in the two-phase region such that reversible adiabatic compression takes place entirely within the two-phase region. The temperature at the end of compression is then only T_c. Many early compressors operated on this principle, known as 'wet compression', but its use has long been

Fig. 3.7.4. Two ways of avoiding, in principle, the superheating loss. (a) by compression in the two-phase region; (b) and (c) by using a refrigerant with a very steep vapour saturation line on T–s coordinates. With evaporation at $-15\,^\circ\mathrm{C}$, R113, R114 and R115 behave like (b); R502 and propane behave like (c).

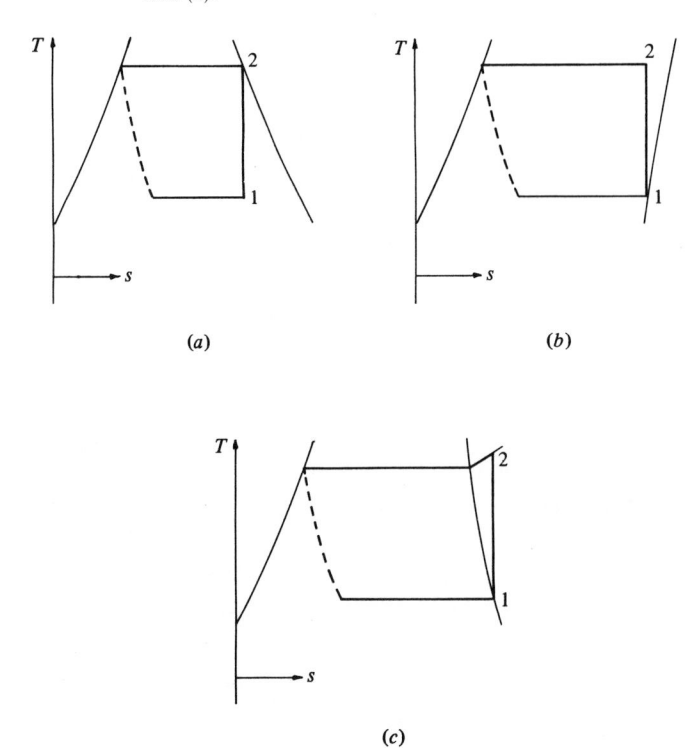

abandoned. It is found in practice, for reasons which will appear later, that the poor performance of a reciprocating compressor when pumping wet vapour far outweighs any theoretical advantage gained. Furthermore, the reduced specific refrigerating effect is accompanied by a reduced volumic refrigerating effect with wet compression, and for a given refrigerating capacity it is necessary to provide a larger machine at greater capital cost.

The other way, shown in Fig. 3.7.4(b), and (c), is to find a refrigerant whose saturated vapour entropy is approximately constant. Isentropic compression would then take the state close to the saturation line, either in the wet region as shown in (b), or with little superheat produced as in (c). Many substances of this kind are known, especially the fluorinated derivatives of ethane such as R114 and R115, and the azeotropic mixture of R115 and R22 known as R502. It may be worth remarking here that symmetry of the T–s diagram is usually associated only with simple molecules like water, ammonia, carbon dioxide and so on. The more complicated the molecular structure, the more unsymmetrical does the diagram become. For a refrigerant which does not superheat much during compression the hatched superheat loss shown in Fig. 3.7.3 is not large.

The deviation from the Carnot cycle caused by superheating is of a fundamentally different nature from the deviation caused by throttling expansion, and in some circumstances it might not be considered a loss at all. The loss caused by throttling expansion is due to an internal irreversibility, a frictional process, which causes a loss of useful refrigerating effect and a loss of work. But the continuation of adiabatic compression beyond point 2C causes an increase of work only if the cooling medium is assumed to be necessarily at a constant temperature. In that event the heat transfer during desuperheating takes place across a temperature difference, and this irreversibility is reflected in the extra work. But if the cooling medium must necessarily rise in temperature, perhaps because there is not enough of it, or it is expensive to pump, an internal régime following the coolant temperature is better than the constant temperature rejection of the Carnot cycle. The Carnot cycle is not then a valid standard for comparison, and the Lorenz cycle should be used, as discussed in section 1.13.

With ammonia as refrigerant it is not possible to use the vapour superheat to achieve a high temperature rise of the cooling water, because the very large specific enthalpy of condensation dominates the conditions and temperature differences in the condenser. With other refrigerants, the specific enthalpy of condensation is sometimes smaller in relation to the heat transfer in desuperheating, particularly when operating near the critical point, and it becomes possible to operate a Lorenz-type heat rejection process.

3.7 Comparison with the Carnot cycle

Example 3.7.1
Evaluate the magnitude of the deviations from the Carnot cycle for the vapour compression cycle of example 3.5.1 with reversible adiabatic compression.

Effect of throttle valve.
In the Carnot cycle, expansion from 3–4C takes place at constant specific entropy:

$$s_{4C} = s_3 = 1.2025 \text{ kJ/kg K}$$

The specific refrigerating effect in the Carnot cycle is:

$$T_e(s_1 - s_{4C}) = 263.15(5.4717 - 1.2025) \text{ kJ/kg}$$
$$= 1123.4 \text{ kJ/kg}$$

compared with 1108.8 kJ/kg in the vapour compression cycle. The loss is thus:

$$(1123.4 - 1108.8) \text{ kJ/kg} = 14.6 \text{ kJ/kg}$$

about 1.3%. In the Carnot cycle work is done in the process 3–4C,

$$h_3 - h_{4C} = h_4 - h_{4C} = T_e(s_4 - s_{4C}) = 14.6 \text{ kJ/kg}$$

and this work is not obtained in the vapour compression cycle.

Extra work in process 2C–2–g, c.
The extra work, shown hatched in Fig. 3.7.3, is:

$$(h_2 - h_{g,c}) - T_c(s_{2C} - s_{g,c})$$
$$(h_2 - h_{g,c}) = (1630.2 - 1467.6) \text{ kJ/kg} = 162.6 \text{ kJ/kg}$$
$$T_c(s_{2C} - s_{g,c}) = 303.15(5.4717 - 4.9793) \text{ kJ/kg}$$
$$= 149.3 \text{ kJ/kg}$$

The extra specific work is therefore:

$$(162.6 - 149.3) \text{ kJ/kg} = 13.3 \text{ kJ/kg}$$

Hence the specific work of compression alone in the Carnot cycle is:

$$(198.6 - 13.3) \text{ kJ/kg} = 185.3 \text{ kJ/kg}$$

The coefficient of performance can be evaluated under two different assumptions:
(a) that the work from the expander is used to help drive the compressor, or
(b) that the work from the expander is wasted.
In case (a) the net work of compression is:

$$(185.3 - 14.6) \text{ kJ/kg} = 170.7 \text{ kJ/kg}$$

and the coefficient of performance is:

1123.4/170.8 = 6.58

In case (b), the work of compression is:

185.3 kJ/kg

and the coefficient of performance is:

1123.4/185.3 = 6.06

Only the first of these is referred to as the Carnot coefficient of performance.

Comparison of results of examples 3.5.1 and 3.7.1

	Vapour compression cycle	Carnot cycle
Specific refrigerating effect/(kJ/kg)	1108.8	1123.4
Specific work of compression/(kJ/kg)	198.6	185.3
Specific work of expansion/(kJ/kg)	Nil	14.6
Net specific work/(kJ/kg)	198.6	170.7
Coefficient of performance (a)	5.58	6.58
Coefficient of performance (b)		6.06

3.8 The effect of evaporating and condensing temperatures on the cycle quantities

The quantities derived so far vary with the evaporating and condensing temperatures, and these variations have important practical consequences. The following discussion is illustrated with reference to ammonia as refrigerant, but the general trends are similar for other refrigerants.

(a) *Specific refrigerating effect*, $(h_1 - h_3)$. At a given condensing temperature h_3 is fixed, and the variation with evaporating temperature depends on the variation of the specific enthalpy of saturated vapour with temperature. Examination of the table of properties of ammonia shows that the value of h_g has a maximum at about 50 °C. Above and below this temperature the change is relatively small, and hence $(h_1 - h_3)$ diminishes only slowly with decreasing evaporating temperature, as shown in Fig. 3.8.1. For other refrigerants the maximum value of h_g may come within the range of normal evaporating temperatures, but again the variation with evaporating temperature is usually small.

With increasing condensing temperature h_3 increases, so that $(h_1 - h_3)$

3.8 Evaporating and condensing temperatures

diminishes with increasing condensing temperature. The relative variation depends on the ratio of the specific heat capacity of the liquid to the specific enthalpy of evaporation. As will be seen, this ratio is lower for ammonia than for the fluorinated refrigerants, and the variation of $(h_1 - h_3)$ is therefore less important for ammonia.

(b) *Volumic refrigerating effect*, $(h_1 - h_3)/v_1$. The numerator of this fraction varies comparatively little, as just shown, but the denominator varies considerably with pressure and hence with evaporating temperature. The volumic refrigerating effect therefore diminishes rapidly with decreasing evaporating temperature, as shown in the second set of curves in Fig. 3.8.1. In practice this means that the refrigerating capacity produced by a compressor of the positive displacement type, with a fixed swept volume

Fig. 3.8.1. The specific refrigerating effect $(h_1 - h_3)$ and the volumic refrigerating effect $(h_1 - h_3)/v_1$ for ammonia as functions of evaporating temperature at three condensing temperatures.

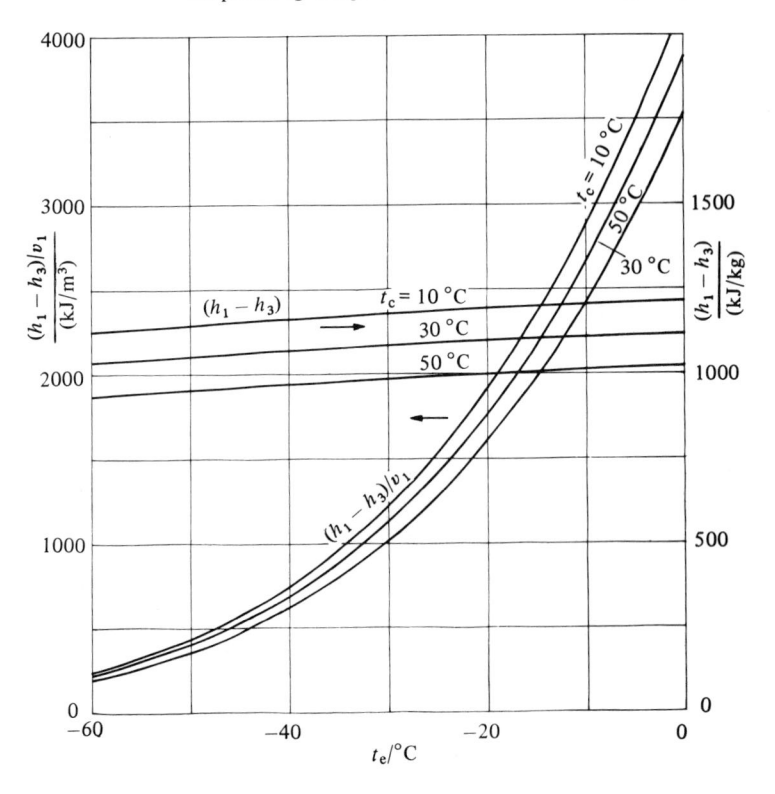

rate, diminishes with decreasing evaporating temperature. Looking at the matter from another point of view, it means that to produce a certain refrigerating capacity, a larger swept volume rate is required for low evaporating temperatures than for high ones. Consequently, plant for low temperatures is larger and more expensive.

(c) *Specific work of isentropic compression*, $(h_2 - h_1)_s$. The variation is shown in Fig. 3.8.2. It increases with decreasing evaporating temperature and with increasing condensing temperature. The reason for this behaviour is evident from an examination of the shape of the lines of constant entropy in the superheat region on a pressure–enthalpy diagram.

(d) *Volumic work of isentropic compression*, $(h_2 - h_1)_s/v_1$. At constant condensing temperature, $(h_2 - h_1)_s$ increases with decreasing evaporating

Fig. 3.8.2. The specific work of isentropic compression $(h_2 - h_1)_s$ and the volumic work of isentropic compression $(h_2 - h_1)_s/v_1$ for ammonia as functions of evaporating temperature at three condensing temperatures.

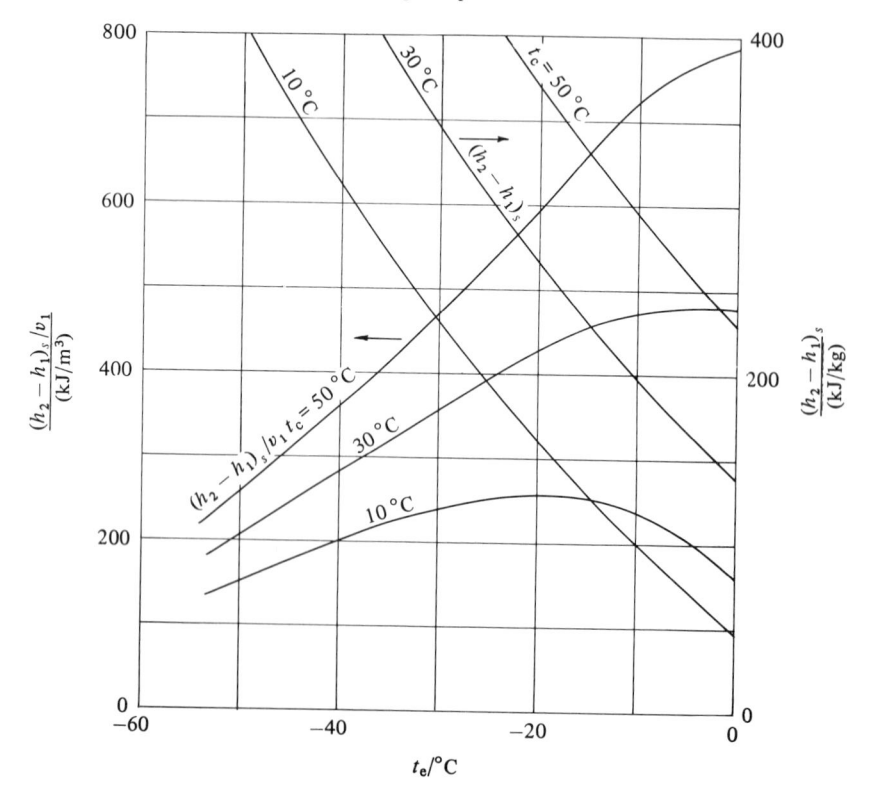

3.8 Evaporating and condensing temperatures

temperature, but so does v_1, and it is seen from the second set of curves in Fig. 3.8.2 that their ratio passes through a maximum.

The volumic work of isentropic compression is also the mean effective pressure acting on the piston of an ideal positive displacement compressor. The power needed to drive such a machine therefore increases with increasing evaporating temperature and thereafter diminishes. This characteristic has an important bearing on the selection of a motor to drive the compressor. The variation for a real machine depends on the type of compressor, and is considered in greater detail in Chapter 4.

Increasing condensing temperature markedly increases the volumic work of isentropic compression, and hence increases the power required to drive a given machine.

(e) *Isentropic coefficient of performance*, $(h_1 - h_3)/(h_2 - h_1)_s$. The numerator of this expression shows only a slow variation with evaporating tempera-

Fig. 3.8.3. The isentropic coefficient of performance for ammonia as a function of evaporating temperature at three condensing temperatures.

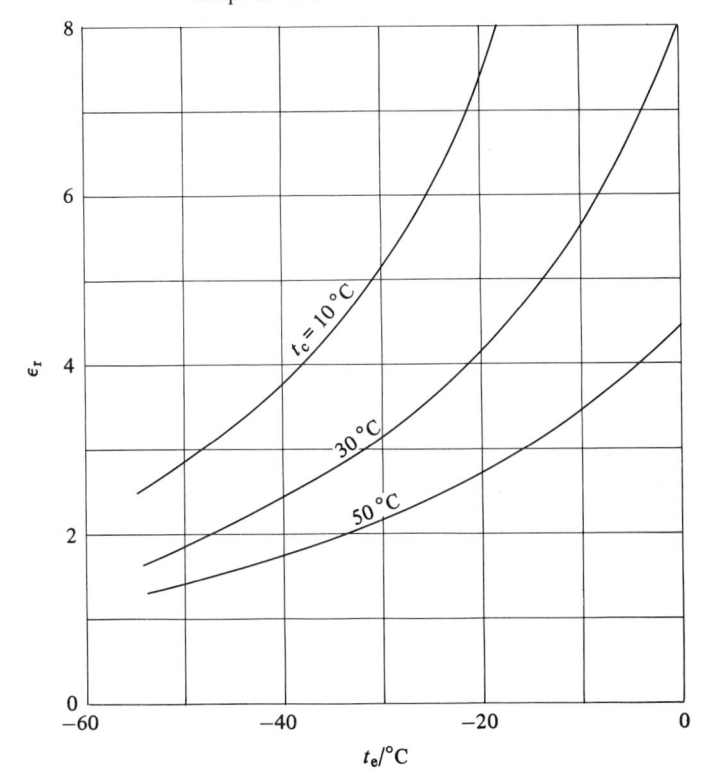

ture, as seen in Fig. 3.8.1. The main variation is caused by that of the denominator. Figure 3.8.3 shows that the isentropic coefficient of performance diminishes quite rapidly with decreasing evaporating temperature. The variation is more marked at lower condensing temperatures.

With increasing condensing temperature the coefficient of performance diminishes. The variation is more marked at the higher evaporating temperatures.

The real coefficient of performance achieved in an actual plant is of course lower than the isentropic value, but it remains true in general that it is more expensive in power to produce a given refrigerating capacity at low evaporating temperatures and high condensing temperatures than at moderate ones.

(f) *The temperature at the end of isentropic compression.* The temperature which the vapour attains inside the compressor cylinder and at which it is discharged has to be considered in practice because high temperatures cause degradation of the lubricating oil with resulting deposits of oleoresins on the valves, and because it enhances the probability of chemical reactions between refrigerant, oil and any moisture which may be present.

Examination of the lines of constant entropy in the superheat region of the pressure–enthalpy diagram shows that decreasing evaporating temperature and increasing condensing temperature both cause an increase in the discharge temperature. This may limit the evaporating temperature which can be attained in single-stage compression, as discussed in Chapter 5.

3.9 The effect of refrigerant properties on the cycle quantities

The quantities discussed so far are listed in Table 3.9.1 for some of the commoner refrigerants. The significance of the differences between them, and the reasons for these, will now be discussed.

(a) *Specific refrigerating effect*, $(h_1 - h_3)$. The specific refrigerating effect determines, through eq. (3.2.11), the mass flow rate of refrigerant required to produce a given refrigerating capacity. The larger the specific refrigerating effect, the smaller the required mass flow rate. Since the densities of liquid refrigerants do not differ by more than a factor of about 2, the specific refrigerating effect is important in determining the sizes of components for handling liquid. For example, the size of orifice needed in the expansion valve is different for different refrigerants mainly on account of the different specific refrigerating effects.

3.9 Refrigerant properties

Table 3.9.1. *Properties of refrigerants*

	Ammonia	Carbon dioxide	R11	R12	R22	R502	Propane
Formula	NH_3	CO_2	CCl_3F	CCl_2F_2	$CHClF_2$	$0.63CHClF_2$ $+0.37C_2ClF_5$	C_3H_8
Molar mass/(kg/kmol)	17.0	44.0	137.4	120.9	86.5	111.6	44.1
Boiling point/°C	−33.3	−78.5†	23.8	−29.8	−40.8	−45.4	−42.1
Critical temperature/°C	132.5	31.0	198.0	112.0	96.0	90.1	96.8
		Saturation cycle quantities −15°/+30 °C					
p_e/bar	2.36	22.9	0.202	1.83	2.96	3.48	2.92
p_c/bar	11.66	72.1(80.0)‡	1.25	7.45	11.92	13.19	10.85
p_c/p_e	4.94	3.15(3.49)	6.19	4.08	4.03	3.78	3.72
(h_1-h_3)/(kJ/kg)	1102	132.0(146.5)	155.4	116.4	162.9	104.4	285
v_1/(m³/kg)	0.509	0.0166	0.762	0.0910	0.0776	0.0500	0.153
$[(h_1-h_3)/v_1]$/(kJ/m³)	2170	7940(8810)	204	1280	2100	2090	1860
$(h_2-h_1)_s$/(kJ/kg)	231	48.6(53.5)	30.9	24.7	34.9	24.0	60.5
$[(h_2-h_1)_s/v_1]$/(kJ/m³)	455	2920(3220)	40.6	272	450	480	394
ε_r	4.77	2.72(2.74)	5.03	4.70	4.66	4.35	4.71
t_2/°C	99	68(77)	44	38	53	37	37
x_4	0.160	0.516(0.463)	0.200	0.266	0.249	0.333	0.280

‡ The figures in brackets for CO_2 are explained in section 3.14.
† Sublimation temperature of solid CO_2 at atmospheric pressure.

The specific refrigerating effect depends mainly on the specific enthalpy of evaporation of the substance, and this in turn depends on the molar mass (molecular weight). The largest enthalpies of evaporation are found for substances with light molecules, such as water and ammonia, whilst the lowest values are found for heavy molecules. There is a rough numerical correlation given by Trouton's rule, according to which the molar entropy of evaporation at the normal atmospheric boiling point of the substance has about the same value for all substances, i.e.

$$Mh_{fg}/T_n \approx 85 \text{ kJ/kmol K} \tag{3.9.1}$$

where M is the molar mass and T_n is the normal boiling point on the absolute scale.

The rule is only approximate, and should not be used instead of tables, unless nothing else is known about the substance. The deviations from Trouton's rule are most marked for liquefied gases at low temperature.

The specific refrigerating effect is less than the specific enthalpy of evaporation by the change of specific enthalpy of saturated liquid between evaporating and condensing temperature, as shown by eq. (3.6.5). This depends on the specific heat capacity of the liquid, which is larger for the lighter molecules such as ammonia. The loss of refrigerating effect as a fraction of the enthalpy of evaporation is numerically the same as the dryness fraction of the two-phase mixture leaving the expansion valve, x_4 in Table 3.9.1.

(b) *Volumic refrigerating effect*, $(h_1 - h_3)/v_1$. The volumic refrigerating effect determines the swept volume rate which is required in a positive displacement compressor to give a certain refrigerating capacity. A high value is usually desirable in order to have a compact compressor, and from the values in the table it is seen that carbon dioxide would require the smallest swept volume rate of all the refrigerants tabulated. On the other hand, R11 has a low volumic refrigerating effect, and it requires a high volume flow rate for a given refrigerating capacity. It is not usually practicable to use R11 with reciprocating compressors on account of this, but it is a useful refrigerant with turbo-compressors.

Because of the differences in volumic refrigerating effects, it is possible to increase the refrigerating capacity of a given plant by charging it with a different refrigerant, e.g. R22 instead of R12. Manufacturers use this principle to increase the range of refrigerating capacities which they can offer from a number of standard compressor bodies. As will be seen, the refrigerant with the higher volumic refrigerating effect will also require

3.9 Refrigerant properties

more power to drive the compressor, so a larger motor must be used. There may also be other modifications required to make the system suitable for a different gas.

The remarks made above about the desirability of a high volumic refrigerating effect do not always apply. For example, in domestic refrigerators with two-pole motors the cylinder size is already small enough, and there is no point in using R22 instead of R12. The latter refrigerant is preferred because of its greater stability to possible reactions with oil and insulating materials, and because of its lower rise in temperature during compression.

A high volumic refrigerating effect at a given evaporating temperature is found to be associated with substances which have high evaporating pressures. The reason for this is again Trouton's rule, coupled with the assumption that the specific volume of the vapour is given approximately by the ideal gas law. Then, approximately, at atmospheric pressure p_{atm}:

$$\frac{h_{fg}}{v_g} = (85 \text{ kJ/kmol K}) \frac{p_{atm}}{\bar{R}}$$

$$= \frac{85}{8.314} \times 10^5 \text{ N/m}^2$$

$$= 1022 \text{ kJ/m}^3 \approx 1000 \text{ kJ/m}^3 \qquad (3.9.2)$$

where $\bar{R} = 8.314$ kJ/kmol K is the molar gas constant, and atmospheric pressure is taken as 1 bar.

Equation (3.9.2) shows that when compared at an evaporating pressure of 1 bar, all substances should have approximately the same value of h_{fg}/v_g, and hence of the volumic refrigerating effect. Since the volumic refrigerating effect rises with evaporating temperature and pressure, as shown by Fig. 3.8.1, the higher the evaporating pressure the higher is the volumic refrigerating effect.

(c) *Specific work of isentropic compression,* $(h_2 - h_1)_s$. This quantity is not itself of special importance in positive displacement compressors, but it has a great influence on the design of turbo-compressors, since it determines the tip speed needed to achieve a given pressure ratio, i.e. for given evaporating and condensing temperatures. Refrigerants with low values of the specific isentropic work require a lower tip speed, and hence a lower rotational speed for a fixed impeller diameter, or a smaller impeller for a fixed rotational speed. It will be seen in Chapter 4 that this fact has an important bearing on the range of refrigerating capacities for which turbo-compressors can be manufactured with satisfactory efficiencies.

As can be seen from Table 3.9.1, the specific work of isentropic compression is highest for the inorganic compounds ammonia and carbon dioxide. There is a very rough correlation with molar mass, but it is not exact enough to be useful.

(d) *Volumic work of isentropic compression*, $(h_2 - h_1)_s/v_1$. This quantity is important for all compressors which produce a more or less constant volume flow rate, for it indicates the power which is needed to drive the compressor. For example, as already remarked, a reciprocating compressor of fixed swept volume rate requires more power when it is working on R22 than when it is working on R12.

The principal factors determining the volumic work are the suction and discharge pressures. Refrigerants of higher vapour pressure, i.e. lower normal boiling points, have larger values of the volumic work of isentropic compression. This corresponds with the findings from Trouton's rule that they also have larger volumic refrigerating effects. By and large, therefore, a refrigerant of higher vapour pressure takes more power for a given swept volume rate, but it produces more refrigerating capacity.

(e) *Isentropic coefficient of performance*, ε_r. The statement in the last sentence would of course be exact if all refrigerants had the same coefficient of performance, but Table 3.9.1 shows that this is not the case. Nevertheless, apart from carbon dioxide which is a rather special case, there are not large differences between the coefficients.

Carbon dioxide is unusual because its critical temperature of 31 °C is only slightly above the condensing temperature for which Table 3.9.1 is calculated. At this condition a relatively large fraction of vapour forms in the expansion valve, and the specific refrigerating effect is much reduced. This suggests that there may be a correlation between the coefficient of performance (for fixed evaporating and condensing temperatures) and the critical temperature of the refrigerant. In Fig. 3.9.1 some relative coefficients of performance are plotted against the reduced condensing temperature, i.e. the ratio of the absolute temperature of condensation to the absolute critical temperature. Two lines can be discriminated, a higher one for substances with fairly simple molecules and a lower one for more complicated molecules, but there are several exceptions. It appears from this that a high coefficient of performance is associated with a high critical temperature and a simple molecule.

When a comparison is made between the coefficients of performance of

3.9 Refrigerant properties

refrigerants at the same reduced evaporating and condensing temperatures, the differences in the isentropic coefficients of performance tend to disappear (Eiseman, 1952), but not entirely. There still remain differences which are probably associated with the shape of the saturated vapour line on the temperature-entropy diagram.

The importance of attaining a good isentropic coefficient of performance needs no emphasis. But it must be remembered that the actual coefficient of performance of a plant depends on other factors, one of which is the volumic refrigerating effect. The matter will be studied in more detail in Chapter 4, but in brief the effect of a high volumic refrigerating effect is to diminish the importance of friction in the compressor, because for the same refrigerating capacity the compressor size required is smaller for the refrigerant with the higher volumic refrigerating effect. As an example of this, the coefficient of performance of a carbon dioxide plant is not so bad, relative to ammonia, as the isentropic coefficients of Table 3.9.1 would lead one to expect.

Fig. 3.9.1. Coefficient of performance for the saturation cycle at -15 to $+30\,°C$ of several refrigerants relative to that of the Carnot cycle, $\varepsilon_C = 5.74$, plotted against reduced condensing temperature.

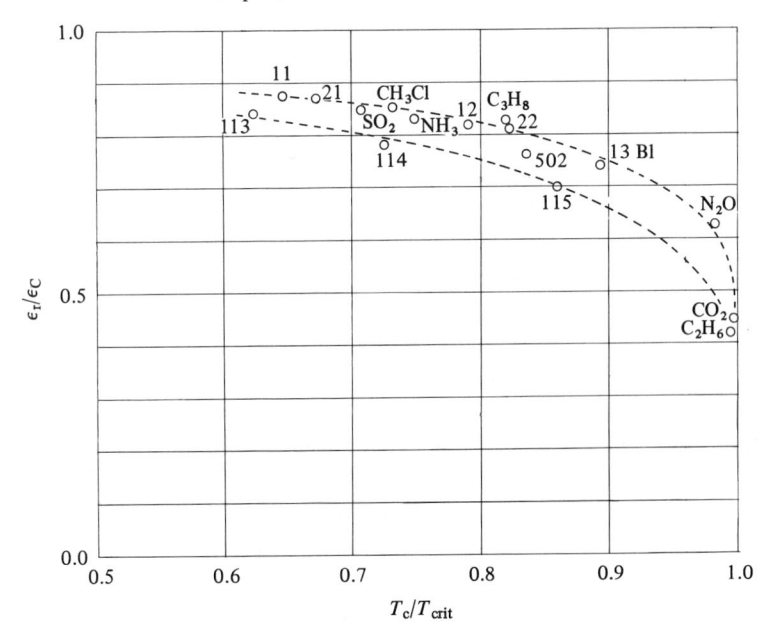

(*f*) *Temperature at the end of isentropic compression.* The rise in temperature of the vapour during compression is a factor of considerable practical importance, for the reasons outlined in section 3.8, and the various refrigerants show great differences between them in this respect. The highest temperature is reached with ammonia, and for this refrigerant a substantial part of the compression process takes place at a temperature above that of condensation and hence above that of the available cooling water. It is practicable therefore, as well as desirable for reasons concerned with the lubricating oil, to cool the cylinders of ammonia by water jackets, and so reduce the work of compression slightly.

For R502, on the other hand, the temperature at the end of compression is only just above condensation temperature, for the cycle considered in Table 3.9.1, and water cooling would be neither necessary or useful. Similar remarks apply to R12, but for R22 the temperature rise is greater and water cooling is often necessary.

At lower evaporating temperatures than those of Table 3.9.1, the temperatures at the end of compression would of course be higher, and some form of cooling may be necessary for most refrigerants. The practical limit is usually considered to be about 150 °C for most refrigerants.

Because of irreversibility during compression, the temperatures attained at the end of *adiabatic* compression would be higher than those of Table 3.9.1.

Temperature rise during compression is particularly important for refrigerants used in hermetic or semi-hermetic systems, in which the windings of the motor are cooled by the suction vapour. The suction vapour is thus first raised in temperature over the windings, and then during compression, so that very high temperatures may be reached at the end. For this reason, and because of the high density of the vapour which improves heat transfer, R502 is much favoured for such plant and has partly displaced R22. R12 continues in use for hermetic systems working at moderate evaporating and condensing temperatures.

The temperature rise of a vapour during compression is associated with its molecular structure. The simplest molecules, having only one atom, such as helium, have the highest temperature rises. Diatomic ones such as oxygen and nitrogen come next, followed by substances like ammonia, sulphur dioxide, etc. Amongst the fluorinated refrigerants, the derivatives of methane show, generally, higher temperature rises than those for the derivatives of ethane, such as R115. Many ethane derivatives in fact become two-phase mixtures when their saturated vapours are compressed isentropically, as shown in Fig. 3.7.4(*b*).

3.10   Subcooled liquid

(*g*) *Vapour pressures and pressure ratio.* The pressures in the evaporator and condenser are shown in Table 3.9.1. They are of course related to the boiling points, a substance of low boiling point having a high vapour pressure. As a general rule, a pressure above atmospheric pressure is desired in the evaporator and low side, in order to prevent the ingress of air and moisture. At the same time the pressure in the condenser and high side should not be so high that special construction and components are required. Carbon dioxide is an exceptional refrigerant in this respect, with its high working pressures. These demand a very heavy form of construction, which is expensive. Cylinder blocks were often solid forgings instead of castings. On the other hand, its high volumic refrigerating effect means that the bore required in the cylinder and pipes is smaller than for the other refrigerants such as ammonia and R22. The working pressures for R11 are low, and the evaporating pressure is always less than atmospheric pressure. As already remarked, this substance is not suitable for use in reciprocating compressors. In turbo-compressors the problem of air leakage inwards is dealt with by special seals and by automatic purging arrangements, or by hermetic construction.

When the desired evaporating temperature is very low, the evaporating pressure is necessarily less than atmospheric pressure for the two common refrigerants ammonia and R22. Nevertheless they are commonly used in this way, but adequate purging arrangements are provided. If a refrigerant of higher vapour pressure is selected, the pressure in the condenser may be too high. It is then necessary to use a cascade system, described in Chapter 5, in which two or more different refrigerants are used.

The pressure ratio is important in reciprocating compressors because of its effect on volumetric efficiency. Table 3.9.1 shows that there are some significant differences between refrigerants, e.g. the high values for ammonia and R11 compared with R12 and R22. As a rough indication of pressure ratio, it tends to increase, for given evaporating and condensing temperature, with the boiling point of the refrigerant.

The higher pressure ratio of ammonia compared with R22 may be noted. Though the boiling point of R22 is lower than that of ammonia, the vapour pressure of ammonia increases with temperature more rapidly. At about 36 °C they have the same vapour pressure, and above that temperature the vapour pressure of ammonia is the higher.

3.10   Subcooled liquid

The state of liquid refrigerant leaving the condenser and entering the expansion valve has so far been assumed as saturated. In practice some degree of subcooling may be acquired, and the point 3 moves to the left of the saturated liquid line on the pressure–enthalpy diagram, as shown in Fig. 3.10.1. The specific refrigerating effect is still given by:

$$h_1 - h_3$$

but h_3 now has a lower value than $h_{f,c}$ the specific enthalpy of saturated liquid at condensing temperature. The increase in specific refrigerating effect:

$$h_3 - h_{f,c}$$

depends on the specific heat of the liquid refrigerant. It is highest for ammonia, which has the highest specific heat capacity of all liquid refrigerants. The proportional increase in the specific refrigerating effect is however:

$$\frac{h_3 - h_{f,c}}{h_1 - h_{f,c}}$$

and this depends on the specific enthalpy of evaporation as well as on the specific heat capacity of the liquid. Because of the large specific enthalpy of evaporation of ammonia, the proportional increase in the specific refrigerating effect for each degree of subcooling is quite small, of the order 0.4%/K. For a refrigerant like R502 on the other hand, the proportional increase is of the order 1.1%/K for $t_e = -15\,°C$, $t_c = 30\,°C$.

The volumic refrigerating effect is of course increased by subcooling in the same way as the specific refrigerating effect. Since the specific work of compression remains the same, the coefficient of performance is improved.

Whether much subcooling can be obtained in practice depends on the design of the condensing system and on the temperature difference between

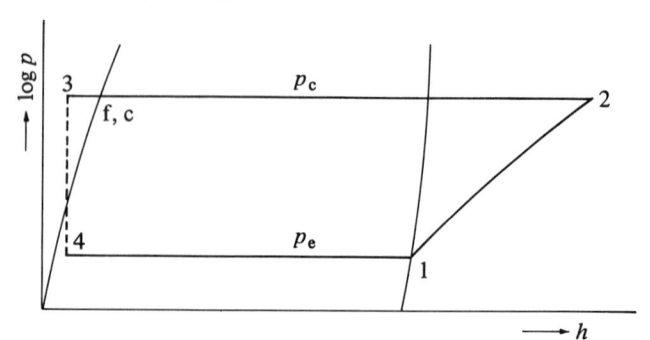

Fig. 3.10.1. Subcooling of liquid from the condenser to state 3 on the p–h diagram.

3.10 Subcooled liquid

condensing refrigerant and the entering cooling medium. Figure 3.10.2 shows a plot of the temperatures of the refrigerant and cooling medium for counterflow of the streams, to a base which represents the heat transfer between one end and any section. If the flow rate of the cooling medium were very large, the outlet temperature t_{w2} would be almost equal to the inlet temperature t_{w1}. This high flow rate is not usually attempted in practice because of the cost of the medium or of pumping it. There is usually a rise in temperature of about 5–10 K for cooling water. The temperature of the cooling medium at inlet is therefore at least this amount below condensing temperature, and in fact low enough to give some subcooling in a counterflow heat exchanger.

The temperature of the refrigerant is shown idealised in the superheat region of Fig. 3.10.2, and it is unlikely that desuperheating takes place as a distinct process. The reason is that even a superheated vapour must condense if it comes in contact with a surface at a temperature below its saturation temperature, i.e. below t_c in this case. With a good heat transfer coefficient on the cooling water side this will happen. It is true that if the water flow rate were restricted so as to make t_{w2} higher than t_c, as shown by the broken line, initial condensation could not take place, but this causes a rather low temperature difference between refrigerant and water at the warm end of the condenser and is expensive in surface area. In practice $(t_c - t_{w2})$ is of the order 2–3 K, called the 'temperature approach'.

It should be noted that to subcool a liquid, heat must be taken from it some distance from the liquid–vapour interface. For if the liquid at the interface itself were cooled, condensation of the vapour would take place on the interfacial surface, and the liquid would remain saturated there. The

Fig. 3.10.2. Temperatures in the condenser with counterflow of water and refrigerant showing the possibility of subcooling to t_3 less than t_c.

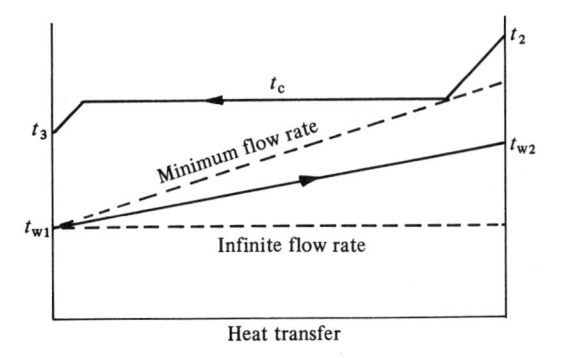

pressure in the container would fall. In a shell-and-tube condenser, for example, in which vapour condenses on the outside of tubes through which water flows, reducing the temperature of the cooling water simply reduces the pressure in the condenser, the liquid draining from the tubes remaining near saturation. (Not quite, because for heat transfer through the condensate film to take place, the temperature of the tube wall and the liquid in contact with it must be somewhat lower than saturation temperature). Under certain conditions of operation it may happen that liquid does not drain freely from the bottom of the condenser and builds up inside. This liquid can then be subcooled, underneath its surface, by transfer of heat to the tubes and the surroundings.

To obtain substantial subcooling it is normally necessary to use a separate heat exchanger, in which the stream of liquid has its temperature reduced by a separate cooling medium, or by the cool entering water before it goes to the condenser. The question then arises whether one would not be better off by using the extra heat transfer surface in the condenser, reducing the condensing pressure and the work of compression. This is a problem in optimisation, the answer to which depends on the type of plant and the refrigerant. As a rule it turns out best, from the point of view of plant performance, to use some of the heat transfer area in a subcooler, at least up to the point where a useful increase in specific refrigerating effect is produced without an undue increase in delivery pressure. Subcooling is more effective with the heavy fluorinated refrigerants than with ammonia. The possibility of obtaining a supply of cold water for the subcooler from another part of a large plant, e.g. feed water to boilers, should always be investigated.

Fig. 3.10.3. The correction terms to obtain the specific enthalpy of subcooled liquid at 3, shown on p–h coordinates.

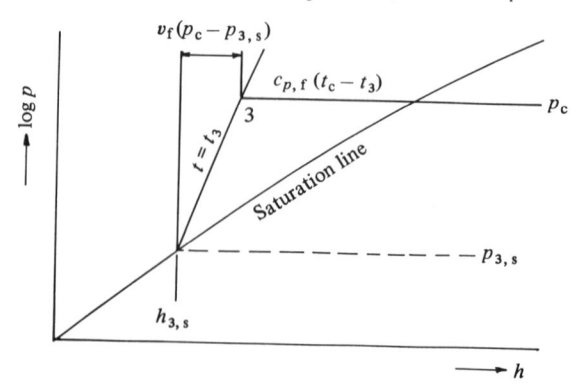

3.10 Subcooled liquid

Substantial subcooling is obtained from heat exchange between liquid and vapour, as described later.

The specific enthalpy of subcooled liquid at state 3 is not usually provided in thermodynamic tables, and it must be estimated. There are two approximate methods for this, corresponding to the two ways in which subcooled liquid can be formed: (a) by cooling saturated liquid at constant pressure, and (b) by compressing saturated liquid at constant temperature. According to the first of these, the specific enthalpy of subcooled liquid is less than that of saturated liquid *at the same pressure* by an amount:

$$h_{f,c} - h_3 = c_{p,f}(t_c - t_3) \tag{3.10.1}$$

where $c_{p,f}$ is the specific heat capacity at constant pressure of the liquid.

According to the other method, the specific enthalpy of subcooled liquid is greater than that of saturated liquid *at the same temperature* by an amount equal to the specific work required to compress it in a steady-flow process from the saturation pressure, $p_{3,s}$ say, corresponding to the temperature t_3, to the condensing pressure p_c, i.e. by:

$$h_3 - h_{3,s} = v_f(p_c - p_{3,s}) \tag{3.10.2}$$

where $h_{3,s}$ is the specific enthalpy of saturated liquid at temperature t_3, pressure $p_{3,s}$, and v_f is the specific volume of liquid, assumed to be constant.

Equation (3.10.1) is not of much value because $c_{p,f}$ is not usually known accurately. Equation 3.10.2 is approximate in that the liquid is assumed incompressible and the internal energy a function of temperature only, but it is the form of correction generally used. The correction terms are illustrated in Fig. 3.10.3

For many refrigerants with moderate vapour pressures the correction term eq. (3.10.2) can be neglected. In this event the specific enthalpy of subcooled liquid is read directly from the saturation table as that of saturated liquid at temperature t_3, regardless of the pressure. This comes to the same thing as assuming in eq. (3.10.1) that the specific heat capacity of subcooled liquid at constant pressure $c_{p,f}$ is the same as that of saturated liquid c_f.

To illustrate the procedure the following example will be worked.

Example 3.10.1
Determine the specific enthalpy of liquid ammonia at a pressure of 11.67 bar and a temperature of 26 °C.

From Tables:

$$h_{3,s} = h_f(26\,°C) = 303.5\,\text{kJ/kg}$$
$$p_{3,s} = p_s(26\,°C) = 10.34\,\text{bar}$$
$$v_f = 0.00166\,\text{m}^3/\text{kg}$$
$$v_f(p_c - p_{3,s}) = (0.00166\,\text{m}^3/\text{kg})(11.67 - 10.34)\,10^5\,\text{N/m}^2$$
$$= 221\,\text{J/kg} = 0.221\,\text{kJ/kg}$$
$$h_3 = (303.5 + 0.2)\,\text{kJ/kg} = 303.7\,\text{kJ/kg}$$

In the above example v_f is taken at 26 °C, but it makes little practical difference whether it is taken at 26 °C or 30 °C, provided that the temperature difference is not large, and that the state is well away from the critical point.

Whether the correction term, 0.2 kJ/kg in example 3.10.1, is important or not must be decided by its value in relation to the differences in specific enthalpy in the cycle, not in relation to the value 303.5 kJ/kg. It is usually negligible for ammonia and medium pressure refrigerants.

When condensation takes place in the neighbourhood of the critical point, as is commonly the case for carbon dioxide, the above approximate method for finding h_3 is not accurate enough. It is necessary to use a table of values for subcooled liquid, or a chart with constant temperature lines in the subcooled region. These are available for carbon dioxide.

The accurate expression, derived using eq. (3.18.9), for the correction term is:

$$h_3 - h_{3,s} = \int_{p_{3,s}}^{p_c} \left[v - T_3\left(\frac{\partial v}{\partial T}\right)_p\right] dp \qquad (3.10.3)$$

in which the integration is carried out at the constant temperature T_3 from pressure $p_{3,s}$ to pressure p_c. The approximation of eq. (3.10.2) results when the coefficient of thermal expansion is neglected and the volume of the liquid is regarded as constant with change of pressure. The specific enthalpy then increases with pressure at constant temperature. Near the critical point, however, the coefficient of thermal expansion becomes large, and since it appears negatively under the integral of (3.10.3) it has the effect of reducing the enthalpy at constant temperature as the pressure increases, as indicated by the isotherm in Fig. 3.14.1.

The error which is caused by using the approximation (3.10.2) under these circumstances is illustrated by the following example.

Example 3.10.2

Tables for carbon dioxide give the specific enthalpy of subcooled 'liquid' at a pressure of 81.06 bar and a temperature of 30 °C as 176.98 kJ/kg. The properties of saturated liquid at 30 °C are: vapour pressure, 72.11 bar; specific enthalpy, 191.21 kJ/kg; specific volume, 1.6861 dm^3/kg. What error arises by using the approximation eq. (3.10.2) instead of the tabular value?

According to eq. (3.10.2) the specific enthalpy of the subcooled liquid is greater than that of the saturated liquid at 30 °C by the amount:

$$\frac{1.686}{1000}(81.06 - 72.11)10^5 \text{ J/kg} = 1.51 \text{ kJ/kg}$$

The specific enthalpy would therefore be:

$$(191.21 + 1.51) \text{ kJ/kg} = 192.72 \text{ kJ/kg}$$

instead of the tabular value 176.98 kJ/kg. Since the specific refrigerating effect for carbon dioxide at $-15 °C / +30 °C$ is 132 kJ/kg (Table 3.9.1), the error, 15.74 kJ/kg, in calculating the effect with subcooling would be quite large.

The decrease of specific enthalpy at constant temperature with increasing pressure near the critical point has important consequences for the performance of refrigerating plant operating under these conditions which will be considered in section 3.14.

3.11 Superheated vapour

The vapour entering the compressor is usually superheated to ensure that it is free from droplets of liquid, and for other reasons which will now be considered. Superheated vapour at point 1 is shown in Fig. 3.11.1 on

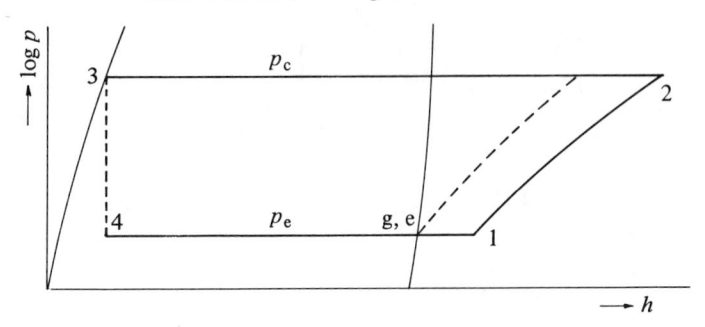

Fig. 3.11.1. Superheating of the vapour leaving the evaporator to state 1 on the $p-h$ diagram.

pressure and enthalpy coordinates and in Fig. 3.11.2 on temperature and entropy coordinates. The specific refrigerating effect is still given by:

$$h_1 - h_3$$

but h_1 is now greater than $h_{g,e}$, the specific enthalpy of saturated vapour at the evaporating temperature. At the same time the specific volume of the vapour is increased by superheat, and the volumic refrigerating effect:

$$\frac{h_1 - h_3}{v_1}$$

may increase or diminish with superheat, depending on the refrigerant. The variation is considered in more detail in section 3.12, but it can be stated here that for ammonia and R22 the volumic refrigerating effect diminishes with superheat, whereas for R12 and R502 it increases.

The specific work of reversible adiabatic compression is increased by superheat. This is indicated in Fig. 3.11.1 by the decreased gradient of the line of constant entropy through point 1 in the superheat region compared with the gradient of the line through point g, e. On the temperature–entropy diagram, Fig. 3.11.2, the increased specific work is shown by the hatched area.

Although the specific work of reversible adiabatic compression is increased by superheat, so is the specific refrigerating effect, and their ratio, ε_r, may increase or diminish. The variation is considered in Section 3.13,

Fig. 3.11.2. Superheating on the T–s diagram. The hatched area shows the increase in specific work of isentropic compression.

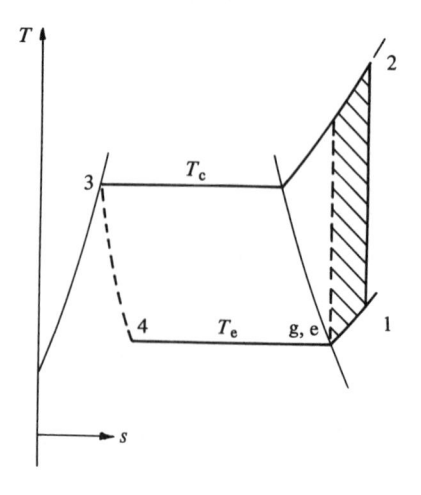

3.11 Superheated vapour

but it may be said here that the variation is in the same direction as the change in the volumic refrigerating effect for the refrigerants mentioned above.

At this stage it is necessary to consider the effect of heat transfer to the refrigerant vapour in the suction pipe from the evaporator to the compressor. The suction pipe usually passes through warm surroundings at least part of the way, and heat transfer to the vapour can take place, causing a rise in its temperature. This is equivalent to wasting a part of the available refrigerating capacity in refrigerating the engine room. By insulation of the suction pipe the waste can be reduced but not entirely eliminated. To distinguish where necessary between refrigerant capacity which is used to refrigerate things which need refrigeration, and that which is used to refrigerate spaces such as engine rooms, the former will be called 'useful' refrigerating capacity, and the latter 'useless'. Corresponding to these capacities there will be useful specific refrigerating effect and useless specific refrigerating effect. The sum of the useful and useless specific refrigerating effects will be called the gross refrigerating effect, i.e. $(h_1 - h_3)$ where h_1 is the specific enthalpy at the compressor inlet.

Superheat which is acquired after the refrigerant leaves the evaporator by refrigerating the outside air is again useless superheat. Useless superheat does not of course increase the useful specific refrigerating effect, though it does increase the specific volume of the vapour, and consequently the useful volumic refrigerating effect diminishes if the superheat is acquired uselessly. Similarly the volumic work of reversible adiabatic compression is increased without any compensating gain by useless superheat.

It is obvious that, as a general rule, not much superheat can be acquired by taking heat directly from the things to be cooled. The bodies or fluids being refrigerated are only 5–10 K warmer than the evaporating refrigerant, and the vapour leaving the evaporator cannot possibly be superheated by more than this by direct heat transfer. In some circumstances, when a fluid is being refrigerated in counterflow with the boiling refrigerant, it may be possible to bring the vapour outlet temperature somewhere near the (warm) fluid inlet temperature. But the number of applications where this would give substantial superheat is relatively small.

By incorporating a counterflow heat exchanger between the liquid going to the expansion valve and the vapour going to the compressor, a substantial useful superheat can be produced, because the superheat is acquired in reducing the temperature and specific enthalpy of the liquid going to the expansion valve, with a consequent gain in specific refrigerat-

The vapour compression system

ing effect. Furthermore, the vapour can be made sufficiently warm that it does not need insulation to prevent heat transfer from the atmosphere as it returns to the compressor.

The circuit diagram for the cycle with heat exchange is shown in Fig. 3.11.3, and the cycle is shown on pressure and enthalpy coordinates in Fig. 3.11.4. Point 6 is shown on the saturation line, but it is not necessarily there, since a small amount of superheat may have been acquired in the evaporator. On the other hand, it could be in the wet region.

Applying the steady-flow equation to the evaporator and valve, the refrigerating capacity is:

$$\dot{Q}_e = \dot{m}(h_6 - h_4) \tag{3.11.1}$$

Without the heat exchanger it would be, for the same state 6 leaving the evaporator,

$$\dot{m}(h_6 - h_3)$$

Fig. 3.11.3. The vapour compression cycle with heat exchange between the liquid going to the expansion valve and the vapour going to the compressor.

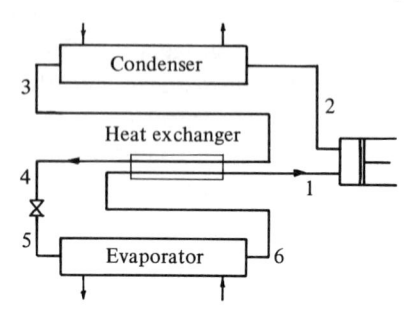

Fig. 3.11.4. Heat exchange on the p–h diagram. In the absence of external heat transfer, the increment $(h_1 - h_6)$ equals the decrease $(h_3 - h_4)$.

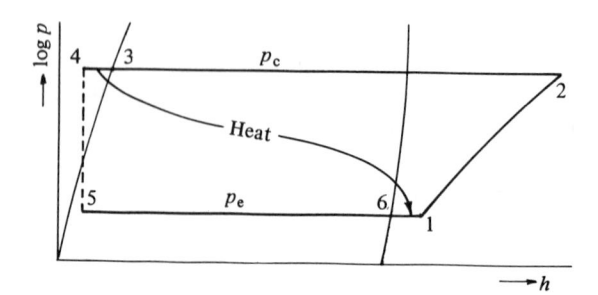

3.11 Superheated vapour

The specific refrigerating effect is thus increased by the amount:

$$h_3 - h_4$$

which is the decrease in specific enthalpy of the liquid in passing through the exchanger.

If the heat transfer to the exchanger from the surroundings is small, then:

$$h_3 - h_4 = h_1 - h_6 \tag{3.11.2}$$

since at steady conditions the flow rates of liquid and vapour are the same. The refrigerating capacity can thus be expressed alternatively as:

$$\dot{Q}_e = \dot{m}(h_1 - h_3) \tag{3.11.3}$$

The same expression results from applying the steady-flow energy equation directly to a control surface surrounding both the evaporator and the heat exchanger.

For calculation of refrigerating capacity, either eq. (3.11.1) or (3.11.3) may be used, whichever is more convenient. For stating the gross refrigerating capacity of compressors, at defined inlet state 1, and defined state 3, then eq. (3.11.3) is naturally used.

For liquid and superheated vapour at constant pressure, the change in specific enthalpy can be approximately represented by the product of a specific heat capacity and a change of temperature. Equation (3.11.2) then becomes:

$$c_{p,f}(t_3 - t_4) = c_p(t_1 - t_6) \tag{3.11.4}$$

where $c_{p,f}$ is the specific heat capacity of the subcooled liquid, and c_p is the specific heat capacity of the superheated vapour, both at constant pressure.

The specific heat capacity of liquid is always greater than the specific heat

Fig. 3.11.5. Temperatures in the heat exchanger. The rise in temperature of the vapour is greater than the drop in temperature of the liquid.

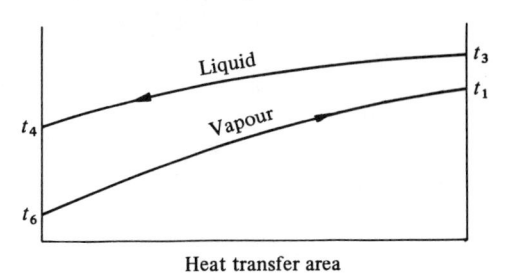

capacity of the superheated vapour. Hence the rise of temperature of the vapour, $(t_1 - t_6)$, is always greater than the drop in temperature of the liquid, $(t_3 - t_4)$. It is possible therefore by making the heat exchanger large enough, to bring the temperature t_1 as near as desired to the temperature t_3; but it is not possible to bring the temperature of the liquid down to $t_6 \approx t_e$, the evaporating temperature. In heat exchanger terminology, the vapour stream is the stream with the lower capacity rate, and it dictates the maximum heat transfer which can be obtained. The conditions in a finite exchanger would be as shown in Fig. 3.11.5. The temperature approach, $(t_4 - t_6)$, at the cold end is necessarily greater than that, $(t_3 - t_1)$, at the warm end.

An interesting possibility of the cycle with heat exchange was recognised by Grindley (1912). Since the temperature of the vapour at 1 can be made nearly equal to $t_3 = t_c$, compression can in principle be carried out isothermally instead of adiabatically. The cycle is shown on temperature and entropy coordinates in Fig. 3.11.6. The specific work of reversible isothermal compression from state 1 to state 2 on the saturation line is less than the specific work of reversible adiabatic compression from state 1 to the pressure p_c. For those refrigerants which show an improvement of the coefficient of performance ε_r with superheating, the Grindley cycle would give a further gain. It is always difficult, however, to achieve

Fig. 3.11.6. The Grindley cycle on the T–s diagram. Heat exchange with liquid raises the temperatue of the vapour from T_6 to $T_1 = T_c$, the condensing temperature. Isothermal compression is shown by 1–2.

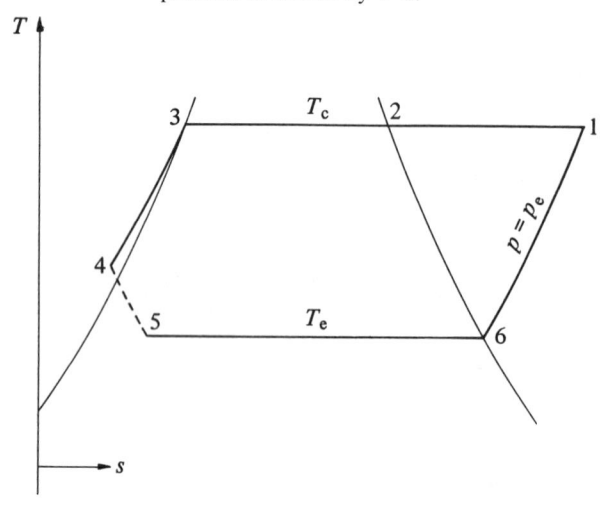

3.11 Superheated vapour

isothermal compression, especially in modern high-speed compressors of either reciprocating or centrifugal type. For these machines the Grindley cycle is therefore merely an interesting academic possibility. In recent years the screw compressor, with injection of cooling oil into the vapour during compression, has been introduced. Test results with these machines show that the delivery temperature is much lower than that for reversible adiabatic compression, but still higher than isothermal. If used with suction to liquid line heat exchange, with substantial superheating, screw compressors could operate on a cycle approaching that of Grindley.

Example 3.11.1

A manufacturer's tables for a certain compressor on R12 give the refrigerating capacity as 9.65 kW and the shaft power as 3.82 kW at the following conditions:

evaporating temperature	$-20\,°C$
condensing temperature	$30\,°C$
temperature of liquid leaving the condenser	$25\,°C$
temperature of vapour entering the compressor	$18\,°C$

Determine the mass flow rate at these conditions and the isentropic efficiency.

Supposing that the compressor operates an evaporator with liquid feed via a thermostatic expansion valve giving a superheat at the outlet of 5 K, and the superheating to 18 °C is done in a heat exchanger, determine the temperature of the liquid reaching the expansion valve and the surface area required in the heat exchanger based on an overall coefficient of heat transfer of 50 W/m² K.

Properties of R12 (Datum: liquid at $0\,°C$, $h_f = 200$ kJ/kg, $s_f = 1$ kJ/kg K exactly):

Saturated liquid

t	p_s	v_f	h_f
°C	bar	dm³/kg	kJ/kg
3	3.402	0.7209	202.78
4	3.512	0.7226	203.71
25	6.516	0.7629	223.65

Superheated vapour. (Saturation temperatures in brackets)

t	$p = 1.509$ bar $(-20\,°C)$		$p = 7.449$ bar $(30\,°C)$	
	h	s	h	s
°C	kJ/kg	kJ/kg K	kJ/kg	kJ/kg K
−15	345.67			
15	363.91	1.6452		
20	367.01	1.6559		
75			396.09	1.6434
80			399.64	1.6536

Using the numbering of Fig. 3.11.3, and interpolating linearly,

h_1(vapour, 1.509 bar, 18 °C) = 365.77 kJ/kg

h_3(liquid, 25 °C) = 223.65 kJ/kg

$h_1 - h_3 = 142.12$ kJ/kg

Hence

$$\dot{m} = \frac{9.65}{142.12}\text{ kg/s} = 0.0679\text{ kg/s}$$

and the specific shaft work is:

$$w = \frac{3.82}{0.0679}\text{ kJ/kg} = 56.26\text{ kJ/kg}$$

s_1 (vapour, 1.509 bar, 18 °C) = 1.6516 kJ/kg K

Interpolating linearly at $p = 7.449$ bar, $s_2 = s_1$ gives:

$t_2 = 79\,°C$

$h_2 = 398.94$ kJ/kg

$(h_2 - h_1)_s = 33.17$ kJ/kg

The isentropic efficiency is then:

$$\eta_{\text{isen}} = \frac{33.17}{56.26} = 0.590$$

h_6(vapour, 1.509 bar, −15 °C) = 345.67 kJ/kg

$h_1 - h_6 = 20.10$ kJ/kg

By (3.11.2) this is also equal to $h_3 - h_4$, hence:

$h_4 = (223.65 - 20.10)$ kJ/kg $= 203.55$ kJ/kg

3.11 Superheated vapour

corresponding to a temperature of:

$$t_4 = 3.8\,°C$$

by interpolation.

The temperatures at each end of the heat exchanger are shown diagrammatically in Fig. 3.11.5. As mentioned earlier, the approach between the streams is closest at the warm end. The temperature profiles must be as shown, plotted against area, provided that the overall coefficient remains constant throughout, because the rate of heat transfer per area and therefore the rate of change of fluid temperature must be lowest at the end where the temperature difference is lowest. Hence the profiles are convex upwards. The logarithmic mean temperature difference is:

$$\frac{18.8 - 7.0}{\ln(18.8/7.0)}\,K = 11.94\,K$$

The heat transfer rate is:

$$(0.0679)(20.10)\,kW = 1.365\,kW$$

and the area required is:

$$\frac{1365}{(50)(11.94)}\,m^2 = 2.29\,m^2$$

In calculating the temperature at 4 the specific enthalpies of the subcooled liquids have been taken as the same as those of saturated liquids at the same temperature. If the subcooling is allowed for by the approximation (3.10.2), which is expected to be valid since the temperature is well below critical, the correction term at 3 would be:

$$\frac{0.7629}{1000}(7.449 - 6.516)10^5\,J/kg = 0.071\,kJ/kg$$

and

$$h_3 = 223.72\,kJ/kg$$

Similarly, the corrections to the specific enthalpies at 3 °C and 4 °C are 0.29 kJ/kg and 0.28 kJ/kg. Taking these into account, the temperature at point 4 becomes 3.6 °C. The error is not large. But the magnitudes of the corrections show that in practice there is not much point in carrying more than one decimal place in the enthalpies unless they are taken into account.

The area required for the heat exchanger in this example shows that the exchanger may be quite a large item of plant and must add to the cost of the

The vapour compression system 228

whole installation. Because the coefficient of heat transfer on the vapour side is much less than that on the liquid side it naturally pays to use a finned surface in the exchanger

If the exchanger were omitted on grounds of cost, the compressor manufacturer's rating would not be obtained as *useful* refrigerating capacity. It would of course always be possible to warm the suction vapour up to 18 °C by not insulating the suction pipe and allowing heat transfer from the atmosphere, but this would not be useful refrigeration: the plant is not intended to refrigerate the ambient air, but a cold space at perhaps −10 °C, and only refrigerating capacity which comes from there can be counted as useful. In order to recover the capacity available in the suction vapour at the lower temperature a heat exchanger is essential.

3.12 Dependence of the volumic refrigerating effect on the suction state

The question which was deferred in the last section will now be considered, i.e. how does superheat alter the volumic refrigerating effect?

Fig. 3.12.1. Diagram of specific enthalpy versus specific volume for ammonia. The specific volume of liquid is so small that the saturated liquid line virtually coincides with the h axis.

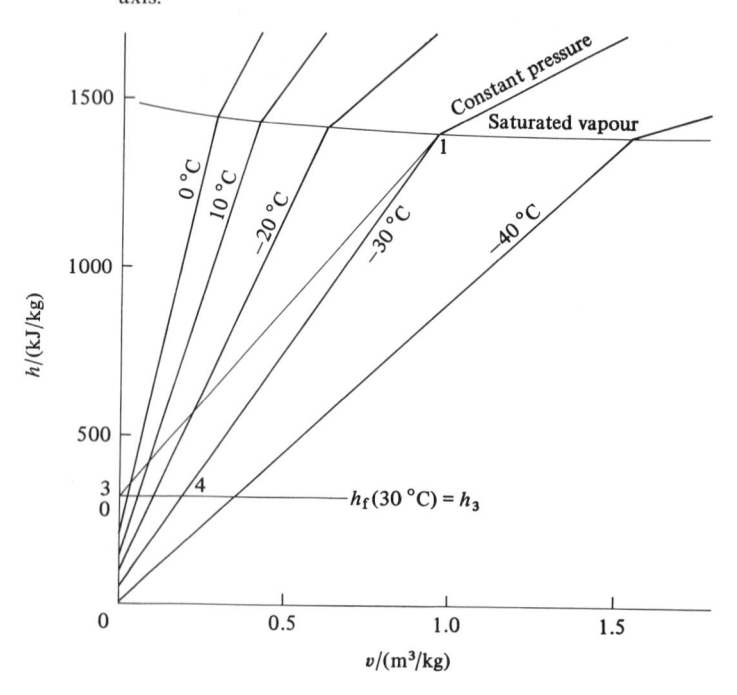

3.12 Volumic refrigerating effect

To make the enquiry complete, it will be supposed that the state of the vapour at the inlet to the compressor, denoted by 1, can vary at constant evaporating pressure from somewhere in the two-phase region, with completely wet compression, to superheated vapour at a temperature equal to the condensing temperature. The temperature, and hence the specific enthalpy, of the liquid leaving the condenser at 3 will be supposed constant. The volumic refrigerating effect considered will be the gross effect, $(h_1 - h_3)/v_1$, assumed to be employed usefully in refrigerating external bodies directly or by using a heat exchanger. The heat exchanger itself need not be considered in the analysis: it is simply a means of recovering, at evaporating temperature, refrigerating effect from the superheated vapour which may be too warm to be useful directly.

The lowest value which the specific enthalpy at state 1 can have is when $h_1 = h_3$, i.e. when the two-phase mixture leaving the expansion valve passes straight to the compressor doing no refrigeration at all. In this trivial (and useless) case, the specific refrigerating effect is zero, and so is the volumic refrigerating effect. Any small increase in h_1 from this condition causes an increase in $(h_1 - h_3)$ and of the ratio $(h_1 - h_3)/v_1$. Thereafter, as state 1 moves towards increasing dryness, $(h_1 - h_3)$ increases, but so does v_1. It is readily shown that for all substances the ratio $(h_1 - h_3)/v_1$ increases with increasing h_1 and v_1, so long as state 1 remains in the two-phase region. This is illustrated in Fig. 3.12.1 on a plot of specific enthalpy against specific volume. The ratio $(h_1 - h_3)/v_1$ is the gradient of the straight line joining point 0 at $h = h_3$, $v = 0$, to point 1. As point 1 moves up the straight line of constant pressure $p = p_e$, the gradient of the straight line 0-1 increases.

When state 1 passes into the superheat region, what happens then depends on how the constant pressure line $p = p_e$ continues there. For a substance like ammonia, for which Fig. 3.12.1 is drawn, the gradient of the line 0-1 diminishes as point 1 moves into the superheat region. Consequently, for ammonia the volumic refrigerating effect diminishes with superheat at the evaporating and condensing temperatures under consideration.

For other refrigerants the variation with superheat is shown in Fig. 3.12.2, at an evaporating temperature of $-15\,°C$ and a condensing temperature of $30\,°C$. The ordinate shows the ratio of the volumic refrigerating effect for superheated vapour to that for saturated vapour at the compressor inlet, state 1. Curves of this type showing the variation of volumic refrigerating effect, and of the isentropic coefficient of performance discussed in the next section, were introduced by Linge (1956).

It may noticed that these curves place the refrigerants, with the exception

The vapour compression system

of carbon dioxide, in an order of behaviour which is the same as their order with respect to temperature rise during isentropic compression. For example, R502 and propane, the highest curves apart from CO_2, have the lowest temperature rises, and ammonia, the lowest curve, has the highest temperature rise.

For ammonia it is clear that superheat is not beneficial to the volumic refrigerating effect, but on the contrary harmful. There is therefore a loss in refrigerating capacity by operating an ammonia plant with substantial superheat, even obtained usefully and of course still more if it is obtained uselessly. In any event the temperature at the end of compression is already high enough for ammonia, even with saturated vapour at inlet, so that high superheats are not used in practice.

This account is incomplete however. It will be seen in Chapter 4 that superheat improves the volumetric efficiency of reciprocating compressors. Consequently, the refrigerating capacity may still increase with superheat if the increase of volumetric efficiency is sufficient to outweigh the decrease in volumic refrigerating effect. This in fact is usually so for the first 10 K of superheat. Consequently, ammonia plant should be operated with some superheat of this order, despite the increased discharge temperature which results.

With R12 and R502 there is an increase of volumic refrigerating effect

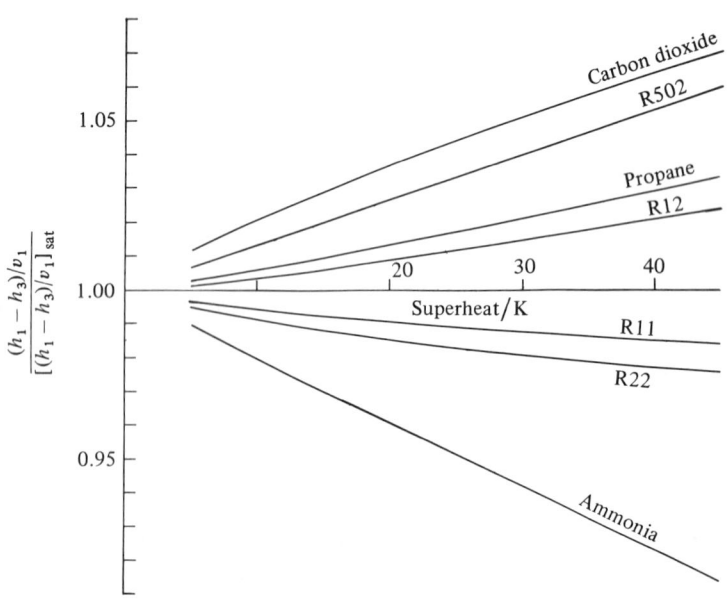

Fig. 3.12.2. The ratio of the volumic refrigerating effect with superheat to the value at saturation for some refrigerants.

3.13 Isentropic coefficient of performance

with superheat which reinforces the increase in the volumetric efficiency. Consequently, the refrigerating capacity of R12 and R502 plants is increased with superheat, and they may be operated with as high suction vapour temperatures as are practicable considering any possible limitations on discharge temperature. Since these refrigerants have a comparatively low rise in temperature during compression, they can usually be operated with as much superheat as can be obtained by the use of a heat exchanger, provided that the pressure ratio is not exceptionally high, and that account is taken of the rise in temperature across the motor when they are used in hermetic compressors.

With R22, the change of volumic refrigerating effect is negative, but there is not a serious reduction with superheat. The increase in volumetric efficiency with superheat is usually dominant therefore. So superheat should be used if practicable, but the rise in temperature of R22 is higher than for R12 and R502 and this places a limitation on the amount of superheat which can be used, especially in hermetic compressors.

R11 shows a reduction of volumic refrigerating effect with superheat, but this is of no practical importance because R11 is used mainly with turbocompressors with nearly saturated suction.

3.13 Dependence of the isentropic coefficient of performance on the suction state

The state at point 1 is again considered to vary from the minimum value when $h_1 = h_3$, up to saturation and then into the superheat region.

Fig. 3.13.1. Sketched diagram of specific enthalpy versus specific entropy. As point 1 moves up the straight line of constant evaporating pressure the isentropic coefficient of performance increases so long as point 2 remains in the two-phase region.

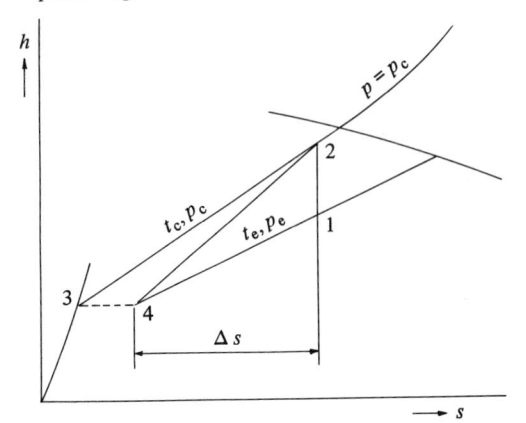

When the specific enthalpy at 1 is the same as at 3, there is zero specific refrigerating effect. But since work is still needed to compress the wet vapour, the coefficient of performance is zero. A small change in h_1 immediately gives some refrigerating effect and raises the coefficient of performance from zero to a small finite value.

The subsequent variation of the coefficient as h_1 increases will be examined, following Ewing (1914), on the diagram of specific enthalpy versus specific entropy, Fig. 3.13.1. The cycle is shown with state 2 at the end of isentropic compression still in the two-phase region. Since only isentropic compression is under consideration here, it will be understood that state 2 refers to the state at the end of isentropic compression, in order to simplify the notation.

The coefficient of performance is:

$$\varepsilon_r = \frac{h_1 - h_3}{h_2 - h_1}$$

$$= \frac{(h_1 - h_3)/\Delta s}{(h_2 - h_1)/\Delta s} \qquad (3.13.1)$$

where:

$$\Delta s = s_1 - s_4 \qquad (3.13.2)$$

Since the isotherm $t = t_e$ is a straight line in the two-phase region, the numerator $(h_1 - h_3)/\Delta s$ remains constant so long as point 1 remains in the two-phase region.

The denominator is:

$$(h_2 - h_1)/\Delta s = (h_2 - h_3)/\Delta s - (h_1 - h_3)/\Delta s$$

the second term of which is constant. The first term is the gradient of the straight line joining point 4 to point 2.

As point 1 moves up the straight line $p = p_e$, and point 2 moves up the straight line $p = p_c$, the gradient of the line 4–2 diminishes, and the denominator of eq. (3.13.1) diminishes. Consequently ε_r continues to increase so long as point 2 remains in the two-phase region. This is true for all substances.

As point 1 moves further up the straight line $p = p_e$, point 2 eventually reaches the saturation line and then moves into the superheat region. Since the gradient of a constant pressure line on h–s coordinates is given by:

$$\left(\frac{\partial h}{\partial s}\right)_p = T \qquad (3.13.3)$$

where T is the absolute thermodynamic temperature, there is no sharp

3.13 Isentropic coefficient of performance

break in the line $p = p_c$ as it crosses the saturation line into the superheat region, although it is then no longer straight. Consequently ε_r continues to increase for a short distance. Eventually, however, as the gradient of the line $p = p_c$ increases, the specific work increases faster than the specific refrigerating effect and the value of ε_r begins to fall. For some refrigerants a maximum value of ε_r is reached with point 1 still in the two-phase region, and thereafter ε_r diminishes.

The condition at this maximum is shown in Fig. 3.13.2, with the straight line 4–2 tangential to the line of constant pressure $p = p_c$. Noting that the gradient of the line 4–2 at this condition is:

$$\frac{h_2 - h_4}{\Delta s} = \left(\frac{\partial h}{\partial s}\right)_p = T_2 \qquad (3.13.4)$$

where T_2 is the absolute thermodynamic temperature at the end of isentropic compression, and that:

$$\frac{h_1 - h_4}{\Delta s} = \left(\frac{\partial h}{\partial s}\right)_p = T_e \qquad (3.13.5)$$

where T_e is the absolute thermodynamic temperature of evaporation, the maximum coefficient of performance is given by:

$$\varepsilon_{r,\max} = \frac{T_e}{T_2 - T_e} \qquad (3.13.6)$$

(Gosney, 1967).

Fig. 3.13.2. Ewing's construction showing a maximum coefficient of performance for some refrigerants when the straight line 4–2 is tangential to the curve of constant condensing pressure in the superheat region.

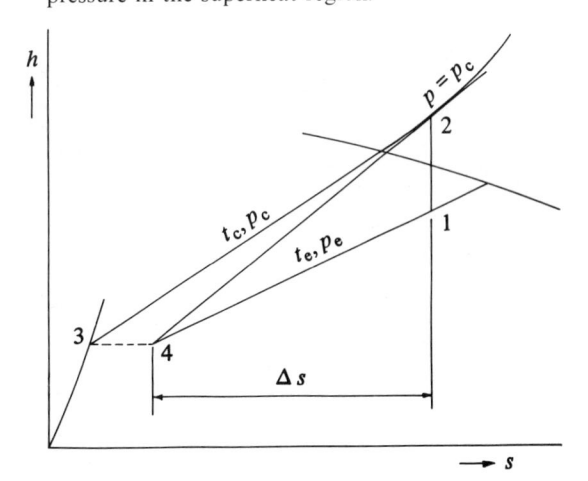

Whether a maximum of the above type occurs whilst the point 1 is still in the two-phase region depends on the properties of the refrigerant. By an extension of the above considerations it may be shown that such a maximum occurs if:

$$\varepsilon_{r,sat} > \frac{T_e}{T_{2,sat} - T_e} \qquad (3.13.7)$$

Table 3.13.1. $t_e = -15\,°C$, $t_c = 30\,°C$

Refrigerant	$\varepsilon_{r,sat}$	$t_{2,sat}$ °C	$\dfrac{T_e}{T_{2,sat} - T_e}$	Maximum?
Ammonia	4.77	99	2.26	Yes
Carbon dioxide	2.72	68	3.11	No
Propane	4.71	37	4.96	No
R11	5.03	44	4.38	Yes
R12	4.70	38	4.87	No
R22	4.66	53	3.80	Yes
R502	4.35	37	4.96	No

Fig. 3.13.3. The ratio of the isentropic coefficient of performance with superheat to the value at saturation for some refrigerants.

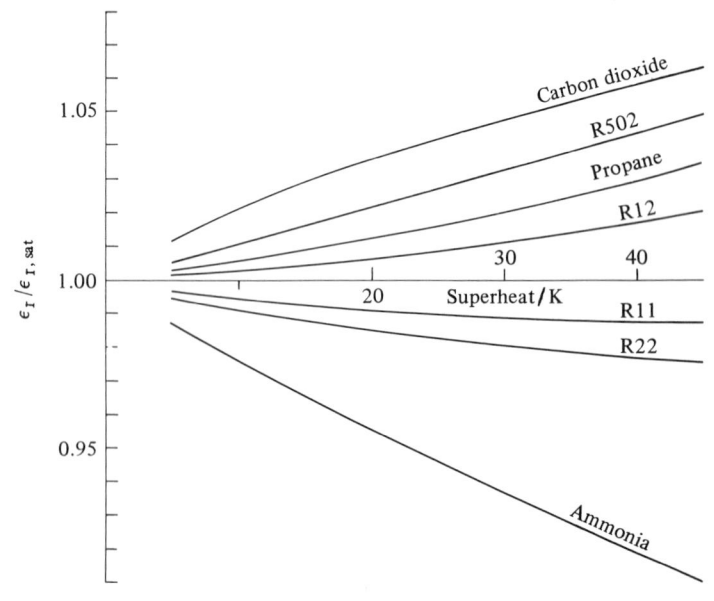

where $\varepsilon_{r,sat}$ is the value of ε_r with point 1 on the saturation line, and $T_{2,sat}$ is the temperature at the end of isentropic compression from point 1 (saturated) to the pressure p_c. Some values for refrigerants at $t_e = -15\,°C$ and $t_c = 30\,°C$ are given in Table 3.13.1. The last column indicates whether a maximum of the type shown in Fig. 3.13.2 occurs.

For those refrigerants, ammonia, R11 and R22, which show a maximum of this type, at the evaporating and condensing temperatures considered, the isentropic coefficient of performance falls from the maximum as point 1 moves up to saturation, and then continues to fall as point 1 moves into the superheat region. For the other refrigerants, which show no maximum, the value of ε_r increases with increasing dryness at point 1, and then continues to increase as point 1 moves into the superheat region. Figure 3.13.3 shows the variation.

It will be noticed that the variation of ε_r with superheat is similar in sense and of the same order as the variation of the volumic refrigeration effect, shown in Fig. 3.12.2. This similarity occurs because the ratio of $(h_1 - h_3)/v_1$ to ε_r is equal to $(h_2 - h_1)/v_1$, the volumic work of isentropic compression, from eq. (3.3.4). It will be shown in Chapter 4 that the volumic work of isentropic compression, which is equal to the mean effective pressure of an ideal compressor, can be expressed as a function of the pressures and is approximately independent of superheat.

It must again be emphasised that the values of ε_r given in Fig. 3.13.3 are based on the gross specific refrigerating effect, i.e. assuming that the superheat can be obtained usefully. This is not normally possible of course without using a heat exchanger between the liquid and suction pipes.

3.14 Operation near the critical point

Because the critical temperature of carbon dioxide is only 31 °C, condensation takes place very near the critical temperature when the temperature of the cooling water is about 20–25 °C. At higher temperatures of the cooling water, condensation is not possible at all when account is taken of the necessary temperature difference between the refrigerant and the water.

The cycle for saturated suction with condensation just below critical temperature is shown in Fig. 3.14.1 on pressure–enthalpy coordinates. This diagram is drawn to the shape corresponding to a linear pressure scale. It is clear that the specific refrigerating effect $(h_1 - h_3)$ is much reduced, for a given evaporating temperature, as the condensing temperature rises and point 3 moves up the saturation line towards the critical point.

When the cooling water temperature is too high to allow condensation, the high-pressure vapour in the 'condenser' is simply reduced in temperature, its final temperature being determined by that of the cooling water and the amount of surface area available for heat transfer. Such a cycle is shown in Fig. 3.14.2. The terminal temperature of the supercritical vapour lies on

Fig. 3.14.1. The pressure-enthalpy diagram for carbon dioxide with a linear pressure scale. Condensation just below critical temperature is shown.

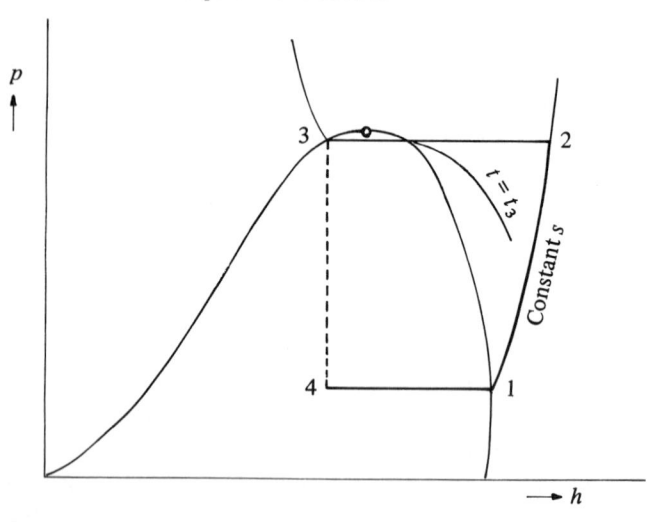

Fig. 3.14.2. The cycle for carbon dioxide when condensation is not possible. By raising the high-side pressure the specific refrigerating effect can be increased from $(h_1 - h_3)$ to $(h_1 - h_{3a})$.

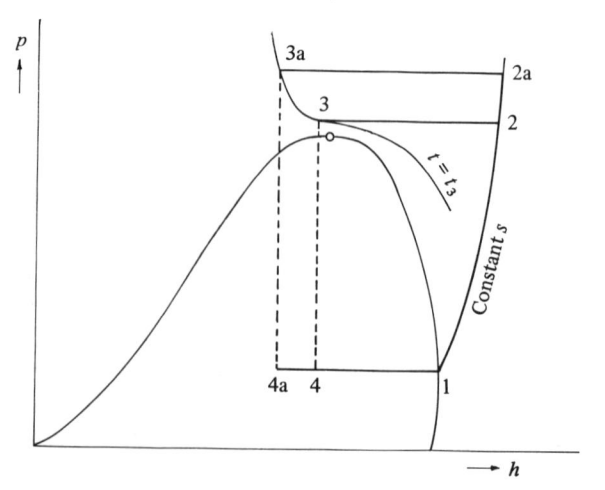

3.14 Operation near critical point

an isotherm t_3, which one may assume to be some degrees above the temperature of the entering cooling water if counterflow is used. The fluid at state 3 entering the expansion valve undergoes a fall in temperature as the pressure falls owing to the Joule–Thomson effect, i.e. because of its deviation from ideal gas behaviour. This drop in temperature takes it down to the saturation line; further reduction of pressure takes the state into the two-phase region and eventually to point 4. The specific refrigerating effect is, as shown in Fig. 3.14.2 drawn approximately to scale for carbon dioxide, rather low: less than half of the specific enthalpy of evaporation at the evaporating temperature.

A peculiarity of operation at this condition, however, is that the refrigerating capacity and the coefficient of performance of the system can both be raised by increasing the pressure in the highside. Thus, maintaining the same terminal temperature t_3, by raising the pressure to give the cycle 1–2a–3a–4a the value of $(h_1 - h_3)$ has been increased to $(h_1 - h_{3a})$. Because of the low gradient of the t_3 isotherm, a useful increase in the specific refrigerating effect can be brought about by a modest rise in pressure. Since conditions at point 1 are not changed, the refrigerating capacity of a machine can be increased by this means.

It should be noted that the curvature of the isotherms in the subcooled and supercritical region renders the usual approximate method for determining specific enthalpy, section 3.10, very inaccurate. It is necessary to use a chart or tables which give values in this region.

Examination of Fig. 3.14.2 shows an increase in $(h_2 - h_1)$ when the delivery pressure is increased, but that owing to the steepness of the line of constant entropy the increase is not so large as the increase in $(h_1 - h_3)$. The result is that the coefficient of performance increases with increasing high-side pressure up to a certain value and then begins to decline as the extra refrigerating effect no longer compensates for the extra work.

Because of the curvature of the isotherms in the subcooled liquid region just below the critical point, this same behaviour of a maximum coefficient of performance is exhibited even when condensation could take place. Referring back to Table 3.9.1, the figures given in parentheses show the improvement which is possible for the conditions of evaporation at $-15\,°C$ and carbon dioxide entering the expansion valve at a temperature of $30\,°C$. The cycle with a high-side pressure above critical pressure gives a slightly better coefficient of performance than the one with condensation.

Several charts have been published which include lines enabling the optimum high-side pressure to be found, e.g. that for carbon dioxide in the book by Macintire (1937). The optimum pressure may also be found by the construction shown in Fig. 3.14.3 on a pressure–enthalpy diagram with a linear pressure scale, (Inokuty, 1928, Tsidzik, Barmin & Veinberg, 1946). The tangent to the isotherm at 3 and the tangent to the line of constant entropy at 2 must intersect at $h = h_1$ when ε_r is a maximum. For, differentiating the equation (3.3.4):

$$\varepsilon_r d(h_2 - h_1) + (h_2 - h_1)d\varepsilon_r = d(h_1 - h_3) \qquad (3.14.1)$$

and putting $d\varepsilon_r = 0$ gives:

$$\varepsilon_{r,\max} = \frac{d(h_1 - h_3)}{d(h_2 - h_1)} \qquad (3.14.2)$$

Fig. 3.14.3. Trial and error construction on linear pressure-enthalpy coordinates to find the high-side pressure which maximises the coefficient of performance. The straight lines from point O at $h = h_1$ are tangential to the curve of constant temperature $t = t_3$ and the curve of constant entropy $s = s_1$, respectively.

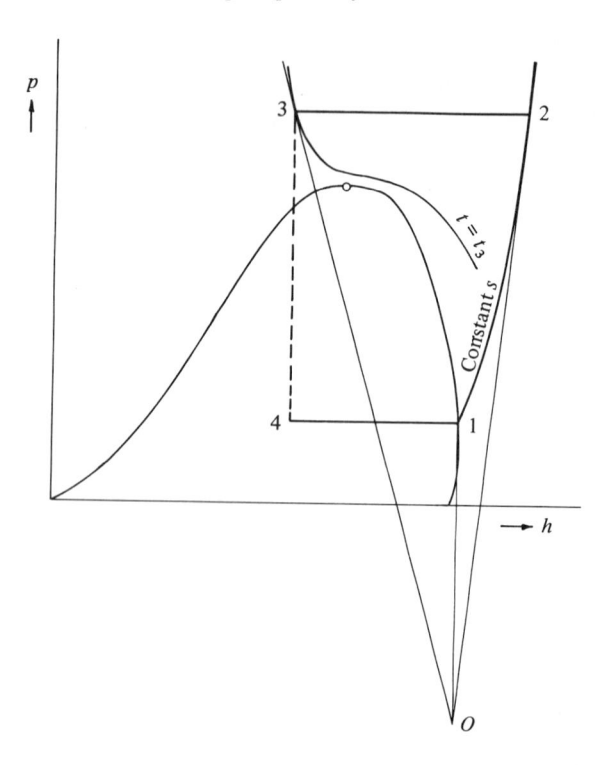

3.14 Operation near critical point

Combining with eq. (3.3.4) gives the condition:

$$\frac{d(h_1 - h_3)}{d(h_2 - h_1)} = \frac{h_1 - h_3}{h_2 - h_1} \qquad (3.14.3)$$

required for a maximum, from which the construction of Fig. 3.14.3 follows.

An approximate value for the optimum pressure can be found by drawing a tangent from a point at evaporating pressure and dryness fraction $x = 0.8$ to the isotherm t_3.

The value of p_2 found thus for maximum ε_r may not maximise the actual coefficient of performance because ε_r is only one of the factors which determine the actual coefficient. Nevertheless, the increase in the specific refrigerating effect gives a real increase in refrigerating capacity. In many situations, such as a ship carrying produce through tropical waters in the days when carbon dioxide was commonly used at sea, the engineer might wish to maximise the refrigerating capacity regardless of power provided that it was within the available motor power.

A question which arises here is how the pressure in the high side can be changed when no condensation takes place. In normal operation, with condensation, the condensing pressure can be raised simply by reducing the flow rate of the cooling water. This raises the temperature of the water leaving, and in order to preserve the mean temperature difference between the fluids the condensing temperature and pressure have to rise.

When condensation is not taking place, reducing the flow rate of the cooling water would simply raise the temperature of the supercritical vapour at the end of cooling, and this would be undesirable because of the increase in h_3 and consequent reduction of refrigerating effect. The pressure in the high side is almost independent of temperature, because it is filled with a single-phase fluid. To raise the pressure it is necessary to increase the mass of fluid in the high side, i.e. to put more carbon dioxide into it. This could be done by temporarily shutting the expansion valve so as to transfer some charge from the evaporator, but this would leave the evaporator short of liquid. The heat transfer coefficient there would fall and the evaporating temperature would fall, thus reducing capacity at the time when it is most wanted. The usual practice is therefore to add some more carbon dioxide to the system from storage cylinders for the duration of the high cooling water temperature. Carbon dioxide systems are not usually provided with liquid receivers on account of the very heavy shells needed to withstand the high pressure.

The above considerations, described mainly for carbon dioxide, have to be taken into account for some other refrigerants when they are used in low temperature cascade systems, treated in Chapter 5.

3.15 The effects of heat transfers and pressure drops between the components

It has been assumed so far that no change of thermodynamic state of the refrigerant takes place between leaving one component and entering the next, though it has been pointed out that the steady-flow energy equations of section 3.2 apply provided that the subscripts are interpreted as the entry and exit states for the component considered. Usually, because of heat transfer and pressure drop, the state of the fluid changes in the connecting pipes, and the effects of this will now be considered.

(a) *Suction pipe.* The most important changes are those in the suction pipe, because they modify the state of the vapour at the entry to the compressor and cause significant changes in performance.

Heat transfer to the vapour in the suction pipe is a useless refrigerating effect, and in general is minimised by insulation of the pipe. The specific volume at the entry to the compressor is increased, and the volume flow rate for a given refrigerating capacity is increased. If this were the only effect, a larger compressor would be needed, or the refrigerating capacity from a given compressor would be reduced. Increase of superheat at the entry to a reciprocating compressor produces an increase of the volumetric efficiency, however, and in practice it is often found that a modest superheat actually improves the refrigerating capacity, even though that superheat is not used usefully, simply because of the effect in improving compressor performance.

When a liquid–suction heat exchanger is used, the vapour leaving the exchanger will already be near to room temperature so that little subsequent heat transfer can take place. Provided that the heat exchanger is placed near to the evaporator, this usually eliminates the need for insulation.

Pressure drop in the suction pipe is always harmful because it increases the specific volume of the vapour and also increases the pressure ratio over the compressor. The volumic refrigerating effect is reduced, and with reciprocating compressors there is a reduction of volumetric efficiency. The specific work of compression is increased and the coefficient of performance reduced accordingly.

The best way to appreciate the effect of a pressure drop is to convert it into a change of saturation temperature. For example, with R22 at a nominal evaporating temperature of $-20\,°C$ say, pressure 2.448 bar, a pressure drop of 0.1 bar corresponds to a change of saturation temperature of about 1.1 K. The compressor then operates as if the evaporating temperature were $-21.1\,°C$ without pressure drop in the suction pipe. The

3.15 Heat transfers and pressure drops

effects of this on the refrigerating capacity and power of the compressor can then be determined from the performance tables or graphs. As a rule a change of about 1 K in saturation temperature is regarded as acceptable.

Pressure drop can be reduced by reducing the velocity of the fluid, i.e. by using a larger pipe size. In practice careful consideration has to be given to vapour velocity with refrigerants R12, R22 and R502 in order to ensure adequate return of the oil from the evaporator to the compressor. A certain minimum velocity is required, of the order of 6 m/s in vertical pipes, and the associated pressure drop has to be accepted. The effects of any suction–liquid heat exchanger and of bends and valves in the pipe on pressure drop must not be forgotten.

Pressure drop is sometimes caused deliberately by a throttling valve as a simple but not very efficient means of reducing the refrigerating capacity which the compressor applies to the evaporator.

The arrangement is shown diagrammatically in Fig. 3.15.1(a) and the cycle is shown on the log p–h diagram in Fig. 3.15.1(b). The reduction in the mass flow rate is caused mainly by the reduced density at the compressor inlet, but there is a further effect caused by the increased pressure ratio across the compressor.

Fig. 3.15.1. (a) Use of a throttling valve in the suction pipe to reduce the refrigerating capacity or to reduce motor power, (b) the cycle on the p–h diagram.

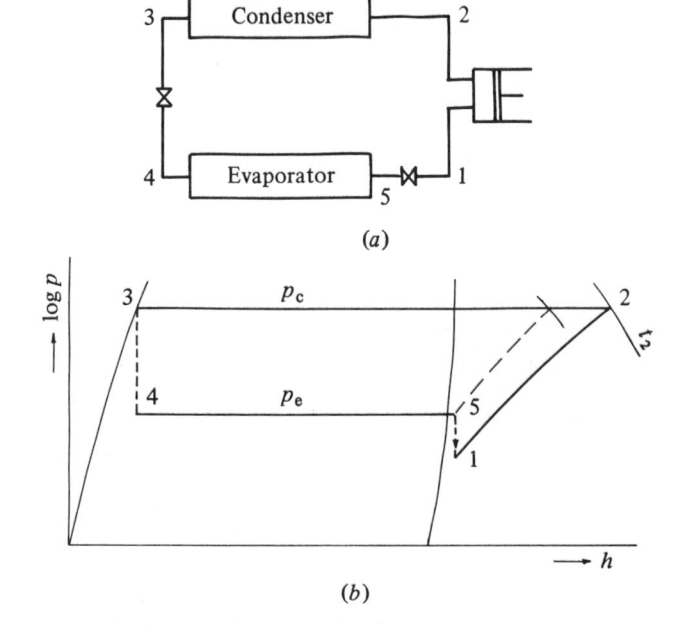

The flow through the throttling valve takes place at constant enthalpy, i.e.:

$$h_5 = h_1$$

and the state at 1 can be located.

Example 3.15.1

Refrigerant 12 boils in an evaporator at a pressure of 1.509 bar (saturation temperature $-20\,°C$) and is superheated by 5 K at the end. It is then throttled to a pressure of 1.237 bar (saturation temperature $-25\,°C$) before compression to the condensing pressure of 6.516 bar (saturation temperature $25\,°C$).

Determine the state after throttling, the approximate reduction in the mass flow rate which is caused by it, and the delivery temperature for isentropic compression.

Properties of R12, superheated vapour

	v	h	s
	dm³/kg	kJ/kg	kJ/kg K
1.509 bar $-15\,°C$	111.48	345.67	1.5784
1.237 bar $-20\,°C$	134.32	343.36	1.5824
1.237 bar $-15\,°C$	137.44	346.31	1.5939
6.516 bar $\;\;\;45\,°C$	29.792	376.00	1.5913
6.516 bar $\;\;\;50\,°C$	30.486	379.53	1.6024

$h_5(1.509\text{ bar}, -15\,°C) = 345.67 \text{ kJ/kg}$

At a pressure of 1.237 bar the specific enthalpy $h_1 = h_5$ is found at a temperature between $-20\,°C$ and $-15\,°C$.

Interpolation gives:

$$t_1 = -16\,°C$$
$$v_1 = 136.76 \text{ dm}^3/\text{kg}$$
$$s_1 = 1.5914 \text{ kJ/kg K}$$

The original specific volume was:

$$v_5 = 111.48 \text{ dm}^3/\text{kg}$$

The reduction in the mass flow rate, neglecting the change in volumetric efficiency, is then:

$$\frac{(1/111.48) - (1/136.76)}{1/111.48} = 0.185 \quad \text{(i.e. } 18.5\%\text{)}$$

3.15 Heat transfers and pressure drops

Interpolating $s_2 = s_1 = 1.5914$ at a pressure of 6.516 bar gives, near enough, $t_2 = 45\,°\mathrm{C}$.

Valves used for suction throttling are usually called evaporator pressure regulating valves when they are constructed so as to respond to the pressure in the evaporator, i.e. to move towards the closed position when the evaporating pressure falls. In this form they are used to prevent the evaporating temperature from falling to too low a value, i.e. to reduce capacity.

They may also be constructed so as to respond to the pressure at the inlet to the compressor, i.e. to move towards the closed position when the suction pressure rises and open again when it falls. In this form they are used to limit the power taken by the compressor driving motor and are called crankcase pressure regulating valves. In both cases the throttling increases the delivery temperature, as will be evident from Fig. 3.15.1(b).

The amount of throttling which can be used when the aim is to reduce capacity depends on how low the suction pressure can be allowed to go and on the delivery temperature. In open-type compressors the point to watch is whether the shaft seal is a balanced one, i.e. whether the pressure forces are balanced so that its operation is independent of the actual values. In hermetic compressors with cooling of the motor by the suction gas, the limitation may be the density of the suction gas for adequate cooling.

(b) *Delivery pipe.* Heat transfer from the warm or hot vapour leaving the compressor to the ambient air causes no change in the coefficient of performance. It simply removes some of the desuperheating load from the condenser.

Pressure drop between the compressor and condenser is harmful because it increases the specific work of compression, and by raising the pressure ratio over the compressor reduces volumetric efficiency.

Water-cooled condensers are usually installed near the compressor, so that the pipe is short. Long delivery pipes are needed when the compressors are in a basement engine room and the condensers are of the air-cooled or evaporative type on the roof.

(c) *Liquid pipe.* Heat transfer from the liquid going from the receiver to the expansion valve provides some useful subcooling and increases the specific refrigerating effect and the refrigerating capacity without any increase in power. It occasionally happens, however, that the heat transfer takes place the other way, to the liquid from the ambient air. This may occur, for example, when a water-cooled condenser is provided with very cold water, so that the condensing temperature is below the temperature of the ambient air in the building through which the liquid pipe passes. There is a danger then that the liquid refrigerant will partially vaporise. This not only reduces the specific refrigerating effect, but also upsets the operation of the expansion valve, which is designed to take liquid alone. As a result the refrigerating capacity of the plant falls rather severely when this happens.

Pressure drop in the liquid pipe has a similar effect in causing partial vaporisation of the liquid refrigerant and a reduction in the refrigerating capacity. The most usual cause of pressure drop in liquid pipes is not, however, fluid friction, but change in height when the liquid flows upwards from a receiver to an expansion valve and evaporator at a higher level. If the liquid is saturated at the bottom, formation of vapour commences immediately as it flows upwards, but in practice a small degree of subcooling is usually present, and vapour would not form until the pressure had fallen to the saturation pressure corresponding to the temperature. The degree of subcooling required to prevent vaporisation within a given lift can be determined using the steady-flow energy equation with the gravity terms and the expression for the specific enthalpy of subcooled liquid, eq. (3.10.2). Suppose subcooled liquid with specific enthalpy h_1 enters the bottom of a pipe at height z_1, and flows adiabatically upwards until it is saturated liquid with specific enthalpy h_2 at a height z_2. Neglecting kinetic energy terms because the liquid is almost incompressible and the pipe section is supposed constant, the steady-flow equation gives:

$$0 = h_2 - h_1 + g(z_2 - z_1) \tag{3.15.1}$$

By eq. (3.10.2), the change of specific enthalpy is approximately:

$$h_1 - h_2 = v_f(p_1 - p_2) \tag{3.15.2}$$

Combining:

$$p_1 - p_2 = \frac{g(z_2 - z_1)}{v_f} \tag{3.15.3}$$

In applying eq. (3.15.3), p_1 is the pressure at the bottom of the pipe (approximately receiver pressure), and p_2 is the saturation pressure corresponding to the temperature of the liquid.

Example 3.15.2
Ammonia liquid leaves a receiver at a pressure of 13.5 bar and a temperature of 34 °C, and flows adiabatically up a vertical pipe. At what height above the inlet will it become saturated liquid?

From the ammonia tables, the saturation temperature at 13.5 bar is 35 °C. The liquid is therefore subcooled by 1 K. The saturation pressure at a

3.15 Heat transfers and pressure drops

temperature of 34 °C is 13.12 bar, and the liquid becomes saturated when the pressure falls to this value.

p_1 = 13.5 bar

p_2 = 13.12 bar

$p_1 - p_2 = 0.38$ bar = 38 000 N/m^2

v_f = 0.0017 m^3/kg $g = 9.81$ m/s^2

Therefore:

$$z_2 - z_1 = \frac{(0.0017)(38\,000)}{9.81} \text{ m} = 6.6 \text{ m}$$

(d) *Expansion valve to evaporator piping.* As a rule the expansion valve is located close to the evaporator, and when possible inside the refrigerated space. Heat transfer to the piping, as to the valve itself, is then included in the refrigerating effect. With liquid chillers the valve has to be mounted outside and some heat transfer from the ambient air may occur, which reduces the refrigerating effect.

Pressure drop in the piping between valve and evaporator is not usually significant when one expansion valve feeds one grid of pipe, but when a distributor is used to apportion liquid between several grids a pressure drop occurs. From the point of view of the coefficient of performance of the cycle, this pressure drop is of no importance, since for a given evaporating temperature the drop in pressure has to take place somewhere, and it makes no difference whether it happens in the valve itself or in a distributor.

In discussing the effects of pressure drop in the evaporator one has to be clear about the conditions being compared. Suppose for instance two systems with identical compressors and components apart from the evaporators, of which one has a low pressure drop relative to the other. This difference could be brought about by different flow arrangements, one having a long pipe, the other having the same length of pipe divided in several parallel paths (commonly called 'circuits'). Supposing the same heat rate to the two evaporators, the conditions of pressure and temperature at the entry to the compressor and expansion valve in the two systems would be identical, because the compressor's refrigerating capacity is the same in each. Apparently pressure drop makes no difference therefore, and the coefficients of performance would be the same. Measurements of pressure in the evaporators would show that the pressure drop between condenser and evaporator outlet takes place mainly in the expansion valve in the system with small drop in the evaporator, whereas in the other system the pressure drop between condenser and evaporator outlet is distributed between the valve and the evaporator.

With a two-phase mixture, however, a change in pressure is necessarily associated with a change in the temperature. Consequently, in the evaporator of higher pressure drop the temperature falls from the outlet of the expansion valve to the outlet of the evaporator. The average temperature difference therefore between the fluids being cooled and the refrigerant is less in the evaporator with the higher pressure drop. To obtain the same heat rate, as assumed above, the fluid being cooled would have to be warmer. Thus, though the two systems compared have the same capacity and coefficient of performance, the one with the low pressure drop in the evaporator is capable of cooling fluid at a lower temperature.

Conversely, if the cooled fluid temperatures and not the heat rates are assumed to be the same, the evaporator with pressure drop would operate at a lower suction pressure than the one without. The capacity of the compressor would be less because of the reduced density of the suction vapour, and the coefficient of performance would be less according to the curves of Fig. 3.8.3. It is in this sense only therefore that pressure drop in the evaporator can be said to be deleterious to the coefficient of performance of a plant.

It is now appreciated that the above account is incomplete in that it neglects the increase of the heat transfer coefficient which is usually associated with higher pressure drop produced by increased loading of a circuit. In fact, an optimum pressure drop can be found at which the increased coefficient of heat transfer compensates for the reduced average temperature difference.

In flooded evaporators with a considerable depth of liquid inside them it is necessary to consider the effect of this on the pressure. The pressure at the bottom is greater than that at the surface where the vapour is leaving. Consequently, it needs a slightly higher temperature to make it boil. For a given design evaporating pressure and temperature, the effect is a reduction of the temperature difference between the refrigerant and the fluid being cooled. It is most pronounced in evaporators for low temperature using refrigerants of the fluorinated hydrocarbon type which have relatively high liquid densities. For example, a depth of 1 m of R22 liquid at a temperature of $-60\,°C$ corresponds to a pressure difference of about 0.14 bar. This in turn corresponds to a change of saturation temperature of about 6.7 K, an important loss of temperature difference between the fluid being cooled and the refrigerant. Under these and similar circumstances special arrangements have to be adopted in the evaporator to ensure that the depth of

3.16 The effect of foreign substances in the circuit

liquid is not excessive. One method in a shell-and-tube evaporator was described in section 2.7. Otherwise the reduced temperature difference has to be allowed for in the design.

The effect of the depth of liquid on the temperature difference is called the 'static head penalty'. It is occasionally of importance in evaporators working at higher temperatures, for example very tall gravity circulation brine chillers working at low temperature difference.

3.16 The effect of foreign substances in the circuit

(*a*) *Non-condensable gases.* A non-condensable gas such as air inside a plant, having leaked in or been left over from assembly operations, collects in the high side of the plant because it cannot pass the liquid seal in the receiver. The mass of gas inside the volume of the high side can be calculated from the ideal gas equation if the specific gas constant or the molar mass of the gas are known.

The partial pressure of the gas is added to that of the refrigerant at the temperature of condensation and so increases the pressure in the condenser and the delivery pressure at the compressor. The specific work of compression is therefore increased and the coefficient of performance is reduced. For reciprocating compressors the increased pressure ratio lowers the volumetric efficiency.

For a given temperature of condensation the temperature of the liquid leaving the condenser is not much changed and the specific refrigerating effect is unaltered. With reference to the actual pressure in the condenser, however, the liquid leaving is subcooled, and an amount of subcooling greater than the normal value is one indication of the presence of a non-condensable gas.

An estimate of the amount of gas is best made when the plant is at standstill. The temperature of the liquid in the receiver is compared with the pressure, as in the following example.

Example 3.16.1

The pressure in the high side of an ammonia plant at rest is 11 bar when the temperature of the ammonia in the receiver is 25 °C. Assuming that the non-condensable gas is air, $R = 287$ J/kg K, and that the volume of the high side of the plant is 1.5 m³, measured from the compressor delivery to the liquid level in the receiver, determine the mass of non-condensable gas in the plant.

From Tables:
saturation pressure at 25 °C is:
10.03 bar
partial pressure of air:
$$= (11.00 - 10.03) \text{ bar} = 0.97 \text{ bar}$$
$$= 97\,000 \text{ N/m}^3$$
mass of air is:
$$m = \frac{pV}{RT} = \frac{(97\,000)(1.5)}{(287)(298)} \text{ kg}$$
$$= 1.7 \text{ kg}$$

If the plant is not at rest this type of calculation becomes inaccurate because the temperature of the liquid–vapour interface, which determines the saturation pressure, is not known accurately. In fact the concentration of the non-condensable gas will be different in the condenser and the receiver. Nevertheless, an estimate has to be made sometimes without shutting down the plant. With a shell-and-tube condenser operating correctly, i.e. without liquid blanketing the bottom tubes, the temperature of the liquid refrigerant leaving may be used as an approximate value of the interface temperature.

(b) *Lubricating oil.* When oil dissolves in a refrigerant it affects its thermodynamic properties. The principal change is a reduction of the vapour pressure by an amount depending on the nature of the oil and the refrigerant and on how much oil dissolves.

So far as ammonia is concerned, it is usually thought that the amount of oil in solution with ammonia liquid is too small to have any significant effect, though there do not seem to be any data available on the point.

With fluorinated hydrocarbons and straight hydrocarbons the amount of oil in solution is much larger. The properties of some refrigerant–oil solutions are given in the literature, e.g. ASHRAE 1980. Jaeger & Löffler (1970) give quite comprehensive data on the vapour pressure, miscibility, density and kinematic viscosity of R10, R12, R13, R13B1 and R22 with five types of oil. In the absence of accurate measured data, an approximate calculation can be made on the assumption that the weak oil–refrigerant solution follows Raoult's law. According to this the partial pressure of the major component in a solution, in this case the refrigerant, is proportional to its mole fraction in the solution. Since the vapour pressure of the oil is negligible, the partial pressure of the refrigerant is the same as the total

3.16 The effect of foreign substances in the circuit

pressure. Raoult's law applies strictly only to ideal solutions in which there is no heat of mixing, but oil and hydrocarbon derivative refrigerants approximate to these.

The mole fraction of a refrigerant R in a solution with oil is defined as:

$$\psi_R = \frac{m_R/M_R}{(m_R/M_R) + (m_{oil}/M_{oil})} \quad (3.16.1)$$

where m_R and m_{oil} are the masses of refrigerant and oil respectively in a sample of the solution, and M_R and M_{oil} are the molar masses (molecular weights). The value of M_{oil} can be obtained from the supplier, but average values are about 300 kg/kmol for naphthenic and 400 kg/kmol for paraffinic oils.

Example 3.16.2

A solution of R12 and an oil having a molar mass of 400 kg/kmol contains 10% by mass of oil. Determine the vapour pressure of the solution at a temperature of $-20\,°C$. The molar mass of R12 is 121 kg/kmol, approximately.

Since masses are proportional to mass fractions, the mole fraction of R12 in the solution is:

$$\psi_R = \frac{0.9/121}{(0.9/121) + (0.1/400)} = 0.967$$

The vapour pressure of pure R12 at $-20\,°C = 1.509$ bar. Hence the vapour pressure of solution at $-20\,°C$:

$$= 0.967(1.509) \text{ bar}$$
$$= 1.459 \text{ bar}$$

If such a solution were present in an evaporator at a temperature of $-20\,°C$ the pressure would therefore be about 3.3% less than if only pure R12 were present. The density of the vapour would be about 3.3% less and the capacity of a compressor having a fixed swept volume rate would also be 3.3% less. Looking at the same thing another way, if the design pressure in the evaporator were 1.509 bar, for a design temperature of $-20\,°C$, the actual temperature for this pressure would be somewhat higher, in this case about $-19\,°C$, thus lowering the available temperature difference between the things being cooled and the refrigerant.

The vapour formed by a refrigerant–oil solution is superheated, because in the above example the pressure of saturated vapour at $-20\,°C$ is 1.509 bar whereas the actual pressure is only 1.459 bar. As refrigerant boils away from a solution the boiling point and the corresponding superheat of the vapour rise. This happens in a dry expansion evaporator, where towards the end there is superheated vapour of the refrigerant in contact with oil now much depleted in refrigerant. The refrigerant dissolved in this oil is eventually carried back to the compressor without producing its refrigerating effect.

The accurate treatment of systems containing oil requires more detailed consideration of the thermodynamic properties of the oil–refrigerant solutions. Properties of oil–R12 solutions have been given by Bambach (1955) and the application of these to cycle calculations by von Cube, Benke & von Sambeck (1958). A method of estimating the thermodynamic properties of mixtures from a minimum number of measurements is provided by Löffler (1970).

3.17 By-passing of hot gas

By-passing of the vapour delivered by the compressor back to the suction side is often used as a means of reducing the refrigerating capacity.

Fig. 3.17.1. (a) By-passing of hot gas from the compressor delivery to suction, (b) the cycle on the p–h diagram, showing throttling from 2 to 6.

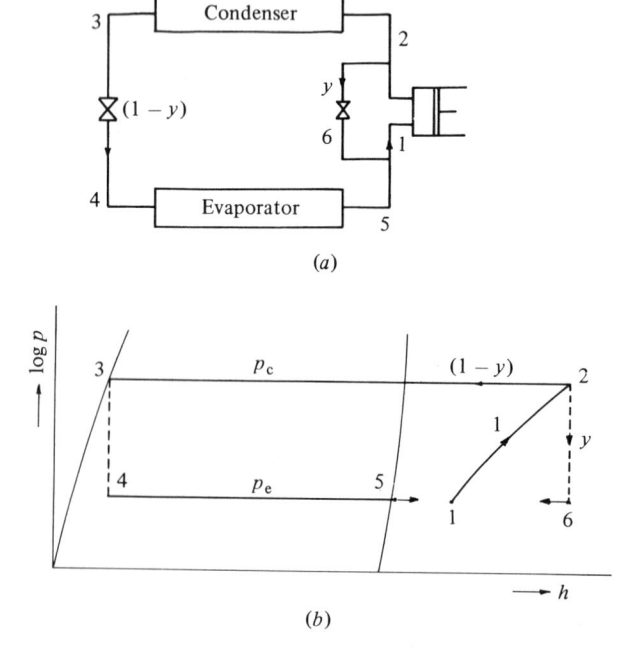

(a)

(b)

3.17 By-passing of hot gas

A valve must of course be interposed to maintain the pressure difference across the compressor, as shown in Fig. 3.17.1(a). The throttling process through the by-pass valve is isenthalpic and:

$$h_2 = h_6 \tag{3.17.1}$$

The temperature of the vapour falls during passage through the valve, but it is still a good deal warmer than the vapour coming from the evaporator with which it mixes to give the state 1 at the beginning of compression.

The steady-flow equation applied to the mixing point gives:

$$\dot{m}_5 h_5 + \dot{m}_6 h_6 = \dot{m}_1 h_1 \tag{3.17.2}$$

Putting $y = \dot{m}_6/\dot{m}_1$ for the fraction of the vapour compressed which is by-passed, $(1 - y) = \dot{m}_5/\dot{m}_1$ for the fraction which goes around the circuit, eq. (3.17.2) with (3.17.1) gives:

$$y = \frac{h_1 - h_5}{h_2 - h_5} \tag{3.17.3}$$

For a given y and given state at 5, this equation defines the state at the beginning of compression when equilibrium is reached, assuming that the law of compression is known. This usually needs a trial and error solution because of the lack of an algebraic relation between h_1 and h_2. Working the other way, however, assuming the state at 2 and isentropic compression, the fraction y can be readily found.

This fraction y is not the whole reduction of capacity. There is a further reduction in the mass flow rate owing to the lower density of the higher temperature vapour entering the compressor. It should be noted that the considerations of section 3.12 are not relevant here. The specific refrigerating effect remains the same, but the mass flow rate through the evaporator is reduced. For reciprocating compressors there may however be an improvement in volumetric efficiency at the higher superheat, as explained in Chapter 4, which offsets the change in density.

Example 3.17.1
A vapour compression system employing R12 and provided with a hot gas by-pass according to Fig. 3.17.1, operates with evaporating and condensing conditions the same as in Example 3.15.1. Supposing the delivery temperature to be the same as in that example, determine the fraction y which is by-passed and the reduction in refrigerating capacity, assuming isentropic compression and neglecting the change in volumetric efficiency.

The properties of R12 given in Example 3.15.1 are used together with the following additional values for superheated vapour.

	v	h	s
	dm³/kg	kJ/kg	kJ/kg K
1.509 bar −10 °C	114.08	348.68	1.5900
1.509 bar −5 °C	116.65	351.70	1.6013

For the delivery state found in Example 3.15.1:

$$s_2 = 1.5914 \, \text{kJ/kg K}$$

and:

$$h_2 = 376.00 \, \text{kJ/kg}$$

For isentropic compression $s_1 = s_2$ and by interpolation at a pressure of 1.509 bar:

$$h_1 = 349.05 \, \text{kJ/kg}$$

From Example 3.15.1,

$$h_5 = 345.67 \, \text{kJ/kg}$$

Hence by eq. (3.17.3):

$$y = \frac{349.05 - 345.67}{376.00 - 345.67} = 0.111$$

The specific volumes are:

$$v_1 = 114.40 \, \text{dm}^3/\text{kg at } s_1 = 1.5914 \, \text{kJ/kg K}$$
$$v_5 = 111.48 \, \text{dm}^3/\text{kg}$$

The total reduction in capacity is then:

$$(0.111)(114.40)/(111.48) = 0.114, \text{ i.e. } 11.4\%$$

Comparing this result with that found in Example 3.15.1 it is seen that for the same delivery temperature hot gas by-pass gives less reduction in capacity than suction throttling. It does not however cause low crankcase pressures with the attendant possibility of drawing air into the system if the pressure falls below atmospheric.

Hot gas by-pass can be applied in the simple form shown in Fig. 3.17.1 when the delivery temperature is not too high. For refrigerants like R12 and R502 at moderate pressure ratios the method can give useful reductions of capacity. When applied to these refrigerants at higher pressure ratios, or to R22 and ammonia, delivery temperatures may become excessive before any useful reduction of capacity is attained. When the plant incorporates a

317 By-passing of hot gas

shell-and-tube condenser, saturated vapour may be by-passed from this, instead of superheated vapour from the delivery pipe, and this reduces the delivery temperature somewhat. In many cases, however, it is still too high and liquid refrigerant can be injected into the suction pipe to desuperheat the vapour before compression, as shown in Fig. 3.17.2(a). The same result can be achieved by injecting the hot gas directly into the low side between the expansion valve and the evaporator (Fig. 3.17.2(b)). In either case some of the refrigerating effect is used to desuperheat the hot gas.

Referring to Fig. 3.17.2(b), let y be the fraction of the compressed vapour which is diverted. Taking a control surface around the evaporator to cut the circuit at points 1, 3 and 5, the refrigerating capacity is given, with $h_5 = h_2$ by:

$$\dot{Q}_e = \dot{m}_1(h_1 - h_3) - y\dot{m}_1(h_2 - h_3) \tag{3.17.4}$$

where $\dot{m}_1$ is the mass flow rate at point 1. The first term on the right of eq. (3.17.4) is the refrigerating capacity without the by-pass; the second term represents a fictitious load, often called a 'dummy load', imposed by the hot gas.

Fig. 3.17.2. (a) Reduction of delivery temperature by the injection of liquid refrigerant into the suction pipe. (b) The same end achieved by injection of hot gas into the evaporator.

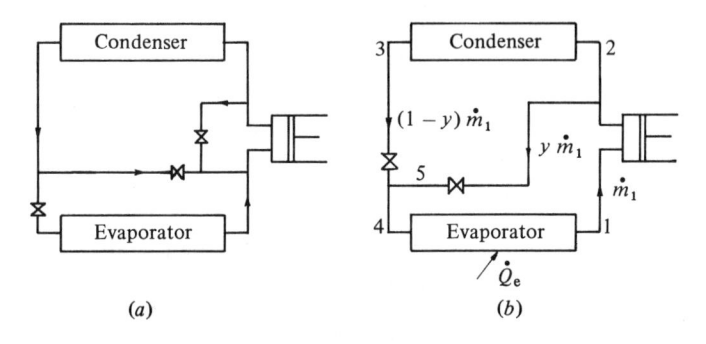

Fig. 3.17.3. A 'run-around' system of zero refrigerating capacity for testing large compressors. An alternative method is to cool the vapour without condensing it. The by-pass valve can then be eliminated.

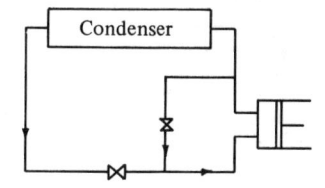

When:

$$y = \frac{h_1 - h_3}{h_2 - h_3} \qquad (3.17.5)$$

the refrigerating capacity is reduced to zero. The evaporator could then be eliminated entirely giving the system shown in Fig. 3.17.3. Such a system has no value as a refrigerator of course, but it is often used to test large compressors because it eliminates the need to provide a large heat transfer rate to the evaporator and greatly reduces the size of condenser needed. The condenser has to be large enough only to condense liquid to desuperheat the hot gas from 2 to 1. This system is sometimes called a 'run around' system.

Another method of reducing capacity by using the hot gas is shown in Fig. 3.17.4, sometimes called 'balance loading', in which a separate heat exchanger is used. Part of the condensation is done by evaporation of refrigerant.

Reduction of refrigerating capacity by by-passing compressed vapour or using refrigerating effect to do the condensation is obviously wasteful since the power taken by the compressor remains the same. Like the use of electrical power for heating, its justification is that it is often convenient, and the various methods are commonly used on small systems where the compressors are not usually provided with devices for capacity reduction.

3.18 The computation of thermodynamic properties

The tables or charts of thermodynamic properties of refrigerants are suitable for performing a small number of calculations, but for the repetition involved in the optimisation of plant designs computers are

Fig. 3.17.4. The use of hot gas from the compressor to provide a 'dummy load' by heat exchange with boiling refrigerant.

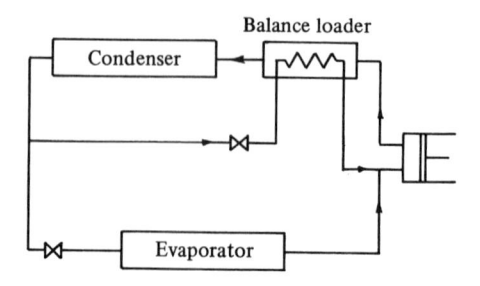

3.18 The computation of thermodynamic properties

commonly used. It is then necessary to incorporate the thermodynamic properties in the program.

In principle only two equations are needed to find all the thermodynamic properties of a substance. These are the equation of state, i.e. the relation between the pressure, temperature and specific volume (or density); and a specific heat capacity at constant volume (or pressure) as a function of temperature at any given constant volume (or pressure). In practice, however, it is impossible to find a single equation of state which applies over the whole of the vapour and liquid regions, and it is necessary to supplement the data by equations representing the saturation vapour pressure and the specific volume (or density) of the saturated liquid as functions of temperature. From these four equations all thermodynamic properties can be derived by using fundamental thermodynamic relations. The methods for doing this are outlined in this section.

Treating the specific entropy s as a function of the absolute temperature T and the specific volume v:

$$ds = \left(\frac{\partial s}{\partial T}\right)_v dT + \left(\frac{\partial s}{\partial v}\right)_T dv \qquad (3.18.1)$$

But:

$$\left(\frac{\partial s}{\partial T}\right)_v = \frac{c_v}{T} \qquad (3.18.2)$$

and:

$$\left(\frac{\partial s}{\partial v}\right)_T = \left(\frac{\partial p}{\partial T}\right)_v \qquad (3.18.3)$$

hence:

$$ds = c_v \frac{dT}{T} + \left(\frac{\partial p}{\partial T}\right)_v dv \qquad (3.18.4)$$

where c_v is the specific heat capacity at constant volume.

Similarly by treating the entropy as a function of absolute temperature and pressure:

$$ds = c_p \frac{dT}{T} - \left(\frac{\partial v}{\partial T}\right)_p dp \qquad (3.18.5)$$

where c_p is the specific heat capacity at constant pressure.

According to the First Law of Thermodynamics:

$$du = T ds - p dv \qquad (3.18.6)$$

where u is the specific internal energy. Substituting for $\mathrm{d}s$ from eq. (3.18.4) gives:

$$\mathrm{d}u = c_v\mathrm{d}T + \left[T\left(\frac{\partial p}{\partial T}\right)_v - p\right]\mathrm{d}v$$

$$= c_v\mathrm{d}T + \frac{T^2}{v}\left(\frac{\partial(pv/T)}{\partial T}\right)_v \mathrm{d}v \qquad (3.18.7)$$

The specific enthalpy h is given by:

$$\mathrm{d}h = T\mathrm{d}s + v\mathrm{d}p \qquad (3.18.8)$$

and on substituting for $\mathrm{d}s$ from eq. (3.18.5):

$$\mathrm{d}h = c_p\mathrm{d}T - \left[T\left(\frac{\partial v}{\partial T}\right)_p - v\right]\mathrm{d}p$$

$$= c_p\mathrm{d}T - \frac{T^2}{p}\left(\frac{\partial(pv/T)}{\partial T}\right)_p \mathrm{d}p \qquad (3.18.9)$$

It will be seen that the right-hand sides of eqs. (3.18.4), (3.18.5), (3.18.7) and (3.18.9) contain (a) a first term which depends on c_v or c_p and thus ultimately on measurements of heat, and (b) a second term which depends on the p–v–T equation only and which is obtainable by differentiation.

To find values of s, u and h, the expressions must be integrated. Because s, u and h are properties, i.e. depend only on the state, the result of the integration must be independent of the path of integration. Consequently the path can be chosen to suit the data available.

Fig. 3.18.1. Path of integration for eq. (3.18.9). The first term is integrated at constant pressure p_0: the second at constant temperature T_1.

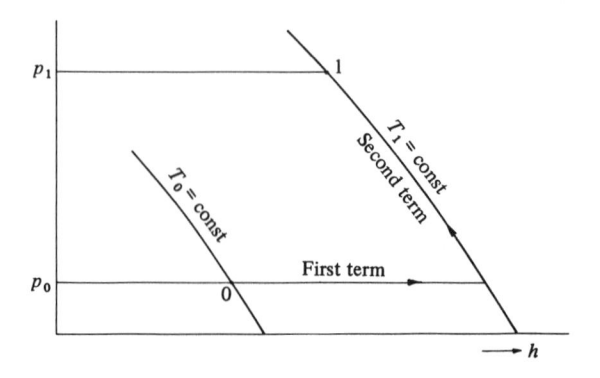

3.18 The computation of thermodynamic properties

Suppose for instance that c_p is known at one pressure p_0 as a function of temperature. The first term in the right-hand side of eq. (3.18.9) can be evaluated for a path at constant pressure p_0 between T_0 and T_1 say. For this path the second term contributes nothing. The second term can be evaluated for a path at constant temperature T_1 from pressure p_0 to pressure p_1. For this path the first term contributes nothing. Thus the specific enthalpy h_1 at p_1, T_1 relative to an initial value h_0 at p_0, T_0 can be found. The path of integration is shown diagrammatically in Fig. 3.18.1. The forms of the data available will now be considered.

Specific heat capacities

The specific heat capacities of vapours depend on both pressure and temperature, but at zero pressure all vapours behave ideally, i.e.

$$\left(\frac{pv}{RT}\right)_{p \to 0} = 1 \qquad (3.18.10)$$

In the ideal gas state the specific heat capacities are functions of temperature only. Consequently, if zero pressure is selected as the path of integration in the above example it is only necessary to have one equation expressing either $c_{v,0}$ or $c_{p,0}$ as a function of temperature. Either may be specified, because at zero pressure:

$$c_{p,0} - c_{v,0} = R \qquad (3.18.11)$$

the specific gas constant. The subscript 0 denotes a value in the ideal gas state, i.e. at zero pressure or density.

The specific heat capacities in the ideal gas state were formerly obtained by measurements at finite pressures and extrapolating to zero pressure. The development of the application of quantum mechanics to molecular models has now reached a stage where it is considered to provide a more accurate determination of $c_{v,0}$ than calorimetric methods, and the values now adopted have been found in this way. The necessary atomic and molecular constants are determined by very accurate spectroscopic methods. The detailed calculations of the values for the fluorinated refrigerants are given by Barho (1965) and for ammonia by Haar (1968).

The graph of $c_{v,0}$ against temperature for an ideal gas shows a number of steps which are connected by curves containing exponential functions. For ease of integration, however, the curve is approximated, over a limited range, by a polynomial in T of the form:

$$c_{v,0} = \sum_{i=0}^{i=n} C_i T^i \qquad (3.18.12)$$

which is easily integrated term by term. The C_i are constants for the gas and as many terms are used as are required to represent $c_{v,0}$ with sufficient accuracy over the range.

Equation of state

At the present time there is no method for establishing an equation of state for a vapour from the basic atomic and molecular constants. It has to be based therefore on measurements of p, v and T.

The earliest equation of state, apart from the ideal gas equation, was that of van der Waals:

$$\left(p + \frac{a}{v^2}\right)(v - b) = RT \qquad (3.18.13)$$

which predicts in a qualitative way the behaviour of real gases but which is of little use for the accurate calculation of properties. More recent forms are given below.

Beattie–Bridgman (BB)

$$p = \frac{RT(1 - \varepsilon)}{v^2}(v + B) - \frac{A}{v^2} \qquad (3.18.14)$$

where $A = A_0(1 - a/v), B = B_0(1 - b/v), \varepsilon = c/vT^3$. $A_0, B_0, a, b,$ and c are constants for a gas.

Benedict–Webb–Rubin (BWR)

$$p = \frac{RT}{v} + \frac{RTB_0 - A_0 - C_0/T^2}{v^2} + \frac{RTb - a}{v^3}$$

$$+ \frac{a\alpha}{v^6} + \frac{c}{T^2 v^3}\left(1 + \frac{\gamma}{v^2}\right)\exp(-\gamma/v^2) \qquad (3.18.15)$$

in which $A_0, B_0, C_0, a, b, c, \alpha,$ and γ are constants for a gas. The BWR equation has been much used for expressing the properties of light hydrocarbons.

Martin–Hou (MH)

$$p = \frac{RT}{v - b} + \sum_{i=2}^{i=5} \frac{A_i + B_i T + C_i \exp(-kT/T_c)}{(v - b)^i} \qquad (3.18.16)$$

in which B_4, C_4 and A_5 are usually taken as zero, the remaining A_i, B_i and C_i being constants for the gas. k may be regarded as another adjustable constant or as a constant for all gases with the value about 4.2. T_c is the absolute critical temperature.

3.18 The computation of thermodynamic properties

The Martin–Hou equation has been used in the Du Pont tabulations of the properties of the fluorinated refrigerants.

The determination of the constants in the BWR and MH equations obviously requires an extensive knowledge of the gas properties. It occasionally happens that one has to perform calculations on gases for which very few properties are known. The Redlich–Kwong equation has been proposed for this as somewhat better than van der Waals:

$$p = \frac{RT}{v-b} - \frac{a}{T^{1/2}v(v+b)} \qquad (3.18.17)$$

where $b = \Omega_b RT_c/p_c$ and $a = \Omega_a R^2 T_c^{2.5}/p_c$, in which T_c and p_c are the critical temperature and pressure, and Ω_b and Ω_a are determined from two known points. When no (p, v, T) points are known, Ω_b is taken as 0.0867 and Ω_a as 0.4278. Great accuracy is naturally not to be expected from the predictions of eq. (3.18.17).

In addition to the above forms of the equation of state, the so-called virial forms have been used by experimenters for many years for expressing their results over a limited range. The virial equations are:

$$\frac{pv}{RT} = 1 + \sum_{i=1}^{i=n} A_i(T)p^i \qquad (3.18.18)$$

and:

$$\frac{pv}{RT} = 1 + \sum_{i=1}^{i=n} B_i(T)\rho^i \qquad (3.18.19)$$

where the density $\rho = 1/v$ is used on the right-hand side for convenience, and the $A_i(T)$ and $B_i(T)$ are functions of temperature only. i takes the values 1, 2, 3...n to as many terms as are required to represent the results with sufficient accuracy. The NEL tables for R12 (see Watson, 1975) are based on eq. (3.18.19) using eight functions B_1 to B_8, which in turn are expressed in terms of 15 constants.

It seems probable that the virial equations require more terms to represent the data with a given accuracy than the BWR or the MH equations. On the other hand they can be programmed somewhat more simply on account of their regular formation.

Vapour pressure
The vapour pressure equation usually takes the form:

$$\ln p = P_1 + \frac{P_2}{T} + P_3 \ln T + P_4 T^n \qquad (3.18.20)$$

where P_1, P_2, etc. and n are constants for the substance. For the mere determination of vapour pressure a simpler equation can be used. But the equation has to be differentiated in order to find the specific enthalpy of evaporation and it is essential to retain all of the terms for accurate work.

Liquid specific volume

Over a limited range the specific volume of the liquid may be assumed to be a linear function of temperature without great error. More accurate formulations often take the form, expressed as density:

$$\rho_\mathrm{f} = \sum_{i=0}^{i=n} D_i(1 - T/T_\mathrm{c})^{i/3} \tag{3.18.21}$$

with as many terms as required.

The use of the above equations for calculating the entropy and enthalpy will now be discussed with particular reference to the use of the virial forms.

Supposing the virial form in pressure, eq. (3.18.18), is available, it is convenient to use eq. (3.18.9) to find h. The first term on the right-hand side is obtained by integrating (3.18.12) and adding RT, i.e.:

$$\int_0^T c_{p,0}\,\mathrm{d}T = \int_0^T (c_{v,0} + R)\,\mathrm{d}T = \sum_{i=0}^{i=n} C_i \frac{T^{i+1}}{i+1} + RT \tag{3.18.22}$$

The second term on the right-hand side is:

$$-T^2 \int_0^p \left(\frac{\partial(pv/T)}{\partial T}\right)_p \frac{\mathrm{d}p}{p} = -RT^2 \int_0^p \sum_{i=1}^{i=n} A'_i(T) p^{i-1}\,\mathrm{d}p$$

$$= -RT^2 \sum_{i=1}^{i=n} A'_i(T) \frac{p^i}{i} \tag{3.18.23}$$

where $A'(T)$ denotes the differential of the function $A(T)$ with respect to T. The integration of $c_{p,0}\,\mathrm{d}T$ was carried out at zero pressure from $T = 0$ to the required final value T. Consequently the enthalpy so obtained is reckoned above a datum of the ideal gas at absolute zero. This can be reduced to the usual datum state of liquid at a specified temperature when the liquid enthalpies have been found.

Using eq. (3.18.5) to determine the entropy, the first term is:

$$\int_0^T c_{p,0} \frac{\mathrm{d}T}{T} = \int_0^T (c_{v,0} + R) \frac{\mathrm{d}T}{T}$$

$$= (C_0 + R)\ln T + \sum_{i=1}^{i=n} C_i \frac{T^i}{i} + \mathrm{const} \tag{3.18.24}$$

3.18 The computation of thermodynamic properties

A constant is used in eq. (3.18.24) instead of a lower limit of integration. This limit cannot be absolute zero because the entropy would go to minus infinity there. The constant can be arbitrarily set to zero since only differences of entropy are needed in calculations. If the datum is subsequently reduced to that of a liquid state it will of course disappear.

The second term in eq. (3.18.5) is:

$$-\left(\frac{\partial v}{\partial T}\right)_p dp = -\frac{1}{p}\left(\frac{\partial (pv)}{\partial T}\right)_p dp$$

$$= -\frac{R}{p}dp - R\sum_{i=1}^{i=n}[TA'_i(T) + A_i(T)]p^{i-1}dp \quad (3.18.25)$$

on using eq. (3.18.18). This then integrates to:

$$-R\ln p - R\sum_{i=1}^{i=n}[TA'_i(T) + A_i(T)]\frac{p^i}{i} + \text{const} \quad (3.18.26)$$

A constant again is used instead of a lower limit of integration for the same reason as before.

When the virial equation in density, eq. (3.18.19), is available it is more convenient to calculate the internal energy and to convert this to enthalpy by:

$$h = u + pv \quad (3.18.27)$$

The first term on the right-hand side of eq. (3.18.7) gives on integration:

$$\int_0^T c_{v,0} dT = \sum_{i=0}^{i=n} C_i \frac{T^{i+1}}{i+1} \quad (3.18.28)$$

on using eq. (3.18.12). The second term gives:

$$T^2 \int_0^v \left(\frac{\partial (pv/T)}{\partial T}\right)_v \frac{dv}{v} = -RT^2 \sum_{i=1}^{i=n} B'_1(T)\frac{\rho^i}{i} \quad (3.18.29)$$

on using eq. (3.18.19) and converting the integration to density instead of specific volume. It will be observed that this has exactly the same form as eq. (3.18.23) with density instead of pressure.

Using the virial equation in density the entropy is most easily found from eq. (3.18.4). The first term on the right gives:

$$\int_0^T c_{v,0}\frac{dT}{T} = C_0 \ln T + \sum_{i=1}^{i=n} C_i \frac{T^i}{i} + \text{const} \quad (3.18.30)$$

The second term on the right gives:

$$\int^T T\left(\frac{\partial p}{\partial T}\right)_v dv = -R\ln\rho - R\sum_{i=1}^{i=n}[TB'_1(T) + B_i(T)]\frac{\rho^i}{i} + \text{const}$$

(3.18.31)

Saturation properties

Values of h_g and s_g for the saturated vapour are found by the above procedure by using the saturation pressure, temperature and specific volume or density. The specific enthalpy of evaporation, h_{fg}, is calculated from Clapeyron's equation:

$$\frac{dp_s}{dT} = \frac{h_{fg}}{T(v_g - v_f)}$$

(3.18.32)

where p_s is the saturation pressure at absolute temperature T and v_g and v_f are the specific volumes of saturated vapour and liquid, respectively. Clapeyron's equation expresses the thermodynamic identity eq. (3.18.3) for a two-phase mixture of a pure liquid and its vapour. The derivative dp_s/dT is found by differentiating the vapour pressure equation (3.18.20). Then:

$$h_f = h_g - h_{fg}$$

(3.18.33)

and

$$s_f = s_g - \frac{h_{fg}}{T}$$

(3.18.34)

When the equation of state is given in the virial form in density it has to be solved for the density corresponding to the desired values of p and T. The usual way of doing this is by iteration, starting with a value of v assuming the ideal gas equation, i.e. $v_{id} = RT/p$. The corresponding density $\rho_{id} = 1/v_{id}$ is then used in the right-hand side of eq. (3.18.9) to obtain a better value of pv/RT and a new v is calculated and substituted back. This process converges very rapidly at low densities. At high densities Newton's method may be quicker, however.

The following example is presented as an illustration of a modern formulation of the basic properties.

Example 3.18.1

The NEL tables for R12 (Watson, 1975) are based on the following formulations.

The specific heat capacity at constant volume in the ideal gas state is

3.18 The computation of thermodynamic properties

given by eq. (3.18.12) with:

$$C_0 = 4.798\,36 \times 10^{-2} \quad \text{kJ/kg K}$$
$$C_1 = 2.381\,54 \times 10^{-3} \quad \text{kJ/kg K}^2$$
$$C_2 = -2.949\,85 \times 10^{-6} \quad \text{kJ/kg K}^3$$
$$C_3 = 1.373\,74 \times 10^{-9} \quad \text{kJ/kg K}^4$$

The equation of state is of the virial form in density, and the functions $B_i(T)$ in eq. (3.18.19) with $n = 8$ have the form:

$$B_i(T) = \sum_{j=0}^{j=3} \frac{A_{ij}}{\theta^j}$$

where $\theta = T/385.15$ K
The coefficients A_{ij} have the values:

$$A_{10} = 4.188\,836\,59 \quad \text{dm}^3/\text{kg}$$
$$A_{11} = -11.485\,800\,3 \quad \text{dm}^3/\text{kg}$$
$$A_{12} = 8.647\,258\,3 \quad \text{dm}^3/\text{kg}$$
$$A_{13} = -3.516\,951\,12 \quad \text{dm}^3/\text{kg}$$
$$A_{20} = -9.660\,655\,02 \quad (\text{dm}^3/\text{kg})^2$$
$$A_{21} = 21.944\,726\,8 \quad (\text{dm}^3/\text{kg})^2$$
$$A_{22} = -10.842\,881\,7 \quad (\text{dm}^3/\text{kg})^2$$
$$A_{30} = 17.992\,098\,9 \quad (\text{dm}^3/\text{kg})^3$$
$$A_{31} = -41.288\,638\,2 \quad (\text{dm}^3/\text{kg})^3$$
$$A_{32} = 23.786\,465\,8 \quad (\text{dm}^3/\text{kg})^3$$
$$A_{40} = 0.048\,526\,637\,2 \quad (\text{dm}^3/\text{kg})^4$$
$$A_{41} = -0.162\,237\,281 \quad (\text{dm}^3/\text{kg})^4$$
$$A_{60} = 4.632\,773\,62 \quad (\text{dm}^3/\text{kg})^6$$
$$A_{61} = -6.168\,879\,11 \quad (\text{dm}^3/\text{kg})^6$$
$$A_{80} = 1.299\,502\,93 \quad (\text{dm}^3/\text{kg})^8$$

The other coefficients A_{23}, A_{33}, A_{42}, etc. are all zero. The saturated vapour pressure is given by:

$$\ln\left(\frac{p_s}{\text{bar}}\right) = 47.492\,146\,03 - \frac{3783.397\,029}{(T/\text{K})} - 5.852\,561\,143 \ln(T/\text{K})$$
$$+ 1.568\,931\,884 \times 10^{-8}(T/\text{K})^3$$

The liquid density is given by an equation of the form (3.18.21) with $n = 6$. The coefficients will not be given here. A value of v_f which is required in the example is:

$$v_f(-40\,°C) = 0.659\,6\ dm^3/kg$$

The specific gas constant for R12 is:

$$R = 0.687\,561\ bar\ dm^3/kg\ K$$

Determine the specific enthalpy of saturated vapour at 30 °C above a datum of saturated liquid at $-40°C$. The outline of the solution will be given with the principal intermediate quantities.

The specific internal energies of the ideal gas are obtained using eq. (3.18.28):

$$64.469\ kJ/kg\ at\ -40\,°C \quad T = 233.15\ K$$
$$99.485\ kJ/kg\ at\ \ \ 30\,°C \quad T = 303.15\ K$$

The saturated vapour pressures are:

$$p_s(-40\,°C) = 0.642\,346\ bar$$
$$p_s(\ \ 30\,°C) = 7.434\,180\ bar$$

To find the density at $-40°C$ of saturated vapour, the first trial value is:

$$\rho_{id} = [0.642\,346/(0.687561)(233.15)]\ kg/dm^3$$
$$= 0.004\,007\ kg/dm^3$$

Using this value in the right-hand side of the equation of state gives after two iterations:

$$\rho_g(-40\,°C) = 0.004\,127\ kg/dm^3$$

Similarly the density at 30 °C is found after six iterations:

$$\rho_g(30\,°C) = 0.042\,077\ kg/dm^3$$

The pressure correction terms in the internal energy expression, eq. (3.18.29), are found to be:

$$-1.283\ kJ/kg\ at\ 233.15\ K, \quad -40\,°C.$$
$$-7.410\ kJ/kg\ at\ 303.15\ K, \quad 30\,°C.$$

The pv terms are:

$$p_s v_g(-40\,°C) = 15.564\ kJ/kg$$
$$p_s v_g(\ \ 30\,°C) = 17.668\ kJ/kg$$

3.18 The computation of thermodynamic properties

Hence:
$$h_g(-40\,°C) = (64.469 - 1.283 + 15.564)\,\text{kJ/kg} = 78.750\,\text{kJ/kg}$$
$$h_g(30\,°C) = (99.485 - 7.410 + 17.668)\,\text{kJ/kg} = 109.743\,\text{kJ/kg}$$

These are relative to a datum of ideal gas at 0 K. By differentiating the vapour pressure equation:

$$dp_s/dT\,(-40\,°C) = 0.030\,226\,7\,\text{bar/K}$$

and using this in Clapeyron's equation with values of

$$T = 233.15\,\text{K}, \quad v_g = (1/0.042\,077)\,\text{dm}^3/\text{kg},$$
$$v_f = 0.659\,6\,\text{dm}^3/\text{kg}$$

gives:
$$h_{fg}(-40\,°C) = 170.290\,\text{kJ/kg}$$
Hence:
$$h_f(-40\,°C) = (78.750 - 170.290)\,\text{kJ/kg} = -91.540\,\text{kJ/kg}$$

above a datum of ideal gas at 0 K. Taking $h_f = 0$ at $-40\,°C$ as datum:

$$h_g(30\,°C) = (109.743 + 91.540)\,\text{kJ/kg} = 201.283\,\text{kJ/kg}$$

which is the value, rounded to two places, given in the NEL tables.

The corresponding value according to the Du Pont tables is 199.618 kJ/kg. Differences of this order are to be expected between tables based on different formulations of the basic data.

There is no need in an actual program to reduce the enthalpies to a datum of liquid at $-40\,°C$ or some other temperature of course unless, in order to save computing time, some data is inserted numerically from the tables.

It appears from the above example that the virial equation in density is not so convenient as the virial equation in pressure because it has to be solved for density at a given T and p. Unfortunately most of the modern formulations are in terms of density or specific volume.

In performing calculations of cycle quantities it is necessary to solve the equation for entropy at a given s and p, e.g. at the end of isentropic compression. This involves two approximative procedures. Selecting a trial temperature, the equation of state is solved by iteration for the density at a given pressure. This temperature and density are then used in the entropy expression to find s. According as the value so found is less or greater than the required known value, the trial temperature is increased or decreased until a fit is obtained. Usually once two neighbouring values are found the final state is fixed by a linear interpolation.

Much of the calculation of cycle quantities can be eliminated if equations are known giving h and s for saturated and superheated vapour directly as functions of pressure and temperature. The properties of ammonia in the NBS Circular 142 (1923) are given in this form. Other equations for the fluorinated refrigerants have been developed, e.g. for R12 and R22 by Gatecliff & Lady (1971).

For approximate computation of cycle quantities the method of section 3.6 is often used.

Although it was stated earlier that two equations are needed to find all of the thermodynamic properties of a substance, it is possible to combine the information in these into one function, called a characteristic function. An example is the Gibbs function or free enthalpy, g, defined as:

$$g = h - Ts \tag{3.18.35}$$

Supposing that this is given as a function of temperature and pressure, $g(T,p)$, its differential is:

$$\mathrm{d}g = \left(\frac{\partial g}{\partial T}\right)_p \mathrm{d}T + \left(\frac{\partial g}{\partial p}\right)_T \mathrm{d}p \tag{3.18.36}$$

But:

$$\mathrm{d}g = \mathrm{d}h - T\mathrm{d}s - s\mathrm{d}T$$
$$= -s\,\mathrm{d}T + v\mathrm{d}p \tag{3.18.37}$$

by eq. (3.18.8). Comparing coefficients in eqs. (3.18.36) and (3.18.37):

$$s = -\left(\frac{\partial g}{\partial T}\right)_p \tag{3.18.38}$$

$$v = \left(\frac{\partial g}{\partial p}\right)_T \tag{3.18.39}$$

Thus having selected p and T say, s and v can be found by differentiation, and hence h from eq. (3.18.35). Another characteristic function is the Helmholtz function or free energy:

$$f = u - Ts \tag{3.18.40}$$

given as a function of T and v.

Modern formulations of the properties of steam are given as $g(T,p)$ or $f(T,v)$, (e.g. Schmidt, 1969), but more than one equation is required to cover the whole region of interest. Few equations are available at present for refrigerants, but a formulation for ammonia is given by Ahrendts, J., and Baehr, H. D. in *V. D. I. Forschungsheft* 596, 1979.

4

Compression processes

4.1 Introduction

The principal types of compressors used in refrigeration have been described in Chapter 2. In this chapter the working of these machines is examined in detail. Reciprocating compressors are treated first because they are the most commonly encountered machines, and because they provide an easier introduction to the principles of compression. Much of the treatment of reciprocating compressors is applicable to positive displacement compressors of the rotary type, but the importance of one of these, the screw compressor, justifies a section devoted to it alone.

Later in the chapter the principal relations for steady-flow compressors are developed, and finally a study is made of the centrifugal compressor, the only steady-flow machine to be used in refrigeration in large numbers.

To begin, the ideal reciprocating compressor is defined and its performance is examined. The differences between the behaviour of a real machine and the ideal machine are then described and discussed.

4.2 The ideal reciprocating compressor

An ideal compressor is defined as one having no clearance volume remaining unswept at the end of delivery, and in which the internal operations, shown in Fig. 4.2.1 as a plot of pressure inside the cylinder against the volume of vapour enclosed, are specified as follows:

Induction, $D-A$, takes place adiabatically at a pressure $p_A = p_1$, the constant pressure in the suction pipe.

Compression, $A-B$, takes place reversibly and adiabatically, i.e. at constant entropy.

Delivery, $B-C$, takes place adiabatically at a pressure $p_B = p_2$, the constant pressure in the delivery pipe.

The volume of vapour induced by the ideal compressor is denoted V_A, and it is equal to the swept volume V_{sw}. The state of the vapour at A is the same as the state of the vapour in the suction pipe, denoted by subscript 1. In particular the specific volume v_A at A is the same as that, v_1, at 1. The mass of vapour inside the cylinder at A is therefore:

$$m_A = V_{sw}/v_1 \tag{4.2.1}$$

This is also the mass of vapour delivered since no vapour is left inside the cylinder at C. The state of the vapour delivered to the delivery pipe is the same as that of the vapour at B.

Although the flow through the compressor is not steady, it is convenient to introduce a mass flow rate $\dot{m}$ and a swept volume rate $\dot{V}_{sw}$. The swept volume rate is defined as:

$$\dot{V}_{sw} = NCV_{sw} \tag{4.2.2}$$

where N is the rotational speed of the machine, and C is the number of cylinders. In rotary compressors C is the number of intervane or interlobe spaces generated per revolution of the shaft to which the speed N applies. The mass flow rate is then given by:

$$\dot{m} = \dot{V}_{sw}/v_1 \tag{4.2.3}$$

The work done on the vapour in compressing it from A to B is given for a reversible adiabatic process by:

$$W_{A-B} = -\int_{V_A}^{V_B} p \, dV)_s \tag{4.2.4}$$

Fig. 4.2.1. An indicator diagram for an ideal reciprocating compressor.

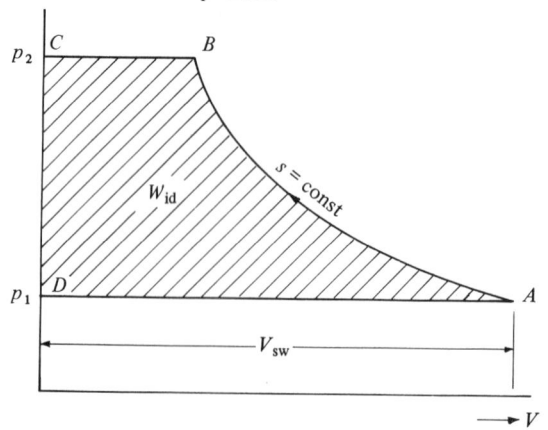

4.2 The ideal reciprocating compressor

where the symbol $)_s$ indicates integration at constant entropy. This is not the whole of the work which the piston has to do. During delivery the piston has to do work $p_2 V_B$ in expelling the vapour from the cylinder, whilst during induction the vapour entering does work $p_1 V_A$ on the piston. The net work which the piston has to do for the whole process is therefore:

$$W_{id} = -\int_{V_A}^{V_B} p\, dV)_s + p_2 V_B - p_1 V_A$$

$$= \int_{p_1}^{p_2} V\, dp)_s \quad (4.2\,5)$$

where the subscript id indicates that it is the work done in the ideal compressor. In Fig. 4.2.1, W_{id} is represented by the area shown hatched.

The specific work, i.e. work divided by the mass of vapour compressed, denoted by lower case w, is given by:

$$w_{id} = \int_{p_1}^{p_2} v\, dp)_s \quad (4.2.6)$$

and the power of the ideal compressor of swept volume rate $\dot{V}_{sw}$ is:

$$\dot{W}_{id} = \dot{m} w_{id} = \frac{\dot{V}_{sw}}{v_1} \int_{p_1}^{p_2} v\, dp)_s \quad (4.2.7)$$

The mean effective pressure is defined as the power divided by the swept volume rate, and is denoted by w_v. From eq. (4.2.7):

$$w_{v,id} = \frac{\dot{W}_{id}}{\dot{V}_{sw}} = \frac{1}{v_1} \int_{p_1}^{p_2} v\, dp)_s \quad (4.2.8)$$

Integration of eqs. (4.2.6) to (4.2.8) requires a knowledge of the relation between p and v during compression at constant entropy. Such a relation can be found from the vapour tables, by following a constant specific entropy, and by interpolation determining values of p and v. This tedious procedure can be eliminated, however, by using the thermodynamic identity:

$$dh = T\, ds + v\, dp \quad (4.2.9)$$

and noting that at constant entropy it becomes:

$$dh)_s = v\, dp)_s \tag{4.2.10}$$

On integration:

$$(h_2 - h_1)_s = \int_{p_1}^{p_2} v\, dp)_s \tag{4.2.11}$$

The expressions for specific work, power and mean effective pressure of the ideal compressor, eqs. (4.2.6) to (4.2.8) respectively, then become:

$$w_{id} = (h_2 - h_1)_s \tag{4.2.12}$$

$$\dot{W}_{id} = \dot{m}(h_2 - h_1)_s = \frac{\dot{V}_{sw}}{v_1}(h_2 - h_1)_s \tag{4.2.13}$$

and:

$$w_{v,id} = (h_2 - h_1)_s / v_1 \tag{4.2.14}$$

where $)_s$ denotes that the states 1 and 2 are taken at the same entropy, and 1 and 2 denote states in the suction and delivery pipes, respectively.

Equation (4.2.12) will be recognised as the quantity which was called the specific work of isentropic compression in section 3.3, and eq. (4.2.13) is the same as eq. (3.3.5). The subscript c used in Chapter 3 to denote compressor is omitted in this chapter since it is understood that only the compressor work is under consideration.

The mean effective pressure, eq. (4.2.14), is the same as the quantity called the volumic work of isentropic compression in Chapter 3.

The variations of $(h_2 - h_1)_s$ and of $(h_2 - h_1)_s/v_1$ have been discussed in Chapter 3 and illustrated in Figs. 3.8.1 and 3.8.2 and in Table 3.9.1.

Although eq. (4.2.14) is the direct way of calculating the mean effective pressure of the ideal compressor, it is often useful to have an algebraic expression for it which can be manipulated for various purposes. An approximation which turns out to be useful is that the relation between pressure and specific volume during compression can be represented by:

$$p v^k = \text{const} \tag{4.2.15}$$

where k is called the index of isentropic compression, and is approximately constant during the process. The integral form of $w_{v,id}$, eq. (4.2.8), may then be integrated to give:

$$w_{v,id} = p_1 \frac{k}{k-1}\left[\left(\frac{p_2}{p_1}\right)^{(k-1)/k} - 1\right] \tag{4.2.16}$$

4.2 The ideal reciprocating compressor

Except for ideal gases with constant specific heat capacities, for which:

$$k = c_p/c_v = \gamma \qquad (4.2.17)$$

where c_p and c_v are the specific heat capacities at constant pressure and constant volume, respectively, no real vapour exhibits a constant value of k, either along one line of constant entropy or between one line of constant entropy and the next. The usefulness of eq. (4.2.16) therefore depends entirely on how closely it can predict $w_{v,id}$ in practice. The following example illustrates the point for ammonia.

Example 4.2.1

Ammonia is compressed in an ideal compressor from the state of saturated vapour at 2.908 bar (saturation temperature $-10\,°C$) to a pressure of 15.54 bar (saturation temperature $40\,°C$). Determine the mean effective pressure by eq. (4.2.14) and by eq. (4.2.16).

From the ammonia tables:

$$p_1 = 2.908 \text{ bar} \qquad v_1 = 0.4185 \text{ m}^3/\text{kg} \qquad h_1 = 1431.6 \text{ kJ/kg}$$
$$s_1 = 5.4717 \text{ kJ/kg K}.$$

By linear interpolation at a pressure $p_2 = 15.54$ bar, for the specific entropy $s_2 = s_1 = 5.4717$ kJ/kg K:

$$v_2 = 0.1132 \text{ m}^3/\text{kg} \qquad h_2 = 1679.2 \text{ kJ/kg}.$$

From eq. (4.2.14):

$$w_{v,id} = \frac{(1679.2 - 1431.6)}{0.4185} = 591.6 \text{ kJ/m}^3 = 5.916 \text{ bar}$$

To use eq. (4.2.16) a value for k must be found. It is assumed for the moment that if k is chosen to fit the initial and final volumes, it will fit the curve in between. Then from eq. (4.2.15),

$$\left(\frac{15.54}{2.908}\right) = \left(\frac{0.4185}{0.1132}\right)^k \qquad k = 1.282$$

Equation (4.2.16) then gives:

$$w_{v,id} = (2.908 \text{ bar})\frac{1.282}{0.282}\left[\left(\frac{15.54}{2.908}\right)^{0.282/1.282} - 1\right]$$

$$= 5.893 \text{ bar}$$

The difference between the two results is about 0.4%.

The principal reason for the (small) difference between the two results in example 4.2.1 is that the curve $pv^k = \text{const}$ does not exactly fit the curve

Compression processes 272

Fig. 4.2.2. Values of the index k in eq. (4.2.15) for ammonia to fit the initial and final states of isentropic compression from pressure p_1 corresponding to an evaporating temperature t_e, to pressure p_2 corresponding to a condensing temperature t_c. The initial state of the vapour is saturated.

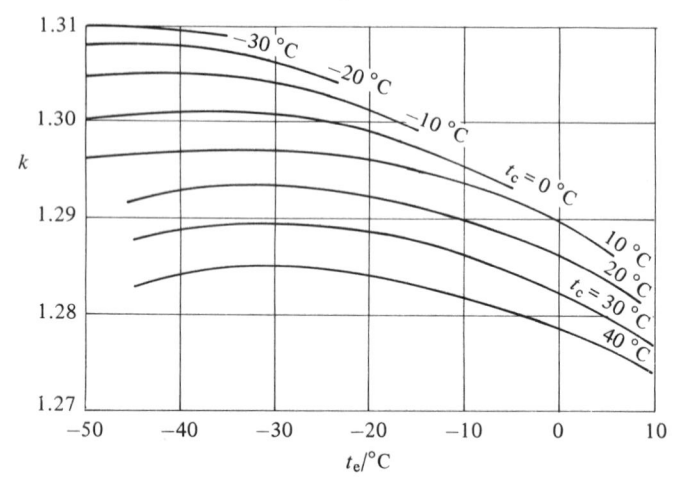

Fig. 4.2.3. Values of the index k for R12: lower curves for initially saturated vapour, upper curves for an initial temperature of 15 °C.

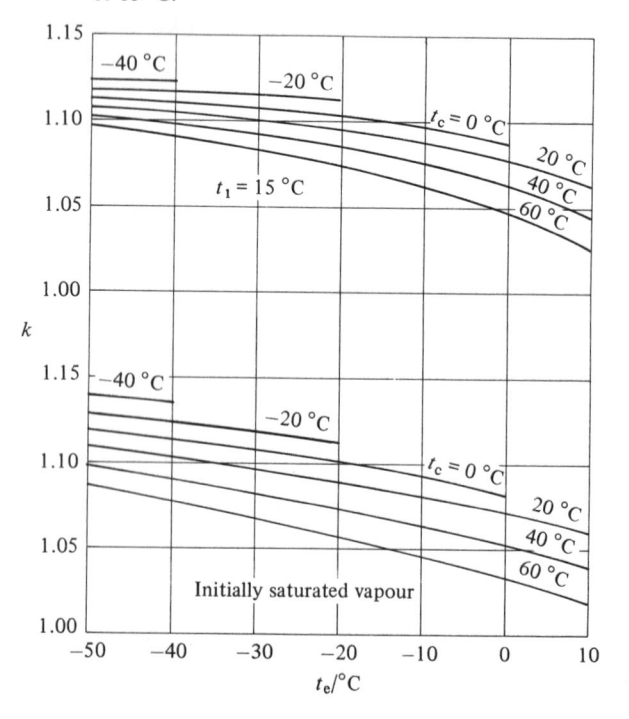

4.2 The ideal reciprocating compressor

$s = $ const when plotted on p–v coordinates, even though the value of k is made to fit the initial and final points. Another reason is that the data given in the thermodynamic tables are not exactly consistent. This applies particularly to the older tables for which the enthalpies and entropies were determined from calorimetric experiments at finite pressure. Most modern tables have better consistency in the superheat region because the entropies and enthalpies are calculated from the values in the ideal gas state by the use of the p–v–T relation.

For most practical purposes, however, the accuracy of eq. (4.2.16) is sufficient. If a value of k had to be determined every time it would be of no great advantage. Fortunately k does not vary much. For approximate purposes it is often taken as 1.3 for ammonia. Some values for various refrigerants are given in Figs. 4.2.2 to 4.2.5.

The integration of $v\mathrm{d}p$ in the form given in eq. (4.2.16) fails in the special case when $k = 1$. Going back to eq. (4.2.8) and using:

$$pv = \mathrm{const} \tag{4.2.18}$$

gives:

$$w_{v,\mathrm{id}} = p_1 \ln(p_2/p_1) \tag{4.2.19}$$

Fig. 4.2.4. Values of the index k for R22: lower curves for initially saturated vapour, upper curves for an initial temperature of 15 °C.

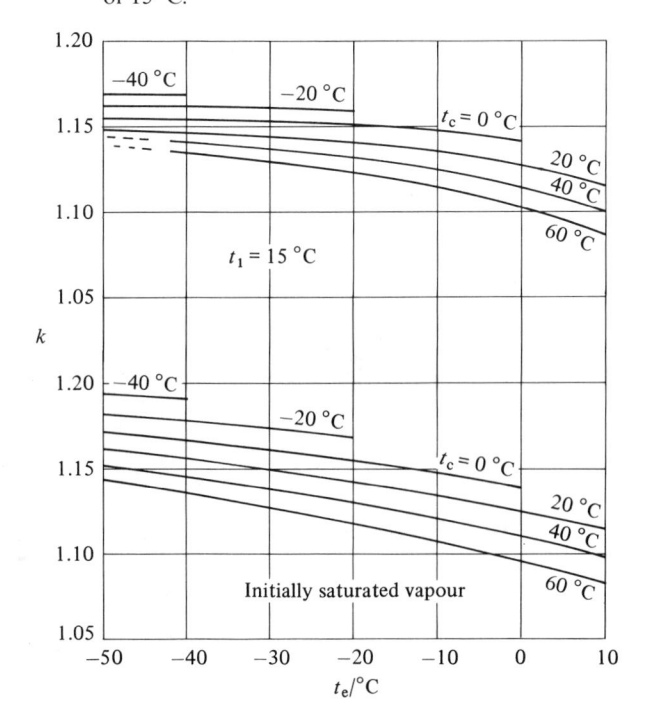

When the mean effective pressure has been determined, the power for a known swept volume rate is found from eq. (4.2.8), as in the following example.

Example 4.2.2
The swept volume rate of an ideal compressor is 0.1 m³/s. Determine the power for the conditions of example 4.2.1.
From example 4.2.1:

$$w_{v,id} = 591.6 \text{ kJ/m}^3$$

Therefore:

$$\dot{W}_{id} = w_{v,id} \dot{V}_{sw} = (591.6)(0.1) \text{ kJ/s} = 59.16 \text{ kW}$$

The same result for example 4.2.2 could have been obtained by the

Fig. 4.2.5. Values of the index k for R502: lower curves for initially saturated vapour, upper curves for an initial temperature of 15 °C.

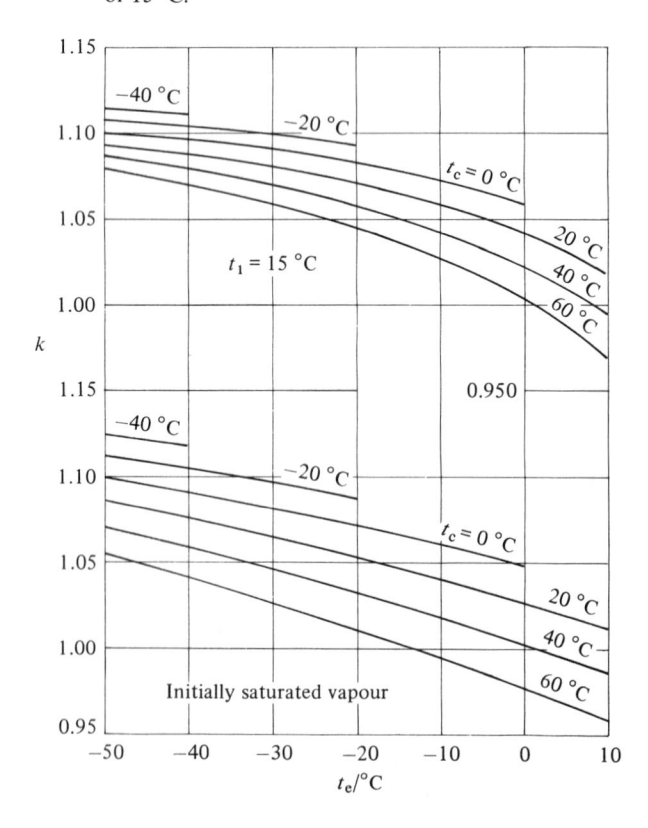

4.2 The ideal reciprocating compressor

methods of Chapter 3, via the specific work of isentropic compression and a mass flow rate. In this chapter, however, the power will be calculated in all cases from the swept volume rate, because this procedure accords better with the requirements in practice. The question of power usually arises in the form of how big a motor is needed to drive a given compressor, not to provide a given mass flow rate. In the simple example above the results are the same, but when real compressors are considered they are not. In Chapter 3 the emphasis was largely on specific work because ideal coefficient of performance was under consideration; in this chapter the emphasis is on mean effective pressure because the aim is to provide a motor big enough to drive a given compressor.

The function of k and the pressure ratio which occurs in eq. (4.2.16) recurs frequently, and it will be convenient to use a symbol for it:

$$Y = \frac{k}{k-1}\left[\left(\frac{p_2}{p_1}\right)^{(k-1)/k} - 1\right] \qquad (4.2.20)$$

Values of Y as a function of pressure ratio for various values of k are given in Fig. 4.2.6. The specific work of isentropic compression may then be written:

$$(h_2 - h_1)_s = p_1 v_1 Y \qquad (4.2.21)$$

Fig. 4.2.6. Values of the function Y defined by eq. (4.2.20).

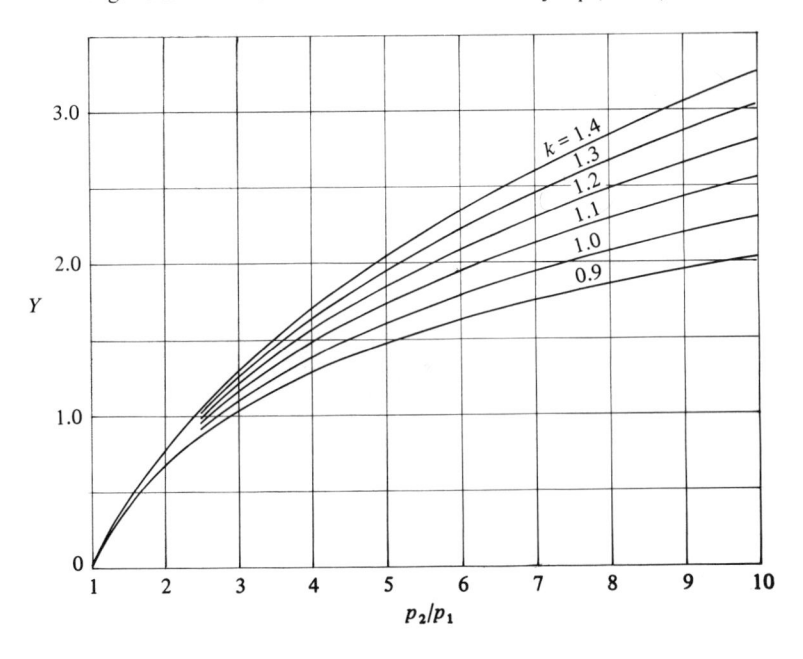

and the volumic work as:

$$\frac{(h_2 - h_1)_s}{v_1} = p_1 Y \qquad (4.2.22)$$

which is a function only of the pressures p_1 and p_2 and the value of k.

Mean effective pressure, although defined as a work divided by a volume, has the same dimensions as pressure and is commonly expressed in units of pressure. In basic SI units

$$1 \text{ J/m}^3 = 1 \text{ N/m}^2 \qquad (4.2.23)$$

When pressure is expressed in bars, one has:

$$1 \text{ bar} = 10^5 \text{ N/m}^2 = 10^5 \text{ J/m}^3 \qquad (4.2.24)$$

4.3 The ideal compressor with clearance

In real compressors it is impossible to make the piston sweep through the whole of the volume enclosed. At the top dead centre position there always remains a small volume of vapour, in the valve ports and in the space between the top of the piston and the cylinder head necessary to accommodate production tolerances, which is inaccessible to the piston. This clearance volume, which will be denoted by V_c, is usually expressed as a fraction of the swept volume, i.e.

$$c = V_c / V_{sw} \qquad (4.3.1)$$

Fig. 4.3.1. An indicator diagram for an ideal compressor with clearance.

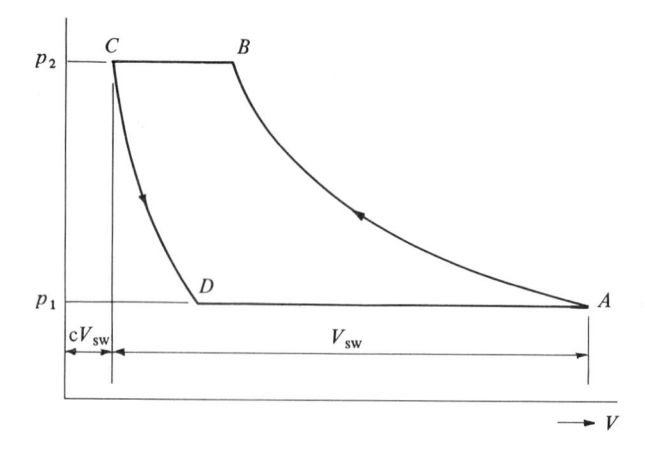

4.3 The ideal compressor with clearance

will be called the clearance fraction. Its value depends on the type and construction of the compressor. In the early machines with very large cylinders it was often as low as 0.005, and this value was attained in some VSA machines. The modern high-speed compressor for pumping the fluorinated hydrocarbon gases has demanded larger port areas for the valves, and with these the clearance fraction is about 0.03–0.04, whilst in small compressors for domestic refrigerators it may rise to 0.06 or more.

The effect of clearance on the diagram for the ideal compressor is shown in Fig. 4.3.1. The volume of vapour V_C at point C expands as the piston moves outwards and the pressure of the vapour inside the cylinder does not fall immediately to p_1 but follows the curve C–D. To extend the definition of the ideal compressor to one with clearance, it will be specified that the process C–D takes place reversibly and adiabatically, i.e. at constant entropy.

The first obvious effect of clearance, shown in Fig. 4.3.1, is to reduce the volume of vapour induced from V_A in Fig. 4.2.1 to $(V_A - V_D)$ in Fig. 4.3.1. The ratio:

$$\eta_{v,\text{idc}} = \frac{V_A - V_D}{V_{sw}} \qquad (4.3.2)$$

will be called the volumetric efficiency of the ideal compressor with clearance. Since:

$$V_A = (1 + c)V_{sw}$$

and:

$$V_D = \left(\frac{V_D}{V_C}\right)_s V_C = c\left(\frac{V_D}{V_C}\right)_s V_{sw}$$

eq. (4.3.2) becomes:

$$\eta_{v,\text{idc}} = 1 + c - c\left(\frac{V_D}{V_C}\right)_s \qquad (4.3.3)$$

where again the subscript s indicates that the states C and D are taken at the same entropy.

Because the mass inside the cylinder at C is the same as that at D, the ratio of the volumes is also the ratio of the specific volumes and:

$$\eta_{v,\text{idc}} = 1 + c - c\left(\frac{v_D}{v_C}\right)_s \qquad (4.3.4)$$

Example 4.3.1
Determine the volumetric efficiency of an ideal compressor with a clearance fraction of 0.04 operating under the conditions of example 4.2.1.

For an ideal compressor the state at D is the same as at A, and the state at C is the same as at B; the specific volumes are as given in Example 4.2.1:

$$v_C = 0.1132 \text{ m}^3/\text{kg}$$
$$v_D = 0.4185 \text{ m}^3/\text{kg}$$

and:

$$\eta_{v,\text{idc}} = 1 + 0.04 - 0.04\left(\frac{0.4185}{0.1132}\right) = 0.892$$

When a value of k the isentropic index is known, the ratio of the specific volumes may be replaced by:

$$\frac{v_D}{v_C} = \left(\frac{p_2}{p_1}\right)^{1/k} \tag{4.3.5}$$

and eq. (4.3.4) becomes:

$$\eta_{v,\text{idc}} = 1 + c - c\left(\frac{p_2}{p_1}\right)^{1/k} \tag{4.3.6}$$

Fig. 4.3.2. Variation of the volumetric efficiency of an ideal compressor having clearance fraction $c = 0.04$ with pressure ratio. Volumetric efficiency falls to zero at the pressure ratios 26 and 69 for $k = 1.0$ and 1.3, respectively.

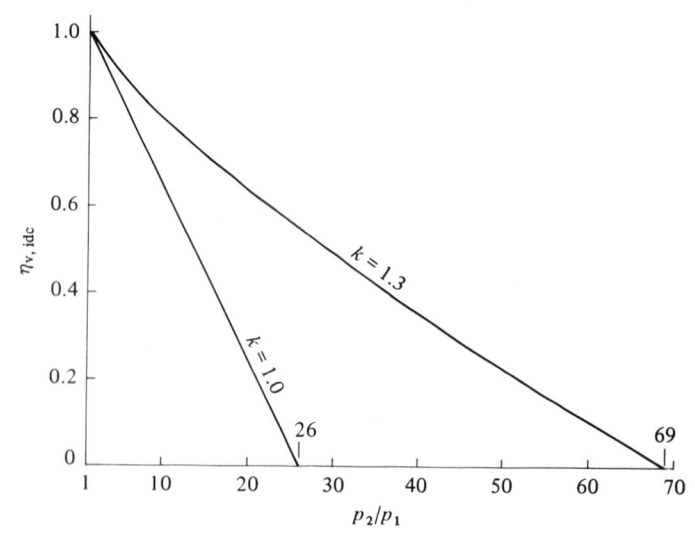

4.3 The ideal compressor with clearance

It is clear that if the value of k is chosen to fit the initial and final specific volumes, as was done in example 4.2.1, the result given by eq. (4.3.6) must be exactly the same as that given by eq. (4.3.4). The value of eq. (4.3.6) is that, for limited accuracy, a value of k can be assumed, or taken from Figs. 4.2.2 to 4.2.5 if greater accuracy is needed, without going through the procedure of following the isentropic process in the tables or on a chart of properties.

An immediate consequence of eq. (4.3.6) is that the volumetric efficiency of an ideal compressor with clearance diminishes with increasing pressure ratio, and eventually falls to zero at a pressure ratio given by:

$$\left(\frac{p_2}{p_1}\right)_{max} = \left(\frac{1+c}{c}\right)^k \qquad (4.3.7)$$

At this condition, the diagram Fig. 4.3.1 degenerates into a single isentropic line, the vapour being alternately compressed and expanded along the same path. The vapour in the clearance space fills the whole of the cylinder volume on re-expansion, and no fresh vapour is taken in. This condition is not normally encountered in refrigerating compressors, except possibly when the compressor itself is used as a vacuum pump to extract air from the system before charging. (This is not regarded as good practice.) p_2 is then the atmospheric pressure, and p_1 is the lowest pressure reached when the flow through the compressor ceases. True vacuum pumps are made with as low a clearance fraction as possible, usually reduced by filling with oil, so that the pressure p_1, called the terminal vacuum, can be very low.

Between the values $\eta_{v,idc} = 1$ at $(p_2/p_1) = 1$, and $\eta_{v,idc} = 0$ at $(p_2/p_1)_{max}$ given by eq. (4.3.7), the variation is as shown in Fig. 4.3.2. The very high pressure ratios are not used in practice, partly because the volumetric efficiency is then so poor, and partly because rather high delivery temperatures are attained. Figure 4.3.3 shows the variations of $\eta_{v,idc}$ for the range of pressure ratios encountered in practice.

The mass flow rate of vapour taken in and delivered by an ideal compressor with clearance is:

$$\dot{m} = \frac{\eta_{v,idc} \dot{V}_{sw}}{v_1} \qquad (4.3.8)$$

and there is a refrigerating capacity associated with this mass flow rate, for given conditions in the cycle, which is given by eq. (3.2.11) with the above value of $\dot{m}$.

The work done by the piston in the ideal compressor with clearance is deduced by noting that the specific work, i.e. the work per mass of vapour induced and delivered, remains unaltered and is still given by eqn. (4.2.12).

This is so because all the internal processes are defined as reversible ones. The work done in compressing the part of the vapour which remains in the clearance space is exactly recovered. The power is therefore given by:

$$\dot{W}_{\text{idc}} = \dot{m}(h_2 - h_1)_s = \eta_{v,\text{idc}} \dot{V}_{\text{sw}} \frac{(h_2 - h_1)_s}{v_1} \tag{4.3.9}$$

on using $\dot{m}$ from eq. (4.3.8). The mean effective pressure is then:

$$w_{v,\text{idc}} = \eta_{v,\text{idc}} \frac{(h_2 - h_1)_s}{v_1} \tag{4.3.10}$$

Thus the power and the mean effective pressure are reduced by the factor $\eta_{v,\text{idc}}$ owing to the effect of clearance. Using the alternative expression for $(h_2 - h_1)_s$ given by eq. (4.2.11), the mean effective pressure may also be written:

$$w_{v,\text{idc}} = \frac{\eta_{v,\text{idc}}}{v_1} \int_{p_1}^{p_2} v \, dp)_s \tag{4.3.11}$$

When it is permissible to use a value of k, the index of isentropic compression and expansion, eq. (4.3.11) becomes:

$$w_{v,\text{idc}} = p_1 \eta_{v,\text{idc}} Y \tag{4.3.12}$$

where Y is defined by eq. (4.2.20).

Fig. 4.3.3. Variation of the volumetric efficiency of an ideal compressor having clearance fraction $c = 0.04$ with pressure ratio over the practical range.

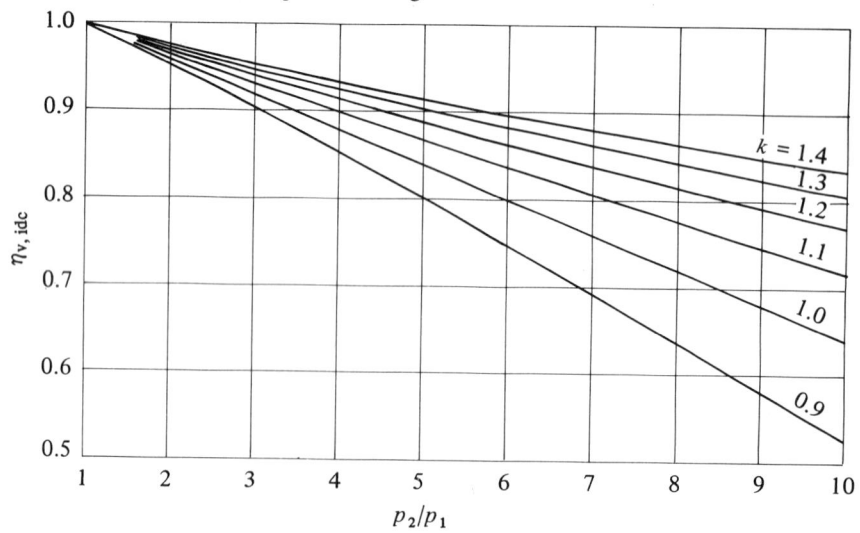

4.3 The ideal compressor with clearance

In the special case of $k=1$, the mean effective pressure is:

$$w_{v,\text{idc}} = p_1 \eta_{v,\text{idc}} \ln(p_2/p_1) \qquad (4.3.13)$$

Example 4.3.2
Determine the mean effective pressure of an ideal compressor with clearance fraction 0.04 for the conditions of example 4.2.1. Find also the power for a swept volume rate of 0.1 m³/s.

From example 4.2.1:

$$w_{v,\text{id}} = 5.916 \text{ bar}$$

From example 4.3.1:

$$\eta_{v,\text{idc}} = 0.892$$

Therefore:

$$w_{v,\text{idc}} = (0.892)(5.916 \text{ bar}) = 5.277 \text{ bar}$$

The power is then:

$$\dot{W} = (0.1 \text{ m}^3/\text{s})(5.277 \text{ bar}) = 52.77 \text{ kW}$$

Fig. 4.3.4. Variation of the mean effective pressure of an ideal compressor having clearance fraction $c = 0.04$, with evaporating temperature, at a fixed delivery pressure of 7.45 bar corresponding to a condensing temperature of 30 °C. The refrigerant is R12.

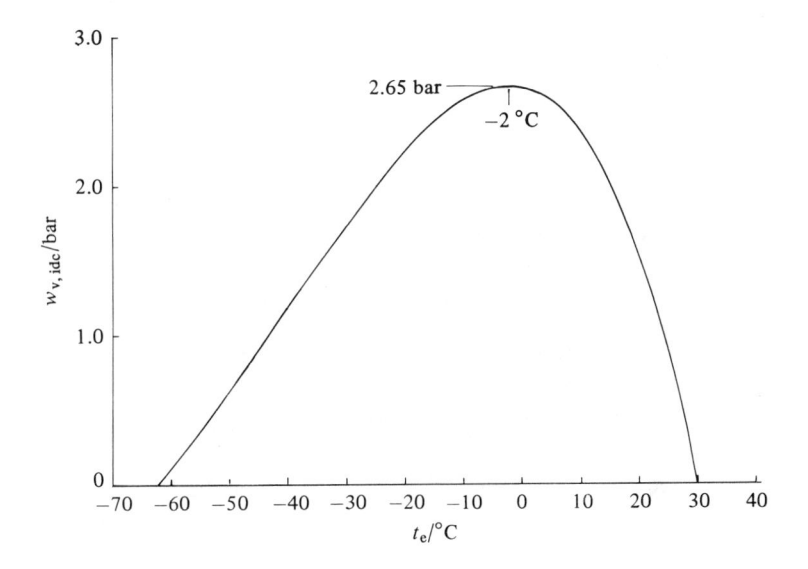

Because the function Y increases with pressure ratio, as shown in Fig. 4.2.6, whilst $\eta_{v,\text{idc}}$ decreases with pressure ratio, as shown in Fig. 4.3.2, the value of the mean effective pressure passes through a maximum as p_1 varies at fixed p_2. There is also a maximum as p_2 varies at fixed p_1. The first of these maxima is of some importance in refrigerating compressors. For a given condensing temperature, which determines p_2, the mean effective pressure increases at first as the evaporating temperature, and with it p_1, diminishes, as for example during a pull down when the plant is first switched on. The mean effective pressure increases to a peak and then begins to fall as the evaporating temperature drops still further. The shape of the curve of mean effective pressure is shown in Fig. 4.3.4 to a base of evaporating temperature. Compressors operating on air-conditioning duties work at evaporating temperatures near the value for peak power, but compressors for lower temperature duties have to operate most of the time at well below peak power. It may not be economic to provide them with a motor capable of delivering the peak power for the few occasions when it may be needed, so that it is common practice on small systems to arrange for the evaporator to be temporarily starved of refrigerant when the suction pressure is high, and enable the compressor to pull it down rapidly through the peak power range. The method of doing this by limit charging of the expansion valve has been described in Chapter 2. It should be noted that the motor must still have enough torque to keep it turning at the peak power. The rapid pull-down simply ensures that it does not run for a long period at a power which would cause overheating.

Other methods of reducing motor size are to cut out some of the cylinders of a multi-cylinder compressor at high suction pressure, or to use a CPRV described in section 2.12.

The determination of the pressure ratios for maximum mean effective pressure is readily done by differentiating eq. (4.3.12) or (4.3.13), holding p_2 and p_1 fixed in turn. The working is tedious but straightforward and will not be given here. The final equations which must be solved are as follows:

p_2 fixed:

$$A(k-1)r^{1/k} + r^{(k-1)/k} - k = 0 \qquad (4.3.14)$$

Table 4.3.1. *Solutions of eqs. (4.3.14) to (4.3.17) for $c = 0.04$*

	$k =$	1.0	1.1	1.2	1.3
p_2 fixed	$p_2/p_1 =$	2.472	2.602	2.731	2.857
p_1 fixed	$p_2/p_1 =$	8.333	10.94	14.51	19.44

4.3 The ideal compressor with clearance

or for $k = 1$:

$$\ln r + Ar - 1 = 0 \tag{4.3.15}$$

p_1 fixed:

$$Akr^{1/k} - Ar^{(2-k)/k} - (k-1) = 0 \tag{4.3.16}$$

or for $k = 1$:

$$A \ln r - (1/r) + A = 0 \tag{4.3.17}$$

where:

$$r = (p_2/p_1) \quad \text{at max } w_{v,\text{idc}}$$

and:

$$A = c/(1+c)$$

Some solutions of these for $c = 0.04$ are given in Table 4.3.1.

Graphs for the determination of maximum power have been given by Thomsen (1951). There is a useful discussion of the matter by Lorge (1968).

Example 4.3.3

A reciprocating compressor has 4 cylinders of bore 130 mm and stroke 100 mm, and a clearance fraction of 0.04. It is directly coupled to a motor, and normally operates with Refrigerant 502 on freezing duties at an evaporating temperature of $-40\,°C$ and a condensing temperature of $35\,°C$. What torque must the motor be capable of providing in order that it will not stall during pull-down? A value of $k = 1.1$ may be assumed.

From Table 4.3.1 the pressure ratio at maximum mean effective pressure is 2.602. From R502 tables, the condensing pressure at $35\,°C$ is 14.90 bar. Assuming that the condensing pressure is regulated at this value, the maximum mean effective pressure occurs at a suction pressure of:

$$p_1 = \frac{14.90 \text{ bar}}{2.602} = 5.726 \text{ bar}$$

at which the evaporating temperature is just under $0\,°C$. The maximum mean effective pressure of the ideal compressor with clearance 0.04 is:

$$w_{v,\text{idc}})_{\max} = (5.726 \text{ bar})\frac{1.1}{0.1}(2.602^{(0.1/1.1)} - 1)$$

$$= 5.721 \text{ bar} = (5.721)\,10^5 \text{ J/m}^3$$

The swept volume is:

$$V_{sw} = \tfrac{1}{4}\pi(0.13)^2(0.1)4 \text{ m}^3 = 0.005309 \text{ m}^3$$

The maximum torque is:

$$T_{max} = \frac{w_{v,idc})_{max} V_{sw}}{2\pi} = \frac{(5.721)(10^5)(0.005309)}{2\pi} \text{N m}$$

$$= 483.4 \text{ N m}$$

In practice a larger torque than this would have to be assumed because of the supplementary effects of the pressure drops through the valves and the bearing friction on the mean effective pressure, as described in section 4.5.

Variation of the clearance fraction

A method of reducing the pumping capacity of a reciprocating compressor which was commonly used at one time was to increase the clearance fraction deliberately by incorporating a chamber (usually called a clearance pocket) placed in communication with the clearance space. Some clearance pockets were of a fixed volume and brought in or out by a shut-off valve, whilst a more elaborate arrangement was a variable pocket permanently in communication with the clearance space. The variation was achieved by moving a piston in the pocket according to the reduction of capacity which was required. The effect is shown in Fig. 4.3.5, where $c'V_{sw}$ is the total clearance with the pocket open. The re-expansion curve now occupies a larger fraction of the stroke and a reduced volume of vapour is taken in.

If the re-expansion process is isentropic, as assumed in the ideal

Fig. 4.3.5. The effect of increasing the clearance fraction from c to c'. Re-expansion is not completed until D' instead of D, and the volumetric efficiency is reduced.

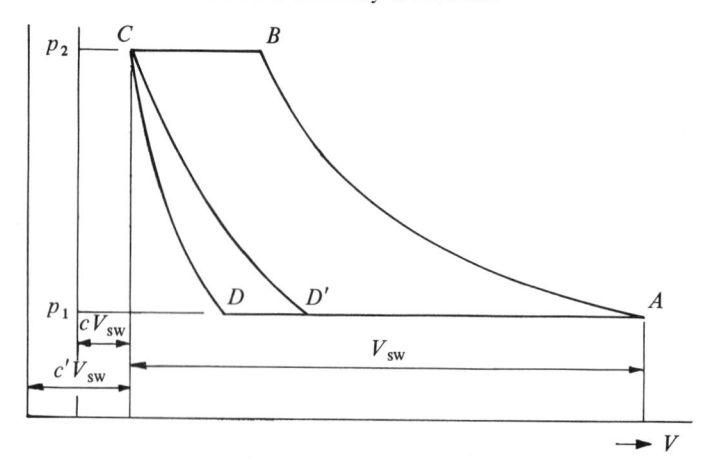

compressor, then the proportional reduction of mean effective pressure is the same as the proportional reduction of mass flow rate. This is true even when the processes are not isentropic, provided that the index of the re-expansion curve is the same as that of the compression curve. But in an actual compressor, there are supplementary amounts of work for pumping the gas into and out of the cylinders, and to overcome resistance in the bearings, etc., and the mean effective pressure does not fall in proportion to the mass flow rate.

Example 4.3.4

An ammonia compressor with a normal clearance fraction of 0.03 operates at an evaporating temperature of $-20\,°C$ and a condensing temperature of $30\,°C$. Assuming ideal compressor behaviour, how much extra clearance has to be introduced to reduce the capacity by 50% ?

A value of $k = 1.289$ is determined from Fig. 4.2.2. From the tables:

$$p_1 = p_e = 1.902 \text{ bar}$$
$$p_2 = p_c = 11.66 \text{ bar}$$
$$\frac{p_2}{p_1} = \frac{11.66}{1.902} = 6.130$$

From eq. (4.3.6):

$$\eta_{v,\text{idc}} = 1 + 0.03 - 0.03(6.130)^{1/1.289}$$
$$= 1 - 0.03(3.082) = 0.9075$$

is the volumetric efficiency with normal clearance. With the clearance fraction increased to c' say, the volumetric efficiency has to be:

$$\eta'_{v,\text{idc}} = 0.9075/2 = 1 - c'(3.082)$$

and:

$$c' = 0.1772$$

The required increase of clearance fraction is therefore: $0.1772 - 0.03 = 0.1472$ of the swept volume.

Another method for reducing capacity is the provision of a partial duty port in the cylinder wall which is connected to the suction pipe at will. Vapour is pushed out of the port back into the suction pipe until the port is covered by the piston. In deciding the position of the port for a given reduction of capacity it is necessary to take account of the effect of clearance.

Example 4.3.5

For the compressor of example 4.3.4 with the normal clearance fraction of 0.03, at what point in the stroke should the partial duty port close in order to reduce the capacity to $\frac{1}{3}$?

With clearance fraction 0.03 the volumetric efficiency of the ideal compressor is 0.9075. To reduce this to $\frac{1}{3}$, the port has to close at the fraction:

$$\tfrac{2}{3}(0.9075) = 0.605$$

of the stroke from bottom dead centre.

These methods of reducing pumping capacity are now more or less obsolete.

4.4 The real compressor: volumetric efficiency

The behaviour of the ideal compressor with clearance goes some way towards describing the behaviour of real compressors, particularly with regard to the way in which volumetric efficiency and mean effective pressure vary with p_1 and p_2. When it comes to calculating values of these quantities for real compressors, however, it is necessary to apply corrections in order to obtain realistic values. It is not the aim of this section to list the values of the corrections obtained from numerous experiments, although a few general indications are given, but to discuss the reasons for departure from ideality and the ways in which the various imperfections affect the results.

Fig. 4.4.1. Indicator diagram for a real compressor shown diagrammatically.

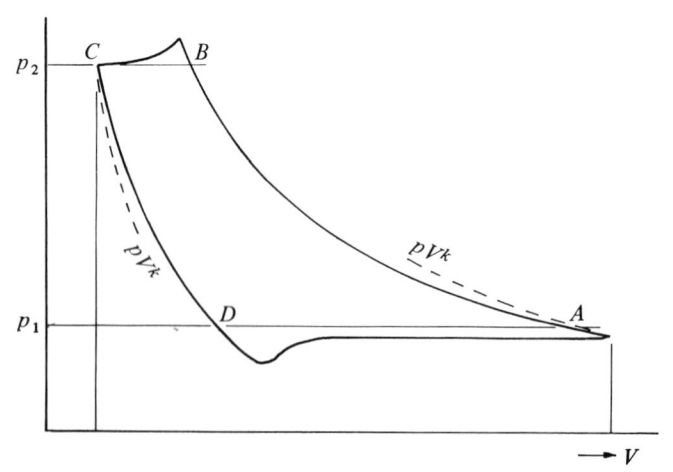

4.4 The real compressor: volumetric efficiency

The difference between the ideal compressor with clearance and the real compressor will be discussed firstly with reference to the indicator diagram for a real machine shown diagrammatically in Fig. 4.4.1. The most obvious differences are that suction and delivery do not take place at the constant pressures p_1 and p_2 respectively. Because the moving parts of the valves have mass, a pressure difference is needed to accelerate them from rest, and some time elapses before a valve is fully open. This appears on the indicator diagram as compression apparently continuing after point B is reached, and expansion apparently continuing after point D is reached. Even when a valve is fully open, a difference of pressure between the vapour inside the cylinder and that outside occurs, because of the flow through the ports.

Examination of the curves for compression and expansion shows that they are not constant entropy ones. They are usually less steep. Suppose for example that the points A and B are connected by the equation:

$$p_1 V_A^n = p_2 V_B^n \qquad (4.4.1)$$

and that n is evaluated from the measured volumes V_A and V_B on the indicator diagram. The value of n so found is invariably less than k, the isentropic index. Similarly the index m, determined from the expansion line, is found to be less than k. The reasons for this are the effects of heat transfer between the vapour and the cylinder walls, and leakage.

Another difference between ideal and real compressors which appears on the indicator diagram is that point A does not always coincide with the piston position at outer dead centre, or, putting it another way, the pressure inside the cylinder at outer dead centre has not yet reached p_1. The effect is hardly noticeable in slow-speed compressors but is evident in high-speed ones.

The volume of vapour apparently taken into the cylinder during suction is:

$$V_A - V_D$$

and the apparent or indicated volumetric efficiency is defined as:

$$\eta_{v,\text{ind}} = \frac{V_A - V_D}{V_{sw}} \qquad (4.4.2)$$

Because the expansion line is found in practice to be less steep than the isentropic line through point C, the indicated volumetric efficiency is less than that of the ideal compressor with the same clearance fraction.

The actual volumetric efficiency is determined by a test in which the mass flow rate $\dot{m}$ is measured at steady conditions, and the specific volume v_1 at the entry to the compressor. The volumetric efficiency is then defined as:

$$\eta_v = \dot{m}v_1/\dot{V}_{sw} \qquad (4.4.3)$$

The efficiency given by this operational definition is termed simply volumetric efficiency in this book. In the literature it is sometimes called the overall, real or actual volumetric efficiency.†

To complete the definition of volumetric efficiency it is necessary to specify the point at which v_1 is to be determined: for example, in the suction pipe before the compressor flange and stop valve, or entering the cylinder just before the suction valve. The distinction becomes very important in connection with hermetic compressors in which the motor windings are cooled by suction vapour passing over them.

The volumetric efficiency determined by tests is found to be less than $\eta_{v,ind}$ found from the indicator diagram. The reasons for this difference are (a) heat transfer between the vapour and the compressor body, (b) the effects of throttling through the valve ports and (c) leakage. These will now be discussed.

(a) *Heat transfer effects.* Heat transfer to the vapour can take place either in the manifold or suction chamber before the vapour enters the cylinder, or from the cylinder walls during the induction stroke. In either case it increases the specific volume above the value v_1 assumed in the definition of the volumetric efficiency, and the mass of vapour trapped inside the cylinder at the beginning of compression is reduced.

Leaving aside hermetic compressors, differences of design have an influence on the opportunity for heat transfer to occur. If the suction manifold is formed in the casting, as in many modern compressors, there is likely to be more heat transfer to the suction vapour than when the suction manifold is separate and bolted to the block.

Inside the cylinder heat transfer can be affected by the direction of flow of the vapour. In uniflow compressors, described in section 2.3, the vapour enters through the top of the piston and comes into contact with cylinder walls which have not immediately before been subjected to hot delivery

---

† A somewhat different efficiency has been used on the Continent in the past with the name, in its German form, 'Liefergrad'. It included the effect of a difference between the actual suction state and that of saturated vapour. Unfortunately the term has often been translated into English simply as volumetric efficiency. The German 'Kältemaschinen Regeln', 5th edition. Müller, Karlsruhe, 1958, accepted its definition as eq. 4.4.3 above, but this has not always been followed in the literature.

4.4 The real compressor: volumetric efficiency

vapour. In return flow machines the cold vapour meets metal which has just been swept by hot vapour and consequently has more opportunity to accept heat. This was thought to be a significant difference at one time, but it is not supported by experimental evidence (Lorentzen, 1949, Bendixen, 1959).

A much more important difference is observed between the behaviour of wet and dry suction vapour. The rate of heat transfer to a wet vapour is much higher than to a dry one, and the observed volumetric efficiency is much lower. This fact is well established by experimental results e.g. Shipley (1910), Wirth (1933) and Smith (1935).

Experiments have further shown that the volumetric efficiency continues to improve with superheat and this needs some explanation. Fig. 4.4.2(a) from Giffen & Newley (1940) shows how the volumetric efficiency of an ammonia compressor increases as the suction state changes from wet to a moderately superheated one. The indicated volumetric efficiency shows only a small increase. Figure 4.4.2(b) from Gosney (1953) gives results for an R12 compressor with dry suction only. Again the volumetric efficiency increases with superheat whilst the indicated volumetric efficiency remains

Fig. 4.4.2. Experimental results for volumetric efficiency. (a) Giffen & Newley: ammonia, single-cylinder VSA compressor of bore 88.9 mm, stroke 101.6 mm, speed 200 rev/min. $A: p_1 = 2.41$ bar, $p_2 = 9.31$ bar; $B: p_1 = 2.41$ bar, $p_2 = 10.55$ bar; $C: p_1 = 2.07$ bar, $p_2 = 10.55$ bar. (b) Gosney: R12, single-cylinder VSA compressor of bore 36.5 mm and stroke 36.5 mm, clearance fraction 0.028, speed 431 rev/min. $p_1 = 1.655$ bar, $p_2 = 7.45$ bar.

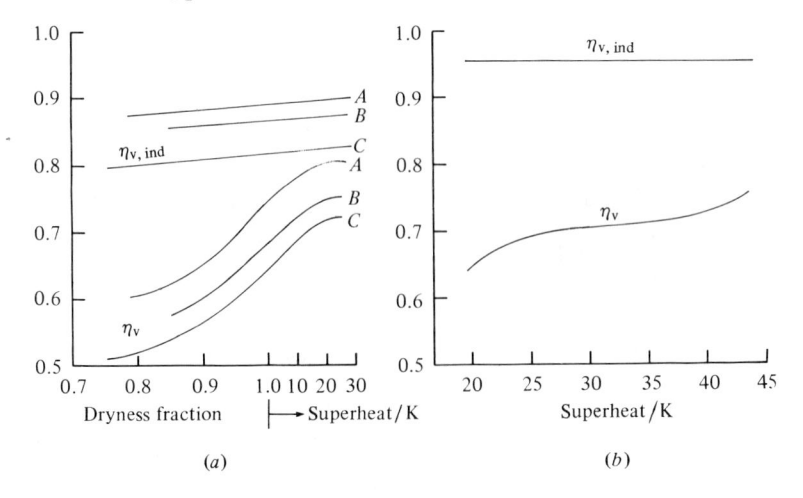

constant. This increase of volumetric efficiency has been found by many investigators.

A possible reason for the improvement with superheat is that superheated vapour may entrain liquid droplets, but this is not thought to be the whole explanation. Wirth (1933) suggested that condensation takes place on the cylinder walls towards the end of the compression stroke. To see how this can occur consider the conditions shown in Fig. 4.4.3 which represents a compression process on a temperature–entropy diagram, with superheated vapour initially at point A and the delivered vapour at point B. The temperature of the cylinder wall T_w is shown somewhere between T_A and T_B as is to be expected. This temperature is not constant of course. It undergoes cyclic variations as heat transfer to and from the metal takes place, but because of the good thermal conductivity of the metal the swings are not expected to be large. Constructing a constant pressure line through the saturated vapour state at T_w, the intersection with the compression line is shown as point X. As compression proceeds from A, the vapour, initially colder than the walls, becomes warmer and the entropy increases. After $T = T_w$ the vapour is warmer than the walls and heat transfer to the walls reduces the entropy. Up to this point only heat transfer to and from dry vapour is involved. As compression proceeds beyond point X however the vapour is now in contact with a surface below its dew point and

Fig. 4.4.3. A compression process on temperature–entropy coordinates showing the possibility of condensation when point X is reached at which the mean wall temperature T_w is equal to the dew point temperature of the vapour.

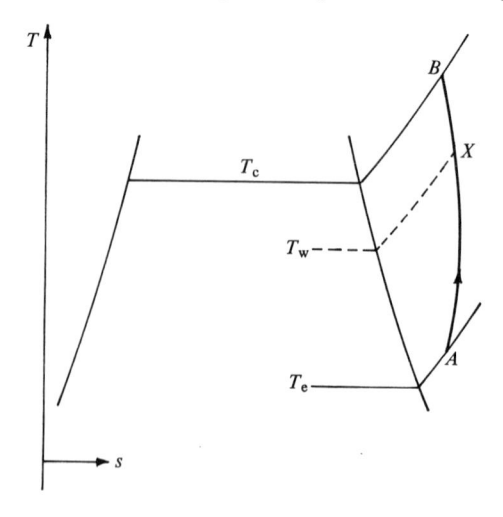

4.4 The real compressor: volumetric efficiency

condensation can occur. The release of enthalpy of evaporation raises the wall temperature locally, whilst the entropy of the cylinder contents diminishes and with it the volume. The observed index of compression n is consequently depressed below the isentropic index k, which is in accord with experimental results.

The condensed liquid remains as a film on the wall and leaks past the piston or it is partly swept up into the clearance space. In this event a part of the charge is left behind in the cylinder as liquid instead of as high pressure vapour in the clearance space. The volumetric efficiency is reduced in either case.

When the vapour is very soluble in the lubricating oil, as with R12 or propane, the effect of condensation can be reinforced by the refrigerant going into solution.

At steady conditions the liquid must ultimately evaporate. If this occurs during the expansion process the effect is to depress the value of the observed index of expansion m. Many tests have shown low values of this index and there is little doubt that this is at least partly explained by evaporation of liquid. A low value of m is evident on the indicator diagram and lowers the value of $\eta_{v,ind}$ but it does not directly alter the difference $\eta_{v,ind} - \eta_v$. For this to be affected, evaporation must continue during the suction stroke, and prevent new vapour from entering. That this is probably so is shown by the fact that the indicated volumetric efficiencies in Fig. 4.4.2 remain nearly constant.

The effect of increasing superheat can now be interpreted as causing an increase of the cylinder wall temperature and delaying the point X at which condensation begins. If the superheat is so high that the wall temperature is above the temperature of saturation corresponding to the delivery pressure, condensation cannot take place at all. Absorption in an oil film is still possible however and this may explain the continued rise of η_v shown in Fig. 4.4.2(b) at values of the superheat at which the wall temperature would be expected to have reached the saturation temperature corresponding to delivery pressure.

Supporting evidence for the occurrence of cyclical condensation comes from direct measurement of the cylinder wall temperature, e.g. by Cheglikov (1963) and by Brown & Kennedy (1971) who also evaluated the actual heat transfer coefficient between the vapour and the wall.

Although cyclical condensation has been discussed above with reference to superheated vapour at suction, it probably occurs with slightly wet vapour and explains part of the rise in volumetric efficiency as the suction state approaches dryness.

Compression processes

The poor volumetric efficiency when working with wet vapour at suction explains why wet compression was abandoned at an early stage in the development of mechanical refrigeration, despite the fact that it has apparent theoretical advantages for some refrigerants as discussed in sections 3.12 and 3.13. Modern compressors operate with enough superheat to give a good volumetric efficiency and, for certain refrigerants, to obtain the benefit of the increased volumic refrigerating effect, taking into account the delivery temperature. This should not as a rule exceed about 150 °C because of possible degradation of the lubricating oil.

(b) *Throttling through the valve ports.* The difference in pressure between the vapour in the suction chamber or manifold and that in the cylinder during the suction stroke reduces the volumetric efficiency by reducing the density of the vapour trapped inside the cylinder at the end of the stroke. This occurs because the vapour flowing through the port undergoes a throttling process to the lower pressure inside the cylinder during which its specific enthalpy remains approximately constant. Later in the stroke the vapour which entered earlier is compressed back to the suction pressure p_1 by the vapour entering behind it. The compression takes place approximately adiabatically and the temperature rises. During the whole induction stroke every particle of vapour undergoes a different process and the state of the vapour cannot be shown except as an average on a thermodynamic diagram. The net effect however is an increase in the specific enthalpy at point A and a reduction of the mass which is enclosed. The consequent reduction of the volumetric efficiency is not evident on the indicator diagram.

(c) *Leakage.* Leakage may take place between the piston and the cylinder wall, and between the valves and their seats. Since the crankcase is normally connected to the suction pressure via an equalising port, the effect of leakage past the piston is basically the same as leakage back through the suction valve.

Leakage to suction occurs during compression, delivery and expansion. Its effect is to lower the index of compression n and to raise the index of expansion m. It thus tends to raise the indicated volumetric efficiency $\eta_{v,ind}$, but of course it lowers the volumetric efficiency η_v.

Leakage through the delivery valve tends to raise the index of compression n and to lower the index of expansion m, thus lowering both $\eta_{v,ind}$ and η_v.

Both kinds of leakage raise the delivery temperature, because the vapour

4.4 The real compressor: volumetric efficiency

which is throttled back to a lower pressure is subsequently re-compressed adiabatically, as explained above in connection with pressure drop through normally open valves. A sudden rise in delivery temperature is a sign of excessive leakage, often caused by a valve breakage.

Leakage across the valves when they are supposed to be closed depends greatly on the condition of the surfaces. It is known that a scratch on a valve or its seat can markedly affect performance. The presence of oil is also an important factor in achieving a good seal, particularly in the very small reed valves used in hermetic compressors.

Leakage past the piston depends on the clearance between the piston and the bore, on the amount of oil on the cylinder walls and on the suction state. The pistons used in small compressors have no rings and the clearance has to be very fine, of the order 0.05 mm or less, for them to work satisfactorily. They also need a good supply of oil. Smith (1961) has investigated the effect of varying the clearance on the performance of compressors with ringless pistons, whilst Kalitenko (1960) has investigated the effect in large compressors with ringed pistons for ammonia. Brown & Pearson (1963) showed that leakage depends on the suction superheat. From a study on a small R12 compressor they concluded that the leakage is associated with a laminar flow of oil containing dissolved refrigerant along the cylinder wall.

Another type of leakage can occur for quite different reasons. This happens when the valves ought to be shut according to the equilibrium of pressure and spring forces acting on them but in fact remain open because of dynamic effects. For example the suction valve may not be shut at point A in Fig. 4.4.1, and as the piston returns vapour is pushed back into the suction pipe. This loss, termed 'blow-by', has been investigated experimentally by Pearson (1958). It becomes important at low absolute suction pressure and is the main reason why reciprocating compressors have poor volumetric efficiency when the absolute pressure drops to about 0.1 bar, irrespective of the pressure ratio.

The importance of the effects (a) and (c) discussed above is very dependent on the rotational speed of the compressor. In a given time so much heat transfer and so much static leakage can occur but the effects on volumetric efficiency are diluted if the speed is raised to increase the mass flow rate through the machine. On the other hand the effects of pressure drop through the ports and of dynamic leakage increase with speed. Because of these opposing tendencies it is found that the volumetric efficiency reaches a maximum at a certain speed, which depends on the design of the compressor and also on the refrigerant. Lorentzen (1951) has shown that the higher the molar mass of the refrigerant the lower is the speed at which the maximum η_v is observed. An approximate relation is:

$$\mathscr{V}_{v,opt}\sqrt{[M/(kg/kmol)]} = 420 \text{ m/s} \qquad (4.4.4)$$

where M is the molar mass of the refrigerant and $\mathscr{V}_{v,opt}$ is the velocity of the vapour through the valve port at the maximum volumetric efficiency.

The application of high-speed computers to the behaviour of valves in compressors has been a feature of compressor development during the last twenty years, and several groups of researchers have built up extensive programs for predicting valve behaviour. The equations originally developed by Costagliola (1950) for ideal gases have been refined and extended to vapours. The results obtained have a bearing not only on the performance of the compressor but also on its mechanical behaviour, for example to predict the life of valves and springs before breakage. An account of these developments, with many references, is given by MacLaren, Kerr & Tramschek (1975).

Suction and delivery pipe effects

Induction 'ramming' by tuning the suction pipe length has long been exploited in high-performance internal combustion engines, and similar effects can be obtained in compressors. The principle is that a suction stroke generates a pulse of rarefaction which travels back along the pipe at the velocity of sound in the vapour. It is reflected at an open end with a change of phase, i.e. as a compression pulse which travels towards the cylinder. Should it arrive just before the suction valve closes it supercharges the cylinder above the normal suction pressure. On the other hand should it arrive back when the suction valve is closed it is reflected without change of phase, travels towards the open end as a compression pulse, and is reflected there as a rarefaction pulse. Should this pulse arrive at the suction valve just before it closes the cylinder is undercharged. Thus, according to the suction pipe length in relation to the engine speed, an improvement or a worsening of performance compared with the no-pipe condition can be obtained.

The theory of the phenomena for a single cylinder air compressor was given by Bannister (1959) who also gave experimental results showing an increase of air flow of 18% over the no-pipe value. A somewhat simpler method of predicting the qualitative behaviour was given by Williams (1959). It is based on summing the pulses arriving at the suction valve which have made one or more double traverses of the pipe. These pulses may have originated during the same suction stroke or during earlier ones.

The relation of pipe length to engine speed is usually expressed as the delay angle Φ defined as:

$$\Phi = (2NL/a)360° \qquad (4.4.5)$$

where N is the rotational speed, L is the length of the suction pipe and a is the velocity of sound in the vapour. The delay angle thus expresses the ratio of the time, $2L/a$, for a pulse to make a double journey along the pipe to the time, $1/N$, of a revolution of the crank.

Assuming that a pulse is attenuated by a factor of one half during its double journey, Williams' method applied to a single cylinder compressor predicts a useful gain when the delay angle lies between 60° and 90°, and a drop in performance when it lies between 180° and 240°. For delay angles greater than 360° the peaks and troughs are repeated at corresponding values of Φ. Translating these angles into pipe lengths for a single cylinder R22 compressor running at $N = 25$ rev/s, with $a = 175$ m/s at an evaporation temperature of -15 °C, extra performance is to be expected with suction pipe lengths which are between 0.58 m to 0.87 m, and lowered performance at lengths which are between 1.75 m to 2.33 m. Effects will also be found at lengths which are multiples 5, 9, 13 etc. of these, but with greater attenuation.

For multi-cylinder compressors account has to be taken of the fact that a pulse arriving at one cylinder may have originated from another. Furthermore a pulse from one cylinder can interfere with another before it has travelled down the pipe and back. Quite apart from the pulse effect a multi-cylinder compressor needs enough volume in the suction manifold to ensure that the high rate of intake of one cylinder at mid-stroke does not starve another which is near the end of the induction stroke.

To obtain reflection with a change of phase the suction pipe must have an open end into a vessel such as a shell-and-tube evaporator or a knock-out pot at the appropriate distance from the compressor. In a hermetic compressor with a case at low-side pressure the required length of pipe can be accommodated inside the case.

Similar results can be obtained by tuning the delivery pipe, a suitable length providing a rarefaction pulse just before closure of the delivery valve.

Compressor valves, unlike those of internal combustion engines, are spring loaded and the pulsations in the pipes modify their movement. This is taken into account in more recent work, as described by MacLaren, Kerr & Tramschek, already cited. Modern work also recognises that the substantial pulses in the pipes are not small amplitude waves as assumed in acoustic theory.

Experimental values of volumetric efficiency

There are numerous reports in the literature giving measured volumetric efficiencies, and other values can be deduced from manufacturers' published ratings which are based at least partly on test results. Owing to the wide variety of designs, refrigerants and rotational speeds, it is not possible to express volumetric efficiency accurately in terms of only one or two parameters. Even for the same machine, caution must be exercised in translating results for one vapour to another. For example, because k for ammonia is greater than k for R12, one would expect the volumetric efficiency of a compressor on ammonia to be higher than that at the same pressure ratio on R12. But in view of the behaviour expressed by eq. (4.4.4), this is not necessarily so. The same compressor tested on ammonia and R12 is not at the same point on the curve of volumetric efficiency versus speed, and the efficiency on R12 may be higher.

Nevertheless, it is sometimes necessary to make an estimate of volumetric efficiency, and for this the following equations may be found useful.

For ammonia compressors of traditional design, Löffler (1941) has analysed the results from many tests, and found a mean value of m during expansion of 1.15, almost independent of speed or pressure ratio, provided that the vapour at entry is superheated by 5 K, and that the jacket water enters at the temperature of the water leaving the condenser. If cooler water is used, the superheat must be 10–15 K at entry. The volumetric efficiency is then given by:

$$\eta_v = 1 + c - c\left(\frac{p_2}{p_1}\right)^{1/1.15} - K\left[\left(\frac{p_2}{p_1}\right)^{0.17} - 1\right] \quad (4.4.6)$$

where:

$$K = \frac{6.95}{(D/\text{cm})^{0.36}(L/\text{cm})^{0.168}(N/(\text{rev/min}))^{0.24}} \quad (4.4.7)$$

and D is the bore, L the stroke, N the rotational speed. (N.B. K is not dimensionless!) The author has found that eq. (4.4.6) predicts the volumetric efficiency of at least one make of VSA compressor reasonably well, but it should be noted that it does not indicate an optimum speed. In fact it gives η_v increasing always with speed, and this cannot be true. It shows that such expressions should not be used outside the range of the tests on which they are based.

Pierre (1952) reported average refrigerating capacities for about 150 compressors on R12, with shaft powers in the range 1–60 kW. With no subcooling leaving the condenser and a temperature of 18 °C at suction, the

4.4 The real compressor: volumetric efficiency

refrigerating capacity is given within ± 10% by:

$$\frac{\dot{Q}_e}{\dot{V}_{sw}} = \left[\left(890 - 6.6\frac{t_c}{°C}\right)\frac{p_1}{\text{bar}} - 180\right] \text{kJ/m}^3 \qquad (4.4.8)$$

By putting the best line through points calculated from this equation, it can be deduced that the average volumetric efficiency is given by:

$$\eta_v = 0.948 - 0.0285(p_2/p_1) \qquad (4.4.9)$$

Fig. 4.4.4. Approximate volumetric efficiencies of some compressors as functions of pressure ratio. (A) A VSA machine like that of Fig. 2.3.3 pumping ammonia. Owing to the relatively low clearance fraction the volumetric efficiency holds up well at high pressure ratios. (B) A modern compressor like that of Fig. 2.3.8 pumping R12 and R22. (C) A small open-type compressor like that of Fig. 2.3.6 pumping R12 and R22. (D) A hermetic compressor like that of Fig. 2.3.10 pumping R12. This curve is shown as a broken line because the volumetric efficiency shown is not the true value: it is based on the temperature and specific volume of the vapour entering the case and is consequently lower than the value based on the temperature of the vapour entering the cylinder.

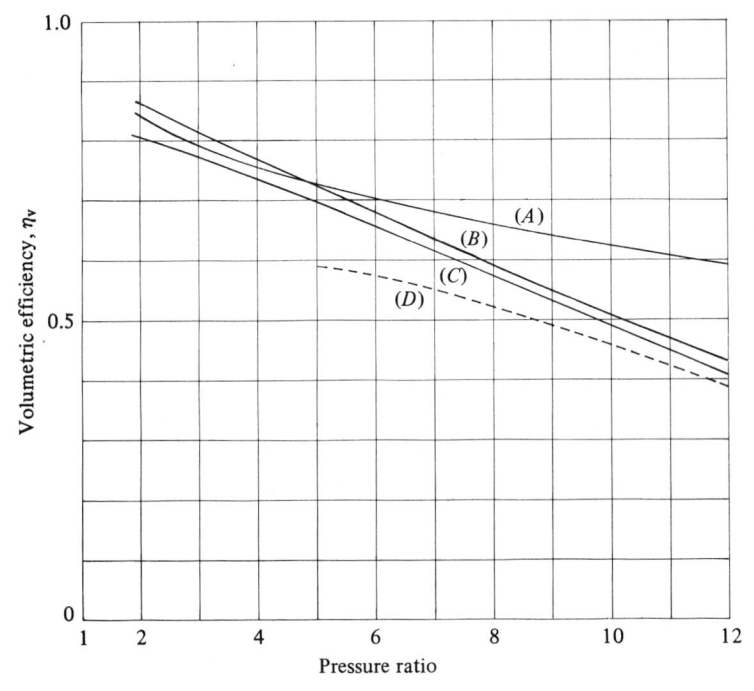

Equation (4.4.9) indicates that the index of the re-expansion line for R12 is approximately unity for these compressors, a result which has been confirmed from indicator diagrams.

The compressors to which eq. (4.4.9) applies are mostly of the type using reed or ring plate valves for suction and delivery, and running at comparatively low piston speeds. Modern multi-cylinder compressors with annular valve rings and unloading gear have higher clearance fractions and run at higher piston speeds. As already remarked, the effect of higher speed is to reduce the importance of heat transfer effects, and there is some evidence that the index of the expansion process is higher than in slow-speed compressors. Results on many of these machines show that the volumetric efficiency falls off with pressure ratio faster than would be expected of an ideal compressor with the same clearance, and indicate the need for a subtractive term similar to that in Löffler's expression.

Approximate volumetric efficiencies of some of the compressors illustrated in Chapter 2 are shown in Fig. 4.4.4 as functions of pressure ratio.

Refrigerating capacity

Once the volumetric efficiency is established for given inlet conditions and delivery pressure, the mass flow rate through the machine is given by eq. (4.4.3). The mass flow rate, which is the effective output of the machine, might then be exhibited on a graph as a function of suction pressure and delivery pressure, or as a function of evaporating temperature and condensing temperature, assuming no pressure drops in the pipes. Manufacturers of compressors however invariably present their compressor performances as refrigerating capacities calculated as:

$$\dot{Q}_e = \dot{m}(h_1 - h_3) \qquad (4.4.10)$$

where $(h_1 - h_3)$ is the difference of specific enthalpy at two specified states of the refrigerant. As a rule h_3 is taken as liquid at condensing pressure with a few degrees of subcooling, and h_1 is taken as superheated vapour at evaporating pressure with either a stated superheat or a stated temperature. Users of such data must acquaint themselves with the basis of the method of rating, and ensure that the conditions are relevant to the proposed application of the compressor. It must always be borne in mind that if the stated suction superheat is not obtained usefully the stated capacity will not be available in the evaporator: part of it will be wasted in refrigerating the plant room, etc.

Suction pipe pressure drop must always be taken into account when using manufacturers' data, though some makers allow for a nominal pressure drop in the capacity tables.

4.4 The real compressor: volumetric efficiency

Fig. 4.4.5. Refrigerating capacity and shaft power as functions of evaporating temperature at four condensing temperatures shown on the curves. The compressor is of similar construction to that of Fig. 2.3.8 and is pumping R22 at a rotational speed of 1450 rev/min. It has four cylinders of nominal bore and stroke 92 mm and 67 mm respectively, giving a swept volume rate of 42.9 dm^3/s. The cylinders are not water cooled.

Below line A compound compression must be used to avoid excessive temperatures. Below line B excessive temperatures will be encountered at part load and compound compression must be used. Above line C compressors should have modified valves which give higher capacity and lower shaft power than those shown here.

The rating conditions are: useful superheat 8.4 K at the compressor suction and 5.5 K liquid subcooling at the expansion valve. (From data supplied by Hall-Thermotank Products Ltd.)

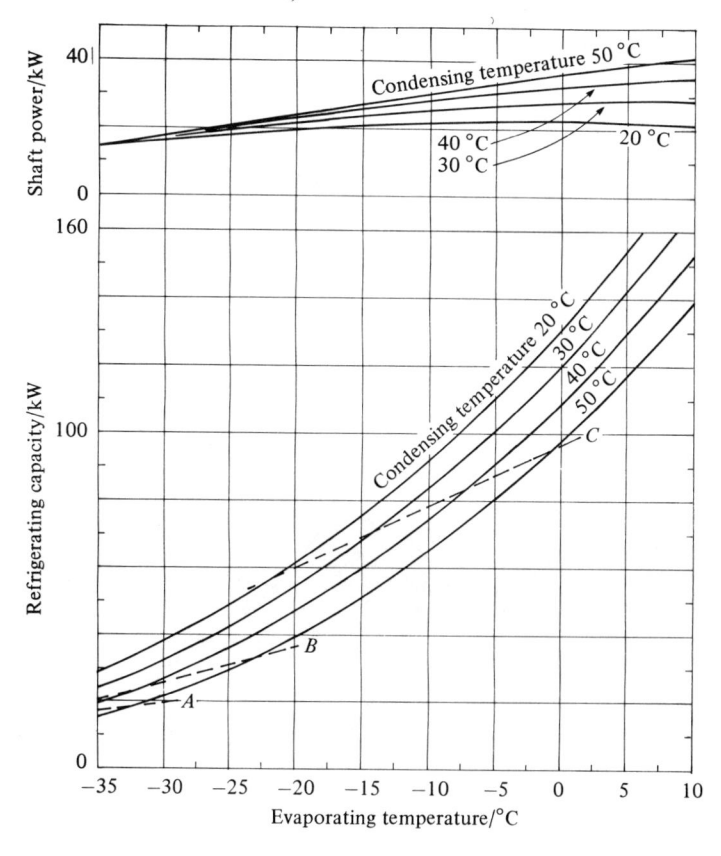

Another question which arises is how representative is the data of the actual product, taking into account variation in manufacturing tolerances. Do the capacity curves represent the average or the minimum performance?

Typical curves of refrigerating capacity for a modern high-speed compressor are shown in Fig. 4.4.5.

4.5 Shaft power and mean effective pressure

The power taken at the shaft of a compressor is composed of several items which will now be discussed in terms of their effects on the mean effective pressure, defined as:

$$w_v = \dot{W}/\dot{V}_{sw} \qquad (4.5.1)$$

where $\dot{W}$ is the shaft power and $\dot{V}_{sw}$ is the swept volume rate.

The largest component of w_v is the mean effective pressure of the indicator diagram, obtained by measuring the area of the actual indicator diagram and dividing the area, in the units of work, by the swept volume. This will be denoted by $w_{v,ind}$. The smaller component, denoted by $w_{v,f}$, represents the work needed to overcome the various resistances in the machine, such as the bearings, sliding of the piston, rubbing at the shaft seal or stuffing box, and the work to drive an internal oil pump. The mean effective pressure at the shaft is then:

$$w_v = w_{v,ind} + w_{v,f} \qquad (4.5.2)$$

Examination of the diagrammatic indicator diagram, Fig. 4.5.1, shows

Fig. 4.5.1. Division of the indicator diagram into areas for the calculation of mean effective pressure.

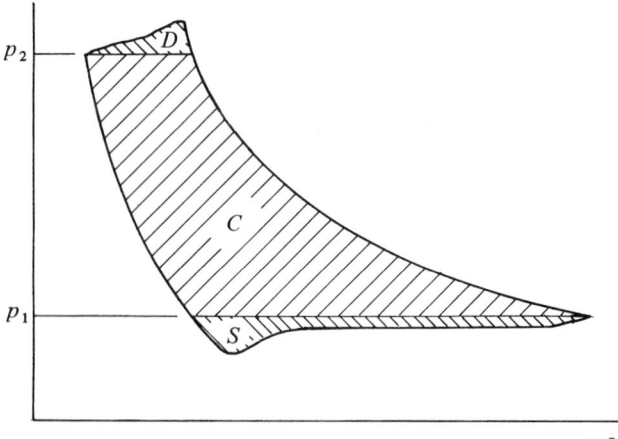

4.5 Shaft power and mean effective pressure

that it can be considered as made up of the three parts: area C, lying between the lines $p_1 = \text{const}$ and $p_2 = \text{const}$; area S, the irreversible work done during the introduction of the vapour into the cylinder; area D, the irreversible work done during the expulsion of the vapour from the cylinder.

Areas S and D used to be known as the 'wiredrawing' areas in connection with steam engines, a descriptive term but not a very accurate one.

Area C is very similar to the area of the indicator diagram of the ideal compressor, but on account of the fact that the observed indices of compression and expansion are less than the isentropic ones, the actual area C is slightly less. When the indices of compression and re-expansion are equal, both n say, the mean effective pressure of the area C is given by a modification of eq. (4.3.12) on replacing k in each of the terms by n:

$$w_{v,C} = p_1 \left[1 + c - c\left(\frac{p_2}{p_1}\right)^{1/n} \right] \frac{n}{n-1} \left[\left(\frac{p_2}{p_1}\right)^{(n-1)/n} - 1 \right] \quad (4.5.3)$$

This expression may be used when definite information about n is available from tests. In the absence of such data it is recommended to approximate $w_{v,C}$ by $w_{v,\text{id}}$ and to use either eq. (4.3.10) or (4.3.12), whichever is convenient. The difference between k and n in modern high-speed compressors is not large, and the suggested procedure will overestimate the power slightly and so be on the conservative side.

Values of the areas S and D may also be expressed as mean effective pressures:

$$w_{v,S} = \frac{\text{Area } S}{V_{\text{sw}}} \quad (4.5.4)$$

$$w_{v,D} = \frac{\text{Area } D}{V_{\text{sw}}} \quad (4.5.5)$$

which can be determined from tests in which indicator diagrams are taken. Using a computer it is possible to estimate values of $w_{v,S}$ and $w_{v,D}$ using the methods outlined by McLaren, Kerr & Tramschek (1975). For preliminary design or approximate estimation of power the following values of $w_{v,S} + w_{v,D}$ may be used: for ammonia, 0.05–0.1 bar (5–10 kJ/m^3); for high-density (fluorinated) refrigerants, 0.1–0.5 bar (10–50 kJ/m^3). The values depend considerably on the rotational speed and on the design of the valves.

Values of the frictional mean effective pressure $w_{v,f}$ are obtained from tests in two ways. The best method is by measurement of the shaft power and the indicated power. A less accurate way is to motor the compressor with the head removed, but this does not provide the loads on the bearings or the side thrust on the piston and the result is usually smaller than that obtained by the first method. For modern compressors the value of $w_{v,f}$ is of the order 0.5–1.0 bar (50–100 kJ/m^3).

This additive method of estimating mean effective pressure and power is to be preferred in practice to methods based on the use of an isentropic efficiency, as defined by eq. (3.3.8) and shown in examples 3.5.2 and 3.5.3. The isentropic efficiency varies greatly with pressure ratio, and cannot be relied on. Furthermore, if the compressor turns out to have a better volumetric efficiency than was originally supposed, the method using isentropic efficiency can underestimate the power, because it bases it on a smaller mass flow rate than is actually attained. The above additive method bases the power on the actual swept volume of the machine, and though there may be uncertainties in the values of $w_{v,S}$, $w_{v,D}$ and $w_{v,f}$, the upper and lower bounds of these can be estimated reasonably well, and their effects on the shaft power determined.

The variation of power is usually shown as a function of evaporating temperature with condensing temperature as a parameter. Some curves for a typical high-speed machine are shown in Fig. 4.4.5. It will be seen that they exhibit the peak power feature which was discussed in section 4.3 for the ideal compressor with clearance.

The method of calculating power explained above will now be used to obtain an expression for the coefficient of performance of a plant employing reciprocating compressors. The shaft power is given by:

$$\dot{W} = \dot{V}_{sw}\left[\eta_{v,\text{ide}}\frac{(h_2 - h_1)_s}{v_1} + w_{v,S} + w_{v,D} + w_{v,f}\right] \quad (4.5.6)$$

The refrigerating capacity is:

$$\dot{Q}_e = \dot{V}_{sw}\eta_v(h_1 - h_3)/v_1 \quad (4.5.7)$$

where $(h_1 - h_3)/v_1$ is the volumic refrigerating effect, using the notation of Chapter 3.

The coefficient of performance is $\dot{Q}_e/\dot{W}$, but it will be more convenient here to use its reciprocal $1/\varepsilon$:

$$\frac{1}{\varepsilon} = \frac{\eta_{v,\text{ide}}(h_2 - h_1)_s}{\eta_v(h_1 - h_3)} + \frac{w_{v,p,f}}{\eta_v(h_1 - h_3)/v_1} \quad (4.5.8)$$

4.5 Shaft power and mean effective pressure

where $w_{v,p,f}$ stands for the sum of the mean effective pressures of pumping and friction. The ratio:

$$\varepsilon_r = \frac{(h_1 - h_3)}{(h_2 - h_1)_s} \tag{4.5.9}$$

was introduced in Chapter 3 as the coefficient of performance of the ideal cycle employing isentropic compression, and was found to be a property of the refrigerant for given evaporating and condensing temperatures and a given suction state 1. With eq. (4.5.9), eq. (4.5.8) becomes:

$$\frac{1}{\varepsilon} = \frac{\eta_{v,idc}}{\eta_v}\frac{1}{\varepsilon_r} + \frac{w_{v,p,f}}{\eta_v(h_1 - h_3)/v_1} \tag{4.5.10}$$

This equation shows the relative importance of the factors which make the real coefficient of performance of a plant less than that, ε_r, of the comparison cycle. The first term of the equation shows the effect of poor volumetric efficiency of a compressor when compared with that of an ideal compressor with the same clearance fraction. The effects of pressure drops through the valve ports, heat transfer and leakage reduce η_v without a corresponding reduction of $\eta_{v,idc}$. The machine takes almost as much power as the ideal compressor with the same clearance, but it delivers a smaller mass flow rate and hence produces a lower refrigerating capacity.

The second term of eq. (4.5.10) shows the importance of having good volumetric efficiency and a high value of the volumic refrigerating effect $(h_1 - h_3)/v_1$. The higher the product of these, i.e. the denominator of the second term, the smaller the effect of given mean effective pressures of pumping and friction on the real coefficient of performance. This explains why, for example, carbon dioxide which appears according to Table 3.9.1 to have such a poor coefficient of performance ε_r, is not so bad in practice as this would indicate. The very large volumic refrigerating effect reduces the second term of eq. (4.5.10) to a small proportion of the first term. The equation has been used by the author (Gosney, 1972) to analyse the differences in performance between R12, R22 and R502 when used in the same compressor.

Example 4.5.1

An ammonia compressor has 8 cylinders of bore 10 cm, stroke 8 cm, and a clearance fraction of 0.04. The speed is 1200 rev/min. Assuming mean effective pressures of pumping of 0.05 bar each at suction and delivery, and 0.5 bar for bearing resistance, etc., estimate the power taken at an evaporating temperature of $-25\,°C$ and a condensing temperature of $25\,°C$.

$$p_1(-25\,°C) = 1.516 \text{ bar}$$
$$p_2(25\,°C) = 10.027 \text{ bar}$$
$$p_2/p_1 = 6.614$$

From Fig. 4.2.2 a value of $k = 1.291$ is estimated. Then:

$$\eta_{v,idc} = 1.04 - (0.04)(6.614)^{1/1.291} = 0.8672$$

$$w_{v,id} = (1.516 \text{ bar})\frac{1.291}{0.291}(6.614^{0.291/1.291} - 1)$$

$$= 3.570 \text{ bar}$$

and:

$$w_{v,ide} = (0.8672)(3.570 \text{ bar}) = 3.10 \text{ bar}$$

Adding the supplementary mean effective pressures gives:

$$w_v = (3.10 + 0.05 + 0.05 + 0.5) \text{ bar} = 3.70 \text{ bar}$$

$$\dot{V}_{sw} = \frac{\pi}{4}(8)(0.1)^2(0.08)\frac{1200}{60} \text{ m}^3/\text{s} = 0.1005 \text{ m}^3/\text{s}$$

The power is then:

$$\dot{W} = (3.70)(10^5)(0.1005) \text{ W} = 37.2 \text{ kW}$$

The value of $w_{v,id}$ could have been found from the tables without resorting to a value for k by using eq. (4.2.14). For saturated ammonia vapour at $-25\,°C$:

$$h_1 = 1411.6 \text{ kJ/kg}$$
$$v_1 = 0.7715 \text{ m}^3/\text{kg}$$
$$s_1 = 5.6978 \text{ kJ/kg K}$$

At a pressure of 10.027 bar corresponding to $25\,°C$ saturation temperature:

$$h_2 = 1688.1 \text{ kJ/kg at } s_2 = s_1 = 5.6978 \text{ kJ/kg K}$$

Hence:

$$w_{v,id} = \frac{1688.1 - 1411.6}{0.7715} \text{ kJ/m}^3 = 358.4 \text{ kJ/m}^3$$

$$= 3.584 \text{ bar}$$

which is in satisfactory agreement with the value already obtained using an isentropic index of compression.

4.6 Scale effects in reciprocating compressors

In common with other types of reciprocating machinery such as internal combustion engines, the rotational speeds of compressors have been increased over the years with the aim of obtaining more refrigerating capacity from a given size of machine. This trend has implications for the design of reciprocating compressors which will now be considered.

Suppose that two geometrically similar compressors having the same swept volume rate are compared: one with speed N and a representative linear dimension L, the other with higher speed rN and representative linear dimension L'. For the same swept volume rate:

$$N L^3 = r N L'^3$$

and:

$$L' = L/r^{1/3} \tag{4.6.1}$$

The masses m and m' of the low- and high-speed machines, respectively, are proportional to volume, and:

$$m' = m/r \tag{4.6.2}$$

The higher speed machine is thus smaller and lighter than the slow-speed one. The saving in cost is not only that of the steel, cast iron, etc., which is not so very great, but the whole series of reduced costs which follow when the size of a product is reduced: area of factory floor, size of machine tools, lifting gear, transport, and so on, are all smaller. The user too, in most instances, will prefer the smaller compressor, because it takes up less of his space and releases some for productive work. Consequently the production of the older type of slow-speed machine with a few cylinders of large bore has ceased almost completely, except for some special purposes.

The modern compressor is not, however, geometrically similar to the old one because of a basic limitation imposed by the piston speed. As the rotational speed is increased from N to rN, with a reduction of stroke from L to $L/r^{1/3}$, the piston speed increases by the factor $r^{2/3}$. Geometrical similarity at the same capacity therefore demands an increase in piston speed, and this turns out to be unacceptable after a certain point. The reason is that the areas S and D in Fig. 4.5.1 increase with the velocity of the vapour through the valve ports, approximately as the square of velocity. If the ratio of the port diameter to the bore is fixed by similarity, then the areas S and D increase with r in the ratio $r^{4/3}$, approximately, giving undesirably high indicated mean effective pressure and power. This places a limit on the piston speed which can be used. The mean piston speed is defined as:

$$\mathscr{V}_p = 2NL \tag{4.6.3}$$

where L is the stroke, and its maximum value with conventional valve designs is about 3 to 4 m/s. This figure has altered surprisingly little during this century. The large slow-speed compressor quoted in Chapter 2 had a mean piston speed of 2 m/s. Improved valve designs enabled this to be increased to 3 m/s, at which it stayed for many years, and some recent designs have allowed an increase to 4 m/s.

One way of preventing the piston speed from rising in proportion to the rotational speed is to reduce the stroke/bore ratio, and this has been done in most modern machines. The stroke is now usually less than the bore of the cylinder. Another way is to reduce the stroke and provide more cylinders to give the desired swept volume rate. The number cannot be increased beyond 12 or 16 in practice, however. This has the curious result, pointed out by Soumerai (1965), that it is not possible to build a machine above a certain swept volume rate for a given rotational speed, within the limit of piston speed mentioned above.

In practice this means that reciprocating compressors are no longer used for very large refrigerating capacities. Many manufacturers extend their range by offering two compressors coupled to one motor, the so-called 'tandem' arrangement, or by providing two or more compressor–motor combinations on a common framework, called 'duplex' or 'multiplex'. One can further choose a high-pressure refrigerant in order to get the most capacity from a machine. But when all this has been done, high-speed reciprocating compressors are excluded for very large capacities, and screw compressors or turbo-compressors must be used.

Example 4.6.1

For a stroke/bore ratio of 0.75, a maximum piston speed of 4 m/s, and a rotational speed of 1450 rev/min, determine the maximum refrigerating capacity for which an ammonia compressor of 8 cylinders could be built, to operate at an evaporating temperature of $-20\,°C$ and a condensing temperature of $30\,°C$. The volumetric efficiency is taken as 0.65.

From eq. (4.6.3) the stroke is:

$$L = \frac{(4)(60)}{2(1450)}\ \text{m} = 0.08276\ \text{m}$$

$$D = \frac{0.08276}{0.75}\ \text{m} = 0.1103\ \text{m}$$

4.7 Screw compressors

and the swept volume rate is:

$$\dot{V}_{sw} = 8\frac{\pi}{4}(0.1103)^2(0.08276)\frac{1450}{60} \text{ m}^3/\text{s} = 0.1529 \text{ m}^3/\text{s}$$

Vapour at the compressor suction will be assumed to be usefully superheated by 5 k. From the ammonia tables:

$v_1 = 0.6385 \text{ m}^3/\text{kg}$

$h_1 = 1430.8 \text{ kJ/kg}$

$h_3 = 322.8 \text{ kJ/kg}$

$h_1 - h_3 = 1108.0 \text{ kJ/kg}$

The refrigerating capacity is therefore:

$$\dot{Q}_e = (0.65)(0.1529)\frac{(1108.0)}{(0.6385)} \text{ kW} = 172.5 \text{ kW}$$

4.7 Screw compressors

The description of the operation of the twin-screw compressor given in Chapter 2 indicates that the 'swept volume' is the volume of the spaces on the male rotor and the flutes on the female rotor. In the usual form of compressor used in refrigeration, the male rotor rotates at input speed N say, and the female rotor rotates at $2N/3$ because there are 4 lobes on the male rotor and 6 flutes on the female rotor, as shown in Fig. 4.7.1. This combination is chosen mainly on the grounds of stiffness of the rotors, and is suitable for most applications. For very high delivery pressures a $6 + 8$ combination may be used, which is even stiffer.

For the usual $4 + 6$ combination, each interlobe space and each flute is

Fig. 4.7.1. Cross-section of the 4-lobed male rotor and the 6-fluted female rotor of a screw compressor.

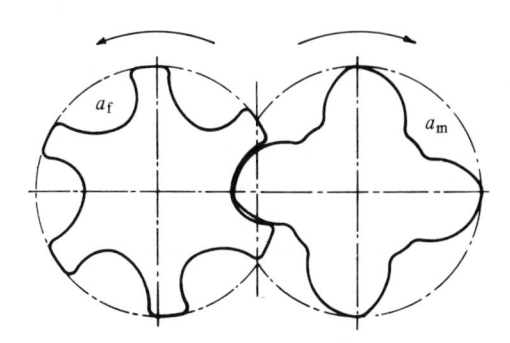

filled with vapour at the point when it is cut off from the suction port by the rotation, and before remeshing has begun. Denoting the cross-sectional areas in Fig. 4.7.1 by a_m and a_f for the male and female rotors respectively, the swept volume rate is given by:

$$\dot{V}_{sw} = N(4a_m + \tfrac{2}{3}6a_f)L = 4N(a_m + a_f)L \tag{4.7.1}$$

where N is the rotational speed of the male rotor, and L is the length of the rotors. Putting:

$$K = \frac{a_m + a_f}{\pi D^2/4} \tag{4.7.2}$$

the swept volume rate is given by:

$$\dot{V}_{sw} = K\pi N D^2 L \tag{4.7.3}$$

where D is the diameter of the rotors. In machines with asymmetric rotors shown in Fig. 4.7.1, K is about 0.155.

Corresponding to this swept volume there is a volumetric efficiency defined by eq. (4.4.3) which is less than unity. Unlike the volumetric efficiency of a reciprocating compressor which is very dependent on pressure ratio because of the effect of clearance, that of a screw machine is almost constant. The clearance volume which remains filled with high-pressure vapour at the end of delivery is very small. From this point of view, therefore, screw compressors can work at high pressure ratios without loss of volumetric efficiency.

The principal factor affecting volumetric efficiency is leakage. In dry machines there is a gap between the rotors and the case, and vapour can leak from one space to one at lower pressure. Because of this leakage, dry machines have a characteristic which is to some extent an aerodynamic one, and the volumetric efficiency depends on the rotational Mach number (see section 4.11) as well as on the ratio of gap width to rotor diameter. Because of the high mass flow rate through a dry machine in relation to the surface area, the heat transfer from the gas is negligible.

In the screw compressor with oil injection, the gaps are filled with oil, and tolerable leakage can be attained at a lower peripheral speed. At the same time, a substantial rate of heat transfer from the vapour can be obtained, and the delivery temperature can be reduced below what it would be in a reciprocating machine at the same conditions.

A peculiar feature of screw compressors, which also applies to multi-vane rotary compressors, is that they have a built-in pressure ratio which is fixed by their geometry, for a particular gas. The opening of the delivery port

4.7 Screw compressors

takes place when the volume of the trapped vapour has been reduced by a fixed ratio, φ say. Corresponding to this built-in volume ratio there is a built-in pressure ratio Π given by:

$$\Pi = \varphi^n \qquad (4.7.4)$$

where n is the index of compression. This of course depends on the gas being compressed and also on the conditions of operation, especially on how much cooling during compression is effected by the injected oil.

When such a compressor operates against a pressure ratio less than Π, the vapour inside the machine is over-compressed, as shown on the indicator diagram in Fig. 4.7.2(a). The vapour compressed to Πp_1 rushes out through the port when it opens and the pressure inside falls rapidly to p_2, at which pressure the rest of the gas is delivered. The unresisted expansion from Πp_1 to p_2 is an irreversible process which lowers the isentropic efficiency. The extra work is represented by the hatched area.

When the pressure ratio is greater than Π, the gas inside the compression spaces has reached only the pressure Πp_1 at the time when the delivery port opens. Gas already in the delivery pipe therefore rushes back through the port and raises the pressure inside the spaces to p_2, at which pressure all of the gas is expelled, as shown in Fig. 4.7.2(b). Again the extra work is represented by the hatched area.

Assuming that the compression process can be satisfactorily represented by the law $pV^n = \text{const}$, the diagram work in both cases is given by:

$$p_1 V_{sw} \frac{n}{n-1} [\Pi^{(n-1)/n} - 1] + p_1 V_{sw} \frac{r - \Pi}{\varphi}$$

Fig. 4.7.2. (a) Indicator diagram showing delivery against a pressure p_2 which is less than the built-in pressure ratio. (b) indicator diagram showing delivery against a pressure p_2 which is greater than the built-in pressure ratio.

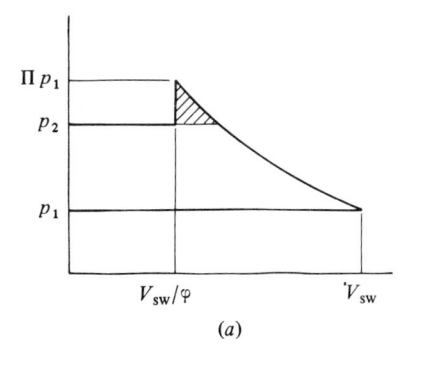

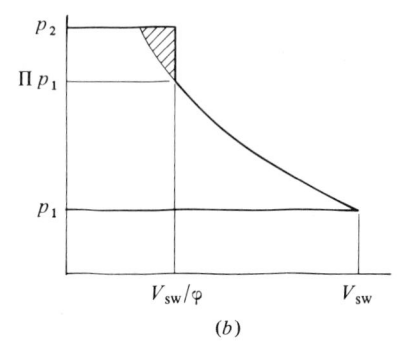

where $r = p_2/p_1$ is the actual pressure ratio. The diagram work which would be done if compression were performed from p_1 to p_2 and delivery took place at p_2 would be:

$$p_1 V_{sw} \frac{n}{n-1} [r^{(n-1)/n} - 1]$$

The ratio of these, which will be called the built-in efficiency, η_φ, represents the unavoidable loss which occurs when a compressor of built-in volume ratio φ operates against a pressure ratio r, even when the internal compression process may be considered reversible. It is given by:

$$\eta_\varphi = \frac{[n/(n-1)][r^{(n-1)/n} - 1]}{[n/(n-1)][\varphi^{(n-1)} - 1] + (r - \varphi^n)/\varphi} \quad (4.7.5)$$

on dividing the above terms and eliminating Π using eq. (4.7.4). For the special case of $n = 1$, $\Pi = \varphi$ and:

$$\eta_\varphi = \frac{\ln r}{\ln \varphi + (r - \varphi)/\varphi} \quad (4.7.6)$$

The variation of this built-in efficiency with r for several values of φ is shown in Fig. 4.7.3 for a value of $n = 1.15$. It will be observed that for a particular value of φ the built-in efficiency does not fall off so rapidly when the

Fig. 4.7.3. Dependence of the built-in efficiency on the pressure ratio for several values of the built-in volume ratio.

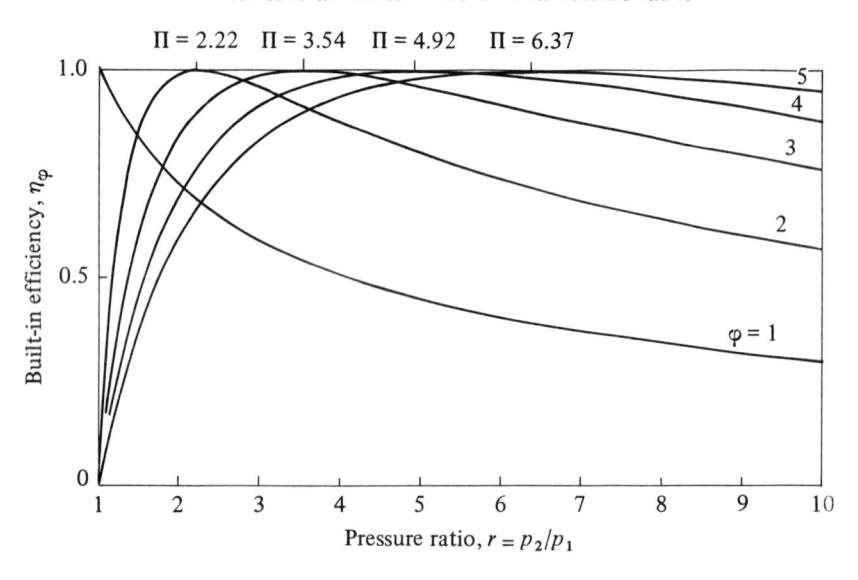

4.7 Screw compressors

pressure ratio exceeds Π as it does for $r < \Pi$. Consequently, accepting that a perfect match of Π and r is not always going to be possible, it is better to choose a built-in volume ratio such that Π is a little lower than the average operating pressure ratio.

This is particularly important if the plant has to do frequent pull-downs, i.e. pulling a warm load down to a low temperature, during which time the pressure ratio tends to be low.

Another point which has to be remembered is that under part load conditions the built-in pressure ratio changes. This was illustrated for a twin-screw compressor in which the built-in ratio is changed by altering the radial port in Fig. 2.4.4. If a substantial period of part load running is expected the built-in ratio should be selected accordingly.

The case of $\varphi = 1$ in Fig. 4.7.3 represents a machine with no internal compression, such as a Root's blower or a liquid pump used on a gas. Obviously it is very inefficient as a gas compressor at pressure ratios above 1.5 say, but may be acceptable for very low ratios.

The dependence of the built-in efficiency on the index of compression is shown in Fig. 4.7.4 for $\varphi = 4$ and two values of n. When compared at the same built-in pressure ratio, however, there is very little difference. The curve for $\varphi = 6.06$ with $n = 1$ coincides with the curve for $\varphi = 4$ with $n = 1.3$.

The built-in efficiency expresses only the losses associated with the

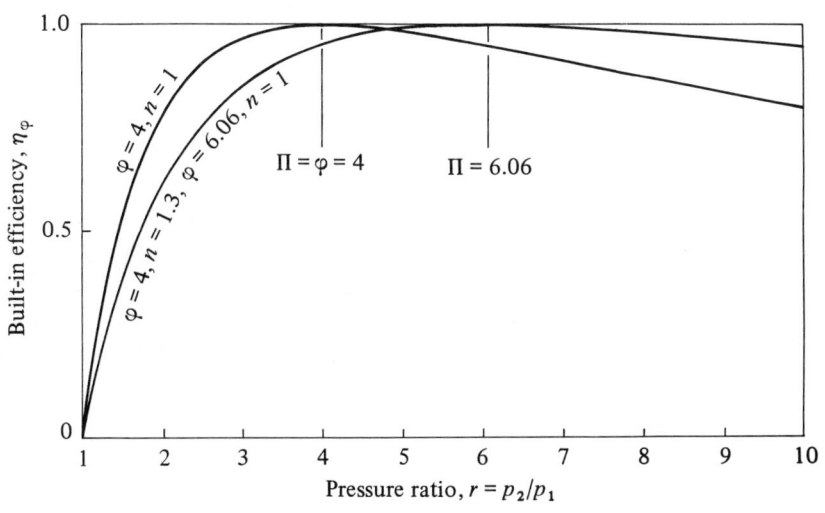

Fig. 4.7.4. The effect of n on the built-in efficiency for the same built-in volume ratio. For the same built-in pressure ratio n has very little effect.

mismatch of Π and r of course. The other losses in the machine have to be taken into account to obtain the actual isentropic efficiency.

If one defines an 'ideal' compressor with a built-in volume ratio φ as one in which the only losses are those associated with over or under compression, the mean effective pressure $w_{v,id}$ is given by:

$$w_{v,id} = p_1 \left[\frac{n}{n-1}(\varphi^{(n-1)} - 1) + \frac{r - \varphi^n}{\varphi} \right]$$

$$= p_1 \left[\frac{\varphi^{(n-1)} - n}{n-1} + \frac{r}{\varphi} \right] \qquad (4.7.7)$$

For the special case of $n = 1$:

$$w_{v,id} = p_1 \left[\ln\varphi + \frac{r - \varphi}{\varphi} \right] \qquad (4.7.8)$$

A compressor with a built-in volume ratio does not exhibit the power maxima with varying p_1 and p_2 like those shown by a reciprocating compressor. Supposing p_1 is kept constant whilst p_2 is varied, differentiation of eq. (4.7.7) gives:

$$\frac{dw_{v,id}}{dp_2} = \frac{1}{\varphi} \qquad (4.7.9)$$

For φ constant, this is constant and always positive, i.e. mean effective pressure always increases with delivery pressure. On the other hand, if p_2 is supposed constant whilst p_1 is varied:

$$\frac{dw_{v,id}}{dp_1} = \frac{1}{n-1}(\varphi^{(n-1)} - n) \qquad (4.7.10)$$

Provided that $n > 1$, this is always positive, always zero or always negative according as $\varphi^{(n-1)}$ is greater than, equal to or less than n.

For $n = 1.15$ say, the criterion is $\varphi = 2.54$. Thus for compressors of low built-in volume ratio the mean effective pressure and power tend to diminish with increasing suction pressure, whilst for those with high built-in volume ratio the mean effective pressure tends to increase with increasing suction pressure. For high suction pressure applications some manufacturers of screw compressors offer a model with a built-in volume ratio of about 2.5 and the power taken by this is almost constant.

The actual mean effective pressure is greater than the 'ideal' one of course because of the pressure drops through the ports and because of friction. These tend to increase with increasing loading and so modify the above results somewhat in that even the machines with low built-in volume ratio show some increase of power with suction pressure increasing.

The peak isentropic efficiency of twin-screw compressors with oil injection currently on the market is about 75%. Efficiencies at other conditions can be estimated by applying the factor η_φ, bearing in mind that at part load the built-in volume ratio differs from that at full load.

Some of the references at the end of section 2.8 contain certain details of performance, but there is a scarcity of information generally in the published literature.

4.8 Steady-flow compression

Whereas the positive displacement compressor operates by inducing a discrete volume of gas, compressing it into a smaller volume and then pushing it out, compression in a turbo-compressor is a continuous process. The rise in pressure of a particle of fluid is produced entirely by forces which are the result of accelerations in a stream of fluid moving at high velocity. The accompanying shear forces ('fluid friction') acting in the fluid introduce a basic difference in the way in which the integral of volume with respect to pressure can be interpreted as work, which will now be discussed.

Neglecting kinetic energy and potential energy terms, the steady-flow energy equation gives:

$$\dot{Q} + \dot{W} = \dot{m}(h_2 - h_1) \tag{4.8.1}$$

where subscripts 1 and 2 denote the inlet and outlet states respectively, $\dot{m}$ is the mass flow rate, $\dot{Q}$ is the rate of heat transfer *to* the control surface and $\dot{W}$ is the power *to* the control surface.

In turbo-compressors the term $\dot{Q}$ can almost always be neglected, because the through-put of vapour is so large in relation to the surface area available for heat transfer that the ratio $\dot{Q}/\dot{m}$ is small in comparison with the change of specific enthalpy. It is true that most turbo-compressors have a cooler for the lubricating oil, but the heat from the oil is largely the result of viscous shear in the bearings. If $\dot{W}$ in eq. (4.8.1) is taken to mean the shaft power minus the power absorbed in the bearings and seals, the assumption that $\dot{Q}$ is zero will be very nearly correct. Then:

$$\dot{W} = \dot{m}(h_2 - h_1) \tag{4.8.2}$$

Equations (4.8.1) and (4.8.2) are applicable whether the process is reversible or not, provided that h_2 is the actual specific enthalpy at delivery. The effect of internal friction is to change both sides of the equations by the same amounts. For example, in eq. (4.8.2), if more power is taken because of internal friction, the fluid would leave with a higher specific enthalpy, which for a dry vapour or gas means a higher temperature.

For the special case of reversible compression, eq. (4.8.2) becomes:

$$\dot{W}_{\text{isen}} = \dot{m}(h_2 - h_1)_s \tag{4.8.3}$$

because a reversible adiabatic process is necessarily isentropic, denoted as before by the symbol $)_s$. Using eq. (4.2.1):

$$\dot{W}_{\text{isen}} = \dot{m} \int_{p_1}^{p_2} v \, dp)_s \tag{4.8.4}$$

Thus for isentropic compression the integral of $v \, dp$ gives the specific work in the steady-flow machine as well as in the ideal reciprocating compressor.

When compression is not reversible a difference appears between the expressions for work in the two machines on account of the different ways in which they operate. For positive displacement machines the sources of irreversibility are the heat transfer within the cylinder and the work of pumping vapour in and out. The irreversibility caused by viscous shearing in the fluid during the actual compression process is small, and one can always say without appreciable error that the contribution made to the whole work during a pressure change dp is given by $V \, dp$, and that the work, apart from mechanical friction, is given by the area enclosed on the indicator diagram.

In a turbo-compressor this is no longer true. The process of compression in a turbo-compressor requires much higher velocities of the fluid than in a reciprocating compressor, with higher gradients of velocity and shear stresses at the boundaries where the work transfer occurs. Whereas in the reciprocating compressor the entire work done by the piston is done in overcoming a normal force equal to the pressure multiplied by the area, in a turbo-compressor work has to be done against the viscous shear forces as well as against the normal pressure forces. In these circumstances, an element of specific work dw is not equal to $v \, dp$, but is greater than it. The ratio:

$$\eta_{\text{ss}} = \frac{v \, dp}{dw} \tag{4.8.5}$$

is called the small stage efficiency of the process. For an adiabatic one:

$$dw = dh \tag{4.8.6}$$

4.8 Steady-flow compression

and the small stage efficiency becomes:

$$\eta_{ss} = \frac{v\,dp}{dh} \qquad (4.8.7)$$

For the whole process from initial state 1 to final state 2, the ratio:

$$\eta_{pol} = \frac{1}{h_2 - h_1} \int_{p_1}^{p_2} v\,dp \qquad (4.8.8)$$

is called the polytropic efficiency. When η_{ss} is constant, $\eta_{pol} = \eta_{ss}$.

The significance of these efficiencies will be discussed with reference to the adiabatic compression processes shown diagrammatically in Fig. 4.8.1. The isentropic process is shown as 1–2′, the real process as 1–2. At each stage of the real process the specific volume is larger than for the isentropic process because of the internal fluid friction. The specific work of the real process $(h_2 - h_1)$ is greater than the specific work of the isentropic process $(h'_2 - h_1) = (h_2 - h_1)_s$. The ratio:

$$\eta_{isen} = \frac{(h_2 - h_1)_s}{(h_2 - h_1)} \qquad (4.8.9)$$

Fig. 4.8.1. Variation of specific volume with pressure during steady-flow adiabatic compression, shown diagrammatically. 1–2 shows a possible real process, 1–2′ shows an ideal isentropic process.

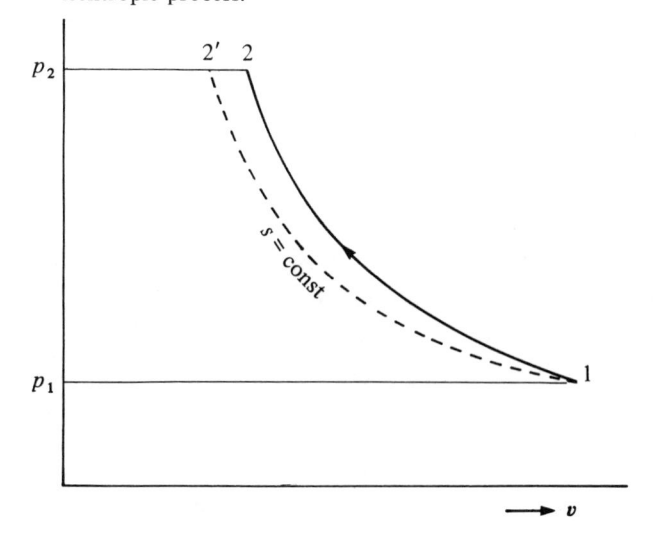

Compression processes 316

has already been introduced and defined in Chapter 3 as the isentropic efficiency.

The specific work of the isentropic process is given by the integral in eq. (4.8.4) as the area to the left of the curve $1-2'$, but the work of the real process is not given by the area to the left of the curve $1-2$. The real specific work is greater than this area by an amount which does not show on the diagram at all, and which is expressed by the polytropic efficiency eq. (4.8 8). Thus the specific work of the real process, $(h_2 - h_1)$, is greater than the specific work of the isentropic process, $(h_2 - h_1)_s$, for two reasons. Firstly because the specific work is greater for the real process than the area to the left of the curve on the $p-v$ diagram; secondly because the area to the left of the real process is greater than the area to the left of the isentropic process. The first of these reasons is concerned with the inherent irreversibility of the process, which is measured by the polytropic efficiency. The second cause comes in because the effects of friction at an early stage of the process increase the specific volume at the later stages and require more work to pump the larger volume through the pressure range.

It will be recalled that, in connection with steam turbines, the 'reheating' caused by irreversibility at an early stage increases the specific volume at subsequent stages, and thereby increases the work delivered. The effect of friction is thus partly offset, and the isentropic efficiency of the whole turbine is greater than the efficiency of a single stage by a factor called the 'reheat factor'. In compressors, however, it is the other way round. The effect of inefficiency in the early stages is to increase the specific volume in later stages and thus increase the work which has to be put in. The isentropic efficiency of the whole machine is less than the efficiency of a single stage.

Finally, it may be remarked that the small stage or polytropic efficiency compares a quantity, $v\,dp$ or integral $v\,dp$, relating to a real process with the specific work in that process, whereas the isentropic efficiency compares a quantity relating to a hypothetical process, $(h_2 - h_1)_s$, with the specific work. The small stage and polytropic efficiencies are thus more fundamental than the isentropic efficiency.

Unfortunately there is a difficulty in applying them to an actual machine. The variation of specific volume with pressure throughout is seldom known accurately, and the integral $v\,dp$ cannot be properly evaluated. For compressors pumping ideal gases with constant specific heat capacities, the difficulty is circumvented by assumptions which will be explained below. For compressors pumping real vapours there is a further difficulty in that the lack of an equation of state and a simple expression for enthalpy blocks

4.8 Steady-flow compression

the direct application of efficiencies obtained from tests. The procedure commonly used in practice is not strictly correct; but since it is widely used in the literature and gives some useful results, it is explained below.

Consider first the case of an ideal gas with constant specific heat capacities. For an element of the process in which the specific enthalpy changes by dh, and the temperature by dT:

$$dh = c_p dT = \frac{c_p}{R} d(pv) \qquad (4.8.10)$$

where c_p is the specific heat capacity of the gas at constant pressure, and R is the specific gas constant. The small stage efficiency is then:

$$\eta_{ss} = \frac{Rv\,dp}{c_p d(pv)} \qquad (4.8.11)$$

Differentiating $d(pv)$ and rearranging gives:

$$\left(\frac{c_p \eta_{ss}}{c_p \eta_{ss} - R}\right)\frac{dv}{v} + \frac{dp}{p} = 0 \qquad (4.8.12)$$

If the small stage efficiency is constant throughout the process, the term in the brackets is constant. Denoting it by n where:

$$n = \frac{c_p \eta_{ss}}{c_p \eta_{ss} - R} \qquad (4.8.13)$$

eq. (4.8.12) may be integrated to give:

$$pv^n = \text{const} \qquad (4.8.14)$$

during the process. The integral of $v\,dp$ can now be evaluated as:

$$\int_{p_1}^{p_2} v\,dp = p_1 v_1 \frac{n}{n-1}\left[\left(\frac{p_2}{p_1}\right)^{(n-1)/n} - 1\right] \qquad (4.8.15)$$

The polytropic efficiency is then calculated from eq. (4.8.8) with:

$$h_2 - h_1 = c_p(T_2 - T_1) \qquad (4.8.16)$$

for an ideal gas with constant c_p.

Expressing the specific gas constant R in terms of the ratio γ of the specific heat capacities:

$$R = c_p - c_v = c_p \frac{\gamma - 1}{\gamma} \qquad (4.8.17)$$

where $\gamma = c_p/c_v$, n may be written:

$$n = \frac{\gamma \eta_{ss}}{\gamma \eta_{ss} - \gamma + 1} \tag{4.8.18}$$

and solved for $\eta_{ss} = \eta_{pol}$

$$\eta_{pol} = \frac{n}{n-1} \frac{\gamma - 1}{\gamma} \tag{4.8.19}$$

These relations are used for ideal gas compressors as follows. From measurements of p_1 and T_1 at the inlet, and of p_2 and T_2 at the outlet, the value of n is calculated from:

$$\frac{T_2}{T_1} = \left(\frac{p_2}{p_1}\right)^{(n-1)/n} \tag{4.8.20}$$

which follows from eq. (4.8.14) on inserting the ideal gas equation. Knowing γ, the polytropic efficiency is found from eq. (4.8.19).

In reverse, to determine the power taken by a compressor for which the polytropic efficiency is known, the value of n is calculated from eq. (4.8.19), and T_2 is then found from eq. (4.8.20). The specific work is then given by eq. (4.8.16), and the power is obtained by multiplying this by the mass flow rate. The following example will illustrate the method.

Example 4.8.1
Air is compressed in steady flow from an absolute pressure of 1 bar and a temperature of 20 °C to an absolute pressure of 4 bar. The small stage efficiency, assumed constant, for the process is 0.8. Determine the specific work, and the shaft power for a mass flow rate of 2 kg/s. For air $\gamma = 1.4$, $c_p = 1.00$ kJ/kg K.
From eq. (4.8.19):

$$0.8 = \frac{n}{n-1} \frac{0.4}{1.4} \quad \text{and} \quad \frac{n}{n-1} = 2.8$$

From eq. (4.8.20):

$$T_2 = (293)(4^{1/2.8})\,\text{K} = 481\,\text{K}$$

From eq. (4.8.16), the specific work is:

$$1.00(481 - 293)\,\text{kJ/kg} = 188\,\text{kJ/kg}$$

The power is therefore:

$$(2)(188)\,\text{kW} = 376\,\text{kW}$$

4.8 Steady-flow compression

For the real vapours used as refrigerants, the derivations of eq. (4.8.13) and eq. (4.8.19) are not valid, because eq. (4.8.10) does not hold. For most vapours it is not possible to express the change in enthalpy simply as a constant multiplying the pv product. Nevertheless, eq. (4.8.19) is widely used in practice for vapours, replacing γ with k the index of isentropic compression. When this approximation is made, however, the value of n should be determined from the test results by going back to eq. (4.8.14) with the specific volumes read from the vapour tables, not by using eq. (4.8.20). The following example illustrates the difference in procedure. The result should be checked to ensure that it satisfies (4.8.8).

Example 4.8.2

Ammonia is compressed in steady flow from an absolute pressure of 2.908 bar, saturated vapour, to an absolute pressure of 6.149 bar, with a (constant) small stage efficiency of 0.8. Determine the specific work and the delivery temperature.

From Fig. 4.2.2 the value of k is taken as 1.293. From eq. (4.8.19), using k instead of γ:

$$0.8 = \frac{n}{n-1} \frac{0.293}{1.293}$$

and:

$$n = 1.395$$

From the ammonia tables:

$$v_1(2.908 \text{ bar, saturated}) = 0.4185 \text{ m}^3/\text{kg}$$
$$h_1(2.908 \text{ bar, saturated}) = 1431.6 \text{ kJ/kg}$$

From eq. (4.8.14):

$$v_2 = 0.4185 \left(\frac{2.908}{6.149}\right)^{1/1.395} \text{ m}^3/\text{kg} = 0.2447 \text{ m}^3/\text{kg}$$

Interpolating in the ammonia tables at $p_2 = 6.149$ bar:

$$h_2 = 1558.2 \text{ kJ/kg}$$
$$t_2 = 50.3 \,^\circ\text{C}$$

The specific work is then:

$$(1558.2 - 1431.6) \text{ kJ/kg} = 126.6 \text{ kJ/kg}$$

To check that this result satisfies the basic definition eq. (4.8.8), using

$n = 1.395$ in eq. (4.8.15) gives:

$$\int_1^2 v\,dp = (2.908)(10^5)(0.4185)\frac{1.395}{0.395}\left[\left(\frac{6.149}{2.908}\right)^{0.395/1.395} - 1\right] \text{J/kg}$$

$$= 101.5 \text{ kJ/kg}$$

and:

$$\eta_{\text{pol}} = 101.5/126.6 = 0.802$$

In this example the approximate procedure using eq. (4.8.19) has turned out to be quite accurate. The reason is that the enthalpy of superheated ammonia vapour is given quite closely by a Callendar-type equation:

$$h = h_0 + B(pv) \tag{4.8.21}$$

in which h_0 and B are constants. It can be shown that when this relation holds eq. (4.8.19) is exact. In a later example, 4.11.2, it will be seen that the method using eq. (4.8.19) is not very accurate for R22. There is in fact no reason to expect it to work for this and similar vapours of high molar mass. Equation (4.8.21) represents exceptional behaviour and only holds for a few simple molecules such as water and ammonia.

A more accurate analysis of polytropic efficiency for real vapours has been given by Schultz (1962), but in most of the literature on turbo-compressors the above simplified method is used. Authors unfortunately do not always state the value of k which they have used in deducing the polytropic efficiency from test results, and since it occurs as $(k-1)$ in eq. (4.8.19) a small error is significant. When k is recorded the same value should be used in converting the efficiency back into specific work rather than values taken from Figs. 4.2.2–4.2.5 or from tables.

Head

An expression much used in connection with turbo-compressors is 'head', and it is not always clear what is meant by it. The term comes from pump technology by an analogy which is not exact. A pump lifting water to a height of 10 m, say, is described as producing a head of (10 m) g where g is the local acceleration of gravity. Since the units of g are m/s² the units of head are:

$$\frac{\text{m}^2}{\text{s}^2} = \frac{\text{J}}{\text{kg}} \tag{4.8.22}$$

i.e. the units of specific work or energy.

4.8 Steady-flow compression

The concept of head is useful for pumps because a pump produces approximately the same head with any fluid and any value of g. Furthermore the head of a pump has a real significance as the output of the pump because the liquid which has been lifted to a certain height has a potential energy which can be recovered completely in principle by allowing it to fall to its original level again.

For a compressible fluid there is no corresponding quantity which unambiguously represents the output of a compressor. Pressure at the delivery flange of a compressor cannot be interpreted as specific work or energy without assuming some relation between pressure and density. According to the relation assumed one can define:

$$\text{isentropic head} = (h_2 - h_1)_s = \int_1^2 v \, dp)_s \qquad (4.8.23)$$

in which the relation between p and v is supposed to be that of the reversible adiabatic process, and:

$$\text{polytropic head} = \int_1^2 v \, dp)_{\text{pol}} \qquad (4.8.24)$$

in which the relation between p and v is supposed to be that of the polytropic process, eq. (4.8.14). The polytropic head is preferred in practice because it is usually held to give a better correlation of performance than the isentropic head.

It should be noted that the head of a pump or compressor, correctly expressed, does not depend on the local value of g. When head is expressed, incorrectly, as a height of a fluid column it does depend on the local g, because this implies cancellation of the local g with the standard acceleration used to define the kilogramme force.

Because head as defined above has the units of specific work the word is often used to express a variety of other quantities with these units. Thus the specific work of a compressor, i.e. the shaft power divided by the mass flow rate, is sometimes called the input head, and the theoretical specific work calculated by change of momentum (eq. (4.10.5)) is sometimes called the Euler head. These uses of the word serve no purpose and simply confuse the distinction between what is put into a compressor and what it does. In the remainder of this chapter the term is used only in the sense of eq. (4.8.23) or (4.8.24).

4.9 Stagnation enthalpy, pressure and temperature

When kinetic energy terms are included the steady-flow energy equation is:

$$\dot{Q} + \dot{W} = \dot{m}(h_2 + \tfrac{1}{2}\mathscr{V}_2^2) - \dot{m}(h_1 + \tfrac{1}{2}\mathscr{V}_1^2) \tag{4.9.1}$$

where $\mathscr{V}_1$ and $\mathscr{V}_2$ are the velocities of the fluid at the entry and exit of the control surface. It is clear that the terms h and $\tfrac{1}{2}\mathscr{V}^2$ always occur together, and it is convenient to give their sum a name. The quantity:

$$h_t = h + \tfrac{1}{2}\mathscr{V}^2 \tag{4.9.2}$$

is called the specific stagnation or total enthalpy. The name stagnation indicates that it is the specific enthalpy which is developed when a stream of fluid is decelerated from velocity $\mathscr{V}$ to rest.

In terms of h_t, eq. (4.9.1) becomes:

$$\dot{Q} + \dot{W} = \dot{m}(h_{t2} - h_{t1}) \tag{4.9.3}$$

and for adiabatic flow:

$$\dot{W} = \dot{m}(h_{t2} - h_{t1}) \tag{4.9.4}$$

Corresponding to the stagnation enthalpy there is a stagnation or total pressure defined by:

$$\int_p^{p_t} v\,dp)_s = h_t - h = \tfrac{1}{2}\mathscr{V}^2 \tag{4.9.5}$$

The stagnation pressure is the pressure developed when the fluid is

Fig. 4.9.1. Relation of the stagnation pressure p_t to the static pressure p shown diagrammatically on pressure–specific enthalpy and pressure–specific volume coordinates.

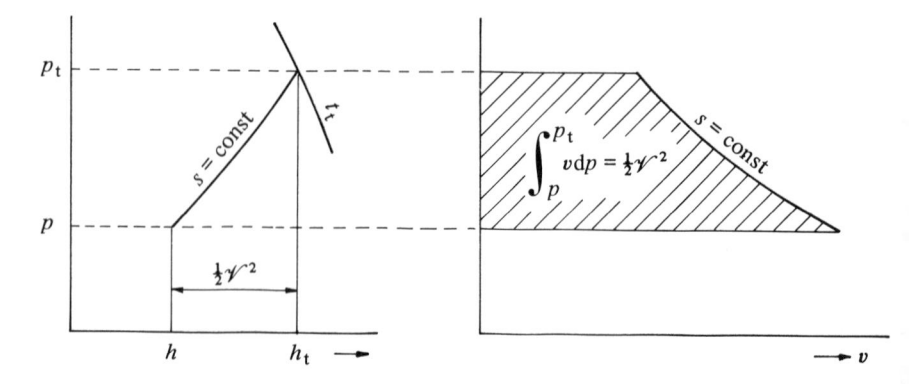

4.9 Stagnation enthalpy, pressure and temperature

decelerated reversibly and adiabatically, i.e. at constant entropy, from velocity $\mathscr{V}$ to rest. The definition is illustrated on pressure–enthalpy and pressure–volume coordinates in Fig. 4.9.1.

Examination of Fig. 4.9.1 shows that there is a temperature t_t, called the stagnation or total temperature, corresponding to the values h_t and p_t.

The stagnation pressure and temperature are approximately the values developed ahead of a bluff body at rest in a moving fluid: not exactly, because the deceleration is not exactly reversible and adiabatic. By making a small hole at the stagnation point, and connecting it to a manometer, the stagnation pressure may be determined, after suitable correction. Similarly, a thermocouple placed in the hole will read the stagnation temperature, after correction.

The true pressure and temperature are the values which would be read by instruments travelling with the fluid at the same velocity. To emphasise the difference, they are sometimes called the static pressure and temperature.

For the special case of an ideal gas with constant heat capacities, the stagnation temperature is obtained by:

$$c_p(t_t - t) = \tfrac{1}{2}\mathscr{V}^2 \qquad (4.9.6)$$

directly, but this method is not applicable to vapours.

For an incompressible fluid, or when the difference between p_t and p is so small that the change in specific volume can be neglected, eq. (4.9.5) becomes:

$$v(p_t - p) = \tfrac{1}{2}\mathscr{V}^2 \qquad (4.9.7)$$

Example 4.9.1

Ammonia vapour flows in a duct at a pressure of 2.908 bar, a temperature of 0 °C and a velocity of 200 m/s. Determine the stagnation pressure and temperature.

From the ammonia tables:

$h(2.908 \text{ bar}, 0\,°\text{C}) = 1456.8 \text{ kJ/kg}$

$s(2.908 \text{ bar}, 0\,°\text{C}) = 5.5656 \text{ kJ/kg K}$

From eq. (4.9.2):

$h_t = 1456.8 \text{ kJ/kg} + \tfrac{1}{2}(200^2) \text{ J/kg} = 1476.8 \text{ kJ/kg}$

By interpolation in the ammonia tables at a pressure of 3.412 bar, the specific enthalpy for a specific entropy of 5.5656 kJ/kg K is: 1477.6 kJ/kg and $t = 10.3\,°\text{C}$.

For the same entropy at a pressure of 3.280 bar, the specific enthalpy is: 1472.4 kJ/kg and $t = 7.7\,°C$.

Interpolating linearly between these, the stagnation pressure is 3.392 bar, and the stagnation temperature is about $9.9\,°C$.

The problem can also be worked using the integral form of the change of specific enthalpy at constant entropy, eq. (4.9.5), and the approximation when a value of k is known, eq. (4.2.21):

$$(h_t - h)_s = p_1 v_1 Y$$

where Y is given by eq. (4.2.20) with (p_t/p) as the pressure ratio. Then:

$$v(2.908\text{ bar}, 0\,°C) = 0.4383\text{ m}^3/\text{kg}$$

and

$$\tfrac{1}{2}(200^2)\text{ J/kg} = (2.908 \cdot 10^5)(0.4383)\frac{1.3}{0.3}\left[\left(\frac{p_t}{p}\right)^{0.3/1.3} - 1\right]\text{J/kg}$$

where an approximate value of $k = 1.3$ has been assumed. Solving for:

$$p_t/p = 1.1663$$

gives:

$$p_t = (1.1663)(2.908\text{ bar}) = 3.393\text{ bar}$$

The isentropic and polytropic efficiencies may be expressed in terms of stagnation quantities instead of static ones. Thus the isentropic efficiency based on stagnation quantities becomes:

$$\eta_{isen} = \frac{(h_{t2} - h_{t1})_s}{(h_{t2} - h_{t1})} \tag{4.9.8}$$

for an adiabatic process. Similarly the numerator of eq. (4.8.8) can be evaluated by integration from p_{t1} to p_{t2}. In most compressors, however, the kinetic energies at the inlet 1 and the exit 2 are not large, and it is customary to express the efficiencies as static values.

4.10 Centrifugal compressors

Almost the only type of steady-flow compressor used in refrigeration is the centrifugal compressor. A general description of the operation has been given in Chapter 2. In this section an elementary treatment of a single-stage machine is presented. For more details the specialist literature, such as the book by Ferguson (1963), should be consulted. Unfortunately, most works treat only ideal gases.

The notation used will be: subscript 1 for the state of the fluid at the inlet flange and subscript 2 for the state at the delivery flange, in accordance with

4.10 Centrifugal compressors

the notation already used. Conditions at the inlet to the impeller will be denoted by subscript i, and conditions at the exit by subscript x.

From the suction flange to the impeller no work is done on the vapour, and the specific stagnation enthalpy remains constant. There may be some acceleration of the vapour owing to the decreasing area for flow, and the static pressure falls a little. Depending on the construction, the vapour approaches the impeller in a more or less axial direction, though in certain circumstances it may have some initial rotation. This applies in multi-stage machines in which the vapour from the preceding stage is still rotating, and also with the method of capacity regulation by inlet guide vanes, mentioned later. For the present axial flow will be assumed.

On entering the impeller the fluid is turned through a right angle into the channels of the impeller, and it flows outwards in these. Work is done on the fluid by the impeller and the specific stagnation enthalpy rises. At the same time the static pressure, stagnation pressure and absolute velocity of the fluid rise. The velocity relative to the impeller usually diminishes. Leaving the impeller, the fluid enters the diffuser in which the absolute velocity is reduced and the static pressure rises. The fluid is then collected into the volute or scroll and flows into the delivery pipe. Between the outlet of the impeller and the delivery pipe no work is done and the specific stagnation enthalpy remains constant. It will be noted from this description that the only change in the stagnation enthalpy occurs across the impeller, where the work is done on the fluid.

Fig. 4.10.1. Velocity diagram at the tip of the impeller of a centrifugal compressor.

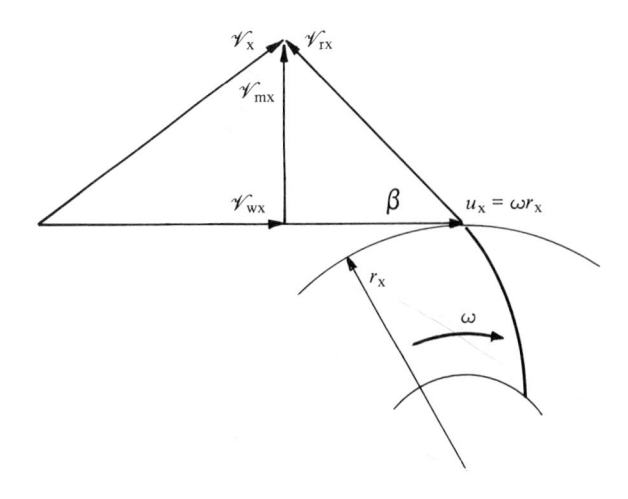

The torque required to turn the impeller is to be found, in principle, by applying the steady-flow momentum equation to a control surface surrounding the impeller. For axial approach to the eye of the impeller, the fluid there has no moment of momentum about the axis. At the periphery of the impeller the specific moment of momentum of the fluid leaving is $r_x \mathscr{V}_{wx}$ where r_x is the radius of the impeller and $\mathscr{V}_{wx}$ is the tangential or whirl component of the absolute velocity of the fluid leaving. For a mass flow rate of $\dot{m}$ the torque T is given by:

$$T = \dot{m} r_x \mathscr{V}_{wx} \tag{4.10.1}$$

and the power is:

$$\dot{W} = \dot{m} \mathscr{V}_{wx} u_x \tag{4.10.2}$$

where $u_x = r_x \omega$ is the peripheral or tip speed of the impeller, and ω is the angular velocity.

The relation between the velocities at the outlet of the impeller is shown in the velocity diagram, Fig. 4.10.1. The vapour leaving the impeller has the velocity $\mathscr{V}_{rx}$ relative to the impeller, and absolute velocity $\mathscr{V}_x$ compounded from $\mathscr{V}_{rx}$ and u_x. The whirl component $\mathscr{V}_{wx}$ is the projection of $\mathscr{V}_x$ on the direction of u_x. Denoting the angle between $\mathscr{V}_{rx}$ and u_x as β, the whirl component is:

$$\mathscr{V}_{wx} = u_x - \mathscr{V}_{mx} \cot \beta = u_x \left(1 - \frac{\mathscr{V}_{mx}}{u_x} \cot \beta \right) \tag{4.10.3}$$

where $\mathscr{V}_{mx}$ is called the meridional component of $\mathscr{V}_x$. The power is then:

$$\dot{W} = \dot{m} u_x^2 \left(1 - \frac{\mathscr{V}_{mx}}{u_x} \cot \beta \right) \tag{4.10.4}$$

and the specific work is:

$$w = \frac{\dot{W}}{\dot{m}} = u_x^2 \left(1 - \frac{\mathscr{V}_{mx}}{u_x} \cot \beta \right) \tag{4.10.5}$$

Introducing the work coefficient μ defined by:

$$\mu = w/u_x^2 = \mathscr{V}_{wx}/u_x \tag{4.10.6}$$

and a flow coefficient φ_x at the tip defined by:

$$\varphi_x = \mathscr{V}_{mx}/u_x \tag{4.10.7}$$

eq. (4.10.5) becomes:

$$\mu = (1 - \varphi_x \cot \beta) \tag{4.10.8}$$

4.10 Centrifugal compressors

The kinetic energy of the fluid leaving the impeller will be needed later. This is given by:

$$\tfrac{1}{2}\mathscr{V}_x^2 = \tfrac{1}{2}(\mathscr{V}_{mx}^2 + \mathscr{V}_{wx}^2) = \tfrac{1}{2}u_x^2(\varphi_x^2 + \mu^2) \tag{4.10.9}$$

The meridional component $\mathscr{V}_{mx}$ is proportional to the volume flow rate from the impeller, and a graph of μ versus φ_x shows how the specific work absorbed by the impeller varies with the flow rate. Figure 4.10.2 shows the variation for the three blade angles: backward curved blades with $\beta = 45°$; radial blades with $\beta = 90°$; and forward curved blades with $\beta = 135°$. It is seen that for a given tip speed u_x, the specific work absorbed by the impeller increases with β. If compression were reversible and adiabatic, the specific work of isentropic compression would increase with β, and so would the pressure ratio developed. Even for real machines, it is confirmed in practice that a higher pressure ratio is produced by forward bladed impellers than by radial and backward bladed ones for a given tip speed.

The variation of shaft power with flow rate is obtained by multiplying the values from Fig. 4.10.2 by the flow coefficient φ_x. The result is shown in Fig. 4.10.3. It is seen that the power taken by an impeller with forward curved blades rises rapidly with flow rate. Such an impeller is said to have an overloading characteristic. Forward bladed impellers are not currently used in refrigerant compressors, but they are used in fans when the maximum pressure ratio for a given tip speed is desired.

For the same tip speed u_x, the absolute velocity of the fluid leaving the impeller increases with β. The kinetic energy has to be recovered in the diffuser to produce a rise in pressure. The diffusion process is usually

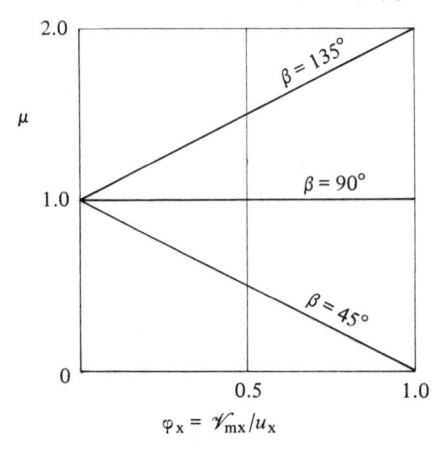

Fig. 4.10.2. Variation of the work coefficient μ from eq. (4.10.8) with the flow coefficient at the tip φ_x, for three values of β.

regarded as one of the least efficient parts of the whole compression process, and consequently backward curved impellers, which require the least pressure rise in the diffuser, have been mostly favoured in the past for refrigerants. Furthermore, with the fluorinated refrigerants, having sonic velocities of the order 100–150 m/s, supersonic flow may be encountered in the diffuser when the absolute velocity leaving the impeller is high. In recent years, however, the design of diffusers has been advanced to the point where radial bladed impellers can be employed.

If the torque in eq. (4.10.1) is understood to comprise the sum of all the moments about the axis of all forces acting on the control surface, including frictional and viscous ones, and if $\mathscr{V}_{wx}$ is the true whirl component, eq. (4.10.1) is then an exact equation. Approximation enters, however, when T is taken as the shaft torque, and when the value of $\mathscr{V}_{wx}$ is inferred from the velocity diagram, assuming that the angle β, the true fluid angle, is the same as β' the angle of the blades, and that $\mathscr{V}_{mx}$ is constant over the outlet area of the impeller.

The difference between the fluid angle and the blade angle results from internal circulation which takes place inside the channels of the impeller. Since a torque has to be applied to turn the impeller, the gas passing through must be resisting this torque, and the only way in which this can happen is that the pressure on the forward facing sides of the blades must be greater than the pressure on the backward facing sides. In other words, the

Fig. 4.10.3. Variation of the product of the work coefficient and the flow coefficient, which is proportional to power, with flow coefficient.

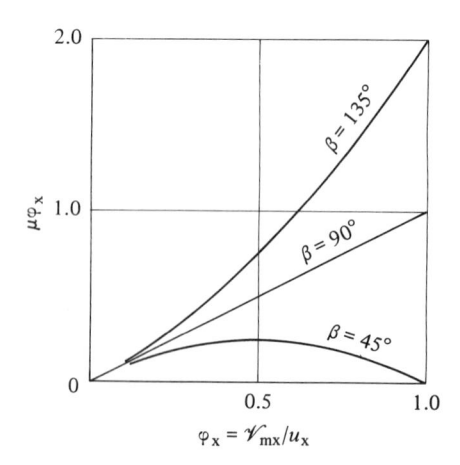

4.10 Centrifugal compressors

fluid piles up on the forward facing sides. The origin of this pressure gradient across the channel is the Coriolis component of acceleration which comes into play when a body moves outwards or inwards along a rotating radius. The pressure gradient across the channel causes an eddy which rotates in a direction opposite to that of the impeller and makes the true fluid angle β less than the blade angle β'. The true whirl component is thus less than the presumed whirl component, $\mathscr{V}'_{wx}$ say. The ratio:

$$\sigma = \mathscr{V}_{wx}/\mathscr{V}'_{wx} \qquad (4.10.10)$$

is called the slip factor. Several theoretical and empirical expressions have been given for it, which are listed and compared by Ferguson (1963).

The slip factor can be raised by increasing the number of blades, but after a point this becomes undesirable because of the increasing losses. The number is therefore a compromise. Impellers for refrigerants are currently using between $\beta'/3$ and $\beta'/4$ blades, where β' is the blade angle in degrees.

The assumption that $\mathscr{V}_{mx}$ is constant over the outlet area of the impeller leads to an error in estimating torque from the mass flow rate, and it is necessary to introduce a correction factor for the velocity profile at the tip of the impeller. Calling this factor λ, the expression for the work coefficient becomes:

$$\mu = \sigma(1 - \lambda \varphi_x \cot \beta') \qquad (4.10.11)$$

It is difficult in practice to separate σ and λ, except by using one of the expressions for σ, and deducing μ from a measurement of the specific work.

Example 4.10.1

An impeller of diameter 28 cm has backward curved blades of $\beta' = 55°$ and an area for flow at the periphery of $0.004\,\text{m}^2$. Determine the specific work and the power when it rotates at 12 000 rev/min and has a flow coefficient $\varphi_x = 0.3$. The specific volume of the vapour leaving the impeller is $0.03\,\text{m}^3/\text{kg}$. A slip factor $\sigma = 0.8$ may be assumed, and the effects of velocity profile may be neglected. $\cot 55° = 0.7002$.

The work coefficient is given by eq. (4.10.11) with $\lambda = 1$:

$$\mu = 0.8[1 - (0.3)(0.7002)] = 0.632$$

The tip speed is:

$$u_x = \pi(0.28)(12\,000/60)\,\text{m/s} = 175.9\,\text{m/s}$$
$$u_x^2 = (175.9)^2\,\text{J/kg} = 30.94\,\text{kJ/kg}$$

The specific work is therefore:

$$w = (0.632)(30.94)\,\text{kJ/kg} = 19.55\,\text{kJ/kg}$$

The meridional component at the exit is:

$$\mathscr{V}_{mx} = (0.3)(175.9)\,\text{m/s} = 52.77\,\text{m/s}$$

and the mass flow rate is:

$$\dot{m} = \frac{(52.77)(0.004)}{0.03}\,\text{kg/s} = 7.036\,\text{kg/s}$$

The power is then:

$$\dot{W} = (7.036)(19.55)\,\text{kW} = 137.6\,\text{kW}$$

It will be seen from the working of the last example that no properties of the fluid are involved in the determination of the work coefficient or the specific work. These depend only on the velocities and the coefficients. The specific volume enters when the mass flow rate is to be determined from the flow coefficient.

The momentum equation, with suitable corrections, gives the work which an impeller can absorb, but it does not indicate how much pressure rise is produced. This depends on the irreversibilities in the machine: a very bad impeller such as an agitator rotating in a closed vessel absorbs work but produces no pressure at all. For a reversible adiabatic compressor, the specific work would be given by eqs. (4.8.3) and (4.8.4) as:

$$(h_2 - h_1)_s = \int_{p_1}^{p_2} v\,\mathrm{d}p)_s \tag{4.10.12}$$

and the final pressure p_2 can be determined by the methods used in example 4.9.1 for finding stagnation pressure.

When compression is adiabatic but not reversible, the specific work is:

$$(h_2 - h_1) = \frac{1}{\eta_{pol}} \int_{p_1}^{p_2} v\,\mathrm{d}p \tag{4.10.13}$$

As already stated, the exact relation between p and v during compression is not known, only the initial and final states. Nevertheless, it is customary to use eq. (4.10.13) on the assumption that the process is represented by eq. (4.8.14), and the value of n is determined to fit the initial p_1, v_1 and the final p_2, v_2. The use of eqs. (4.10.12) and (4.10.13) is shown in the following example.

4.10 Centrifugal compressors

Example 4.10.2
The specific work to an impeller is 16 kJ/kg when it is inducing saturated vapour of R12 at a pressure of 3.086 bar (saturation temperature 0 °C). Determine the delivery pressure developed if (a) compression is reversible and adiabatic, and (b) the polytropic efficiency is 0.75.

(a) It is convenient to use the integral form of the change of specific enthalpy at constant entropy, eq. (4.2.21), when only small differences are involved which are difficult to determine accurately from tables. From Fig. 4.2.3 a value of k is estimated as 1.06, the final pressure not yet being known. From the tables for R12:

$$v_1 = 0.05539 \, \text{m}^3/\text{kg}$$

Equating the given specific work to the change of specific enthalpy:

$$(16)(1000) \, \text{J/kg}$$
$$= (3.086)(10^5)(0.05539)\frac{1.06}{0.06}\left[\left(\frac{p_2}{p_1}\right)^{0.06/1.06} - 1\right] \text{J/kg}$$

Solving for the pressure ratio:

$$p_2/p_1 = 2.489$$

and

$$p_2 = (2.489)(3.086 \, \text{bar}) = 7.681 \, \text{bar}$$

(b) The polytropic head is by eq. (4.10.13):

$$\int_{p_1}^{p_2} v \, dp = (0.75)(16) \, \text{kJ/kg} = 12 \, \text{kJ/kg}$$

Assuming that the polytropic efficiency and the small stage efficiency are equal, eq. (4.8.19) gives:

$$\frac{n}{n-1} = 0.75 \frac{1.06}{0.06} = 13.25 \quad \text{and} \quad n = 1.0816$$

Therefore:

$$(12)(10^3) \, \text{J/kg}$$
$$= (3.086)(10^5)(0.05539)(13.25)\left[\left(\frac{p_2}{p_1}\right)^{1/13.25} - 1\right] \text{J/kg}$$

Solving for the pressure ratio:

$$p_2/p_1 = 1.982$$

and

$$p_2 = (1.982)(3.086 \text{ bar}) = 6.116 \text{ bar}$$

As already remarked, the use of eq. (4.8.19) to determine n needs some justification since it was derived for an ideal gas. A check will therefore be made to see if the definition of polytropic efficiency is satisfied. From the R12 tables:

$$h_1 = 351.48 \text{ kJ/kg}$$
$$h_2 = (351.48 + 16) \text{ kJ/kg} = 367.48 \text{ kJ/kg}$$

At a pressure of 6.116 bar this h_2 locates a state of the vapour delivered. Interpolating in the R12 tables one finds:

$$t_2 = 32.1\,°\text{C}$$
$$v_2 = 30.076 \text{ dm}^3/\text{kg}$$

The value of n is then found to be 1.1201, i.e. higher than that given by eq. (4.8.19).

The corresponding polytropic head is:

$$3.086(10^5)(0.05539)\frac{1.1201}{0.1201}\left[\left(\frac{6.116}{3.086}\right)^{0.1201/1.1201} - 1\right] \text{J/kg}$$

$$= 12.13 \text{ kJ/kg}$$

Since this is close to the given value of 12 kJ/kg, the method has worked in this instance, despite the fact that it does not give the correct value of n.

The ability of a compressor to produce pressure may be given more directly in the form of head coefficients defined as:

isentropic head coefficient $\qquad \psi_{\text{isen}} = \dfrac{(h_2 - h_1)_s}{u_x^2}$ \hfill (4.10.14)

polytropic head coefficient $\qquad \psi_{\text{pol}} = \dfrac{1}{u_x^2}\int_{p_1}^{p_2} v\, dp$ \hfill (4.10.15)

The head coefficients are related to the corresponding efficiencies and the work coefficient by:

$$\mu = \frac{\psi_{\text{isen}}}{\eta_{\text{isen}}} = \frac{\psi_{\text{pol}}}{\eta_{\text{pol}}} \tag{4.10.16}$$

4.11 Generalised characteristics

Example 4.10.3
The tip speed of an impeller is 150 m/s, and the isentropic head coefficient $\psi_s = 0.55$ when inducing saturated R12 vapour at a pressure of 3.086 bar. Determine the pressure developed.

$$u_x^2 = (150)^2 \text{ J/kg} = 22.5 \text{ kJ/kg}$$

From eq. (4.10.14):

$$(h_2 - h_1)_s = (0.55)(22.5) \text{ kJ/kg} = 12.37 \text{ kJ/kg}$$

Using the procedure of example (4.10.2) with $k = 1.06$:

$$(12.37)(10^3) \text{ J/kg}$$

$$= (3.086)(10^5)(0.05539)\frac{1.06}{0.06}\left[\left(\frac{p_2}{p_1}\right)^{0.06/1.06} - 1\right] \text{ J/kg}$$

Solving for the pressure ratio:

$$p_2/p_1 = 2.032$$

and

$$p_2 = (2.032)(3.086 \text{ bar}) = 6.271 \text{ bar}$$

If instead of the isentropic head coefficient in example 4.10.3, the polytropic head coefficient had been given, it would not be possible from this information alone to find the pressure developed. It is then necessary to know either the work coefficient or the polytropic efficiency.

4.11 Generalised characteristics

Any two of the quantities μ, η_{pol} and ψ_{pol} (or of μ, η_{isen} and ψ_{isen}), at all conditions of operation of a compressor would give all the information required to calculate the pressure developed and the power taken at any flow rate. The aim of compressor testing is to obtain this knowledge for a particular machine, with the intention of using it in published rating curves for a variety of conditions of operation, and of extrapolating it to other refrigerants and other sizes of similar machines. The values of μ, η and ψ depend on a number of variables, however, such as the dimensions of the machine, the properties of the fluid, the flow rate and so on. To eliminate the need to test at every conceivable combination of these, a correlation is sought using the principles of dimensional analysis between (μ, η, ψ) and the dimensionless groups containing the independent variables.

Several types of correlation have been proposed and used, but the method which appears to correlate best the data for refrigerant compressors is based on the dimensionless groups:

flow coefficient $\quad \varphi_x = \mathscr{V}_{mx}/u_x \quad$ (4.11.1)

rotational Reynolds number $\quad (Re)_{rot} = 2u_x r_x/v_x \quad$ (4.11.2)

where v_x is the kinematic viscosity of the fluid leaving the impeller:

rotational Mach number $\quad (Ma)_{rot} = u_x/\mathscr{V}_{son,1} \quad$ (4.11.3)

where $\mathscr{V}_{son,1}$ is the sonic velocity in the vapour entering the compressor.

The rotational Reynolds and Mach numbers are not true Reynolds and Mach numbers because they are based not on the velocity of the fluid but on that of the impeller tip.

Of these groups the flow coefficient is of primary importance. For a given compressor running at a given speed with the same vapour, the variation of the rotational Reynolds and Mach numbers is small. Under these conditions the generalised characteristics of the machine would be plots of

Fig. 4.11.1. Generalised characteristics of a centrifugal compressor showing correlation when tested on air and on R11. Rotational Mach numbers: 0.473 on air, 1.197 on R11. Axial entry, and blade angle at impeller tip $\beta' = 56.5°$. The polytropic efficiency has a maximum for both air and R11 at $\varphi_x = 0.306$. (After Wiesner, 1960). The sloping line shows μ according to eq. (4.10.8).

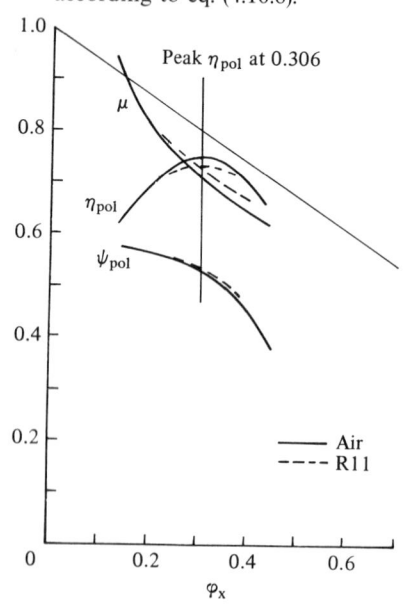

4.11 Generalised characteristics

μ, η and ψ against the flow coefficient φ_x. Experience has shown that the best correlation is obtained when the polytropic values rather than the isentropic values are used. Such generalised characteristics are shown in Fig. 4.11.1.

A disadvantage of using φ_x as one of the dimensionless groups is that it is not measured directly in tests, but has to be inferred from the other measured quantities. This is done as follows. Suppose that the measured quantities are $\dot{m}$ the mass flow rate, N the rotational speed, and the inlet and outlet states. p_1, t_1 and p_2, t_2. The dimensions of the impeller are also known. From the states at 1 and 2, the change of specific enthalpy is found, and the work coefficient from:

$$\mu = (h_2 - h_1)/u_x^2 \qquad (4.11.4)$$

Since the only change of stagnation enthalpy occurs across the impeller, the leaving specific enthalpy h_2, neglecting kinetic energy at 2, is equal to the specific stagnation enthalpy leaving the impeller. Hence:

$$h_2 = h_1 + \mu u_x^2 = h_x + \tfrac{1}{2}\mathscr{V}_x^2 \qquad (4.11.5)$$

The specific kinetic energy is given by eq. (4.10.9), therefore:

$$h_x - h_1 = u_x^2(\mu - \tfrac{1}{2}\mu^2 - \tfrac{1}{2}\varphi_x^2) \qquad (4.11.6)$$

This equation alone is not enough to determine φ_x since h_x is still unknown. Another relation is obtained, following a suggestion of Wiesner & Caswell (1959) that the flow through well designed impellers is very nearly reversible and adiabatic, most of the irreversibilities occurring in the diffuser, by putting $s_x = s_1$. This relation implies for a given h_x and s_x, a pressure p_x and a specific volume v_x at the exit from the impeller. The flow coefficient is then:

$$\varphi_x = \dot{m}v_x/a_x u_x \qquad (4.11.7)$$

where a_x is the area for flow at the periphery of the impeller. The procedure to find φ_x is therefore to assume a trial value, usually between 0.1 and 0.5, and find a trial h_x from (4.11.6). From refrigerant tables, or a chart, v_x for the trial h_x at $s_x = s_1$ is determined and a new value of φ_x is calculated from (4.11.7). The calculation is then repeated using the new φ_x as the starting value. Convergence is quite rapid, as will be seen in the next example.

When the generalised characteristics are given and it is required to convert these to pressure ratios, mass flow rates and powers, the procedure is reversed.

Example 4.11.1

A single-stage centrifugal compressor has an impeller of diameter 170 mm and a flow area at the periphery of 41 cm². At a rotational speed of 20 400 rev/min when pumping R12 initially saturated at 2.191 bar ($-10\,°C$ saturation) to a final pressure of 6.695 bar (26 °C saturation) the inlet volume flow rate is 0.32 m³/s and the shaft power is 114 kW.

Determine the work coefficient, polytropic efficiency, polytropic head coefficient and flow coefficient.

Properties of R12 vapour

Saturated at 2.191 bar ($-10\,°C$)

v_g dm³/kg	h_g kJ/kg	s_g kJ/kg K
76.65	347.13	1.5600

Superheated at 6.695 bar
(*Saturation temperature* 26 °C)

t °C	v dm³/kg	h kJ/kg	s kJ/kg K
30	26.74	364.97	1.5542
35	27.47	368.58	1.5660
40	28.18	372.17	1.5776
45	28.88	375.73	1.5888

The mass flow rate is:

$$\dot{m} = [(0.32)/(0.07665)] \text{ kg/s} = 4.175 \text{ kg/s}$$

and the specific work:

$$w = 114/4.175 = 27.31 \text{ kJ/kg}$$

The tip speed of the impeller is:

$$u_x = [(20\,400)(0.17)\pi/60]\,\text{m/s} = 181.6\,\text{m/s}$$

and:

$$u_x^2 = (181.6)^2 \text{ J/kg} = 32.98 \text{ kJ/kg}$$

The work coefficient is:

$$\mu = 27.31/32.98 = 0.828$$

4.11 Generalised characteristics

For adiabatic flow the final specific enthalpy is:

$$h_2 = (347.13 + 27.31) \text{ kJ/kg} = 374.44 \text{ kJ/kg}$$

Interpolating for this value of h at $p = 6.695$ bar:

$$v_2 = 28.63 \text{ dm}^3/\text{kg}$$

which corresponds to an average index of compression of:

$$n = 1.1342$$

giving:

$$\int v \, dp$$

$$= (2.191)10^5 (0.07665) \frac{1.1342}{0.1342} \left[\left(\frac{6.695}{2.191} \right)^{0.1342/1.1342} - 1 \right] \text{J/kg}$$

$$= 20.06 \text{ kJ/kg}$$

The polytropic efficiency is therefore:

$$\eta_{\text{pol}} = 20.06/27.31 = 0.734$$

and the polytropic head coefficient is:

$$\psi_{\text{pol}} = (0.828)(0.734) = 0.608$$

Equation (4.11.6) gives:

$$h_x - h_1 = 32.98[0.828 - \frac{1}{2}(0.828^2) - \frac{1}{2}\varphi_x^2] \text{ kJ/kg}$$

$$= 16.00 - 16.49 \varphi_x^2$$

Equation (4.11.7) gives:

$$\varphi_x = \frac{4.175 \, v_x (10^3)}{(4.1)(181.6)} \text{ kg/m}^3 = 5.607 \, v_x \text{ kg/m}^3$$

These equations with the condition that at the impeller outlet $s_x = s_1$ determine φ_x. The solution may be obtained by trial and error in the refrigerant tables, which is rather laborious. The procedure suggested here is to determine, from the tables, a value for the isentropic index and to use the integral form, (4.2.11), for $(h_x - h_1)_s$.

Interpolating for $s = 1.5600$ kJ/kg K at $p = 6.695$ bar gives:

$$v = 27.10 \text{ dm}^3/\text{kg}$$

corresponding to an average value of

$$k = 1.0743$$

Substituting the volume ratio for the pressure ratio:

$$\frac{p_x}{p_1} = \left(\frac{v_1}{v_x}\right)^k$$

in the usual form of the integral gives:

$$(h_x - h_1)_s$$

$$= (2.191)10^5(0.07665)\frac{1.0743}{0.0743}\left[\left(\frac{0.07665}{v_x\,\text{kg/m}^3}\right)^{0.0743} - 1\right] \text{J/kg}$$

$$= 242.8\left[\left(\frac{0.07665}{v_x\,\text{kg/m}^3}\right)^{0.0743} - 1\right] \text{kJ/kg}$$

Taking a trial value of $\varphi_x = 0.2$ in (4.11.6) gives:

$$h_x - h_1 = 16.00 - 16.49(0.2)^2 = 15.34 \text{ kJ/kg}$$

Substituting this in $(h_x - h_1)_s$ and solving for v_x gives:

$$v_x = 0.03361 \text{ m}^3/\text{kg}$$

giving:

$$\varphi_x = (5.607)(0.03361) = 0.188$$

from (4.11.7), which does not quite agree with the trial value, but is in fact very close to the true value. A second iteration with $\varphi_x = 0.188$ gives the result:

$$\varphi_x = 0.1877$$

which satisfies all the equations.

Wiesner & Caswell (1959) have given an expression for the direct determination of φ_x, but it incorporates ideal gas behaviour and it is not clear what accuracy it would have with vapours.

Example 4.11.2

The compressor of example 4.11.1 compresses R22 vapour initially saturated at 4.976 bar (0 °C saturation) to a final pressure of 14.601 bar (38 °C saturation). Assuming that at the same flow coefficient the same values of μ and η_{pol} as in the previous example hold, determine the required rotational speed, the power and the inlet volume flow rate.

4.11 Generalised characteristics

Properties of R22 vapour.
Saturated at 4.976 bar (0 °C)

v_g	h_g	s_g
dm³/kg	kJ/kg	kJ/kg K
47.14	405.36	1.7518

Superheated at 14.601 bar (Saturation temperature 38 °C)

t	v	h	s
°C	dm³/kg	kJ/kg	kJ/kg K
55	17.82	431.57	1.7501
60	18.33	435.92	1.7632
65	18.83	440.22	
70	19.31	444.48	

Superheated at 11.919 bar (Saturation temperature 30 °C)

t	v	h	s
°C	dm³/kg	kJ/kg	kJ/kg K
40	21.03	423.16	1.7400
45	21.64	427.38	1.7534

Following example 4.8.2 using eq. (4.8.19), the value of the isentropic index is first found. Interpolating for $s = 1.7518$ kJ/kg K at $p = 14.601$ bar:

$$v = 17.891 \text{ dm}^3/\text{kg}$$

giving:

$$k = 1.1111$$

(4.8.19) then gives:

$$0.734 = \frac{n}{n-1} \frac{0.1111}{1.1111}$$

and:

$$\frac{n}{n-1} = 7.3407 \quad n = 1.1577$$

Hence

$$v_2 = 47.14 \left(\frac{4.976}{14.601}\right)^{1/1.1577} = 18.60 \text{ dm}^3/\text{kg}$$

Compression processes

Interpolating for this value at $p = 14.601$ bar:

$$h_2 = 438.24 \text{ kJ/kg}$$
$$h_1 = 405.36$$
$$w = h_2 - h_1 = 32.88 \text{ kJ/kg}$$

To check whether this satisfies the basic definition of polytropic efficiency,

$$\int_1^2 v\, dp = (4.976)10^5(0.04714)(7.3407)\left[\left(\frac{14.601}{4.976}\right)^{1/7.3407} - 1\right] \text{J/kg}$$
$$= 27.20 \text{ kJ/kg}$$

and by (4.8.8):

$$\eta_{\text{pol}} = 27.20/32.88 = 0.827$$

The agreement with the starting value of 0.734 is not satisfactory and indicates that with R22 the use of (4.8.19) can cause serious error. It is necessary in this case to search for a final state which satisfies the requirements.

At a temperature of 68 °C, $p = 14.601$ bar;

$$h_2 = 442.78 \text{ kJ/kg}$$

giving

$$w = h_2 - h_1 = 37.42 \text{ kJ/kg}.$$

The corresponding value

$$v_2 = 19.12 \text{ dm}^3/\text{kg}$$

gives the compression index:

$$n = 1.1929$$

and:

$$\int_1^2 v\, dp$$
$$= (4.976)10^5(0.04714)\frac{1.1929}{0.1929}\left[\left(\frac{14.601}{4.976}\right)^{0.1929/1.1929} - 1\right] \text{J/kg}$$
$$= 27.58 \text{ kJ/kg}$$

4.11 Generalised characteristics

and:

$$\eta_{pol} = 27.58/37.42 = 0.737$$

which is in reasonable agreement with the starting value. Hence:

$$u_x^2 = (37.42/0.828) \text{ kJ/kg} = 45\,190 \text{ J/kg}$$
$$u_x = 212.6 \text{ m/s}$$
$$N = [212.6/\pi(0.17)]\text{rev/s} = 23\,885 \text{ rev/min}$$

Equation (4.11.6) gives, with $\varphi_x = 0.188$:

$$h_x - h_1 = 45.19[0.828 - \tfrac{1}{2}(0.828)^2 - \tfrac{1}{2}(0.188)^2] \text{ kJ/kg}$$
$$= 21.12 \text{ kJ/kg}$$
$$h_x = 426.48 \text{ kJ/kg}$$

Inspection in the refrigerant tables or on an accurate chart will show that the pressure at which this value of h_x coincides with $s_x = s_1 = 1.7518$ kJ/kg K is very nearly 11.919 bar.

At $p = 11.919$ bar, $s_x = 1.7518$ kJ/kg K gives:

$$h_x = 426.88 \text{ kJ/kg}$$
$$v_x = 21.57 \text{ dm}^3/\text{kg}$$

Accepting this as being close enough, (4.11.7) gives:

$$\dot{m} = \frac{(0.188)(4.1)212.6}{21.57} \text{ kg/s} = 7.597 \text{ kg/s}$$

and:

$$\dot{W} = (7.597)(37.42) \text{ kW} = 284 \text{ kW}$$
$$\dot{V}_1 = (7.597)(0.04714) \text{ m}^3/\text{s} = 0.358 \text{ m}^3/\text{s}$$

The foregoing examples are intended to show the principles involved in correlating turbo-compressor performance in terms of the polytropic efficiency and the flow coefficient at the outlet of the impeller. In practice a number of short cuts can be devised, as well as further refinements such as taking account of an efficiency for the flow through the impeller, of a drop in pressure with some irrecoverable loss between the inlet duct and the inlet to the impeller, and so on. The emphasis has been placed on the non-ideal gas behaviour of most refrigerant vapours and how this is to be accounted for.

4.12 Effects of Reynolds and Mach numbers

The generalised characteristics of Fig. 4.11.1 strictly apply at only one value of the rotational Reynolds number, and a family of such curves ought to be produced with $(Re)_{rot}$ as a parameter. It is found, however, that the rotational Reynolds number has relatively little effect; its main one is to reduce the polytropic efficiency slightly at lower $(Re)_{rot}$. Wiesner & Caswell have given correction factors.

The rotational Mach number also has effects on the characteristics, but according to some results of Wiesner (1960), even a wide variation of $(Ma)_{rot}$ has comparatively little effect. The differences between the results of tests of the same compressor on air and on R113 were traced rather to the different velocity profiles at the impeller tip. The polytropic head coefficient was almost identical for air and R113, the slight differences appearing in the work coefficients and the polytropic efficiencies.

On the other hand, the true Mach number at points inside the compressor has important effects if it exceeds or even approaches unity. The most critical position is at the entry to the impeller, since shocks there would have a bad effect on the flow inside the impeller channels. It is usual to limit the Mach number there, based on the velocity of the fluid relative to the impeller, to about 0.9. The fluid velocity relative to the impeller is compounded of the absolute velocity of the approaching fluid and the velocity of the tips of the inlet vanes. For a given volume flow rate and a given rotational speed, the absolute velocity of the fluid can be reduced by increasing the diameter of the eye, but this increases the velocity of the tips of the inlet vanes. The relative velocity is a minimum when the angle which it makes with the plane of the impeller is about 34°, as the following analysis shows.

The entrance to the impeller, sometimes called the eye of the impeller and

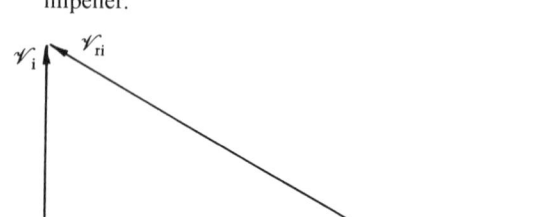

Fig. 4.12.1. Velocity diagram at the tip of the inducer or eye of the impeller.

$\mathscr{V}_i$ $\mathscr{V}_{ri}$

α

$u_i = \omega r_i$

4.12 Effects of Reynolds and Mach numbers

sometimes the inducer, is an annular area. The full area of the inlet duct is not available for flow because of the area taken by the hub, so that the fluid is accelerated as it enters the impeller. The blades of the impeller are turned in the direction of rotation to guide the fluid into the channels. Assuming that the absolute axial velocity $\mathscr{V}_i$ is uniform across the annular area for flow, the highest velocity relative to the impeller occurs at the tip of the inlet blades. The velocity diagram at this point is shown in Fig. 4.12.1, in which the absolute velocity $\mathscr{V}_i$ is at right angles to the plane of rotation, and u_i the velocity of the tips of the inlet blades is in the plane of rotation. For a volume flow rate $\dot{V}_i$, which is nearly the same as $\dot{V}_1$ at the inlet flange, and rotational speed N, then:

$$\mathscr{V}_i = \frac{\dot{V}_i}{\pi(r_i^2 - r_h^2)} \tag{4.12.1}$$

and:

$$u_i = 2\pi N r_i \tag{4.12.2}$$

where $2r_h$ is the diameter of the hub and $2r_i$ is the diameter of the inducer or eye. The square of the relative velocity $\mathscr{V}_{ri}$ is then:

$$\mathscr{V}_{ri}^2 = \frac{\dot{V}_i^2}{\pi^2(r_i^2 - r_h^2)^2} + 4\pi^2 N^2 r_i^2 \tag{4.12.3}$$

Differentiating this with respect to r_i^2 and equating to zero gives:

$$(r_i^2 - r_h^2) = \frac{\dot{V}_i^{2/3}}{2^{1/3}\pi^{4/3} N^{2/3}} \tag{4.12.4}$$

Substitution in eq. (4.12.1) gives the absolute axial velocity of the fluid for the condition of minimum relative velocity:

$$\mathscr{V}_i = (2\pi \dot{V}_i N^2)^{1/3} \tag{4.12.5}$$

which is independent of the diameter of the hub.

Introducing:

$$\rho = r_h/r_i \tag{4.12.6}$$

into eq. (4.12.4) and solving for r_i:

$$r_i = \frac{1}{(1-\rho^2)^{1/2}} \left(\frac{\dot{V}_i}{2^{1/2}\pi^2 N}\right)^{1/3} \tag{4.12.7}$$

and using eq. (4.12.2),

$$\cot \alpha = \frac{u_i}{\mathscr{V}_i} = \frac{2^{1/2}}{(1-\rho^2)^{1/2}} \tag{4.12.8}$$

where α is the angle between $\mathscr{V}_{ri}$ and the plane of rotation. If there were no hub, $\rho = 0$ and:

$$\cot \alpha = 2^{1/2} \quad \text{with} \quad \alpha = 35.3° \tag{4.12.9}$$

but with an average value of $\rho = \tfrac{1}{3}$ say:

$$\cot \alpha = \frac{3}{2} \quad \text{with} \quad \alpha = 33.7° \tag{4.12.10}$$

This is the angle at which the inlet blades should be set for normal operation at the design point.

The minimum relative velocity $\mathscr{V}_{ri}$ is given by:

$$\mathscr{V}_{ri} = \mathscr{V}_i \operatorname{cosec} \alpha = (2\pi \dot{V}_i N^2)^{1/3} \left(\frac{3-\rho^2}{1-\rho^2}\right)^{1/2} \tag{4.12.11}$$

and the Mach number is:

$$(Ma)_i = \frac{\mathscr{V}_{ri}}{\mathscr{V}_{son,i}} \tag{4.12.12}$$

where $\mathscr{V}_{son,i}$ is the sonic velocity at the inlet to the impeller. Sonic velocity is given by the expression:

$$\mathscr{V}_{son,i} = (kp_i v_i)^{1/2} \tag{4.12.13}$$

where k is the isentropic index for an infinitesimal process. It is taken from Fig. 4.2.2 etc. at $t_c = t_e$.

The above analysis is based on an axial velocity of approach. In the event of the fluid having some rotation at the inlet, as when it comes from a previous stage, or when stationary pre-whirl vanes are used to reduce the flow rate, the effects have to be allowed for.†

As already stated, the relative velocity diminishes whilst the fluid passes through the impeller, and with the rise in temperature the sonic velocity increases. The Mach number, therefore, relative to the impeller, diminishes towards the tip. The absolute velocity on the other hand increases, and may become supersonic. This does not matter so long as the fluid does not come into contact with anything stationary, but when it does so on meeting vanes, if any, in the diffuser, a shock results. Compressors of traditional design usually avoided running at this condition, but some modern machines of compact design and high speed do in fact operate with shock in the diffuser (Jassniker, 1962). In other designs diffusion from supersonic to

---

† Centrifugal compressors for air often employ fixed pre-whirl vanes to reduce the inlet Mach number.

subsonic flow may take place in the free vortex in the vaneless space between the impeller and the diffuser.

Example 4.12.1
The radius of the inlet blade tips for the compressor of example 4.11.1 is 4.5 cm, and the radius of the hub is 2.5 cm. Determine the Mach number at the inlet blade tips for the conditions of example 4.11.1.
The flow area at inlet to the impeller is:

$$(4.5^2 - 2.5^2) \text{ cm}^2 = 43.98 \text{ cm}^2$$

Neglecting changes in density between the inlet and the impeller entry:

$$\dot{V}_i = \dot{V}_1 = 0.32 \text{ m}^3/\text{s}$$

and the absolute velocity is:

$$\mathscr{V}_i = \frac{(0.32)(10^3)}{43.98} \text{ m/s} = 72.76 \text{ m/s}$$

$$u_i = \frac{\pi(0.09)(20\,400)}{60} \text{ m/s} = 96.13 \text{ m/s}$$

Assuming axial flow with no whirl component,

$$\mathscr{V}_{ri} = (72.76^2 + 96.13^2)^{1/2} \text{ m/s} = 120.6 \text{ m/s}$$

Taking k as 1.1 at $t_e = t_c = -10\,°\text{C}$ in Fig. 4.2.3 the sonic velocity is:

$$[(1.1)(2.191)10^5(0.07665)]^{1/2} \text{ m/s} = 135.9 \text{ m/s}$$

and the Mach number is:

$$120.6/135.9 = 0.89$$

4.13 Peak performance characteristics

The generalised characteristics of Fig. 4.11.1 show that the polytropic efficiency increases with the flow coefficient to a maximum and thereafter decreases. Away from this peak the efficiency is too low to be acceptable, and the design should be aimed at operating somewhere near the peak efficiency. In making preliminary calculations it would be helpful if one knew in advance at what value of the flow coefficient the maximum polytropic efficiency could be obtained. According to Stepanoff (1955) there is a correlation between the polytropic head coefficient and the flow coefficient at maximum efficiency for all impellers having the same blade angle β'. The relation is:

$$\psi_{\text{pol}} = 0.69(1 - \varphi_x \cot \beta') \qquad (4.13.1)$$

Wiesner & Caswell (1959) and Wiesner (1960) have tested this relation against information obtained from a large number of impellers pumping various gases and vapours, and have shown that it represents the peaks of the polytropic efficiency curves reasonably well.

The maximum polytropic efficiency for impellers of all blade angles appears to lie within a comparatively narrow range of the flow coefficient $\varphi_x = 0.30$ to 0.35, which is independent of the rotational Mach number. It is possible therefore to select a flow coefficient to give maximum efficiency at the design point, and to predict the head coefficient in advance.

The same type of information regarding peak efficiency may be presented in terms of a modified flow coefficient defined by:

$$\varphi_{mod} = \dot{V}_1/ND^3 \qquad (4.13.2)$$

where $\dot{V}_1$ is the volume flow rate at the inlet and $D = 2r_x$ is the outer diameter of the impeller. Figure 4.13.1 gives the maximum efficiency as a function of the modified flow coefficient for one series of compressors operating at low rotational Mach numbers (Anderson, 1962). (Note that this is not an efficiency curve for one compressor, but the locus of the peak efficiencies of a series of machines against the modified flow coefficient at which these are attained.) Since the modified flow coefficient is somewhat easier to handle than the coefficient φ_x, it will be used to explore the

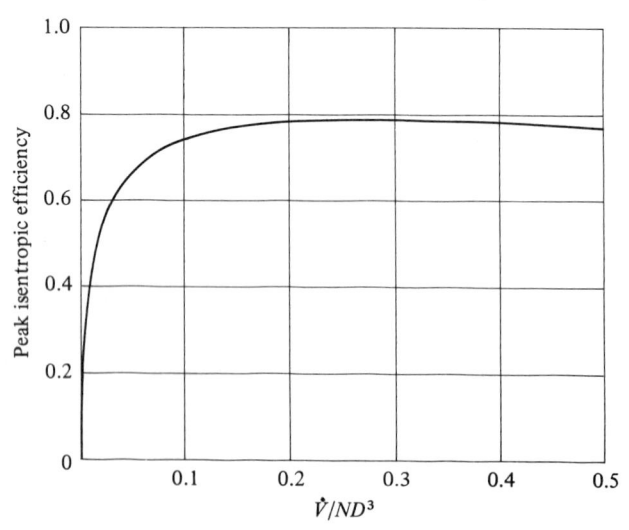

Fig. 4.13.1. Peak performance characteristic of a range of centrifugal compressors (after Anderson, 1962).

4.13 Peak performance characteristics

implications of the peak efficiency for the application of centrifugal compressors in refrigeration cycles.

It will be assumed that the evaporating and condensing temperatures are 5 °C and 40 °C, respectively, with saturated suction vapour. The maximum rotational speed available for the compressor will be taken as 3000 rev/min, corresponding to direct drive from a two-pole motor on 50 Hz supply, neglecting slip. An isentropic head coefficient, $\psi_{isen} = 0.55$ will be assumed, and to preserve the efficiency the modified flow coefficient will be not less than 0.05. Table 4.13.1 shows the results of calculations on these assumptions. Column (1) shows the change of specific enthalpy required for isentropic compression, and column (2) gives the impeller tip speed u_x from eq (4.10.14). Column (3) gives the impeller diameter needed for this tip speed at 50 rev/s, and column (4) is the minimum volume flow rate at inlet to ensure that φ_{mod} is at least 0.05 for good efficiency. Columns (5) and (6) give the specific volume and the specific refrigerating effect, and column (7) gives the minimum refrigerating capacity for the flow coefficient to be at least 0.05.

Several things appear from this table. The tip speeds are very high, and that for ammonia would probably be ruled out on considerations relating to the stresses in the impeller. The others would be allowable, however. The refrigerating capacities are also very large and, apart from R11, would find few applications. There are considerable differences between the refrigerants in respect of these minimum capacities. This means that if a plant of a given rotational speed for a given refrigerating capacity is required, it is not possible to chose the refrigerant at will. For example, if the capacity required is 2500 kW at 50 rev/s, only R11 could do it in single stage with good efficiency.

By increasing the rotational speed the minimum refrigerating capacity comes down. Examination of eq. (4.13.2) shows that, holding ND constant to keep the tip speed constant, the minimum volume flow rate to give

Table 4.13.1

	$(h_2 - h_1)_s$	u_x	D	$\dot{V}_1$	v_1	$(h_1 - h_3)$	$\dot{Q}_e$
	(kJ/kg) (1)	(m/s) (2)	(m) (3)	(m³/s) (4)	(m³/kg) (5)	(kJ/kg) (6)	(kW) (7)
NH₃	156.50	533	3.40	97.9	0.243	1076.7	434 000
R11	21.85	199	1.27	5.11	0.332	156.7	2 410
R12	17.23	177	1.13	3.58	0.0475	115.1	8 660
R22	23.90	208	1.33	5.84	0.0404	157.5	22 800
R13B1	9.67	133	0.844	1.504	0.0129	61.76	7 180

acceptable φ_{mod} is inversely proportional to the square of the rotational speed. By raising this to 24 000 rev/min = 400 rev/s say, the figures in column (7) are reduced by a factor of 64. This would bring all the refrigerants except ammonia into a range of capacity suitable for small to medium sized air conditioning installations. It does not however change the tip speeds required, which are determined solely by the required isentropic heads and the assumed head coefficient.

The head coefficient assumed above corresponds to that for a backward curved impeller. A radial-bladed impeller has a higher head coefficient, and this reduces somewhat the required tip speed. A forward bladed impeller would have a still higher head coefficient, but the efficiency tends to be less on account of the extra duty imposed on the least efficient part of the compressor, i.e. the diffuser. Forward bladed impellers are not currently used in refrigeration.

Another way of reducing the minimum capacity whilst preserving acceptable efficiency is to split the compression into stages, using two or more impellers, usually on a common shaft. The tip speed is then reduced because the isentropic head per stage is reduced, and the diameter for a given rotational speed is reduced. There is some sacrifice in efficiency because of the greater loss in multiple entries to the impellers and in the transfer passages bringing vapour from the periphery to the eye of the next impeller. There are compensating advantages, however, in that the refrigeration cycle can be improved by introducing desuperheating of vapour between the stages and precooling of the liquid. These benefits are described in Chapter 5.

In recent years the trend in the design of compressors for air-conditioning has been away from staged operation at low rotational speeds (commonly using R11 as refrigerant), towards single-stage operation with higher rotational speeds achieved by gearing or by frequency changing of the electric supply, coupled with radial bladed impellers. Industrial machines have not on the whole followed this trend. Versatility for pumping a range of gases and vapours is regarded as important and this is more easily obtained with conservative designs. Such compressors are often driven by steam or by internal combustion prime movers, and a choice of rotational speed is available to suit different gases.

4.14 Capacity reduction

One of the merits of the centrifugal compressor compared with the reciprocating compressor is its ability to achieve a smooth reduction of flow rate. In some designs this may be down to 10% of full load.

4.14 Capacity reduction

In the small to medium sized compressors which are made in large numbers for air-conditioning applications a commonly used method is by pre-rotation of the suction vapour. A set of adjustable guide vanes in the inlet duct deflects the stream in the direction of the impeller rotation. The relative velocity of the vapour with respect to the impeller is reduced and with it the volume flow rate. The effect is shown on the velocity diagram at inlet in Fig. 4.14.1. The net result is an almost proportional reduction of power with flow rate and with refrigerating capacity. At low flow rates the

Fig. 4.14.1. Velocity diagram at the tip of the inducer, showing the reduction of the relative velocity $\mathscr{V}_{ri}$ with pre-whirl, compared with that shown in Fig. 4.12.1 for axial entry.

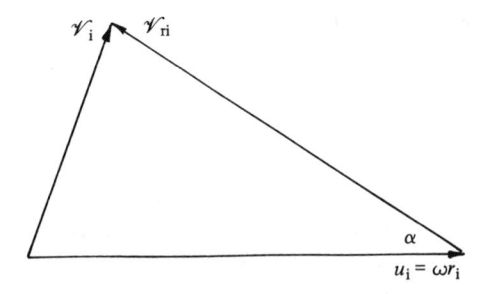

Fig. 4.14.2. Refrigerating capacity of a typical three-stage turbo-compressor pumping R11, (a) as a function of evaporating temperature at a constant condensing temperature of 43 °C and (b) as a function of condensing temperature at a constant evaporating temperature of 4.5 °C. The parameter on the curves is the percentage of rated speed. Point P is the full load point at rated speed on both curves.

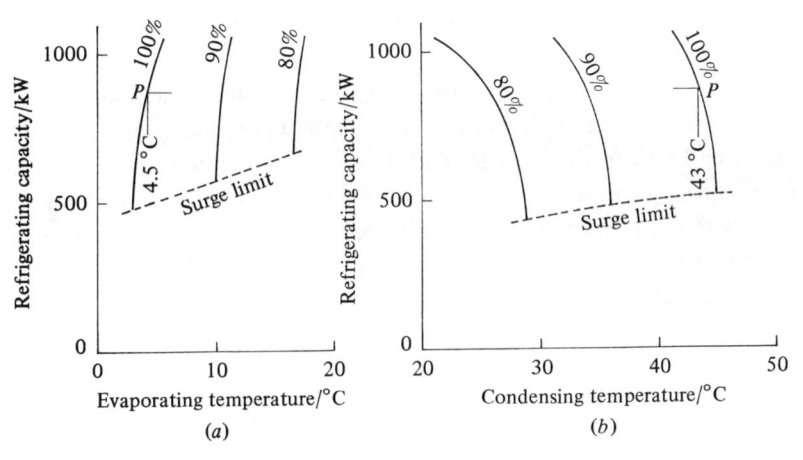

diffuser is not well matched to the impeller and it may be necessary to adjust the width of the diffuser channels at the same time. Another method used on compressors of this type is the movement of the vanes in the diffuser. By rotating these the area for flow is reduced and the static pressure at the outlet of the impeller is increased. This is turn causes a reduction of the flow rate.

In the large industrial type of compressor the principal methods of capacity reduction are (a) speed reduction, (b) suction throttling, (c) raising the delivery pressure by altering the flow rate of the condenser water, or directly by a throttle valve and (d) by-passing vapour from delivery to suction. These possibilities will be discussed with reference to Fig. 4.14.2 which shows the characteristics of a turbo-compressor of the industrial type. A curve in Fig. 4.14.2(*a*), for a constant condensing temperature, may be compared with one of the curves of Fig. 4.4.5 for a reciprocating compressor. It is seen that the shape is quite different. For the reciprocating compressor the shape of the capacity curve is largely determined by the change in the density of the suction vapour, whereas for the turbo-compressor the very steep line is caused by the change in the volume flow rate as well as by the change in density. At a given speed a large reduction of capacity can be brought about by a small drop in evaporating temperature. This may be a real drop or one simulated by introducing a pressure drop in the suction pipe with a valve. At a given evaporating temperature a large reduction of capacity can be brought about by a small change of speed. Figure. 4.14.2(*b*) shows the effect of increasing the condensing temperature. Again it may be a real increase obtained by throttling the cooling water flow rate or a simulated one produced by pressure drop in the delivery pipe.

As a rule method (a) gives a larger reduction of shaft power for the same reduction of capacity than (b) or (c), but it is not usually available with drive by electric motor.

Whichever of the three methods is used the available reduction is limited by the onset of 'surging', which is an instability of operation caused by mismatch between the impeller and diffuser at low flow rates. Capacities below the surge limit can be obtained only by by-passing hot gas from the delivery to the suction, as described in section 3.17. The control system has therefore to make provision for the changeover from methods (a), (b) or (c) to method (d) at low loads.

5

Multi-stage vapour compression systems

5.1 Introduction

In many applications of refrigerating plant such as air-conditioning, ice making, chilled storage of foods, etc., the difference in temperature between the condenser and the evaporator is not greater than about 40 K, and the simple vapour compression system treated so far in this book is quite adequate for this purpose. There are other circumstances however in which a much larger temperature difference is wanted. These are mainly of two kinds: either because an unusually low temperature is required in the evaporator, or because an unusually high temperature is required in the condenser. Examples of the first kind occur in food freezing, where an evaporating temperature of $-40\,°C$ is common with a condensing temperature of $30-40\,°C$, in chemical industry, where temperatures down to $-100\,°C$ are often desired, and in the liquefaction of natural gas at $-161\,°C$. Examples of the second kind occur when the refrigerator is used as a heat pump with a condensing temperature of perhaps $70\,°C$, and in the conditioning of high-speed aircraft which have to reject heat at the stagnation temperature corresponding to the air speed, about $130\,°C$ at Mach 2 in a (static) air temperature of $-50\,°C$.

As the difference in temperature between the condenser and the evaporator increases, several factors cause reductions of the refrigerating capacity and of the coefficient of performance which eventually become unacceptable. At the same time operating problems of cooling or lubrication arise or become more acute.

The answer to many of these problems is the adoption of one of the two main types of multi-stage system: (a) staged (compound) compression, or the (b) cascade system.

In staged compression the same refrigerant is used throughout the system but it is compressed in stages. In a cascade system two or more closed circuits each containing different refrigerants are used, the condenser of one circuit being cooled by the evaporator of another. Frequently the two methods are combined, since a cascade system may incorporate multi-stage compression in each circuit. An example of this kind is described in Chapter 8 for the liquefaction of natural gas.

5.2 Compression in stages

Compression in stages may be performed in physically distinct machines working in series, so that each stage compresses the vapour over a part of the overall pressure ratio and provides only a part of the overall specific work of compression. The compressors are then called the first stage, second stage, etc. compressors, or, when there are only two, the low-stage and the high-stage compressors. In refrigerating plant practice the first-stage machine is often called a 'booster' from the fact that at one time many old single-stage plants were converted to two-stage operation by the addition of a low-stage compressor to deal with additional duties such as freezing of foodstuffs at a lower temperature than had formerly been demanded.

Staged compression may also be performed in one machine, with some of the cylinders used for the first stage, others for the second, and so on. For machines with a single crank and piston rod several ingenious arrangements have been adopted in the past, requiring the use of compound pistons with up to three different diameters giving annular compression spaces, though such machines are not common nowadays. Reciprocating compressors in which staged compression can be carried out by any of these means are called 'internally compounded' or, for short, compound compressors.

Similar remarks apply to turbo-compressors: the stages may be carried by a single shaft, a common arrangement in small machines, or they may be distinct machines. Large plants often have several machines in series, each machine containing several stages.

The reasons for using, and the benefits obtainable from, staged compression differ greatly between one type of system and another. The following treatment covers some of the main points in connection with three different types of compressor.

Reciprocating compressors

One reason which is common to most applications is the need to overcome the reduction in volumetric efficiency which occurs at high

5.2 Compression in stages

pressure ratios, i.e. at unusually low evaporating temperatures or unusually high condensing temperatures. The reduction in volumetric efficiency occurs, as explained in Chapter 4, because of the re-expansion of the vapour in the clearance space. The following example illustrates the limitation.

Example 5.2.1
The volumetric efficiency of a compressor on R22 is given approximately by:

$$\eta_v = 1.09 - 0.116 r^{0.7}$$

where r is the pressure ratio. Determine the minimum evaporating temperature which could be attained when the condensing temperature is 40 °C.

The volumetric efficiency falls to zero when:

$$r = \left(\frac{1.09}{0.116}\right)^{1/0.7} = 24.54$$

At a condensing temperature of 40 °C, the saturation vapour pressure of R22 is 15.34 bar. The volumetric efficiency falls to zero therefore when the suction pressure is:

$$p_1 = 15.34/24.54 = 0.625 \text{ bar}$$

corresponding to an evaporating temperature of:

$$t_e = -50.5 \,°\text{C}$$

This is a limiting value of course at which no mass flow rate is produced, and consequently no refrigerating capacity.

Even for evaporating temperatures higher than the limiting value, it may be beneficial to compress in stages in order to obtain more refrigerating capacity from a given total swept volume. An approximate break-even point occurs when the total swept volume rates of the two stages equal the swept volume rate of a single-stage machine to give the same capacity. When the volumetric efficiency is given by an expression of the form:

$$\eta_v = A - Br^{1/n} \tag{5.2.1}$$

assuming equal pressure ratios r in the two stages and compression according to pV^n = constant, the break-even point occurs at an overall pressure ratio R:

$$R = r^2 = (A/B)^{2n/3} \tag{5.2.2}$$

In practice a number of other reasons may decide the evaporating temperature below which staged compression is adopted. For some refrigerants an important one is the high delivery temperature which occurs at high pressure ratios. Delivery temperature can always be lowered by going to wet compression of course, but it has been seen in Chapter 3 that this practice does not produce the maximum refrigerating capacity for some refrigerants, and in Chapter 4 that it has bad effects on the volumetric efficiency. To obtain the full benefits of compression in the superheat region may mean that the final delivery temperature will be too high considering the chemical stability of the refrigerant and the lubricating oil. By withdrawing the vapour between the stages it can be desuperheated before entry to the next stage, by methods to be explained, and the final delivery temperature can be reduced.

An important advantage which results from staged compression is that the refrigeration cycle can be improved by expansion of the liquid in stages, as explained in Section 5.4, with a consequent increase in the refrigerating capacity and in the coefficient of performance.

Furthermore, when two or more suction pressures are available it is possible to carry refrigeration loads at several temperatures.

Turbo-compressors

The limitations of a single impeller were discussed in Chapter 4, where it was seen that the change of specific enthalpy which can be produced is related to the size of the machine if peak efficiency is to be maintained. Compression in two or more stages offers a way out of the difficulties there outlined. Again advantage may be taken of the staged compression to improve the cycle by desuperheating between stages and by expansion of the liquid in stages.

Rotary compressors

Some types of positive displacement rotary compressor, particularly the screw compressor, do not exhibit the pronounced decrease of volumetric efficiency with increasing pressure ratio which is usual in reciprocating compressors. Moreover, by the injection of oil during the compression process the delivery temperature can be prevented from rising excessively with pressure ratio. Screw compressors can often perform in single-stage duties which would require two stages of reciprocating compressors, and in fact are commonly so used. This practice, however, forfeits the benefits which staged compression can give by incorporating staged expansion of liquid. Whether these benefits would be worth the extra cost of plant involved depends on the circumstances, and needs to be worked out in particular cases.

In many circumstances the adoption of staged compression depends on a number of factors rather than on one hard reason. Much depends on the sizes of plant available in a manufacturer's range and whether the refrigerating capacity required is such that more than one machine would be needed in any event, whether different temperatures are wanted, and whether standby capacity is desired.

5.3 Desuperheating between stages

Desuperheating means reducing the temperature of the vapour delivered by one stage before it enters the next. When used in air compressors this practice is usually called intercooling, but this term will not be employed here because it is not strictly accurate when applied to some methods of desuperheating; and because the term intercooling has been used by some manufacturers to denote reduction of the temperature of the liquid between stages.

Desuperheating by direct heat transfer to cooling water or air has a limited use in refrigerant compressors because there is often no water available at a low enough temperature to provide effective desuperheating. The situation is quite different to that in air compressors, where the air enters the first stage at approximately the same temperature as that of the

Fig. 5.3.1. Desuperheating of the vapour between stages by heat transfer to water or air. (*a*) Diagrammatic arrangement. (*b*) Representation on the temperature–entropy diagram showing isentropic compression and desuperheating to approximately condensing temperature.

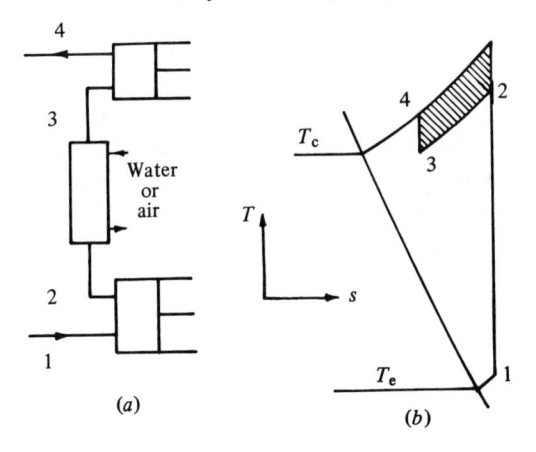

available cooling water. In an ammonia compressor the vapour entering the first stage is usually much below ambient temperature, and the most that can be expected is to desuperheat the vapour from the first stage to condensing temperature approximately. On occasions a small amount of especially cool water may be available for desuperheating, but this cannot usually be relied on.

For refrigerants like R12 and R502 which are commonly used with considerable superheat at entry to the compressor, desuperheating by water or by the ambient air may be all that is required. These refrigerants have only moderate rises of temperature during compression and high delivery temperatures are not to be expected. In many installations sufficient desuperheating is obtained in the piping between stages by direct heat transfer to the atmosphere.

Figure 5.3.1(a) shows the desuperheating arrangement in diagrammatic form, and the states of the vapour are shown on T–s coordinates in Fig. 5.3.1(b). The temperature t_3 after desuperheating is shown as approximately the same as the condensing temperature. With the intermediate pressure as shown, the delivery temperature t_2 from the first stage is higher than the delivery temperature t_4 from the second stage. By decreasing the intermediate pressure, t_2 would be reduced and t_4 would be increased.

If the first object of desuperheating is to reduce the maximum temperature to which the vapour is subjected, it is clear that this is best attained when the delivery temperatures from the two stages are equal, for a given state at 1, p_1 and t_1, and a given final pressure p_4. The intermediate pressure to give equal delivery temperatures can be determined from the refrigerant tables by inspection.

Example 5.3.1

Ammonia is compressed in two stages from an initial state of saturated vapour at $-30\,°\text{C}$ to a final pressure of 10.03 bar. The vapour is desuperheated between stages to a temperature of 25 °C. Assuming that the temperatures of the vapour after compression in each stage are the same as those for isentropic compression, determine the intermediate pressure which would give equal delivery temperatures.

At state 1, $p_1 = 1.195$ bar, $s_1 = 5.7801$ kJ/kg K. By inspection in the ammonia tables, at an intermediate pressure:

$$p_2 = 5.345 \text{ bar} \qquad s_2 = s_1 = 5.7801 \text{ kJ/kg K} \qquad t_2 = 70.7\,°\text{C}$$

Desuperheating at this pressure to 25 °C gives:

$$s_3 = 5.4341 \text{ kJ/kg K}$$

5.3 Desuperheating between stages

At the final pressure 10.03 bar

$$s_4 = s_3 = 5.4341 \text{ kJ/kg K} \qquad t_4 = 71.0\,°\text{C}$$

Thus an intermediate pressure of very nearly 5.345 bar satisfies the requirement.

The pressure ratios in the stages are:

    low-pressure stage $5.345/1.195 = 4.473$

    high-pressure stage $10.03/5.345 = 1.876$

Example 5.3.1 shows that if the intermediate pressure is chosen for minimum delivery temperature the pressure ratios in the stages are very different. In extreme cases this would result in a very low volumetric efficiency in the first stage. On the other hand, if the pressure ratios were made more nearly equal to give a better volumetric efficiency in the first stage, which is desirable because the first-stage machine is the larger machine, the delivery temperature leaving the second stage would be quite high, about 108 °C in the above example.

Desuperheating by direct heat transfer gives some saving in the specific work of compression, represented for the case of reversible adiabatic compression by the hatched area in Fig. 5.3.1(b). The shaft power and the swept volume rate in the second stage are also reduced.

A much more effective method of desuperheating is by the injection of liquid refrigerant into the interstage vapour, as shown diagrammatically in Fig. 5.3.2(a) and represented on a pressure–enthalpy diagram in Fig. 5.3.2(b). By injecting sufficient liquid refrigerant, the interstage vapour can obviously be desuperheated to saturation, or even made wet if desired.

Fig. 5.3.2. Desuperheating by injection of liquid refrigerant. (a) Circuit diagram. (b) Representation of states on the pressure–enthalpy diagram showing desuperheating to saturation.

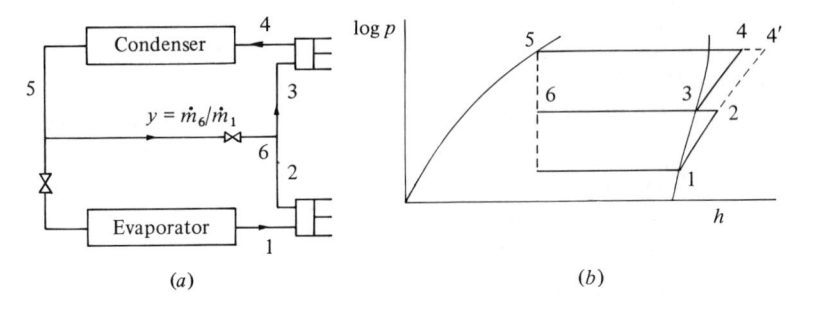

The question which then arises is how does this affect the coefficient of performance of the whole cycle, considering that the liquid refrigerant used for desuperheating is apparently wasted since it produces no refrigeration in the evaporator.

To investigate the point the cycle 1–2–3–4–5 in Fig. 5.3.2(b) will be compared with the cycle 1–2–4′–5 in which there is no desuperheating. The first stages of these are identical, with reversible adiabatic compression from 1–2, and for comparison it will be assumed that the mass flow rates are the same, thus giving identical refrigerating capacities in the evaporator.

The second stages differ, however, in that a larger mass flow of vapour, because of the added refrigerant, is compressed in the cycle with desuperheating. The extra mass flow rate is found by drawing an enthalpy balance around the mixing point. Supposing unit mass flow rate through the evaporator and first stage, and y the mass rate of injected liquid, the enthalpy balance around the mixing point gives:

$$h_2 + yh_5 = (1+y)h_3 \qquad (5.3.1)$$

and:

$$y = \frac{h_2 - h_3}{h_3 - h_5} \qquad (5.3.2)$$

$$1 + y = \frac{h_2 - h_5}{h_3 - h_5} \qquad (5.3.3)$$

The volume flow rate through the second stage for the cycle 1–2–3–4–5 is proportional to $(1+y)v_3$, and for the cycle 1–2–4′–5 it is proportional to v_2, where v_3 and v_2 are the specific volumes at 3 and 2. The volume flow rate through the second stage for the cycle with desuperheating is less than that for the cycle without desuperheating if:

$$(1+y)v_3 < v_2 \qquad (5.3.4)$$

that is if:

$$\frac{h_2 - h_5}{v_2} < \frac{h_3 - h_5}{v_3} \qquad (5.3.5)$$

on using (5.3.3) and re-arranging.

Provided that (5.3.5) holds, the volume flow rate to the second stage is less when the vapour is desuperheated than when it is not, despite the increase in the mass flow rate. Since the second stage in the two cases compared works over the same pressure ratio, the power taken by the second stage is less when (5.3.5) holds. It is true that the index of compression is slightly different in the two cases, but in fact this has very little effect on the power, the main influence being the volume rate of vapour which has to be compressed.

5.3 Desuperheating between stages

The inequality (5.3.5) will be recognised as the criterion, discussed in Chapter 4, for whether the volumic refrigerating effect decreases or increases with superheat. The behaviour of several refrigerants was shown in Fig. 3.12.2. Ammonia clearly satisfies inequality (5.3.5) at the conditions used for drawing the diagram, and consequently it is worth while using liquid ammonia to desuperheat the vapour from the first stage. For R12 and R502 inequality (5.3.5) is not satisfied, and it is not beneficial to use liquid for desuperheating with these refrigerants, but the opposite, especially bearing in mind that these refrigerants have moderate delivery temperatures, and the need for considerable superheat at suction to improve volumetric efficiency. R22 behaves marginally like ammonia, and if liquid is used for desuperheating it does not cause any striking loss of performance.

Even with ammonia and R22, the effect of superheat on volumetric efficiency should be considered, and generally desuperheating should not be carried right down to saturation, still less into the two-phase region.

The difference between refrigerants just noted is a very real one, and to emphasise it the result may be explained as follows. When liquid ammonia is injected into superheated vapour, the resulting desuperheated vapour consists of the original part at a lower temperature and the additional part representing the evaporated injected refrigerant. The decrease in volume of the original part caused by the reduction of temperature is greater, for ammonia, than the extra volume of the evaporated liquid, and consequently the total volume of the vapour diminishes. For R12 and R502, on the other hand, the volume of the evaporated liquid is greater than the decrease in volume of the original vapour, and the total volume increases.

It should be noted that, with ammonia, not only is the power taken by the second stage reduced by desuperheating, but so also is the capital cost of the second-stage machine, on account of its smaller size.

Example 5.3.2
Ammonia at a pressure of 1.195 bar ($t_e = -30\,°C$) with 5 K useful superheat, is compressed in a first stage of compression to 4.294 bar. It is desuperheated to a superheat of 5 K by injection of liquid coming to the injection valve at 30 °C. Determine the volume flow rate at entry to the second stage of compression for a refrigerating capacity of 60 kW, and compare this with the volume flow rate when there is no desuperheating.

The notation of Fig. 5.3.2 will be used, and to estimate the temperature leaving the first stage, isentropic compression will be assumed.

$h_1(1.195\text{ bar}, -25\,°C) = 1416.0\text{ kJ/kg}$

$s_1(1.195\text{ bar}, -25\,°C) = 5.8278\text{ kJ/kg K}$

Multi-stage vapour compression systems

$s_2 = s_1$ gives:

$h_2 = 1591.6 \text{ kJ/kg}$

$v_2 = 0.3689 \text{ m}^3/\text{kg}$

$h_3(4.294 \text{ bar}, 5\,°\text{C}) = 1456.5 \text{ kJ/kg}$

$v_3(4.294 \text{ bar}, 5\,°\text{C}) = 0.2966 \text{ m}^3/\text{kg}$

$h_5(\text{liquid at } 30\,°\text{C}) = 322.8 \text{ kJ/kg}$

The mass flow rate through the first stage is:

$$\frac{60}{(1416.0 - 322.8)} \text{ kg/s} = 0.05488 \text{ kg/s}$$

(5.3.3) gives:

$$1 + y = \frac{1591.6 - 322.8}{1456.5 - 322.8} = 1.1192$$

and the mass flow rate through the second stage is:

$(0.05488)(1.1192) \text{ kg/s} = 0.06142 \text{ kg/s}$

The volume flow rate at 3 is therefore:

$(0.06142)(0.2966) \text{ m}^3/\text{s} = 0.0182 \text{ m}^3/\text{s}$

If no desuperheating had been performed, the volume flow rate would have been:

$(0.05488)(0.3689) \text{ m}^3/\text{s} = 0.0202 \text{ m}^3/\text{s}$

an increase of 11% over the value for desuperheating. (The final delivery temperature would be prohibitive in this example without desuperheating if the condensing temperature were much above 30 °C.)

The assumption of isentropic compression used in this example is a useful one for estimating delivery temperature from the first stage, but better information from tests should be used when this is available. For water-cooled compressors the rate of heat transfer to the jacket water is of the order 12–15% of the shaft power taken by the machine. For air-cooled compressors the heat transfer to the surrounding air is a much lower fraction of the power and is often taken as zero for the purposes of estimating delivery temperature.

With hermetic and semi-hermetic compressors the electrical power should be used in estimating delivery temperature unless test figures are known. The isentropic temperature rise should not be used of course,

5.4 Multi-stage expansion

because of the unknown rise in temperature of the vapour passing over the motor before it enters the cylinder.

Screw compressors have a considerable rate of heat transfer to the oil in the barrel, and it is often possible to dispense with desuperheating without attaining excessive temperatures. It should not be forgotten, however, that for ammonia desuperheating by liquid injection has a beneficial effect on the coefficient of performance.

The injection valve shown in Fig. 5.3.2 may be a thermostatic expansion valve of the type described in Chapter 2, regulating the flow of liquid refrigerant to give a desired superheat at point 3. It may alternatively be made responsive to the actual delivery temperature from the second-stage compressor, thus using only enough liquid to prevent an excessive temperature there.

5.4 Multi-stage expansion

When two or more different suction pressures are available, as in a multi-stage compression system, advantage may be taken of these to reduce the irreversibility associated with the throttling process through the expansion valve by withdrawing the flashed vapour at each stage and compressing it from a higher pressure than the lowest suction pressure. The system was called at one time the Windhausen system, after its inventors (Windhausen & Windhausen 1901).

The reduction in temperature of the liquid refrigerant as it passes through the expansion valve is brought about by the transfer of enthalpy from the liquid to provide the enthalpy of evaporation of the flash vapour. At any

Fig. 5.4.1. Precooling by expansion to an intermediate pressure with separation of liquid and vapour in a flash-chamber. (a) Circuit diagram. (b) Representation on the pressure–enthalpy diagram, drawn as for R22. The mixing of states 2 and 7 gives state 3.

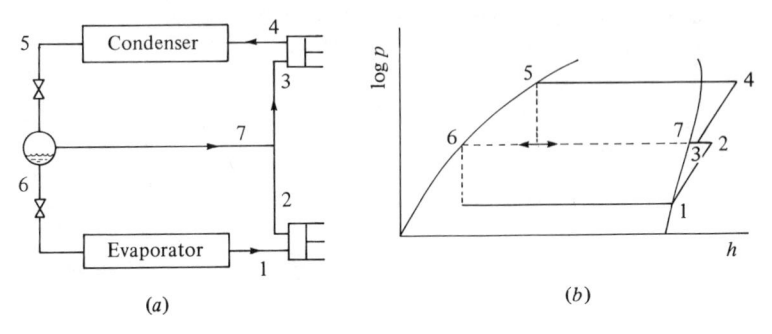

stage during the expansion the vapour formed down to that point has performed its function and it passes through the remainder of the throttling process and the evaporator as a passenger. Nevertheless, in the simple system it has to be compressed from the lowest pressure. In the Windhausen system it is compressed from one or more higher pressures with a resulting saving in work. Ideally an infinite number of intermediate pressures would be used, but in practice a useful improvement in refrigerating capacity and in coefficient of performance can be obtained when only one intermediate pressure is available.

This method of reducing the irreversibility of a part of a cycle has some analogies with the method of regenerative feed heating used in steam power plants. The irreversibility associated with raising the temperature of the cool feed water by heat transfer from high furnace temperature is reduced in these plants by using steam bled from the turbines at intermediate pressures to heat the feed water.

The basic components of the system, with two compressors in series and two expansion valves are shown in Fig. 5.4.1(a), and the states of the refrigerant around the cycle are shown on pressure–enthalpy coordinates in Fig. 5.4.1(b). The states at 1 and 7 are shown as saturated, but of course state 1 may be superheated depending on the method of feeding the evaporator, and state 7 may be slightly superheated or even wet if some priming occurs in the separator vessel.

The analysis of the system requires the determination of the mass flow rates at the several points, and these are determined (a) for the evaporator by applying the steady-flow energy equation when the refrigerating capacity is known, as described in Chapter 3, and (b) for the second-stage compressor by applying the steady-flow energy equation to the separator vessel, which is usually assumed to be adiabatic. The state 3 at entry to the second-stage compressor is found by applying the energy equation to the mixing of streams 7 and 2.

For a refrigeration capacity $\dot{Q}_e$, the mass flow rate through the evaporator and first-stage compressor is:

$$\dot{m}_6 = \dot{m}_1 = \frac{\dot{Q}_e}{h_1 - h_6} \qquad (5.4.1)$$

For the separator vessel:

$$\dot{m}_5 = \dot{m}_6 + \dot{m}_7 \qquad (5.4.2)$$

and:

$$\dot{m}_5 h_5 = \dot{m}_6 h_6 + \dot{m}_7 h_7 \qquad (5.4.3)$$

5.4 Multi-stage expansion

if the separator is adiabatic. Hence:

$$\frac{\dot{m}_7}{\dot{m}_1} = \frac{h_5 - h_6}{h_7 - h_5} \tag{5.4.4}$$

on using (5.4.1) and (5.4.2).
The mass flow rate through the second-stage compressor is:

$$\dot{m}_3 = \dot{m}_1 + \dot{m}_7 \tag{5.4.5}$$

and the state at the point 3 is fixed by the known intermediate pressure, and the specific enthalpy given by:

$$\dot{m}_3 h_3 = \dot{m}_1 h_2 + \dot{m}_7 h_7 \tag{5.4.6}$$

The state at 2 leaving the first-stage compressor may be determined by assuming isentropic compression unless better information is available.

In the event of the separator not being quite adiabatic, a heat transfer rate to it, $\dot{Q}_x$ say, calculated from the surface area, thickness of insulation and the temperature difference, may be inserted on the left-hand side of (5.4.3).

The explanation of the benefit of staged expansion given above was related to a reduction of the work of compression for a given mass flow rate of liquid entering the first expansion valve. Another way of looking at the matter, however, is to regard the method as a means of increasing the specific refrigerating effect of the liquid arriving at the evaporator, with a consequent increase of the refrigerating capacity available from a given volume flow rate in the first-stage compressor. This is a most important consideration in practice, not only because a smaller first-stage machine can be used to provide a demanded capacity, with an associated lower cost, but a further improvement in the coefficient of performance follows on account of the reduced proportional effects of friction and gas pumping, as outlined in Chapter 4.

Table 5.4.1

	$(h_1 - h_6)/(h_1 - h_5)$
Ammonia	1.155
R11	1.212
R12	1.316
R22	1.280
R502	1.447
Carbon dioxide, (state 5, saturated liquid)	3.723
Carbon dioxide, (state 5, $t = 30\,°C, p = 81$ bar)	1.728

This improvement of the specific refrigerating effect is given by referring to Fig. 5.4.1(b) as the ratio:

$$\frac{h_1 - h_6}{h_1 - h_5}$$

and it differs considerably between the refrigerants. Table 5.4.1 shows some values for an evaporating temperature of $-40\,°C$, an intermediate temperature in the separator of $-5\,°C$, and a condensing temperature of $30\,°C$. Saturated states are assumed at points 1 and 7, though, as already explained, there will be some superheat at state 1, especially for R12 and R502. Two conditions are given for carbon dioxide, one for saturated liquid at point 5, and one when advantage has been taken of increased pressure to improve the specific refrigerating effect as explained in Chapter 3.

The large value for R502 is noteworthy. It results from the fact that R502 has an unusually low specific enthalpy of evaporation combined with a specific heat capacity of the liquid about the same as that of R22. This combination gives R502 a rather poor coefficient of performance in single stage, but makes the benefit from staged expansion greater than for the other substances except carbon dioxide.

The importance which is attached in practice to the improvement of the specific refrigerating effect has led to the naming of the separator vessel as a 'precooler', which emphasises its role as a method of improving refrigerat-

Fig. 5.4.2. Liquid subcooling by evaporation of refrigerant at intermediate pressure: an alternative arrangement to that of Fig. 5.4.1, with float controlled expansion valve.

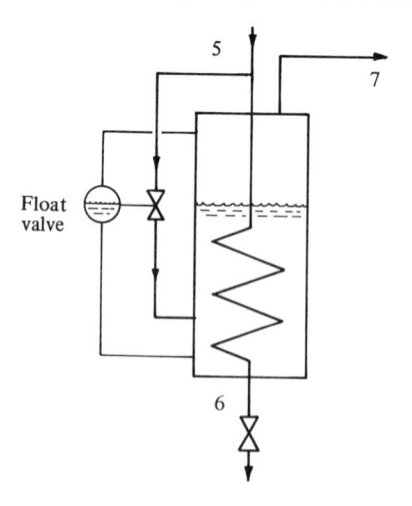

5.4 Multi-stage expansion

ing capacity rather than coefficient of performance. Some manufacturers have used the name 'economiser', but this is not used here since the term has been overworked in a number of engineering applications.

The circuit of Fig. 5.4.1 is essentially that used in applications where it is possible to place the separator close to the evaporator. When this is not possible a number of operating troubles are likely to occur, because the liquid leaving the separator is saturated and tends to flash into vapour if a pressure drop or heat transfer from the surroundings is encountered. As a result the second expansion valve at the evaporator may not be able to pass the required mass flow rate of refrigerant and the evaporator will be underfed.

The same end is often achieved by different means, therefore. Instead of a separator, a heat exchanger in which the main stream of the liquid is subcooled by evaporation of part of the refrigerant at intermediate pressure is used, as shown in Fig. 5.4.2. The liquid passes through the subcooling coil at the full condensing pressure and no flashing can occur on the way to the evaporator. Only one main expansion valve is required, and since this has the full pressure difference across it this valve can be smaller than the valves used in the system of Fig. 5.4.1. The subsidiary expansion valve used to feed the shell of the subcooler is a much smaller one. The disadvantage of the system is of course the cost of the subcooling coil, and the loss which occurs because the liquid leaving the subcooling coil at 6 in Fig. 5.4.2 is inevitably warmer than the liquid leaving the separator in Fig. 5.4.1, assuming that the

Fig. 5.4.3. Subcooler used in small systems with dry expansion at intermediate pressure controlled by a thermostatic expansion valve with the bulb at point 7. Placing the bulb at point 3 instead would cause more liquid feed through the subcooler and provide desuperheating as well.

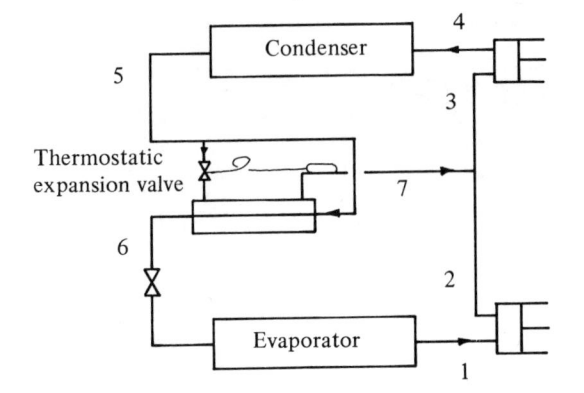

intermediate pressure is the same in each case. The temperature approach attained in the subcooling coil, i.e. $(t_6 - t_i)$, where t_i is the saturation temperature corresponding to the intermediate pressure, may be of the order 2–10 K, depending on the area provided for heat transfer, and because of this the specific refrigerating effect is somewhat lower than for the circuit of Fig. 5.4.1.

In small plants the rather expensive shell shown in Fig. 5.4.2 is dispensed with and a double-tube annular heat exchanger is used with refrigerant feed to the evaporative side by a thermostatic expansion valve, as indicated diagrammatically in Fig. 5.4.3. Since the thermostatic valve requires an operating superheat of about 5 K it is important that the heat exchanger work in counterflow (as shown), otherwise it will be impossible to reduce the temperature approach $(t_6 - t_i)$ below 5 K, and in fact will require a very large heat transfer surface to attain 6 or 7 K even.

5.5 Combined liquid cooling and desuperheating

The mixing of the vapour from the separator or subcooler with the vapour from the first stage of compression causes some desuperheating, but the resulting drop in temperature entering the second stage will not usually be sufficient for those refrigerants which have high delivery temperatures, such as ammonia or R22. By placing the bulb of the thermostatic expansion valve in Fig. 5.4.3 at point 3 after mixing instead of at point 7, the superheat at point 3 is regulated to about 5 K. The expansion valve in such a case acts as a combined feed for the subcooler and an injection valve for desuperheat-

Fig. 5.5.1. Combined liquid subcooling and desuperheating in the form usual in ammonia plants. (a) Circuit diagram. (b) Representation on the pressure–enthalpy diagram approximately to scale for ammonia at $-40°/-5°/30\,°C$. Some superheat is shown at points 1 and 3.

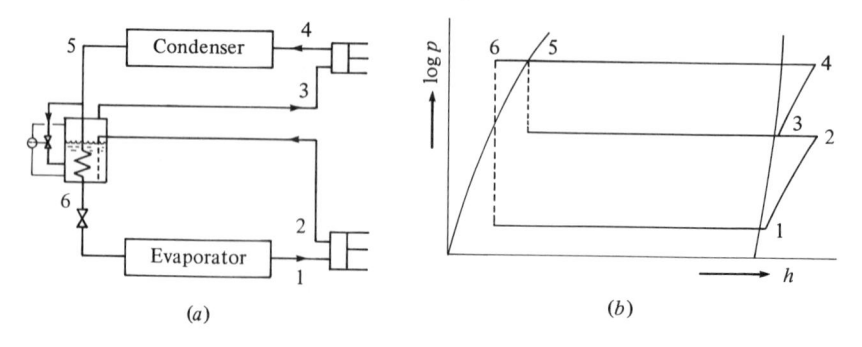

5.5 Combined liquid cooling and desuperheating

ing. The refrigerant state at point 7 will then be slightly wet because the vapour is carrying the liquid for the desuperheating process.

With the expansion valve bulb located at 3 the state there is controlled, and the analysis of the cycle is most readily performed by drawing the enthalpy balance around a control surface passing through points 5, 6, 2 and 3, leaving the unknown state at 7 entirely inside the surface. Then:

$$\dot{m}_3 h_5 + \dot{m}_1 h_2 = \dot{m}_1 h_6 + \dot{m}_3 h_3 \tag{5.5.1}$$

for adiabatic conditions, noting that $\dot{m}_5 = \dot{m}_3$ and $\dot{m}_6 = \dot{m}_1$. When a calculated heat transfer rate $\dot{Q}_x$ from the surroundings is to be included it goes on the left-hand side of (5.5.1). For the adiabatic case:

$$\frac{\dot{m}_3}{\dot{m}_1} = \frac{h_2 - h_6}{h_3 - h_5} \tag{5.5.2}$$

enables the mass flow rate through the second stage to be found when that through the first stage, $\dot{m}_1$, is known.

In ammonia systems the subcooling and desuperheating are often carried out as shown in Fig. 5.5.1. The vapour from the first-stage compressor is bubbled from a perforated pipe through the liquid in the subcooler, care being taken in the piping arrangements to ensure that liquid cannot flow back when the plant is not running from the shell to the delivery side of the first-stage compressor.

The analysis of the circuit of Fig. 5.5.1, the state at 3 being assumed or known, is exactly the same as that already given, eqs. (5.5.1) and (5.5.2).

Example 5.5.1
An ammonia system like Fig. 5.5.1 operates at the following conditions:

evaporating temperature	$-35\,°C$	$p = 0.932$ bar
intermediate temperature	$-2\,°C$	$p = 3.983$ bar
condensing temperature	$30\,°C$	$p = 11.66$ bar
temperature of liquid at 5	$29\,°C$	
temperature of liquid at 6	$1\,°C$	
superheat at 1 and 3	5 K	

Neglecting heat transfer from the surroundings and pressure drops in the piping, determine the volume flow rates at entry to the compressors for a refrigerating capacity of 100 kW. Assume isentropic compression in the first stage for the purpose of determining h_2.

From the ammonia tables:

$h_1(0.932 \text{ bar}, -30\,°C) = 1408.1 \text{ kJ/kg}$

$s_1(0.932 \text{ bar}, -30\,°C) = 5.9144 \text{ kJ/kg K}$

$v_1(0.932 \text{ bar}, -30\,°C) = 1.245 \text{ m}^3/\text{kg}$

$h_2(3.983 \text{ bar}, s = 5.9144 \text{ kJ/kg K}) = 1608.8 \text{ kJ/kg}$

$h_3(3.983 \text{ bar}, 3\,°C) = 1454.2 \text{ kJ/kg}$

$v_3(3.983 \text{ bar}, 3\,°C) = 0.3185 \text{ m}^3/\text{kg}$

$h_5(\text{liquid}, 29\,°C) = 318.0 \text{ kJ/kg}$

$h_6(\text{liquid}, 1\,°C) = 185.7 \text{ kJ/kg}$

From the heat and enthalpy balance on the evaporator:

$$\dot{m}_1 = \frac{100}{1408.1 - 185.7} \text{ kg/s} = 0.08181 \text{ kg/s}$$

The volume flow rate at entry to the first stage is then:

$$\dot{V}_1 = (0.08181)(1.245) \text{ m}^3/\text{s} = 0.102 \text{ m}^3/\text{s}$$

From (5.5.2):

$$\frac{\dot{m}_3}{\dot{m}_1} = \frac{1608.8 - 185.7}{1454.2 - 318.0} = 1.253$$

The mass flow rate through the second stage is:

$$\dot{m}_3 = (0.08181)(1.253) \text{ kg/s} = 0.1025 \text{ kg/s}$$

and the volume flow rate at the entry to the second stage is:

$$\dot{V}_3 = (0.1025)(0.3185) \text{ m}^3/\text{s} = 0.0326 \text{ m}^3/\text{s}.$$

Another method of treating the cycle of Fig. 5.5.1 is based on the idea that exactly the same heat and work transfers would be produced if the whole system were replaced by two independent cycles in series, that is by a cascade system (section 5.11) in which the two circuits contain the same refrigerant. In the low-temperature circuit the specific refrigerating effect is $(h_1 - h_6)$ and the refrigerating capacity is $\dot{m}_1(h_1 - h_6)$. To desuperheat and condense the vapour in the low-temperature circuit, a heat transfer rate of $\dot{m}_1(h_2 - h_6)$ is required, and this is regarded as a load on the high temperature circuit which operates with the specific refrigerating effect $(h_3 - h_5)$ and the refrigerating capacity $\dot{m}_3(h_3 - h_5)$. Consequently:

$$\dot{m}_3(h_3 - h_5) = \dot{m}_1(h_2 - h_6) \tag{5.5.3}$$

in agreement with eq. (5.5.2).

5.6 The choice of intermediate pressure

A refrigerating load at the intermediate temperature can be accommodated by including it on the right-hand side of eq. (5.5.3) because it simply adds to the load on the high-temperature circuit.

This form of calculation is often more convenient in practice than the earlier method because the refrigerating capacities and powers of the compressors are usually available in tables or graphs as functions of evaporating temperature. The low stage is treated as a single-stage system with the intermediate temperature as its condensing temperature, and the high stage is treated as a single-stage system with the intermediate temperature as its evaporating temperature. An example based on this method is given in section 5.7.

5.6 The choice of intermediate pressure

It is well known that for air compressors with intercooling to the initial temperature the minimum theoretical work is required when the pressure ratios in the stages are equal. This condition also gives equal delivery temperatures.

There is no reason to expect this rule to hold for compressors in multistage refrigeration systems, because the temperatures at entry to the stages are not usually equal, and the mass flow rates are different. Nevertheless, several investigators have found that a condition for minimum work exists somewhere near the point of equal pressure ratios.

The first extensive calculations were those of Behringer (1928) for two-stage ammonia cycles with liquid subcooling and desuperheating to saturation temperature. Liquid leaving the condenser was assumed to be already subcooled by 5 K. Over the range of evaporating temperatures from $-35\,°C$ to $-10\,°C$, an optimum intermediate temperature was found at approximately:

$$t_{opt} = t_m + 5 \text{ K} \tag{5.6.1}$$

where t_m is the temperature corresponding to the pressure for equal pressure ratios in the stages. Behringer did not in fact propose equation (5.6.1), but it represents his results quite closely.

The minima on Behringer's curves are not very pronounced, and appreciable deviations are allowable without increasing the power unduly. His findings have been widely used in ammonia practice, especially because making the pressure ratio in the first stage somewhat greater than that in the second helps to reduce the delivery temperature from the second stage.

In a later investigation for ideal cycles Czaplinski (1959) claimed to establish an optimum at an absolute intermediate temperature:

$$T_{opt} = (T_e T_c)^{1/2} \tag{5.6.2}$$

where T_e and T_c are the absolute temperatures of evaporation and condensation, respectively.

Calculations for more realistic cycles, allowing for compressor and motor efficiencies, have been presented by Baumann & Blass (1961). They found that provided there is the same clearance fraction in the machines, an optimum occurred at a point quite close to equal pressure ratios in the stages.

Serdakov (1961) reported experimental work on a two-stage ammonia system at an evaporating temperature of $-32.5\,°C$ and a condensing temperature of $22.5\,°C$. A maximum coefficient of performance was found at an intermediate temperature of $-2\,°C$, which is quite close to the value given by (5.6.1).

In practice the above considerations are often of academic value only, since the intermediate temperature is determined by other factors, for example the need for refrigeration at some intermediate temperature as described in section 5.8, or by the capacities of the machines available as described next.

5.7 The balance between the stages

It has been assumed so far that the intermediate pressure was known and that the problem was to determine the mass flow rates through the stages and from these find the required swept volume rates. In practice the matter is often presented the other way round, the swept volume rates being known and intermediate pressure unknown. This is especially so in a compound compressor. For example in an eight cylinder machine six of the cylinders may be given to the first stage and two to the second, a ratio of swept volume rates of 3. The intermediate pressure when the machine is running will then assume a value such that the mass flow rates satisfy the enthalpy balance around the interstage arrangements whatever they happen to be. For example, in the cycle of Fig. 5.5.1 equation (5.5.2) must hold, and this implies:

$$\frac{h_2 - h_6}{h_3 - h_5} = \frac{\dot{V}_{swH}\,\eta_{vH}\,v_1}{\dot{V}_{swL}\,\eta_{vL}\,v_3} \tag{5.7.1}$$

where $\dot{V}_{sw}$ and η_v are the swept volume rates and the volumetric efficiencies with subscripts L and H for the first and second stages, and v_1, v_3 are the specific volumes at the points 1 and 3.

An approximate method of determining the pressure which satisfies this equation has been given by Dauser (1941) and for R22 Soumerai (1953) has plotted functions of the refrigerant properties which enable the solution to be determined.

5.7 The balance between the stages

In general it is necessary to use a graphical solution or a trial and error method which can be performed by a computer. The graphical method will be outlined to illustrate the principle.

Re-arranging eq. (5.7.1) as:

$$\frac{\dot{V}_{swL}(h_2 - h_6)\eta_{vL}}{v_1} = \frac{\dot{V}_{swH}(h_3 - h_5)\eta_{vH}}{v_3} \qquad (5.7.2)$$

the right-hand side will be recognised as the refrigerating capacity of the second-stage compressor at an evaporating temperature equal to the intermediate temperature. The left-hand side of eq. (5.7.2) is the rate of heat transfer required to condense the vapour delivered by the first-stage compressor to a condenser at the intermediate temperature.

Using the refrigerating capacity curves for the two stages, the heat transfer rate required to condense the vapour from the first stage is calculated and plotted to a base of intermediate temperature together with the refrigerating capacity of the second stage. The point of intersection gives the balance condition. The following example illustrates the procedure.

Example 5.7.1

The first stage of a compound ammonia compressor has a swept volume rate of $0.129 \, \text{m}^3/\text{s}$. At an evaporating temperature of $-50\,°\text{C}$ the refrigerating capacity is given by the following table at various delivery pressures corresponding to the intermediate temperature t_i. The figures are based on 5 K superheat at inlet to the stage and liquid to the expansion valve at the intermediate temperature.

$t_i/°\text{C}$	-22	-20	-18	-16
$\dot{Q}_e/\text{kW}$	40.8	40.1	38.8	36.6

The second stage has a swept volume rate of $(0.129/3) \, \text{m}^3/\text{s} = 0.043 \, \text{m}^3/\text{s}$. The refrigerating capacity at a condensing temperature of $40\,°\text{C}$ and various suction pressures corresponding to the intermediate temperatures t_i is as follows. The figures are based on a superheat of 5 K at entry to the stage and liquid at $40\,°\text{C}$ to the expansion valve.

$t_i/°\text{C}$	-22	-20	-18	-16
$\dot{Q}_e/\text{kW}$	38.3	44.0	49.9	56.2

Determine the intermediate pressure when the stages are coupled for an evaporating temperature of $-50\,°\text{C}$ and a condensing temperature of $40\,°\text{C}$, assuming subcooling of liquid to the intermediate temperature and desuperheating of the interstage vapour to a superheat of 5 K.

To determine the specific enthalpy of the vapour from the first stage isentropic compression will be assumed. With the subscripts referring to Fig. 5.5.1:

$$h_1 (0.4088 \text{ bar}, -45\,°\text{C}) = 1383 \quad \text{kJ/kg}$$
$$s_1 (0.4088 \text{ bar}, -45\,°\text{C}) = 6.2034 \text{ kJ/kg K}$$

For the several intermediate pressures, the values of h_6 and h_2 are determined, and from these the ratio $(h_2 - h_6)/(h_1 - h_6)$ is calculated. Multiplying the values of $\dot{Q}_e$ for the first stage by this ratio gives the heat transfer rate, $\dot{Q}_i$ say, required to condense the vapour delivered by the first stage at intermediate pressure. Values of $\dot{Q}_i$ and of $\dot{Q}_e$ for the second stage

t_i °C	h_2 kJ/kg	h_6 kJ/kg	$\dfrac{h_2 - h_6}{h_1 - h_6}$	$\dot{Q}_i$ kW
−22	1573	81	1.146	46.8
−20	1587	90	1.158	46.4
−18	1601	99	1.170	45.4
−16	1615	108	1.182	43.3

are plotted against intermediate temperature in Fig. 5.7.1. The curves intersect at a temperature of $-19.3\,°\text{C}$, corresponding to a pressure of 1.965 bar at which the stages will be in balance. The refrigerating capacity at this condition is obtained by interpolating in the table for the refrigerating capacity of the first stage above, and is found to be 39.6 kW.

From the graph of Fig. 5.7.1 it will be noticed that the heat rate $\dot{Q}_i$ changes slowly with intermediate temperature when compared with the refrigerating capacity of the second stage. Hence a *rough* estimate of intermediate temperature can be made by neglecting the variation of the first-stage capacity and assuming a multiplying factor of about 1.15. This rough estimate will give some guide as to where to begin the accurate calculations.

The assumption of isentropic compression in the first stage is the usual one when better information about delivery temperature is not available, and fortunately it is not critical. Another method is to add the measured power to the first stage to the refrigerating capacity and to subtract the measured heat rate to the compressor jacket water: as indicated in section 5.3.

For a screw compressor the delivery temperature is usually well below the value given by isentropic compression, with a maximum value of about 100 °C, and the calculations should be adjusted accordingly.

5.7 The balance between the stages

When the liquid at point 6 is not subcooled to the intermediate temperature the figures for the capacity of the first stage should be modified by reducing them by the factor:

$$\frac{h_1 - h_6}{h_1 - h_{f,i}}$$

where $h_{f,i}$ is the specific enthalpy of liquid at the intermediate temperature t_i corresponding to the intermediate pressure.

In the practical operation of two-stage refrigerating plant observation of the intermediate pressure can give a useful indication of how the stages are behaving, because the pressure is quite sensitive to variations in the relative

Fig. 5.7.1. Graph for examples 5.7.1 and 5.8.1. Balance is attained where the refrigerating capacity of the second stage equals the heat rate needed to condense the vapour from the first stage.

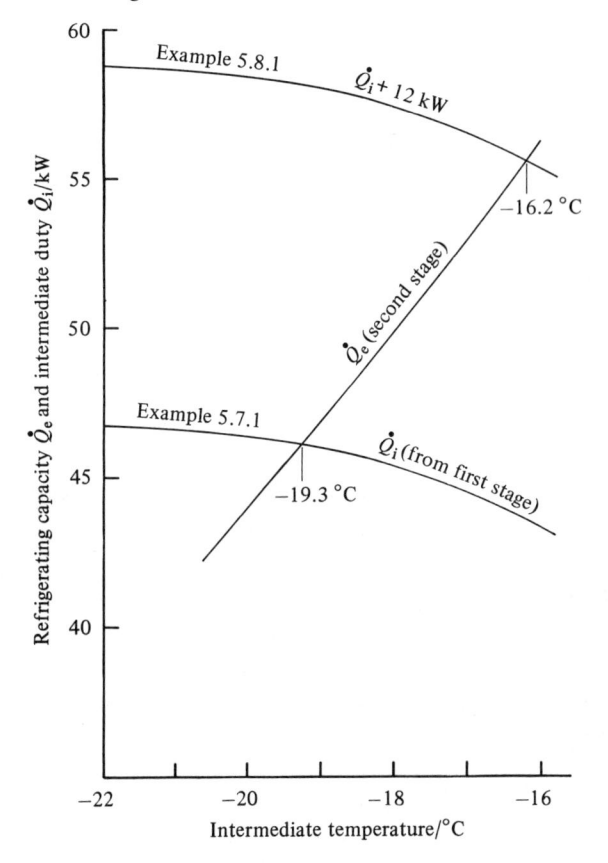

pumping capacities. Supposing that because of valve wear the capacity of the first stage diminishes, the mass flow rate to the second stage diminishes and the second stage can pump this reduced amount at a lower intermediate pressure. In Fig. 5.7.1 the line for $\dot{Q}_i$ is moved down and the intersection point falls. On the other hand, if the capacity of the second stage diminishes because of wear the intermediate pressure goes up, corresponding to a lowering of the line $\dot{Q}_e$ (second stage) in Fig. 5.7.1. Physically this means that the density of the vapour to the second stage has to rise to compensate for the reduced volume flow rate which the stage can handle.

5.8 Refrigeration at different temperatures

When two or more suction pressures are available it is possible to operate evaporators at different temperatures. It is true that evaporators at different temperatures may be worked by one compressor by throttling the vapour from the warmer one through a back-pressure valve, but this practice is not economical for large plants.

The piping arrangements for the warmer evaporator can be of several types. The simplest method suitable for a circuit like that of Fig. 5.4.3 is to feed the evaporator with liquid from point 5 through a separate expansion valve and to take the vapour from the evaporator into the intermediate pressure suction at point 3. Another method for larger plants like that of Fig. 5.5.1 would be to pump liquid refrigerant from the subcooler–desuperheater drum at the intermediate temperature to remote evaporators and to return the two-phase mixture to the drum exactly as in a normal pumped circulation system.

The calculations may be done either on a mass flow basis as in example 5.5.1, adding the mass flow rate of vapour formed by the intermediate load to the mass flow rate through the second stage; or on a capacity basis like example 5.7.1, adding the intermediate load to the load $\dot{Q}_i$ coming from the first stage.

Example 5.8.1

Referring to the plant in example 5.7.1, it is desired to carry a load of 12 kW at the intermediate temperature provided that this would be less than $-15\,°\text{C}$. Would the plant be suitable without modification of the swept volume rates?

In Fig. 5.7.1 12 kW have been added to the $\dot{Q}_i$ ordinates to give the upper curve. This intersects the capacity curve of the second stage at $-16.2\,°\text{C}$. This would be the evaporating temperature for the 12 kW load and would be satisfactory.

In this example the intermediate load is small relative to the load on the first stage, and consequently it brings little disturbance when introduced. In many circumstances, however, when the first stage and the intermediate loads are about equal, and when both can vary from 0 to 100%, careful control of the plant is needed to ensure that the rated powers of the motor or motors are not exceeded. Capacity modulation is usually desirable on both stages and careful sizing of the motors is required.

Evaporation at different temperatures is most often used in plants such as dairies with refrigeration duties at several temperatures, e.g. about 0 °C for milk cooling, $-15\,°C$ for ice cream making and perhaps $-40\,°C$ for the ice cream hardening rooms. The possibility of using several evaporators at different temperatures should also be considered when a fluid is to be cooled over a large temperature range. By this means a useful reduction in power can be obtained, and the process made to approach that envisaged in the Lorenz cycle treated in section 1.13.

5.9 Dual-effect compression

The benefits of liquid pre-cooling described in section 5.4 can be obtained even when two distinct stages of compression are not available by using a method of supercharging a compressor invented by Voorhees (1905) and called by him *multiple effect compression*. Voorhees envisaged the possibility of taking in vapour at several pressures, but in practice only one pressure higher than evaporating pressure was used and so the name dual-effect compression is more appropriate. Its main application in the past was in carbon dioxide systems, to which it was applied by W. Stokes and H. Brier (H. Brier, 1914), for compensating the greatly reduced refrigerating effect when operating near the critical point. For some years the merits of the system were hotly debated, but eventually the method became standard practice on carbon dioxide machines for marine use. With the decline of carbon dioxide for this purpose dual-effect compression became virtually obsolete. It has now been revived for screw compressors. Because of the oil injection and the very low clearance, these can be used over high pressure ratios, i.e. at low evaporating temperatures, without encountering excessive delivery temperatures or suffering from low volumetric efficiency. Thus a screw compressor can be used in single-stage where a reciprocating compressor would require two stages. But this forfeits the benefits of liquid pre-cooling. Hence dual-effect compression can be used with advantage.

The principle of the method will first be described with reference to a traditional compressor, shown in Fig. 5.9.1, with normal suction and delivery valves in the head. Ports are provided in the cylinder wall near the outer end of the piston stroke which are uncovered by the piston. These

Multi-stage vapour compression systems

ports are connected when uncovered to vapour at an intermediate pressure, p_2 say, which rushes in and raises the pressure inside the cylinder from the first suction pressure p_1 to pressure p_2, approximately. The mixture of vapour already inside the cylinder when the ports were uncovered, and the vapour which entered through the ports, is then compressed by the piston and delivered to the condenser.

The process is shown on a hypothetical indicator diagram in Fig. 5.9.1. Without pressure drop through the ports the pressure would rise instantaneously as shown by the full line, but in practice the shape indicated by the broken line would be expected. In the following treatment the effect of this will be neglected.

States inside the cylinder are denoted by A, B, C as shown. The states of the streams entering are denoted by 1 and 2, and that of the stream leaving by 3.

Applying the First Law of Thermodynamics to a system comprising the vapour which eventually ends up in the cylinder at B, and neglecting heat transfer:

$$0 = m_B u_B - m_A u_A - (m_B - m_A)u_2 - (m_B - m_A)p_2 v_2 \quad (5.9.1)$$

Fig. 5.9.1. Principle of a dual-effect compressor. Between A and B an extra charge of vapour is taken in through ports in the cylinder wall opened and closed by the piston.

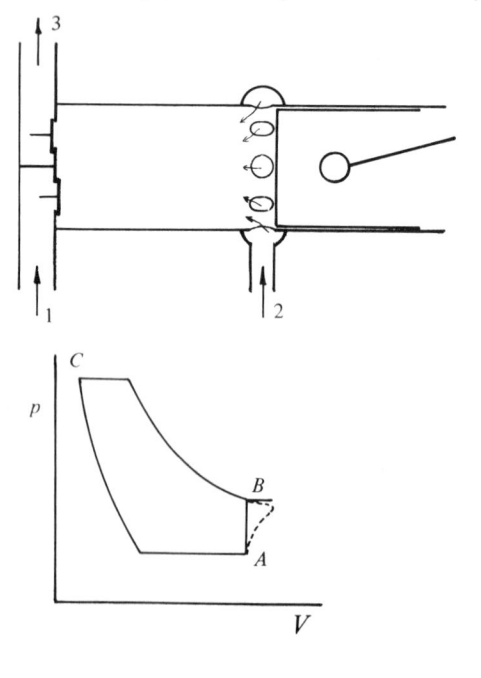

5.9 Dual-effect compression

where m denotes mass, u specific internal energy, v specific volume and p pressure.

In this equation $m_B u_B$ is the internal energy of the vapour finally within the cylinder, and $m_A u_A + (m_B - m_A)u_2$ is the internal energy of the same mass of vapour initially, partly inside the cylinder and partly in the pipe. The term $(m_B - m_A)p_2 v_2$ is the work done (by the fluid behind) in pushing the fluid mass $(m_B - m_A)$ into the cylinder.

Replacing internal energies by enthalpies:

$$mu = mh - pV \tag{5.9.2}$$

dividing by:

$$V = m_A v_A = m_B v_B \tag{5.9.3}$$

and rearranging the unknowns (state B) on the left-hand side, one obtains:

$$\frac{h_B - h_2}{v_B} = \frac{h_A - h_2}{v_A} + (p_B - p_A) \tag{5.9.4}$$

This equation relating properties at B to those at A and 2 was first obtained in this form by Sparks (1935).

Assuming that the states at A and 2 are known, the value of the right-hand side of eq. (5.9.4) can be calculated. Suppose this value is R. Then the determination of the state B involves finding a pair of values h_B and v_B in the refrigerant tables at the pressure p_B which satisfy:

$$\frac{h_B - h_2}{v_B} = R \tag{5.9.5}$$

This may be done by plotting the quantity $(h - h_2)/v$ to a base of temperature or superheat, but the line which results for most refrigerants is very nearly straight. Consequently linear interpolation can be used, as will be shown in example 5.9.1.

Having located the state B the mass inside the cylinder there is:

$$m_B = V/v_B \tag{5.9.6}$$

The mass initially inside the cylinder at A is:

$$m_A = V/v_A \tag{5.9.7}$$

and the mass $(m_B - m_A)$ which flows in can be found.

Example 5.9.1

In a dual-effect carbon dioxide compressor the volume enclosed by the piston when the ports open and close is $0.65\,\mathrm{dm}^3$. The vapour inside the cylinder when the ports open is saturated at $-20\,°C$, 19.71 bar. The vapour entering at the higher pressure is saturated at $-5\,°C$, 30.46 bar. Assuming adiabatic conditions, and that the pressure inside the cylinder rises to 30.46 bar by the time the ports close, determine the mass of vapour which flows in through the dual-effect ports.

Properties of carbon dioxide. (Adapted from, Newitt, Pai, Kuloor & Hugill, 1956.) (Datum: liquid at $-40\,°C$)

p bar	t °C	h kJ/kg	v dm^3/kg
19.71	-20	323.7	19.47 (saturation)
30.46	-5	320.5	12.14 (saturation)
30.46	0	325.3	12.69
30.46	10	342.1	13.80

The right-hand side of eq. (5.9.4) is:

$$R = \frac{323.7 - 320.5}{0.01947}\,\text{kJ/m}^3 + (30.46 - 19.71)10^2\,\text{kJ/m}^3$$

$$= 1239\,\text{kJ/m}^3$$

For superheated vapour at 30.46 bar:
at 0 °C:

$$[(325.3 - 320.5)/0.01269]\,\text{kJ/m}^3 = 378\,\text{kJ/m}^3$$

at 10 °C:

$$[(342.1 - 320.5)/0.0138]\,\text{kJ/m}^3 = 1565\,\text{kJ/m}^3$$

Interpolating the value 1239 between these linearly gives the temperature t_B as:

$$t_B = \frac{1239 - 378}{1565 - 378}\,10\,\text{K} + 0\,°C = 7.25\,°C$$

at which the specific volume v_B is:

$$v_B = 13.49\,\text{dm}^3/\text{kg}$$

by linear interpolation. The mass of vapour originally inside the cylinder is:

$$m_A = (0.650/19.47)\,\text{kg} = 0.0334\,\text{kg}$$

The mass of vapour inside when the ports close is:

$$m_B = (0.650/13.49)\,\text{kg} = 0.0482\,\text{kg}$$

5.9 Dual-effect compression

and the mass which flows in is therefore:

$$m_B - m_A = (0.0482 - 0.0334)\,\text{kg} = 0.0148\,\text{kg}$$

In applying the above method to an actual compressor it is necessary to make some assumption about the state at A. In some cases the mass of vapour m_1 say, which flows in from the evaporator will be known from tests or from a volumetric efficiency, and the mass m_A can be found from:

$$m_A = m_1 + m_c \tag{5.9.8}$$

where m_c is the mass of vapour left in the clearance space at the end of delivery.

For a reciprocating compressor it must be remembered that the volume $V_A = V_B$ is less than the sum of the swept volume and the clearance volume because of the width of the dual-effect ports.

The steady-flow equation applied to the whole compressor gives, allowing a heat rate $\dot{Q}_c$ to coolant:

$$\dot{W} = (\dot{m}_2 + \dot{m}_1)h_3 - \dot{m}_1 h_1 - \dot{m}_2 h_2 + \dot{Q}_c \tag{5.9.9}$$

$\dot{Q}_c$ is approximately zero for an air-cooled high-speed compressor. For the reasons given in section 4.5, however, it is safer to base the

Fig. 5.9.2. The relation between the state at A and that at B shown on a pressure–enthalpy diagram. The pressure scale must be linear. (a) The case of $h_2 < h_A$. (b) The special case when the entropy at B is the same as at A.

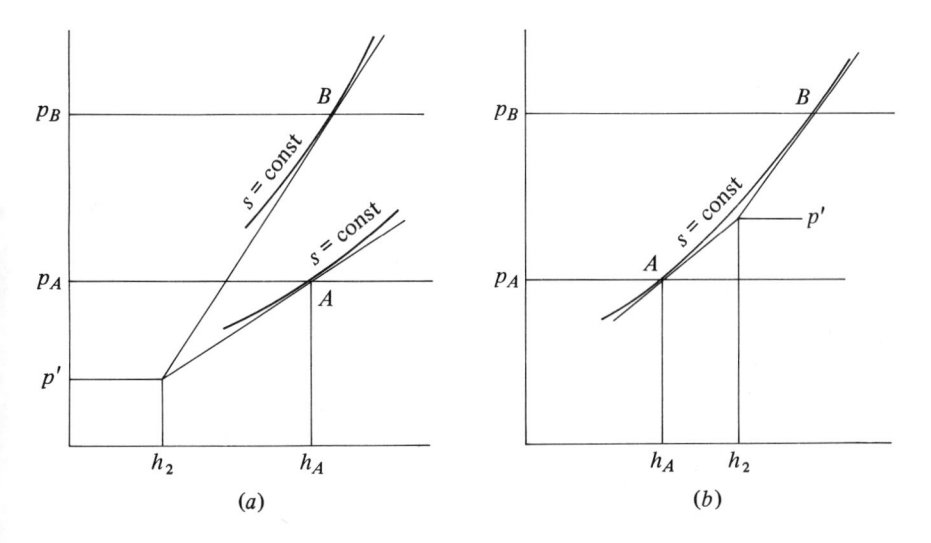

calculation of power on the area of the indicator diagram with suitable supplements for friction and pressure drops. The mean effective pressure of the diagram of Fig. 5.9.1 is given by:

$$w_v = (1+c-e)p_2 \frac{n}{n-1}\left[\left(\frac{p_3}{p_2}\right)^{(n-1)/n} - 1\right] + (1+c-e)(p_2 - p_1)$$

$$- c\left(\frac{p_3}{p_1}\right)^{1/n} p_1 \frac{n}{n-1}\left[\left(\frac{p_3}{p_1}\right)^{(n-1)/n} - 1\right] \quad (5.9.10)$$

where c is the clearance fraction, e is the fraction of the swept volume occupied by the dual effect ports and n is the index of compression, assumed the same as that for expansion. The pressures p_1 and p_2, which are those usually measured, are assumed the same as p_A and p_B, respectively. As explained in section 4.5, the index k of isentropic compression may be safely used in place of n to obtain a conservative estimate of the power.

The thermodynamic treatment leading to eq. (5.9.4) gives the state of the fluid at B but it does not show very clearly on a diagram of properties where B is located with respect to states A and 2. The following construction shows this on a pressure–enthalpy diagram with a linear pressure scale. Referring to Fig. 5.9.2(a) which shows a part of the superheated vapour region, a tangent is drawn at the point A to the line of constant entropy there. This tangent cuts the line $h = h_2$ at a pressure p' say. From the point so obtained, (p', h_2), a line is drawn tangential to a line of constant entropy at pressure p_B, and the tangent point gives the required state B. For at B:

$$v_B = \left(\frac{\partial h}{\partial p}\right)_s = \frac{h_B - h_2}{p_B - p'} \quad (5.9.11)$$

Therefore:

$$\frac{h_B - h_2}{v_B} = p_B - p' \quad (5.9.12)$$

Similarly:

$$\frac{h_A - h_2}{v_A} = p_A - p' \quad (5.9.13)$$

Subtracting eq. (5.9.13) from (5.9.12) gives eq. (5.9.4), which is to be satisfied.

If h_2 is greater than or equal to h_A the same construction applies, but p' then lies above or on the line $p = p_A$.

At one particular value of h_2 the entropy at B has the same value as at A, as shown in Fig. 5.9.2(b). For this to occur the value of h_2 has to be somewhat greater than h_A. This is approximately the case for compressors

5.9 Dual-effect compression

working on ammonia and the fluorinated refrigerants. Consequently, for practical calculations one can often estimate the specific volume at state B by taking an isentropic compression from state A, without having to go through the calculation of example 5.9.1 each time.

Screw compressors

In a reciprocating compressor the dual-effect port has necessarily to be placed at outer dead centre, because the alternating movement of the piston would uncover it prematurely if it were placed near mid stroke. In a screw compressor this limitation does not apply, and the process can take place somewhat as shown in Fig. 5.9.3. At point B an interlobe space is opened to the intermediate pressure. Vapour rushes in, but because the interlobe space is diminishing some of this has to be pushed out again from C to D. At D the interlobe space is disconnected from the intermediate pressure and normal compression continues. It will be seen that this position of the port offers advantages compared with a position at the beginning of compression. On the other hand, for compressors fitted with slide valves for capacity regulation, the dual effect is put out of action as the slide valve opens.

The thermodynamics of the process is the same as that given earlier, and the specific volume at C, which is the same as the specific volume at D, is found in the same way.

It is clear from the diagram that the use of a dual-effect port raises the built-in pressure ratio of the compressor, for if the delivery port is opened at the volume V_E the pressure inside the cylinder has to be higher there than if compression had been completed along the broken line.

Fig. 5.9.3. Dual effect in a screw compressor. At B an interlobe space is opened to a higher pressure, and closed at D.

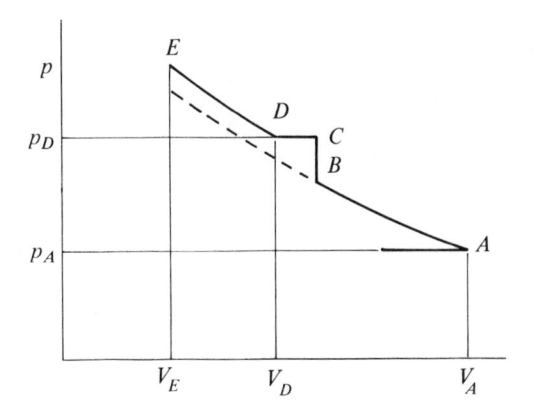

Application to pre-cooling

The above analysis of the filling process presupposes a known value of the pressure p_2, the intermediate pressure. When dual-effect compression is used for liquid pre-cooling this pressure is not known in advance. Its value depends on a balance between the flash chamber and the compressor such that the rate at which flash vapour is produced equals the rate at which it can flow in through the dual-effect ports. If these rates are not the same the intermediate pressure adjusts itself until they are.

Referring to Fig. 5.9.4 where the compressor may be understood as either a reciprocating or a screw machine, an enthalpy balance on the pre-cooler gives the rates of flow of refrigerant at 4 and 6 in the ratio:

$$\frac{\dot{m}_4}{\dot{m}_6} = \frac{h_2 - h_6}{h_2 - h_4} \tag{5.9.14}$$

For a screw compressor the clearance fraction is small and it will be sufficiently accurate to assume that this ratio is equal to the ratio of the mass at D to the mass at A in Fig. 5.9.3. Hence the condition of balance is that:

$$\frac{h_2 - h_6}{h_2 - h_4} = \frac{V_D}{V_A} \frac{v_A}{v_D} \tag{5.9.15}$$

where V_D/V_A is a ratio fixed by the position of the dual-effect port, and v_A and v_D are the specific volumes of the vapour at A and D, respectively.

For a reciprocating compressor V_D/V_A equals unity, but eq. (5.9.15) is not exact because not all of the mass inside the cylinder at D is delivered on account of the effect of clearance. Nevertheless it will be assumed for the purpose of the discussion that eq. (5.9.15) holds.

Fig. 5.9.4. The use of dual effect compression for pre-cooling.

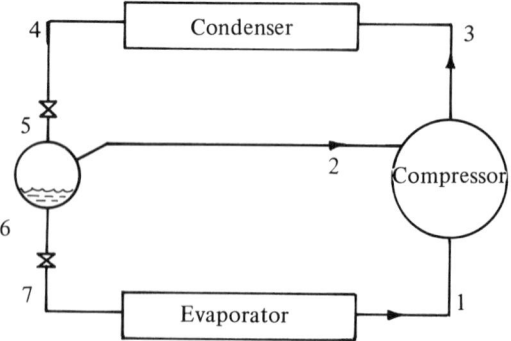

5.9 Dual-effect compression

Figure 5.9.5 shows the balance points corresponding to eq. (5.9.15) for several refrigerants at specified conditions. Three values of V_D/V_A are shown.

The ratio of the refrigerating capacity with pre-cooling to that without is given by the ratio:

$$\frac{h_1 - h_6}{h_1 - h_4}$$

with reference to the numbered points in Fig. 5.9.4, because the mass flow rate through the evaporator is the same in either case. For the refrigerants shown in Fig. 5.9.5 the values of h_1 and h_2 are not very different. Hence the ordinates of Fig. 5.9.5 at the balance points also express the improvement in refrigerating capacity with and without pre-cooling.

It is clear that the maximum improvement in capacity is found when the

Fig. 5.9.5. The balance points for common refrigerants, evaporating at $-40\,°\mathrm{C}$, liquid to valve at $30\,°\mathrm{C}$, for three values of V_D/V_A. The broken lines are not the compressor mass ratio lines, but merely indicate balance points with the same V_D/V_A values.

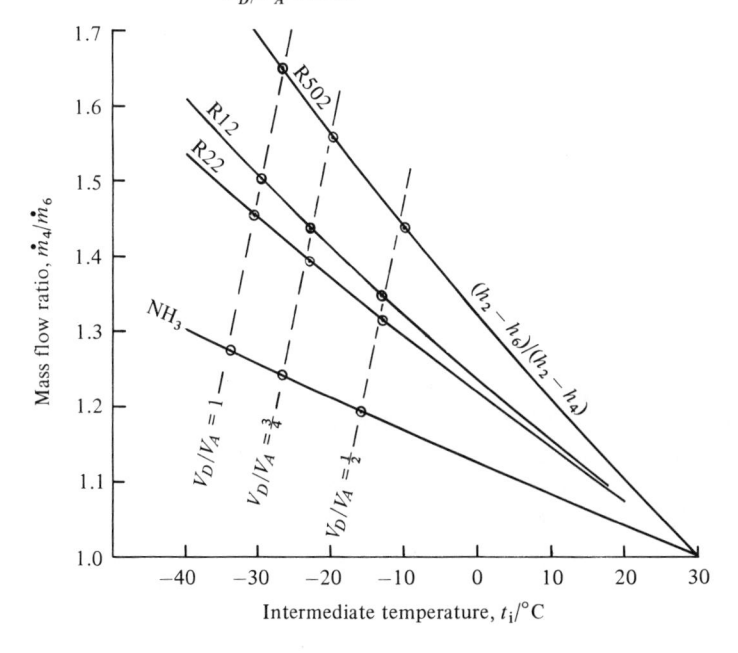

dual-effect port closes as early as possible, i.e. when V_A/V_D is as near unity as possible. The system then balances at an intermediate pressure which is lower than the balance pressure for a ratio V_D/V_A less than unity.

On the other hand there is little improvement in the coefficient of performance when the intermediate pressure is low, because the flashed vapour has to be compressed over nearly the whole pressure ratio and there is very little saving in work. If the intermediate pressure were very high there would again be little saving of work because there would not be much flashed vapour to compress. There is in fact an optimum intermediate pressure for maximum coefficient of performance. For R22 this occurs at an intermediate temperature of about $-10\,°C$, for the conditions of Fig. 5.9.5, for the theoretical coefficient of performance. The optimum for the actual coefficient of performance has to be worked out in detail taking into account the effects of friction and the irreversibilities associated with the dual-effect process. These tend to move the optimum towards a lower intermediate pressure, mainly because of the fact that the same friction power is spread over a larger refrigerating capacity there.

In a screw compressor there is a certain amount of choice about the position of the dual-effect port. In a twin-screw compressor there are two ports, one for the male rotor spaces and the other for the female rotor spaces. They are situated in the common axial plane of the rotors.

The choice is limited however by the geometry of the machine. Since the ports are opened by the axial movement of the helical sealing line in the casing, and because the ports are uncovered at one edge and covered at the other, they are open for a little more than one quarter of a revolution on the male side and one sixth on the female side, assuming the usual 4/6 arrangement. This means in practice that V_D/V_A cannot be made equal to unity, to give the largest increase of capacity. On the other hand, if the value of V_D/V_A is chosen to give a substantial improvement in the coefficient of performance the open positions of the dual-effect ports straddle the 80% swept volume line at which the capacity regulating slide valve opens, and as soon as the slide valve opens the intermediate pressure can flow to the suction space via the interlobe space. The pressure in the separator falls to suction pressure and the pre-cooling effect is lost. Furthermore, with the circuit of Fig. 5.9.4 when the pre-cooler pressure falls to suction pressure there is no pressure difference to drive the refrigerant through the second expansion valve. Special system arrangements have to be provided for this.

With reciprocating compressors pre-cooling by dual effect has been used for ammonia, but its most successful application was with carbon dioxide to offset the serious reduction in capacity which occurs at high cooling water

5.10 The manufacture of solid carbon dioxide

temperatures. The increase of capacity is very striking with carbon dioxide. Assuming evaporation at $-20\,°C$. with saturated liquid at $30\,°C$ to the expansion valve, the system balances at an intermediate pressure of about $4\,°C$ with an increase of capacity of over 75%. With optimum high side pressure, as described in section 5.14, the increase is somewhat less.

The dual-effect method of treating intermediate pressure vapour can of course be used with a subcooler like that of Fig. 5.4.2 instead of a separator/pre-cooler.

5.10 The manufacture of solid carbon dioxide

When liquid carbon dioxide from a cylinder under pressure is allowed to escape into the atmosphere, i.e. to undergo a throttling expansion at constant enthalpy, it emerges as a two-phase mixture of solid, in the form of snow, and vapour. This behaviour occurs because normal atmospheric pressure is lower than the pressure at the triple point of carbon dioxide, 5.18 bar at $-56.6\,°C$. Below this pressure liquid cannot exist as a stable phase. The triple point is represented by a line on the pressure–enthalpy diagram shown in Fig. 5.10.1. This diagram is plotted with a logarithmic pressure scale in order to open it up at the lower pressures. The

Fig. 5.10.1. Phase boundary lines for carbon dioxide on the pressure–enthalpy diagram. The pressure is plotted to a logarithmic scale.

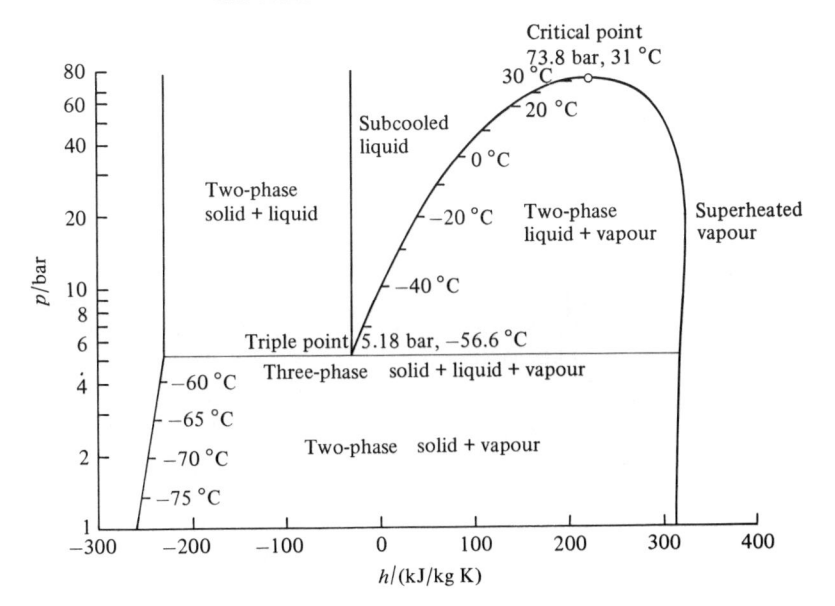

boundary curves for saturated liquid and vapour in equilibrium and for saturated solid and vapour in equilibrium are shown. The boundary curves for solid and liquid in equilibrium are shown as vertical lines because the pressure required to change the equilibrium temperature is very high.

Throttling expansion is the usual method for making solid carbon dioxide, but it is clear that to obtain a good yield the carbon dioxide before throttling should have as low a specific enthalpy as possible. In a simple system, with no additional refrigeration, this state of minimum specific enthalpy would be liquid at a temperature somewhat above that of the available cooling water, say 25 °C and a pressure of about 64 bar. When the temperature of the cooling water is near the critical temperature the pressure could be increased in order to reduce the specific enthalpy, as described in section 3.14.

If the gas to be processed arrives at atmospheric pressure, several stages of compression are needed to bring the gas up to its condensing temperature or the necessary supercritical pressure when the cooling water temperature is high. An opportunity is thus provided to stage the expansion of the carbon dioxide with the withdrawal of flash vapour and the compression of this from intermediate pressures.

Figure 5.10.2(a) shows a typical circuit. For simplicity only two stages of compression are shown, though in practice more would be used as a rule. Figure 5.10.2(b) shows the states on the pressure–enthalpy diagram.

Beginning at state point 6, liquid expands at constant enthalpy through

Fig. 5.10.2. The manufacture of solid carbon dioxide. (a) Circuit diagram, showing only two stages of compression for simplicity. (b) Representation on the pressure–enthalpy diagram.

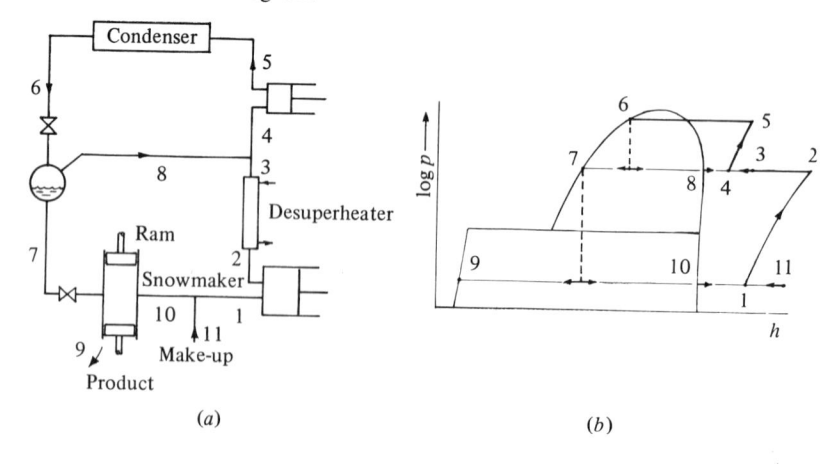

5.10 The manufacture of solid carbon dioxide

the first expansion valve into a flash chamber or pre-cooler in which liquid 7 separates and vapour 8 goes off to the second stage compressor. The liquid 7 undergoes a second isenthalpic expansion bringing the state into the solid–vapour region. This operation takes place in the snowmaker as a batch process, several snowmakers being operated from one condensing system. While snow is being formed in the snowmaker the vapour 10, called the 'revert gas', is being given off. It returns via a collecting vessel, not shown, to the first stage compressor after mixing with the make-up gas. When the snowmaker is full a hydraulic ram comes down and compresses the snow into a block of density about 1450 kg/m^3, subsequently expelling the block from the bottom.

The analysis of the system follows closely the methods for multi-stage systems already treated. Indicating mass flow rates and specific enthalpies at the various points by subscript numbers to correspond, the steady-flow energy equation applied to a control surface around the flash chamber gives:

$$\dot{m}_6 h_6 = \dot{m}_7 h_7 + \dot{m}_8 h_8 = \dot{m}_7 h_7 + \dot{m}_6 h_8 - \dot{m}_7 h_8 \tag{5.10.1}$$

from which:

$$\frac{\dot{m}_7}{\dot{m}_6} = \frac{h_8 - h_6}{h_8 - h_7} \tag{5.10.2}$$

Assuming that enough snowmakers are operating to approximate to quasi-steady conditions, the mass and enthalpy balances for the snowmakers give:

$$\dot{m}_7 h_7 = \dot{m}_9 h_9 + \dot{m}_{10} h_{10} = \dot{m}_9 h_9 + \dot{m}_7 h_{10} - \dot{m}_9 h_{10} \tag{5.10.3}$$

and:

$$\frac{\dot{m}_9}{\dot{m}_7} = \frac{h_{10} - h_7}{h_{10} - h_9} \tag{5.10.4}$$

Supposing that the mass flow rate $\dot{m}_9$ of the snow is known, the flow rate of the make-up gas $\dot{m}_{11}$ is the same, and the mass flow rate through the first stage compressor $\dot{m}_1 = \dot{m}_7$ is then given by eq. (5.10.4). The mass flow rate through the second-stage compressor is $\dot{m}_4 = \dot{m}_6$, given by eq. (5.10.2) in terms of $\dot{m}_7$.

Having determined the mass flow rates, the required swept volumes and the powers can be found.

The above concerns a simple form of the process. Many more complicated ones have been considered, particularly suitable for high cooling water temperatures, with staged cooling and compression of the gas in the supercritical region. Several of these processes are considered by Stickney (1932), who gives optimum conditions of operation.

Dual-effect compression has also been applied to the manufacture of solid carbon dioxide. In one method described by H. Brier & J. H. Brier (1933) a triple-effect compressor was used, treating gas at three different suction pressures.

The performance of the whole plant can be much improved by using auxiliary refrigeration from a vapour compression or absorption plant to condense the carbon dioxide at a lower temperature.

The solid carbon dioxide produced by the above methods consists of compressed snow, which is white and opaque. Vahl (1961) has proposed a method involving heat transfer to a refrigerated surface which produces clear carbon dioxide ice.

Example 5.10.1

A plant for making solid carbon dioxide is generally similar to that of Fig. 5.10.2 except that the condensation of the carbon dioxide is performed at a temperature of $-15\,^\circ\text{C}$, 22.92 bar, by an auxiliary ammonia refrigerating system whose coefficient of performance is 2.9. The make-up vapour enters at atmospheric pressure (1.013 bar) and a temperature of $20\,^\circ\text{C}$, and mixes with the saturated revert vapour from the snowmakers before compression in the first stage. The vapour delivered by the first-stage compressor at a pressure of 5.551 bar is desuperheated by direct water cooling to $30\,^\circ\text{C}$ before mixing with the saturated vapour from the liquid precooler. The vapour from the second-stage compressor is also desuperheated to $30\,^\circ\text{C}$ before further desuperheating and condensation by the ammonia plant.

Assuming isentropic efficiencies of 0.7 for the carbon dioxide compressors, determine the total shaft power required for a production rate of solid carbon dioxide at atmospheric pressure, $-78.5\,^\circ\text{C}$, of 1 kg/s.

Properties of carbon dioxide. (Adapted from Newitt et al., 1956[†]) (Datum: liquid at $-40\,^\circ\text{C}$)

Saturation

p bar	t_s $^\circ$C	h_i kJ/kg	h_f kJ/kg	h_g kJ/kg
1.013	-78.5	-259.26		311.84
5.551	-55.0		-27.97	317.74
22.92	-15.0		50.16	323.22

5.10 The manufacture of solid carbon dioxide

Superheated vapour

	1.013 bar		5.551 bar		22.92 bar	
t	h	s	h	s	h	s
°C	kJ/kg	kJ/kg K	kJ/kg	kJ/kg K	kJ/kg	kJ/kg K
−20	359.35	1.9439				
−10	367.64	1.9758				
10			379.01	1.7018		
20	392.81	2.0666	388.07	1.7334		
30			397.13	1.7637	378.60	1.4484
90			452.13	1.9293		
100			461.49	1.9545		
110					462.55	1.6943
120					472.59	1.7205

† The properties of the superheated vapour were found by second-degree interpolation in Newitt's tables, which give values at rather widely spaced integral pressures in atmospheres. The second differences in the entropy are quite erratic. The more recent tables of Vukalovich & Altunin (1968) could not be used because they do not go below 0 °C in the superheat region.

Using the numbering of Fig. 5.10.2, the specific enthalpies are:

h_6(liquid, $-15\,°C) = 50.16$ kJ/kg

h_7(liquid, $-55\,°C) = -27.97$ kJ/kg

h_8(saturated vapour, $-55\,°C) = 317.74$ kJ/kg

h_9(solid, $-78.5\,°C) = -259.26$ kJ/kg

h_{10}(saturated vapour, $-78.5\,°C) = 311.84$ kJ/kg

Equation (5.10.2) gives:

$$\frac{\dot{m}_7}{\dot{m}_6} = \frac{(317.74 - 50.16)}{(317.74 + 27.97)} = 0.7740$$

Equation (5.10.4) gives:

$$\frac{\dot{m}_9}{\dot{m}_7} = \frac{(311.84 + 27.97)}{(311.84 + 259.26)} = 0.5950$$

Putting $\dot{m}_9 = 1$ kg/s

$\dot{m}_7 = 1.6807$ kg/s

$\dot{m}_6 = 2.1714$ kg/s

By differences:

$\dot{m}_8 = 0.4907$ kg/s

$\dot{m}_{10} = 0.6807$ kg/s

At steady conditions the rate of make-up $\dot{m}_{11} = \dot{m}_9$. $h_{11}(1.013 \text{ bar}, 20\,°\text{C}) = 392.81$ kJ/kg, and on mixing with the revert vapour:

$$h_1 = \frac{392.81 + (0.6807)(311.84)}{1.6807} \text{ kJ/kg} = 360.02 \text{ kJ/kg}$$

Interpolating linearly between values in the table at 1.013 bar gives:

$t_1 = -19.2\,°\text{C}$

$s_1 = 1.9465$ kJ/kg K

For isentropic compression $s_2 = s_1$, and by interpolation at $p = 5.551$ bar:

$t_2 = 96.8\,°\text{C}$

$h_2 = 458.52$ kJ/kg

The power taken by the first-stage compressor is therefore:

$$\dot{W}_{LP} = \frac{(1.6807)(458.52 - 360.02)}{0.7} \text{ kW} = 236 \text{ kW}$$

After desuperheating to $t_3 = 30\,°\text{C}$:

$h_3(\text{vapour}, 5.551 \text{ bar}, 30\,°\text{C}) = 397.13$ kJ/kg

and after mixing with vapour 8:

$$h_4 = \frac{(1.6807)(397.13) + (0.4907)(317.74)}{2.1714} \text{ kJ/kg} = 379.19 \text{ kJ/kg}$$

Interpolating in the table at 5.551 bar gives:

$t_4 = 10.2\,°\text{C}$

$s_4 = 1.7024$ kJ/kg K

After isentropic compression to 22.92 bar, $s_5 = s_4$ gives:

$t_5 = 113.1\,°\text{C}$

$h_5 = 465.66$ kJ/kg

The power taken by the second-stage compressor is therefore:

$$\dot{W}_{HP} = \frac{(2.1714)(465.65 - 379.19)}{0.7} \text{ kW} = 268 \text{ kW.}$$

After desuperheating to $30\,°\text{C}$:

$h(\text{vapour}, 22.92 \text{ bar}, 30\,°\text{C}) = 378.60$ kJ/kg and the heat transfer rate to the ammonia plant is:

2.1714(378.6 − 50.16) kW = 713.2 kW.

This requires a power to the ammonia plant of:

$$\frac{713.2}{2.9}\,\text{kW} = 246\,\text{kW}$$

The total shaft power is therefore:

(236 + 268 + 246) kW = 750 kW

An actual plant of the type of example 5.10.1 would probably have other heat exchangers which have been omitted for simplicity. In particular the revert gas from the snowmakers might be used to subcool the liquid carbon dioxide.

5.11 Cascade systems

In a cascade system separate refrigerating cycles are arranged in series, as shown diagrammatically in Fig. 5.11.1. Boiling of refrigerant in circuit B causes the condensation of refrigerant in circuit A. The refrigerant in A then boils in the evaporator of its circuit to produce a still lower temperature. The intermediate heat exchanger between the circuits is usually called a 'cascade condenser'.

It is advantageous to use a cascade system when the difference between the temperature at which heat is ultimately rejected and the temperature at which refrigeration is required is so large that a refrigerant with suitable properties at both temperatures cannot be found. For example, because the

Fig. 5.11.1. A simple cascade system with two circuits.

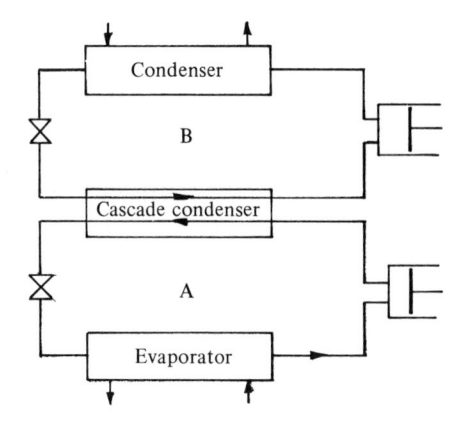

vapour pressure curves of substances are of roughly similar shape it is not possible to find a refrigerant which has a suitably high pressure in the evaporator and a reasonably low pressure in the condenser when the difference in temperature is large. This may present constructional difficulties. Furthermore, the volumic refrigerating effect, as discussed in section 3.9, is closely related to the vapour pressure, and all substances have approximately the same volumic refrigerating effect at the same evaporating pressure. Even if a very low evaporating pressure can be accommodated, therefore, it will inevitably be associated with a low volumic refrigerating effect and will require a large volume flow rate of the vapour.

It was shown in section 3.9 that the isentropic coefficient of performance is related to the critical temperature of the refrigerant, and that the condensing temperature should be well below this for good performance. Unfortunately, for low temperature working, the ratio of the absolute temperature at the normal boiling point to the absolute critical temperature (known as the Guldberg number) is approximately the same for most substances, about 0.6. Consequently, if a refrigerant of high vapour pressure is selected for the evaporator, i.e. a refrigerant of low boiling point, in order to give a good volumic refrigerating effect, it will have a low critical temperature. The plant will still be capable of working in the manner described for carbon dioxide in section 3.14, but the coefficient of performance will not be very good.

The pressures which can be accommodated in a system must be considered in relation to the type of equipment which is available and to the size of the plant. A system can of course be designed for any pressure in principle, high or low, but refrigerating plants are not constructed in this way. They are assembled from more or less standard components available on the market. For example, R22 would be usable in a staged compression system with turbo-compressors down to an evaporating temperature of $-100\,°C$ or even lower, at which its volumic refrigerating effect is only about 25 kJ/m^3 assuming liquid precooling to $-40\,°C$. But this would be possible only for systems above a certain capacity and subject to availability of the machines. Using reciprocating compressors R22 has been used down to about $-75\,°C$ at an absolute suction pressure of 0.15 bar. Nevertheless there are operating difficulties when working at such low suction pressures, such as inward leakage of air which has to be removed eventually, and which in the case of flammable refrigerants such as propane would form explosive mixtures internally. With reciprocating compressors, valve operation is impaired when the suction pressure is below about

5.11 Cascade systems

0.2 bar absolute, unless the valves are specially designed for the work. Some shaft seals may be unusable because of the reversal of the direction of pressure difference. All of these difficulties can be allowed for in design but they make for non-standard machines.

If a refrigerant such as R13 with a high vapour pressure is selected to give good volumic refrigerating effect and a high pressure in the evaporator, there will be problems at the warm end of the system because of its low critical temperature, 29 °C, and high condensing pressure, 36 bar at 25 °C for example. It has in fact been used in single-stage systems at this condition but with non-standard high side equipment.

It will be clear from the above that although there is no alternative to a cascade system below a certain evaporating temperature of the order −100 °C, there is an area where either staged or cascade systems may be used, depending on the material available and the size of plant.

A point in favour of the cascade system for reciprocating compressors is that it simplifies the problem of oil return compared with staged compression systems in which arrangements have to be made to share the oil between compressors. On the other hand this point has little force if the system is small enough for a compound compressor to be used, which will have a cost advantage because of its one driving motor and control system compared with the two sets necessary in the cascade system.

The cascade system introduces a necessary difference of temperature between the condensing refrigerant in the cold circuit and the evaporating temperature in the warmer circuit, which causes a loss which is not present in the staged compression system.

An important point of design in the low-temperature circuit of a cascade system when using a refrigerant of very high vapour pressure is to ensure either that the vessels, etc. are strong enough to take the pressure at ambient temperature when the plant is at standstill, or that the pressure rise is limited. The usual way of doing this in small systems is to make the total volume of the low-temperature circuit large enough to ensure that the refrigerant 'dries out' on warming up before ambient temperature is reached. In order that the volume be not too large, the amount of liquid in the circuit is made as small as possible by using dry expansion or spray-type evaporators, by eliminating the liquid receiver and using a high side float control, and by making the liquid lines as small and as short as possible. When all this has been done it is often found that the total volume is still insufficient and it becomes necessary to include an expansion vessel in the circuit.

Example 5.11.1

The charge of R13 in the low-temperature circuit of a cascade system is 30 kg, and the circuit is designed for a gauge pressure of 25 bar. Determine the volume of the circuit if this pressure is not to be exceeded at standstill when the temperature may rise to 30 °C.

From the table of properties of R13 the specific volume of vapour at an absolute pressure of 26 bar and a temperature of 30 °C is:

$$v = 0.00768 \text{ m}^3/\text{kg}$$

The required volume is therefore:

$$30(0.00768) \text{ m}^3 = 0.23 \text{ m}^3$$

In larger systems it may be cheaper to provide a separate vessel capable of taking the full standing pressure into which the charge is pumped before shut down.

The properties of some refrigerants which have been used in the low-temperature circuits of cascade systems are shown in Table 5.11.1. At still lower temperatures ethylene and methane would come into consideration, as will be described in Chapter 8 concerning natural gas. The inclusion of carbon dioxide may seem strange considering that its triple point is -56.6 °C and that it is solid at the evaporating temperature of -70 °C assumed in the table. Nevertheless it has been used as a refrigerant below -56.6 °C as a suspension of crystals in a non-freezing liquid. An interesting system of this kind is described by Cervenka & Cvejn (1967).

The symbols used and the quantities given are the same in Table 5.11.1 as in Table 3.9.1.

The analysis and design of cascade systems for given duties involves no principles other than those treated so far. There is one point which requires consideration however. It was seen in section 3.11 that for some refrigerants such as R12 and R502 superheating of the suction vapour, provided that it is obtained usefully, increases the volumic refrigerating effect and the isentropic coefficient of performance; but for others such as ammonia the opposite effects occur. At low temperatures, however, superheating, even when obtained usefully, is detrimental to the volumic refrigerating effect of all the refrigerants of Table 5.11.1. This is shown in Fig. 5.11.2 which is similar to Fig. 3.12.2 but for the conditions likely in a low-temperature cascade cycle. Apparently, therefore, only enough superheat should be used to ensure a dry suction and a good volumetric efficiency.

This conclusion may not be valid in practice for several reasons. At low

5.11 Cascade systems

Table 5.11.1. *Properties of refrigerants*

	Nitrous oxide	Ethane	R13	Carbon dioxide	R13B1
Formula	N_2O	C_2H_6	$CClF_3$	CO_2	$CBrF_3$
Molar mass/(kg/kmol)	44.0	30.1	104.5	44.0	148.9
Boiling point/°C	−88.3	−88.6	−81.4	−78.5†	−57.8
Critical point/°C	36.1	32.4	28.9	31.0	67.0
		Saturation cycle quantities, −70°C / −30°C			
p_e/bar	2.85	2.48	1.79	1.98	0.54
p_c/bar	13.42	10.61	8.40	14.29	3.22
p_c/p_e	4.70	4.27	4.68	7.21	5.92
$(h_1 - h_3)/(\text{kJ/kg})$	165	198	104	165	95
$v_1/(\text{m}^3/\text{kg})$	0.128	0.217	0.0842	0.186	0.204
$[(h_1 - h_3)/v_1]/(\text{kJ/m}^3)$	1290	912	1239	886	466
$(h_2 - h_1)_s/(\text{kJ/kg})$	46	50	25.5	50.5	22
$[(h_2 - h_1)_s/v_1]/(\text{kJ/m}^3)$	361	230	303	271	108
ε_r	3.57	3.97	4.09	3.27	4.31
t_2'/°C	36	−3	−16	52	−11

†Sublimation temperature at atmospheric pressure.

Fig. 5.11.2. Variation of volumic refrigerating effect with superheat for some refrigerants at an evaporating temperature of $-70\,°C$ and a condensing temperature of $-30\,°C$.

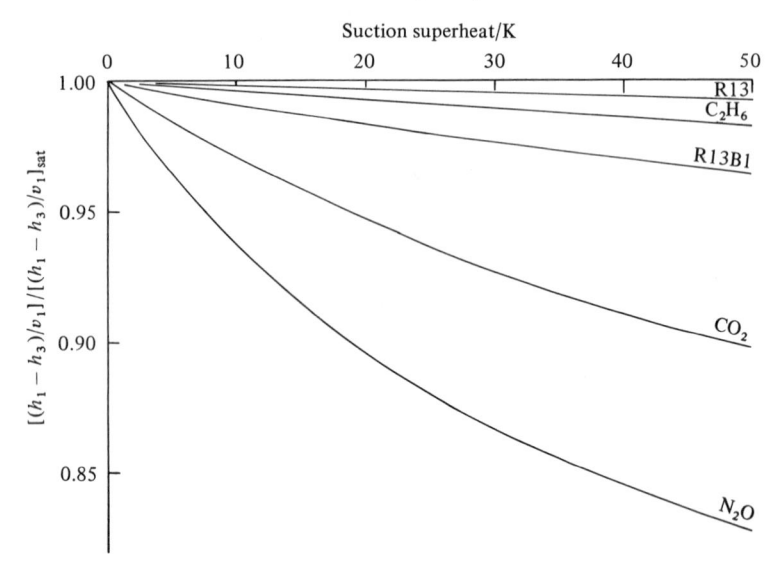

Fig. 5.11.3. A two-circuit cascade system with heat exchange between the low temperature suction vapour and the liquid in the same circuit in H.E.1, and with the liquid in the next (warmer) circuit in H.E.2.

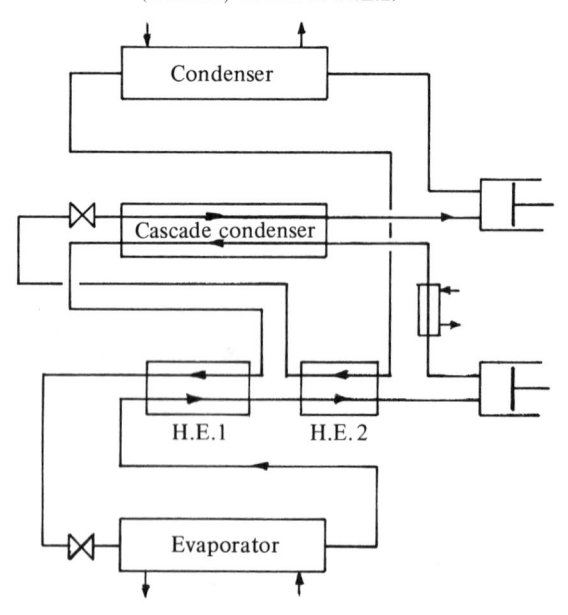

5.11 Cascade systems

temperatures superheat may easily be acquired non-usefully owing to inadequate insulation of the pipework and heat transfer inside the compressor. This is especially so when the density of the vapour is low. Such superheat gives no increase in refrigerating effect but increases the work of compression and requires a larger swept volume in the compressor. Recognising this, it may be better to use a liquid to suction heat exchanger near the evaporator and dispense with insulation for the suction pipe. Furthermore the compressor then operates at normal temperature and does not need special low-temperature materials incorporated nor special lubricating oil.

It was shown in section 3.11 that heat exhange between the liquid and suction vapour can raise the temperature of the vapour as near to condensing temperature as desired, but that it is not possible at the same time to reduce the liquid temperature to the evaporating temperature. This remains true of each cycle in a cascade system, but there may be scope for further heat exchange if desired by using the suction vapour in one cycle to lower the temperature of the liquid in the next warmer cycle. The method is shown in Fig. 5.11.3 in which H.E.1 indicates the heat exchanger in the cold cycle and H.E.2 indicates the heat exchanger in which the vapour to the low temperature cycle compressor is further superheated. The warmer suction to the low compressor may mean that desuperheating of the vapour

Fig. 5.11.4. Three-circuit cascade system in which the vapour delivered by the low-stage compressor is warmer than the evaporating temperature in the high-stage.

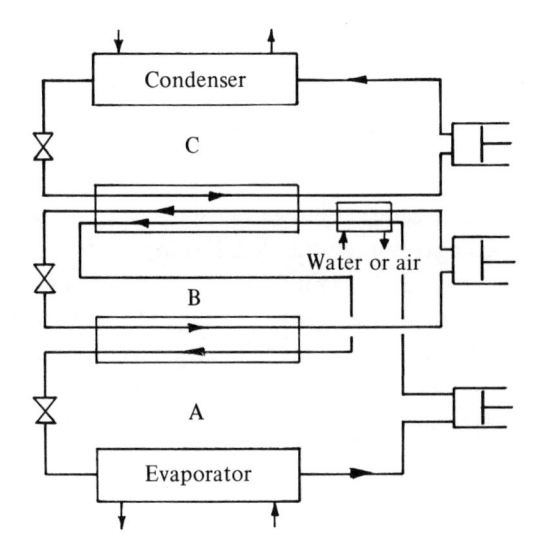

delivered by this compressor can be partly achieved by water. This would reduce the load on the cascade condenser which would have to be carried by the warmer circuit.

This illustrates a general rule for designing cascade systems: that the heat transfer required in the lower stages should be carried as high up the cascade as possible. There is no point in using refrigeration at a low temperature to cool something which could equally well be done by refrigeration at a higher temperature, or indeed by cooling water at ambient temperature. Another example is shown in Fig. 5.11.4. The vapour delivered by the compressor in stage A may be warmer not only than the evaporating temperature in stage B but warmer even than the evaporating temperature in stage C. Instead therefore of using some of the refrigerating capacity in stage B to provide the desuperheating before condensation, with the consequent expenditure of work in the compressor for stage B, it is more economical to transfer part of the desuperheating load to the warmer evaporator in stage C, where less work is required to pump it finally up to the condensing temperature. Again advantage should be taken of desuperheating by water if the temperature is high enough.

Cascade systems can be used like staged compression systems to carry loads at the intermediate temperatures. The principal application is in cooling a process stream through a large temperature range, as is needed in the liquefaction of natural gas, treated in section 8.7.

Cascade systems may of course incorporate an absorption system as one of the circuits. Such systems are not common, but they have been used to condense carbon dioxide at a lower temperature than ambient for the manufacture of solid, as described already in section 5.10.

In many circumstances multi-stage compression is used in each cycle of the cascade system. When this is done advantage should be taken of liquid precooling or subcooling at the intermediate pressures.

The auto-cascade

A variation of the cascade system was proposed by Ruhemann in 1946, in which only one compressor is used. Consider for illustration a two-cycle cascade system employing R13 and R22. Figure 5.11.5 displays the vapour pressure of these substances on a Dühring plot which will be explained in the next chapter. For the moment it may be regarded simply as a plot of vapour pressure against temperature with a distorted pressure scale to make the lines straight.

For an evaporating temperature of $-65\,°C$ in the R13 circuit the pressure is 2.25 bar, and for a condensing temperature of $-15\,°C$ in the R13

5.11 Cascade systems

circuit the pressure is 13.2 bar. To condense the R13, R22 has to boil at a temperature less than $-15\,°C$. If its evaporating temperature is selected as $-22\,°C$ the suction pressure in the R22 circuit is the same as that in the R13 circuit. Furthermore, if the condensing pressure in the R22 circuit is

Fig. 5.11.5. Vapour pressure, to a distorted scale, versus temperature for R13 and R22 to illustrate the working of the auto-cascade.

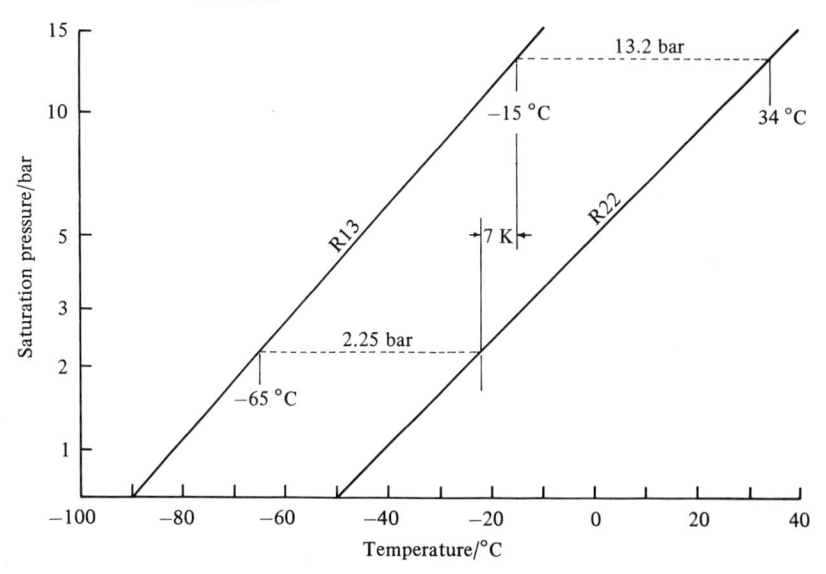

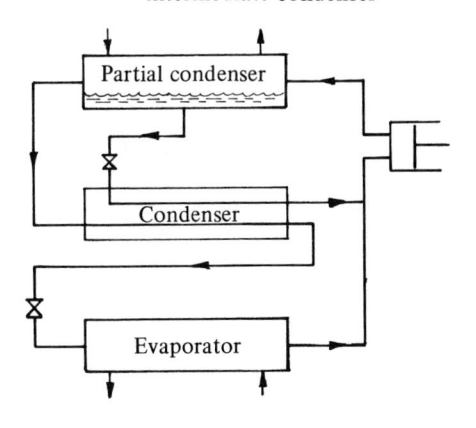

Fig. 5.11.6. Diagrammatic arrangement of Ruhemann's auto-cascade. Boiling of the lower vapour pressure component serves to condense the component of higher vapour pressure in the intermediate condenser

arranged to be 34 °C the delivery pressure in the two circuits is the same. The vapours can then be mixed at suction and compressed together provided that they can be separated afterwards.

The separation can be brought about by partial condensation using a circuit as shown in Fig. 5.11.6. When the mixed vapours pass into the condenser the temperature is too high for the R13 to condense at the pressure 13.2 bar, but the R22 can condense provided that its partial pressure in the mixture is high enough. The condensed R22 is then separated from the R13 vapour and is throttled to the pressure in the low side at which it boils at -22 °C. Its refrigerating effect is then used to condense the R13 vapour -15 °C, and this liquid goes on to the evaporator where it boils at -65 °C.

Such a plant cannot work in practice exactly as described because when the mixture of R13 and R22 enters the condenser both vapours condense to some extent but in different proportions. The condensed liquid is in fact a solution rich in R22 but still containing some R13, and the vapour is rich in R13 but still containing some R22. The vapour pressure of a solution depends on its composition as well as on its temperature and it boils over a range of temperature at constant pressure. These relations are treated in Chapter 6 in connection with absorption systems. The analysis of the system of Fig. 5.11.6 is based on these same principles.

The auto-cascade has not been used in small systems to any extent. Chaikovsky & Kuznetsov (1963) reported tests on a small one employing R22 and R12. It has turned out to have an important application in the liquefaction of natural gas which will be described in section 8.7.

6

The vapour absorption system

6.1 Introduction

The general mode of operation of the vapour absorption system has been described in Chapter 1 with reference to the system which uses ammonia as the refrigerant and water as the absorbent. In recent years another type of system using water as the refrigerant and a solution of lithium bromide, LiBr, in water as the absorbent has come into use, principally for producing chilled water for air-conditioning plants.

Although the lithium bromide system is not so versatile as the ammonia–water system in that it is restricted to evaporating temperatures above 0 °C, and although it has appeared much more recently, it is proposed to treat it first because the calculation of it is easier. The main reason for this is that the vapour produced when aqueous solutions of lithium bromide are boiled is virtually pure water vapour, and the rectification process, required in ammonia–water systems, is absent.

6.2 Composition of mixtures

Calculation of absorption refrigerators requires some knowledge of the thermodynamics of solutions and of how their properties depend on the composition.

Fig. 6.2.1. The mixing of two solutions displayed on a linear scale of mass fraction.

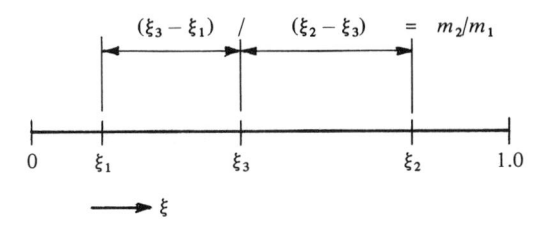

$(\xi_3 - \xi_1) / (\xi_2 - \xi_3) = m_2/m_1$

$0 \quad \xi_1 \quad \xi_3 \quad \xi_2 \quad 1.0$

$\longrightarrow \xi$

401

Composition is expressed as the mass fraction of one of the components. The symbol ξ will be used here to denote the mass fraction of lithium bromide, i.e. in a solution containing mass m_L of LiBr and m_W of H_2O, the mass fraction of LiBr is defined as:

$$\xi = \frac{m_L}{m_L + m_W} \qquad (6.2.1)$$

and the mass fraction of H_2O is:

$$1 - \xi = \frac{m_W}{m_L + m_W} \qquad (6.2.2)$$

Mass fractions can be represented on the base of a diagram of properties, the scale going from 0 to 1, as shown in Fig 6.2.1. Mixing two masses m_1 and m_2 of solutions having the mass fractions ξ_1 and ξ_2, respectively, produces a solution having the intermediate mass fraction ξ_3, say. From the constancy of the mass of LiBr before and after:

$$(m_1 + m_2)\xi_3 = m_1\xi_1 + m_2\xi_2 \qquad (6.2.3)$$

and:

$$\frac{m_2}{m_1} = \frac{\xi_3 - \xi_1}{\xi_2 - \xi_3} \qquad (6.2.4)$$

as displayed in Fig. 6.2.1.

Another quantity which specifies the composition of a mixture is the mole fraction, denoted by ψ. In a mixture containing masses m_L and m_W of LiBr and H_2O, respectively, the numbers of moles of each are:

$$n_L = m_L/M_L \qquad (6.2.5)$$

and:

$$n_W = m_W/M_W \qquad (6.2.6)$$

where M_L and M_W are the molar masses ('molecular weight') of LiBr and H_2O, respectively. The mole fractions of each component are then:

$$\psi = \frac{n_L}{n_L + n_W} \qquad (6.2.7)$$

and:

$$1 - \psi = \frac{n_W}{n_L + n_W} \qquad (6.2.8)$$

where it is understood again that ψ without subscript stands for the mole fraction of LiBr.

6.3 Vapour pressure of solutions

The mole fraction is used in texts on physical chemistry because it is more fundamentally related to the properties of solutions than the mass fraction. For the analysis of cycles in this chapter, only the mass fraction will be used.

It should be noted that ξ stands, by convention, for the mass fraction of the anhydrous substance. Lithium bromide can form with water the crystals $LiBr.5H_2O$, $LiBr.3H_2O$, $LiBr.2H_2O$ and $LiBr.H_2O$, and it would be impossibly confusing if the mass of water in each of these was included. When making up solutions in practice to a desired ξ, it is necessary to know the amount of water in the salt actually used.

6.3 Vapour pressure of solutions

When a salt like lithium bromide is dissolved in water, the boiling point of the water at a given pressure is raised. If on the other hand the temperature of the solution is held constant, the effect of the dissolved salt is to reduce the vapour pressure of the solution below that of the pure solvent.

In general, the solute itself may have some vapour pressure which is added to the (reduced) vapour pressure of the solvent to give the total vapour pressure over the solution, and this is the case with solutions of ammonia in water. The vapour pressure of pure lithium bromide is so small compared with that of water however, at the temperatures under consideration here, that it can be entirely neglected. The vapour which is in equilibrium with the solution is therefore pure water vapour, and the total pressure is equal to the reduced vapour pressure of the water.

Equilibrium between any liquid and a vapour is regarded as a dynamic equilibrium in which the rate of passage of molecules across the interface from liquid to vapour equals the rate of passage from vapour to liquid. If the number of molecules in unit volume of the liquid is reduced because of the presence of a solute it is to be expected that the number of molecules in unit volume of the vapour will also be reduced, i.e. the pressure will be reduced. When a strict proportionality exists between the mole fraction of a component in solution and its partial pressure, the solution is called ideal. *If the solution of lithium bromide in water were ideal, the vapour pressure of the solution would be given by:*

$$p = (1 - \psi)p_W \tag{6.3.1}$$

where $(1 - \psi)$ is the mole fraction of the water in the solution and p_W is the vapour pressure of pure water at the same temperature as the solution.

Equation (6.3.1) is usually known as Raoult's Law, but it is not an exact general relation. It has a status similar to that of the equation of state of an ideal gas, i.e. it is a definition of ideal behaviour which is approached by real substances under certain conditions. Real gases approach ideal gas behaviour as the pressure approaches zero. Real solutions approach Raoult's Law as the mole fraction of the component contributing the vapour pressure tends towards unity.

Strong aqueous solutions of lithium bromide in fact deviate strongly from eq. (6.3.1), as shown by the following example.

Example 6.3.1

The measured vapour pressure of a solution of mass fraction $\xi = 0.5$ of LiBr in H_2O at 25 °C is 8.5 mbar (Pennington, 1955). Compare this value with that deduced from eq. (6.3.1).

Molar masses: LiBr 86.8 kg/kmol, H_2O 18.0 kg/kmol. The mole fraction of water in the solution is:

$$1 - \psi = \frac{0.5/18.0}{(0.5/18.0) + (0.5/86.8)} = 0.828$$

Vapour pressure of water at 25 °C = 31.7 mbar. Therefore, according to eq. (6.3.1):

$$p = (0.828)(31.7) = 26.2 \text{ mbar}$$

Fig. 6.3.1. Dühring plot of the vapour pressure of solutions of lithium bromide in water. Based on the data of (Pennington, 1955.)

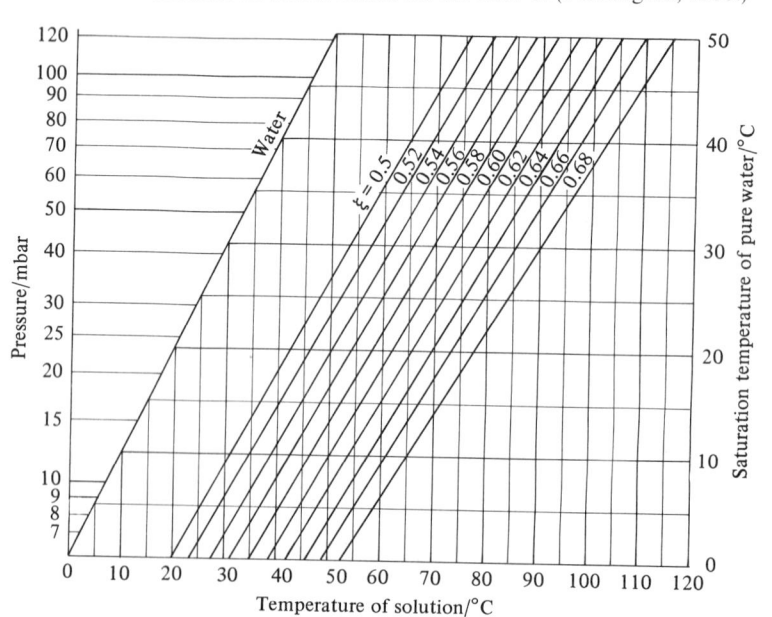

6.3 Vapour pressure of solutions

instead of 8.5. The ratio is $(8.5)/(26.2) = 0.324$, known as the activity coefficient.

The presentation of vapour pressure of solutions requires a plot (or tables) containing the three variables, temperature, pressure and mass fraction. Any pair of these may be used as the coordinates of a chart with the other as a parameter. For many purposes the most convenient choice is a plot of pressure against temperature for various mass fractions, as shown in Fig. 6.3.1, based on the data of Pennington (1955), for the range of mass fractions and pressures usually encountered.

When vapour pressure is plotted on a linear scale against temperature, as shown in Fig. 1.3.1, strongly curved lines result which make interpolation difficult, and which also make the representation of low-pressure values inaccurate. Several methods of plotting to give approximately straight lines have been suggested. For example, the logarithm of the pressure against temperature (linear) gives a better pressure scale, but not straight lines. Another plot frequently used is $\log p$ against $1/T$, where T is the absolute temperature, which gives nearly but not quite straight lines. A better result is obtained by plotting $\log p$ against $1/(T-C)$, where C is a constant determined for the substance. Both of these plots have the disadvantage that the temperature scale is not uniform, and to obtain an accurate reading of temperature the graph paper has to be specially drawn. The same difficulty applies to another plot known as the Cox plot in which a straight line is drawn on the chart with a logarithmic pressure scale to represent a reference substance such as water. The temperature scale is marked off as the corresponding saturation temperatures for water. The resulting grid then gives very nearly straight lines when data for other related substances are plotted.

In refrigeration plant the actual pressure itself is for many purposes not of great importance, but the saturation temperature of the refrigerant is. Pressure gauges for refrigeration are frequently marked with temperature scales, and one often speaks of the evaporation and condenser gauge readings as temperatures. A plot which gives saturation temperature of the refrigerant to a linear scale is therefore useful, as shown in Fig. 6.3.1. It is called a Dühring plot and is based on the fact that the saturation temperatures (or boiling points) *at the same pressure*, of two substances not too dissimilar, give a straight line when plotted against each other. The basic scales are therefore simply linear temperature scales which can be plotted on ordinary graph paper. In Fig. 6.3.1 an ordinate of a point, on the right-hand linear scale of temperature, shows the temperature of water which has the same vapour pressure as a solution at the temperature given by the abscissa of the point. Naturally, the line for pure water passes through points having the same values on the ordinate and abscissa.

Corresponding to the values of water temperature, a scale of pressure can be laid out on the ordinate, and this is shown on the left-hand side. This scale is of course non-uniform, and cannot be interpolated accurately. But if accuracy is needed, it is easy to read the temperature from the linear scale and look up the vapour pressure of water in the steam tables. In fact, as will be seen, it is not usually necessary to know the pressure accurately, because the right-hand ordinate scale gives the temperatures in the evaporator and condenser directly.

Example 6.3.2
Determine the equilibrium vapour pressure of a solution of mass fraction 0.56 at 60 °C.

From the left-hand scale the vapour pressure may be estimated as about 31 or 32 mbar.

From the right-hand scale, the temperature of pure water which has the same vapour pressure is, very nearly, 25 °C. From steam tables, this corresponds to a vapour pressure of 31.7 mbar.

6.4   The cycle of operations

The vapour pressure of a solution, as of all liquids, is a statement of an equilibrium situation, i.e. one in which nothing happens. If however the solution of example 6.3.1, having a vapour pressure of 8.5 mbar, were placed in contact with water vapour at a pressure of, say, 7 mbar, inside a closed space, the vapour and liquid could not be in equilibrium. Liquid would have to evaporate until equilibrium was restored by the combined effects of

Fig. 6.4.1. The principle of cooling by absorption. Boiling of the water in the left-hand vessel at 5 °C is caused by the absorption of water vapour in the solution at 25 °C in the right-hand vessel.

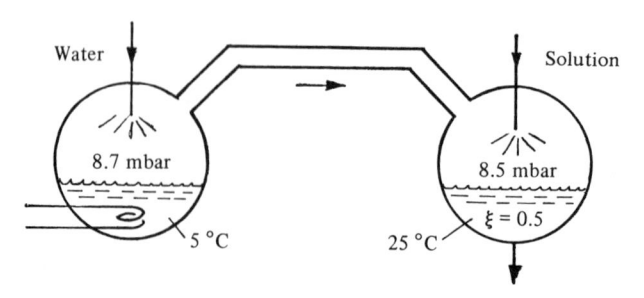

6.4 The cycle of operations

increasing pressure of the vapour and a drop in the temperature of the liquid due to evaporation. Conversely, if the solution were placed in contact with water vapour at a pressure of, say, 10 mbar, vapour would have to condense on the solution until equilibrium was restored by the combined effects of a decrease in the pressure of the vapour and a rise in the temperature of the liquid due to condensation.

Pure water at 7 °C has a vapour pressure of about 10 mbar, and consequently a solution of mass fraction 0.5 at 25 °C can draw water vapour from the surface of the pure water and cause evaporation there. Provided that a fresh supply of solution at 25 °C is continuously available, the process could go on indefinitely, and the water would evaporate until its vapour pressure dropped to a little over 8.5 mbar, say 8.7 mbar at a temperature of 5 °C, Fig. 6.4.1. The difference between 8.7 and 8.5 would be made up by the pressure drop due to the flow in the pipe connecting the vessels, and by the slight departure from equilibrium arising from the fact that the process is a dynamic and not a static one. The water evaporating at 5 °C would be able to take its enthalpy of evaporation from bodies or fluids at a higher temperature, i.e. it would be able to refrigerate them. It is clear that the solution of lithium bromide acts like a compressor in drawing vapour away from the water surface and in causing the water temperature to fall until it can perform some useful refrigeration. This is the principle of the absorption refrigerator. The vessel in which the vapour from the evaporator is dissolved is known as the absorber.

In a continuous process the solution supplied to the absorber must be stronger, in LiBr, than the solution leaving. The mass fractions of these strong and weak solutions will be denoted by ξ_s and ξ_w. An important quantity in the calculation of an absorption system is the mass flow rate of the strong solution which is needed to absorb unit mass flow rate of vapour from the evaporator. This quantity will be called the circulation factor, and denoted by λ. It is determined by the two solution concentrations from the fact that all the LiBr which enters with the strong solution leaves with the weak solution. Supposing unit mass flow rate from the evaporator, the flow rates of the solutions are λ and $(\lambda + 1)$, and the mass balance for LiBr gives:

$$\lambda \xi_s = (\lambda + 1)\xi_w \tag{6.4.1}$$

hence:

$$\lambda = \frac{\xi_w}{\xi_s - \xi_w} \tag{6.4.2}$$

The vapour which dissolves in the absorber changes from vapour to liquid in solution, and consequently gives out heat. As a result, the solution must be cooled by external means, or it would become too warm to absorb vapour at the desired evaporator pressure.

Figure 6.4.1 shows the essential elements of a refrigeration process but not of a refrigeration cycle. It would be possible but expensive to supply new strong solution continuously, and to feed fresh water to the evaporator; but this would be undesirable on account of the accumulation of the hardness salts and scaling as in a steam boiler. It is better simply to boil the solution to make new strong solution and condense the vapour to make distilled water to supply the evaporator, as shown in the circuit diagram Fig. 6.4.2.

Before the solution is boiled, however, it must be raised to a higher

Fig. 6.4.2. The components of a simple absorption system using a solution of lithium bromide in water. Water vapour is drawn from the evaporator at the mass flow rate $\dot{m}$ and absorbed by strong solution coming from the boiler at the rate $\lambda \dot{m}$, giving $(\lambda + 1)\dot{m}$ of weak solution which is pumped to the boiler. Heat is given out in the absorber and condenser at the rates $\dot{Q}_a$ and $\dot{Q}_{con}$ respectively, whilst heat is accepted in the boiler and evaporator at the rates $\dot{Q}_b$ and $\dot{Q}_e$, respectively. The evaluation of these rates is treated in section 6.6.

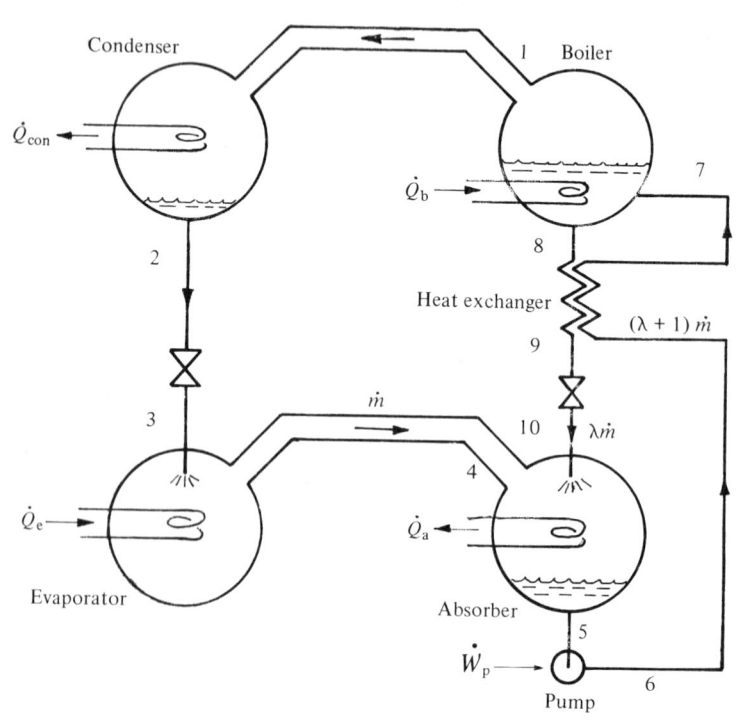

6.4 The cycle of operations

pressure by a pump, so that the vapour which is generated can be condensed by the available cooling water. For example, if cooling water is available at 25 °C, condensation will be able to occur at, say, 30 °C allowing a difference of temperature for heat transfer, and the pressure inside the condenser must be 42.4 mbar or more. (This pressure also exists in the boiler, apart from any pressure drop between the two vessels.) Thus, in the absorption system, liquid is pumped from lower to higher pressure, instead of vapour as in the compression system. The liquid has a much smaller volume than an equal mass of vapour, and hence the work taken by the pump is only a small fraction of the work which would be required by a compressor to pump the vapour from the evaporator up to the condensing pressure. On the other hand a large heat transfer is required in the boiler.

The difference in pressure between the boiler and condenser, i.e. the high side, and the evaporator and absorber, i.e. the low side, must be maintained by expansion valves or restrictions in the pipes feeding the water to the evaporator and the solution to the absorber. This difference is rather small in the system using lithium bromide and water, of the order 34 mbar for the conditions cited above. This can be balanced by a water column of only 35 cm height. Instead of a throttle valve, therefore, a U-tube may be used in which the pressure difference is maintained by a column of water. Otherwise the nozzles which spray the water into the evaporator and the solution into the absorber may be sized to serve as the restrictors, and this is the practice in some large systems.

The solution leaving the boiler is of course hot, whilst that leaving the absorber is relatively cool. The hot solution is not capable of absorbing vapour at the absorber pressure until it is cooled to absorber temperature, and part at least of this cooling can be done by heat transfer to the weak solution in the heat exchanger shown in Fig. 6.4.2. There is a snag however in applying this effect further, and in bringing the strong solution right down to absorber temperature. Under some conditions of operation the solution is so strong leaving the boiler that it would deposit crystals if cooled to absorber temperature. For example, a solution of mass fraction 0.66 is saturated at 32 °C, and if it were cooled below this temperature crystals would block the heat exchanger. Even were the bulk solution not cooled so far in the heat exchanger, parts of the solution in the boundary layer would become chilled when they encountered the pipes carrying the cooling water in the absorber and would deposit crystals there. It is necessary with these very strong solutions to mix them first with the weak solution already in the absorber and to recirculate the mixture by a pump. This practice has the additional advantage of providing better heat transfer

The vapour absorption system

conditions by the continued forced circulation of solution at a high rate over the cooling pipes.

At the low pressures inside the apparatus the specific volume of water vapour is very large, e.g. 147 m^3/kg for saturated vapour at 5 °C, and large vapour pipes are needed to prevent excessive pressure drops. To eliminate these, the condenser and boiler may be combined inside the same vessel, and so may the evaporator and absorber. One manufacturer has taken this a step further by placing the four major components inside a single shell divided into higher and lower pressure regions by a diaphragm, as shown in Fig. 6.4.3.

Since both the pressures inside the system are well below atmospheric pressure, there is a possibility of air leaking in, and this must be removed by a purger. This is a two-stage ejector of the type used in condensers for steam power plant, and working on the principle described in section 1.7.

The boiler may be heated by steam or by hot water, and this is the usual practice in large systems with refrigerating capacities up to 3.5 MW. Small systems are usually heated directly by oil or gas, and employ the principle of the bubble pump (see section 6.19) to return the weak solution to the boiler. They are hermetically sealed and require no purging.

Fig. 6.4.3. Arrangement of the single shell version of the lithium bromide in water absorption system, with typical operating temperatures.

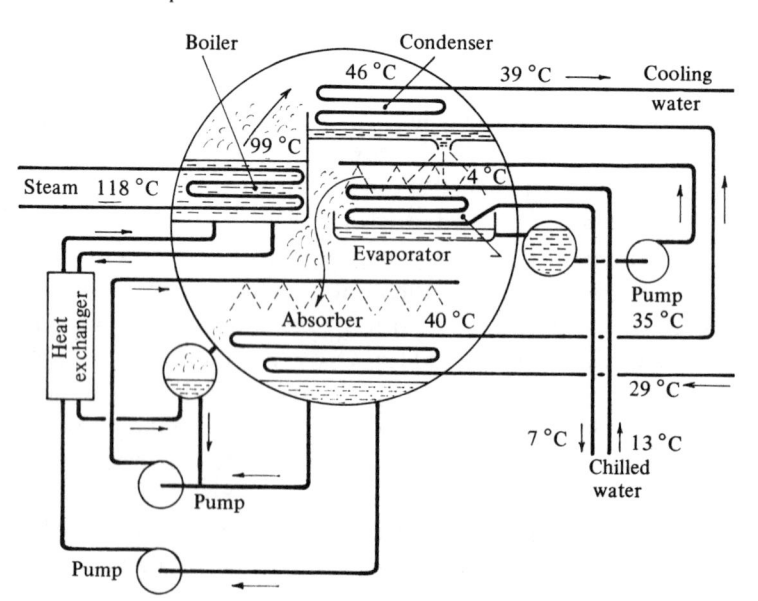

6.5 Enthalpy of liquid and vapour

In the system shown in Fig. 6.4.3, primary refrigerant water evaporates around the tubes carrying the water to be chilled, exactly as in a vapour compression system. The tubes are not however totally immersed in the primary refrigerant, as in the usual type of flooded shell-and-tube evaporator shown in Fig. 2.7.1(c), because at the low pressure in the evaporator the hydrostatic head is significant. For the water to boil on the bottom tubes the pressure in the vapour space would have to be so much less, with a consequent lowering of the absorbing pressure. The refrigerant water is therefore sprayed over the tubes carrying the chilled water, wetting them adequately for good heat transfer. Provided that there is vapour space between the tubes, pressure does not build up at the bottom. The water draining from the tubes is recirculated by the pump shown in Fig. 6.4.3.

Because there must be a difference of temperature between the outflowing chilled water and the primary refrigerant water, it is not possible with this arrangement to produce chilled water at much less than 5 °C. By circulating the primary refrigerant directly to the external circuit, colder water can be obtained. It would normally be pumped to a pressure above atmospheric before circulating it to the external cooler batteries, etc., and when returned would be flashed down to evaporator pressure again. Provided that the external circuit is tight this works reasonably well. But if direct contact air washers are supplied, the water picks up dirt and dissolves air and these are carried back to the evaporator where they have to be eliminated.

6.5 Enthalpy of liquid and vapour

When anhydrous lithium bromide and water at the same temperature mix to form a solution adiabatically there is a noticeable rise in temperature. If the experiment is performed in such a way as to maintain the temperature constant, there is a heat transfer from the substances whilst dissolving. This heat transfer, expressed per unit of mass of solution formed, is called the integral heat of solution. In the old convention used in chemistry, heat of solution was reckoned positive when from the solution. In modern terminology the change of enthalpy during mixing is stated, $(\Delta h)_i$. This has the same value as the integral heat of solution, but opposite sign.

The specific enthalpy of a solution formed at constant temperature from the pure components is then given by:

$$h = (1 - \xi)h_W + \xi h_L + (\Delta h)_i \tag{6.5.1}$$

where h_W and h_L denote the specific enthalpies of the pure components water and lithium bromide. This relation is exhibited graphically in Fig. 6.5.1. A mixing process in which no heat transfer occurred whilst the temperature remained constant would be represented by the straight line joining h_W and h_L. This would be found for an ideal solution, i.e. one which obeyed eq. (6.3.1) over the whole range of mass fractions. The specific enthalpies of the actual solutions are found by setting off the values $(\Delta h)_i$ from the straight line at each mass fraction.

For the temperature at which the curve is drawn on Fig. 6.5.1, there is a mass fraction (of saturated solution) above which the solution cannot exist. Points in the region to the right of this crystallisation curve represent two-phase mixtures of saturated solution and crystals up to the mass fraction corresponding to the composition of the crystals, and thereafter mixtures of crystals and anhydrous salt. The isotherm would therefore show changes of direction before it reached the point h_L representing the specific enthalpy of the anhydrous salt.

The integral enthalpy of solution is involved when the solution is made up from its pure components. It is to be distinguished from the differential enthalpy of solution which is involved when a small amount of one of the components is added to a large amount of an already existing solution. It has a different value for the two components. The negative of the differential enthalpy of solution of the water, i.e. the heat given out per unit mass of

Fig. 6.5.1. Specific enthalpy of a solution at constant temperature. Mixing without heat of solution would be represented by the broken line joining h_W and h_L. The heat of solution, when positive, i.e. given out, depresses the actual specific enthalpy below the broken line.

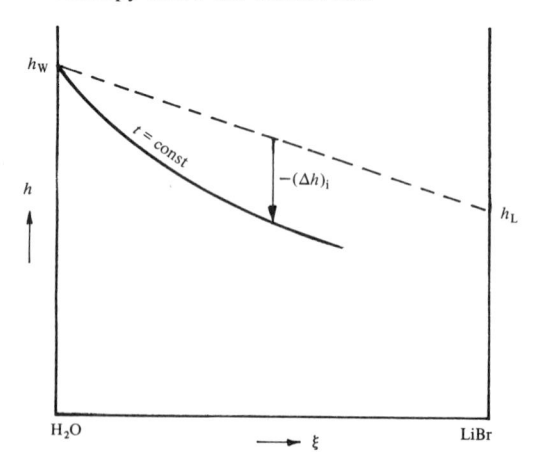

6.5 Enthalpy of liquid and vapour

water when it is added to an amount of solution large enough for the mass fraction to remain constant, is also known as the heat of dilution.

The integral heat of solution of $LiBr-H_2O$ mixtures has been measured by Lange & Schwartz (1928) at a temperature of 25 °C. Values have also been computed by Löwer (1960) from the vapour pressure data and correlated with those of Lange & Schwartz. Figure 6.5.2 is based on Löwer's values.

The specific enthalpy of a pure substance implies only one datum state at which the specific enthalpy is arbitrarily made zero for convenience. Equation (6.5.1) shows that for a solution, two datum states are needed, one

Fig. 6.5.2. Specific enthalpy of solutions of lithium bromide in water. Datum states: 0 kJ/kg at 0 °C for liquid water and solid LiBr. The broken line shows the limit of solubility.

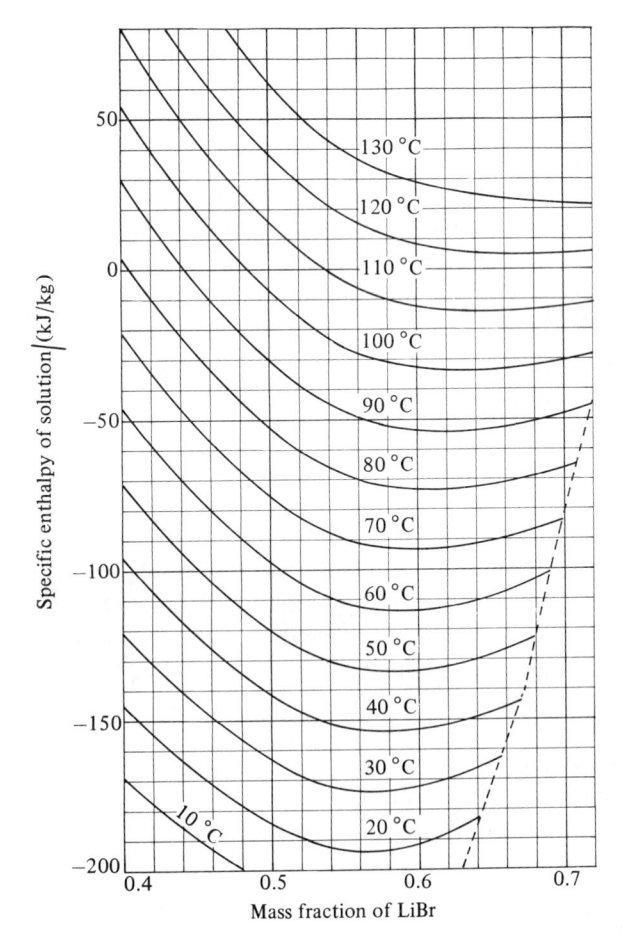

for each pure component, which are quite independent. Figure 6.5.2 is based on the specific enthalpies of liquid water and solid anhydrous lithium bromide each being zero at 0 °C. But other choices are possible, and it is always advisable to ascertain the datum states when using a chart of enthalpies. This applied especially when it is necessary to take data for one of the substances from other sources. For example, in treating the lithium bromide system the enthalpies of vapour will be taken from the usual steam tables, calculated above liquid water at 0 °C (strictly the triple point 0.01 °C, but this difference can be neglected). It is naturally essential that the chart values should be reckoned above the same datum for the water component.

The properties of the pure water substance circulating in the condenser and evaporator may be taken from steam tables. For practical purposes the specific enthalpy of pure liquid water within the temperature range of interest may be approximated by:

$$h(H_2O \text{ liquid}) = 4.19 \frac{(t - t_0)}{K} \text{kJ/kg} \qquad (6.5.2)$$

where t is the temperature, and t_0 is the datum temperature, 0 °C.

The water vapour produced in the boiler is superheated vapour because at a given pressure it is generated by a solution boiling at a higher temperature. For example, referring to Fig. 6.3.1, a solution of mass fraction 0.68 boils at 90 °C when the pressure is the same as the saturation pressure of water at 30 °C, i.e. 42.4 mbar. The vapour is therefore given off at a temperature of 90 °C and is thus superheated by 60 K. The specific enthalpy may be read from steam tables as 2669 kJ/kg. Many abridged steam tables do not give properties of superheated vapour at close intervals of pressure, but fortunately at low pressures the enthalpy is almost independent of pressure, and the pressure can be disregarded. The specific enthalpy is then read at the required temperature in the nearest pressure column. With sufficient accuracy, the specific enthalpy of the vapour leaving the boiler is given by:

$$h(H_2O \text{ vapour at low pressure}) = 2501 + 1.88 \frac{(t - t_0)}{K} \text{kJ/kg}$$

$$(6.5.3)$$

6.6 Steady-flow analysis

Using subscripts according to the notation of Fig. 6.4.2 to indicate the specific enthalpies at the various points in the cycle, the steady-flow energy equation will be applied to each component in turn. As usual, changes of kinetic energy and height will be neglected. The mass flow rate of

6.6 Steady-flow analysis

pure water through the condenser and evaporator will be denoted by $\dot{m}$, the rate of flow of strong solution from the boiler by $\lambda\dot{m}$, and the rate of flow of weak solution from the absorber by $(\lambda + 1)\dot{m}$.

(a) *Condenser.* The heat transfer rate outwards $\dot{Q}_{con}$, is given by:

$$\dot{Q}_{con} = \dot{m}(h_1 - h_2) \tag{6.6.1}$$

(b) *Expansion valve or restrictor.*

$$h_3 = h_2 \tag{6.6.2}$$

(c) *Evaporator.* The refrigerating capacity, i.e. the heat transfer rate to the evaporator, $\dot{Q}_e$, is given by:

$$\begin{aligned}\dot{Q}_e &= \dot{m}(h_4 - h_3) \\ &= \dot{m}(h_4 - h_2)\end{aligned} \tag{6.6.3}$$

on using eq. (6.6.2).

Equations (6.6.1)–(6.6.3) are exactly the same as those stated for the vapour compression system, eqs. (3.2.5)–(3.2.11).

(d) *Solution valve or restrictor.*

$$h_9 = h_{10} \tag{6.6.4}$$

(e) *Absorber.* The heat transfer rate from the absorber, $\dot{Q}_a$, is given by:

$$\begin{aligned}\dot{Q}_a &= \dot{m}[h_4 + \lambda h_{10} - (\lambda + 1)h_5] \\ &= \dot{m}[h_4 + \lambda h_9 - (\lambda + 1)h_5]\end{aligned} \tag{6.6.5a}$$

on using eq. (6.6.4).

Equation (6.6.5a) may be rearranged as:

$$\dot{Q}_e = \dot{m}[(h_4 - h_5) + \lambda(h_9 - h_5)] \tag{6.6.5b}$$

which shows more clearly how the absorber heat is made up. The first term $(h_4 - h_5)$ represents the enthalpy of evaporation which is given out when the vapour at 4 is changed into liquid at 5, plus a small contribution from the difference between the enthalpy of water and the enthalpy of the solution which is less than that of pure water at the same temperature. The second term $\lambda(h_9 - h_5)$ represents mainly the heat needed to cool the solution from the temperature at which it enters the absorber to the temperature at which it leaves, with a small contribution from the enthalpy of solution because the concentrations are different.

If a circulating pump is used as shown in Fig. 6.4.3, some of the power to this appears in the absorber balance and should be added to the right-hand side of eq. (6.6.5). The amount is usually small compared with the enthalpy changes, however. The same applies to a circulating pump used in the evaporator.

(*f*) *Solution pump.* The steady-flow energy equation applied to the main solution pump gives the power input, assuming adiabatic operation:

$$\dot{W}_p = (\lambda + 1)\dot{m}(h_6 - h_5) \qquad (6.6.6)$$

but this equation is not useful because the difference between the specific enthalpies of subcooled solution at 6 and saturated solution at 5, both at the same temperature very nearly, is not distinguished on the chart. It will be recalled that the same point occurs in connection with the feed pump work in the analysis of a steam power cycle. Pump power in practice is calculated from the pump indicator diagram. For the ideal pump:

$$\dot{W}_p = (\lambda + 1)\dot{m}(p_6 - p_5)v_{sol} \qquad (6.6.7)$$

where v_{sol} is the specific volume of the solution at 5, approximately $0.00055 \, \text{m}^3/\text{kg}$ at $\xi = 0.66, 40\,°\text{C}$.

Compared with the heat rates in the other parts of the plant, the pump power is very small, but it must be estimated in order to size the driving motor. The pump efficiency must be applied to $\dot{W}_p$ as found from eq. (6.6.7).

(*g*) *Heat exchanger.* Assuming that it is externally adiabatic, i.e. neglecting heat transfer to the surroundings, the heat transfer rate between the two streams is:

$$\dot{Q}_x = \lambda \dot{m}(h_8 - h_9) = (\lambda + 1)\dot{m}(h_7 - h_6) \qquad (6.6.8)$$

The state at 8 leaving the boiler is usually known, and the state at 9 is assumed for design purposes. As already mentioned, it is not possible to cool the solution to near absorber temperature on account of crystallisation. Having decided on what temperature is permissible at 9, the surface area required in the exchanger can be determined.

Assuming the states at 8, 9 and 6, eq. (6.6.8) may be used to determine the state at 7.

(*h*) *Boiler.* The rate of heat transfer $\dot{Q}_b$ to the boiling solution is given by:

$$\dot{Q}_b = \dot{m}[h_1 + \lambda h_8 - (\lambda + 1)h_7] \qquad (6.6.9a)$$
$$= \dot{m}[(h_1 - h_7) + \lambda(h_8 - h_7)] \qquad (6.6.9b)$$

6.6 Steady-flow analysis

If there are no heat losses from the heat exchanger, the boiler heat rate may also be obtained by drawing the control surface around the combined boiler and heat exchanger to give:

$$\dot{Q}_b = \dot{m}[h_1 + \lambda h_9 - (\lambda + 1)h_6] \quad (6.6.10a)$$
$$= \dot{m}[(h_1 - h_6) + \lambda(h_9 - h_6)] \quad (6.6.10b)$$

which saves the evaluation of h_7 from the heat exchanger balance.

Equation (6.6.10b) shows how the boiler heat is made up. If enthalpy of solution is neglected, being only about 10% of the whole, the first term $(h_1 - h_6)$ is the heat needed to generate steam at 1 from water at 6. Considering the absorber temperature as more or less fixed, then this term is approximately constant. The second term $\lambda(h_9 - h_6)$ is the heat needed to raise the temperature of solution from the temperature at 6 to the temperature at which it is supplied again to the absorber. If the heat exchange were complete, so that $t_9 = t_6$, then this term would be nearly zero. When the difference between the solution concentrations is small, and λ is large, this second term has an important effect on the heat needed, considering that heat exchange cannot be complete for practical reasons.

The coefficient which expresses the efficiency of a heat operated refrigerator has been defined in Chapter 1 and termed the heat ratio ζ. Neglecting the small work transfers to the pumps, it is given for the above cycle by:

$$\zeta = \dot{Q}_e/\dot{Q}_b \quad (6.6.11)$$

The use of the steady-flow equations is demonstrated in the following example.

Example 6.6.1

A lithium bromide system operates at the following conditions:

temperature of:

evaporation	5 °C	($p_e = 8.72$ mbar)
condensation	50 °C	($p_c = 123.3$ mbar)
boiling	110 °C	
absorption	40 °C	

Assuming equilibrium states leaving the boiler and evaporator, no pressure drops, and complete heat exchange, i.e. the strong solution leaves the exchanger at 40 °C, determine the heat transfers for unit mass flow rate of water to the evaporator, and the heat ratio.

From the assumption of equilibrium (saturation) states leaving the boiler and absorber, the mass fractions of the strong and weak solutions are determined from Fig. 6.3.1 as:

$\xi_w(40\,°C, 8.72\text{ mbar}) = 0.578$

$\xi_s(110\,°C, 123.3\text{ mbar}) = 0.660$

The circulation factor λ is, by eq. (6.4.2):

$$\lambda = \frac{0.578}{0.660 - 0.578} = 7.05$$

The specific enthalpies may then be listed:

h_1 (steam tables or eq. (6.5.3), 110 °C) = 2706 kJ/kg

h_2 (steam tables or eq. (6.5.2), 50 °C) = 209 kJ/kg

h_4 (steam tables or eq. (6.5.3), 5 °C) = 2510 kJ/kg

h_5 (Fig. 6.5.2), $\xi = 0.578$, 40 °C) = -154 kJ/kg

$h_6 \approx h_5$

h_8(Fig. 6.5.2, $\xi = 0.66$, 110 °C) = -13 kJ/kg

h_9 (Fig. 6.5.2, $\xi = 0.66$, 40 °C) = -146 kJ/kg

The heat transfers will be evaluated per unit mass of water to the evaporator. These will be denoted by lower case $q = \dot{Q}/\dot{m}$.
Condenser (eq. (6.6.1)):

$q_{con} = (h_1 - h_2) = (2706 - 209)\text{ kJ/kg} = 2497\text{ kJ/kg}$

Evaporator (eq. (6.6.3)):

$q_e = (h_4 - h_2) = (2510 - 209)\text{ kJ/kg} = 2301\text{ kJ/kg}$

Absorber (eq. (6.6.5b)):

$q_a = (h_4 - h_5) + \lambda(h_9 - h_5)$
$= [(2510 + 154) + 7.05(-146 + 154)]\text{ kJ/kg}$
$= (2664 + 56)\text{ kJ/kg} = 2720\text{ kJ/kg}$

Heat exchanger (eq. (6.6.8)):

$q_x = \lambda(h_8 - h_9) = [7.05(-13 + 146)]\text{ kJ/kg} = 938\text{ kJ/kg}$

Boiler (eq. (6.6.10b)):

$q_b = (h_1 - h_6) + \lambda(h_9 - h_6)$
$= [(2706 + 154) + 7.05(-146 + 154)]\text{ kJ/kg}$
$= (2860 + 56)\text{ kJ/kg} = 2916\text{ kJ/kg}$

6.7 The effect of inefficiencies in the cycle

As a check on the arithmetic it may be noted that the sum of the evaporator and boiler heats must equal the sum of the condenser and absorber heats: 5217 kJ/kg. The heat ratio is (eq. (6.6.11)):

$$\zeta = 2301/2916 = 0.789$$

6.7 The effect of inefficiencies in the cycle

The actual heat ratio for a lithium bromide system operating under the conditions of example 6.6.1 would be about 0.6 instead of 0.79. The main reason for this difference is that the strong solution from the boiler cannot be cooled to 40 °C in practice because it is then too close to the crystallisation line for safety at all possible absorber temperatures. Furthermore, the actual difference between the mass fractions is less than in the example, and this increases the circulation factor and aggravates the effects of inefficiency in the exchanger.

The reasons for the departure of the actual concentrations from the theoretical ones are mainly the following.

(i) Pressure drops between the evaporator and the absorber, and between the boiler and the condenser. The absorber pressure is slightly less than the evaporator pressure, and the equilibrium mass fraction for a given absorber temperature is increased. Similarly, because the boiler pressure is greater than the condenser pressure, the equilibrium mass fraction for a given boiler temperature is reduced.

(ii) Hydrostatic head in the boiler. The pressure of the water at the bottom exceeds that in the vapour space and the equilibrium mass fraction is reduced.

(iii) Departures from equilibrium. The solution leaving the absorber is not in practice completely saturated with respect to water vapour at the absorber temperature. This would require infinite surface for heat and mass transfer, but there has to be a finite difference in practice between the actual pressure and the saturation pressure, analogous to the temperature approach in a heat exchanger. Similarly, in the boiler equilibrium may not be attained.

The effect on heat ratio is illustrated by the following example.

Example 6.7.1

Assuming that the effects (i) to (iii) above change the mass fractions in example 6.6.1 from 0.66 to 0.65, and 0.578 to 0.6, and that the temperature of the strong solution leaving the heat exchanger is 70 °C, all other temperatures remaining the same, determine the new heat ratio and the heat exchanger load.

The new circulation factor is:

$$\lambda = \frac{0.60}{0.65 - 0.60} = 12.0$$

The new specific enthalpies are:

h_5 (Fig. 6.5.2, $\xi = 0.60$, 40 °C) = -153 kJ/kg

$h_6 \approx h_5$

h_8(Fig. 6.5.2, $\xi = 0.65$, 110 °C) = -14 kJ/kg

h_9 (Fig. 6.5.2, $\xi = 0.65$, 70 °C) = -90 kJ/kg

The new boiler heat is:

$$\begin{aligned} q_b &= (h_1 - h_6) + \lambda(h_9 - h_6) \\ &= [(2706 + 153) + 12.0(-90 + 153)] \text{ kJ/kg} \\ &= (2859 + 756) \text{ kJ/kg} = 3615 \text{ kJ/kg} \end{aligned}$$

The evaporator heat remains the same, and the new heat ratio is:

$$\zeta = 2301/3615 = 0.636$$

The new heat transfer in the heat exchanger is:

$$q_x = \lambda(h_8 - h_9) = 12.0(-14 + 90) \text{ kJ/kg} = 912 \text{ kJ/kg}$$

Example 6.7.1 shows the strong effect of imperfect heat exchange on heat ratio. The term $\lambda(h_9 - h_6)$ has gone up from 56 kJ/kg in example 6.6.1 to 756 kJ/kg in example 6.7.1. The other term in the boiler heat has hardly changed. It also shows that even to achieve cooling of the strong solution to 70 °C, the heat exchanger has to be a rather large piece of equipment, handling 912 kJ/kg, which is approximately one quarter of the heat transfer to the solution in the boiler.

6.8 Solutions of ammonia in water

The composition of ammonia in water solutions (aqua-ammonia for short) is expressed by the mass fraction exactly as for solutions of lithium bromide in water. But, following the usual practice, the symbol ξ will stand for the mass fraction of ammonia, i.e. the refrigerant, and the adjectives strong and weak will mean strong and weak in ammonia, i.e. in the refrigerant. When treating lithium bromide solutions strong and weak meant strong and weak in the lithium bromide, i.e. in the absorbent.

Liquid ammonia and water are completely miscible in all proportions,

6.8 Solutions of ammonia in water

and hence can form solutions of all mass fractions from 0 to 1, at normal temperatures. At low temperatures, the mass fraction in the solution is limited by the deposition of solid phases, either pure water ice, pure ammonia ice, or the compounds $NH_3 \cdot H_2O$ and $2NH_3 \cdot H_2O$. There are

Fig. 6.8.1. The eutectic points of ammonia–water solutions.

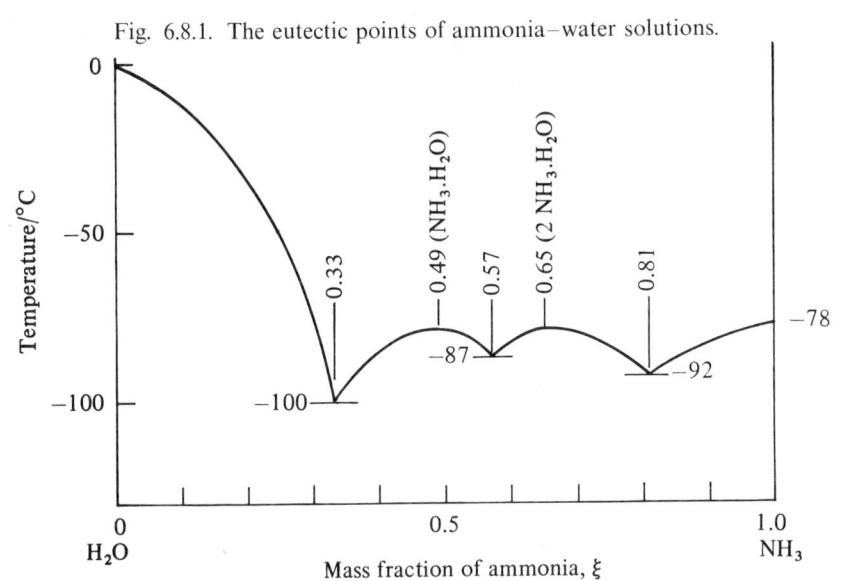

Fig. 6.8.2. Dühring plot of the vapour pressure of ammonia-water solutions. (Based on the data of Wucherer, 1932).

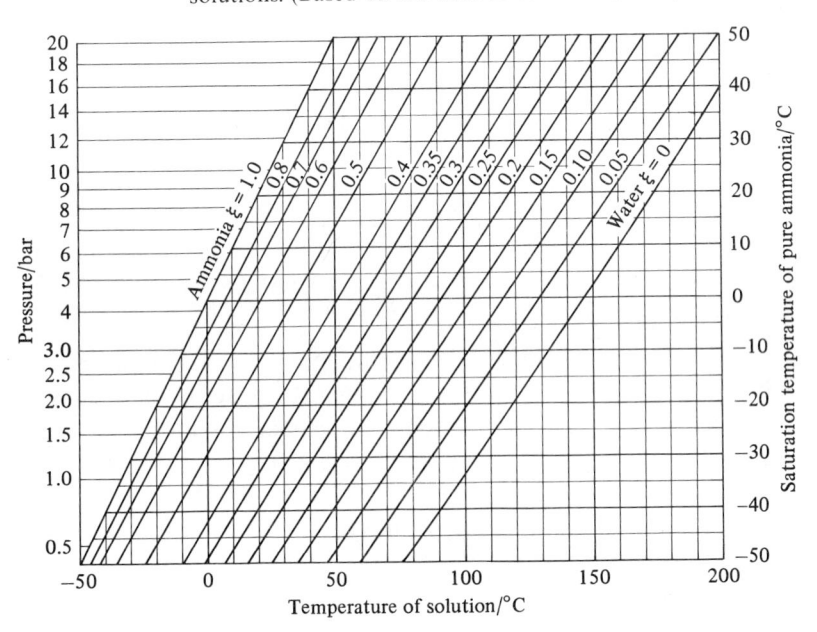

three eutectic points as shown in Fig. 6.8.1, corresponding to equilibrium between solution and two of these solid phases. It is perhaps worth noting that water in the evaporator of an ammonia plant, whether of absorption or compression type, cannot block the expansion valve by freezing there, as it can do in a Refrigerant 12 plant. A small fraction of water in the ammonia simply lowers the freezing point below the value $-78\,°C$ for pure ammonia.

The effect of ammonia in solution in water is to lower the vapour pressure of the water. At the same time the effect of the water in solution in the ammonia is to lower the vapour pressure of the ammonia. The total pressure over a solution is thus made up of contributions from both ammonia and water, but it always lies between the vapour pressures of pure water and pure ammonia. This is not true of all solutions. For example, Refrigerant 502 is a mixture of R22 and R115, with a vapour pressure higher than that of either R22 or R115, indicated by the fact that its boiling point at 1 atm is $-45.4\,°C$, lower than that of R22, $-40.8\,°C$, and of R115 $-39.1\,°C$.

The vapour pressure of aqua-ammonia as determined by Wucherer (1932) is presented on a Dühring plot in Fig. 6.8.2. Again the Dühring plot is convenient because, taking ammonia as the reference substance, the (distorted) vertical scale of pressure gives the evaporating and condensing temperatures of the pure ammonia to a linear scale. Thus a solution of mass fraction 0.4 at $40\,°C$ has a vapour pressure of approximately 3 bar reading from the left-hand pressure scale. Reading from the right-hand scale the saturation temperature of ammonia at the pressure is $-9\,°C$ and the pressure may be read accurately from ammonia tables as 3.029 bar. An absorber at $40\,°C$ from which solution of mass fraction 0.4 was leaving could therefore absorb vapour from an evaporator in which ammonia was boiling at $-9\,°C$, neglecting pressure drop and departure from equilibrium.

Solutions of ammonia and water, like those of lithium bromide in water, show considerable departure from ideality, as is shown by the following example.

Example 6.8.1
Compare the vapour pressure of a solution of $\xi = 0.4$ at $40\,°C$, 3.029 bar, with that predicted by Raoult's Law (eq. (6.3.1)) applied to each component.

Molar masses: water 18.0 kg/kmol, ammonia 17.0 kg/kmol. The mole fraction of ammonia is:

$$\psi = \frac{0.4/17.0}{(0.4/17.0) + (0.6/18.0)} = 0.414$$

6.9 Composition of the vapour

Vapour pressure of ammonia at 40 °C = 15.54 bar. Therefore by eq. (6.3.1), pressure of ammonia in vapour

$$= (0.414)(15.54) \text{ bar} = 6.43 \text{ bar}$$

Vapour pressure of water at 40 °C = 0.07375 bar. Therefore by eq. (6.3.1), pressure of water in vapour

$$= (1 - 0.414)(0.07375) \text{ bar} = 0.0432 \text{ bar}$$

The total pressure would be:

$$(6.43 + 0.04) \text{ bar} = 6.47 \text{ bar}$$

instead of 3.029 bar.

Water in ammonia thus behaves, qualitatively, like lithium bromide in water in giving a vapour pressure much less than that predicted by ideal behaviour. The significance of this for refrigerant–absorbent combinations will be examined later, in section 6.22.

In treating lithium bromide solutions it was pointed out that the lithium bromide is virtually involatile and that the vapour given off when a solution is boiled is pure water vapour. This is not the case with aqua-ammonia, however, and the vapour pressure of the water is not at all negligible. Consequently it is not sufficient in an aqua-ammonia absorption refrigerator simply to boil the solution to obtain pure ammonia again; the arising vapour has to be separated from the associated water vapour by the process of rectification. It will be shown that this implies that more vapour has to be boiled off than is passed to the condenser, and the extra heat needed for this purpose makes the heat ratio of the aqua-ammonia system inferior to that of the lithium bromide one.

6.9   Composition of the vapour

Raoult's law of ideal solution behaviour predicts the vapour composition as well as the vapour pressure, but again it does not agree with that found by experiment, as the following example shows.

Example 6.9.1

Determine the mass fraction of ammonia in the vapour given off by solution of mass fraction 0.4 at a temperature of 40 °C, assuming ideal solution behaviour.

Using the pressure contributions from example 6.8.1, the specific volume of ammonia vapour at a pressure of 6.43 bar and a temperature of 40 °C is, by interpolation in the ammonia tables, 0.2267 m^3/kg. Similarly, the

The vapour absorption system

specific volume of water vapour at 40 °C and a pressure of 0.0432 bar is 33.3 m³/kg. The mass fraction of water vapour in the mixture is therefore:

$$\frac{1/33.3}{(1/0.2276)+(1/33.3)} = 0.0068$$

and the mass fraction of ammonia is $1 - 0.0068 = 0.9932$.

Experimental results by Wucherer (1932) indicate that the measured mass fraction of water in the vapour is 0.012 at the above conditions.

Figure 6.9.1 shows the results of Wucherer plotted against pressure on a linear saturation temperature of ammonia scale, with the mass fraction in the liquid as a parameter. It will be convenient from now on to distinguish the mass fraction of ammonia in the liquid as ξ' and the mass fraction of ammonia in the vapour as ξ''.

6.10 Bubble and dew points: distillation

Figure 6.9.1 is a convenient way of presenting numerical information, but for the purpose of understanding the distillation process a

Fig. 6.9.1. Mass fraction ξ'' of ammonia in the vapour which is in equilibrium with a solution of ammonia in water having mass fraction ξ' as a function of pressure and saturation temperature of pure ammonia.

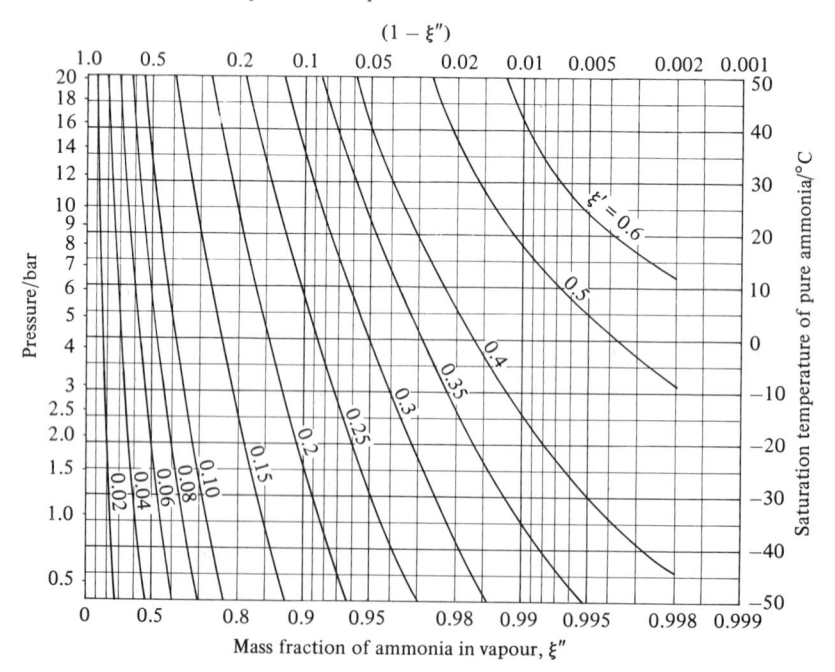

6.10 Bubble and dew points: distillation

schematic diagram showing the boiling temperatures of solutions at a given constant pressure against the composition of the liquid is more useful. In the boiler of an absorption system the pressure is constant throughout (and approximately equal to condensing pressure), and in Fig. 6.10.1 the boiling points of solutions at constant pressure are shown diagrammatically as a function of mass fraction of ammonia in the liquid by the line $W - P' - A$. It will be noted that the boiling points lie between those of the pure components t_W and t_A. A solution of composition shown as ζ' on being warmed to temperature t_P at P' is just on the point of boiling, whence this point is known as the bubble point, and the curve $W - P' - A$ as the bubble point curve.

The mass fraction of ammonia in the vapour first produced is not ξ' however but ξ'' shown as point P'', i.e. the vapour is richer in ammonia than the solution.

Point P'' may also be approached from above by cooling some vapour from a temperature higher than t_P until condensation just begins, i.e. dew forms. Point P'' is therefore a dew point for vapour of composition ξ'', whence the curve $W - P'' - A$ joining all such points is called the dew point curve.

Liquid at P' is in equilibrium with vapour at P'', i.e. they can exist in

Fig. 6.10.1. Diagrammatic plot of the bubble points of a solution of ammonia in water, shown by the lower curve. The upper curve is the dew point line. The composition of vapour in equilibrium with solution P' is given by P''.

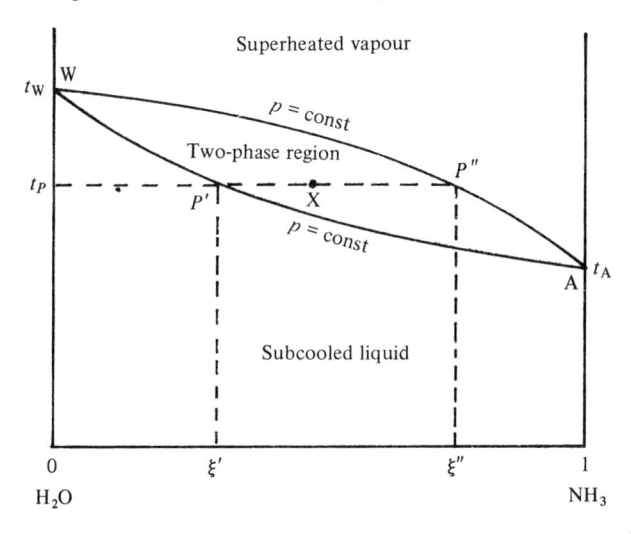

equilibrium with each other in any proportions indefinitely so long as the temperature and pressure remain the same. A pure liquid and its saturated vapour can also exist in equilibrium in any proportions indefinitely, but in that case it is only necessary to specify either temperature or pressure because the one implies the other. Because the solution and vapour have two components there is one extra degree of freedom compared with a pure liquid.

Below the curve $W - P' - A$ all solutions having pressure p for which the curves are drawn are below their bubble point, and this region therefore represents subcooled solution. Above curve $W - P'' - A$ all vapours having pressure p for which the curves are drawn are above their dew point, and this region represents superheated vapour.

Between the bubble and dew curves lies a two-phase region. The point X for example represents a two-phase mixture of liquid of mass fraction ξ', and equilibrium vapour of mass fraction ξ'', both at the same temperature t_p. The dryness fraction or quality of the mixture at X is given by:

$$x = \frac{\xi_X - \xi'}{\xi'' - \xi'} \tag{6.10.1}$$

For different pressures, different pairs of curves must be drawn, corresponding to higher boiling points for higher pressures and vice versa.

When vapour is formed from an ammonia–water solution the first fraction of vapour to come off contains more ammonia than the solution. Consequently the solution is weakened in ammonia and the bubble point temperature rises. What happens subsequently depends on the conditions.

Fig. 6.10.2. Boiling a solution of ammonia in water at constant pressure when the vapour is not allowed to escape.

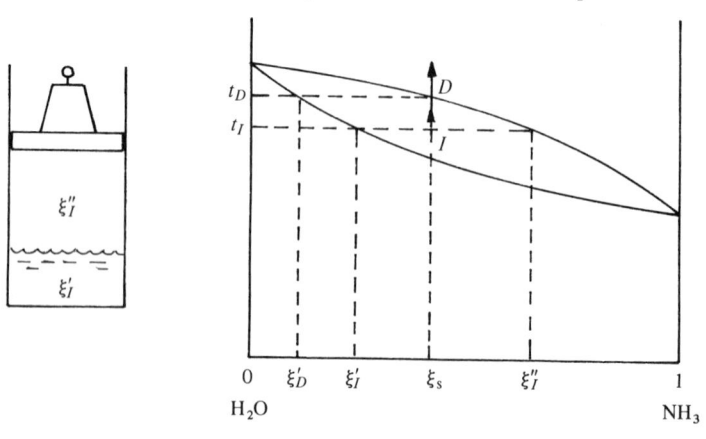

6.10 Bubble and dew points: distillation

Suppose that the solution of composition ξ_s is enclosed in a cylinder by means of a piston loaded with a weight so that the pressure remains constant, Fig. 6.10.2. As vaporisation continues the temperature rises, but the mass fraction of the total contents of the cylinder must stay constant at ξ_s, since neither component can escape. As indicated in Fig. 6.10.2 the state of the system changes at constant mass fraction ξ_s and rising temperature to point D. An intermediate state is represented by I at temperature t_I, where the liquid has the composition ξ'_I and the vapour ξ''_I. The last traces of liquid finally disappear when the temperature t_D is reached, and the liquid composition just before it vanishes is given by ξ'_D. Further heating raises the temperature of the vapour at constant pressure and produces superheated vapour, of the same composition as the original liquid. Conversely t_D is a dew point temperature for vapour of composition ξ_s.

On the other hand, if the solution is boiled in a vessel with an outlet, the pressure still remaining constant, the vapour is removed as it forms. Referring to Fig. 6.10.3, the first portion of vapour removed has the mass fraction ξ''_s. But the liquid is thereby weakened to ξ'_2 so that the next portion to come over has the mass fraction ξ''_2 say, the next ξ''_3 and so on. Whilst the solution is boiled from an initial mass fraction ξ'_s to a final mass fraction ξ'_w, the vapour given off becomes weaker from ξ''_s to ξ''_w, and if this vapour is totally condensed the resulting condensate will have a mass fraction intermediate between ξ''_s and ξ''_w. This is the principle of a batch

Fig. 6.10.3. Boiling a solution of ammonia in water when the vapour is allowed to escape. Solution of mass fraction ξ' gives off stronger vapour, ξ'', and thereby is weakened. If the vapour is totally condensed the resulting liquid has a composition somewhere between ξ''_s and ξ''_w.

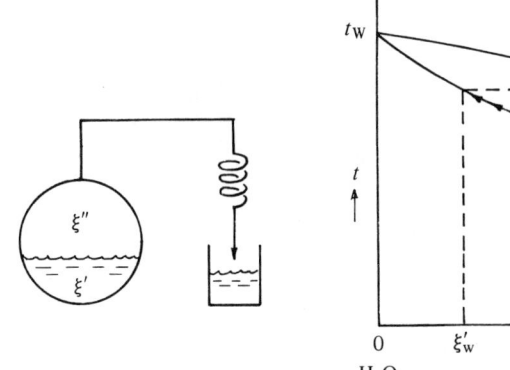

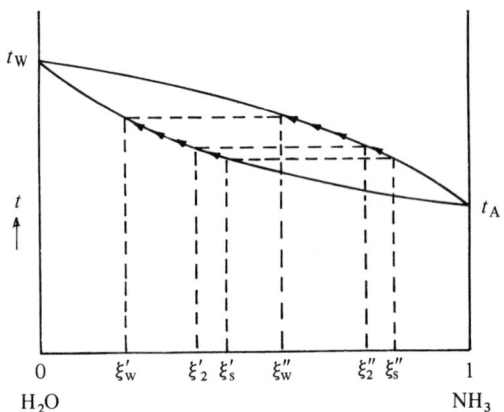

distillation process. The total distillate is necessarily weaker than the first fraction to come over, ξ_s''.

The vapour given off can be made much stronger in a continuous distillation process by arranging counterflow of the incoming solution, the *feed*, with the arising vapour, in direct contact with each other so that heat and mass transfer can occur. This happens in the lower part of a distillation column, known as the *stripping* section, shown diagrammatically in Fig. 6.10.4. The arrangements for ensuring good contact between the descending stream of liquid and the ascending stream of vapour may take the form of packing which is wetted by the solution, or of trays on which the liquid stands whilst the vapour is bubbled through it.

Figure 6.10.3 shows that as the solution becomes weaker on boiling its temperature rises. Hence at any section of the column the vapour ascending, which has originated from warmer solution at a lower level, is warmer than the solution which it meets descending from a higher level. The arising vapour is therefore partially condensed and the descending liquid partially evaporated. The condensate is weaker than the vapour from which it comes and the vapour is consequently enriched. The vapour given off is stronger than the liquid from which it comes so that the descending solution is weakened. The result is a net transfer of ammonia from the descending solution to the ascending vapour, and of water from the

Fig. 6.10.4. A stripping process in which by counterflow of ascending vapour and descending liquid the ascending vapour is gradually strengthened towards ξ_s'' and the descending liquid is gradually weakened.

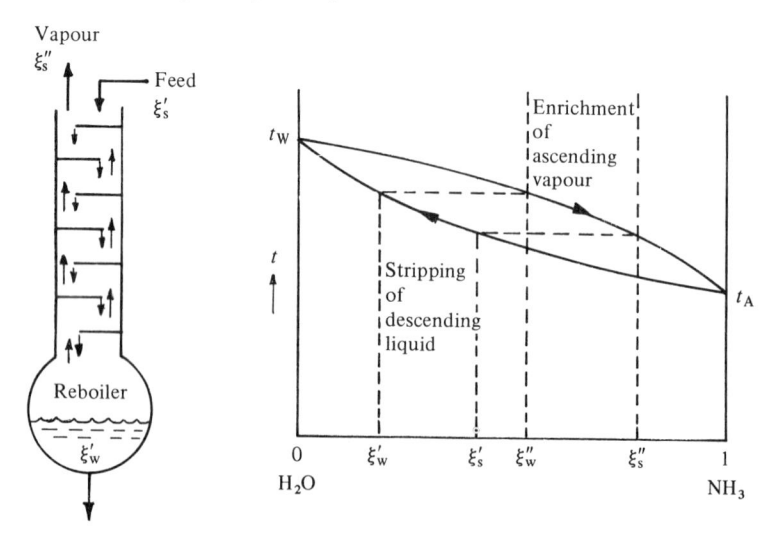

6.10 Bubble and dew points: distillation

ascending vapour to the descending solution. In the limit, with infinite transfer surface, the vapour at the top could be brought into equilibrium with the strong solution entering there, but in practice a certain approach to equilibrium is all that is possible.

By a similar counterflow process, this strong vapour can be enriched to any desired degree in the *rectifying* section of the column, Fig. 6.10.5, which is placed above the stripping section. The descending stream in this consists of a condensed part of the almost pure vapour of composition ξ_r'' leaving the column at the top. The descending stream is called the *reflux*, and the condenser which produces it the *reflux condenser*. Conditions in the rectifying section are similar to those in the stripping section. At any section the ascending vapour is warmer than the descending liquid, and a net transfer of ammonia takes place from the liquid to the vapour whilst a transfer of water takes place from the vapour to the liquid. The vapour is gradually enriched towards the top.

The whole apparatus, usually called the *generator* in absorption refrigerators, for recovering ammonia from the strong solution coming from the absorber therefore consists of a boiler (usually called the *reboiler* because it reboils the liquid descending from the column), a stripping section (also sometimes called an *analyser*), and the rectifying column with reflux condenser. In chemical engineering practice the exhausted solution leaving the reboiler is called the *bottoms*.

Fig. 6.10.5. A rectification process in which by counterflow of ascending vapour and descending liquid from the reflux condenser the ascending vapour is enriched towards $\xi = \xi_r''$.

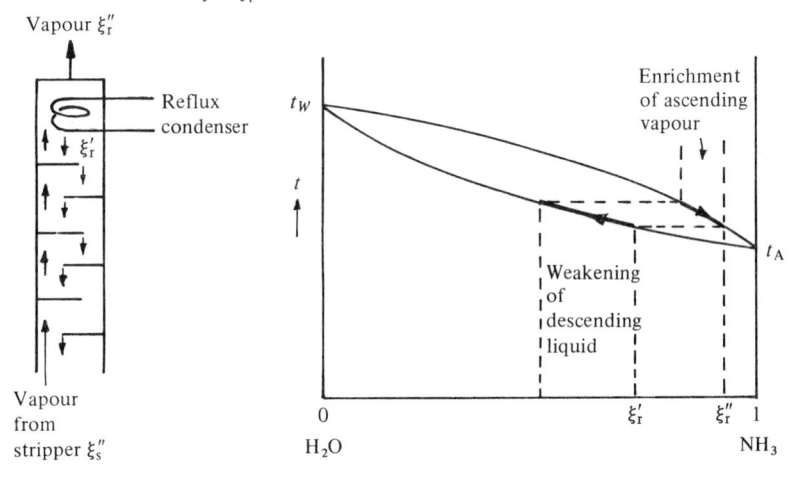

The vapour absorption system 430

6.11 Pressures and temperatures in the aqua-ammonia cycle

The bubble point and dew point curves of Fig. 6.10.1 are drawn for one fixed value of the pressure. In the absorption system there are two pressures: (a) in the evaporator and absorber, (b) in the condenser and generator. The pressure in the evaporator is determined as in the compression system by the temperature desired there, which must be somewhat less than that of the fluids to be cooled; and the pressure in the condenser is determined by the temperature desired there, which must be somewhat higher than that of the cooling medium.

In a real plant there will of course be small differences of pressure between the evaporator and the absorber, and between the condenser and the generator, because of the fact that flow takes place through pipes between them. In the following treatment these differences will be neglected. They must not be neglected in practice however, especially for low-temperature plant, because a small difference of pressure in the absorber can make a significant difference to the equilibrium solution strength.

In Fig. 6.11.1, bubble point and dew point curves are drawn (not to scale) for the pressure $p_e = p_a$ in the evaporator and absorber, and $p_c = p_b$ in the generator and condenser. Typical state points on the diagram are numbered to correspond with the circuit diagram shown in Fig. 6.11.2. It is supposed that the ammonia after rectification is very nearly pure.

Fig. 6.11.1. Pressures and temperatures in an ammonia–water absorption system. The points are numbered to correspond with those in Fig. 6.11.2.

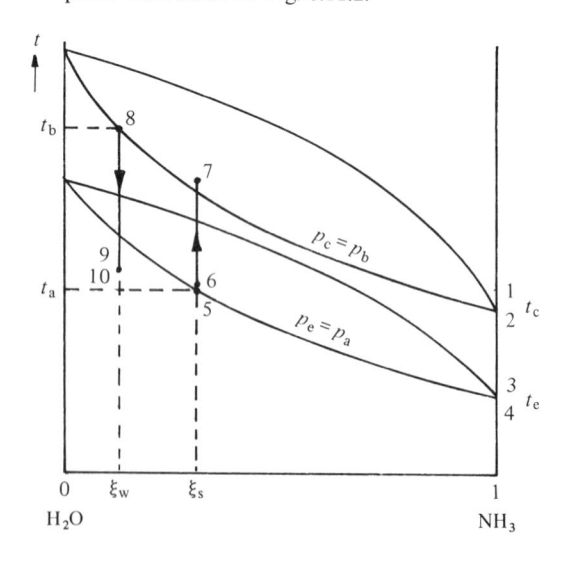

6.11 Pressures and temperatures in aqua-ammonia cycle

Since the condenser and absorber are cooled by the same cooling medium, usually water, the temperature in the absorber is approximately the same as that in the condenser, depending on how the cooling water is used. If it is passed through them in parallel these temperatures may be very nearly the same, but if it is passed through them in series, as is usually done, one will be higher than the other. In the following description, it will be assumed that $t_c > t_a$, as will be the case when the cooling water is passed first through the absorber.

The temperature t_b of the liquid in the reboiler is determined by the temperature of the available heating medium.

Beginning at point 1, which represents nearly pure vapour leaving the generator, at condensing pressure p_c and condensing temperature t_c, the vapour is condensed to give saturated liquid at point 2. This liquid passes through the expansion valve to point 3 which represents a two-phase mixture at temperature t_e. In the evaporator, saturated or slightly superheated vapour at point 4 is produced, pressure p_e, temperature $t_e = t_4$.

Fig. 6.11.2. Components of an ammonia–water absorption system, essentially the same as those of Fig. 6.4.2 but with a stripping and rectification column on the boiler.

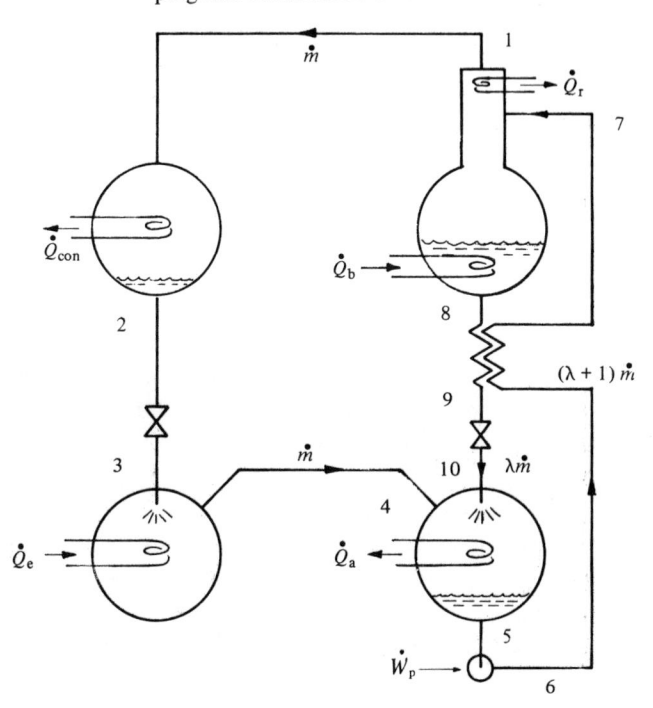

So far the operations are concerned with pure ammonia and are exactly the same as those in the compression refrigerator.

The vapour 4 meets in the absorber a weak solution of mass fraction ξ_w at the state 10. This solution originated in the boiler as liquid, 8, at its bubble point, at the temperature $t_8 = t_b$ and pressure p_b. These values fix the equilibrium composition ξ_w. The weak solution is subsequently reduced in temperature in the heat exchanger to a temperature t_9 which is slightly above absorber temperature. It is possible in principle to cool the weak liquid down to t_a by using a large enough heat exchanger, but economical design usually dictates a certain approach of a few degrees. The weak solution at 9 is at pressure p_b and is therefore very much subcooled. After passing through the valve to pressure p_a at 10 it is still subcooled because 10 lies below the bubble curve for p_a. It can therefore absorb the vapour at 4 and produce strong solution of mass fraction ξ_s at point 5, in equilibrium on the bubble point curve at temperature t_a and pressure p_a. These values fix the equilibrium composition ξ_s. In a real plant this equilibrium is not quite achieved, but for simplicity it is supposed here that it is.

The solution 5 passes through the pump to point 6. In compressing a liquid there is only a very small change in temperature, and hence points 5 and 6 are almost coincident. But the state 6 is of course at generator pressure $p_6 = p_c$, and with respect to the upper bubble point curve at this pressure 6 is now subcooled solution, well below its bubble point temperature. In the heat exchanger the temperature rises to give point 7 at the same mass fraction. The location of this state 7 depends on the amount of heat exchanger surface provided, but it cannot reach the temperature t_8 because the mass flow rate of the strong solution is greater than the mass flow rate of the weak solution and their specific heat capacities are not much different. In heat exchanger terminology, the capacity rate of the weak solution stream is less than that of the strong solution stream, and the weak solution is thus the controlling one. It can be cooled to t_6 without violating the enthalpy balance, but the strong solution cannot be warmed up to t_8. State 7 usually lies somewhere near its bubble point, but it is often in the two-phase region, as it is shown on Fig. 6.11.1.

6.12 Enthalpy of aqua-ammonia

The specific enthalpy of solutions of ammonia in water is found by the same procedure as already described for lithium bromide in water in section 6.5, i.e. by measurement of the integral enthalpy of solution for various mass fractions. Calculating h by eq. (6.5.1), the specific enthalpy can

6.12 Enthalpy of aqua-ammonia

be plotted as a field of isotherms against the mass fraction. This is shown in Fig. 6.12.1. based on the values determined by Zinner (1934).

These isotherms give the specific enthalpy of the solution at the temperature marked without regard to the pressure. When treating the enthalpy of pure liquid refrigerants, pressure has been found not to make a significant difference, except for substances near their critical point, and the specific enthalpies of saturated and subcooled liquid have been treated as

Fig. 6.12.1. Specific enthalpy of solutions of ammonia in water. Datum states: 0 kJ/kg for liquid water at 0 °C, 0 kJ/kg for liquid ammonia at -40 °C. The lines of constant pressure are labelled with the saturation temperatures of pure ammonia at the corresponding pressures. (Based on the data of Zinner, 1934.)

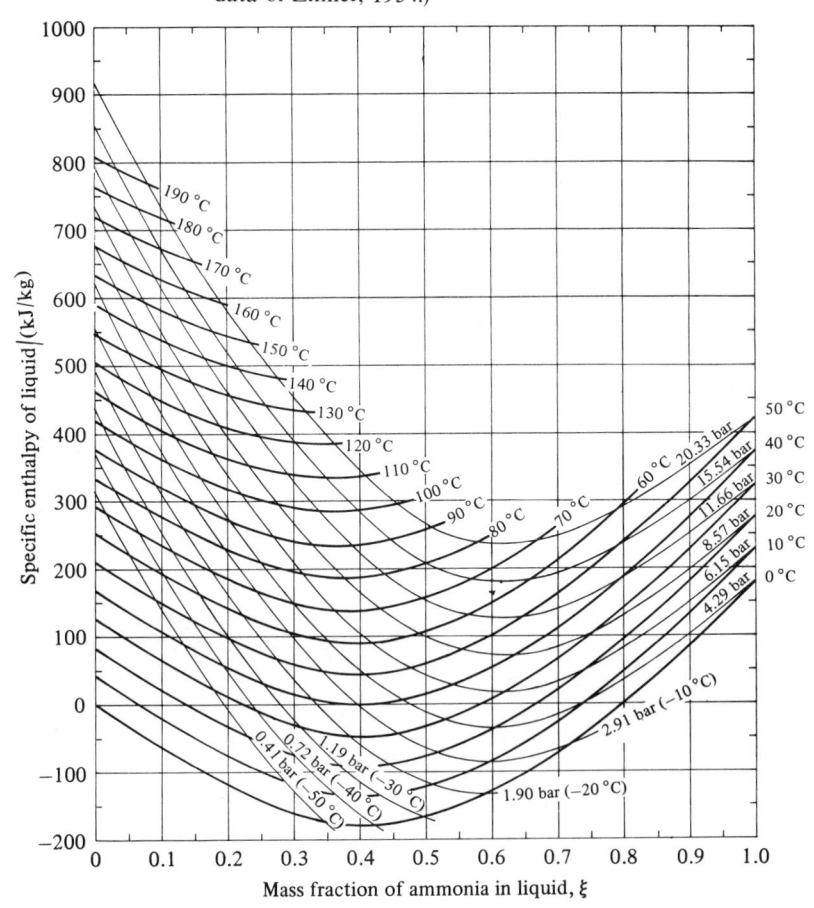

equal. This is also implied in Fig. 6.12.1, with the further justification that the values are not known to an accuracy which would warrant distinguishing between liquid at its bubble point and liquid of the same composition and temperature under a higher pressure. Consequently the specific enthalpy of any solution can be read from Fig. 6.12.1.

It is convenient for the study of the absorption cycle to show the equilibrium data of Fig. 6.8.2 on the enthalpy composition diagram as a family of lines of constant pressure. This is shown diagrammatically in Fig. 6.12.2 with two lines of constant pressure representing pressures in the absorber p_a and in the boiler p_b. As these isobars approach the $\xi = 1$ axis, they meet there the isotherms t_e and t_c for the evaporating and condensing temperatures. Towards the $\xi = 0$ axis, they become much steeper than the isotherms and ultimately reach the axis at values of h equal to those of pure water boiling at these pressures. Unlike the values of specific enthalpy given by the isotherms, the values given by the isobars are *equilibrium values at the bubble points only*.

The lines of constant pressure in Fig. 6.12.1 are drawn not for equal intervals of pressure but for even values of the saturation temperature of ammonia. For example, for an evaporating temperature of $-10\,°C$ the evaporating pressure is 2.908 bar and a line at this pressure is provided. This practice has the advantage of giving a family of pressure lines directly related to the evaporating and condensing temperatures of approximately

Fig. 6.12.2. Skeleton diagram of lines from Fig. 6.12.1 showing lines of constant evaporating and condensing pressures, p_e and p_c, and lines of constant evaporating and condensing temperature, t_e and t_c.

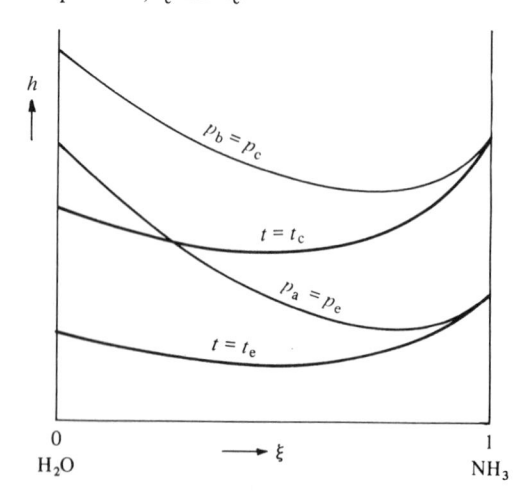

6.12 Enthalpy of aqua-ammonia

equal spacing on the diagram without changing the plotting interval. (Examination of the $h-\xi$ charts cited below will show that the pressure interval has to be changed arbitrarily in order to preserve the spacing.) Interpolation for intermediate evaporating and condensing temperatures can be readily done by eye.

Vapour

When it comes to the enthalpy of the equilibrium vapour produced by solutions of ammonia in water at their bubble point, the state of knowledge is far from complete. There are in fact at present no direct measurements of these enthalpies, or of any related thermodynamic quantities such as the specific volumes which would enable the enthalpy to be determined with confidence. It is usually assumed that ammonia and water vapours mix without enthalpy of mixing, implying that no chemical compounds are formed in the vapour phase as they are in the liquid. The specific enthalpy of vapour is then given by:

$$h'' = \xi'' h''_A + (1 - \xi'') h''_W \qquad (6.12.1)$$

where h''_A and h''_W are the specific enthalpies of pure ammonia and water vapours at the temperature of the mixture. The difficulty which then arises is: at what pressures are these enthalpies to be evaluated? If the components were ideal gases, like oxygen and nitrogen in the atmosphere, there would be no problem because the enthalpies of ideal gases are independent of pressure.

It is clear that the terms in eq. (6.12.1) are not to be evaluated at the total pressure of the mixture, because one of the components, the water, does not exist as a pure vapour at the temperature and total pressure of the mixture. For example, the equilibrium vapour produced by a solution of mass fraction 0.4 at a pressure of 3.029 bar is at a temperature of 40 °C. The vapour pressure of pure water at 40 °C is only 0.07375 bar, so that water vapour does not exist, except perhaps as a metastable supersaturated vapour, at a pressure of 3.029 bar. On the other hand the ammonia vapour would be superheated at the pressure 3.029 bar (saturation temperature $-9\,°C$) and temperature 40 °C.

The procedure adopted here, which is admitted to be approximate only, is to evaluate h''_A and h''_W at the pressures

$$p_A = \psi p \qquad (6.12.2)$$

$$p_W = (1 - \psi) p \qquad (6.12.3)$$

where ψ is the mole fraction of ammonia in the mixture and p is the total pressure. In an ideal gas these would be the partial pressures of the components, defined as the pressures which each would contribute to the total if each occupied the space alone. But in a real gas mixture the partial pressures are to be determined using the vapour tables at the partial densities of the components, and these are not known. As a check that this method represents the mixture reasonably well, however, the mass fraction of water in the vapour has been back-checked and compared with the experimental values of Fig. 6.9.1. The following example illustrates the method.

Example 6.12.1
Determine the specific enthalpy of the equilibrium vapour produced by solution of mass fraction 0.4 at a pressure of 8.571 bar.
From Fig. 6.8.2 the temperature is 75 °C, and from Fig. 6.9.1 the mass fraction of ammonia in the vapour is 0.970. By a calculation similar to example 6.8.1, the mole fraction of ammonia is:

$$\psi = 0.972$$

and of water vapour:

$$1 - \psi = 0.028$$

The pressures p_A and p_W are then:

$$p_A = (0.972)(8.571) \text{ bar} = 8.331 \text{ bar}$$

$$p_W = (0.028)(8.571) \text{ bar} = 0.240 \text{ bar}$$

From the ammonia tables at 75 °C, 8.331 bar:

$$h''_A = 1609 \text{ kJ/kg}$$

From the steam tables at 75 °C, 0.240 bar:

$$h''_W = 2638 \text{ kJ/kg}$$

By eq. (6.12.1) the specific enthalpy of the mixture is:

$$h'' = [(0.970)(1609) + (0.030)(2638)] \text{ kJ/kg}$$

$$= 1640 \text{ kJ/kg}$$

The assumed partial pressures are consistent with the known composition data from the following check. Specific volume of ammonia at 75 °C, 8.331 bar is:

$$v''_A = 0.1882 \text{ m}^3/\text{kg}$$

6.12 Enthalpy of aqua-ammonia

Specific volume of water vapour at 75 °C, 0.240 bar is:

$$v''_W = 6.662 \text{ m}^3/\text{kg}$$

The mass fraction of water vapour in the mixture is then:

$$1 - \xi'' = \frac{1/6.662}{(1/0.1882) + (1/6.662)} = 0.0275$$

and of ammonia:

$$\xi'' = 0.9725$$

as against 0.03 and 0.97 from the experimental data.

Fig. 6.12.3. Specific enthalpy of vapour in equilibrium with a solution of ammonia in water. Datum states the same as for Fig. 6.12.1. The lower family of curves shows the specific enthalpy plotted against the composition of the solution. The upper family shows the same values plotted against the composition of the vapour.

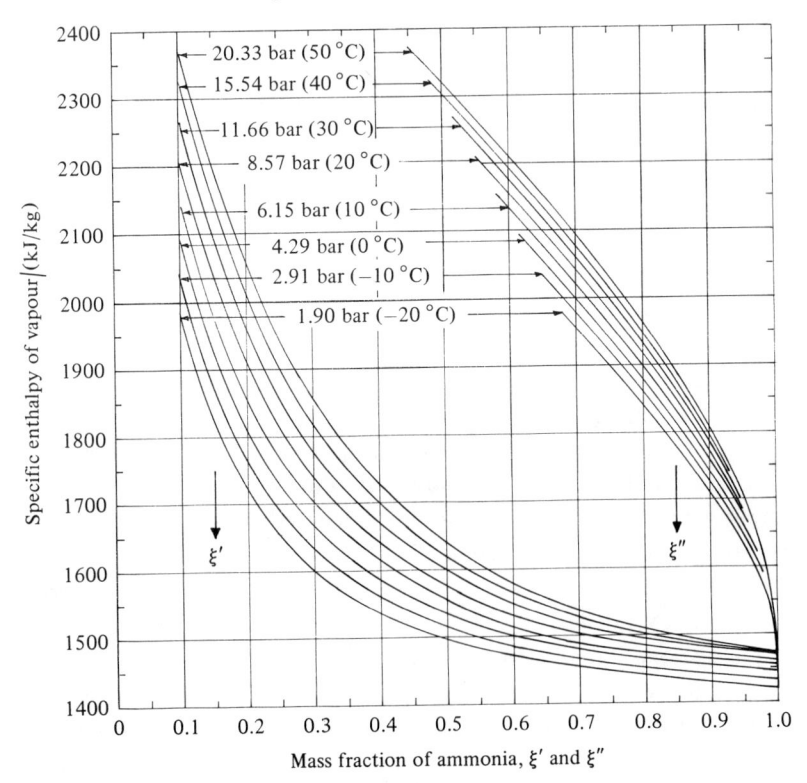

The above calculation could be refined further by trial and error to find values of p_A and p_W which would give closer approximations to ξ''. But this is not in fact very profitable because within the small pressure variations required to put ξ'' correct, the specific enthalpies h_A'' and h_W'' do not vary much.

Values of the specific enthalpy of the equilibrium vapour calculated by the above method are presented in Fig. 6.12.3. In the left-hand family of curves they are plotted against the mass fraction of ammonia in the liquid, ξ', for the same set of pressures as in Fig. 6.12.1. In the right-hand family of curves the enthalpies are plotted against the mass fraction of ammonia in the equilibrium vapour, ξ''. Figure 6.12.3 thus incorporates the data of Fig. 6.9.1, so that one can determine by a horizontal line joining corresponding points in the two families the equilibrium values of ξ'', but not of course so accurately as from Fig. 6.9.1, owing to the small scale on the right-hand side.

A note on the properties of aqua-ammonia

The properties of aqua-ammonia are not known so completely or to the same accuracy as those of important pure substances such as water, ammonia or carbon dioxide. Much of the data rests on a comparatively small number of investigations, which have not always agreed, and the absence of any experimental data at all in the vapour phase has already been mentioned. It is possible therefore that some differences will be found between the data of Figs. 6.8.2, 6.9.1, 6.12.1 and 6.12.3 and that from some charts and tables in current circulation.

The principal compilations and charts in use at present are the following. Merkel & Bošnjaković (1929) first gave a systematic exposition of the calculation of absorption refrigeration on the enthalpy–composition diagram, and they presented tables and charts based on earlier measurements of vapour pressure, and the enthalpies of solution measured by Hilde Mollier (1909). A revised chart incorporating later measurements of Wucherer, and enthalpies of solution by Zinner, which were substantially larger than those obtained by Mollier, appeared in a folder with the second volume of Bošnjaković's book *Technische Thermodynamik* (1937). This chart is still in circulation on the Continent, and is currently available with the book by Niebergall (1959). It is in technical metric units, specific enthalpy (kcal/kg) and pressure (kgf/cm^2).

Another way of presenting the same data on the coordinates of temperature and pressure was introduced by Tandberg & Widell (1937), and Bäckström presented them on a conventional plot of logarithm of pressure versus specific enthalpy (Bäckström & Emblik, 1965).

6.12 Enthalpy of aqua-ammonia

A chart incorporating Wucherer's and Zinner's data in British units was given by Jennings & Shannon (1938).

Independent compilations and reviews of the data have been made by Scatchard, Epstein, Warburton & Cody (1947), and by Macriss, Eakin, Ellington & Huebler (1964), both in British units. Scatchard *et al.* based their work on the vapour pressures of Perman (1901, 1903) and Wucherer (1932), the enthalpies of Zinner (1934), and the vapour compositions of Wucherer, and derived the enthalpy of the vapour by a thermodynamic method based on empirical formulae for the free energy of the vapour. Their results are given as tables with the arguments temperature and mass fraction of ammonia in liquid, subsequently presented as a chart of temperature versus mass fraction by Kohloss & Scott (1950).

Macriss *et al.* reviewed the whole of the available data, including new measurements of vapour pressure by Pierre (1958), and incorporated their own measurements of dew point and enthalpy of liquid. They calculated the enthalpies of vapour by the method which is now usual for pure substances. Starting from the ideal gas state at a low pressure p_0 in which the enthalpies of ammonia and water vapours are accurately known as a function of temperature, the enthalpy is corrected for pressure by the addition of the second term of the right-hand side of eq. (3.18.9) integrated from pressure p_0 at the constant temperature of the mixture. Equation (3.18.9) can be readily integrated when an equation of state is known, but in the absence of the $p-v-T$ data for aqua-ammonia vapour, it has to be estimated using generalised compressibility charts obtained for other substances. Macriss *et al.* presented their results in tabular form and as a chart of specific enthalpy versus mass fraction. This chart is reproduced in the current edition of the ASHRAE Handbook of Fundamentals (1977).

It should be noted that the datum state for the liquid ammonia in the tables of Scatchard *et al.*, and in the tables and chart of Macriss *et al.* is 32 °F (0 °C), and is thus not compatible with the standard ammonia tables in British Units, e.g. NBS Circular 142, which are based on a datum of $-40\,°F$. If data for pure ammonia are taken from the NBS tables and used in conjunction with the aqua-ammonia chart in the current ASHRAE Handbook, they should be corrected by the subtraction of 77.9 Btu/lb, or the values of the chart should be corrected by the addition of 77.9 (ξ) Btu/lb.

Schulz (1972) has produced formulations of the properties of aqua-ammonia in terms of the characteristic function free enthalpy g mentioned at the end of section 3.18. Naturally it is now dependent on composition as well as on T and p. The use of these formulations demands considerable iteration, particularly to determine the conditions of phase equilibrium, which can only be performed by a computer.

6.13 The complete enthalpy–composition diagram

The data in Figs. 6.12.1 and 6.12.3 may be combined on one diagram as shown schematically in Fig. 6.13.1. Several authors including Bošnjaković and Macriss *et al.* have produced their data in this form, but it has the disadvantage that a large part of the centre of the diagram is wasted and it imposes an unnecessarily small scale for specific enthalpy.

The complete diagram has considerable value, however, as an aid to the understanding of the absorption cycle. This development is mainly due to Merkel & Bošnjaković (1929) and to Bošnjaković (1937) who has applied the same principles to many problems concerning mixtures.

Figure 6.13.1 shows only one isobar for clarity, in both the liquid and vapour phases, and the interpretation of points and regions on the diagram will now be considered.

The junction of an isobar and an isotherm in the liquid region, such as point P' marked in Fig. 6.13.1, represents liquid of composition ξ' at its bubble point at the pressure p, temperature t, and having the specific enthalpy h', and corresponds to point P' in Fig. 6.10.1. The vapour with which this liquid can exist in equilibrium is represented by point P'' having the composition ξ'' and specific enthalpy h''. This point is found by

Fig. 6.13.1. Diagram showing how the data of Fig. 6.12.1 and Fig. 6.12.3 may be combined on a single chart. Point P' represents liquid at its bubble point in equilibrium with vapour at P''. Point X represents a two-phase mixture of the phases P' and P''.

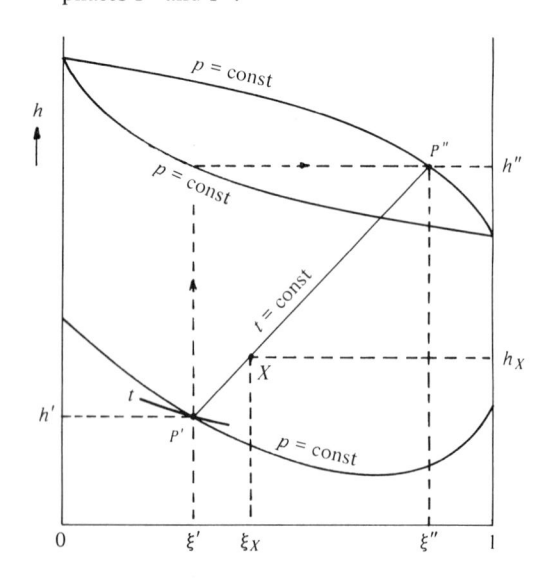

6.13 The complete enthalpy–composition diagram

following $\xi' = $ const to intersect the constant pressure line $p = $ const in the left-hand family of vapour curves, and then going across at constant h to intersect the constant pressure line $p = $ const in the right-hand family of curves, as indicated by the broken line. (Or, more accurately, the value of ξ'' could be read from Fig. 6.9.1.) The point P'' corresponds to P'' in Fig. 6.10.1. The straight line which joins $P' - P''$ is an isotherm at temperature t, since the liquid and its equilibrium vapour are both at the same temperature.

A point such as X on this isotherm represents a two-phase mixture, corresponding to point X in Fig. 6.10.1. The specific enthalpy at X is given by:

$$h_X = (1 - x)h' + xh'' \qquad (6.13.1)$$

where x is the dryness fraction (quality) at X. From eq. (6.13.1), x is given by:

$$x = \frac{h_X - h'}{h'' - h'} \qquad (6.13.2)$$

Since the line $P' - X - P''$ is straight, x is also given by:

$$x = \frac{\xi_X - \xi'}{\xi'' - \xi'} \qquad (6.13.3)$$

Fig. 6.13.2. The interpretation of arbitrary points L and V on the h–ξ diagram. L may represent liquid at its bubble point, sub-cooled liquid or a two-phase mixture, depending on the pressure. V may represent vapour at its dew point, superheated vapour or a two-phase mixture.

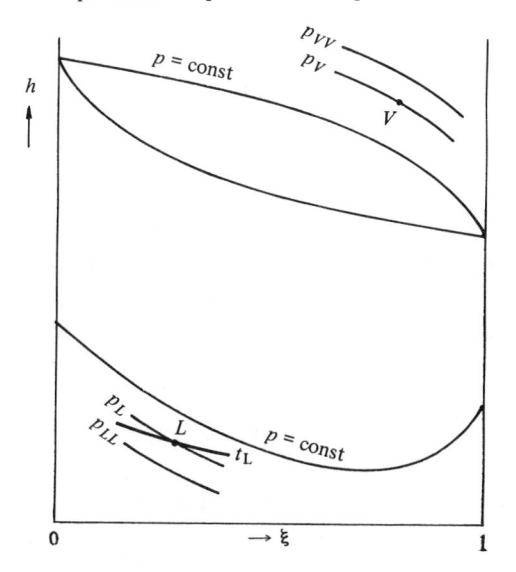

It is not practicable to show isotherms in the two-phase region on published charts because there is a different set of them for each pressure.

The region above the line $p=$ const in the vapour phase represents superheated vapour of mixtures which are at the pressure p. But points such as V, Fig. 6.13.2, though superheated when considered with respect to the pressure p, will also lie on the saturation lines for some higher pressure p_V say, and in the two-phase region for a still higher pressure, p_{VV} say. Consequently, the mere location of a point on the diagram does not itself indicate the state unambiguously: it is necessary to know the pressure to interpret the significance of a point given simply by the coordinates h and ξ.

In the liquid region, the region underneath the curve of $p=$ const represents subcooled liquid for all solutions which are at the pressure p. But points such as L, though subcooled when considered with respect to the pressure p, also lie on the saturation line for some lower pressure p_L. Furthermore, if the pressure at L is actually p_{LL} less than p_L the value on the constant pressure line passing through the point, then L represents a two-phase mixture because it lies above the line $p_{LL}=$ const representing its pressure. Its temperature is not then t_L, the value on the isotherm passing through L, because it is then to be considered as lying on the two-phase isotherm for some lower concentration.

Example 6.13.1

The coordinates of a point in Fig. 6.12.1 are: $\xi = 0.3$, $h = 60$ kJ/kg, and the pressure is (a) 2.91 bar (b) 1.90 bar. What are the states (a) and (b)?

Case (a). Liquid at its bubble point at a pressure of 2.91 bar has a specific enthalpy of 100 kJ/kg at $\xi = 0.3$. State (a) therefore has a lower specific enthalpy than this and hence represents subcooled liquid. Since the isotherms are drawn for subcooled or saturated liquid, the temperature can be determined as 50 °C.

Case (b). Liquid at its bubble point at a pressure of 1.90 bar has a specific enthalpy of about 35 kJ/kg at $\xi = 0.3$. State (b) has a higher specific enthalpy than this and hence represents a two-phase mixture. The isotherms in Fig. 6.12.1 do not apply to two-phase mixtures so it is not possible to read the temperature for state (b) from this diagram directly.

Example 6.13.2

Some *liquid* has a mass fraction of 0.25 and a temperature of 80 °C. Can its state be determined from Fig. 6.12.1?

No. All that can be said is that the specific enthalpy is 210 kJ/kg, since it is known to be liquid. The pressure is indeterminate. The liquid might be at its

6.13 The complete enthalpy–composition diagram

bubble point at a pressure of 4.29 bar, or subcooled at, for example, 8.57 bar. Both states have the same specific enthalpy on the chart.

The point which was unresolved in example 6.13.1(b) will now be considered. The question is, knowing the specific enthalpy h_X, pressure p_X and mass fraction ξ_X of a point X in the two-phase region, how is the temperature to be determined? A graphical trial and error method on the complete h–ξ chart will first be shown, illustrated in Fig. 6.13.3. The starting assumption is that the vapour is the equilibrium vapour for solution at mass fraction ξ_X at its bubble point pressure p_X with the specific enthalpy h_1'' mass fraction ξ_1'', point 1″. From this point a straight line is drawn through point X to cut the line of constant pressure p_X in the liquid region at h_2', ξ_2', point 2′. For this bubble point state a new equilibrium vapour state, h_2'', is determined, point 2″, and another straight line is drawn back from this point through X to cut the line p_X in the liquid region again, giving new values h_3', ξ_3', and point 3′. The procedure is repeated until it converges. Provided that X is not far into the two-phase region, this will happen after only two or three trials.

Fig. 6.13.3. Graphical method of finding the isotherm passing through an arbitrary point X in the two-phase region. Following the direction of the arrows from X, rapid convergence is found towards a line passing through X whose ends represent equilibrium states on the bubble and dew point curves.

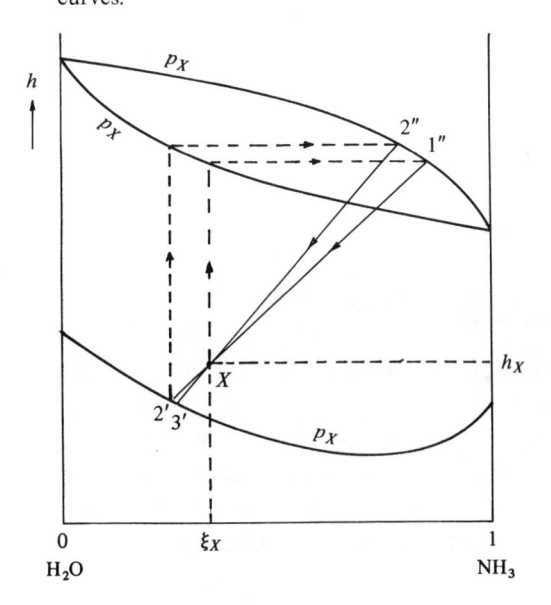

A trial and error procedure is needed if the problem is to be solved by calculation. The points (h', ξ') and (h'', ξ'') at each end of the isotherm passing through X have to satisfy:

$$\frac{h'' - h_X}{\xi'' - \xi_X} = \frac{h_X - h'}{\xi_X - \xi'} \tag{6.13.4}$$

To save working, it is best to choose a starting point (h'_1, ξ'_1) on the bubble point line which is somewhat weaker than ξ_X. Values of (h''_1, ξ''_1) are then determined from the chart, Fig. 6.12.3, and using eq. (6.13.4) new values (h'_2, ξ'_2) are calculated. The process is repeated as necessary.

Example 6.13.3
Determine the temperature and dryness in example 6.13.1(b). A first estimate $\xi'_1 = 0.295$ will be made, giving $h'_1 = 42 \text{ kJ/kg}$, $h''_1 = 1605 \text{ kJ/kg}$ $\xi''_1 = 0.965$. Using eq. (6.13.4), new estimates of h'_2, ξ'_2 are found from:

$$\frac{1605 - 60}{0.965 - 0.30} = \frac{60 - h'_2}{0.3 - \xi'_2}$$

or:

$$h'_2 = 2323\xi'_2 - 637$$

Values which satisfy this equation on the line $p = 1.90$ bar are:

$$h'_2 = 44 \text{ kJ/kg} \qquad \xi'_2 = 0.293$$

and are so little different from the assumed starting values that, within the accuracy of the charts, they may be accepted as correct. Thus the state X is made up of:

liquid $h' = 44 \text{ kJ/kg}$ $\qquad \xi' = 0.293$
vapour $h'' = 1606 \text{ kJ/kg}$ $\qquad \xi'' = 0.964$

and the dryness fraction is:

$$\frac{60 - 44}{1606 - 44} = 0.0102$$

The temperature is estimated from Fig. 6.12.1 as about 46 °C.

6.14 Steady-flow analysis of the aqua-ammonia cycle

Basing the analysis on a flow rate of ammonia through the condenser and evaporator, $\dot{m}$, the steady-flow energy equations for the condenser, expansion valve, absorber, pump, solution valve and heat exchanger, are exactly the same as those developed in section 6.6 for the

6.14 Steady-flow analysis of aqua-ammonia cycle

lithium bromide system, eqs. (6.6.1)–(6.6.8). The expression for the circulation factor is different, however, owing to the convention for the meaning of strong and weak solution. Denoting now solution strong in ammonia, i.e. the refrigerant, by the mass fraction ξ_s, and solution weak in ammonia by the mass fraction ξ_w, a balance on the mass of ammonia entering and leaving the absorber gives:

$$1 + \lambda \xi_w = (\lambda + 1)\xi_s \qquad (6.14.1\text{a})$$

and

$$\lambda = \frac{1 - \xi_s}{\xi_s - \xi_w} \qquad (6.14.1\text{b})$$

It is assumed here that the ammonia circulating in the condenser and evaporator is pure. The effects of residual water in solution will be considered later.

When it comes to the generator, i.e. the combination of reboiler, stripper and rectifier, eq. (6.6.9) applied to a control surface drawn around the whole gives only the net heat transfer. Since it has been shown that there is a heat transfer from the reflux condenser, $\dot{Q}_r$, which did not occur in the lithium bromide system, eq. (6.6.9) must be written:

$$\dot{Q}_b - \dot{Q}_r = \dot{m}[h_1 + \lambda h_8 - (\lambda + 1)h_7] \qquad (6.14.2\text{a})$$

$$= \dot{m}[(h_1 - h_7) + \lambda(h_8 - h_7)] \qquad (6.14.2\text{b})$$

Or, if the control surface is taken to include the heat exchanger, adiabatic with respect to the surroundings:

$$\dot{Q}_b - \dot{Q}_r = \dot{m}[h_1 + \lambda h_9 - (\lambda + 1)h_6] \qquad (6.14.3\text{a})$$

$$= \dot{m}[(h_1 - h_6) + \lambda(h_9 - h_6)] \qquad (6.14.3\text{b})$$

In these equations the subscripts denote the states at the points numbered in Fig. 6.11.2, which are the same as in Fig. 6.4.2.

Example 6.14.1

An aqua-ammonia system operates at the following conditions:

temperatures of:		
evaporation	$-20\,°C$	$(p_e = 1.90 \text{ bar})$
condensation	$40\,°C$	$(p_c = 15.54 \text{ bar})$
absorption	$30\,°C$	
weak solution leaving boiler	$170\,°C$	
weak solution leaving heat exchanger	$40\,°C$	

The vapour absorption system

Assuming equilibrium conditions at absorber and boiler exits, complete rectification, no pressure drops, saturated liquid leaving condenser and saturated vapour leaving the evaporator, determine the heat transfers in the condenser, evaporator, absorber, generator and heat exchanger, and the state of the mixture leaving the heat exchanger.

The heat transfers will be evaluated for unit mass of ammonia to the condenser, i.e. $q = \dot{Q}/\dot{m}$ for each component of the system.

From the equilibrium at the absorber exit:

$$\xi_s(30\,°C, 1.90\,\text{bar}) = 0.38$$

and from the reboiler exit:

$$\xi_w(170\,°C, 15.54\,\text{bar}) = 0.10$$

Hence:

$$\lambda = \frac{1 - 0.38}{0.38 - 0.10} = 2.214$$

Specific enthalpies:

$h_1(\xi = 1, 15.54\,\text{bar}, \text{saturated vapour}) = 1472\,\text{kJ/kg}$

$h_2(\xi = 1, 40\,°C, \text{saturated liquid}) = 372\,\text{kJ/kg}$

$h_3 \approx h_2$

$h_4(\xi = 1, 1.90\,\text{bar}, \text{saturated vapour}) = 1419\,\text{kJ/kg}$

$h_5(\xi = 0.38, 30\,°C) = -46\,\text{kJ/kg}$

$h_6 \approx h_5$

$h_8(\xi = 0.10, 170\,°C) = 673\,\text{kJ/kg}$

$h_9(\xi = 0.10, 40\,°C) = 107\,\text{kJ/kg}$

Condenser:

$$q_{\text{con}} = h_1 - h_2 = (1472 - 372)\,\text{kJ/kg} = 1100\,\text{kJ/kg}$$

Evaporator:

$$q_e = h_4 - h_2 = (1419 - 372)\,\text{kJ/kg} = 1047\,\text{kJ/kg}$$

Absorber:

$$q_a = (h_4 - h_5) + \lambda(h_9 - h_5)$$
$$= [(1419 + 46) + 2.214(107 + 46)]\,\text{kJ/kg} = 2025\,\text{kJ/kg}$$

6.14 Steady-flow analysis of aqua-ammonia cycle

Generator:

$$q_b - q_r = (h_1 - h_6) + \lambda(h_9 - h_6)$$
$$= [(1472 + 46) + 2.214(107 + 46)] \text{ kJ/kg} = 1857 \text{ kJ/kg}$$

Heat exchanger:

$$q_x = \lambda(h_8 - h_9) = 2.214(673 - 107) \text{ kJ/kg} = 1253 \text{ kJ/kg}$$

$$h_7 = h_6 + \frac{\lambda(h_8 - h_9)}{(\lambda + 1)} = \left(-46 + \frac{1253}{3.214}\right) \text{ kJ/kg} = 344 \text{ kJ/kg}$$

The specific enthalpy of liquid of mass fraction 0.38 at its bubble point at a pressure of 15.54 bar, the same as at point 7, is 304 kJ/kg. Therefore the state at 7 is a two-phase mixture. Following the method of example 6.13.3, the properties of the phases are found to be:

liquid $h' = 318$ kJ/kg $\quad \xi' = 0.365$

vapour $h'' = 1735$ kJ/kg $\quad \xi'' = 0.938$

The temperature can be estimated as about 107 °C.

It is clear that the heat transfer to the reboiler, being the heat which drives the system and which has to be paid for, cannot be determined from an enthalpy balance drawn around the whole generator. It is necessary to consider the rectifier separately from the reboiler and stripping column, and this is done in the next section.

For the estimation of the heat transfers above it has again been assumed, as in section 6.6, that $h_6 \approx h_5$. The pump power is given by eq. (6.6.7). The specific volume of aqua-ammonia v_{sol} is given sufficiently accurately for this purpose by:

$$v_{sol} = \frac{0.001}{1 - 0.35\xi} \text{ m}^3/\text{kg} \qquad (6.14.4)$$

Departure from equilibrium at the absorber exit occurs in practice. It may be expressed in two different ways. The liquid leaving, at 30 °C in the above example, will not have attained the equilibrium mass fraction 0.38 corresponding to this temperature and the absorber pressure of 1.90 bar. Suppose that it reaches a mass fraction of only 0.36. At 30 °C this means an equilibrium pressure of 1.63 bar. There is thus a departure from equilibrium of $(1.90 - 1.63)$ bar $= 0.27$ bar.

On the other hand at the actual mass fraction of 0.36 and actual pressure of 1.90 bar, the bubble point temperature would be 34 °C, whereas the actual temperature is 30 °C. The departure from equilibrium can thus be expressed as a subcooling of 4 K.

In calculating actual systems the effect of pressure drops between the components must also be taken into account.

6.15 The rectifier

Consider a control surface drawn across any section of the column above the feed point to enclose the top of the column and the reflux condenser and to cut the (assumed) pure ammonia vapour pipe at 1, as indicated in Fig. 6.15.1. Descending solution crosses the control surface at a rate $L\dot{m}$ say, and ascending vapour crosses at the rate $V\dot{m}$ say. Let the specific enthalpies and mass fractions be h', h'', ξ' and ξ'' for the liquid and vapour, respectively. Note that $'$ and $''$ do not here indicate equilibrium values. As already pointed out, the vapour is warmer than the liquid. The following mass and enthalpy balances then apply, on cancelling $\dot{m}$ from all equations:

total mass $\quad V = 1 + L$ (6.15.1)

ammonia mass $\quad V\xi'' = 1 + L\xi'$ (6.15.2)

enthalpy $\quad Vh'' = h_1 + Lh' + q_r$ (6.15.3)

Fig. 6.15.1. Control surface drawn to include the reflux condenser and to cut the vapour stream leaving the top of the rectifying column and the ascending vapour and descending liquid streams at any level above the feed point.

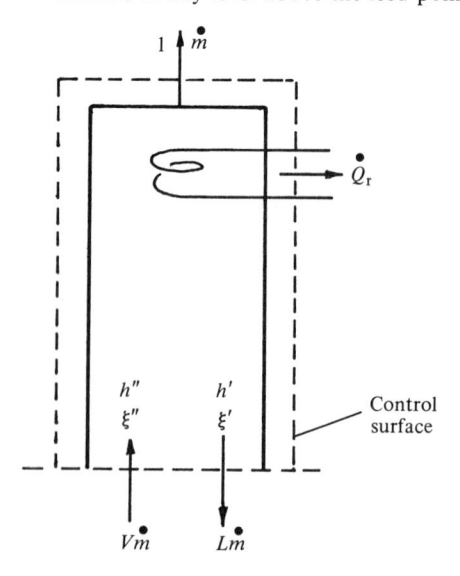

6.15 The rectifier

where $q_r = \dot{Q}_r/\dot{m}$. Solving for L from (6.15.1) and (6.15.2):

$$L = \frac{1-\xi''}{\xi''-\xi'} \tag{6.15.4}$$

Equation (6.15.3) may be written, using eq. (6.15.1).,

$$h_1 + q_r = (1+L)h'' - Lh' = h'' + L(h''-h') \tag{6.15.5}$$

Substituting for L from eq. (6.15.4):

$$h_1 + q_r = h'' + \frac{1-\xi''}{\xi''-\xi'}(h''-h') \tag{6.15.6}$$

Inspection of Fig. 6.15.2 shows that the right-hand side is the enthalpy ordinate at the point R where a straight line joining h', ξ', and h'', ξ'', produced cuts the axis $\xi = 1$. According to the left-hand side the ordinate of R has the value $h_1 + q_r$. Since h_1 and q_r have the same values whatever section

Fig. 6.15.2. Graphical illustration of eq. (6.15.6). The ordinate of point R has a value equal to the right-hand side of the equation, and hence gives the value of $h_1 + q_r$, from which q_r can be found. Since the control surface can cut the column at any arbitrary section, point R must be fixed for all sections of the column above the feed point. It is called the rectifier pole. Note that the states L and V are not necessarily in equilibrium with each other: they simply satisfy the mass and enthalpy balances.

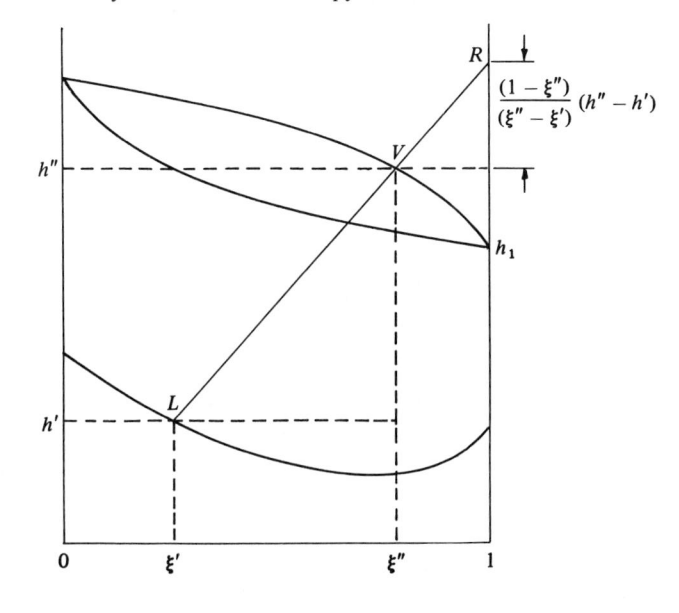

The vapour absorption system

Fig. 6.15.3. To ensure that the ascending vapour is warmer than the descending liquid at every section, the straight line R–V–L must be steeper than the local two-phase isotherm.

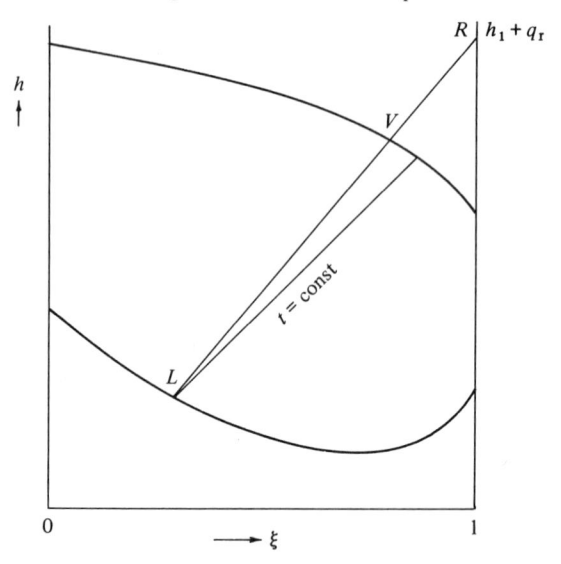

Fig. 6.15.4. The isotherms in the two-phase region begin as verticals at the water end, swing away from the vertical and then swing back to the vertical at the ammonia end. For ammonia in water it is found that provided that the actual rectifier pole R is higher than R_{min} obtained by continuing the isotherm through the feed state 7, the condition of warmer vapour than liquid is satisfied at all higher levels in the column.

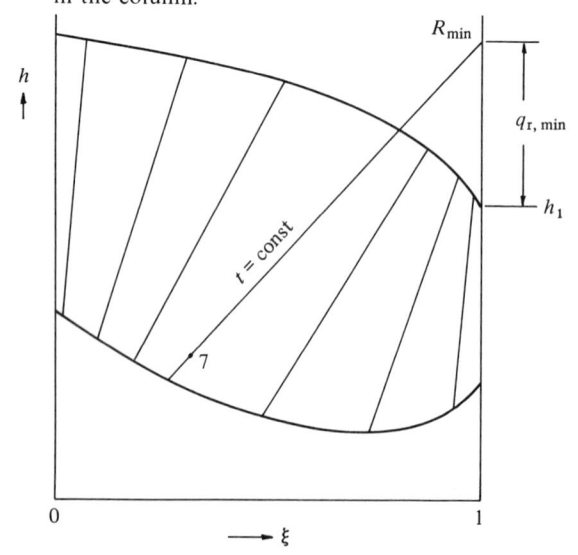

6.15 The rectifier

across the column is chosen as the boundary of the control surface, it follows that the position of R is fixed for all cross-sections of the column above the feed point. The point R is called the pole of the rectifier. Starting at any cross-section, the conditions at cross-sections higher up the column are obtained by rotating the straight line anticlockwise about R.

For rectification to proceed it is essential for the vapour at any cross-section to be warmer than the liquid. Consequently the straight line through R must always be steeper than the isotherm in the two-phase region through h', ξ', as shown in Fig. 6.15.3. The pole must therefore be sufficiently high on the $\xi = 1$ axis to ensure this at every cross-section, and this requirement fixes the minimum amount of reflux and of the rectifier heat.

This minimum value depends on the way in which the gradient of the two-phase isotherms changes, shown diagrammatically in Fig. 6.15.4. Obviously at $\xi = 0$ and $\xi = 1$ the isotherms are vertical. For solutions of ammonia in water, it is usually found that the condition that the vapour must always be warmer than the liquid is satisfied if the straight line through R is made steeper than the two-phase isotherm through the feed state 7. The position of the pole R can then be determined for the limiting case, and the minimum rectifier heat $q_{r,\min}$ can be read from the chart or found from eq. (6.15.6).

The condition of minimum reflux is an ideal one, which cannot be obtained in practice with a finite contact area in the column. The actual reflux heat may be expressed by a rectifier efficiency η_r:

$$\eta_r = \frac{q_{r,\min}}{q_r} \tag{6.15.7}$$

and values of η_r can be found for various arrangements of contact surface.

Example 6.15.1

For the conditions of example 6.14.1, determine the position of the rectifier pole for complete rectification at minimum reflux, the minimum rectifier heat transfer, the minimum heat transfer to the reboiler, and the maximum heat ratio.

From the composition of the two-phase mixture at 7:

$$L = \frac{1 - \xi''}{\xi'' - \xi'} = \frac{1 - 0.938}{0.938 - 0.365} = 0.1082$$

The right-hand side of eq. (6.15.5) is then:

$$h'' + L(h'' - h') = [1735 + 0.1082(1735 - 318)] \text{ kJ/kg}$$
$$= 1888 \text{ kJ/kg}$$

The vapour absorption system

and this is the position of the rectifier pole for minimum reflux on the $\xi = 1$ axis. The rectifier heat for minimum reflux is

$$q_{r,min} = h_R - h_1 = (1888 - 1472) \text{ kJ/kg} = 416 \text{ kJ/kg}$$

From example 6.14.1,

$$q_b - q_r = 1857 \text{ kJ/kg}$$

therefore the minimum reboiler heat is

$$q_{b,min} = (1857 + 416) \text{ kJ/kg} = 2273 \text{ kJ/kg}$$

From example 6.14.1 the specific refrigerating effect is

$$q_e = 1047 \text{ kJ/kg}$$

and the maximum heat ratio is therefore

$$\zeta_{max} = q_e/q_{b,min} = 1047/2273 = 0.461$$

In the above discussion the reflux condenser was taken as situated inside the rectifying section of the column. In many absorption refrigeration plants the reflux is brought back to the top of the column from the main condenser at a rate which can be controlled to give the required purity of the product vapour at point 1. This saves the cost of a separate heat exchanger, but of course requires increased surface in the main condenser. It also gives better control, because the control action obtained by modulating the rate of flow of the reflux is more rapid than that obtained by modulating the rate of flow of cooling water through a separate reflux condenser, with the associated thermal lags.

Example 6.15.2

If the rectifier efficiency is 0.7, determine the heat transfer to the reboiler and the heat ratio for the conditions of example 6.15.1.

The actual rectifier heat is

$$q_r = q_{r,min}/\eta_r = 416/0.7 = 594 \text{ kJ/kg}$$

and the reboiler heat becomes

$$q_b = (1857 + 594) \text{ kJ/kg} = 2451 \text{ kJ/kg}$$

The actual heat ratio is

$$\zeta = 1047/2451 = 0.427$$

In example 6.14.1 the solution leaving the heat exchanger was found to be a two-phase mixture, and this is often the case in aqua-ammonia absorption systems. It may happen however that when the circulation ratio λ is small and the heat exchanger is relatively inefficient the state of the solution at point 7 may be subcooled. In this event point 7 on the $h-\xi$ diagram has to lie on the two-phase isotherm produced. To determine the minimum height of the rectifier pole one has to find the two-phase isotherm which passes through point 7 by trial and error on the complete $h-\xi$ diagram. Provided that the subcooling is not very great, however, it is usually sufficiently accurate to draw a parallel through point 7 to the two-phase isotherm for saturated liquid of the same composition ξ_s.

6.16 The effects of incomplete rectification

In practice rectification is not complete and a small amount of water vapour remains mixed with the ammonia leaving the top of the column. Complete rectification would demand complete equality of temperature and partial pressures at the top of the column, which is not possible in a finite apparatus. It is not economical after a certain point to purify the ammonia further, because potentials for heat and mass transfer diminish towards the top of the column and each step of rectification requires an increasing contact area. Values of ξ_1 in practice are about 0.995–0.998.

With these values the effect of incomplete rectification on the minimum reflux is not large, since the isotherm through the feed state cuts the axis $\xi_1 = 1$ and the ordinate $\xi_1 = 0.998$, say, at very nearly the same point. There is, however, a noticeable difference of temperature between vapour delivered at $\xi_1 = 1$ and $\xi_1 = 0.998$. For a generator pressure of 12 bar, for example, the impure vapour has a temperature of 46 °C compared with 31 °C for pure vapour at that pressure, i.e. it is apparently superheated by 15 K. This difference in temperature can be used to control the reflux rate to give the desired degree of purity.

In the condenser the effect of the small amount of water is small, because it condenses with the ammonia, in which it is completely miscible, and passes straight through in solution.

In the evaporator the effect of water is more serious because it tends to accumulate there. The boiling solution gives off a vapour which is stronger in ammonia and the water is left behind in the evaporator. What happens then depends on the design of the evaporator. With a dry expansion or 'once through' type the velocity of the vapour may be sufficient to entrain the water and carry it back to the absorber. The situation is then very similar to that of lubricating oil in solution with Refrigerant 12. Most large plants however use shell-and-tube evaporators, in which there is a large bulk of ammonia boiling and there is not sufficient velocity immediately above the surface to carry the water droplets off. In these circumstances it is necessary to arrange a continual bleed of liquid from the evaporator to the absorber. The rate of this is determined by a simple mass balance on the amount of water coming in and the amount by which the mass fraction of water in the bulk liquid in the evaporator can be allowed to rise. The vapour leaving the evaporator can be assumed to be nearly anhydrous.

Water in the boiling solution in the evaporator raises the boiling temperature corresponding to a given pressure or lowers the pressure needed for a given evaporating temperature. If water were allowed to accumulate indefinitely the working of the plant would stop. The reduced temperature difference, corresponding to the amount of water allowed to remain in the evaporator, must be taken into account in the design.

Example 6.16.1
The composition of the two-phase mixture entering a flooded evaporator from the expansion valve is $\xi = 0.995$. Determine the rate of flow at which liquid from the evaporator must be purged to the absorber if the amount of water in the evaporator is to be maintained at not more than 2%.

The liquid in the evaporator is so strong that the vapour boiled from it may be assumed to be 100% pure. For unit mass entering the evaporator from the expansion valve, let the fraction z be purged directly to the absorber. The fraction z has the composition $\xi = 0.98$, and a balance on the water entering and leaving gives

$$(1 - 0.995) = (1 - 0.98)z$$

and

$$z = 0.005/0.02 = 0.25$$

the remaining 0.75 being evaporated and returned to the absorber as vapour.

Of the incoming flow, 0.25 is of course an undesirably large fraction of liquid to waste without doing any refrigeration, and it shows the importance in practice of obtaining pure vapour from the rectifier. If this could be pushed up to 0.999, z would fall to 0.05, which is reasonable. If such purity is not possible for some reason or another, it will be necessary to allow the amount of water in the evaporator to rise somewhat.

It is possible in theory to recover the refrigerating effect of the purged liquid by making it pass in heat exchange with the liquid coming from the

6.17 Further study of distillation: Keesom's method

condenser to the expansion valve. But the capital cost may be prohibitive, and in any event this eliminates the possibility of using the condensed liquid to superheat the suction vapour, as described in section 6.18.

6.17 Further study of distillation: Keesom's method

The equations developed in section 6.15 enable the rectifier and boiler heat rates to be determined when some simplifying assumptions are made. In this section a more detailed treatment is given with the aid of an algebra for the mixing of streams suggested by Keesom (1930) which has been applied to the aqua-ammonia absorption system by Ruhemann (1947). This method does not achieve results which cannot be obtained by the method of section 6.15, but it provides a more concise treatment which is easier to remember.

The adiabatic mixing of a stream 1 having mass flow rate $\dot{m}_1$, mass fraction ξ_1, specific enthalpy h_1, with a stream 2 ($\dot{m}_2, \xi_2, h_2$) to give a stream 3 ($\dot{m}_3, \xi_3, h_3$) is governed by the three equations:

$$\dot{m}_1 + \dot{m}_2 = \dot{m}_3 \tag{6.17.1}$$

$$\dot{m}_1 \xi_1 + \dot{m}_2 \xi_2 = \dot{m}_3 \xi_3 \tag{6.17.2}$$

$$\dot{m}_1 h_1 + \dot{m}_2 h_2 = \dot{m}_3 h_3 \tag{6.17.3}$$

representing the total mass balance, individual component balance and the

Fig. 6.17.1. Adiabatic mixing of two streams on the $h-\xi$ diagram is represented by addition of the streams at the top. Subtraction of two streams is shown in the lower line.

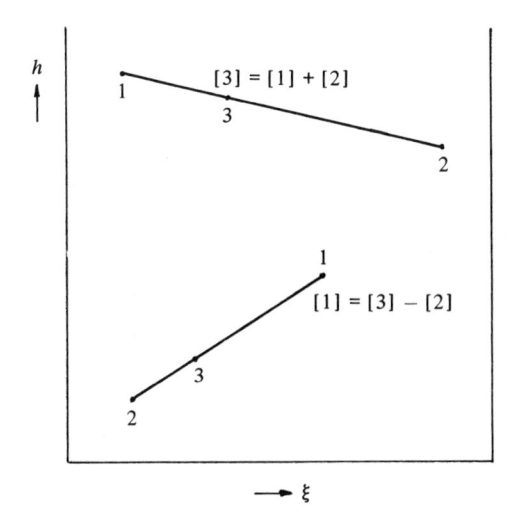

enthalpy balance, respectively. Symbolically one can write:

$$[1] + [2] = [3] \tag{6.17.4}$$

where [1] denotes the stream $(\dot{m}_1, \xi_1, h_1)$, etc., and in which mixing of the streams is regarded as addition. Keesom called the triplet $(\dot{m}, \xi, h)$ a 'phase', but in view of the well established use of the word to mean something different it will be called simply a 'stream'.

On the h–ξ diagram the point 3 representing the result of mixing the streams represented by the points 1 and 2 lies on the straight line joining 1 to 2 at a position determined by the ratio of the mass rates, as shown in Fig. 6.17.1. This is the rule of addition for the streams.

Equation (6.17.4) may also be written as:

$$[1] = [3] - [2] \tag{6.17.5}$$

which gives the rule for subtracting two streams. The resulting point 1 on the h–ξ diagram is obtained on the line 2 to 3 produced beyond 3, as shown in Fig. 6.17.1.

When mixing takes place with heat transfer at the rate $\dot{Q}$, say, one can write:

$$[1] + [2] + [\dot{Q}] = [3] \tag{6.17.6}$$

in which it is understood that the $\dot{Q}$ is taken into the third of the eqs. (6.17.1)

Fig. 6.17.2. The addition of streams with heat transfer.

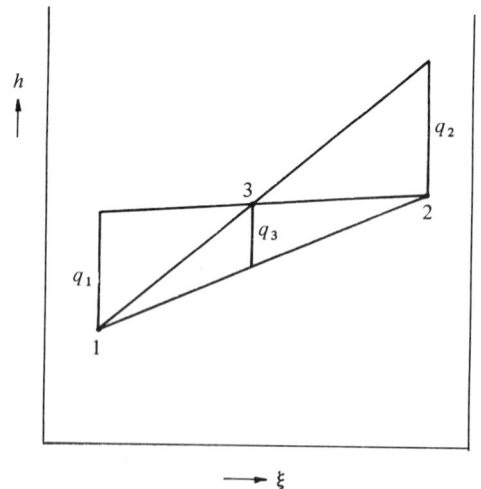

6.17 Further study of distillation: Keesom's method

to (6.17.3) only. The $[\dot{Q}]$ may be grouped with any enthalpy term to give:

$$[1 + q_1] + [2] = [3] \tag{6.17.7}$$
$$[1] + [2 + q_2] = [3] \tag{6.17.8}$$
$$[1] + [2] = [3 - q_3] \tag{6.17.9}$$

where:

$$\dot{m}_1 q_1 = \dot{m}_2 q_2 = \dot{m}_3 q_3 = \dot{Q} \tag{6.17.10}$$

The stream $[1 + q_1]$ is represented by a point on the h–ξ diagram at the composition ξ_1 with the specific enthalpy $(h_1 + q_1)$, and according to eq. (6.17.7) this point is in line with the points 2 and 3, as shown in Fig. 6.17.2. Similarly the points $1, (2 + q_2)$ and 3 are in line, as are $1, 2$ and $(3 - q_3)$.

Keesom's method can be explained in orthodox mathematical terms by noting that the stream, [1] say, (Keesom's 'phase') is in fact a column matrix with the elements:

$$\begin{bmatrix} \dot{m}_1 \\ \dot{m}_1 \xi_1 \\ \dot{m}_1 h_1 \end{bmatrix}$$

Using the rule for matrix addition by adding corresponding terms gives the matrix representing the right-hand sides of eqs. (6.17.1) to (6.17.3). The notation [1], [2] and [3] is then simply an abbreviated way of writing the stream matrices. The heat transfer rate is the matrix:

$$[\dot{Q}] = \begin{bmatrix} 0 \\ 0 \\ \dot{Q} \end{bmatrix} = \begin{bmatrix} 0 \\ 0 \\ \dot{m}_1 q_1 \end{bmatrix} = \ldots$$

with two zero elements indicating that the heat transfer has to appear only in the enthalpy equation.

Applying the method to a control surface enclosing the whole of the generator, i.e. reboiler, stripping and rectifying columns and reflux condenser, the equation is:

$$[7] + [\dot{Q}_b] = [1] + [8] + [\dot{Q}_r] \tag{6.17.11}$$

with the notation of Fig. 6.11.2. On grouping the heat rates with streams:

$$[7] = [1 + q_{r,1}] + [8 - q_{b,8}] \tag{6.17.12}$$

in which $q_{r,1}$ denotes the rectifier heat referred to unit mass rate of stream 1, and $q_{b,8}$ denotes the boiler heat referred to unit mass rate of stream 8. On the h–ξ diagram the points 7, $(1 + q_{r,1})$ and $(8 - q_{b,8})$ are in line as shown in Fig. 6.17.3. The boiler heat referred to unit mass rate at 1 is given by:

$$[7] = [1 + q_{r,1} - q_{b,1}] + [8] \tag{6.17.13}$$

and the points 8, 7 and $(1 + q_{r,1} - q_{b,1})$ are in line.

The point R at the ordinate $(h_1 + q_{r,1})$ at the composition ξ_1 which is now taken as less than unity (for incomplete rectification) is the rectifier pole. The point S at the ordinate $(h_8 - q_{b,8})$ is called the stripping pole. The overall balance of mass, components and enthalpy for the generator is expressed by the fact that the feed state 7 must lie on the straight line joining the rectifier pole and the stripping pole. This line is the main operating line for the whole column.

For the control surface shown in Fig. 6.15.1 with a section cutting the rectifier column at any level above the feed point and below the reflux condenser:

$$[V] = [1] + [L] + [\dot{Q}_r] \tag{6.17.14}$$

or:

$$[V] - [L] = [1 + q_{r,1}] \tag{6.17.15}$$

The right-hand side of this equation is constant and is represented by the point R, the rectifying pole. Hence the points representing V and L must lie on a straight line passing through R, whatever section in the column is

Fig. 6.17.3. The operating line S–7–R for the whole generator expresses the overall balances of mass and heat and enthalpy.

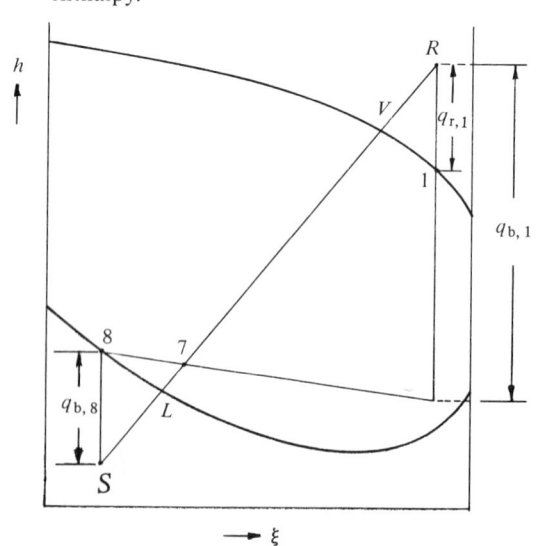

6.17 Further study of distillation: Keesom's method

taken. The line joining L, V and R is a cross-section line for the rectifier.

A similar treatment for the boiler and any part of the stripping column below the feed level gives:

$$[L] - [V] = [8 - q_{b.8}] \tag{6.17.16}$$

Fig. 6.17.4. Going up the rectification column from the feed point 7 the cross-section lines rotate about the pole R. Going down the stripping column, the cross-section lines rotate about the pole S.

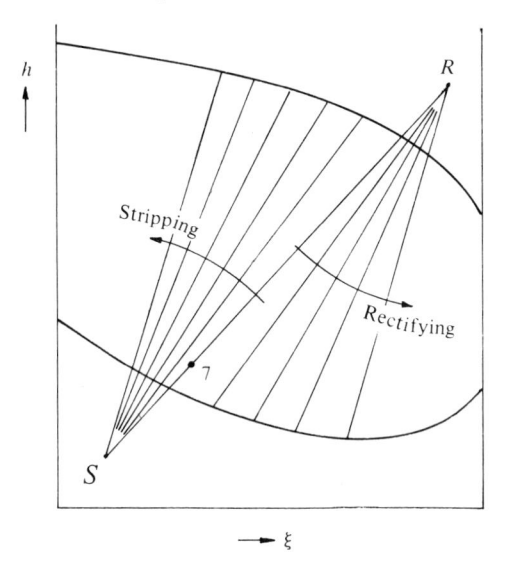

Fig. 6.17.5. An equilibrium stage or theoretical plate is conceived as an adiabatic mixing box in which the entering streams V_a and L_b give two leaving streams V_b and L_a which are in equilibrium with each other.

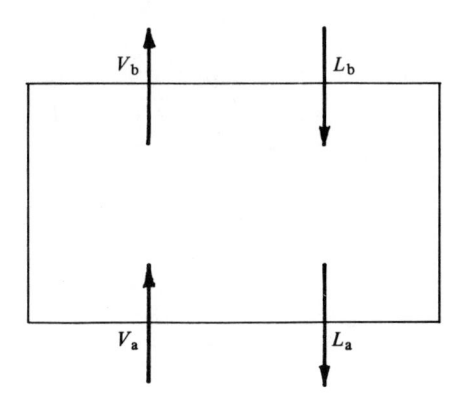

The vapour absorption system   460

Fig. 6.17.6. The relation between the states of an equilibrium stage in a rectifying column shown diagrammatically on the h–ξ diagram.

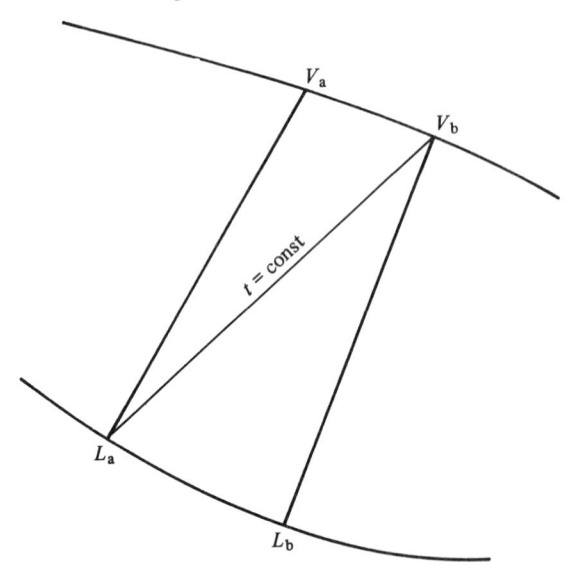

Fig. 6.17.7. The Ponchon–Savarit construction for finding the number of theoretical plates required in a rectifying column.

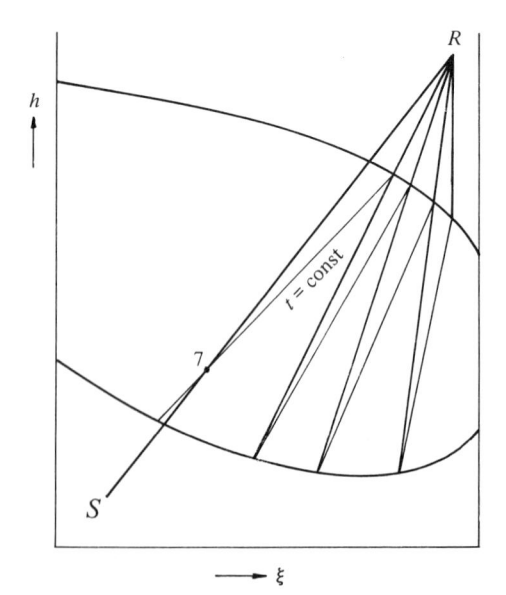

6.17 Further study of distillation: Keesom's method

which shows that the points V, L and the stripping pole S must be in line. This is a cross-section line for the stripping column.

For stripping and rectification to take place, the cross-section lines must be steeper at every section than the isotherms in the two-phase region, i.e. the vapour must everywhere be warmer than the liquid. The rotation of the lines about the poles is illustrated in Fig. 6.17.4.

A useful concept for distillation columns and similar apparatus involving mass transfer is that of the equilibrium stage, also called a theoretical plate. Referring to Fig. 6.17.5, adiabatic mixing occurs between the entering streams V_a and L_b such that the leaving streams V_b and L_a are in equilibrium with each other, i.e. have the same temperature and have compositions represented by the ends of the two-phase isotherm on the bubble point and dew point curves. Consequently the relation between the four states is as shown on the $h-\xi$ diagram in Fig. 6.17.6.

The number of equilibrium stages to achieve a desired purity at the top of the rectifying column can be found by the construction shown in Fig. 6.17.7, known as the Ponchon–Savarit method. Starting from a section immediately above the feed point, the number of equilibrium stages needed can be counted, shown as three in this case. A similar construction applies to the stripping section to find the number of stages needed to reach the desired composition of liquid at the bottom.

An approximation to the equilibrium stage is the bubble-cap plate. A bubble-cap is shown diagrammatically in Fig. 6.17.8. Vapour rising from a

Fig. 6.17.8. Diagrammatic arrangement of a bubble-cap. The rising vapour is forced under the surface of the liquid standing on a plate. Liquid overflows to a lower plate. In actual construction there would be many such bubble-caps or 'mushrooms' on each plate.

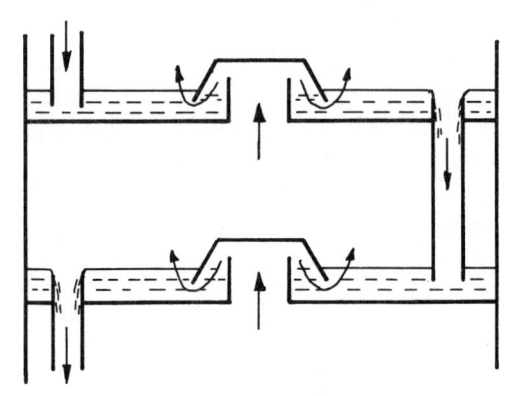

lower plate has to bubble through the liquid standing on the plate and passes up to the next higher plate. With enough contact between the phases, the vapour leaving could be in equilibrium with the liquid leaving, but this does not happen in practice. A plate efficiency is defined by the ratio:

$$\eta_{plate} = N_{theor}/N \qquad (6.17.17)$$

where N_{theor} is the number of equilibrium stages required, according to the construction of Fig. 6.17.7, and N is the actual number of stages.

An actual plate in a distillation column consists of many bubble-caps like Fig. 6.17.8.

Another form of construction is the packed column in which liquid trickles down over the surface of glass or ceramic spheres or rings, various types of which are available commercially. The vapour rises through the interstices. The performance of packed columns is determined experimentally and expressed as a height equivalent to one equilibrium stage.

6.18 Vapour to liquid heat exchange

Since the vapour from the evaporator is eventually warmed up to absorber temperature in the absorber itself, it is clearly advantageous to warm it up before it gets there by using it to subcool the liquid from the condenser in a counterflow heat exchanger. The calculation for the extra specific refrigerating effect is exactly the same as that given in Chapter 3 for a heat exchanger in a vapour compression cycle.

It was found there that a heat exchanger is advantageous for some refrigerants and not for others, and that ammonia is one of the others for which a heat exchanger is not beneficial, apart from the effect of increasing volumetric efficiency. In the absorption system, however, the heat exchanger is always beneficial, because the vapour will be warmed up to absorber temperature in any event.

Whether it is used in practice depends on the expected return as a saving of boiler heat compared with the cost of the exchanger. Because of the low heat transfer coefficient on the vapour side, especially at low evaporating pressure, the exchanger is rather bulky and expensive, and in many cases does not justify its cost.

6.19 The Platen–Munters system

In this method of using the absorption cycle, invented by Baltzar von Platen and Carl Munters in the 1920s, the solution pump and the expansion valves are eliminated by arranging for the whole circuit to be at the same pressure. The system can thus be hermetically sealed and, without

6.19 The Platen–Munters system

any moving parts, it becomes very suitable for use in small refrigerators for domestic use. It is also valuable when no supply of electric power to drive a compression system is available.

The equalisation of pressure is obtained by introducing hydrogen gas into the evaporator and absorber. The liquid ammonia in the evaporator does not then boil, because the total pressure on it is greater than its vapour pressure and no bubbles can form below the surface of the liquid. But the

Fig. 6.19.1. The arrangement of the components in one form of the Platen–Munters absorption system.

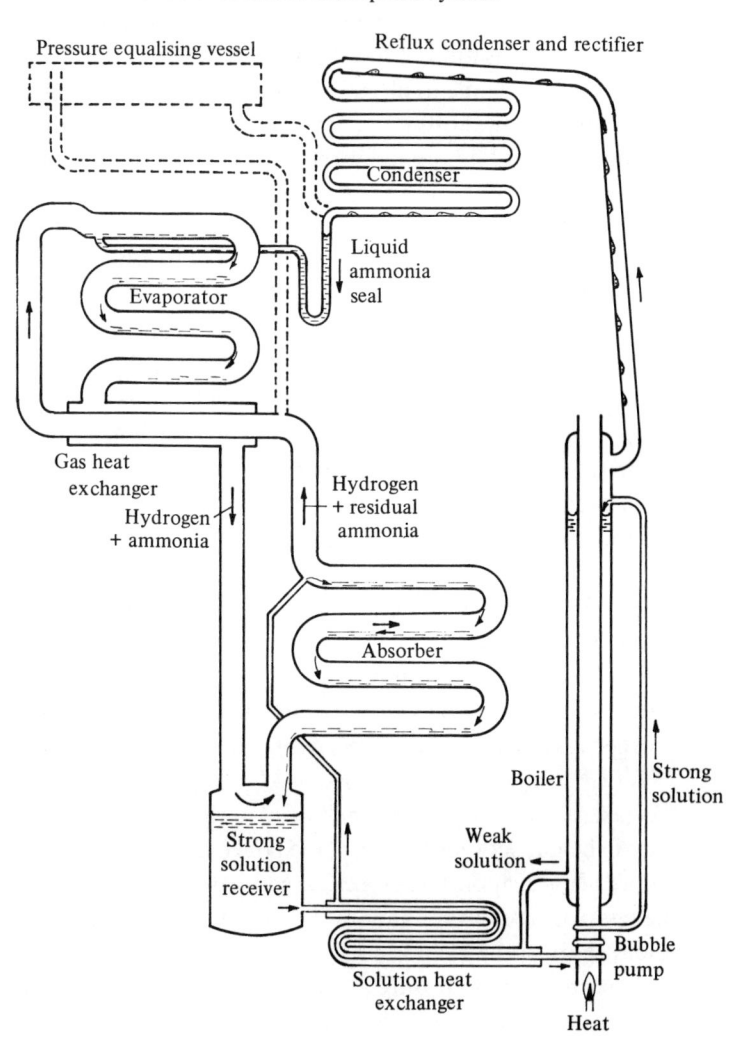

The vapour absorption system

ammonia evaporates into the hydrogen gas, just as a pool of water evaporates into the atmosphere, provided that the hydrogen is not saturated with ammonia. The vapour is carried away from the surface by diffusion, and the Platen–Munters refrigerator is therefore sometimes called the diffusion–absorption system. In the absorber the ammonia is washed out of the hydrogen by the solution and the hydrogen returns to the evaporator. Figure 6.19.1 shows the arrangement of the components of the system in one of its forms.

The tall vertical boiler acts as a combined boiler and stripping section of the generator, the strong solution feed coming in at the top. A strong vapour is delivered to a combined rectifier section and reflux condenser in which partial condensation of the vapour takes place in counterflow with the reflux, and enrichment of the vapour occurs. This vapour is then condensed and it flows to the evaporator via a U-tube liquid seal. In the evaporator the liquid ammonia evaporates into the stream of hydrogen. The hydrogen plus ammonia stream leaving the evaporator passes through a gas heat exchanger which cools the hydrogen coming from the warm absorber. In the absorber the ascending current of hydrogen and ammonia meets descending weak solution which dissolves much of the ammonia, and the hydrogen passes back to the evaporator, the strong liquid going to the boiler. It is carried to the top of the boiler by the action of the bubble-pump, which works on the principle of the coffee percolator. Heat applied to the pump coil causes the formation of bubbles, and the density of the strong solution in the vertical riser is reduced so that the solution is forced to the top by the static head of the solution in the absorber vessel.

The circulations in the system are produced solely by gravity and density differences as follows:

(i) Hydrogen circulates between the absorber and the evaporator because of the greater density of the ammonia-rich gas column, i.e. the one descending from the evaporator.

(ii) Strong liquid rises to the top of the boiler because the liquid head in the absorber vessel (strong liquid receiver) is greater than the head of liquid plus vapour bubbles after the pump.

(iii) Weak solution flows from the boiler to the absorber because of the difference in height between the top of the boiler and the top of the absorber.

A feature of the evaporative process as compared with a boiling process is that the temperature of the evaporating liquid does not remain constant. The coldest part of the evaporator is the end where the hydrogen enters, because here the partial pressure of ammonia in the hydrogen is least. This effect may be used to provide two temperatures in a refrigerator, for freezing

6.20 Heating and cooling media for absorption systems

and ice making say, and for chilled storage of food. Calculation of the evaporating temperature is not so straightforward as in a standard absorption system. The situation is similar to that of the water on the wick of a wet bulb thermometer. If there were no heat transfer from external things, the liquid ammonia would come to a wet bulb temperature for ammonia in hydrogen. This depends on the temperature of the gas and the partial pressure of the vapour in it. The effect of external heat transfer is to raise the evaporating temperature, in the same way that radiation to the wick of a wet bulb thermometer raises its temperature. The temperature which results is an equilibrium one at which the rate of heat transfer from the exterior balances the rate at which enthalpy is carried away by the diffusion of vapour and convection from the surface. As evaporation proceeds and the hydrogen becomes richer in ammonia vapour, the partial pressure of ammonia increases and the potential for diffusion decreases, so the temperature of the evaporating ammonia rises.

Figure 6.19.1 shows a vessel above the evaporator, connected to it and also to the condenser. Through this vessel pressure equalisation between the condenser and evaporator can occur. Suppose that at steady operating conditions the pressures are balanced with a liquid seal in the U-tube. Conditions then change because the ambient temperature around the condenser rises and the condensing pressure increases. If the pressure equalising connection were not there, the increased condensing pressure would blow out the liquid from the U-tube and uncondensed ammonia would pass to the evaporator, raising the partial pressure there and raising the evaporating temperature. With the pressure equalising connection, the pressures remain the same in condenser and evaporator and the liquid seal remains. The uncondensed ammonia in the condenser passes to the equalising vessel and displaces hydrogen from there into the evaporator, raising the total pressure to balance the increased pressure in the condenser. Naturally some ammonia is displaced with the hydrogen, but only at the beginning. There is not a continuous leakage of ammonia vapour from the condenser.

The Platen–Munters system was at one time widely used for domestic refrigeration, but it has largely gone out now except for use where there is no electricity, or in places such as hotel bedrooms where its silent operation is an advantage.

6.20 Heating and cooling media for absorption systems

Early ammonia absorption machines were heated directly by coal, but heating by steam was eventually used because of the better control obtainable. Large lithium bromide systems are usually driven by steam or hot water, but many small ones are directly fired by gas or oil. Small domestic refrigerators of the Platen–Munters type may be heated by gas, paraffin (kerosene), or by electricity.

The temperature of the available heating medium is a basic design parameter because it fixes the mass fraction of the solution leaving the boiler for a given condensing temperature and pressure. The solution leaving the boiler is at its bubble point, and the higher the temperature the weaker in refrigerant it is. In order to work a practical system there must be a difference between the mass fractions of the strong and weak solutions. For an aqua-ammonia system the minimum economic difference is about 0.08. For given condensing and absorbing temperatures the boiler temperature then determines the lowest temperature which can be attained in the evaporator, as shown by the following example.

Example 6.20.1
For a temperature of solution of 110 °C leaving the reboiler of an aqua-ammonia plant, operating at absorbing and condensing temperatures of 30 °C and 40 °C, respectively, what is the lowest practical evaporating temperature?

From Fig. 6.8.2., the equilibrium mass fraction of ammonia in the solution leaving the boiler is:

$$\zeta_w(110\,°C, 15.54\text{ bar}) = 0.35$$

The minimum mass fraction in the strong solution is therefore:

$$\zeta_s = 0.35 + 0.08 = 0.43$$

The equilibrium pressure in the absorber for a temperature of 30 °C and mass fraction 0.43 corresponds to a saturation temperature of ammonia of -12.5 °C, and this is the lowest practical evaporating temperature.

To achieve a lower evaporating temperature, the temperature of the heating medium would have to be raised. If this is not possible, the lower temperatures can be reached by the use of a two-stage system, but this of course is much more expensive.

Absorption systems dissipate more than twice as much heat as vapour compression systems, but it is not normally economic to provide them with twice as much cooling water. The same water has to be used for cooling both the absorber and the condenser by passing it through them in series. The usual practice is to pass the water through the absorber first, because provided that the temperature of the heating medium is high enough, a lower absorber temperature usually gives better performance than a lower condensing temperature. The question should however be decided by calculation for each design. There is also the effect on capital cost to consider, remembering that the condenser and generator shells are very thick and expensive, and a reduction of these can be effected by passing the cooling water through the condenser first, without affecting the thickness of the evaporator and absorber shells which are decided by the pressure of ammonia at standstill temperature.

The water for cooling large systems is usually brought back to temperature in a cooling tower. For compression plant these are designed for a range, i.e. temperature of water entering minus temperature of water leaving, of 5–6 K, with an approach, i.e. temperature of water leaving minus wet bulb temperature of the air, of about 4 K. For absorption plant one could provide towers of twice the contact area and preserve the range with twice the water flow rate, but this is too expensive in practice. It is better to receive a reduced flow of warmer water at the top of the tower and to take advantage of the increased potential for heat and mass transfer. When this is done the cooling towers for an absorption plant need be only about 50% larger than for a compression plant of the same duty.

6.21 Performance of absorption systems

The heat ratio of an absorption system, like the coefficient of performance of a compression system, depends on the operating conditions. As a rule, however, the heat ratio does not fall off so rapidly with falling evaporating temperature as does the coefficient of performance of a compression system. An aqua-ammonia absorption system having a refrigerating capacity of about 100 kW would have a heat ratio of 0.55 at an evaporating temperature of 0 °C, a condensing and absorber temperature of 35 °C, and a boiler temperature of 130 °C. At an evaporating temperature of −20 °C the heat ratio would fall to about 0.4, i.e. 73% of the value at 0 °C. For an ammonia compression machine of similar size with condensation at 35 °C, the coefficient of performance referred to electrical power to the motor would be about 4.1 with evaporation at 0 °C, and 1.9 with evaporation at −20 °C, i.e. 46% of the value at 0 °C.

Comparisons of operating costs between the two systems tend therefore to be more favourable to the absorption system at lower evaporating temperatures. So does the comparison of capital cost. Especially is this so when a high-temperature heating medium is available, since it is often possible to achieve in a single-stage absorption system an evaporating temperature which would require two stages of compression.

The heat ratio of an absorption system is not directly comparable with the coefficient of performance of a compression system, because they are basically two different ratios. Work is more valuable than heat and costs more to generate. To bring the two ratios to a comparable basis one must take into account the heat transfer in the steam plant at the power station where the electrical work used to drive the motor of the compression machine is produced. Assuming an average efficiency of electrical work generation and transmission of 0.3 say, the equivalent heat ratio for the compression machine at $0\,°C$ evaporating temperature would be $(0.3)(4.1) = 1.23$. So far as the use of fuel goes, this is the figure to compare with the heat ratio 0.55 of the absorption system.

Even so, the heat ratio of the absorption system does not come up to the heat ratio of the combined compression system and power station. This happens because there is more thermodynamic irreversibility in the operation of the absorption system. In the first place the absorption system cannot utilise the high temperature available from the combustion of fuel, as can the steam cycle in the power station. The large difference of temperature between the temperature in the combustion chamber and the temperature in the reboiler of the absorption machine is wasted, because a heat engine could be operated between them to deliver work. The thermal efficiency of low-pressure steam cycles is low for the same reason. Secondly, there are irreversibilities in the absorption system, principally owing to departures from equilibrium during the absorption and regeneration processes. On the other hand, there are irreversibilities in the compression system, the throttling process through the expansion valve, the compression process, and the conversion of electrical work to mechanical work in the motor.

Expressions were given in section 1.12 for the coefficient of performance and heat ratio of ideal refrigerators. When the absorption system, taking into account the actual reboiler temperature, and the compression system are compared with reference to these values, the difference between them is not so large as that cited above.

Applying eq. (1.12.6) for the work operated refrigerator with:

$$T_1 = (-20 + 273.15)\,K = 253.15\,K$$
$$T_2 = (35 + 273.15)\,K = 308.15\,K$$
$$\varepsilon_C = 253.15/55 = 4.60$$

The actual compression refrigerator cited above achieves $1.9/4.60 = 0.41$ of this.

6.22 Other refrigerant–absorbent combinations

Applying eq. (1.12.10) for the heat operated refrigerator with T_1 and T_2 as before and:

$$T_3 = (130 + 273.15)\,\text{K} = 403.15\,\text{K}$$

$$\zeta_C = \frac{253.15}{55}\,\frac{95}{403.15} = 1.08$$

The actual absorption refrigerator cited above achieves $0.4/1.08 = 0.37$ of this.

As the difference between the temperatures of condensation and evaporation increases, the absorption system becomes relatively better with regard to the ideal.

The large irreversibility associated with the difference of temperature between the combustion chamber where the fuel is burnt and the reboiler of the absorption system can be considerably reduced by combining the refrigerator with a high-pressure steam power cycle. The steam is generated at high pressure and expanded in turbines (designed for working at high exhaust pressure) before being condensed in the heating pipes in the reboiler of the absorption system. By this means useful power can be developed. The scheme is excellent in principle, but is not always applicable because there may not be much demand for power, or it may be difficult to match the outputs of turbines and refrigerator to fluctuating demands.

It should go without saying that when electrical power is used to drive an absorption system, as in some small domestic refrigerators, the result is very wasteful in fuel. But the largest irreversibility occurs then in the electric heater, not in the absorption system. Using electric power to provide heating of any kind, though it may be convenient, is a thermodynamic sin more appropriate to a primitive age of technology than to an advanced one.

6.22 Other refrigerant–absorbent combinations

Only ammonia–water and water–lithium bromide are used at present to any significant extent. For temperatures below those obtainable by lithium bromide and water, 5 °C say, ammonia and water has to be used, although it has several disadvantages. Firstly there are those associated with the use of ammonia itself, the same as in the compression system. Secondly there are the unfavourable properties of the solution, in particular the large heat of solution which has to be provided in the reboiler over and above the enthalpy of evaporation; and the volatility of the water which requires an elaborate rectification process which increases the reboiler heat by the amount rejected in the reflux condenser.

A large number of other combinations have been examined, some of which show promise, but which have not yet been adopted commercially. A number of properties have to be taken into account when judging the suitability of a refrigerant–absorbent combination. One which is generally considered of first rate importance is a large negative deviation from Raoult's Law, as exhibited by both ammonia–water and water–lithium bromide solutions. It means that the vapour pressure of the solution is much less than that of the refrigerant, and consequently a greater difference in temperature between the absorber and evaporator can be obtained for a given solution strength. It also implies that the average difference in equilibrium vapour pressure in the absorber, which is the driving potential for absorption, can be large and so cause vigorous absorption which proceeds very nearly to equilibrium at the absorber outlet.

On the other hand it has to be remembered that the conditions in the boiler are reversed. A large negative deviation from Raoult's Law means that the refrigerant is more difficult to drive out of solution, i.e. that a higher temperature is needed. This does not matter when the system is driven by a high-temperature fluid such as steam, or by direct combustion, but it has to be considered when low-grade heat, e.g. from solar panels, is to be used. Ideally one would like to have a large negative deviation from Raoult's Law in the absorber, and ideal solution behaviour or even a positive deviation in the boiler, but no solutions exhibit this kind of behaviour to any marked extent.

Furthermore a large negative deviation from Raoult's Law is associated with a large enthalpy of solution, and this is undesirable as already noticed. It represents heat which has to be supplied at boiler temperature and then simply rejected at the absorber temperature without doing anything for the refrigerating effect.

Low volatility of the absorbent is of course important in order that rectification can be eliminated. Consequently much interest at present centres around absorbents which are solids, like lithium bromide, or liquids which have high boiling points. Of the first kind, methanol may be mentioned as refrigerant with lithium bromide or lithium bromide with zinc bromide in solution as the absorbent. Of the second kind, R22 as the refrigerant with dibutylphthalate (a chemical used in the plastics industry under the name DBP) as absorbent has been tried.

Further information about unusual combinations of substances is given by Niebergall (1949, 1959), Dannies (1951) and Badylkes and Danilov (1966). Several references to various mixtures are given in the current ASHRAE Handbook, Fundamentals, 1977.

6.23 Resorption systems

With complete rectification of the vapour leaving the generator of an absorption system the refrigerant condenses in the condenser at a constant temperature and boils in the evaporator at a constant temperature exactly as in the vapour compression system. These isothermal processes are well matched to an external régime in which things to be cooled are at a fixed temperature and the fluid to which heat is rejected is at a constant temperature.

It often happens that heat is to be accepted from bodies or fluids undergoing a fall of temperature and that heat is to be rejected to fluids undergoing a rise in temperature. In these circumstances one would like the refrigerant to follow these changes because the better the matching of the internal and external régimes the more efficient is the plant, as was discussed in section 1.13.

A variation of temperature of the refrigerant whilst accepting or rejecting heat can be arranged in an absorption system by using the two-component solution as the refrigerant. As was seen in section 6.10, the equilibrium temperature of the solution rises whilst the more volatile component boils away at constant pressure. Similarly the equilibrium temperature of the solution falls whilst the more volatile component is absorbed at constant pressure.

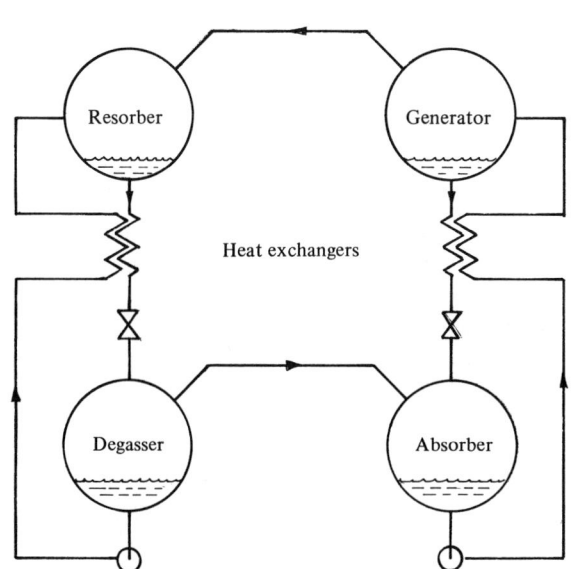

Fig. 6.23.1. The components of a simple resorption system.

These ideas are used in the resorption system as described by Altenkirch (1913), and shown diagrammatically in Fig. 6.23.1 in its simplest form.

Referring to the figure, unrectified vapour leaves the generator and passes to a vessel called the resorber in which the vapour is absorbed by a fairly strong solution, i.e. stronger than the solutions circulating between the generator and the absorber. Absorption of the vapour takes place to give a still stronger solution. During this process heat is rejected to cooling water and the temperature of the solution falls over a range which is to be matched with that of the rise in temperature of the cooling water. The resorber corresponds to the condenser in the normal system.

The solution from the resorber passes in counterflow heat exchange with cold solution and is throttled through the expansion valve into a vessel called the degasser which corresponds to the evaporator in the normal system. In the degasser the solution boils and its temperature rises over a range which is to be matched with that of the fluid being cooled. The vapour formed goes to the absorber in the usual way. The weakened solution remaining is pumped through the heat exchanger back to the resorber.

To achieve the full benefits of the resorption system some care is needed in the design of the resorber and the degasser. The simple arrangement for bulk boiling in the degasser shown in Fig. 6.23.1 would not suffice because it does not provide for separation of the liquid at different temperatures as it gives off vapour. A counterflow arrangement is required in which some approach to equilibrium can be made at each stage, rather like that in the stripping and rectifying columns. The same principle can be extended to the absorber, the cooling fluid for which rises in temperature.

A number of resorption systems were built in the 1930s by G. Maiuri using ammonia and water, which achieved remarkable heat ratios of about unity, but the method has not been generally used. Further developments have been made by Dannies.

According to Altenkirch, Osenbrück proposed the resorption principle for use with a compressor. In Fig. 6.23.1 the combination of absorber, pump, heat exchanger and generator performs exactly the same function as a mechanical compressor, i.e. takes in low-pressure vapour and delivers high-pressure vapour. So far as is known no systems of this type have been built, but somewhat similar principles are employed in the mixed refrigerant cascade systems for the liquefaction of natural gas which are described in Chapter 8.

When the resorber and degasser operate with gradients of temperature along them as described above, a further improvement can be introduced. The solution going to the degasser can be cooled before entering its

6.23 Resorption systems

expansion valve by heat transfer in counterflow to the boiling solution in the degasser. Thus a part of the refrigeration effect is used to subcool the liquid to the expansion valve. In a circuit using a pure fluid this would be pointless because the extra refrigeration obtained by subcooling the liquid is exactly equal to the heat transfer from the boiling refrigerant. This is still true in a solution cycle, but with an important difference. The subcooling of the solution is performed by heat transfer from boiling solution at the warm end of the degasser which is thus recovered at the cold end. The average boiling temperature is lowered. In the compression cycle of Osenbrück this means that for the same expenditure of work in the compressor a lower average degassing temperature can be obtained, or alternatively for the same average degassing temperature the suction pressure can be higher and the compression work reduced. Application of the same method in the resorber enables either the average temperature of resorption to be raised, or for the same temperature of resorption the compression work can be reduced.

In an absorption–resorption system like that of Fig. 6.23.1, the same principle can be introduced into the generator and absorber. In such a machine it is possible to achieve a complete match between the temperatures of the fluids in the external and internal régimes, according to the requirements of the Lorenz cycle, treated in section 1.13.

Finally it may be remarked that the combination of a generator and an absorber can be used for power production by using the high-pressure vapour to drive an expansion engine or turbine which exhausts to the low pressure in the absorber. This is then equivalent to applying the resorption principle to the boiler and condenser part of an ordinary Rankine vapour cycle. Compared with the pure fluid system, however, it confers the ability of accepting and rejecting heat over temperature ranges which match those in the external régime.

7

Ideal gas cycles

7.1 The ideal gas state

The refrigerant employed in the systems described in this chapter is assumed to be in the ideal gas state, or as near to it as required to enable calculations to be made using the ideal gas equation. Before discussing the cycles themselves, a few remarks will be made on the general question of reducing the temperature of a gas by expansion.

A substance is said to be in the ideal gas state when it has the equation of state:

$$pv = RT \qquad (7.1.1)$$

in which p is the pressure, v is the specific volume, T is the absolute thermodynamic temperature and R is a constant which is characteristic of the gas. No substance actually exists in the ideal gas state at a finite pressure, but as the pressure tends towards zero, the ratio pv/T tends to the value R. For most practical purposes the gases oxygen, nitrogen, air, hydrogen, helium and the other rare gases may be treated as ideal gases when the pressure is only a few atmospheres and the temperature is not too far below normal ambient temperature. For brevity a substance in the ideal gas state will be called simply an ideal gas.

It may be shown from the definition of an ideal gas and the Second Law of Thermodynamics that the specific internal energy u and the specific enthalpy h of an ideal gas are functions only of the thermodynamic temperature and are independent of pressure. For infinitesimal changes:

$$du = c_v dT \qquad (7.1.2)$$

and:

$$dh = du + d(pv)$$
$$= c_v dT + R dT$$
$$= c_p dT \qquad (7.1.3)$$

474

7.1 The ideal gas state

where c_v is the specific heat capacity at constant volume and c_p is the specific heat capacity at constant pressure. Clearly:

$$c_p - c_v = R \tag{7.1.4}$$

The specific heat capacities themselves are functions of temperature, but over the temperature ranges to be considered in this chapter the variation will be neglected. Equations (7.1.2) and (7.1.3) may then be integrated to give:

$$u_2 - u_1 = c_v(T_2 - T_1) \tag{7.1.5}$$

$$h_2 - h_1 = c_p(T_2 - T_1) \tag{7.1.6}$$

For approximate calculations on air the values may be taken as:

$$c_v = 0.718 \text{ kJ/kg K} \tag{7.1.7}$$

$$c_p = 1.005 \text{ kJ/kg K} \tag{7.1.8}$$

$$R = 0.287 \text{ kJ/kg K} \tag{7.1.9}$$

To reduce the temperature of an ideal gas it is therefore necessary to reduce its internal energy and enthalpy. It is not possible to do this by expanding an ideal gas in steady flow from a high pressure to a low one through a throttling valve, because in such a process carried out adiabatically there is no change in specific enthalpy. The cooling effect which takes place with real gases, known as the Joule–Thomson effect, is a measure of the departure from the ideal gas state. It becomes significant as a means of cooling only at high pressures, as explained in Chapter 8.

One way of reducing internal energy and enthalpy is to allow the gas to do work in an expansion process, e.g. by letting it push on a piston in a cylinder, insulated so that heat transfer to the gas is largely prevented. The work done by the gas, W, is equal to the decrease in internal energy of the mass m of gas, and:

$$W = mc_v(T_1 - T_2) \tag{7.1.10}$$

When the expansion is reversible and adiabatic, the final specific volume and temperature are given by the expressions:

$$p_1 v_1^\gamma = p_2 v_2^\gamma \tag{7.1.11}$$

$$T_1 v_1^{\gamma-1} = T_2 v_2^{\gamma-1} \tag{7.1.12}$$

$$T_1/p_1^{(\gamma-1)/\gamma} = T_2/p_2^{(\gamma-1)/\gamma} \tag{7.1.13}$$

where:

$$\gamma = c_p/c_v \tag{7.1.14}$$

Ideal gas cycles

The adiabatic expansion may also be carried out in steady flow, using a turbine as the means for accepting work from the gas. Applying the steady-flow energy equation (3.2.3), in the absence of changes of kinetic and potential energies, for adiabatic flow, the shaft power delivered by the turbine is then:

$$\dot{W} = \dot{m} c_p (T_1 - T_2) \qquad (7.1.15)$$

where $\dot{m}$ is the mass flow rate.

The relations between the temperature, pressure and specific volume for the case of reversible and adiabatic expansion in steady flow, are the same as eqs. (7.1.11), (7.1.12) and (7.1.13).

Another way of reducing the temperature of a gas is to allow it to expand in a nozzle so as to acquire a high velocity. Assuming the initial velocity to be zero, the steady-flow energy equation gives, for adiabatic flow:

$$h_1 = h_2 + \tfrac{1}{2} \mathscr{V}_2^2 \qquad (7.1.16)$$

where $\mathscr{V}_2$ is the velocity acquired after the expansion. Hence, using (7.1.6):

$$T_1 - T_2 = \tfrac{1}{2} \mathscr{V}_2^2 / c_p \qquad (7.1.17)$$

for adiabatic flow. If the expansion is reversible as well, the relations between temperature, pressure and specific volume are the same as eqs. (7.1.11), (7.1.12) and (7.1.13) again.

At first sight this seems a very simple way of obtaining a low temperature, but there is a snag, unfortunately. If a thermometer bulb is placed in the stream of high-velocity gas at the temperature T_2, the thermometer does not indicate the temperature T_2. Its reading in fact is observed to be quite close to T_1. The reason is that the high-velocity stream approaching the bulb is decelerated in the boundary layer around it and comes back to zero velocity and the temperature T_1 again. It does not come back to T_1 exactly because of heat conduction and viscous effects in the boundary layer, but it is near enough to render the high-velocity stream useless for cooling anything. (An exceptional instance occurs in vortex flow, described later, where some useful reduction of temperature is found.) The stagnation or total temperature T_{t2} of an ideal gas with constant c_p is defined as:

$$T_{t2} = T_2 + \frac{\mathscr{V}_2^2}{2c_p} \qquad (7.1.18)$$

and for the adiabatic flow through the nozzle it is clearly equal to T_1. The temperature actually attained by a thermometer placed in the stream is

7.2 The constant pressure cycle

called the recovery temperature, T_r, and the ratio:

$$\frac{T_r - T_2}{T_{t2} - T_2} \qquad (7.1.19)$$

is called the recovery factor. The actual temperature T of a gas moving at speed is often called the 'static' temperature to distinguish it from stagnation temperature.

Recovery of temperature also occurs in a throttling process, and is the reason why no temperature drop is observed for an ideal gas. The acceleration of the gas into the jet formed in an orifice undoubtedly results in a lowering of the static temperature, but the subsequent deceleration to a low velocity accompanied by frictional dissipation of kinetic energy causes the temperature to rise again to its initial value. It is important to observe that the above remarks concern steady flow. If a thermometer is placed in a jet of air issuing from a pressure vessel, a lowering of temperature is in fact observed. This low temperature is not however caused by the expansion of the gas into the jet, but by the expansion of the gas inside the vessel doing work in pushing the gas in front of it out against the atmosphere. Suppose the vessel initially charged with gas at pressure p_1, temperature T_1. When the vessel is partly empty, the gas remaining inside has done work and as a result its temperature has fallen. The temperature T_2 inside the vessel when the pressure has fallen to p_2 is approximately given by eq. (7.1.13), insofar as the expansion is reversible and adiabatic, and a thermometer placed in the gas in the vessel, or in the issuing jet, would read this temperature. This method of reduction of temperature by the expansion of compressed gas from a cylinder against the atmosphere has some historical importance, since it was used by Louis Cailletet in 1877 to produce the first evidence of oxygen liquefaction, as described in section 8.1. He used oxygen compressed to about 300 bar and cooled first to $-29\,°C$. On sudden release of the pressure in the vessel, a mist of liquid oxygen was observed inside. In more recent times the method has been used by Sir Francis Simon to make small amounts of liquid helium in the laboratory.

7.2 The constant pressure cycle

The circuit diagram of a refrigerating system which uses the lowering of temperature produced by the expansion of air doing work on a piston or in a turbine is shown in Fig. 7.2.1. Air from the cold room at 1, temperature T_1, pressure p_1, is compressed to pressure p_2, the temperature rising during the process by an amount depending on the way in which compression is performed. The air at 2 enters an air cooler in which it is cooled at approximately constant pressure to temperature T_3, a little above the cooling water temperature. The air at 3 is then expanded in a turbine or other work-producing device, and leaves at a temperature T_4 low enough to refrigerate the cold room. In the cold room heat transfer to the air raises the temperature to T_1 again.

The work delivered by the turbine or expander can be used to provide part of the work required to drive the compressor. The two machines may then be directly coupled to each other. The remaining work required must be supplied from an external supply.

In the cycle as shown in Fig. 7.2.1, the pressure p_1 is atmospheric pressure, the pressure in the cold space. No low-temperature heat exchanger is required because the low-temperature air comes into direct contact with the goods, etc., being refrigerated. This form of the cycle is known as the 'open' type. It corresponds almost exactly to a reversed open type gas turbine plant, in which air at atmospheric pressure is compressed, raised in temperature by burning fuel in it, expanded in a turbine to atmospheric pressure, and finally discharged to the atmosphere.

In its earliest forms the constant pressure air cycle, usually called simply the 'cold air cycle', incorporated reciprocating machinery, and it played an important part in the development of marine refrigeration. In its later forms as used in aircraft and for some other purposes, turbo-machinery was used.

Fig. 7.2.1. Diagrammatic arrangement of an open type cold air system using turbo-machinery.

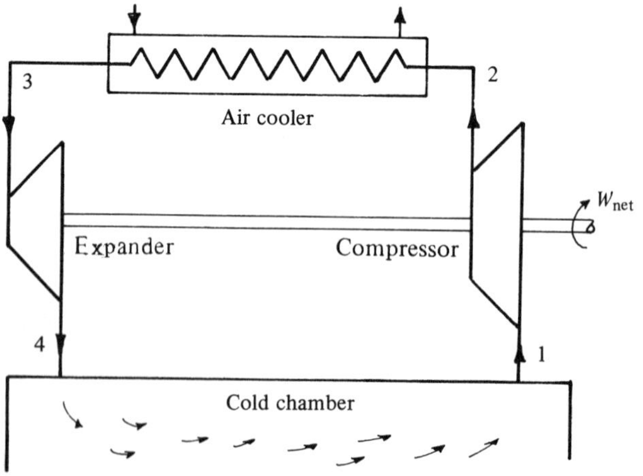

7.2 The constant pressure cycle

The analysis which follows is based on the assumption that the processes are carried out in steady flow with turbo-type machinery. The introduction of reciprocating machinery into the cycle requires a somewhat more detailed treatment, particularly of the expander and the effect of incomplete expansion, which will not be pursued here. It will be assumed that compression and expansion are adiabatic, which is substantially true for turbo-machinery. Kinetic and potential energy changes will be neglected, and temperature changes of the air in the interconnecting pipework will be neglected. They are readily taken into account when necessary. Applying the steady-flow energy equation to each component part of the cycle, using eq. (7.1.6): rate of heat transfer to air in cold room:

$$\dot{Q}_{in} = \dot{m}(h_1 - h_4) = \dot{m}c_p(T_1 - T_4) \tag{7.2.1}$$

Power *to* compressor:

$$\dot{W}_c = \dot{m}(h_2 - h_1) = \dot{m}c_p(T_2 - T_1) \tag{7.2.2}$$

Rate of heat transfer *from* air in cooler:

$$\dot{Q}_{out} = \dot{m}(h_2 - h_3) = \dot{m}c_p(T_2 - T_3) \tag{7.2.3}$$

Power delivered *by* expander:

$$\dot{W}_{exp} = \dot{m}(h_3 - h_4) = \dot{m}c_p(T_3 - T_4) \tag{7.2.4}$$

The net power to be delivered from an external source is then:

$$\dot{W}_{net} = \dot{W}_c - \dot{W}_{exp} = \dot{m}c_p(T_2 - T_1) - \dot{m}c_p(T_3 - T_4) \tag{7.2.5}$$

The coefficient of performance (c.o.p.) is therefore:

$$\varepsilon = \frac{\dot{Q}_{in}}{\dot{W}_{net}} = \frac{T_1 - T_4}{(T_2 - T_1) - (T_3 - T_4)} \tag{7.2.6}$$

Equation (7.2.6) above gives the c.o.p. subject only to adiabatic compression and expansion. It will now be supposed that in addition compression and expansion are reversible, i.e. constant entropy processes, and also that there are no pressure drops in the air cooler, cold room, or piping, i.e. $p_2 = p_3$, and $p_4 = p_1$. Then from eq. (7.1.13):

$$\left(\frac{p_2}{p_1}\right)^{(\gamma-1)/\gamma} = \frac{T_2}{T_1} = \frac{T_3}{T_4} = \frac{T_2 - T_3}{T_1 - T_4} = \rho \tag{7.2.7}$$

The c.o.p. with these assumptions will be denoted by ε_r. From eq. (7.2.6) and (7.2.7):

$$\varepsilon_r = \frac{1}{\rho - 1} \tag{7.2.8}$$

$$= \frac{T_1}{T_2 - T_1} = \frac{T_4}{T_3 - T_4} \tag{7.2.9}$$

The cycle with these assumptions is shown in Fig. 7.2.2 on T–s coordinates. The expressions for ε_r show that the c.o.p. is the same as that of a Carnot cycle operating between T_1 and T_2, or between T_3 and T_4. Now if the object of the refrigerating plant is to maintain a cold space at temperature T_1, and transfer heat to cooling water at temperature T_3, approximately, this could in principle be done by means of a Carnot cycle operating between T_1 and T_3. The c.o.p. of this Carnot cycle would be:

$$\varepsilon_C = \frac{T_1}{T_3 - T_1} = \frac{1}{(T_3/T_4) - 1} \tag{7.2.10}$$

and this is clearly larger than ε_r as given by eq. (7.2.8). The difference between them is illustrated in the following example.

Fig. 7.2.2. The cycle for the system of Fig. 7.2.1 assuming reversible adiabatic compression and expansion, shown on temperature–entropy coordinates.

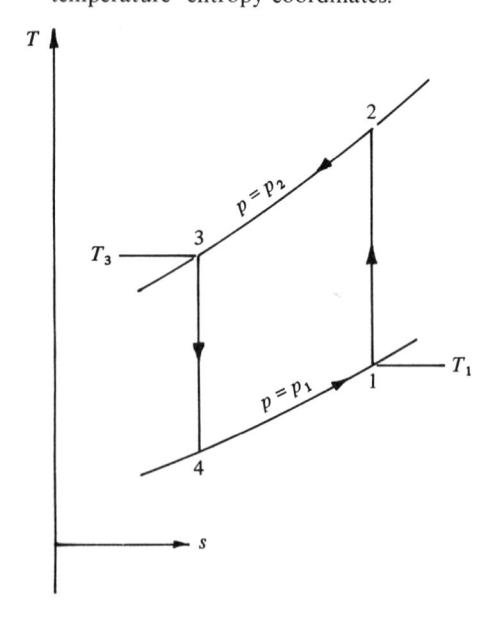

7.2 The constant pressure cycle

Example 7.2.1
The temperature required in the cold space is: $t_1 \approx -15\,°\text{C}$, $T_1 = 258\,\text{K}$. The pressures are: $p_1 = 1$ bar, $p_2 = p_3 = 4$ bar. Cooling water is available at a temperature which enables the high pressure air to be cooled to $t_3 \approx 30\,°\text{C}$, $T_3 = 303\,\text{K}$. For air $\gamma = 1.4$, $(\gamma - 1)/\gamma = 0.2857$. Then:

$$\rho = \left(\frac{p_2}{p_1}\right)^{(\gamma-1)/\gamma} = (4)^{0.2857} = 1.486$$

$$T_2 = \rho T_1 = 1.486 \times 258\,\text{K} = 383.4\,\text{K} \qquad t_2 \approx 110\,°\text{C}$$

$$T_4 = T_3/\rho = (303/1.486)\,\text{K} = 203.9\,\text{K} \qquad t_4 \approx -69\,°\text{C}$$

$$\varepsilon_r = \frac{1}{\rho - 1} = 1/0.486 = 2.058$$

For a Carnot cycle operating between $T_1 = 258\,\text{K}$ and $T_3 = 303\,\text{K}$,

$$\varepsilon_C = \frac{258}{303 - 258} = 5.733$$

Now, considered from the point of view of their internal operation, the constant pressure air cycle with reversible adiabatic compression and

Fig. 7.2.3. A 'thin cycle' operating at a lower pressure ratio than that shown in Fig. 7.2.2.

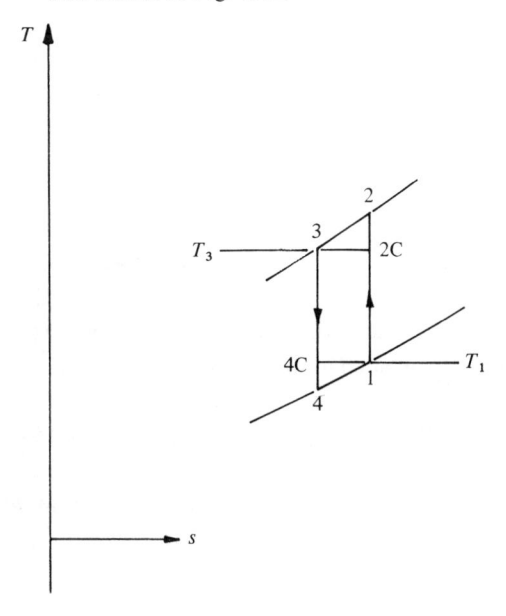

expansion and the Carnot cycle are both reversible, and both are 'ideal' in a sense. The reason why the c.o.p. of the air cycle is lower than that of the Carnot cycle is not then because of its internal imperfections, but because it is being used in an unsuitable application. The air cannot accept heat at constant pressure without rising in temperature, and consequently it is necessary to deliver air to the cold space at a much lower temperature than is really necessary, i.e. at $-69\,°\mathrm{C}$ in the above example, to maintain the space at $-15\,°\mathrm{C}$. Similarly, because air cannot reject heat at constant pressure without falling in temperature, it is necessary to deliver air to the air cooler at a much higher temperature than is really necessary, i.e. $110\,°\mathrm{C}$ in the above example.

By contrast, in the Carnot cycle the processes of taking in and rejecting heat are carried out without a change of temperature. But to do this with an ideal gas requires work to be done in the process, which leads to complication of the cycle and practical difficulties in carrying out an isothermal process with simultaneous heat and work transfer.

The choice of $p_2 = 4$ bar in the above example was arbitrary. By using a lower value of p_2, ε_r can be improved. For example with:

$$p_2 = 2\,\text{bar} \qquad p_2/p_1 = 2$$
$$\rho = (2)^{0.2857} = 1.219$$
$$\varepsilon_r = \frac{1}{1.219 - 1} = 4.566$$

a considerable improvement on 2.058. In fact as p_2 is reduced still further, ε_r approaches ε_C in the limit when ρ approaches T_3/T_1, T_2 approaches T_3, and T_4 approaches T_1. A 'thin' cycle approaching this limit is illustrated in Fig. 7.2.3 on T–s coordinates. The cycle 1–2–3–4 approaches the Carnot cycle $1-T_{2C}-T_3-T_{4C}$ as the entropy width of the cycle diminishes.

As this condition is approached the specific refrigerating effect of the air, $c_p(T_1 - T_4)$, diminishes towards zero, and the mass flow rate of air needed for a given refrigerating capacity rises rapidly towards infinity, with a consequent rapid increase in the size of the plant. Nevertheless, it seems at first sight that by using such a 'thin' cycle with a relatively low value of p_2 and a small specific refrigerating effect, a reasonable c.o.p. could be obtained provided that sufficient air can be moved around the system. Unfortunately, when account is taken of inefficiency of the expander, this encouraging prospect disappears. It is then found that p_2/p_1 cannot be reduced below a certain value because the refrigerating effect vanishes entirely.

7.3 The effect of compressor and expander inefficiency

For adiabatic compression and expansion between fixed pressures the effect of internal irreversibility is to raise the temperature at the end of the process above that which would be obtained in a reversible process. This is illustrated in Fig. 7.3.1. The states at the end of reversible adiabatic processes are now distinguished by a prime, 2' and 4', whilst the actual states are 2 and 4. The isentropic efficiencies of the compressor and expander are defined by:

$$\eta_c = \frac{(h'_2 - h_1)_s}{h_2 - h_1} = \frac{(T'_2 - T_1)_s}{T_2 - T_1} \qquad (7.3.1)$$

and:

$$\eta_e = \frac{h_3 - h_4}{(h_3 - h'_4)_s} = \frac{T_3 - T_4}{(T_3 - T'_4)_s} \qquad (7.3.2)$$

respectively, where it is understood that states 1 and 2' have the same entropy, and 3 and 4' have the same entropy.

It is clear that for a given pressure ratio $p_2/p_1 = p_3/p_4$, the specific refrigerating effect is reduced because of the expander inefficiency from $c_p(T_1 - T'_4)$ to $c_p(T_1 - T_4)$. In the following analysis, T_1 and T_3 will be

Fig. 7.3.1. Modification of the cycle caused by inefficiency of the compressor and expander.

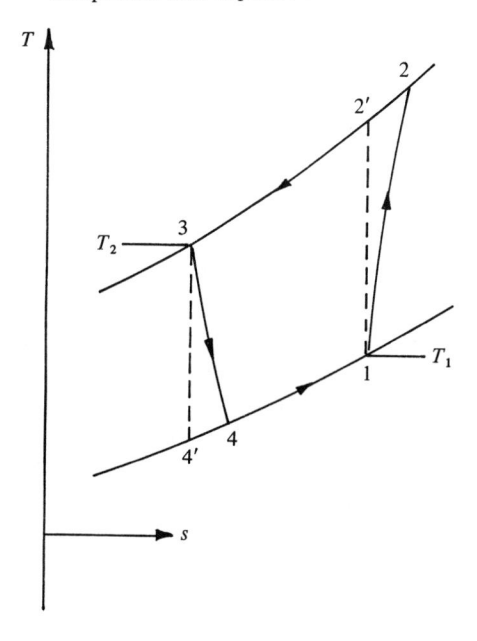

regarded as fixed by the external régime. They are demands which the plant has to meet: T_1 the temperature in the cold room, and T_3 a little above the temperature of the available cooling water. For convenience the ratio T_3/T_1 will be denoted by:

$$\frac{T_3}{T_1} = \theta \tag{7.3.3}$$

It will also be convenient to retain:

$$\rho = \frac{T_2'}{T_1} = \frac{T_3}{T_4'} = \left(\frac{p_2}{p_1}\right)^{(\gamma-1)/\gamma} \tag{7.3.4}$$

a function of the pressure ratio only. Then:

$$\begin{aligned} T_1 - T_4 &= (T_3 - T_4) - (T_3 - T_1) \\ &= \eta_e(T_3 - T_4') - (T_3 - T_1) \\ &= \eta_e \theta T_1(1 - 1/\rho) - T_1(\theta - 1) \end{aligned} \tag{7.3.5}$$

$$T_2 - T_1 = \frac{T_2' - T_1}{\eta_c} = \frac{T_1}{\eta_c}(\rho - 1) \tag{7.3.6}$$

$$T_3 - T_4 = \eta_e(T_3 - T_4') = \eta_e \theta T_1(1 - 1/\rho) \tag{7.3.7}$$

The c.o.p. is therefore, from eq. (7.2.6):

$$\varepsilon = \frac{\eta_e \theta(1 - 1/\rho) - (\theta - 1)}{[(\rho - 1)/\eta_c] - \eta_e \theta(1 - 1/\rho)} \tag{7.3.8}$$

A plot of ε against ρ is shown in Fig. 7.3.2 for various values of η. For simplicity it is assumed that $\eta_c = \eta_e$. The assumed constant conditions are $T_1 = 258$ K, $T_3 = 303$ K, as in example 7.2.1, giving $\theta = 1.174$. For $\eta = 1$, the c.o.p. is the same as that given by eq. (7.2.8), and as already pointed out this value increases towards $\varepsilon_C = 5.73$ as ρ is reduced to approach θ.

The strong effect of inefficiency in the components is shown by the curve for $\eta = 0.99$, with a maximum attainable ε of only 2.92, approximately. For $\eta = 0.8$ the maximum ε is only 0.45.

The value of ρ at which maximum ε occurs is readily found by differentiating eq. (7.3.8). The details will be omitted, but the result is that the value of ρ for maximum ε is given by the root ($> \theta$) of the equation:

$$\frac{1 - \theta(1 - \eta)}{\eta \theta}\rho^2 - 2\rho + 1 + \eta(\theta - 1) = 0 \tag{7.3.9}$$

for the case of $\eta_c = \eta_e = \eta$.

7.3 Effects of compressor and expander inefficiency

The value of ε is zero when $T_4 = T_1$, i.e. when

$$0 = \eta_e \theta(1 - 1/\rho) - (\theta - 1)$$

from eq. (7.3.5). Solving for ρ:

$$\rho_{\varepsilon=0} = \frac{\eta_e \theta}{1 - \theta(1 - \eta_e)} \qquad (7.3.10)$$

Thus, to sum up the results of this section, for any given value of expander efficiency, there is a certain minimum ratio ρ which is greater than θ, and a corresponding pressure ratio p_2/p_1 at which the system will just begin to work at all. Above this minimum value the c.o.p. rises to a maximum and then decreases. The value of the maximum c.o.p. is very sensitive to the value of the expander and compressor efficiency, as shown by the curves of Fig. 7.3.2.

Fig. 7.3.2. Coefficient of performance of air cycles versus ρ, the ratio of the absolute temperature after and before isentropic compression for the pressure ratio (p_2/p_1), for various isentropic efficiencies of the compressor and expander, assumed equal: $\eta_e = \eta_c = \eta$. $T_1 = 258$ K, $T_3 = 303$ K.

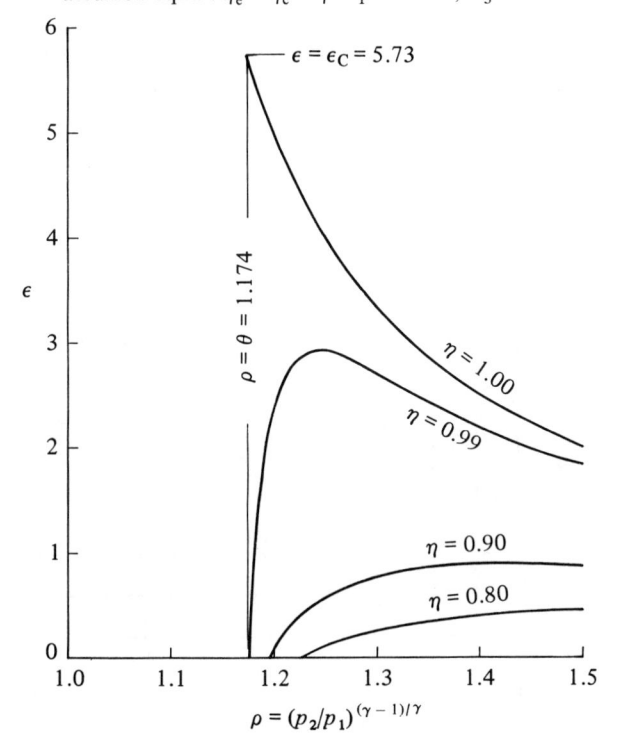

Ideal gas cycles 486

7.4 The constant pressure cycle with internal heat exchange

The effect of inefficiency in the expander can be offset to some extent by incorporating a heat exchanger in the system to warm up the air coming from the cold space by cooling down the air going to the expander, as shown in the circuit diagram of Fig. 7.4.1 and on temperature–entropy coordinates in Fig. 7.4.2.

Air leaves the space at T_1 and the air cooler at T_3. These temperatures are determined by the external conditions, T_1 being the desired temperature in the cold space, and T_3 the temperature fixed by the cooling water available. These streams flow in a counterflow exchanger, with the same mass flow rates and the same specific heat capacities. Hence their temperature change is the same:

$$T_{1a} - T_1 = T_3 - T_{3a} \qquad (7.4.1)$$

In an infinite exchanger T_{1a} would be equal to T_3 and T_{3a} would be equal to T_1. What is actually achieved in a real heat exchanger is expressed by its efficiency defined as:

$$\eta_x = \frac{T_{1a} - T_1}{T_3 - T_1} = \frac{T_3 - T_{3a}}{T_3 - T_1}, \qquad (7.4.2)$$

whose value depends on the amount of heat transfer surface provided. The expression for the c.o.p. in terms of η_x is rather cumbersome, however, and it

Fig. 7.4.1. Modification of the system by introducing a heat exchanger between the cold air going to the compressor and the warm air going to the expander.

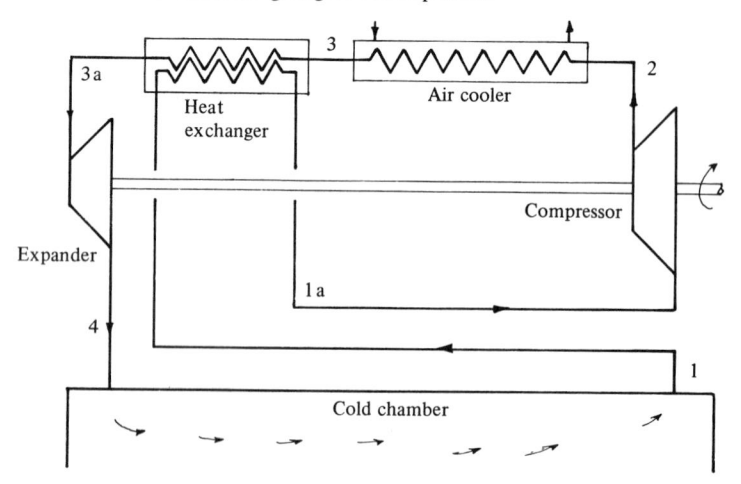

7.4 Constant pressure cycle with internal heat exchange

is convenient to introduce the ratios:

$$\alpha = T_{1a}/T_1 = 1 + \eta_x(\theta - 1) \tag{7.4.3}$$

and:

$$\beta = T_{3a}/T_1 = \theta - \eta_x(\theta - 1) \tag{7.4.4}$$

where:

$$\theta = T_3/T_1 \tag{7.4.5}$$

regarded as fixed by the external conditions. As before, the pressure ratio will be expressed through ρ defined as:

$$\rho = \frac{T'_2}{T_{1a}} = \frac{T_{3a}}{T'_4} = \left(\frac{p_2}{p_1}\right)^{(\gamma-1)/\gamma} \tag{7.4.6}$$

Pressure drops in the pipe work and heat exchangers are neglected.

The refrigerating capacity is:

$$\dot{Q}_{in} = \dot{m}c_p(T_1 - T_4) \tag{7.4.7}$$

where:

$$\begin{aligned} T_1 - T_4 &= (T_{3a} - T_4) - (T_{3a} - T_1) \\ &= \eta_e \beta T_1 (1 - 1/\rho) - T_1(\beta - 1) \end{aligned} \tag{7.4.8}$$

Fig. 7.4.2. The cycle for the system of Fig. 7.4.1 shown on temperature–entropy coordinates.

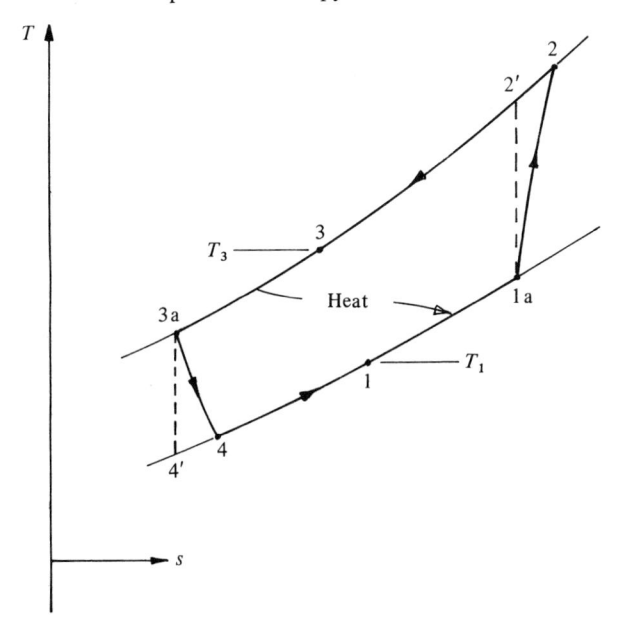

Ideal gas cycles

The net power is:

$$\dot{W}_{net} = \dot{m}c_p[(T_2 - T_{1a}) - (T_{3a} - T_4)] \tag{7.4.9}$$

where:

$$(T_2 - T_{1a}) - (T_{3a} - T_4) = \frac{\alpha T_1(\rho - 1)}{\eta_c} - \eta_e \beta T_1(1 - 1/\rho) \tag{7.4.10}$$

The c.o.p. is therefore:

$$\varepsilon = \frac{\eta_e \beta(1 - 1/\rho) - (\beta - 1)}{[\alpha(\rho - 1)/\eta_c] - \eta_e \beta(1 - 1/\rho)}$$

$$= \frac{(\eta_e \beta/\rho) - (\beta - 1)/(\rho - 1)}{(\alpha/\eta_c) - (\eta_e \beta/\rho)} \tag{7.4.11}$$

Fig. 7.4.3. Coefficient of performance of air cycles with and without heat exchange: $T_1 = 258$ K, $T_3 = 303$ K. A for no heat exchanger, $\eta_e = \eta_c = 1$, and B no heat exchanger, $\eta = 0.9$, are reproduced from Fig. 7.3.2 for comparison. C Perfect heat exchanger, $\eta_e = \eta_c = \eta = 0.9$, D efficiency of heat exchanger 0.933 corresponding to an approach of 3 K at each end, $\eta_e = \eta_c = \eta = 0.9$.

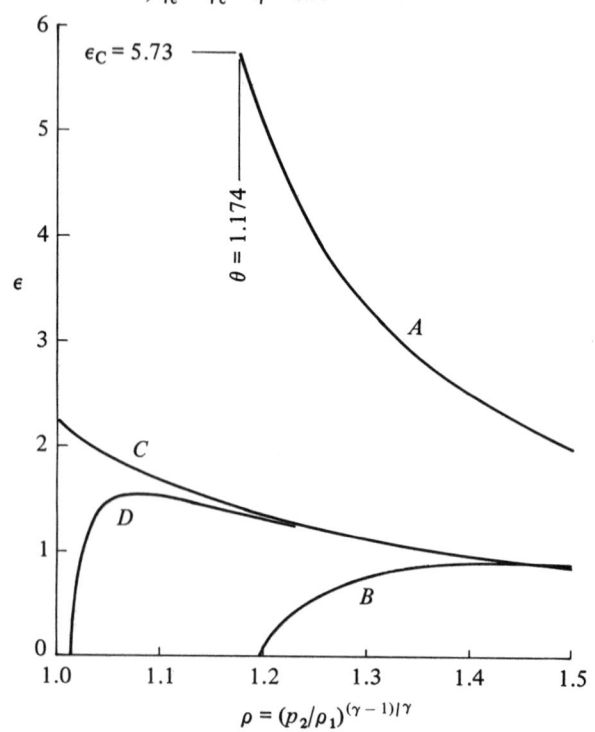

$\epsilon_C = 5.73$

$\theta = 1.174$

$\rho = (p_2/p_1)^{(\gamma - 1)/\gamma}$

7.4 Constant pressure cycle with internal heat exchange

The effect of the heat exchanger on the c.o.p. is shown in Fig. 7.4.3 for the same value of θ as in Fig. 7.3.2 and for $\eta_e = \eta_c = 0.9$. Two cases are illustrated: one (C) for an infinite heat exchanger with $\eta_x = 1$, and the other (D) for a finite exchanger with an approach between the terminal temperatures of 3 K, i.e.

$$T_3 - T_{1a} = T_{3a} - T_1 = 3 \text{ K}$$

corresponding to an efficiency of:

$$\eta_x = 42/45 = 0.933$$

Comparing the values with those of Fig. 7.3.2, it is seen that the c.o.p. can be enhanced by heat exchange. Furthermore the maximum value of ε occurs at a much lower value of ρ, and hence at a lower pressure ratio, than without heat exchange. In the case illustrated with $\eta_x = 0.933$, the maximum ε occurs at:

$$\rho = 1.072$$

corresponding to a pressure ratio of only:

$$p_3/p_2 = (1.072)^{3.5} = 1.276$$

For a given η_x the value of ρ for maximum ε may be found by differentiating eq. (7.4.11). The result is that ρ for maximum ε is the root (> 1) of the equation:

$$\frac{\alpha(1 + \eta\beta - \beta)}{\eta}\rho^2 - 2\alpha\beta\rho + \eta\beta(\beta + \frac{\alpha}{\eta} - 1) = 0 \qquad (7.4.12)$$

for $\eta_e = \eta_c = \eta$.

For the constant pressure cycle without heat exchange the minimum pressure ratio must be such that ρ is greater than θ. At pressure ratios lower than this the temperature at the end of compression could not attain the temperature required to reject heat to the cooling water. It is now seen that with a heat exchanger pressure ratios giving $\rho < \theta$ are usable. In fact with a long enough heat exchanger the pressure ratio can be as low as desired.

As the pressure ratio is reduced the temperature T_4 approaches T_1, and the specific refrigerating effect is reduced. Hence it is necessary to circulate a larger mass flow rate, and the machinery required becomes more bulky. In Fig. 7.4.4 the results from Fig. 7.3.2 and Fig. 7.4.3 are shown replotted to a base of temperature rise in the cold space, i.e. $T_1 - T_4$. For the same value of this temperature rise, and therefore the same mass flow rate of air for a given capacity, the superiority of the cycle with heat exchange is evident.

Ideal gas cycles

The cycles discussed up to now have been open at the cold end, but they may be closed there with a pressure above atmospheric. This reduces the volume flow rate of gas to be circulated for a given refrigerating capacity, and enables the system to be used in circumstances where an open cycle would be ruled out. It does however entrain an extra temperature difference to be taken into account.

When a heat exchanger is used which raises the temperature of the gas undergoing compression to approximately ambient temperature, the compressor is operating under conditions similar to those of an ordinary air compressor. For these it is usual to express the efficiency as an isothermal efficiency, i.e. the reversible isothermal process is regarded as the ideal, because it is possible in principle to reject heat to the surroundings. The specific work for a reversible isothermal compression process is:

$$w_T = RT \ln(p_2/p_1) \quad (7.4.13)$$

and the isothermal efficiency is defined as:

$$\eta_T = w_T/w \quad (7.4.14)$$

Fig. 7.4.4. The data of Fig. 7.4.3 replotted to a base of temperature change of the air within the refrigerated chamber. The letters on the curves have the same meaning as in that figure.

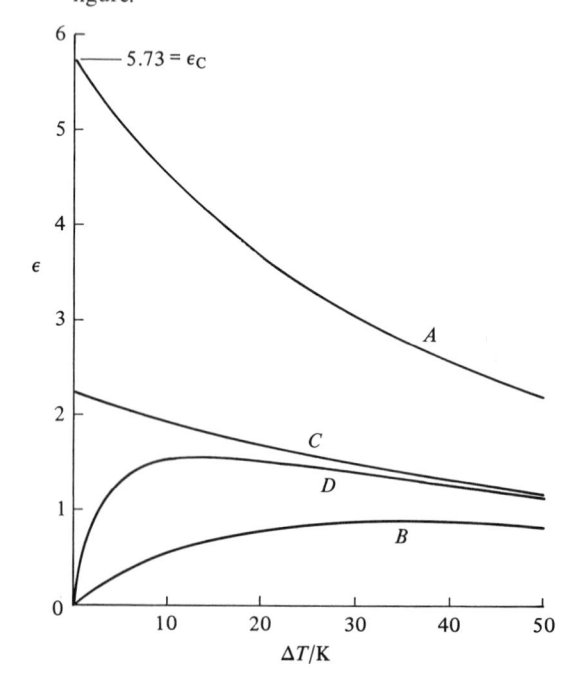

7.4 Constant pressure cycle with internal heat exchange 491

where w is the actual specific work. The isothermal efficiency may be referred to the indicated, shaft or electrical power at the driving motor terminals, and it must be specified which is intended.

The following example illustrates one current use of an ideal gas cooling system with an internal heat exchanger. It is used to condense the boil-off gas from a liquefied natural gas tank. A reciprocating compressor is used with two stages of expansion turbines. The details are taken from Brown (1971) with some modification for the purpose of the example.

Example 7.4.1

Figure 7.4.5 shows the circuit diagram of a helium gas cycle. The absolute pressure at the suction of the compressor is 6 bar and there are pressure drops of 0.5 bar for each stream in the heat exchanger and 1 bar in the high-

Fig. 7.4.5. Circuit diagram of an ideal gas cycle using helium for reliquefying the boil-off from a liquefied natural gas tank.

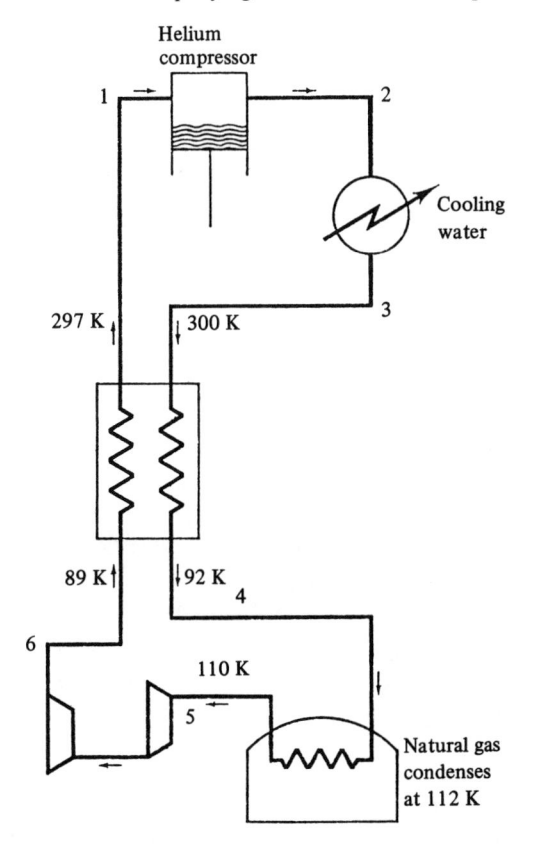

pressure stream between the outlet from the heat exchanger and the entry to the first turbine. Other pressure drops and extraneous heat transfers may be neglected. The polytropic efficiency of the turbine is 0.8 and the shaft isothermal efficiency of the compressor is 0.6.

Determine the power required to drive the compressor and the power delivered by the turbines for a refrigerating capacity of 60 kW.

For helium, $R = 2.08\,\text{kJ/kg K}$, $c_p = 5.20\,\text{kJ/kg K}$, $\gamma = 5/3$.

The relation between polytropic efficiency and the index of compression is given by eq. (4.8.19). For an expansion process a corresponding equation can be derived:

$$\eta_{\text{pol}} = \frac{(n-1)/n}{(\gamma-1)/\gamma}$$

Then for the turbines:

$$0.8 = \frac{(n-1)/n}{2/5}$$

$$\frac{n-1}{n} = 0.3200 \text{ and } n = 1.4706$$

Hence, with $p_6 = 6.5\,\text{bar}$:

$$p_5 = \left(\frac{110}{89}\right)^{1/0.32} 6.5\,\text{bar} = 12.60\,\text{bar}$$

$$p_4 = 13.60\,\text{bar} \qquad p_3 = p_2 = 14.10\,\text{bar}$$

The mass flow rate of helium is:

$$\dot{m} = \frac{60}{5.20(110-92)}\,\text{kg/s} = 0.641\,\text{kg/s}$$

and the power delivered by the turbines is:

$$\dot{W}_t = 0.641(5.20)(110-89)\,\text{kW} = 70.0\,\text{kW}$$

The shaft power taken by the compressor is:

$$\dot{W}_c = \frac{0.641(2.08)(297)}{0.6}\ln\left(\frac{14.10}{6}\right)\,\text{kW} = 564\,\text{kW}$$

The turbine power could be used to help to drive the compressor but it was not considered worth while in this application and it is dissipated by a brake.

The position of the heat exchanger in the above example may seem unusual when compared with the arrangement shown in Fig. 7.4.1. In

7.4 Constant pressure cycle with internal heat exchange

Fig. 7.4.5 the refrigerating effect has been transferred to the high-pressure stream for practical reasons concerned with the long pipes connecting the compressor and turbine group to the pipe coil for condensing the natural gas in the tank. The heat exchanger shown combines this function, of transferring the refrigerating effect, with that of the heat exchanger in Fig. 7.4.1 which is to raise the temperature of the gas at entry to the compressor.

Fig. 7.5.1. (a) The system of Fig. 7.4.1 with the cold chamber at a pressure less than that of the atmosphere, so that the system is open at the warm end. (b) The 'Odessa' system with atmospheric pressure at both cold and warm ends.

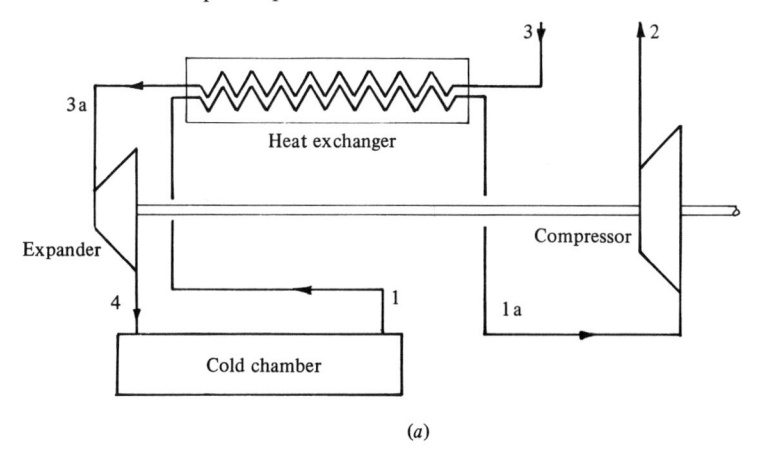

(a)

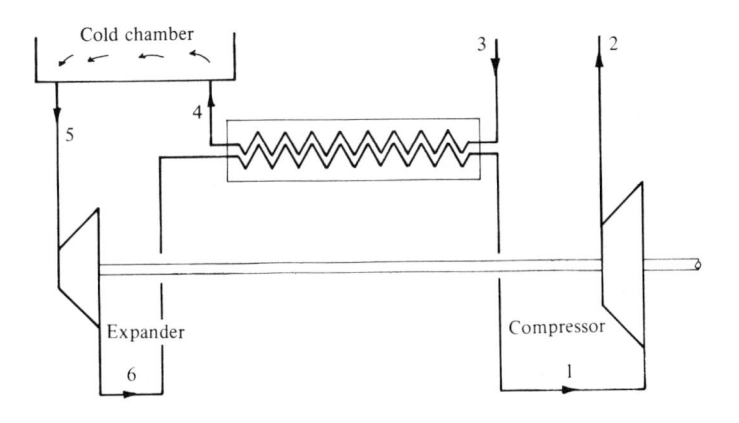

(b)

Ideal gas cycles 494

7.5 The constant pressure cycle with sub-atmospheric pressures

In the form of the constant pressure system described so far the heat transfer to the cold air takes place at atmospheric pressure. The system may therefore be described as 'open' at the cold end. If desired it can be made 'open' at the warm end by operating the cold end at a pressure less than atmospheric pressure. This is shown in Fig. 7.5.1(a).

An interesting variant of this system by which it is made open at both cold and warm ends was developed in Odessa by Martinovsky & Dubinsky (1964). In their system the refrigerating effect of the sub-atmospheric stream is transferred to the atmospheric pressure stream in a heat exchanger, which forms an extension of the heat exchanger already discussed in the last section. Referring to Figs. 7.5.1(b) and 7.5.2, air at atmospheric pressure and temperature is taken in at 3 and cooled in the heat exchanger down to state 4. From 4 to 5 the air accepts heat, doing useful refrigeration, and at 5 it enters the turbine where it expands to a pressure less than atmospheric pressure, emerging at 6 with a reduced temperature. It then passes through the heat exchanger, leaving at state 1 with approximately ambient temperature, and is compressed to atmospheric pressure in the compressor. It is discharged as warm air at 2.

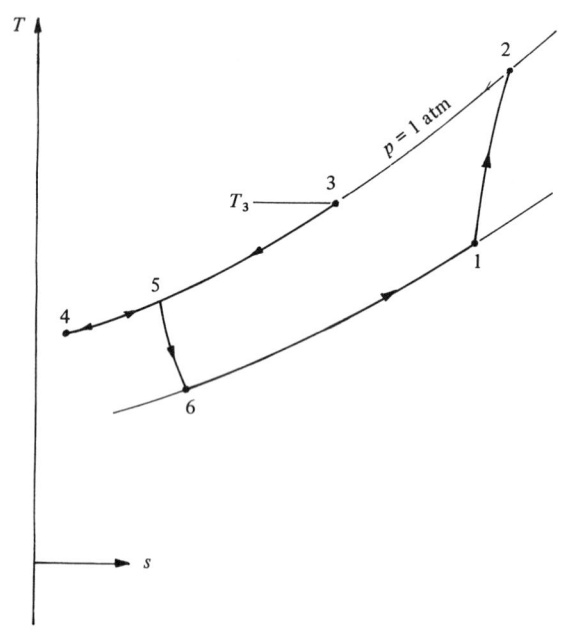

Fig. 7.5.2. Air states for the cycle of Fig. 7.5.1(b) shown on temperature–entropy coordinates.

7.6 Aircraft cooling systems

Because the system may be operated with quite small pressure ratios, it is possible to use axial flow compressors and turbines having a very high isentropic efficiency, of the order 0.9 over these small pressure ratios.

It is important in all these cycles to employ a heat exchanger of high efficiency. A convenient way of obtaining high exchanger efficiency is to use regenerators, shown in Fig. 7.5.3. These are vessels packed with metal foil or gauze, through which the cold and warm gas streams are passed alternatively via a two-way valve arrangement. Because the heat transfer surface does not now have to perform the task of physically separating the two streams and withstand a difference in pressure, it can be made very large, and the cross-sectional area for flow can be made large to minimise the pressure drops.

7.6 Aircraft cooling systems

In jet aircraft, compressed air is available from the compressors for the main engines. This, coupled with the simplicity and reliability of the air system, has made it attractive as a means of cooling the cabins of high-speed aircraft.

For low-speed aircraft flying at moderate altitudes no artificial cooling in flight is usually required, because the outside air temperature is low enough to prevent cabin temperatures from rising too much. In a high-speed aircraft, however, this cooling is not available, because the surface temperature of the aircraft is the recovery temperature corresponding to the speed, and this is near to the stagnation temperature. Even at high altitude

Fig. 7.5.3. Regenerators used instead of the heat exchanger in the 'Odessa' system. The numbers correspond with those in Fig. 7.5.1(b). The upper regenerator is in its warm phase, falling in temperature; the lower one is in its cold phase, rising in temperature.

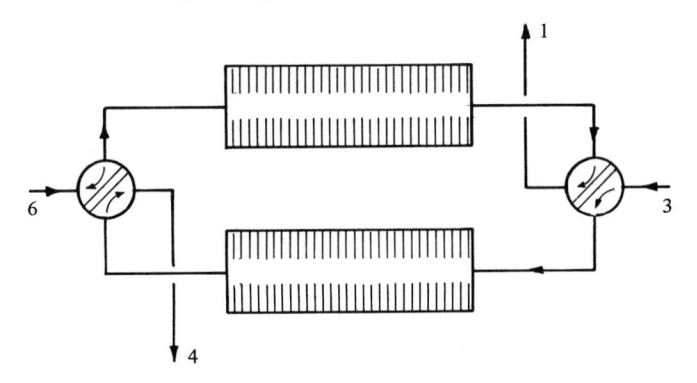

Ideal gas cycles 496

where the air temperature is quite low, the stagnation temperature at speeds of the order Mach one and above is far too high to be tolerable.

A simple form of system for producing cool air is shown in Fig. 7.6.1. Compressed air at 1, still warm, is reduced in temperature by heat transfer with outside air at 4. The temperature of this outside air is effectively its stagnation temperature, because the air is brought to rest in the boundary layer on the heat transfer surface. The stagnated outside air is often described as 'ram' air. So far as heat transfer is concerned, it behaves like stationary air with the stagnation temperature of the outside stream relative to the aircraft. The compressed air leaving the heat exchanger at 2 has a temperature still above the outside stagnation temperature therefore. It then passes through an expansion turbine, delivering work, in which the

Fig. 7.6.1. A simple system for producing cooled air on board aircraft, using compressed air from the main engines.

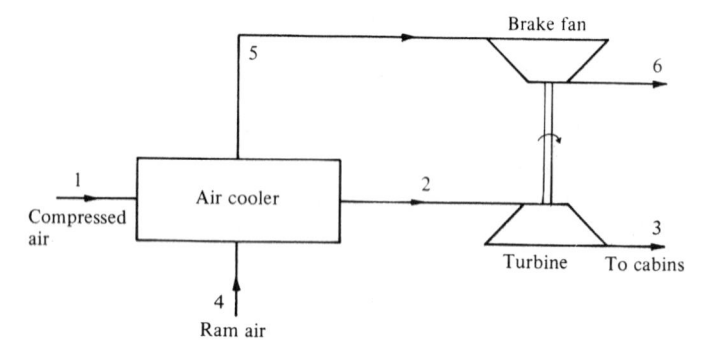

Fig. 7.6.2. A 'bootstrap' system for use when the temperature of the ram air is too high to ensure useful cooling with the simple system.

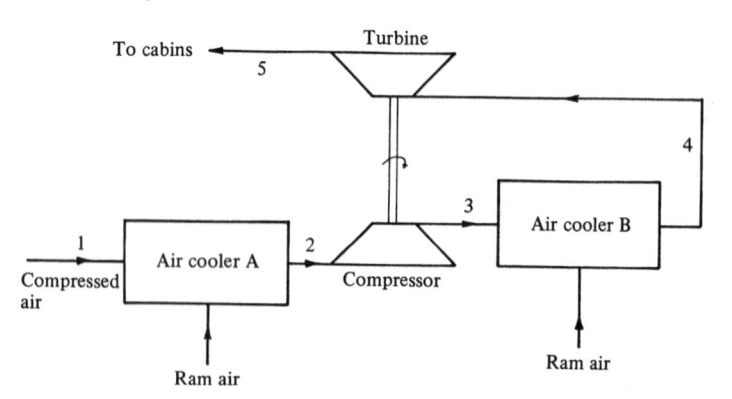

7.7 The Stirling cycle

pressure falls to the cabin pressure and the temperature falls accordingly.

The power delivered by the expansion turbine is taken by a fan or compressor which helps to pump ram air through the heat exchanger and so reduce the drag on the aircraft caused by this resistance.

If the stagnation temperature of the ram air is so high that with the pressure ratio available across the expander the temperature of the air entering the cabins is not low enough, a more complicated scheme known as the 'bootstrap' system may be used. This is shown in Fig. 7.6.2.

The compressed air at 1 is cooled in heat exchanger A to state 2 as before by ram air. It is then further compressed to state 3. The power for this compressor comes from the subsequent expansion process. After compression the air is again cooled by ram air in a second heat exchanger B, to state 4. Its temperature is now a little above stagnation temperature, but it has a higher pressure before expansion than in the simple system. Consequently the temperature at point 5 leaving the expansion turbine is lower.

Further information on aircraft cooling by expansion of air is given by Fiedler (1965).

7.7 The Stirling cycle

It has been pointed out in connection with the constant pressure cycle that if the fluids or bodies being refrigerated are at a constant

Fig. 7.7.1. The Carnot cycle using an ideal gas shown on (a) temperature–entropy coordinates, and (b) pressure–volume coordinates. The pressure–volume diagram is drawn approximately to scale for 1 gramme of air with $p_1 = 1$ bar, $p_3/p_1 = 10$, $T_1 = 258$ K, the temperature of heat acceptance, and $T_2 = 303$ K, the temperature of heat rejection.

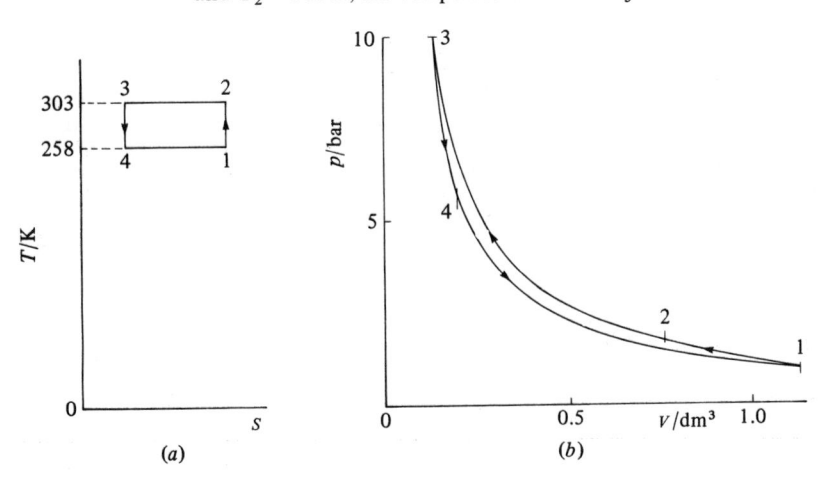

Ideal gas cycles 498

temperature, or nearly so, the temperature of the refrigerant accepting heat should also be constant, in order to obtain a high c.o.p. Similarly, if the cooling fluid is at a constant temperature, the temperature of the refrigerant rejecting heat should be constant. With an ideal gas as the refrigerant, the desired isothermal processes can only be carried out if work is done by or on the gas at the same time as heat is accepted or rejected respectively.

One ideal cycle employing such isothermal processes is the Carnot cycle, in which the cycle is completed by two reversible adiabatic processes, as shown on $T-S$ and $p-V$ coordinates in Fig. 7.7.1. The trouble with the Carnot cycle as a practical mode of operation is that the volumetric refrigerating capacity and the mean effective pressure are very small. The effect of friction between piston and cylinder thus becomes dominant and gives a poor c.o.p. In reverse, the analogous trouble with the Carnot cycle as a practical mode of operation for an engine, is that the ideal indicated mean effective pressure is very small in relation to the maximum pressure of the

Fig. 7.7.2. The Stirling cycle using an ideal gas shown on (*a*) temperature–entropy coordinates, and (*b*) pressure–volume coordinates. The pressure–volume diagram is drawn approximately to scale for the same T_1 and T_2 as Fig. 7.7.1.(*b*), and it shows how the 'weak' parts of the Carnot cycle have been cut off.

It should be noted that the bases of these diagrams are not specific entropy and specific volume. During the processes 1–2 and 3–4 the gas is not in a homogeneous state, and specific quantities have no real significance except as mass averages. Nevertheless it is possible to ascribe an entropy and a volume to a given mass of gas which is not homogeneous.

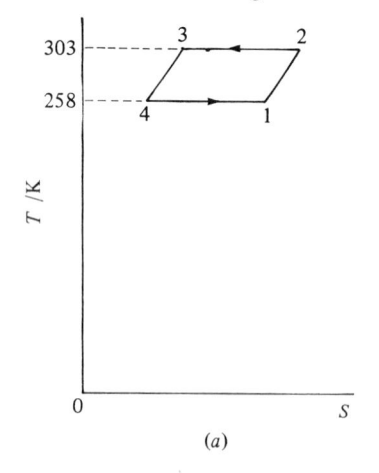

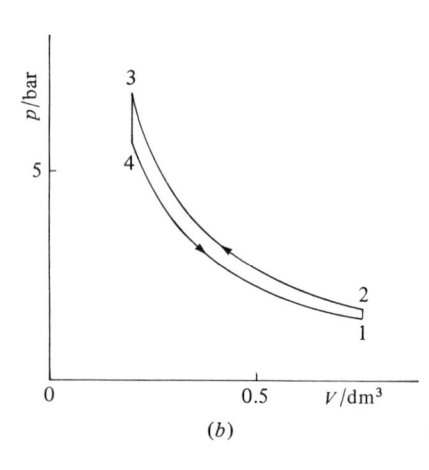

7.7 The Stirling cycle

cycle. For a reasonable value of the maximum pressure, the indicated mean effective pressure is less than the frictional mean effective pressure, so that no power can be delivered at all.

The Stirling cycle provides a way around this difficulty. The isothermal processes are connected by constant volume processes to make a complete cycle, as shown in Fig. 7.7.2 on temperature–entropy and pressure–volume coordinates. But a constant volume process requires heat transfer to the gas while it rises in temperature from T_1 to T_2, and heat transfer from it while it falls in temperature from T_2 to T_1. The essential idea in the Stirling cycle is to use the heat transfer from the gas falling in temperature to provide that for the gas rising in temperature. In principle this could be done by arranging two cylinders in thermal communication, with two cycles being carried on out of phase, so that the gas in one cylinder rises in temperature

Fig. 7.7.3. Diagrammatic arrangement of a Stirling refrigerator. The motions required of the pistons are shown below in full lines and numbered to correspond with Fig. 7.7.2. The approximately sinusoidal motions produced by simple crank mechanisms are shown by the dotted lines.

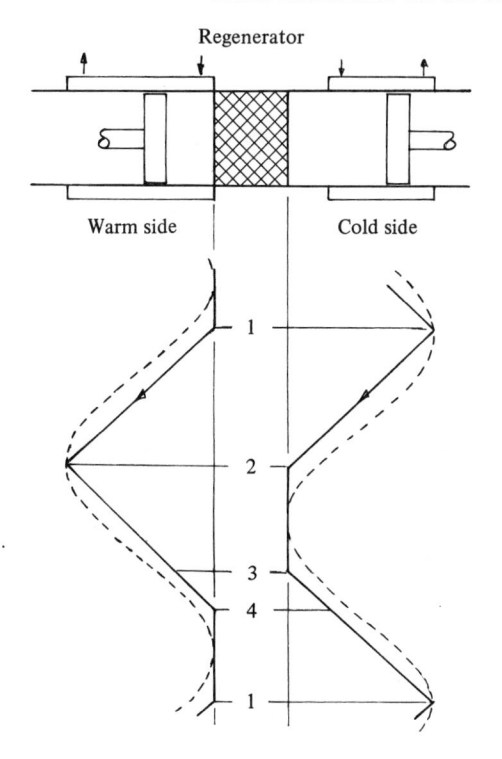

Ideal gas cycles

at the same time as the gas in the other falls in temperature. In the machine proposed by Stirling, however, a regenerator is used to store the change of internal energy of the gas during a constant volume process with falling temperature and give it back to the gas during the next constant volume process with rising temperature. The regenerator is similar to that described in the last section, a mass of gauze or wire mesh which is capable of accepting and rejecting heat to the gas flowing through it, whilst maintaining a temperature gradient along it.

A method of carrying out the Stirling cycle is illustrated in Fig. 7.7.3. Two opposed pistons are coupled to move in a manner to be described. Between them is the regenerator. Arrangements are made around the right-hand cylinder for the gas to accept heat from a fluid which is to be refrigerated, and arrangements are made around the left-hand cylinder for the gas to reject heat to a cooling fluid. Below the diagram of the cylinders the motions of the pistons are plotted, with numbers corresponding to the points marked in Fig. 7.7.2.

Beginning at 1, when the cold gas on the right has finished isothermal expansion accepting heat from the jacket, both pistons then move to the left at the same speed, displacing gas through the warm regenerator at constant enclosed volume. The gas is raised in temperature before entering the left-hand cylinder, and the regenerator becomes colder. When the gas is all transferred to the left-hand cylinder its temperature is T_2, high enough to reject heat to the coolant fluid. From 2–3 the right-hand piston remains stationary while the left-hand one moves inward. The gas is compressed isothermally rejecting heat to the jacket. At 3 the right-hand piston moves to the right at the same speed as the left-hand piston, and the gas is transferred through the cold regenerator to the right-hand side. The regenerator becomes warmer as a result. At 4 the gas is in the right-hand side and the left-hand piston remains stationary while the right-hand one continues to move to the right. The gas expands isothermally at T_1, accepting heat from the jacket until point 1 is reached again.

The rather peculiar motion of the piston required by the above description is not exactly achieved in practice. Instead an approximation to it is obtained by coupling the pistons through connecting rods and cranks which are out of phase. Their resulting motions, approximately sinusoidal, are shown by the broken lines in Fig. 7.7.3. The actual indicator diagram then becomes a smooth closed curve instead of the diagram with corners shown in Fig. 7.7.2. Furthermore, the isothermal processes are not strictly so in practice.

The cycle described above was originally invented by R. and J. Stirling in

7.7 The Stirling cycle

Fig. 7.7.4. Simplified cross-section of a gas refrigerating machine, designed for the liquefaction of air. 1 – main piston moving in cylinder 2; 4–warm side working space; 5 – cold side working space; 6 – two parallel connecting rods with cranks 7; 8 – crankshaft; 9 – displacer rod linked to connecting rod 10 and crank 11; 12 – ports; 13 – cooler in which heat is rejected from the working gas; 14 – regenerator; 15 – freezer; 16 – displacer piston with cap 17; 18 – condenser for the air to be liquefied, with annular channel 19, tapping 20, insulating screening cover 21 and mantle 22; 23 – aperture for entry of air; 24 – plates of the ice separator, joined by the tubular structure 25 to the freezer; 26 – gas tight shaft seal; 27 – gas cylinder supplying working gas; 28 – supply pipe with one-way valve 29. (Reproduced by permission from Philips Technical Review.)

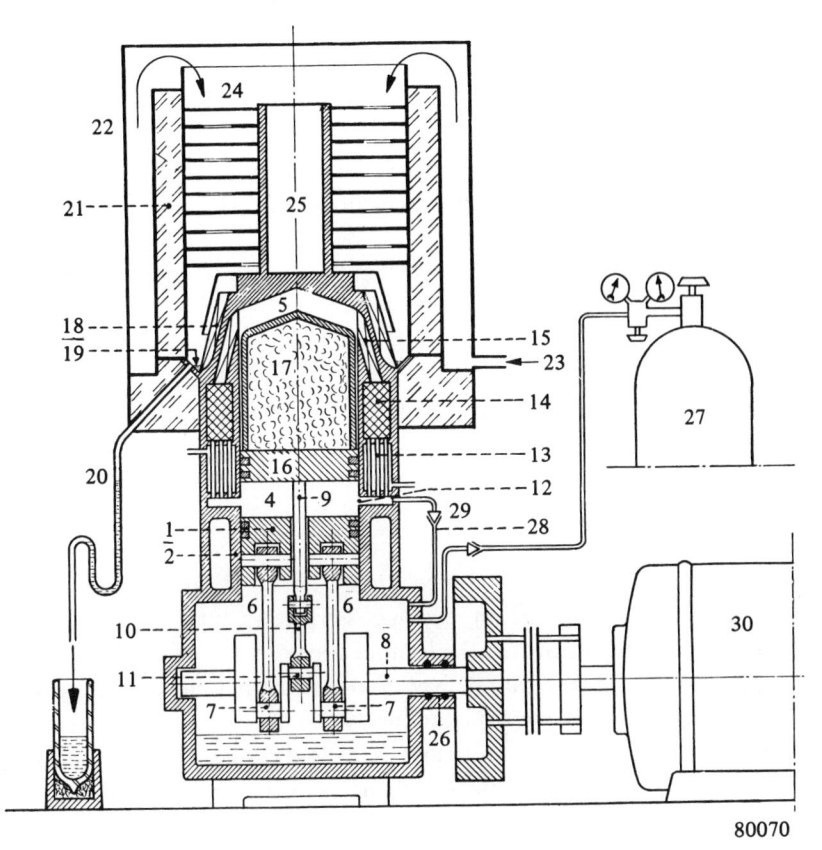

Ideal gas cycles 502

1827 as a heat engine. Its application to refrigeration was made by A. C. Kirk in the 1860s with some success, though few machines seem to have been built (Kirk, 1874). As a practical machine it passed into oblivion as inventors turned their attention to the simpler constant pressure cycle, and to the vapour compression cycle which offered more hope of successful development. It was revived in recent times by J. W. L. Köhler and has been found to have advantages for refrigeration duties at low temperature of the order 150 K and below. Its development at the Philips Research Laboratories into a machine for the liquefaction of air has been described by Köhler & Jonkers (1954). A simplified cross-section is shown in Fig. 7.7.4. It is constructed somewhat differently to the arrangement of Fig. 7.7.3, with a main piston 1 which does the actual compression and expansion of the gas, and a displacer piston 16 which transfers the gas from the warm space 4 to the cold space 5 and back again. During the transfer the pressure is the same on the two sides of the displacer, so that direct leakage of gas from the warm to the cold space is reduced.

The main piston in Fig. 7.7.4 is shown as sealed by rings, but in recent machines a rolling diaphragm seal is used. This is a flexible rubber sock sealed to the piston and the cylinder and backed by oil on the atmospheric side so that the working gas is hermetically sealed. The oil can be supplied under pressure to drive the piston.

By means of a two-stage arrangement much lower temperatures of the order 12 K can be produced.

The working gas in Stirling cycle machines for low temperatures is usually either hydrogen or helium.

There are many possible variations on the theme of the Stirling cycle. A description of these and of the development and applications of the cycle both as an engine and as a refrigerator is given by Walker (1973).

7.8 The vortex tube

In 1931 George Ranque in France observed low temperatures in the rotating flow of air in cyclone separators, and he devised an arrangement, since known as the Ranque tube or vortex tube, which when supplied with compressed air delivers two streams at atmospheric pressure, one cold and the other warm (Ranque 1933). After passing unnoticed for some years, the tube was investigated by Hilsch (1946), who recommended optimum dimensions and measured the performance. Since then many investigators have studied the device and have found means of improving its performance a little. Though the general principle of operation is fairly clear, a convincing quantitative theory of its performance has not yet appeared.

7.8 The vortex tube

The tube can be made either counterflow or uniflow. The counterflow version is shown in Fig. 7.8.1(a). A nozzle arranged tangentially at one end of a tube is supplied with compressed air. The tube at that end is partially closed by a diaphragm with a central orifice approximately half the tube diameter. At the other end of the tube a valve restricts the exit. When the valve is partly open a stream of cold air leaves through the orifice and a stream of warm air leaves through the valve. The temperature of the cold stream and its flow rate depend on the setting of the valve. When the valve is fully closed all the air comes out through the orifice and no reduction of temperature is found. As the valve is opened the flow rate of the cold stream rises and its temperature falls to a minimum value when the flow rate is about one quarter or one third of the total. Further closure of the valve causes the temperature of the cold stream to rise again.

The uniflow design is shown in Fig. 7.8.1(b). The cold stream comes out at the same end as the warm stream, as a central core of air which is separated by a special arrangement of orifice and valve. The uniflow tube is not quite as efficient as the counterflow type.

Fig. 7.8.1. Two forms of the Ranque–Hilsch tube: (a) counterflow and (b) uniflow.

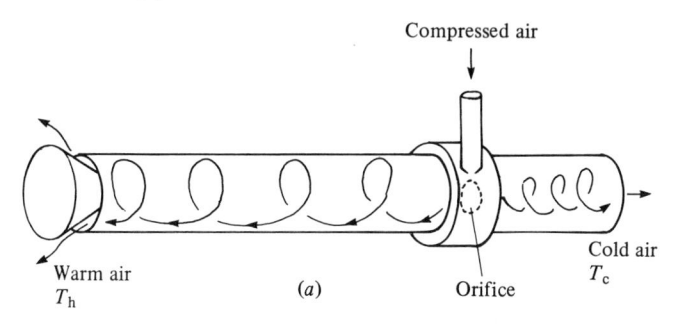

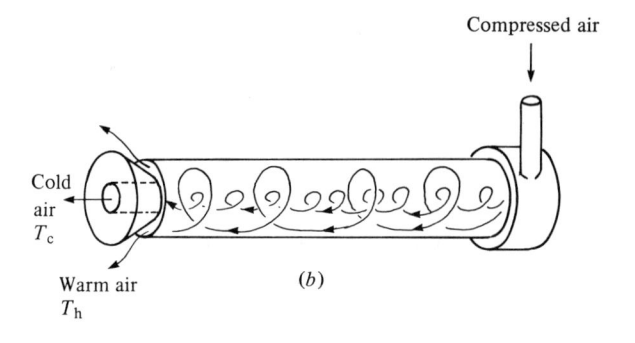

Ideal gas cycles

The vortex tube partly solves the problem mentioned in section 7.1 of how to use expansion in a nozzle to produce useful refrigeration, in other words to produce a stream with a low stagnation temperature. The drop in static temperature takes place in the nozzle as the air expands from the relatively high pressure at inlet to the pressure inside the tube. The air acquires kinetic energy and its temperature falls in accordance with eq. (7.1.17). Entering the tube tangentially therefore is a high-velocity jet at a low temperature, but the stagnation temperature of the jet is still the same as at the entrance to the nozzle, and a thermometer placed in the jet would show no drop of temperature. What happens subsequently in the tube is the removal of kinetic energy from a part of the total flow, which becomes the cold stream, and the dissipation of this kinetic energy by fluid 'friction' in the remainder of the flow which becomes the warm stream.

A mechanical analogue of the process would be an impulse turbine coupled to an inefficient fan. Part of the high-velocity jet does work on the turbine: its kinetic energy falls and so does its stagnation temperature. The work developed by the turbine is used to drive the fan and to raise the stagnation temperature of the remainder of the flow. In the uniflow tube a central core of fluid is decelerated by viscous shear stresses between it and an outer annulus of fluid. The central core loses momentum and kinetic energy. The outer annulus has work done on it by the central core, and by dissipative processes the work causes a rise in the stagnation enthalpy and temperature.

The division of the flow into a central core and an annulus is more a conceptual device than a physical reality of course. In fact there is a continuous gradient of velocity along the radius. But the principle holds, that there is a flow of momentum outwards which reduces the stagnation temperature of layers near the axis. The situation is complicated by the effects of heat conduction and convection, and this has made the development of a quantitative theory difficult.

The above description refers strictly to the uniflow tube. In the counterflow tube a further complication occurs, in that at some point the layers nearer the axis and moving initially towards the valve reverse their direction and flow out through the orifice. There is no fundamental difference in the principle of operation, however.

The performance of the vortex tube is expressed by the mass flow rate of the cold stream, $\dot{m}_c$, and its temperature, T_c, when operating at a given inlet pressure p_1 and temperature T_1. As compared with an ideal device which expanded all of the inlet air at the rate $\dot{m}_1$, say, reversibly and adiabatically to the pressure at the outlet, p_o, usually atmospheric pressure, the isentropic

7.8 The vortex tube

efficiency of the device considered as an expander would be:

$$\eta_e = \frac{\dot{m}_c(T_1 - T_c)}{\dot{m}_1(T_1 - T'_c)} \tag{7.8.1}$$

where T'_c is the temperature after isentropic expansion to pressure p_o. From eq. (7.1.13), putting $p_2 = p_o$, $T_2 = T'_c$:

$$T_1 - T'_c = T_1\left(1 - \frac{1}{\rho}\right) \tag{7.8.2}$$

where:

$$\rho = \left(\frac{p_1}{p_o}\right)^{(\gamma-1)/\gamma} \tag{7.8.3}$$

The ratio:

$$\frac{T_1 - T_c}{T_1 - T'_c} = \varphi \tag{7.8.4}$$

may be called the isentropic temperature drop efficiency. Denoting the cold fraction by μ:

$$\mu = \dot{m}_c/\dot{m}_1 \tag{7.8.5}$$

then:

$$\eta_e = \mu\varphi \tag{7.8.6}$$

Tubes may be designed to give a maximum φ when it is desired to reach as low a temperature as possible, or to give maximum η_e when the largest cooling capacity is wanted.

Hilsch recommended:

for maximum φ $d_o = 0.45\,D$ (7.8.7)

for maximum η_e $d_o = 0.6\,D$ (7.8.8)

where d_o is the diameter of the orifice in a counterflow tube, and D is the diameter of the tube. The other dimensions for best results are approximately:

$$d_n = \tfrac{1}{4}D \tag{7.8.9}$$
$$L = 50\,D \tag{7.8.10}$$

where d_n is the diameter of the nozzle throat and L is the length of the tube on the warm side.

Some of Hilsch's results for three orifice diameters and several inlet

Ideal gas cycles 506

pressures are given in Fig. 7.8.2. The maximum temperature drop is about 55 K at an inlet gauge pressure of 9.81 bar and an inlet temperature of 20 °C. Assuming an atmospheric pressure of 1 bar, approximately, the value of $T_1 - T'_e$ would be:

$$293 \left(1 - \frac{1}{10.81^{0.286}} \right) K = 145 \, K$$

and:

$$\varphi_{max} = 55/145 = 0.38$$

In a larger tube of $D = 17.6$ mm Hilsch obtained $\varphi_{max} = 0.47$.

These values of the isentropic temperature drop efficiency have not been much improved, though with larger tubes some improvement is possible on account of the higher efficiency of larger nozzles. By cooling the warm side of the tube Martynov & Brodyansky (1964) found some improvement. It is also possible, if very cold air is needed, to tap off part of the cold stream near the axis.

The upper parts of the graphs in Fig. 7.8.2 show the temperature of the

Fig. 7.8.2. Hilsch's results for a counterflow vortex tube of 4.6 mm diameter, length 300 mm with a nozzle diameter of 1.1 mm. $t_1 = 20\,°C$; atmospheric pressure at outlet not stated. Cold orifice diameter (a) 1.4 mm (b) 1.8 mm (c) 2.2 mm. The mass rate of air flow through the nozzle is given by $0.214.10^{-3} (\Delta p/\text{bar})$ kg/s where Δp is the gauge pressure.

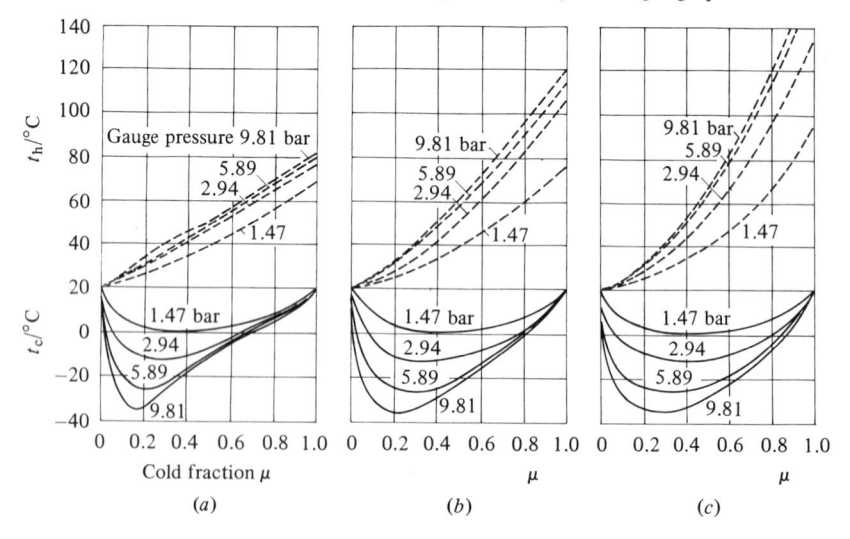

7.8 The vortex tube

warm stream. For adiabatic operation of the tube these are related to the temperature of the cold stream and the cold fraction by the steady-flow equation:

$$\mu T_c + (1 - \mu)T_h = T_1 \tag{7.8.11}$$

after cancelling the specific heat capacity, assumed the same for each stream. In terms of temperature differences:

$$\mu(T_1 - T_c) = (1 - \mu)(T_h - T_1) \tag{7.8.12}$$

When $\mu = 1$ and all the air leaves by the cold side, the process is simply an irreversible throttling process, with no change of specific enthalpy and no change of temperature, $T_c = T_1$. According to eq. (7.8.12), T_h then becomes indeterminate. The values of T_n at $\mu = 1$ in Fig. 7.8.2 must therefore be regarded as limiting values.

As compared with an expansion turbine for reducing the temperature of air, the vortex tube must be regarded as a poor performer. In the example quoted above $\varphi = 0.38$ was obtained at $\mu = 0.22$, giving $\eta_e = 0.084$. With a larger cold stream the value of η_e could be increased to about 0.2. But an expansion engine or turbine could do much better than this. The vortex tube has a role therefore only in circumstances where compressed air is available and when cheapness and light weight are desired. It has found some applications as a laboratory means of producing small refrigerating capacities at quite low temperature, and also for cooling suits for use in hot working conditions. One or two firms specialise in the manufacture of tubes for these purposes.

The length of the vortex tube can be reduced below the 50 diameters recommended by Hilsch, as shown by Parulekar (1961), who has also given results for a variety of diaphragms and nozzles.

8

Cryogenic engineering

8.1 The field of low temperatures

The name 'cryogenics' has been given in recent years to the techniques of reaching and using very low temperatures. 'Very low' with regard to temperature means different things to different people. To the physicist it means temperatures close to 0 K at which the energy retained by a substance at absolute zero has significant effects on its properties. To engineers it may simply mean temperatures below those which have been commonly obtained in the past by vapour compression or absorption machines; or temperatures at which the common material of construction, mild steel, becomes unusable on account of embrittlement.

Nevertheless there is some agreement that cryogenics is concerned with temperatures below about 120 K, approximately, which is a little above the normal boiling point of liquefied natural gas. Below this temperature there is a wide range of industrial and laboratory processes which have some common features. The object of many cryogenic operations is not so much the provision of a service like refrigeration but the manufacture of a product such as liquid nitrogen or gaseous oxygen. Many of the processes have developed out of the concepts of those originally used for making liquid air; and it will be seen that heat transfer between fluids inside the process plays a much more important part than it does in orthodox refrigeration.

Since the temperatures quoted in this chapter are almost all negative Celsius ones they will be given exclusively on the absolute scale ($0\,°C = 273.15$ K).

The refrigeration industry was based on the discoveries of the late eighteenth and early nineteenth centuries that many gases could be liquefied at ambient temperature by compression. The lower temperatures which then became available by boiling the liquefied gases under reduced

8.1 The field of low temperatures

pressure were applied in attempts to liquefy other gases. By 1845 Michael Faraday had liquefied by compression and cooling to a temperature of 163 K (obtained with solid carbon dioxide and ether) all of the then known gases with the exception of oxygen, nitrogen, hydrogen, carbon monoxide, nitric oxide and methane. All attempts to liquefy these failed. In Vienna, Johannes Natterer showed that even at a pressure of 3600 atm with cooling to 195 K, oxygen, nitrogen and hydrogen were still gaseous. For some time the view was held that these gases were 'permanent', until the experiments of

Fig. 8.1.1. Cailletet's apparatus of 1877. The hydraulic press on the right pumps water at a pressure up to 1000 atm into the space surrounding tube T on the left, by working the lever L and pump P and, for the higher pressures towards the end, the wheel V operating a piston in cylinder dP. The gas in the tube at P on the left is compressed by the rise of mercury through the open bottom end of T. The gas is surrounded by a freezing mixture or by boiling sulphur dioxide in the tube M. The pressure can be suddenly released by the valve V'.

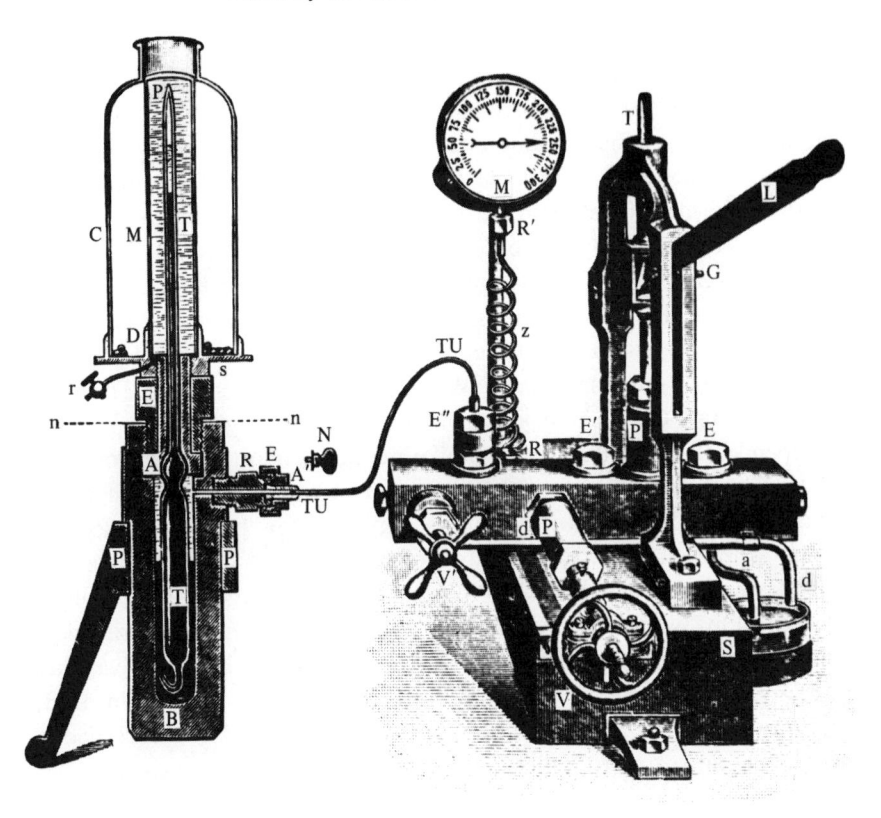

Thomas Andrews in 1861–9 on carbon dioxide showed the relevance of the critical temperature. Above this temperature it is not possible for liquid and vapour to exist in equilibrium, and it then appeared that the failure to liquefy the permanent gases had been caused by their very low critical temperatures. Renewed efforts were made, and in December 1877 evidence of the liquefaction of oxygen was obtained almost simultaneously by Louis Cailletet in France and Raoul Pictet in Geneva. Cailletet compressed oxygen to 300 atm and at the same time cooled it to 244 K by boiling sulphur dioxide. On allowing the gas to expand, doing work against the atmosphere, a mist of liquid became visible. Cailletet's apparatus is shown in Fig. 8.1.1. Pictet as a pioneer of refrigeration already had powerful means of cooling at his disposal in the form of cascade systems using sulphur dioxide and carbon dioxide. He also compressed oxygen, cooled it and then allowed it to expand suddenly. Neither Cailletet nor Pictet could make liquid oxygen in sufficient quantity to see it in a container. This was first achieved by Sigmund von Wroblewski and Karel Olszewski in Poland in 1883, by compression and cooling with ethylene at 137 K under reduced pressure. Air was liquefied by Wroblewski in 1885. He found that oxygen and nitrogen are completely miscible as liquids and not, as had been suspected, only partially so.

By the early 1890s liquid air, oxygen and nitrogen were being produced in laboratories in sufficient amounts for research on their properties. At Leiden, Kamerlingh Onnes developed the cascade refrigerator using methyl chloride, ethylene and nitrogen as refrigerants for liquefying air. His machine was eventually capable of producing it at the rate of 14 dm^3/h and for many years supplied the requirements of the Leiden laboratory.

Meanwhile the possible industrial applications of liquid air had been noticed, mainly with a view to separating from it oxygen, for which there was a small but growing demand. William Hampson in England, Tripler in the USA and Carl Linde in Germany were working on the problem independently and all arrived at similar methods almost simultaneously in 1895. They were based on the reduction of temperature which a gas undergoes on being throttled from a high to a low pressure, called the Joule–Thomson effect. The reduction of temperature is small for air, about 1/4 K per atmosphere, but when it is combined with regenerative cooling, suggested by William Siemens in 1857 as an improvement in Gorrie's cold air machine, a small fraction of the air can be liquefied. Ernest Solvay had already tried regenerative cooling with an expansion engine in 1885, substantially in the form described in section 7.4, but had not succeeded in liquefying air owing to faulty design and heat gains to the expander.

8.1 The field of low temperatures

Linde and Hampson's patent applications are dated within a few days of each other, but Linde was the first to produce liquid air in quantity for sale. His apparatus, shown in Fig. 8.1.2, was already quite advanced, and incorporated two-stage compression and expansion of the air. Linde used a

Fig. 8.1.2. An early Linde machine for liquefying air. Dry CO_2-free air is compressed in the first stage, e, of a Whitehead compressor to about 20 atm, and in the second stage, d, to about 200 atm. Oil and water are separated in f. The high-pressure air is then cooled by a freezing mixture in g, and passes through the central tube of the heat exchanger. It is throttled to the intermediate pressure in the valve a and a part of the gas goes back through the heat exchanger to the suction of the second-stage compressor. The remainder goes on to the second throttling valve b where it expands to atmospheric pressure. A part of the air is liquefied and collects in c whence it can be withdrawn via the tap h. The unliquefied air returns through the heat exchanger and leaves the apparatus. These machines were made with outputs of liquid air of 2 to 300 dm^3/h.

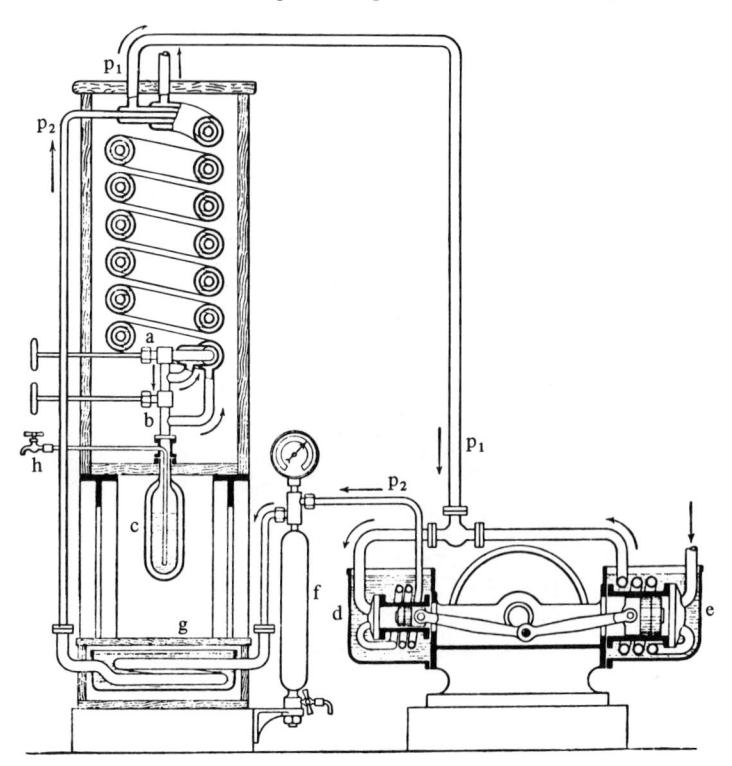

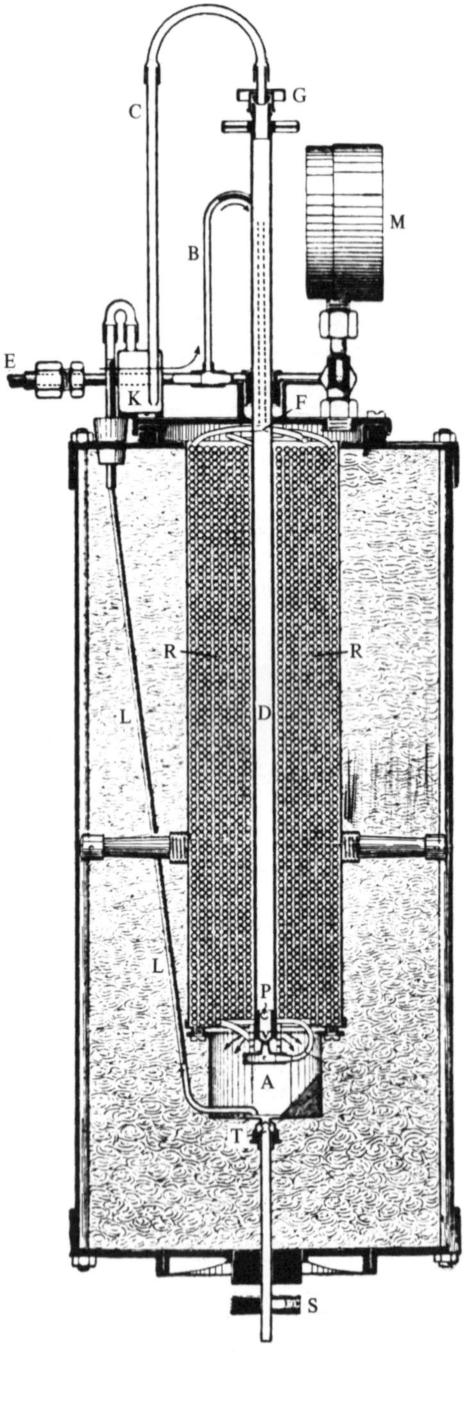
G
C
M
B
E
K
F
R
R
L
D
L
P
A
T
S

8.1 The field of low temperatures

tubular heat exchanger in which the expanded air was used to cool the compressed air coming to the valve. This exchanger was rather massive and the apparatus took a long time to cool down. Hampson, whose apparatus is shown in Fig. 8.1.3, introduced a heat exchanger in which the high-pressure coil is wound inside a space through which the low-pressure revert gas passes. The Hampson heat exchanger has since been commonly used for many systems in which one of the gases is at a low pressure.

The simple Linde–Hampson process is rather inefficient, for reasons which will be seen later. It can be improved by precooling the high-pressure gas by refrigerating plant, or by using some of the high-pressure gas to produce refrigeration by expansion in an engine. Expansion engines were introduced into liquefaction plants by Georges Claude in France and they are now common features of most liquefaction systems. The subsequent development of liquefaction methods is treated later in the chapter.

Back in the laboratory, attempts to reach lower temperatures were continuing. An important step made by James Dewar in 1892 was the

Fig. 8.1.3. Hampson's apparatus for making liquid air on a small scale in laboratories. Air at a pressure of 170–80 atm, dried and freed of carbon dioxide, enters through the tube E, with a branch to the pressure gauge M, and through B to the header F, which feeds 4–6 thin walled copper tubes of 2–3 mm internal diameter, 15–25 m long. These tubes are wound into very tight helices almost completely filling the space R. At the bottom these tubes debouch into the header shown above the letter A, from which the gas expands upwards through an orifice regulated by the valve cone V on the hollow spindle D controlled by the ebonite wheel G. The fraction of the air liquefied collects in A, from which it can be withdrawn downwards through the tap T and hollow spindle with ebonite wheel S. The unliquefied air passes upwards in the space R around the tubes carrying the high-pressure gas and leaves to the atmosphere at the top. The level of liquid in A is indicated by the glycerine manometer K (working on the principle described in section 2.11) connected to the bottom of the liquid in A by the tube L, and to the pressure after expansion via the hole P, the hollow spindle D and the tube C.

The models were made with capacities of 2 and 4 dm^3/h, with about 6% of the compressed air liquefied. Owing to the effectiveness and low mass of the Hampson heat exchanger, liquid air was produced within about 12 min of starting, and full output was attained in another 15 min.

invention of the vacuum flask for holding very cold liquids with little loss. By 1898 all the old permanent gases and fluorine, which had meanwhile been isolated, had been liquefied with the exception of hydrogen. Pictet had claimed to produce a mist of hydrogen but this claim was not generally conceded. It was known that the method of cascade refrigeration with nitrogen or oxygen in the last stage was not capable of liquefying hydrogen because its critical temperature was understood from calculations based on its equation of state to be about 30 K, whereas nitrogen freezes at 63 K and the lowest temperature obtainable by boiling oxygen under reduced pressure is about 60 K. It was also known that the Linde–Hampson process would not work for hydrogen because when compressed and throttled at ambient temperature hydrogen becomes warmer instead of colder. Olszewski however had found that below a temperature of 190 K, called the inversion temperature, the temperature of compressed hydrogen would fall on throttling. By cooling hydrogen with liquid air to below 190 K and then using the Linde–Hampson process, Dewar succeeded in liquefying hydrogen in 1898 in litre quantities, and in 1899 he solidified it by evaporating the liquid under reduced pressure.

In the meantime new gases were being discovered in the atmosphere, beginning with the finding of argon by Lord Rayleigh and William Ramsay in 1894. This gas was liquefied without difficulty since its boiling point lies between those of oxygen and nitrogen. In 1895 Ramsay extracted from the mineral cleveite an element, helium, whose existence had been surmised by Norman Lockyer from his observations of the spectrum of the solar chromosphere in 1878. The subsequent discovery of helium in the atmosphere, and of the other rare gases neon, krypton and xenon, by separation from liquid air was one of the first striking applications of low temperatures to another field of research. Without it they could hardly have been suspected at the time since they seemed to be chemically inert.† The discovery of argon and the rare gases added a new group to the periodic table, a group which turned out to be very suggestive in relating chemical properties to the electronic structure of the atom.

Krypton and xenon can be liquefied by liquid air, and neon by liquid hydrogen, but for some years helium defied all attempts to liquefy it. Its critical temperature 5.2 K is below the freezing point of hydrogen, 13.8 K, so that it cannot be liquefied by compression and cooling by boiling hydrogen. Fortunately its inversion temperature of about 40 K is high enough for it to

† Within the last twenty years several compounds of the noble gases, particularly of xenon, have been prepared.

8.1 The field of low temperatures

be cooled by boiling hydrogen to a temperature at which the Linde–Hampson method can work. It was eventually liquefied by Kamerlingh Onnes in 1908.

Within a short time it appeared that liquid helium had the most unexpected properties. Attempts to solidify it by low temperature failed and it now appears that helium remains a liquid (at atmospheric pressure) right down to absolute zero. It can be solidified by pressure, however, as indicated by the phase diagram in Fig. 8.1.4. The specific heat capacity of liquid helium instead of diminishing as expected with falling temperature, goes towards infinity at a temperature of 2.2 K and becomes finite again below that. This point is known as the lambda point from the shape of the curve of specific heat capacity with temperature. Below 2.2 K helium exists as a different liquid phase known as helium II, with zero viscosity and a thermal conductivity higher than that of silver. Helium II is called a 'superfluid' and it is recognised as a fourth state of aggregation, neither solid, liquid nor gas. Its investigation has thrown a completely new light on the nature of matter.

Not only helium but other materials turned out to have remarkable properties at low temperatures. The most spectacular of these is the superconductivity of metals, discovered by Onnes in 1911. He found that at a temperature of 4.1 K for mercury and 7.2 K for lead their electrical resistivity became immeasurably small: it is impossible to tell experimentally whether it is zero or not. Only recently has it been possible to exploit this discovery industrially.

Fig. 8.1.4. The phase diagram of helium. The lower curve shows the equilibrium between the normal liquid, helium I, and vapour, terminating at the critical point (C.P.).

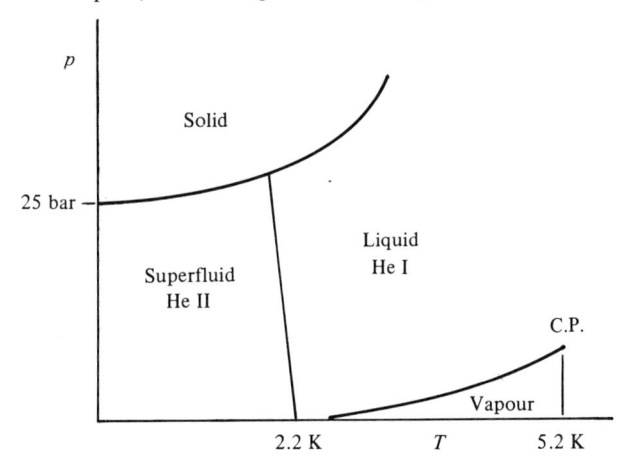

Helium rapidly assumed a position of crucial importance for low-temperature research. Fortunately the need to find a replacement for hydrogen in airships during and after World War I stimulated a search for better sources than the rarer minerals, and it was found in some natural gases in the USA in concentrations up to about 2%. These gases are still the main source of helium. The liquefaction of helium was for many years a rather hit and miss laboratory operation, but with the advent of the Collins liquefier, described in section 8.5, liquid helium became readily available in the 1950s.

By evaporation of helium under reduced pressure Onnes had reached a temperature of 0.83 K in 1922. It seemed at the time that this would be the limit unless some new method could be found. In 1926, however, William F. Giauque and Peter Debye independently suggested the method of adiabatic demagnetisation. When a gas is compressed isothermally its entropy is reduced and subsequent reversible adiabatic expansion lowers the temperature. Giauque and Debye suggested an analogous operation in which isothermal magnetisation of a paramagnetic salt such as gadolinium sulphate replaced the compression of the gas and sudden demagnetisation replaced the adiabatic expansion. By this method Giauque attained a temperature of 0.25 K in 1933 and subsequently temperatures of 0.01 K have been reached.

Fig. 8.1.5. The curves in the T–s plane represent some intensive parameter by which the entropy can be changed at constant temperature, for example pressure of a gas or the magnetisation of a substance. If the curves for this parameter constant are as shown in (a), absolute zero can apparently be attained in a finite number of operations by a sequence of 'compression' and adiabatic 'expansions'. If they are as shown in (b) attainment of absolute zero is not possible in a finite number of operations.

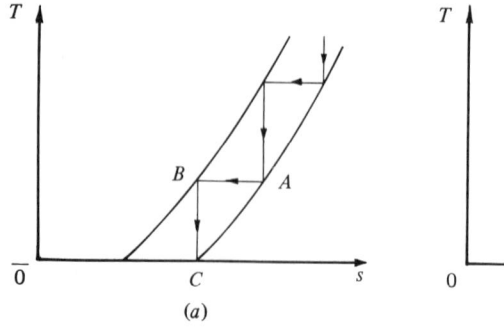

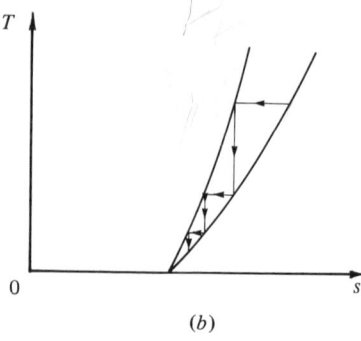

8.1 The field of low temperatures

Magnetisation is the alignment of electronic spins but protons and neutrons also have spins which can be aligned by a magnetic field. A similar method applied to these nuclear spins by Francis Simon and Nicholas Kurti in 1956 produced for a brief instant a temperature of 16×10^{-6} K. This is not a macroscopic temperature, however, but only the temperature associated with the nuclear spins. The macroscopic temperature of the electrons and lattice has only in recent years been lowered to 30×10^{-3} K in a specimen of copper, and liquid helium has been maintained at a temperature of 70×10^{-3} K.

The approach to absolute zero continues, but it is not possible to attain it. Absolute zero happens to be so called because of the way in which the absolute scale is defined by eq. (1.12.3). It was chosen in this way because such a scale corresponds with the ideal gas scale and with ordinary measurements of temperature by thermometers. It could equally well have been defined so as to make the difference of temperature proportional to the logarithm of the ratio of Q_1 and Q_2 in eq. (1.12.3). Such a scale would not go to zero as Q_1 tends to zero but to minus infinity. For low temperatures this scale gives a more realistic picture of events. The range of temperature, called a decade, between 0.1 K and 1 K has the same significance as the decades 1 K to 10 K, 10 K to 100 K and so on.

The non-attainability of the absolute zero is closely connected with the Third Law of Thermodynamics, according to which the entropy of all substances in their most ordered state, crystalline in effect, tends to the same value at 0 K. For many purposes it is convenient to call this value arbitrarily zero, but this is not essential. Referring to Fig. 8.1.5(a) which shows some lines of a constant parameter on temperature–entropy coordinates, a reduction of temperature is possible by isothermal compression followed by reversible adiabatic expansion. If the entropies are different at absolute zero there is no reason why a path ending $A–B–C$ should not be chosen to reach zero. On the other hand if the lines for some constant parameter corresponding to pressure, which may be the magnetisation of a paramagnetic salt, run as shown in Fig. 8.1.5(b) it is clear that the absolute zero will not be attained in a finite number of operations.

The field of research opened up by nuclear refrigeration has required means of providing continuous refrigeration of the heat sink, at temperatures about 0.1 K, which forms the starting point for further descent. Two methods are of interest as examples of systems which work continuously in cycles, one of them a non-flow cycle and the other a steady flow cycle.

The non-flow cycle is the method of adiabatic demagnetisation as developed by J. G. Daunt and C. V. Heer. It incorporates a device called a heat switch or thermal valve. The heat switch is a thin rod of lead which has a superconducting transition temperature of 7.2 K in the absence of a magnetic field. At a temperature below this the application of a magnetic field removes the superconductivity. At the same time the lead becomes a much better conductor of heat, by a factor of a thousand or so. The lead can be made into a thermal insulator or a thermal conductor by removing or applying the field. Using the heat switch, the paramagnetic salt in the working cell can be put into thermal communication with liquid helium whilst it is being magnetised to remove the heat liberated. It is then isolated from the liquid helium and demagnetised, whereupon its temperature falls. By means of the other heat switch it can be put into thermal communication with objects to be cooled. The apparatus has a refrigerating capacity of 7×10^{-6} W at a temperature of 0.2 K. The method of cooling using adiabatic demagnetisation has now been largely superseded by the dilution refrigerator, described below.

The dilution refrigerator is a steady-flow system which works on the enthalpy change which occurs when helium-3 diffuses into helium-4. Helium exists as the isotopes helium-4, the more abundant one with relative atomic mass 4, and helium-3 the rarer one with relative atomic mass 3. Helium-3 forms only 2×10^{-7} of natural helium but it can be made in small quantities by nuclear reactions. The liquids helium-3 and helium-4 are miscible in all proportions above a temperature of 0.87 K, giving a normal liquid or a superfluid, depending on the composition. Below 0.87 K miscibility is limited as indicated by the phase diagram Fig. 8.1.6. A point X

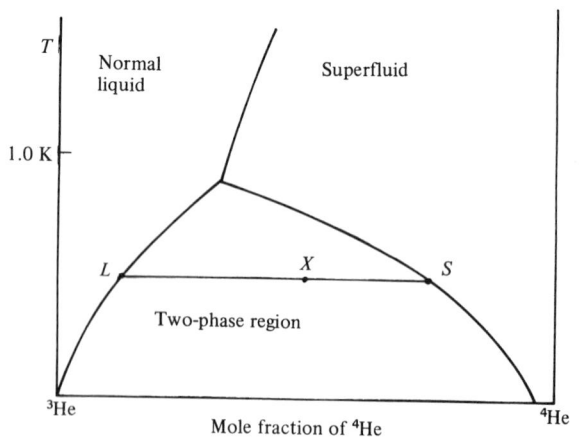

Fig. 8.1.6. The miscibility diagram for mixtures of helium-3 and helium-4.

8.1 The field of low temperatures

represents a mixture of a helium-3-rich layer of composition L floating on a denser helium-4-rich layer of composition S. The liquid L is normal but S is a superfluid. When the equilibrium is disturbed by heating the phase L, helium-3 passes into solution in S with absorption of heat. The process of helium-3 diffusing into a superfluid is in some respects like the evaporation of a liquid into a vapour, but there is an important practical difference. The density of saturated helium vapour at a temperature of 0.1 K is so minute that a practical vapour compression system could not be made to work, whereas the partial density of helium-3 in the liquid is some millions of times greater than the density of the vapour.

When the helium-3 atoms have dissolved in the superfluid helium-4-rich layer they can move quite freely under the influence of a concentration gradient.

The idea of using the phenomenon to produce refrigeration was suggested by Heinz London in 1951, but only since about 1966 have dilution refrigerators been available commercially.

The system is shown diagrammatically in Fig. 8.1.7. Beginning at point 1, helium-3 vapour is compressed by rotary and diffusion pumps to a pressure of about 40 mbar, at which it condenses to a liquid at point 3 by contact with a bath of helium-4 (ordinary helium) at a temperature of about 1.3 K.

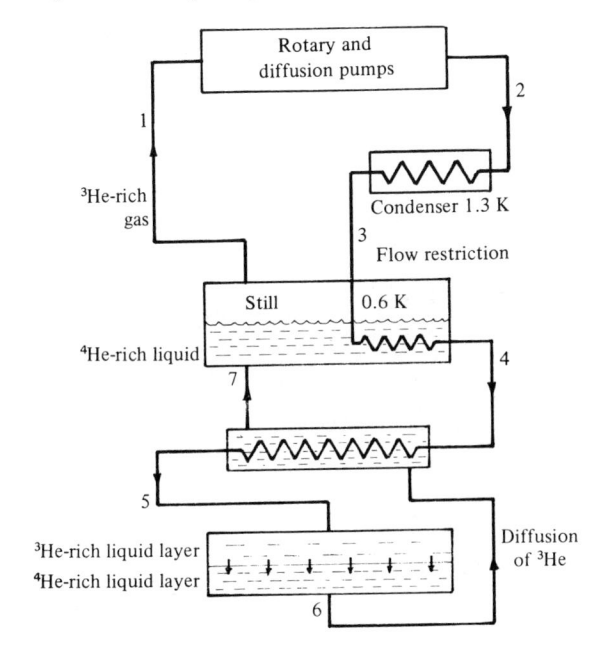

Fig. 8.1.7. The principle of operation of a dilution refrigerator.

After passing through a flow restrictor the helium-3 liquid is further cooled to point 4 by contact with the evaporator or still at a temperature of about 0.6 K. In the heat exchanger the liquid is further cooled to point 5 and it then enters the dilution chamber. There it forms an upper layer. The helium-3 diffuses into the helium-4-rich lower layer and in doing so absorbs heat. The specific refrigerating effect is approximately 0.03 J/g of helium-3 at a temperature of 0.1 K, but diminishing rapidly at lower temperatures. The helium-3 moves from point 6 to point 7 through the helium-4-rich superfluid and into the evaporator or still where the concentration is lower. Application of heat there vaporises helium-3 more rapidly than the less volatile helium-4 and the vapour is taken away by the pumps.

With no heat input to the dilution chamber a temperature of 0.005 K can be attained with the more elaborate models, and about 0.025 K with a smaller and simpler one, shown in Fig. 8.1.8. At a temperature of 0.1 K

Fig. 8.1.8. A small dilution refrigerator, excluding the pumps, vacuum system and condenser. The base temperature, with no heat input to the mixing chamber is 0.025 K. The refrigerating capacity at 0.1 K is 15×10^{-6} W. (By courtesy of Oxford Instruments Ltd.)

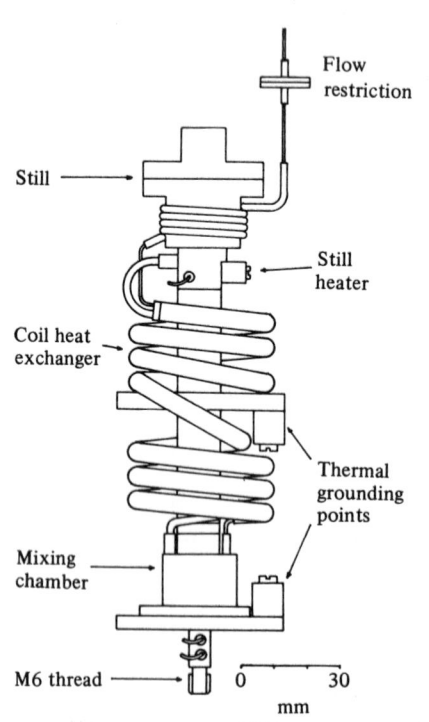

8.1 The field of low temperatures

refrigerating capacities up to 1 mW are available with the largest model. The refrigerating capacities of systems such as these are minute by ordinary standards. A heat rate of 1 mW for example is about the same as that by natural convection in air at ambient temperature from a surface of 1 cm^2 area at a temperature difference of 1 K, or in other terms as the heat rate from burning 1 gramme of fuel oil in a year. Yet they are sufficient for experimental work at these low temperatures.

In fact at very low temperatures all energy changes become small when compared with those at ambient temperature. The specific refrigerating effect of helium-3 quoted above may be compared with those given in Table 3.9.1.

Ultra-low temperatures in the millikelvin and microkelvin ranges have not yet found industrial uses, but considering how liquid hydrogen and helium have now become common instead of rare commodities, it seems likely that it is only a matter of time before uses are found for still lower temperatures.

Cryogenic engineering

The principal large scale uses of cryogenic techniques today are in the separation of gases, especially those of the atmosphere and cracked petroleum gases, and the liquefaction of natural gas. Sections 8.6 and 8.8 are devoted to these processes.

The products of the gas separation industry have found numerous applications. Liquid nitrogen in particular has become the universal cooling medium for temperatures above 77 K, on account of its low cost and relative safety. Considerable amounts are used for the freezing of foodstuffs and for the transport of frozen food, but it is also used in the metals industry for many purposes such as cryo-quenching and cryo-annealing (Darlington, 1969). Cryo-quenching is an interesting application of the Leidenfrost phenomenon. Quenching a hot metal in liquid nitrogen gives a slower and more controlled cooling rate than quenching in water or oil owing to the insulating layer of vapour which is formed, and this is advantageous for thin metal parts. Other applications are based on the fact that rubber and plastic parts become brittle at 77 K and can easily be trimmed or cleaned up. The same technique is used for breaking scrap metal. Copper, for example, which does not become brittle, can be readily separated from iron and steel parts which do.

Liquid nitrogen is produced by plants which also make liquid oxygen, but it is not a waste product. The nitrogen gas itself is almost free because, as will be seen, the work of separating nitrogen from the atmosphere is small.

But the production of liquid nitrogen requires the expenditure of considerable work.

Liquid oxygen was first introduced commercially as an explosive with charcoal, but was sold later mainly for convenience of transport to factories using it for welding and flame cutting. Its use as the oxidant in rocket motors dates from the experiments of R. H. Goddard in the USA in the 1920s and the work of Wernher von Braun in Germany in the 1930s and 40s on the V-2 rockets. With each rocket carrying about $5\frac{1}{2}$ t of liquid oxygen the total demands for the German V-2 campaign became considerable, about 13 000 t/year. Other oxidants such as fluorine (boiling point 86 K) can be used, but they are mostly unpleasant liquids and dangerous to handle, and liquid oxygen has remained the favourite. The Saturn 5 rockets for the Apollo space programme used liquid oxygen in all stages, with kerosene as the fuel in the first stage and liquid hydrogen as the fuel in the second and third stages.

The demand for liquid hydrogen (boiling point 20 K) for the rocket programme has brought in train the construction of plants to make it with outputs of 100 t/day or more with corresponding storage and transport facilities. Liquid hydrogen at 20 K is carried around by rail like any other industrial chemical.

The general availability of liquefied gases at low temperatures has stimulated their use for many purposes. One of these is the production of very high vacua by the method known as 'cryo-pumping'.

The vapour pressure of (solid) nitrogen at a temperature of 20 K is about 1.3×10^{-8} N/m², which represents a high vacuum. Consequently if nitrogen gas is exposed to a surface at this temperature, refrigerated by liquid hydrogen say, it will condense until the pressure comes down to this equilibrium value. This pressure is much below that which can be attained by mechanical or diffusion pumps, although these may be used during the pull-down. Given a condensing surface of sufficient area, high pumping rates can be attained. Cryo-pumping is applied to producing the high vacua needed for the deposition of thin films and is also used in particle accelerators.

Even at moderate vacua cryo-pumping may be competitive with diffusion pumps when large flow rates are required. This has found application in low-density wind tunnels for reaching high Mach numbers on small models about 10 mm long.

Cryo-pumping is also used for space simulation. As usually understood this means simulation of the pressure some hundred kilometers above the earth's surface, of the order 10^{-3} N/m². At this pressure convective heat

8.1 The field of low temperatures

transfer is negligible and heat balances for spacecraft can be realised under laboratory conditions.

Going down to liquid helium temperature (≈ 4 K) an important application is the amplification of the feeble radio signals received from satellites. The signal strength may be only about 10^{-13} W and it is essential to reduce the background noise generated thermally in the materials of the amplifier. Masers (standing for 'microwave amplification by stimulated emission of radiation') operate at about 4 K cooled by liquid helium (Daglish, 1966). Parametric amplifiers can work at higher temperatures of about 20 K. The receivers for satellite communications were originally maintained by regular deliveries of liquid helium but the trend has been towards self-contained closed cycle refrigerators.

Another device requiring liquid helium is the extremely sensitive bolometer used in infra-red detection systems for homing missiles on to heat sources from engine exhausts. These are based on superconducting materials which show a gradual change of resistance instead of a sudden drop at a transition temperature.

Considerably more refrigerating capacity is required for bubble chambers used for research on elementary particles. When the particle passes through a superheated liquid, i.e. metastable liquid at a temperature above its boiling point corresponding to its pressure, the particle provides a nucleus for vapour to form, and the track of the particle is rendered visible. In order that the whole mass of liquid does not boil, the superheated state must last for only a short time, of the order 20 ms. This is secured by a piston which suddenly reduces the pressure over the liquid and then restores it. The bubble chamber at CERN contains about 1 m^3 of liquid hydrogen or deuterium at a temperature of about 20 K. The refrigerating duty for pull-down and filling the chamber is about 7 kW, and during normal operation to make up the losses, the refrigerating duty is about 4 kW. The refrigeration is provided by a closed circuit of the Claude type, described in section 8.5, using hydrogen as refrigerant.

With temperatures of about 4 K becoming available on an industrial scale, attention has been turned to the exploitation of superconductivity. Following its discovery in 1911 it seemed to offer many possibilities for reducing the size of electric conductors, but early findings were disappointing. It was found that most pure metals (now called Type-1 superconductors) lost their superconductivity in a magnetic field. This seemed to rule out their use for motors and electromagnets. It was subsequently discovered however that niobium and certain intermetallic compounds behaved quite differently in that they allowed current and flux to penetrate into the bulk material and could carry current without losing their superconductivity. These are the Type-2 materials. They have as a rule higher transition temperatures than those of Type-1. For niobium it is 9.2 K, and for compounds of niobium with zirconium, titanium, tin or germanium transition temperatures up to 25 K can be obtained. At liquid helium temperature they can carry a large current and support a high flux density without losing their superconductivity. The manufacture of superconductors requires a new technology which cannot be described here. A superconducting 'wire' for magnets or motors is in fact a complex assembly of many superconducting filaments of, say, niobium stannide Nb_3Sn, in a bronze matrix, incorporating pure copper strands to act as a shunt in the event of reversion to the normal conducting state because of refrigeration failure or overload.

The principal use of superconductors up to now has been for making powerful electromagnets. In ordinary electromagnets the magnetic flux density is limited to about 2.5 T by considerations of size and cooling of the windings. Superconducting windings have enabled very powerful magnets with flux densities up to 12 T to be made in a very much smaller size. These small magnets are available commercially and find many uses. Larger superconducting magnets are specially designed and built for high-energy particle accelerators.

Superconducting electrical machines are now under development in several countries. Appleton (1977) describes the development of a homopolar motor with an output of 2.5 MW in which the field windings work at the temperature of liquid helium. Other uses of superconductivity envisaged are for the underground transmission of electric power and for the storage of electric energy. Foner & Schwartz (1974) describe recent developments and possibilities.

It will not be possible in this chapter to treat all current aspects of cryogenic engineering. The plan adopted is to treat the principal techniques for obtaining low temperatures and to follow this with accounts of the two major cryogenic industries, the separation of gases and the liquefaction of natural gas.

Excellent general accounts of the field of low temperatures are given by Din & Cockett (1960) and by Mendelssohn (1977). Zemansky (1968) provides a detailed introduction to low-temperature physics (chapters 14 and 15 are the relevant ones for cryogenics, superfluidity, etc.). White (1979) and Rose-Innes (1964) give useful accounts with much practical advice on low-temperature techniques in the laboratory. Technical aspects of cryogenic engineering are treated in Haselden (1971).

Safety

All cryo-products are dangerous to some degree unless they are properly handled. Obvious dangers are the risk of frost-bite, the generation of pressure in closed containers by rise of temperature, the creation of suffocating, toxic, flammable or explosive atmospheres by evaporation, and the powerful oxidising nature of some products such as liquid oxygen and liquid fluorine.

Users should be informed on the general precautions to be taken and be aware of the hazards associated with individual substances. A book such as the '*Cryogenics Safety Manual*' published by the Institution of Chemical Engineers for the British Cryogenics Council should be studied before engaging in work with these substances.

8.2 The minimum work required to make cryo-products

The object of many cryogenic operations is the manufacture of a product such as liquid air, liquefied natural gas, etc., rather than the acceptance of heat at a low temperature, which is the object of ordinary refrigerating plant. The Carnot and Lorenz criteria were introduced in Chapter 1 for the comparison of the efficiency of refrigerating plant with what is ideally possible. In this section another criterion, the exergy, is developed for evaluating the efficiencies of plants which have a product as their output.

Manufacture of any product has to take place in an environment at an approximately constant pressure, $p_0 \approx 1$ atm, and an approximately constant absolute temperature, T_0 say, of the available water or air for cooling or heating. It will be assumed that heat transfer to or from the environment at this temperature can take place without restriction. Any body of matter which is in equilibrium with the environment cannot undergo any change and consequently cannot deliver any work. It is then said to be in the 'dead state'. Excluding for the present the possibilities of mixing or chemical reaction with the environment, equilibrium implies equality of pressure and temperature with the environment. There is no possibility, for example, of obtaining work from a volume of air at the same pressure and temperature as the rest of the immediate environment. On the other hand a reservoir of compressed air is not in equilibrium when the valve is opened and the air can deliver work by expanding to atmospheric pressure through an engine or turbine. The maximum work which a body of matter can deliver by coming into its 'dead state' in equilibrium with the environment in a steady-flow process is called its 'exergy'.

Suppose that a steady stream of fluid is available at pressure p, absolute temperature T, giving the state shown on the T–s diagram in Fig. 8.2.1. The maximum work is obtained from this stream when it comes into equilibrium with the environment by reversible processes. This implies not only the absence of friction, turbulence, etc., internally but also the absence of any difference of temperature between the fluid and the environment whilst heat transfer takes place. Otherwise it would be possible to obtain still more work by using this difference of temperature to operate a heat engine.

The only operations which can meet these conditions are reversible adiabatic expansion to some intermediate pressure, p' say, followed by reversible isothermal expansion at temperature T_0 from pressure p' to pressure p_0.

The specific work available from these processes is given by the steady-flow equation, neglecting kinetic and potential energies, as:

in the first expansion $h - h'$

in the second expansion $h' - h_0 + T_0(s_0 - s)$

where h' is the specific enthalpy at the end of the first expansion, and h_0, s_0 are the specific enthalpy and entropy respectively of the fluid when it is finally in equilibrium with the environment at p_0, T_0. The total work delivered is then:

$$b = (h - h_0) - T_0(s - s_0) \qquad (8.2.1)$$

Regarding p_0 and T_0 as fixed, the quantity b is a property of the initial state (p, T): it is called, referred to unit mass of fluid, the specific exergy. The name

Fig. 8.2.1. A reversible way of going from an initial state (T, p, h) to the dead state 0 in equilibrium with the environment, shown on temperature and entropy coordinates.

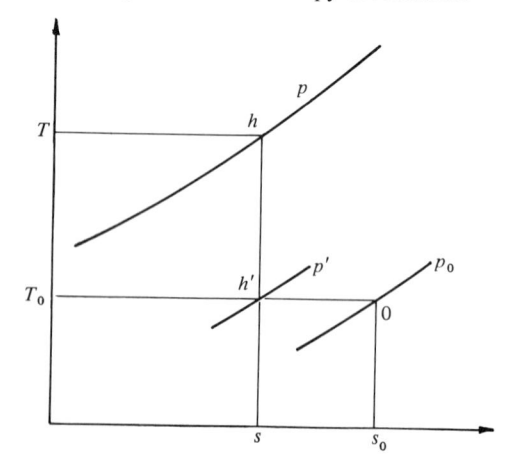

8.2 The minimum work required to make cryo-products

exergy was introduced by Zoran Rant, but the quantity b has also been known under other names such as 'availability in steady flow' (Keenan) and 'technical work ability' (Bošnjaković).

The exergy measures the maximum work which can be obtained when a steady flow of a fluid comes into equilibrium with the environment with no heat transfer other than that with the environment. Conversely it measures the minimum work which is needed to produce the stream of fluid from an initial state in which the fluid is in equilibrium with the environment. If the specific work w of a process is taken as positive when it is done on the fluid, then:

$$w > b \tag{8.2.2}$$

If the fluid is not in equilibrium with the environment initially but at a state denoted by subscript 1, and the result of the process is to produce a state denoted by subscript 2, then:

$$w > b_2 - b_1 \tag{8.2.3}$$

The ratio:

$$\eta_{ex} = \frac{b_2 - b_1}{w} \tag{8.2.4}$$

is the exergetic efficiency of a process in which work is expended on the fluid. When work is delivered by a fluid the right-hand side of eq. (8.2.4) is inverted to make η_{ex} less than unity, following the usual practice for defining the isentropic efficiency of a compressor or an expander.

It must be emphasised that the inequality (8.2.3) holds only for the condition that heat transfer takes place solely with the environment at temperature T_0. If other heat transfers are admitted, for example from a source of heat at a temperature T higher than T_0, then work could be available by operating a heat engine between T and T_0 and the net w for the process could be less than $b_2 - b_1$. Furthermore it is understood that kinetic and potential energies are ignored. For the development of the full inequality, which expresses the Second Law of Thermodynamics under steady-flow conditions in the same way that the steady-flow energy equation expresses the First Law, books on thermodynamics should be consulted. This development is not needed here.

Example 8.2.1

Determine the specific exergy of solid carbon dioxide at a pressure of 1 atm = 1.013 bar and a temperature of $-78.5\,°C$ with reference to an environment at 1 atm and a temperature of 293 K.

The properties of carbon dioxide from example 5.10.1 are used with the additional property: specific entropy of solid at 1 atm, $-78.5\,°C$ = 1.2016 kJ/kg K.

For the gas at $p_0 = 1$ atm, $T_0 = 293$ K:

$h_0 = 392.81$ kJ/kg $s_0 = 2.0666$ kJ/kg K

For the solid at $p = 1$ atm, $T = -78.5\,°C$:

$h = -259.26$ kJ/kg $s = -1.2016$ kJ/kg K

Hence:

$b = [(-259.26 - 392.81) - 293(-1.2016 - 2.0666)]$ kJ/kg

$= 306$ kJ/kg

The shaft work for the manufacture of 1 kg of solid was found to be 750 kJ/kg in example 5.10.1. The exergetic efficiency of that process is therefore:

$306/750 = 0.41$

Table 8.2.1. *Exergies of liquids at their normal boiling points, $T_n (b_0 = 0$ at $p_0 = 1$ atm $T_0 = 300$ K)*

	T_n	b
	K	kJ/kg
Helium	4.224	6820
Hydrogen, normal	20.39	12040
Parahydrogen†	20.268	12600
Parahydrogen ‡	20.268	14300
Neon	27.09	1340
Nitrogen	77.4	770
Air	78.8	740
Argon	87.3	480
Oxygen	90.2	630
Methane	111.7	1090
Ethylene	169.4	430
Ethane	184.5	350

† With reference to parahydrogen in the datum state.
‡ With reference to normal hydrogen in the datum state.(See note at the end of section 8.4.)

8.2 The minimum work required to make cryo-products

The specific exergies of some liquids at their normal boiling points are given in Table 8.2.1. It will be noted that these are all positive, i.e. work is required to manufacture them. Conversely they are capable of delivering work in a process in which they come to equilibrium with the environment, or to certain other states.

Example 8.2.2
Liquid methane at $-161.5\,°C$ is converted into gas at a pressure of 10 bar and a temperature of 25 °C at the rate of 2000 t/day. What power is available from the process in an environment at a temperature of 293 K?

Properties of methane (Datum: liquid at $-161.5\,°C$): superheated vapour at 10 bar, 25 °C:

$$h = 893 \text{ kJ/kg} \qquad s = 5.489 \text{ kJ/kg K}$$
$$b_2 - b_1 = [(893 - 0) - 293(5.489 - 0)] \text{ kJ/kg}$$
$$= -715 \text{ kJ/kg}$$

Hence, assuming heat transfer only with the environment, the specific work must satisfy:

$$w > -715 \text{ kJ/kg}$$

i.e. the specific work *delivered* must be less than 715 kJ/kg. The maximum power available is then:

$$\frac{(2000)(1000)}{(24)(3600)} 715 \text{ kW} = 16\,600 \text{kW}$$

Negative exergies can occur at sub-atmospheric pressure when the work available from the difference in temperature between the fluid and the environment may be insufficient to compress it up to p_0. It must be remembered that mixing with the environment was specifically excluded earlier. If mixing is allowed, then even the air inside a vessel at sub-atmospheric pressure can provide a source of work by allowing the atmosphere to expand into it through a turbine.

The exergy of a liquefied gas may be represented by an area on the temperature–entropy diagram as shown in Fig. 8.2.2. Supposing the gas initially at the pressure p_0 and temperature T_0 of the environment, the first term in the expression for the exergy, eq. (8.2.1), is $(h_f - h_0)$ where h subscript f denotes saturated liquid. It is represented by the area under the line $0 - G - F$ down to absolute zero of temperature. The second term is $T_0(s_f - s_0)$ and this is represented by the area under the line $0 - H$. Hence the exergy is represented by the area shown hatched.

Cryogenic engineering 530

A reversible way of going from 0 to F would be compression of the gas reversibly and isothermally from 0 to H, followed by reversible adiabatic expansion from H to F. Unfortunately such a process is not possible in practice because the pressure required at point H is enormous. Practical operations have to be confined to some maximum pressure less than this value. Because of this limitation and the practical impossibility of carrying out the reversible method of liquefaction, a number of different methods can be used, none of which attains complete reversibility. The principal methods are described in sections 8.3–8.5.

Separation of products

Separation of the components of mixtures is an important application of cryogenics. Some typical processes are described in section 8.6. The following exergetic treatment may be read in conjunction with that section.

It is convenient to work in molar quantities. A bar over the usual letter for a specific property will denote a molar property. Amount of substance will be denoted by n and the mole flow rate by $\dot{n}$.

When a stream of a fluid mixture, which will be characterised by subscript 1, is separated in a steady-flow process into streams characterised by subscript 2, the minimum power required is:

$$\dot{W} = \sum \dot{n}_2 \bar{b}_2 - \dot{n}_1 \bar{b}_1 \qquad (8.2.5)$$

where the summation is taken over the outflowing streams. Kinetic and potential energies are ignored here and in the following treatment.

Fig. 8.2.2. Representation of the exergy of a liquid at F, normal boiling temperature T_n, with respect to a dead state at 0.

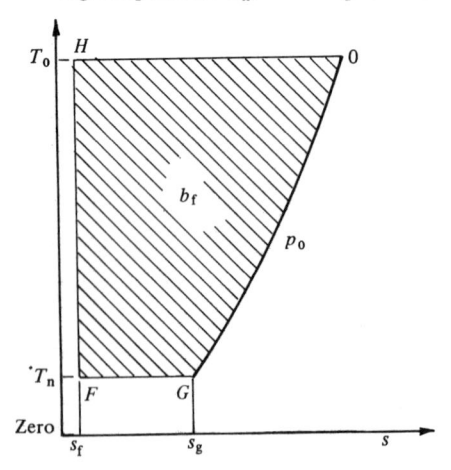

8.2 The minimum work required to make cryo-products 531

In applying this equation it is necessary to know the differences between the molar enthalpies and entropies of different substances. With regard to the enthalpies no difficulty in principle arises because they can be determined in steady-flow calorimetric experiments. With regard to the entropies however there is a fundamental difficulty because the entropy changes can only be measured in reversible processes which cannot be realised in actual experiments.

The way around this difficulty is provided by the Third Law of Thermodynamics, according to which the entropy of each pure chemical species can be assigned the value zero at the absolute zero of temperature. Consequently it is possible to attribute absolute entropies to substances and to use these to find the change of exergy in chemical reactions.

When it is known that no chemical reaction is involved, and that the fluids can be regarded as being in the ideal gas state, the matter can be approached more simply by means of the Gibbs–Dalton law for the entropy of ideal gas mixtures.

The molar entropy of an ideal gas with respect to a datum state $\bar{s}_0$ at pressure p_0 and absolute temperature T_0 is given by:

$$\bar{s} - \bar{s}_0 = \int_{T_0}^{T} \bar{c}_p \frac{dT}{T} - \bar{R} \ln \frac{p}{p_0} \qquad (8.2.6)$$

where $\bar{c}_p$ is the molar heat capacity at constant pressure, which may depend on temperature, and $\bar{R}$ is the molar gas constant.

According to the Gibbs–Dalton law the same expression applies to a component gas in a mixture when p is taken as the partial pressure of the component gas defined by:

$$p = \psi P \qquad (8.2.7)$$

in which ψ is the mole fraction of the gas in the mixture and P is the total pressure. The entropy of the mixture is the sum of the entropies of the components.

The change of exergy on separating a mixture of ideal gases into its components will be derived. The separated gases will be supposed to be at the same pressure P as the mixture and at the same temperature.

Since the enthalpy of an ideal gas depends only on the temperature, the enthalpy contributions to the exergy difference are all zero.

Taking the total pressure and the temperature of the mixture as the datum state for entropy, the entropies of the separated gases are all zero.

The sum of the initial entropies of the gases in 1 mole of mixture is, by eq.

Cryogenic engineering 532

(8.2.6) and (8.2.7):

$$\bar{s}_1 = -\bar{R} \sum \ln \psi$$
$$= \bar{R} \sum \ln \frac{1}{\psi} \qquad (8.2.8)$$

where the summation is taken over all the gases in the mixture. At an environmental temperature T_0, therefore, the change of exergy on separating 1 mole of mixture into its components is:

$$-\bar{b}_1 = \bar{R} T_0 \sum \ln \frac{1}{\psi} \qquad (8.2.9)$$

since all of the $\bar{b}_2$ are zero with the datum chosen. It will be seen that the actual temperatures of the mixture and of the separated gases, provided they are equal, do not come into the expression.

Example 8.2.3
Determine the minimum work needed to separate 1 kg of a mixture of nitrogen and oxygen in the mole proportions 0.79/0.21, respectively, into pure nitrogen and pure oxygen at the same pressure and temperature as the mixture, in an environment at 298 K. The molar masses are: nitrogen 28.0 kg/kmol; oxygen, 32.0 kg/kmol. $\bar{R} = 8314$ J/kmol.

The minimum work is:

$$8314(298)[0.79 \ln(1/0.79) + 0.21 \ln(1/0.21)] \text{ J/kmol}$$
$$= 1273 \text{ kJ/kmol}$$

The molar mass of the mixture is:

$$[0.79(28.0) + 0.21(32.0)] \text{ kg/kmol} = 28.84 \text{ kg/kmol}$$

and the minimum specific work is:

$$(1273/28.84) \text{ kJ/kg} = 44.14 \text{ kJ/kg}$$

Equation (8.2.9) may also be derived by adding the isothermal work needed to compress each component of the mixture reversibly from its partial pressure to the total pressure.

When only one gas in a mixture is to be obtained pure, leaving the others mixed, the change of exergy may be calculated by similar methods. To separate a gas A from a mixture in which its mole fraction is ψ_A, the result is:

$$-\bar{b}_1 = \bar{R} T_0 \left[\psi_A \ln \frac{1}{\psi_A} + (1 - \psi_A) \ln \frac{1}{1 - \psi_A} \right] \qquad (8.2.10)$$

8.2 The minimum work required to make cryo-products

Again this expression may be deduced by calculating the isothermal work to compress the gas A from its partial pressure to the total pressure and the isothermal work to compress the remaining gases as a whole from $(P - p_A)$ to the total pressure.

Example 8.2.4
Determine the minimum work needed to obtain 1 kg of argon gas from the atmosphere in which it is present in the mole fraction 0.0093. $T_0 = 298$ K. The molar mass of argon is 40.0 kg/kmol.

The minimum work per unit amount of air is:

$$8314(298)[0.0093 \ln(1/0.0093) + 0.9907 \ln(1/0.9907)] \text{ J/kmol}$$
$$= 130.7 \text{ kJ/kmol}$$

The mass of argon in 1 kmol of air is:

$$0.0093(40) \text{ kg} = 0.372 \text{ kg}$$

and the minimum work per unit mass of argon is:

$$(130.7/0.372) \text{ kJ/kg} = 351 \text{ kJ/kg}$$

If the product is wanted as a liquid, the exergy of liquefaction has to be added to the exergy of separation. Referring to Table 8.2.1 the total for argon becomes $(351 + 480)$ kJ/kg $= 831$ kJ/kg. In this case the exergy of separation is of the same order as the exergy of liquefaction. For nitrogen on the other hand the exergy of separation from air is quite small compared with that required for liquefaction.

Equation (8.2.10) is a special case of the more general one when the products are impure. Suppose that a mixture of ideal gases A, B, C, ..., having mole fractions $\psi_{A1}, \psi_{B1}, \ldots$, enters an apparatus at the mole rate $\dot{n}_1$, and emerges as streams 2, 3, ... at the same pressure and temperature as stream 1, stream 2 having the composition $\psi_{A2}, \psi_{B2}, \ldots$, and so on for the other streams.

Taking the datum pressure and temperature for entropy as P and T as before, the contributions to the entropy from the temperature all vanish, as do the contributions to the exergy via the enthalpy. The rate of flow of exergy into the apparatus is then:

$$\dot{n}_1 \bar{b}_1 = -\bar{R} T_0 \sum \ln \frac{1}{\psi_{A,B,\ldots}} \qquad (8.2.11)$$

Cryogenic engineering 534

The rate of flow of exergy out of the apparatus is:

$$\dot{n}_2 \bar{b}_2 + \dot{n}_3 \bar{b}_3 + \ldots$$

$$= -\bar{R}T_0 \left[\dot{n}_2 \sum \ln \frac{1}{\psi_{A,B,\ldots}} + \dot{n}_3 \sum \ln \frac{1}{\psi_{A,B,\ldots}} + \ldots \right] \qquad (8.2.12)$$

in which each term in the brackets on the right is summed over the components, and the whole right-hand side is summed over the streams. The rate of increase of exergy is then given by the difference between the right-hand sides of eqs. (8.2.12) and (8.2.11).

For given values of $\dot{n}_1, \psi_{A1}, \psi_{B1}, \ldots$, the values of the $\dot{n}$ and the ψ in the emerging streams are not all independent, since they have to satisfy the mole balances. In general, if there are i components and j emerging streams there are $(i+j)$ equations available, of which i express the facts that the mole flow rate of each component in and out is the same, and j express the facts that in each stream $\sum \psi = 1$.

The number of unknowns is $(j+ij)$, in which j is the number of the $\dot{n}$ for the emerging streams and ij is the number of the mole fractions. The difference $(j+ij) - (i+j) = i(j-1)$ is the number of degrees of freedom, i.e. the number of the outgoing variables which can be specified independently, leaving $(i+j)$ to be determined from the mole balances.

Example 8.2.5
Supposing the data given in example 8.2.3, determine the minimum work needed to separate 1 kg of mixture into nitrogen and oxygen, each 0.99 (mol) pure.

Let subscript 2 denote the nitrogen-rich stream and subscript 3 the oxygen-rich stream. From the mole balances:

$$\frac{\dot{n}_2}{\dot{n}_1} = \frac{0.79 - 0.01}{0.99 - 0.01} = 0.796$$

and:

$$\frac{\dot{n}_3}{\dot{n}_1} = 0.204$$

by difference. The exergy entering per unit amount of mixture is:

$$-\bar{R}T_0[0.79\ln(1/0.79) + 0.21\ln(1/0.21)] = -0.514\bar{R}T_0$$

The exergy leaving per unit amount of mixture is:

$$-0.796\,\bar{R}T_0[0.99\ln(1/0.99) + 0.01\ln(1/0.01)]$$
$$-0.204\,\bar{R}T_0[0.01\ln(1/0.01) + 0.99\ln(1/0.99)]$$
$$= -0.0560\,\bar{R}T_0$$

8.3 The Joule–Thomson effect: real gases

The increase of exergy is then:

8314(298)(−0.056 + 0.514) J/kmol = 1135 kJ/kmol

and the minimum specific work is:

(1135/28.84) kJ/kg = 39.4 kJ/kg

As is to be expected this value is somewhat less than that found in example 8.2.3.

8.3 The Joule–Thomson effect: real gases

As explained in Chapter 7, an ideal gas, i.e. one having the equation of state $pv = RT$, undergoes no change of temperature when it expands without doing work. The ideal gas state, however, is an abstraction which is approached as the pressure tends to zero. A real gas at a finite pressure undergoes a change of temperature when it is throttled. An example of this has been met in the case of carbon dioxide when the temperature of the

Fig. 8.3.1. Adiabatic throttling through an orifice.

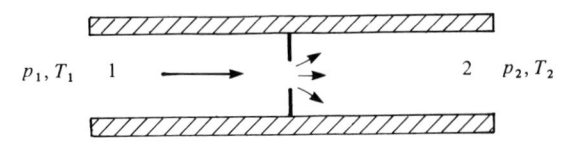

Fig. 8.3.2. States attained by adiabatic throttling from a state 1, shown diagrammatically on pressure–temperature coordinates.

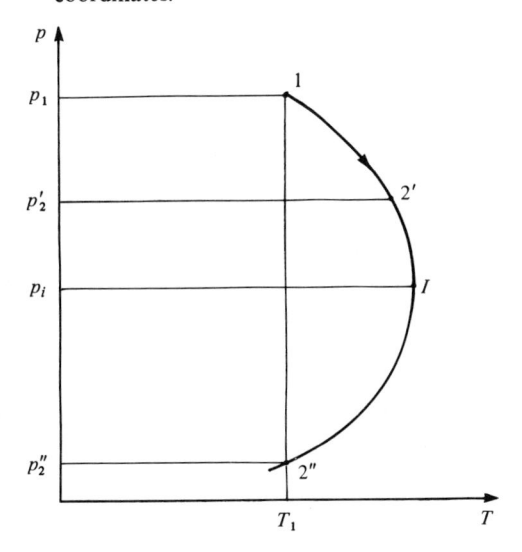

cooling water is too high to allow condensation (section 3.14) and in fact the drop of temperature through the expansion valve of the normal vapour compression system may be regarded as an extreme example of departure from ideality.

In both of these examples the temperature falls, but for every fluid there are certain states from which the temperature rises when the fluid expands without doing work. The matter was first investigated by J. P. Joule and W. Thomson (Lord Kelvin) in their celebrated 'porous plug' experiments. The porous plug was simply a restriction through which the gas was throttled more or less adiabatically from a higher to a lower pressure. It is represented in Fig. 8.3.1 by an orifice. Provided that the velocities at 1 and 2 are small enough for kinetic terms to be negligible, and that the process is adiabatic, the steady-flow energy equation gives:

$$h_1 = h_2 \qquad (8.3.1)$$

where h is the specific enthalpy. Holding p_1 constant at a high value, and T_1 constant, then h_1 is fixed. At various values of p_2 the temperature T_2 can be measured. The pairs of readings p_2, T_2 when plotted on pressure–temperature coordinates give a line of constant specific enthalpy, an isenthalp, as shown in Fig. 8.3.2.

The shape of the isenthalp shows that a drop in pressure from p_1 to p_2' say is accompanied by a rise in temperature. A drop from p_1 to p_i produces a still larger rise, whilst a drop from p_1 to p_2'' would bring the gas back to its initial temperature T_1.

Since the specific enthalpy is a property of a gas, any other starting point lying on the same isenthalp would give final states on the same curve. In particular when the initial pressure is less than p_i, with the temperature chosen so that the state lies on the curve, the result of a drop in pressure is always a drop in temperature.

The point I where the isenthalp reaches a maximum temperature is called an inversion point. For an infinitesimal change in pressure the ratio:

$$\mu = \left(\frac{\partial T}{\partial p}\right)_h \qquad (8.3.2)$$

is called the Joule–Thomson coefficient. It is positive when p is less than p_i, meaning that a small drop in pressure is accompanied by a small drop in temperature, whilst it is negative when p is greater than p_i, indicating a small rise in temperature with a small drop in pressure.

By altering the initial specific enthalpy in the experiment, i.e. by altering p_1 and T_1, a number of isenthalps can be measured and plotted on the

8.3 The Joule–Thomson effect: real gases

diagram, as shown in Fig. 8.3.3. The inversion points on each of the isenthalps are joined by a broken line which is called the inversion locus. At the high-temperature end the inversion locus comes down to zero pressure at a temperature $T_{i,max}$, which is called the maximum inversion temperature. Above this temperature no isenthalp shows the phenomenon of inversion, and adiabatic throttling is always accompanied by a rise of temperature. At the left-hand side the inversion locus joins the pressure–temperature curve for the liquid and vapour in equilibrium.

Below the inversion locus the value of μ is everywhere positive, indicating a reduction of temperature on throttling. Outside the locus it is everywhere negative, indicating a rise of temperature.

Table 8.3.1 gives some maximum inversion temperatures. These are only

Table 8.3.1. *Maximum inversion temperatures*

	$T_{i,max}$
	K
Carbon dioxide	1500
Argon	600
Nitrogen	620
Air	660
Hydrogen	200
Helium	40

Fir. 8.3.3. Isenthalps for air on the pressure–temperature diagram showing the maximum inversion temperature. The position of the inversion locus is approximate only.

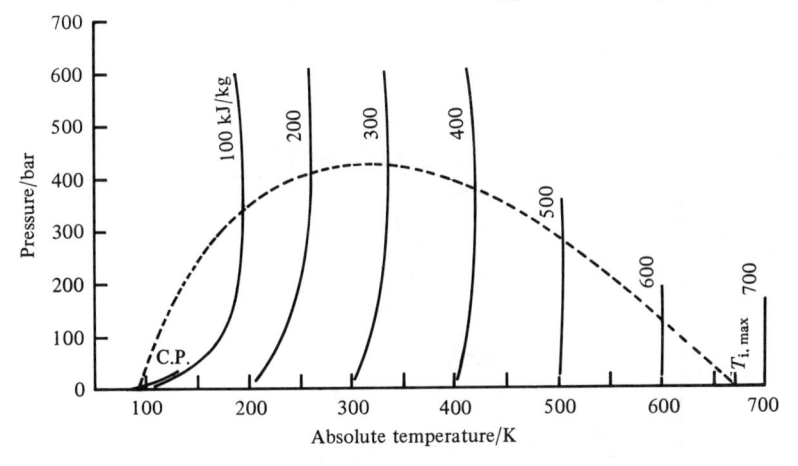

approximate since the exact point of inversion is somewhat difficult to determine experimentally.

It will be observed that neither hydrogen nor helium can be reduced in temperature by adiabatic throttling at normal ambient temperature.

Joule–Thomson experiments enable lines to be drawn, such as those in Fig. 8.3.3, passing through states having the same specific enthalpy. But they give neither the values of h nor the differences in the values of h between the lines. In order to label the lines, further information, such as the specific heat capacities, is required.

The shape of the inversion locus, however, depends only on the equation of state. The general expression for a differential change of enthalpy found in section 3.18 is:

$$\mathrm{d}h = c_p \mathrm{d}T - \left[T\left(\frac{\partial v}{\partial T}\right)_p - v \right] \mathrm{d}p \qquad (8.3.3)$$

Fig. 8.3.4. The Joule–Thomson effect and the inversion phenomenon shown on the temperature–entropy diagram.

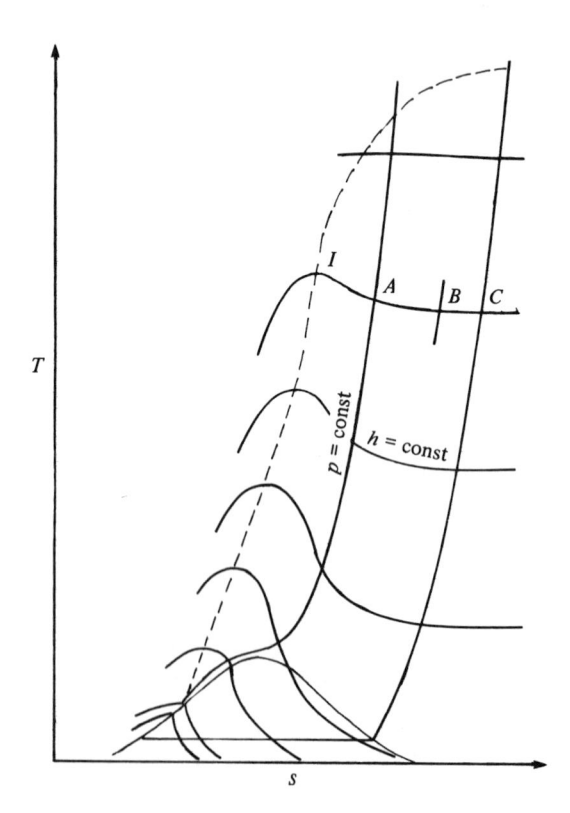

8.3 The Joule–Thomson effect: real gases

For an infinitesimal throttling process $dh = 0$ and:

$$\mu = \left(\frac{\partial T}{\partial p}\right)_h = \frac{1}{c_p}\left[T\left(\frac{\partial v}{\partial T}\right)_p - v\right] \qquad (8.3.4)$$

Along the inversion locus $\mu = 0$ and:

$$T\left(\frac{\partial v}{\partial T}\right)_p - v = 0 \qquad (8.3.5)$$

which may also be written:

$$\frac{T^2}{p}\left[\frac{\partial(pv/T)}{\partial T}\right]_p = 0 \qquad (8.3.6)$$

Elimination of v between eqs. (8.3.5) or (8.3.6) and an equation of state $f(p, v, T) = 0$ gives the equation of the inversion locus on the p–T coordinates of Fig. 8.3.3.

Fig. 8.3.5. Pressure–enthalpy diagram with a logarithmic pressure scale, showing the Joule–Thomson effect and inversion.

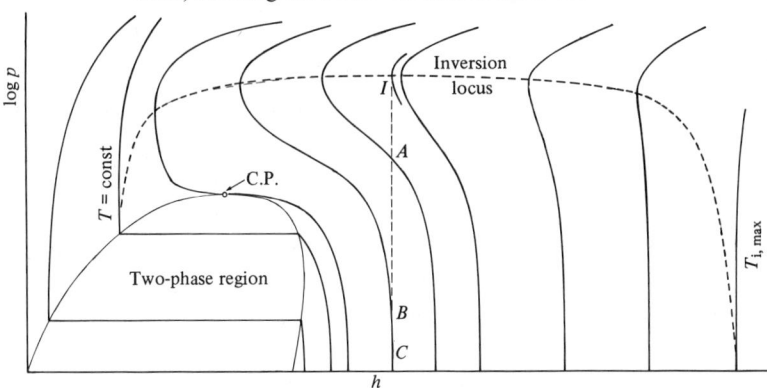

Fig. 8.3.6. Pressure–enthalpy diagram showing how the increased Joule–Thomson effect at low temperatures is related to the increase of specific heat capacity with pressure.

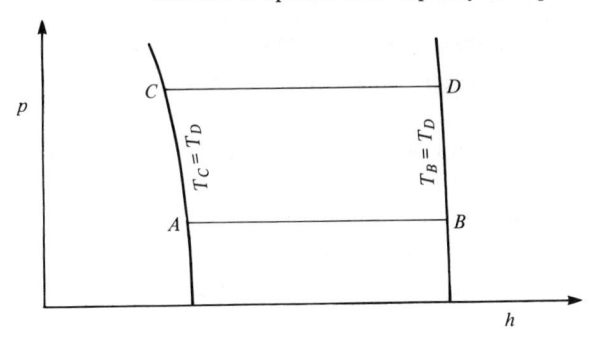

The Joule–Thomson effect may also be exhibited on the temperature–entropy diagram, as in Fig. 8.3.4. The lines of constant enthalpy are seen to become nearly horizontal at the right-hand side, i.e. at low pressure, indicating the approach to ideal gas behaviour. Throttling at low pressure produces negligible change of temperature, as for example from B to C. From a point such as A, however, a drop of temperature occurs on throttling to B or C, and the maximum drop of temperature, for an initial state on a fixed isenthalp, occurs when the starting point is at I. At higher temperatures the peaks in the isenthalps move to the right and finally disappear at the maximum inversion temperature.

Figure 8.3.5 shows the inversion locus on the pressure–enthalpy diagram with pressure to a logarithmic scale. The isotherms at low pressure tend towards the vertical, indicating again the approach to ideal gas behaviour. The points A, B, C and I correspond to the points in Fig. 8.3.4 with the same letters, and the same remarks hold as for the temperature–entropy diagram. The inversion locus comes down towards zero pressure (not shown on the diagram because the scale is logarithmic) at the maximum inversion temperature. At the left-hand side the inversion locus joins the saturated liquid line.

It will be recognised from Figs. 8.3.4 and 8.3.5 that the reduction of temperature accompanying a given drop in pressure increases as the initial temperature falls because the isotherms in the neighbourhood of the inversion locus become more strongly curved at lower temperatures. This fact is connected with the variation of the specific heat capacity at constant pressure. Consider the two points, A and B, inside the inversion locus in Fig. 8.3.6. The mean specific heat capacity at the constant pressure $p_A = p_B$ is $(h_B - h_A)/(T_B - T_A)$. At a higher pressure the points C and D on the same isotherms are more widely spaced and the mean specific heat capacity between C and D is greater than that between A and B. Thus the fact that the Joule–Thomson effect increases at lower temperatures is directly related to the increase with pressure of the specific heat capacity at constant pressure.

8.4 Regenerative cooling: the Linde–Hampson process

The Joule–Thomson effect could be used for refrigeration by carrying out a cycle of essentially the same operations as the vapour compression cycle, shown in Fig. 8.4.1, which is numbered to correspond to Fig. 3.1.1. Gas at ambient temperature and pressure at point 1 is compressed, using several stages of compression, and finally cooled back to ambient temperature at 3. It is then throttled to 4 during which process its temperature falls. From 4 to 1 it accepts heat at constant pressure.

8.4 Regenerative cooling: the Linde–Hampson process

The specific refrigerating effect is:

$$q_e = h_1 - h_3 \tag{8.4.1}$$

exactly as for the vapour compression cycle in eq. (3.2.11). This refrigerating effect is very small, however, even when the pressure is quite high. By compressing air almost up to its inversion pressure, 400 bar say, the specific refrigerating effect is only about 45 kJ/kg with a temperature at point 4 of 45 K below ambient temperature. The cycle is not a very useful one for refrigeration at this temperature therefore, since the vapour compression system can do much better and does not require such high pressures.

Suppose now that a heat exchanger is interposed between the compressed gas and the low-pressure gas, as shown in Fig. 8.4.2. The specific refrigerating effect is now:

$$q_e = h_6 - h_4 \tag{8.4.2}$$

Fig. 8.4.1. (a) Circuit of a refrigerator working on the Joule–Thomson effect. (b) The cycle on the pressure–enthalpy diagram. Compression in several stages is represented diagrammatically by the single compressor in (a).

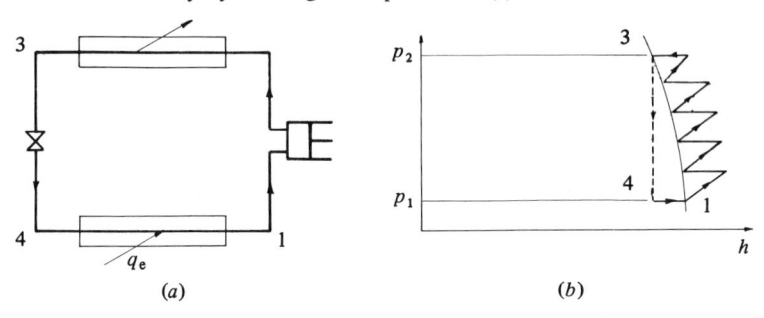

Fig. 8.4.2. (a) Circuit incorporating a heat exchanger. (b) The cycle on the pressure–entropy diagram. The heat exchanger does not alter the specific refrigerating effect but it makes it available over the much lower temperature range T_5 to T_6.

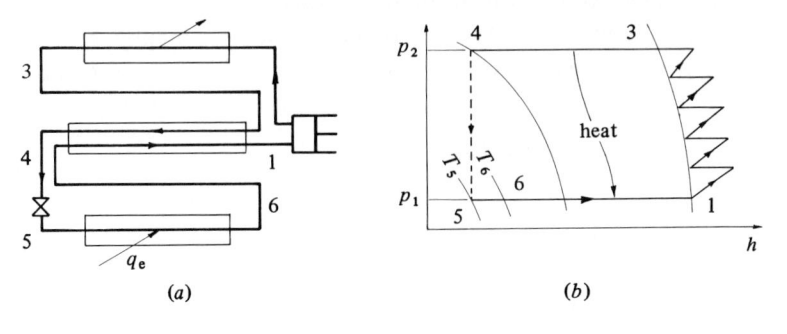

but assuming that the heat exchanger is adiabatic to the surroundings it is also given by:

$$q_e = h_1 - h_3 \tag{8.4.3}$$

exactly the same as without the heat exchanger.

The heat exchanger does not change the specific refrigerating effect, but it makes a fundamental difference in that the refrigerating effect is now available at temperatures which are much lower than ambient temperature, namely from T_5 to T_6 shown on the pressure–enthalpy diagram in Fig. 8.4.2(b).

If the refrigerating capacity produced by such a system is balanced by a heat transfer rate from the outside, a steady condition of operation can be attained. But if it is not, the refrigerating capacity is available to cool down the apparatus itself. Provided that the heat exchanger is long enough to make T_1 almost equal to T_3, and that heat transfer from the outside is made small by sufficient insulation, the reduction of temperature can go on until the state at 5 lies in the two-phase region. This is the principle of the Linde–Hampson liquefier. In the following description the method of numbering is altered from that used so far. Point 1 denoting high-pressure gas at ambient temperature is from now on common to all processes described.

The apparatus for the process is shown diagrammatically in Fig. 8.4.3(a). From now on the compressors will not be shown since they are common to all systems ultimately, and because small liquefiers may be supplied from a cylinder of compressed gas. It is assumed that gas at a suitably high pressure and at approximately ambient temperature is available at point 1 at the inlet to the apparatus. The high-pressure gas flows through a heat exchanger against a stream of returning gas and its temperature is reduced to T_2. On being throttled through a restriction a two-phase mixture forms which is separated into liquid at 4 and a stream of saturated vapour at 5. This stream goes back through the heat exchanger and emerges at a temperature T_6 which is close to T_1. The stream of high-pressure gas will be called for short the forward stream, and the stream of gas returning through the exchanger will be called the revert stream.

Supposing unit mass flow rate of the forward stream at point 1, and the fraction λ of this which is liquefied at 4, the steady-flow equation applied to the apparatus gives:

$$h_1 = \lambda h_4 + (1 - \lambda)h_6 \tag{8.4.4a}$$

and:

$$\lambda = \frac{h_6 - h_1}{h_6 - h_4} \tag{8.4.4b}$$

8.4 Regenerative cooling: the Linde–Hampson process

The specific enthalpy at 6 is fixed by the temperature and pressure at that point. Assuming atmospheric pressure and a temperature $T_6 = T_1$ for the moment, the value of h_6 is fixed. The specific enthalpy at 4 is that of saturated liquid at atmospheric pressure and is fixed. The ratio λ depends only on h_1 therefore, and the lower this is the larger the fraction liquefied. For a given T_1, h_1 can be reduced by raising the pressure, but only up to the inversion pressure for that temperature. Thereafter h_1 increases with increasing pressure. Hence the maximum yield is obtained when the point 1 lies on the inversion locus. For air the inversion pressure at ambient temperature is about 400 bar. It is not in fact economical to go as high as this and the process has been commonly worked at about 200 bar.

Figure 8.4.3(b) shows the states on the $p-h$ diagram. In contrast with that of Fig. 8.4.2(b), the change of specific enthalpy of the forward and revert streams in the exchanger are not now equal, because the mass flow rates are different.

Fig. 8.4.3. (a) The simple Linde–Hampson process for liquefying air. (b) The process on the pressure–enthalpy diagram. State 6 is shown at a temperature lower than T_1 owing to imperfect heat exchange.

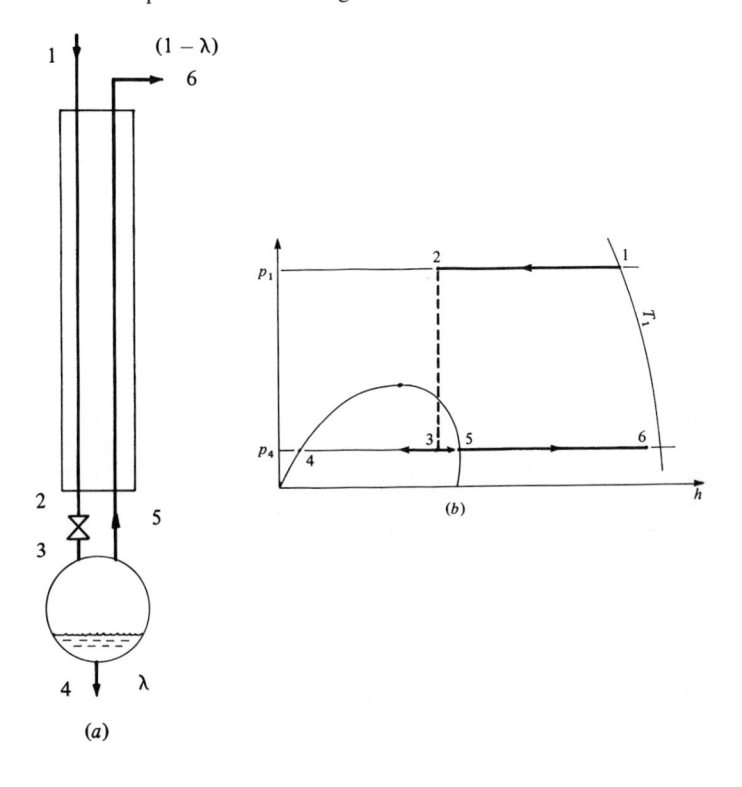

The specific refrigerating effect is $(h_6 - h_1)$. It is not increased by the heat exchanger, merely transferred to a lower temperature. Although it is small compared with the values in vapour compression systems, it drives the whole process, providing the refrigeration to liquefy the gas and to counter heat transfer from the outside. Naturally this heat transfer is reduced as much as possible by insulation.

When air is being liquefied a complication enters because the nitrogen is more volatile than the oxygen. When the split into two phases occurs after the valve, the liquid is richer in oxygen than normal air. The revert gas is correspondingly enriched in nitrogen. To avoid this difficulty the examples in this section are based on the liquefaction of nitrogen with the properties given in Table 8.4.1.

Table 8.4.1. *Properties of nitrogen*

T	$p = 1.013$ bar		$p = 40.52$ bar		$p = 202.6$ bar	
	h	s	h	s	h	s
K	kJ/kg	kJ/kg K	kJ/kg	kJ/kg K	kJ/kg	kJ/kg K
77.4 sat. liq.	29.4	0.418				
77.4 sat.vap.	228.7	2.994				
70			17.2	0.189	30.4	0.107
80	231.5	3.030	37.5	0.459	49.6	0.364
90	242.3	3.157	57.6	0.696	68.4	0.586
100	253.0	3.270	78.4	0.915	87.2	0.784
110	263.6	3.371	101.1	1.131	106.3	0.966
120	274.2	3.463	126.4	1.352	124.5	1.124
130	284.8	3.548	185.6	1.820	144.0	1.282
140	295.3	3.626	243.3	2.253	164.3	1.432
150	305.8	3.698	265.3	2.406	184.2	1.569
160	316.2	3.766	282.7	2.518	204.0	1.697
170	326.7	3.829	298.0	2.611	223.7	1.816
180	337.1	3.889	312.1	2.692	243.2	1.928
190	347.6	3.945	325.5	2.764	262.3	2.031
200	358.0	3.999	338.3	2.830	280.7	2.125
220	378.9	4.098	362.9	2.947	315.3	2.291
240	399.7	4.189	386.6	3.050	347.2	2.429
260	420.5	4.272	409.5	3.142	376.9	2.548
280	441.3	4.349	432.1	3.225	404.8	2.652
300	462.1	4.421	454.3	3.302	431.5	2.744

Datum: $h = 0$ and $s = 0$ for saturated liquid at 63.15 K (triple point).
Abstracted from Technical Note 129, January 1962, National Bureau of Standards.

8.4 Regenerative cooling: the Linde–Hampson process

Example 8.4.1
Liquid nitrogen is made at a pressure of 1 atm = 1.013 bar in the apparatus of Fig. 8.4.3. The high pressure is 200 atm = 202.6 bar, and the temperature $T_1 = 300$ K. Assuming no heat gains to the apparatus and that the revert gas leaves at 300 K, determine the fraction liquefied.

From Table 8.4.1:

h_6(1.013 bar, 300 K) = 462.1 kJ/kg
h_1(202.6 bar, 300 K) = 431.5 kJ/kg
h_4(liquid, 77.4 K) = 29.4 kJ/kg

Hence:

$$\lambda = \frac{462.1 - 431.5}{462.1 - 29.4} = 0.0707$$

The small specific refrigerating effect obtainable by throttling of a compressed gas, only 30.6 kJ/kg in example 8.4.1, means that it is of vital importance not to waste any by allowing the revert gas to escape at a temperature much below T_1. Figure 8.4.4 shows the yield fraction calcu-

Fig. 8.4.4. The yield fraction for the Linde–Hampson process for a pressure of 202.6 bar (200 atm) and $T_1 = 300$ K as a function of the leaving revert gas temperature T_6. The corresponding heat exchanger efficiencies are marked on the curve.

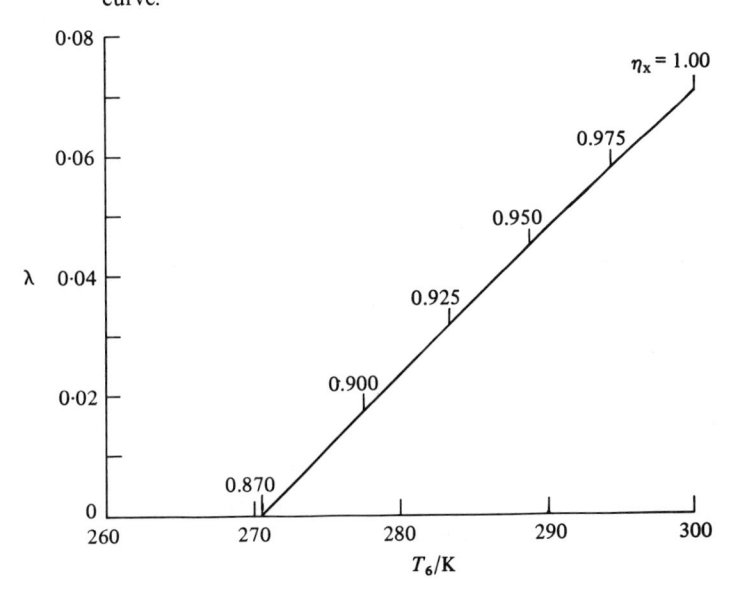

lated as in example 8.4.1 with different assumed values of T_6. It shows that no yield at all is obtained unless the revert gas can be warmed up to at least 271 K. The difference between the actual temperature of the revert gas T_6 and the temperature T_1 to which it could be raised in a very long exchanger is called the 'approach'. In practice an approach of about 2 K has to be obtained for economical operation.

Another method of expressing the performance of a heat exchanger is by its efficiency. This is defined as:

$$\eta_x = \frac{h_6 - h_5}{h_6(\text{max}) - h_5} \quad (8.4.5)$$

where $h_6(\text{max})$ denotes the specific enthalpy of the revert gas at the temperature T_1. The numerator then gives the actual heat transfer per unit mass and the denominator gives the heat transfer per unit mass which could be achieved in principle by a long enough exchanger. When the specific heat capacity of the gas is constant the efficiency can be expressed in terms of the temperatures:

$$\eta_x = \frac{T_6 - T_5}{T_1 - T_5} \quad (8.4.6)$$

The efficiencies of the heat exchanger corresponding to the several values of T_6 are marked on the curve in Fig. 8.4.4. The yield falls to zero when the efficiency drops to 0.87.

The possibility which exists in the heat exchanger of the Linde–Hampson apparatus of raising T_6 to as near as desired to T_1 depends on the fact that the product of the mass flow rate and specific heat capacity of the revert stream is less than the corresponding product for the forward stream. The overall enthalpy balance for the exchanger gives:

$$\dot{m}_1(h_1 - h_2) = \dot{m}_6(h_6 - h_5) \quad (8.4.7)$$

where $\dot{m}_1$ and $\dot{m}_6$ are the mass flow rates of the forward and revert streams. Taking mean specific heat capacities (at constant pressure) for the forward stream $c_{p,F}$ and for the revert stream $c_{p,R}$, then:

$$\dot{m}_1 c_{p,F}(T_1 - T_2) = \dot{m}_6 c_{p,R}(T_6 - T_5) \quad (8.4.8)$$

The mass flow rate of the revert stream is less than that of the forward stream because some of the $\dot{m}_1$ has been abstracted as liquid; and $c_{p,R}$ is lower than $c_{p,F}$, as can be seen by inspection of Table 8.4.1. Consequently the range of temperature $(T_6 - T_5)$ of the revert stream is greater than that, $(T_1 - T_2)$, of the forward stream. Thus T_6 can be made to approach T_1 as near as desired, but T_2 cannot approach T_5.

8.4 Regenerative cooling: the Linde–Hampson process

The products $\dot{m}c_p$ for the two streams are called their capacity rates, and they play a highly important part in considering the possibilities of applying heat exchangers in all situations. The efficiency is defined for the stream having the lower capacity rate, which happens to be the revert stream in the Linde–Hampson apparatus.

The temperatures in the heat exchanger may be displayed graphically to a base of specific enthalpy of one of the streams. They are calculated as in the following example.

Example 8.4.2
Determine the temperature of the forward stream leaving the heat exchanger in example 8.4.1 and plot the temperature profiles to a base of specific enthalpy of the revert stream.

From example 8.4.1, $\lambda = 0.0707$ and $\dot{m}_6 = 0.9293\,\dot{m}_1$. From Table 8.4.1:

$$h_5(1.013\,\text{bar}, 77.4\,\text{K}) = 228.7\,\text{kJ/kg}$$

Hence in eq. (8.4.7):

$$h_2 = [431.5 - 0.9293(462.1 - 228.7)]\,\text{kJ/kg}$$
$$= 214.6\,\text{kJ/kg}$$

Interpolating in Table 8.4.1 at a pressure of 202.6 bar:

$$T_2 = 165.4\,\text{K}$$

Fig. 8.4.5. Temperatures in the heat exchanger of the simple Linde–Hampson process to a base of specific enthalpy of the revert gas h_R considered as a variable drawn to scale for the conditions of example 8.4.1.

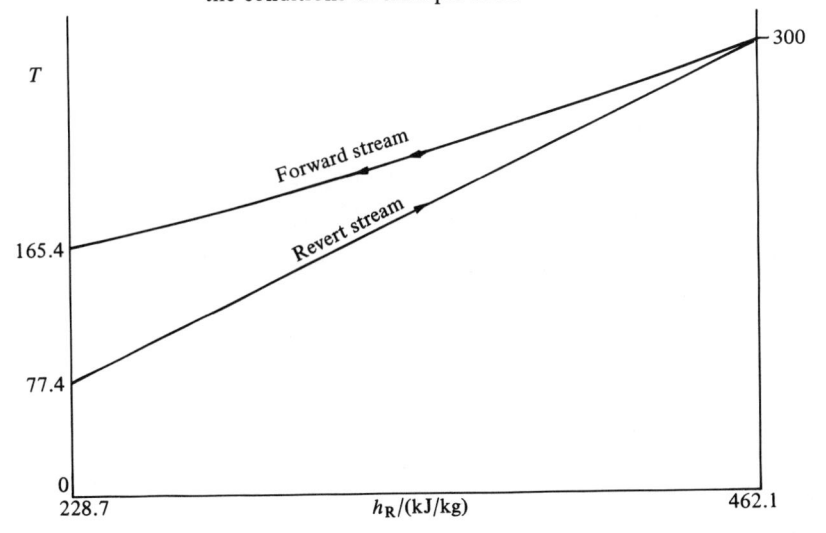

The temperatures at points in the exchanger are found in the same way. Taking a sample point in the forward stream at $T = 200$ K, say, with:

$$h(202.6 \text{ bar}, 200 \text{ K}) = 280.7 \text{ kJ/kg}$$

the specific enthalpy at the corresponding point in the revert stream is:

$$h = \left(426.1 - \frac{431.5 - 280.7}{0.9293}\right) \text{kJ/kg} = 299.8 \text{ kJ/kg}$$

Interpolation in the table at a pressure of 1.013 bar gives:

$$T = 144.3 \text{ K}$$

The results are shown in Fig. 8.4.5 plotted to a base of specific enthalpy of the revert stream. The range of the revert stream is greater than that of the forward stream and there is a difference of temperature of 88 K at the cold end of the exchanger.

An exchanger like that of Fig. 8.4.5 is said to be unbalanced. In an unbalanced exchanger no amount of surface area for heat transfer can reduce the temperature difference at one end, the cold end in this case. Because work can be obtained by using a heat engine across a temperature difference, there is a loss of exergy in an unbalanced heat exchanger, as shown in the following example.

Example 8.4.3
Determine the losses of exergy, expressed per unit mass of liquid formed, in the heat exchanger and expansion valve for the conditions of example 8.4.1.
In addition to the specific enthalpies already cited, the following specific entropies are taken from Table 8.4.1, interpolated where necessary:

$$s_1(202.6 \text{ bar}, 300 \text{ K}) = 2.744 \text{ kJ/kg K}$$
$$s_1(202.6 \text{ bar}, 165.4 \text{ K}) = 1.761 \text{ kJ/kg K}$$
$$s_4(\text{liquid}, 77.4 \text{ K}) = 0.418 \text{ kJ/kg K}$$
$$s_5(1.013 \text{ bar}, 77.4 \text{ K}) = 2.994 \text{ kJ/kg K}$$

The specific exergies are found using eq. (8.2.1) with $T_0 = 300$ K, $p_0 = 1.013$ bar:

$$b_1 = 472.5 \text{ kJ/kg}$$
$$b_2 = 550.5 \text{ kJ/kg}$$
$$b_4 = 768.2 \text{ kJ/kg}$$
$$b_5 = 194.7 \text{ kJ/kg}$$
$$b_6 = 0$$

8.4 Regenerative cooling: the Linde–Hampson process

The specific exergy at 3 is the sum of the exergies of the liquid and vapour since they are at the same temperature:

$$b_3 = [0.0707(768.2) + 0.9293(194.7)] \text{ kJ/kg} = 235.2 \text{ kJ/kg}$$

Heat exchanger. The exergy gained 1–2 is:

$$(b_2 - b_1) = (550.5 - 472.5) \text{ kJ/kg} = 78.0 \text{ kJ/kg}$$

The exergy lost 5–6 is:

$$(b_5 - b_6) = 0.9293(194.7) \text{ kJ/kg} = 180.9 \text{ kJ/kg}$$

The net loss of exergy is then 102.9 kJ/kg. This is for unit mass of gas at 1. For unit mass of liquid delivered it is:

$$(102.9/0.0707) \text{ kJ/kg} = 1455 \text{ kJ/kg}$$

Expansion valve

$$b_2 - b_3 = (550.5 - 235.2) \text{ kJ/kg} = 315.3 \text{ kJ/kg}$$

and per mass of liquid delivered the loss is:

$$(315.3/0.0707) \text{ kJ/kg} = 4460 \text{ kJ/kg}$$

The exergetic efficiency of the process is:

$$768.2(0.0707)/472.5 = 0.115$$

It appears from this example that about one quarter of the loss of exergy in the Linde–Hampson process occurs in the heat exchanger, the remaining three-quarters being caused by the throttling process. These losses are inherent in the process: nothing can be done about them without changing it. In particular it should be noted that the exergy loss in the heat exchanger is there even though the exchanger itself is a perfect one with an efficiency of unity. It is caused solely by the fact that the exchanger is unbalanced. Consequently the aim in the design of cryogenic systems is to arrange matters so that the exchangers are balanced as far as possible, i.e. have equal capacity rates in the streams.

The effect of pressure drops in the exchanger will be briefly considered. Equation (8.4.4b) shows that the yield fraction depends only on the states at 1, 4 and 6 and is apparently independent of pressure drops. For the high-pressure stream this is true because it makes no difference whether the gas is throttled through an orifice or in a pipe: both processes are irreversible in any event.

For the revert stream which is discharged at atmospheric pressure, a pressure drop would mean that the liquid was formed at a pressure above atmospheric pressure. If this liquid is subsequently discharged at atmospheric pressure some of it would have to flash into vapour to maintain equilibrium. The refrigerating effect of this vapour would not be recovered. Consequently the ultimate yield would be less.

To determine the overall efficiency of the Linde–Hampson process the work of compression has to be considered. When treating vapour compression systems in Chapter 3 the standard of reversible isentropic compression was adopted because the vapour at entry to the compressor was in most cases below ambient temperature. No heat could therefore be rejected except towards the end of the process. In the present circumstances the gas is usually at about ambient temperature and reversible isothermal compression is regarded as the ideal. Because of the high pressure needed for the process, several stages of compression are necessary and cooling of the gas between stages enables a stepwise approach to isothermal compression to be made, as indicated in the Fig. 8.4.1(b).

The specific work of reversible isothermal compression from a state 0 to a pressure p_1 is given by:

$$w_T = (h_1 - h_0) - T_0(s_1 - s_0) \tag{8.4.9}$$

where s_1 is the specific entropy at the pressure p_1 and temperature T_0. If p_0 and T_0 are the environment pressure and temperature, this is exactly the same as the change of specific exergy of the gas. The isothermal efficiency is then defined as:

$$\eta_T = w_T/w \tag{8.4.10}$$

where w is the actual specific work of compression. This may be the shaft work, the indicated work or the electrical work at the motor terminals, provided that it is stated with the efficiency. The isothermal efficiency of large gas compressors of the reciprocating type is about 0.6 referred to electrical work.

When the gas is well above its critical temperature the process of isothermal compression is represented quite well by the relation $pv = \text{const}$. The specific work of reversible compression is then given by:

$$w_T = RT_0 \ln \frac{p_1}{p_0} \tag{8.4.11}$$

Example 8.4.4

Determine the work of reversible isothermal compression for the conditions of example 8.4.1 and the exergetic efficiency of the whole process

8.4 Regenerative cooling: the Linde–Hampson process

referred to $T_0 = 300$ K, $p_0 = 1.013$ bar.

$s_0(1.013 \text{ bar}, 300 \text{ K}) = 4.421$ kJ/kg K

$s_1(202.6 \text{ bar}, 300 \text{ K}) = 2.744$ kJ/kg K

Hence:

$$w_T = [431.5 - 462.1 - 300(2.744 - 4.421)] \text{ kJ/kg}$$

$$= 472.5 \text{ kJ/kg}$$

which is the same as b_1 already found in example 8.4.3. Using the approximate eq. (8.4.11) with $R = 297$ J/kg K for nitrogen gives:

$$w_T = 297(300) \ln(200) \text{ J/kg} = 472 \text{ kJ/kg}$$

The work per unit mass of liquid is:

$(472.5/0.0707)$ kJ/kg $= 6683$ kJ/kg

Fig. 8.4.6. The dual pressure process with throttling in two stages and compression of a part of the gas from intermediate pressure.

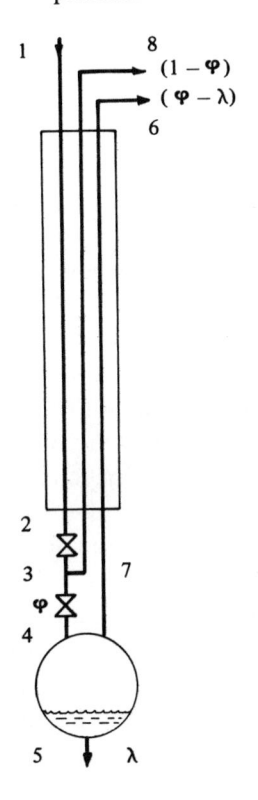

The specific exergy of the liquid nitrogen at 77.4 K was found to be 768.2 and the exergetic efficiency is:

$(768.2/6683) = 0.115$

This is of course the same as the exergetic efficiency found in example 8.4.3 because there is no loss of exergy in the compressor.

The apparatus of Fig. 8.4.3 is used in simple liquefiers with atmospheric pressure downstream of the expansion valve. At an early stage of the development, however, Linde realised that whereas the Joule–Thomson effect is approximately proportional to the pressure difference, the specific work of compression is approximately proportional to the pressure ratio. It is economical therefore to expand the air in two stages as shown in Fig. 8.4.6 with an intermediate pressure stream and a low-pressure stream going back through the exchanger. This practice corresponds to the expansion in series in the vapour compression system discussed in Chapter 5.

Referring to Fig. 8.4.6, suppose the fraction λ of the forward stream at 1 is liquefied and drawn off. Let φ be the fraction of the forward stream which goes on to the second expansion valve. There is then a revert stream $(1 - \varphi)$ at intermediate pressure and another revert stream $(\varphi - \lambda)$ at the low pressure. The steady-flow equation gives:

$$h_1 = \lambda h_5 + (1 - \varphi)h_8 + (\varphi - \lambda)h_6 \qquad (8.4.12a)$$

and:

$$\lambda = \frac{h_6 - h_1 - (1 - \varphi)(h_6 - h_8)}{h_6 - h_5} \qquad (8.4.12b)$$

This fraction is clearly less than the fraction liquefied in the simple apparatus because the numerator is less than $h_6 - h_1$. The intermediate revert stream in fact carries away a part of the refrigerating effect because it still has some pressure with a potential for a drop in temperature on expansion. Up to an intermediate pressure of about 10 bar for air this loss is relatively small and the fraction liquefied is about the same as in the simple apparatus. At higher intermediate pressure the fraction λ declines, but it is accompanied by a significant reduction of the work of compression since the intermediate revert stream has to be compressed over a much smaller pressure ratio.

The optimisation of the dual pressure process requires much calculation and will not be pursued here. Results are given in the books by Hausen (1957) and Ruhemann (1949). Practical operating conditions for plants like Fig. 8.4.6 liquefying air are: high pressure 200 bar, intermediate pressure

8.4 Regenerative cooling: the Linde–Hampson process

40–50 bar, $\varphi = 0.2$. The actual specific work is about 5440 kJ/kg corresponding to an exergetic efficiency of about 0.14.

Precooling

It is obvious that an increased fraction of liquid can be obtained if the specific enthalpy of the high-pressure gas can be reduced by external means such as cooling by a vapour compression plant. But some consideration needs to be given to the place at which the cooling is applied. It would not be best, for example, to reduce the temperature of the gas at point 1 in Fig. 8.4.3 because it would not then be possible to use the forward stream to raise the temperature of the revert stream to ambient temperature, and some of the Joule–Thomson refrigerating effect would be wasted. The external refrigeration has to be applied at some point down the exchanger as shown in Fig. 8.4.7.

Applying the steady-flow equation to a control surface around the lower heat exchanger, valve and separator in Fig. 8.4.7 gives:

$$h_3 = \lambda h_6 + (1 - \lambda)h_8 \tag{8.4.13a}$$

Fig. 8.4.7. Cooling by auxiliary means applied to the high-pressure stream between 2 and 3.

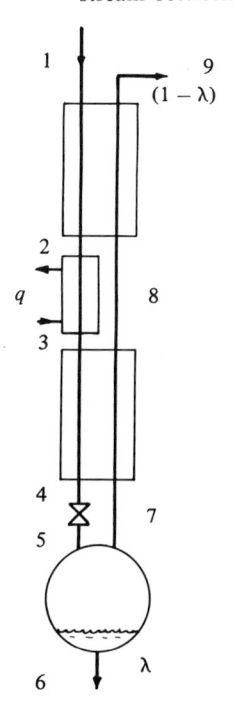

and:

$$\lambda = \frac{h_8 - h_3}{h_8 - h_6} \tag{8.4.13b}$$

where λ is the fraction liquefied. Assuming that the temperatures at 3 and 8 are specified, λ can be found. The external refrigerating effect per unit mass of the forward stream, q say, is given by a balance on the whole apparatus:

$$q = \lambda(h_9 - h_6) - (h_9 - h_1) \tag{8.4.14}$$

and the state at 2 is found from:

$$q = (h_2 - h_3) \tag{8.4.15}$$

The use of eq. (8.4.13b) to find λ requires an assumption about the temperature at 8. Because the revert stream has a lower capacity rate than the high-pressure stream it is possible to bring its temperature at 8 within any desired approach of T_3, exactly as was found for the temperature at the warm end of the exchanger in the simple process. The same holds for the upper exchanger in Fig 8.4.7. The actual temperature approaches $(T_1 - T_9)$ and $(T_3 - T_8)$ depend on the surface areas provided in the exchangers, but they must be made small for economical operation.

The effect of the external cooling on the temperature profiles in the exchangers is shown in Fig. 8.4.8, similar to Fig. 8.4.5 for the simple process,

Fig. 8.4.8. Temperatures in the heat exchanger with auxiliary cooling from 2 to 3.

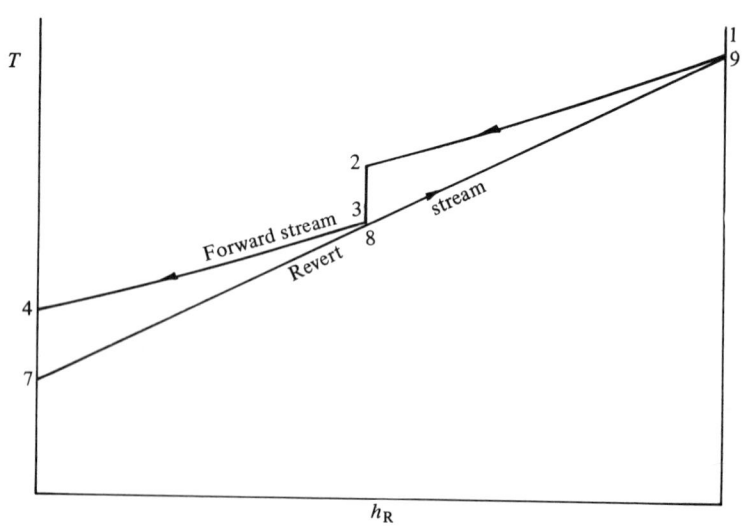

8.4 Regenerative cooling: the Linde–Hampson process

as a step in the temperature plot for the high-pressure stream. The temperature difference at the cold end has been reduced and the loss of exergy in the exchanger has been reduced. The temperature change across the valve is reduced and the loss of exergy in the valve is less. At the same time work has had to be expended in driving the external refrigerating machine. The advantage of precooling lies in the fact that more exergy is saved in the liquefaction process than has to be expended in the refrigerating machine.

The dual pressure process with precooling by an ammonia machine was for many years one of the two principal methods of making liquid air. It was subsequently overtaken by the Heylandt process described in section 8.5, which is marginally more efficient.

External refrigeration also allows gases whose maximum inversion temperature is below ambient temperature to be liquefied. Thus hydrogen with a maximum inversion temperature of about 200 K can be precooled by liquid nitrogen at 77 K and liquefied, as also can neon. Helium can be

Fig. 8.4.9. Auxiliary cooling by, for example, liquid nitrogen, when the cold nitrogen gas is taken back through the upper heat exchanger.

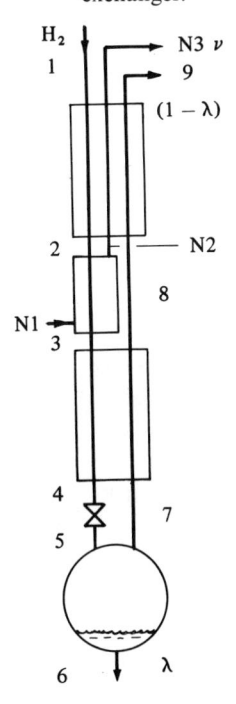

Cryogenic engineering

liquefied by using a final precooling stage of hydrogen boiling at 20.4 K at atmospheric pressure, or preferably below 20.4 K under reduced pressure.

When another fluid such as nitrogen is used for precooling, the refrigerating effect of the nitrogen gas is recovered by passing it back through the heat exchanger at the warm end, as shown in Fig. 8.4.9.

Example 8.4.5

Hydrogen is liquefied in an apparatus like Fig. 8.4.9 with cooling by nitrogen boiling at 77.4 K. The hydrogen gas enters at a pressure of 101.3 bar = 100 atm and a temperature of 300 K. Liquid is withdrawn at 1.013 bar = 1 atm. The temperature approaches are as follows:

$$T_1 - T_9 = T_1 - T_{N3} = 5 \text{ K}$$
$$T_3 - T_8 = 5 \text{ K}$$
$$T_3 - T_{N1} = 2.6 \text{ K}$$

Determine the fraction liquefied, the mass of liquid nitrogen needed per unit mass of hydrogen liquid produced, and the temperature of the high-pressure hydrogen gas at the cold end of the upper heat exchanger.

The properties of hydrogen are given in Table 8.4.2.

$$h_8(1.013 \text{ bar}, 75 \text{ K}) = 1301 \text{ kJ/kg}$$
$$h_3(101.3 \text{ bar}, 80 \text{ K}) = 1192 \text{ kJ/kg}$$
$$h_6(\text{liquid}, 20.4 \text{ K}) = 271 \text{ kJ/kg}$$

Table 8.4.2. *Specific enthalpy*/(kJ/kg) of normal hydrogen

T K	$p = 1.013$ bar h kJ/kg	$p = 101.3$ bar h kJ/kg
20.4 sat. liq.	271	
20.4 sat. vap.	718	
70	1248	1040
80	1354	1192
90	1463	1334
100	1574	1471
280	3941	3984
300	4227	4275

8.4 Regenerative cooling: the Linde–Hampson process

Hence:

$$\lambda = \frac{1301 - 1192}{1301 - 271} = 0.1058$$

$h_9(1.013 \text{ bar}, 295 \text{ K}) = 4155 \text{ kJ/kg}$

$h_1(101.3 \text{ bar}, 300 \text{ K}) = 4275 \text{ kJ/kg}$

Values for nitrogen from Table 8.4.1 are:

$h_{N3}(1.013 \text{ bar}, 295 \text{ K}) = 456.9 \text{ kJ/kg}$

$h_{N1}(\text{liquid}, 77.4 \text{ K}) \quad = 29.4 \text{ kJ/kg}$

$h_{N2}(1.013 \text{ bar}, 77.4 \text{ K}) = 228.7 \text{ kJ/kg}$

Let v be the mass flow rate of nitrogen per unit mass of high-pressure hydrogen. Then the steady-flow equation applied to a control surface cutting points 1, 9, 8, 3, N1 and N3 gives:

$$h_1 - h_3 = (1 - \lambda)(h_9 - h_8) + v(h_{N3} - h_{N1}) \tag{8.4.16}$$

Inserting values and solving for v:

$v = 1.242$

The mass of nitrogen required to liquefy 1 kg of hydrogen is:

$(1.242/0.1058) \text{ kg} = 11.74 \text{ kg}$

The steady-flow equation applied to the nitrogen evaporator gives:

$$h_2 - h_3 = v(h_{N2} - h_{N1}) \tag{8.4.17}$$

Inserting values and solving for h_2:

$h_2 = 1440 \text{ kJ/kg}$

Interpolating in the table at 101.3 bar:

$T_2 = 97.7 \text{ K}$

A feature peculiar to hydrogen liquefiers is an arrangement to catalyse the transformation from ortho- to para-hydrogen, the enthalpy change of which is not included in the above example. At ambient temperature hydrogen consists of a mixture of about 0.75 mole fraction of ortho-hydrogen and 0.25 of para-hydrogen. The difference between these is the relative orientation of the proton spins in the molecule. The mixture 0.75/0.25 is called normal hydrogen. Freshly made liquid hydrogen has this composition, but at 20 K the equilibrium composition is 0.998 para-hydrogen, i.e. nearly pure para-hydrogen. The change is exothermic with a specific enthalpy of about 700 kJ/kg per unit mass of ortho-hydrogen converted. Freshly made liquid thus liberates about 525 kJ/kg in a change which proceeds slowly over days or weeks. This quantity is larger than the specific enthalpy of evaporation and consequently there is a rapid rate of boil-off even when the liquid is perfectly insulated. Hydrogen liquefiers therefore incorporate means of accelerating the transformation by a ferric hydroxide gel catalyst so that the liquid produced is almost pure para-hydrogen. This transformation increases the net work of liquefaction and the exergy of para-hydrogen is greater than that of normal hydrogen.

8.5 The use of expansion machines: the Claude process

It has been seen that the Joule–Thomson method of reducing temperature requires considerable pressures to produce a fairly small change, e.g. nearly 50 bar to achieve a drop of 10 K with air. The same drop in temperature can be produced by very much smaller pressure if the expansion is carried out reversibly in a process in which work is delivered. For air at a temperature of 300 K initially, for example, a drop of 10 K can be achieved with a pressure of only 1.13 bar expanded to 1 bar.

In principle it is possible to liquefy air and similar gases by direct expansion from an initial state at ambient temperature and a suitable pressure, as shown in Fig. 8.5.1. For air a pressure of about 180 bar initially

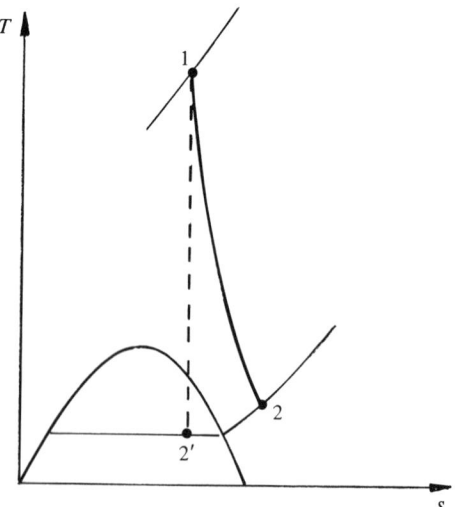

Fig. 8.5.1. Isentropic expansion 1 to 2', and real expansion 1 to 2, on the temperature–entropy diagram.

8.5 The use of expansion machines: the Claude process

would produce a liquid fraction of 0.1 on expanding isentropically to a pressure of 1 bar. Because isentropic expansion is only a limiting case and in practice there would be an increase in entropy, the state at the end of expansion in a cylinder would in fact be saturated or superheated vapour, and no liquid would be produced. However it would seem that by precooling the compressed air with revert gas it would be possible to liquefy directly on expansion, and this was in fact achieved by Georges Claude in his early experiments. But in practice this method is never used. Whether expansion engines or turbines are employed, the accelerated heat transfer which the presence of liquid inside them causes, as compared with a dry gas, reduces the yield of liquid considerably. The Joule–Thomson throttling valve on the other hand is compact and can easily be insulated. The liquid

Fig. 8.5.2. The Claude process for liquefying air. Part of the high pressure stream at 2 is diverted through an engine with reduction of temperature. The cold gas so produced is returned to the revert stream at 8.

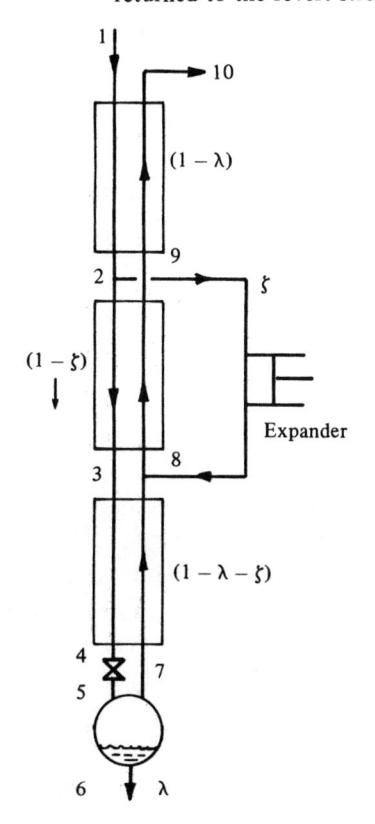

passing through it has little time to gain enthalpy from heat transfer and it can be discharged directly into a vacuum flask with little loss. Consequently the final stage of almost all liquefiers is still a throttling valve.

The combination of an expansion machine with a final throttling process, which is usually called the Claude system, is shown in its most general form in Fig. 8.5.2. The gas under pressure enters at 1 and is first cooled by heat transfer to the revert gas. A fraction ζ of the flow is diverted at 2 through the expansion machine and undergoes a drop in pressure and in temperature, being returned to the revert stream at 8. Cooling of the high-pressure gas continues from 2 to 3 and then from 3 to 4 before entering the throttling valve where a fraction λ of the original high-pressure gas is liquefied and can be drawn off.

The enthalpy balance for the apparatus at steady conditions and assuming adiabatic operation is:

$$h_1 = \lambda h_6 + (1 - \lambda) h_{10} + \zeta(h_2 - h_8) \tag{8.5.1}$$

and:

$$\lambda = \frac{h_{10} - h_1}{h_{10} - h_6} + \zeta \frac{h_2 - h_8}{h_{10} - h_6} \tag{8.5.2}$$

The first term on the right-hand side is the same as the fraction which would be liquefied in the Linde–Hampson apparatus with the same states at 1, 6 and 10 (corresponding to states 1, 4 and 6 in Fig. 8.4.3): the second term gives the extra fraction which is liquefied as a result of using the expansion machine.

At first sight it appears that by increasing ζ up to its maximum value of 1 the fraction λ would also increase, but this is clearly ridiculous because when $\zeta = 1$ there is nothing left to liquefy. The fallacy in the argument lies in the assumption that h_{10} can be chosen at will so as to be only arbitrarily less than the value $h_{10}(\max)$ when the revert gas at 10 has the same temperature as the ingoing stream at 1, i.e. perfect recuperation. This is always possible in the Linde–Hampson process, provided that the process never goes too far outside the inversion locus, because the capacity rate of the revert stream is less than that of the forward stream. In the Claude process this is not necessarily so. Above a certain fraction ζ, it is impossible to obtain perfect recuperation, as will be seen. One cannot simply select a value of ζ and assume h_{10} to find λ without considering the temperature approaches in the heat exchangers.

This type of problem is a very common one in cryogenic systems and it is desirable to employ a systematic approach which will give the answer

8.5 The use of expansion machines: the Claude process

without guesswork and which will be applicable to more complicated systems. The principles of the method will be outlined for the system of Fig. 8.5.2.

As constants for the problem it will be supposed that the high pressure, the low pressure, and the temperature at point 1 are known or assumed. The temperature of the gas at the withdrawal point 2 is assumed to be fixed for the problem and also the temperature to which it is expanded in the machine, T_8. To avoid irreversibilities caused by mixing streams at different temperatures it will be supposed that the expanded stream is returned to the revert stream at the correct point, i.e. where the temperature is the same as that of the expanded stream.

For the whole apparatus to work the temperature of the revert stream at all points in the apparatus must be less than, or in the limit equal to, that of the forward stream. As a rule this means checking temperature differences at the junctions of the heat exchangers, although as will be seen later it is advisable to ensure that no temperature pinches occur inside them. Disregarding this possibility here, the constraints which have to apply are:

$$T_{10} \leqslant T_1 \tag{8.5.3}$$

$$T_9 \leqslant T_2 \tag{8.5.4}$$

$$T_8 \leqslant T_3 \tag{8.5.5}$$

$$T_7 \leqslant T_4 \tag{8.5.6}$$

The temperatures T_1, T_2, T_8 and T_7 are known or assumed. The first pair of inequalities are therefore applied in the form:

$$h_{10} \leqslant h_{10}(\text{max}) \tag{8.5.7}$$

$$h_9 \leqslant h_9(\text{max}) \tag{8.5.8}$$

where $h_{10}(\text{max})$ and $h_9(\text{max})$ are the specific enthalpies of the low-pressure gas at temperatures T_1 and T_2, respectively. The second pair of inequalities are applied in the form:

$$h_3(\text{min}) \leqslant h_3 \tag{8.5.9}$$

$$h_4(\text{min}) \leqslant h_4 \tag{8.5.10}$$

where $h_3(\text{min})$ and $h_4(\text{min})$ are the specific enthalpies of the high-pressure gas at temperatures T_8 and T_7, respectively.

The procedure is then to apply the steady-flow equation to control surfaces cutting the points where the constraints apply. The result will be a set of linear inequalities in λ and ζ which will determine the permissible field of these quantities.

From eq. (8.5.1):

$$h_{10} = \frac{h_1 - \lambda h_6 - \zeta(h_2 - h_8)}{1 - \lambda} \qquad (8.5.11)$$

Hence using inequality (8.5.7):

$$\frac{h_1 - \lambda h_6 - \zeta(h_2 - h_8)}{1 - \lambda} \leqslant h_{10}(\text{max}) \qquad (8.5.12)$$

and on re-arranging:

$$\zeta \geqq \frac{\lambda[h_{10}(\text{max}) - h_6] - [h_{10}(\text{max}) - h_1]}{h_2 - h_8} \qquad (8.5.13)$$

This relation defines a region on a plot of ζ versus λ within which the inequality (8.5.13) holds and outside which it does not. This region will be called a permissible region, and the region on the other side of the line the forbidden region. When the equality sign in (8.5.13) holds the temperature approach between streams 1 and 10 is zero, i.e. there is perfect recuperation.

For a control surface cutting the high-pressure stream just above the take-off point for the expanded stream, the same relation as (8.5.13) will be obtained with h_2 replacing h_1 and $h_9(\text{max})$ replacing $h_{10}(\text{max})$, i.e.

$$\zeta \geqslant \frac{\lambda[h_9(\text{max}) - h_6] - (h_9(\text{max}) - h_2]}{h_2 - h_8} \qquad (8.5.14)$$

For a control surface cutting 3 and 8 below the point of return of the expanded gas:

$$(1 - \zeta)h_3 = \lambda h_6 + (1 - \zeta - \lambda)h_8 \qquad (8.5.15)$$

and:

$$h_3 = \frac{\lambda h_6 + (1 - \zeta - \lambda)h_8}{1 - \zeta} \qquad (8.5.16)$$

Using inequality (8.5.9):

$$\frac{\lambda h_6 + (1 - \zeta - \lambda)h_8}{1 - \zeta} \geqslant h_3(\text{min}) \qquad (8.5.17)$$

and on re-arranging:

$$\zeta \leqslant 1 - \frac{h_8 - h_6}{h_8 - h_3(\text{min})} \lambda \qquad (8.5.18)$$

For a control surface cutting points 4 and 7 the same relation applied with h_7 replacing h_8 and $h_4(\text{min})$ replacing $h_3(\text{min})$, i.e.

$$\zeta \leqslant 1 - \frac{h_7 - h_6}{h_7 - h_4(\text{min})} \lambda \qquad (8.5.19)$$

8.5 The use of expansion machines: the Claude process

The use of these inequalities will be shown by an example in which for illustration a slightly simplified version of Fig. 8.5.2 is used. To avoid the complications arising from the partial separation of air during liquefaction, nitrogen will be assumed as the fluid.

Example 8.5.1
Nitrogen is liquefied in the apparatus of Fig. 8.5.2 with the lower heat exchanger absent, i.e. $h_4 = h_3$ and $h_8 = h_7$. The high-pressure gas enters at 40.52 bar (40 atm), 300 K, and the low-pressure is 1.013 bar (1 atm). The gas is withdrawn at 2 where the temperature is 180 K and is expanded to produce saturated vapour at 8. Investigate the permissible field of λ and ζ and find the maximum value of λ.

Listing the specific enthalpies at the principal points:

h_1(40 atm, 300 K) = 454.3 kJ/kg

h_2(40 atm, 180 K) = 312.1 kJ/kg

h_8(1 atm, 77.4 K, vapour) = h_7 = 228.7 kJ/kg

h_6(1 atm, 77.4 K, liquid) = 29.4 kJ/kg

The limiting values are:

h_{10}(max)(1 atm, 300 K) = 462.1 kJ/kg

h_9 (max)(1 atm, 180 K) = 337.1 kJ/kg

h_3 (min)(40 atm, 77.4 K) = h_4(min) = 32.2 kJ/kg

Substituting values in inequality (8.5.13) gives:

$\zeta \geqslant 5.188 \lambda - 0.0935$

The boundary of this relation is plotted in Fig. 8.5.3. The forbidden side of the line is shown hatched. The permissible region gives the values of λ and ζ for which the temperature at point 10 is less than 300 K. Substituting values in inequality (8.5.14) gives:

$\zeta \geqslant 3.689 \lambda - 0.300$

The boundary of this relation is shown in Fig. 8.5.3 with the forbidden side hatched. Within the permissible region the temperature at 9 is less than 180 K. However, it is seen that this consideration is not restrictive here since the permissible region embraces that already found above.

From inequality (8.5.19) with $h_7 = h_8$, etc.:

$\zeta \leqslant 1 - 1.014\lambda$

This boundary is also shown in Fig. 8.5.3. Points below the line indicate conditions where the temperature of the nitrogen just before the valve is higher than 77.4 K.

It is clear that λ and ζ can lie anywhere inside the quadrilateral defined by the origin O and the points B, C, and D without going outside the constraints.

The maximum yield occurs at C where:

$$5.188\lambda - 0.0935 = 1 - 1.014\lambda$$

i.e.

$$\lambda = 0.176 \quad \zeta = 0.821.$$

At this condition there is perfect recuperation at point 10 and also zero temperature difference between streams 3 and 8. Almost the whole fraction of the stream at 3 is liquefied, the flash vapour formed in the valve amounting to only $(1 - 0.176 - 0.821)$, i.e. 0.003 of the original stream at 1.

Fig. 8.5.3. Plot of the diverted fraction ζ versus the yield fraction λ for example 8.5.1. The hatched sides of the lines show conditions at which the revert stream would be warmer than the forward stream.

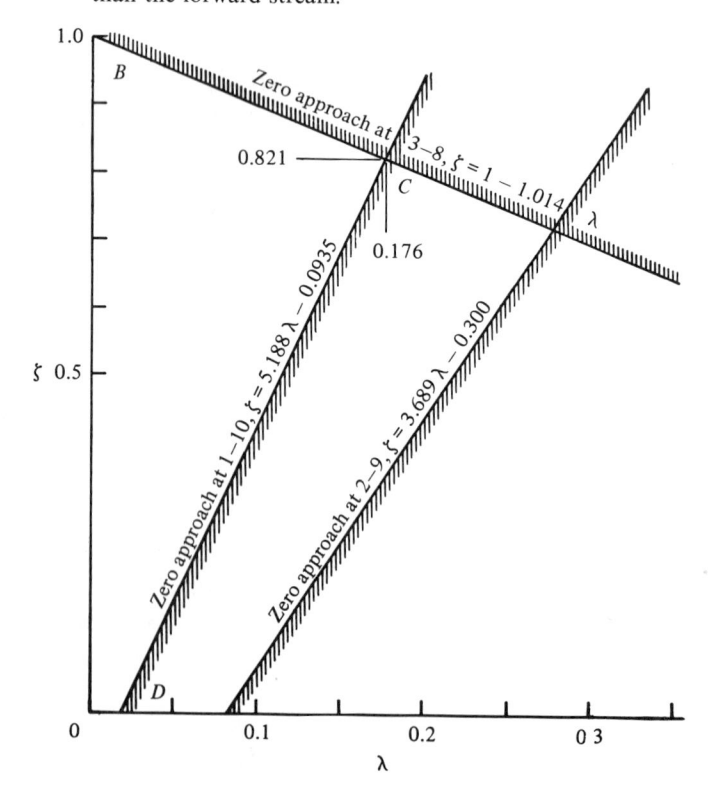

8.5 The use of expansion machines: the Claude process

In practice of course one would have to work somewhere inside the permissible quadrilateral $OBCD$ in order to allow finite temperature approaches in the heat exchangers.

Referring back to eq. (8.5.2), the reason why λ cannot be increased indefinitely simply by increasing ζ is that above $\zeta = 0.821$ in example 8.5.1 the operating point is forced off the line representing perfect recuperation by the fact that the temperature difference between 3 and 8 has vanished: hence h_{10} has to diminish.

If the work from the expander is not recovered and used to help to drive the compressor, then the point of maximum λ is obviously also the point of maximum economy of operation. This applies to many small systems. For large systems, however, the power may be worth recovering and then one is interested in the ratio:

$$y = \frac{\lambda}{w_c - \zeta w_e} \qquad (8.5.20)$$

where w_c is the specific work of compression and w_e is the specific work of expansion, both regarded as positive. Equation (8.5.20) implies the linear

Fig. 8.5.4. Plot of the diverted fraction ζ versus the yield fraction λ showing that the minimum net specific work is obtained when the yield fraction has its highest value.

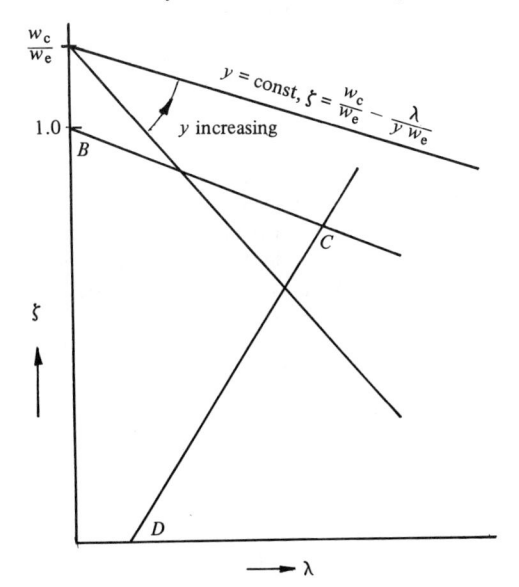

relation:

$$\zeta = \frac{w_c}{w_e} - \frac{\lambda}{yw_e} \qquad (8.5.21)$$

between λ and ζ for a constant value of y. Hence on the plot of λ versus ζ straight lines of constant y radiate from the point w_c/w_e on the ζ-axis with the gradient diminishing as y increases, as shown in Fig. 8.5.4. Since w_c/w_e is greater than unity because the temperature at entry to the expander is less than that at entry to the compressor, and because the efficiencies tend to increase w_c but to reduce w_e, it is seen that the maximum value of y occurs at the maximum permissible value of λ at point C.

In a variation of the Claude arrangement due to Heylandt, the first heat exchanger in Fig. 8.5.2 is omitted and a higher pressure is used, about 200 bar for air. It can be shown by the method of example 8.5.1 that this permits a larger fraction to be liquefied, although at the expense of more compression work.

When the method of example 8.5.1 is applied to the Heylandt process it is found that the limiting temperature approaches occur between points 2 and 9 (or 1 and 10 since the first heat exchanger is absent) and points 3 and 8, leaving necessarily a finite approach between 4 and 7. Thus it is apparently possible to balance the upper heat exchanger exactly. This is not quite true however when the variation of the specific heat capacity with temperature of the forward stream is taken into account. Because c_p increases as the temperature falls, whereas that of the low-pressure stream is almost constant or if anything increases with temperature, the plots of temperatures in the upper heat exchanger are somewhat as shown in

Fig. 8.5.5. Temperature possibilities in a heat exchanger when the warm stream has a specific heat capacity decreasing with temperature increasing. An arbitrarily small approach can be obtained at either end or somewhere in between, but not everywhere simultaneously. h_R is the specific enthalpy of the revert gas considered as a variable increasing from h_8 at the cold end of the exchanger to h_9 at the other end.

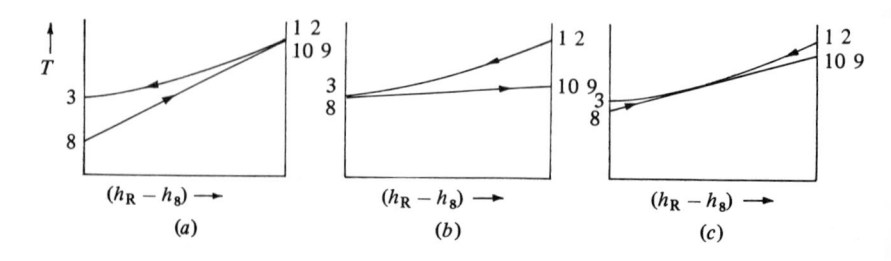

8.5 The use of expansion machines: the Claude process

Fig. 8.5.5. One can have an arbitrarily small approach at one end or the other, shown in (a) and (b), but not at both simultaneously. Somewhere in the exchanger an arbitrarily small approach can be obtained, as in (c), leaving necessarily finite approaches at the ends.

This occurrence, where the overall enthalpy balance for the exchanger apparently shows the possibility of infinitesimal approaches at both ends, but which is not possible when intermediate sections are considered, is often called a 'temperature nip' or 'pinch point'. It is often found in cryogenic plant and must be watched for. The result of not noticing one would be the design of heat exchangers for terminal temperature differences which could not possibly be obtained.

The temperature of the gas leaving the expansion machine depends on the isentropic efficiency and on the rate of heat transfer to the expanding gas. For an expansion machine the isentropic efficiency is defined as:

$$\eta_{\text{isen}} = \frac{w_e}{(h_2 - h_8)_s} \qquad (8.5.22)$$

using the nomenclature of Fig. 8.5.2 for h_2 and h_8, where the subscript s denotes that the entropy at 8 is the same as that at 2. For an adiabatic machine:

$$w_e = h_2 - h_8 \qquad (8.5.23)$$

the actual change of specific enthalpy, but because of heat transfer to the gas:

$$w_e > h_2 - h_8 \qquad (8.5.24)$$

and consequently:

$$\frac{h_2 - h_8}{(h_2 - h_8)_s} < \eta_{\text{isen}} \qquad (8.5.25)$$

Calculation of the temperature leaving the expander requires the determination of the isentropic enthalpy drop for states which end in the two-phase region.

Example 8.5.2

Determine the ratio $(h_2 - h_8)/(h_2 - h_8)_s$ which is implied by the temperatures at entry and exit of the expander in example 8.5.1.

The entropies are:

$s_2(40\,\text{atm},\ 180\,\text{K}) = 2.692\,\text{kJ/kg}$

$s_{g,8}(1\,\text{atm, sat. vapour}) = 2.994\,\text{kJ/kg}$

$s_{f,8}(1\,\text{atm, sat. liquid}) = 0.418\,\text{kJ/kg}$

The dryness fraction after isentropic expansion to a pressure of 1 atm is then:

$$x'_8 = \frac{2.692 - 0.418}{2.994 - 0.418} = 0.883$$

The specific enthalpy is then:

$$h'_8 = [29.4 + 0.883(228.7 - 29.4)] \text{ kJ/kg}$$
$$= 205.4 \text{ kJ/kg}$$

and:

$$(h_2 - h_8)_s = (312.1 - 205.4) \text{ kJ/kg} = 106.7 \text{ kJ/kg}$$

The actual specific enthalpy change is:

$$(312.1 - 228.7) \text{ kJ/kg} = 83.4 \text{ kJ/kg}$$

giving a ratio:

$$83.4/106.7 = 0.782$$

According to eq. (8.5.25) the isentropic efficiency would have to be higher than this. Such a value would be higher than that actually attained in practice.

For the Linde–Hampson process it was seen in section 8.4 that pressure drop in the forward stream did not affect the yield fraction. For the same pressure at point 1 the work of compression is the same so that the exergetic efficiency is unaltered.

In systems using expansion machines on the other hand, pressure drop reduces the yield fraction by reducing the pressure ratio and the temperature drop across the expansion machines. There is a loss of exergy of the forward stream which is not compensated elsewhere. If the work from the expansion machines is recovered there is a further loss in that this work is reduced.

Both the Claude and Heylandt methods have a better exergetic efficiency than the simple Linde–Hampson method. This is associated with the better balancing of the heat exchange part of the process, resulting in a smaller loss of exergy. The reduction more than outweighs the loss of exergy in the expansion machine.

When compared with the Linde dual pressure process with precooling by a vapour compression plant the efficiencies are not very different. By a small margin the Heylandt system appears to be the best with a specific work requirement of about 2900 kJ/kg when liquefying air, corresponding to an

8.5 The use of expansion machines: the Claude process

exergetic efficiency of about 0.25. It gives a simpler system by dispensing with the ammonia plant and, adapted to the production of liquid oxygen and nitrogen, is still widely used.

Liquid air itself as an end product is not produced in large quantities today. Modern methods of liquefaction are associated with a separation process which is described in section 8.6.

Processes of the Claude type employing two-stage expansion have been proposed. Several variations of these are treated by Rabes (1930), but they have been little used in large plants.

Multi-stage expansion is commonly employed in hydrogen and helium liquefiers, however. The best known of these is the Collins liquefier, already mentioned in section 8.1, which is manufactured in quite large numbers for use in laboratories.

The principal advantage of using more than one expansion machine is that it offers the possibility of more complete balancing of the heat exchange process than do the Claude and Heylandt methods. How this is so is illustrated on temperature and entropy coordinates in Fig. 8.5.6. A flow of compressed gas, of which the fraction ultimately liquefied is λ say, enters at 1. The fraction $(1 - \lambda)$ leaves the apparatus at 0, and with make-up gas is recompressed to 1. If the fraction diverted through the first expansion machine 1–3 is made equal to λ, the remaining forward stream 1–2 has the

Fig. 8.5.6. Temperature–entropy diagram showing how approximate reversibility can be obtained with balancing of the heat exchangers, by diverting appropriate fractions of the high-pressure stream through expansion machines and returning the exhaust to the revert stream.

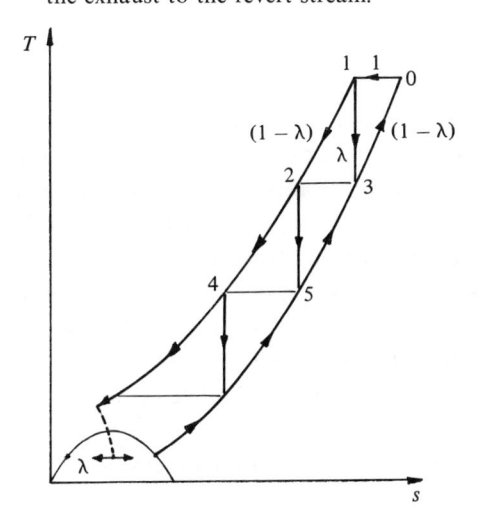

mass flow rate $(1 - \lambda)$. The mass flow rates of the forward and revert streams in the first exchanger are thus equal. Since at ambient temperature the specific heat capacities of the two streams are very nearly equal the exchanger is balanced.

In the next exchanger the specific heat capacity of the compressed gas is slightly larger than that of the revert gas. To balance this exchanger, the mass flow rate 2–4 has to be somewhat less than that in 5–3. Hence the fraction diverted through the expansion engine 2–5 has to be somewhat smaller than λ of the original stream.

By diverting appropriate fractions through the expansion machines each exchanger except the last can be balanced, apart from the existence of pinch points inside them. But since the range of temperature in each exchanger is small the pinch points do not embarass the design.

The last exchanger is not balanced, but again the exergy loss is small because of the small temperature range. In it the compressed gas is reduced to a temperature such that Joule–Thomson expansion takes the state into the two-phase region.

The above description was based on isentropic expansion processes as shown in Fig. 8.5.6. In practice of course there is an increase in entropy in each expansion on account of irreversibility and of heat transfer.

Multi-stage expansion also enables the maximum pressure to be reduced because one engine does not have to produce the whole of the refrigeration for the process. This again reduces the loss of exergy in the expansion valve.

The type of process just described is clearly very suitable for the liquefaction of helium which has a very small liquefaction step even at atmospheric pressure.

Expansion machines

The first expansion engines used by Claude were rather similar to reciprocating steam engines, but his later designs were single acting with poppet valves. With his process the whole cylinder operated at low temperature, and special lubricants such as pentane or petroleum ether had to be used. The Heylandt high-pressure process on the other hand operates with the inlet gas at ambient temperature, the exhaust being at low temperature. By providing a non-conducting extension to the piston it is possible to maintain the piston seals and the rod gland at ambient temperature so that lubrication problems are avoided. The work done in overcoming friction goes off to ambient temperature instead of into the expanding gas.

The Collins helium liquefier employs a reciprocating engine of unusual

8.5 The use of expansion machines: the Claude process 571

Fig. 8.5.7. An expansion turbine used for the liquefaction of helium. (With acknowledgment to B.O.C. Ltd)

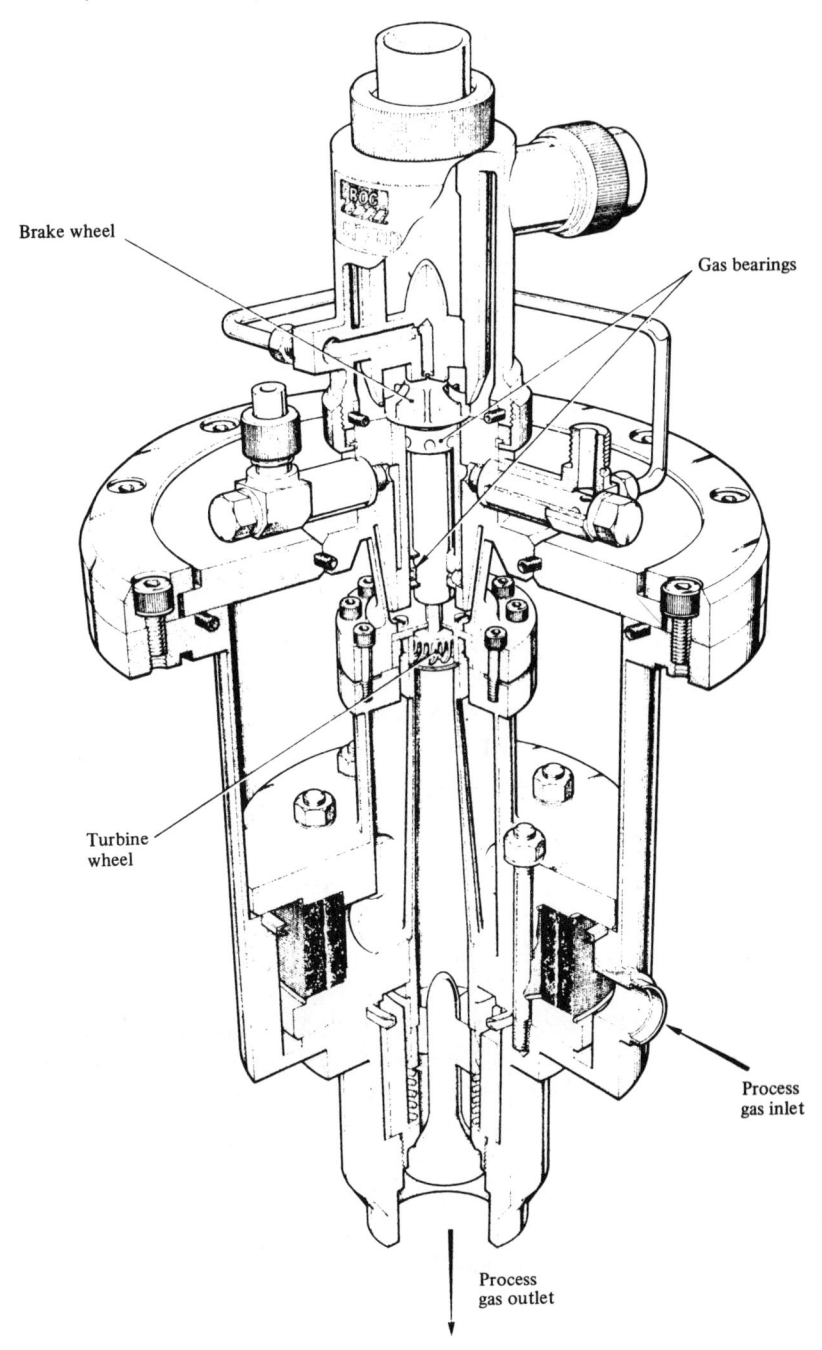

Brake wheel

Gas bearings

Turbine wheel

Process gas inlet

Process gas outlet

Cryogenic engineering 572

design in that the inlet and exhaust valves are at the rod end of the cylinder. The piston rod is in tension always and it can be made sufficiently thin to be flexible, allowing the piston to be self-aligning in the bore and making a good fit possible, and also reducing heat conduction along the rod.

Many industrial processes are intended for gas separation by distillation and for these the pressures employed are lower. Single-stage turbines of the radial flow type are widely used for these duties. They are also used for smaller plants for helium and hydrogen liquefaction. Figure 8.5.7 shows one of these.

An interesting account of the development of expansion machines and processes employing them is given by Collins & Cannaday (1958). Detailed drawings of several types of machines are given by Usyukin (1965). The design and analysis of expanders are treated by Blackford, Halford & Tantam in Haselden (1971).

Expansion processes in which work is developed are used in the modern form of the Stirling machine, as described in section 7.7.

8.6 The separation of gases

The separation of oxygen and nitrogen from the atmosphere is the largest industrial use of temperatures below 100 K, and the principles involved will be treated in this section.

Before the application of low temperatures for this purpose, oxygen was produced by various chemical methods, such as Brin's process in which barium peroxide was formed and then decomposed, but its industrial uses were few. It was used for breathing apparatus, and in the oxyhydrogen flame for making the brilliant 'limelight' and for welding special metals, but it was hardly used in general industry. The availability of oxygen from the atmosphere in the early years of this century stimulated the use of oxyhydrogen and later oxyacetylene welding for steel. For this a purity of about 98% was needed. The later development of flame-cutting required a purity of about 99%. Welding oxygen is usually supplied now for all purposes at about 99.3% purity. Originally the oxygen was delivered to works as compressed gas, but by about 1920 it was being transported in liquid form for large users. The nitrogen left over found applications in the Frank–Caro process for making cyanamide and the Haber process for the synthesis of ammonia.

Argon, 0.93% of the atmosphere, found a use in filling the envelopes of electric lamps and later for argon-arc welding. The trace gases neon and krypton were employed in gas-discharge tubes for advertising purposes from the 1920s.

8.6 The separation of gases

The largest separation plants today produce oxygen for use in steel making. This so-called 'tonnage oxygen' is produced as gas of purity down to 99% at or near the place of consumption in units with outputs of 50 to 1000 t/day. In open-hearth furnaces it gives higher production rates than the ordinary air blast as well as a better quality of steel. In processes specially adapted to the use of an oxygen blast, such as the Linz–Donawitz process (named after the towns in Austria where it was developed), very high production rates can be attained, typically about 150 t of steel in less than an hour with the use of about 10 t of oxygen. Oxygen is also used in electric furnaces for making special steels.

Tonnage oxygen is also used in the Lurgi process for making town gas from petroleum and in the manufacture of ethylene oxide.

Since World War II, liquid oxygen has been used in considerable quantities for rocket propulsion.

The commercial success of air separation suggested the use of low-temperature techniques for separating other gas mixtures. There was a growing demand for hydrogen for the manufacture of margarine by hydrogenation of oils into solid fats, for airships, and again for the synthesis of ammonia. Extraction of hydrogen from water-gas by condensing the other components was the first application of low-temperature techniques to gases other than air.

Ethylene (boiling point 169 K) has for some decades been an important base chemical on account of its numerous reactions and its ability to polymerise. Unfortunately it does not occur naturally to any great extent. One of the first large-scale sources was coke-oven gas from which ethylene was separated by low-temperature condensation, but most of it now comes from refinery gases.

Deuterium, heavy hydrogen, is important for the manufacture of heavy water, used as a moderator in the nuclear power industry. It was traditionally obtained by electrolysis of ordinary water. Since heavy hydrogen occurs only as 0.14% of ordinary hydrogen the process is difficult. Pure deuterium however boils at 23.66 K as against pure normal hydrogen at 20.39 K and this difference is sufficient to enable a practical separation process to be operated.

Helium is separated from natural gas by condensation of the methane and nitrogen. With its very low boiling point of 4.2 K and its unusual properties at low temperatures, helium is the only liquid available for low-temperature research.

Gas separation processes today are many and varied, but the main principles will be shown by a discussion of the separation of oxygen and nitrogen from the air. The composition of the atmosphere is given in Table 8.6.1, from which it is seen that argon is also an important constituent. To avoid complicating matters this will be neglected here, and the air treated as a simple mixture of nitrogen and oxygen in the mole proportions 0.79/0.21.

The principles of gas separation are the same as those of distillation treated in Chapter 6. There is one fundamental difference, however, between the distillation processes treated there and the ones under consideration here. In the generator of the absorption system the heat transfer can be supplied by external means such as steam or combustion at a temperature higher than ambient temperature, and heat can eventually be rejected to the environment by direct transfer. In low-temperature distillation such external heating is not permissible because the corresponding rejection of heat to the environment cannot be done by direct transfer: it needs refrigeration to pump the heat up to the ambient temperature and this requires a large expenditure of work. Consequently low-temperature distillation apparatus is, as nearly as possible, externally adiabatic. The refrigerating effect, whether produced by Joule–Thomson throttling or resisted expansion, is applied solely to making low-temperature products and to countering the heat gains to the apparatus.

It will clearly make a difference to the refrigeration requirements if the

Table 8.6.1. *Composition of the atmosphere*

	Mole fraction in dry air	Boiling point K
Nitrogen	0.7809	77.36
Oxygen	0.2095	90.18
Argon	0.0093	87.3
Carbon dioxide	0.0003	194.7 (sublimes)
	Mole fraction $\times 10^6$	
Neon	18.2	27.09
Helium	5.2	4.22
Krypton	1.1	121.4
Xenon	0.09	164.1
Ozone	0.03	160.7

The principal variable components are: water vapour up to about 0.06 in warm humid climates, sulphur dioxide up to 1×10^{-6} (1 p.p.m.) and hydrocarbons up to 8×10^{-6} in industrial atmospheres. Carbon dioxide, average 335×10^{-6} in 1978, increases about 1 p.p.m. each year.

8.6 The separation of gases

separated materials are wanted in liquid or in gaseous form. It was seen in section 8.2 that the exergy of separation of nitrogen, for example, is small compared with the exergy of liquefaction. Consequently much more work is needed when one or more of the products is delivered as liquid. In practice this implies that processes for gaseous products work at pressures lower than for liquid products.

In the oxygen and nitrogen solution the boiling points show that nitrogen is the more volatile of the two. It corresponds to the ammonia in the aqua-ammonia solutions considered in Chapter 6. The composition there was expressed by the mass fraction of ammonia. Following the usual practice for gas separation the composition will be expressed here by the mole fraction of nitrogen, denoted by ψ without a subscript, the oxygen mole fraction being understood as $(1 - \psi)$. A prime, as in ψ', will denote a liquid state at the bubble point (saturated liquid) and a double prime, as in ψ'', will denote a vapour state at the dew point (saturated vapour).

Fig. 8.6.1. Solutions of oxygen and nitrogen: mole fractions of nitrogen in liquid and vapour in equilibrium, plotted from the data of Din (1960).

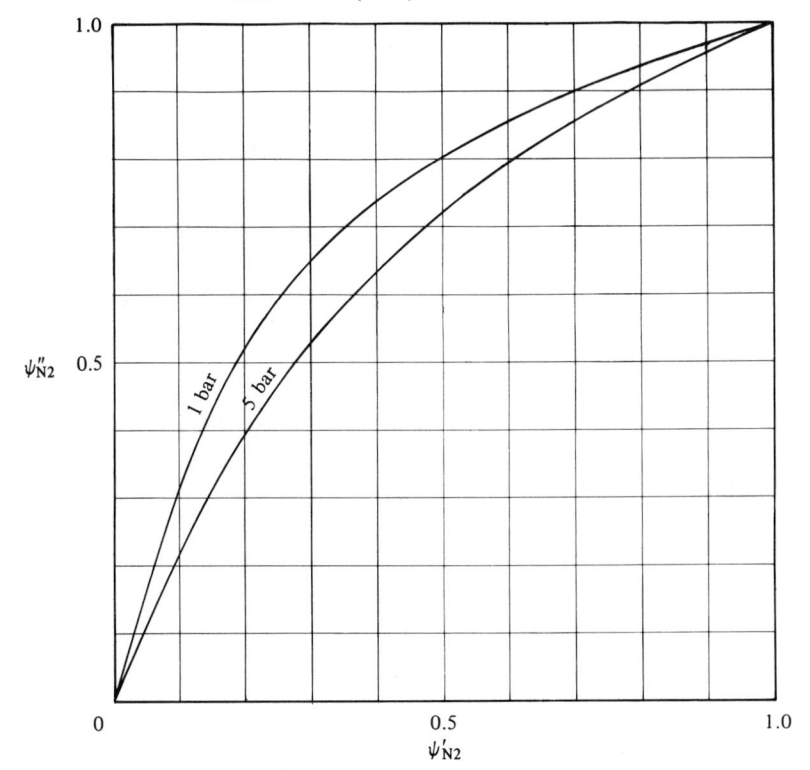

Cryogenic engineering 576

The relation between the mole fraction of nitrogen in the saturated liquid and the mole fraction in the saturated vapour is exhibited for two pressures in Fig. 8.6.1, based on the data of Din (1960). The boiling point versus composition diagram is given in Fig. 8.6.2.

For the application of the steady-flow energy equation values of the enthalpies must be known. In accordance with the use of mole fractions, molar enthalpies will be used, indicated by $\bar{h}$. Figure 8.6.3 shows the molar enthalpy versus composition diagram with two lines of constant pressure. A point on this diagram must be interpreted according to the principles discussed in Chapter 6. The lines of constant temperature in the liquid region give the molar enthalpies of solutions at the stated temperatures, the effect of pressure being disregarded. The lines of constant pressure give the molar enthalpies at the bubble points only. The plots of vapour enthalpy against liquid composition which were shown in Fig. 6.12.3 have been omitted here for clarity. The curves can be plotted if required by reference to Fig. 8.6.1.

The analysis of the plant is carried out using the same methods as

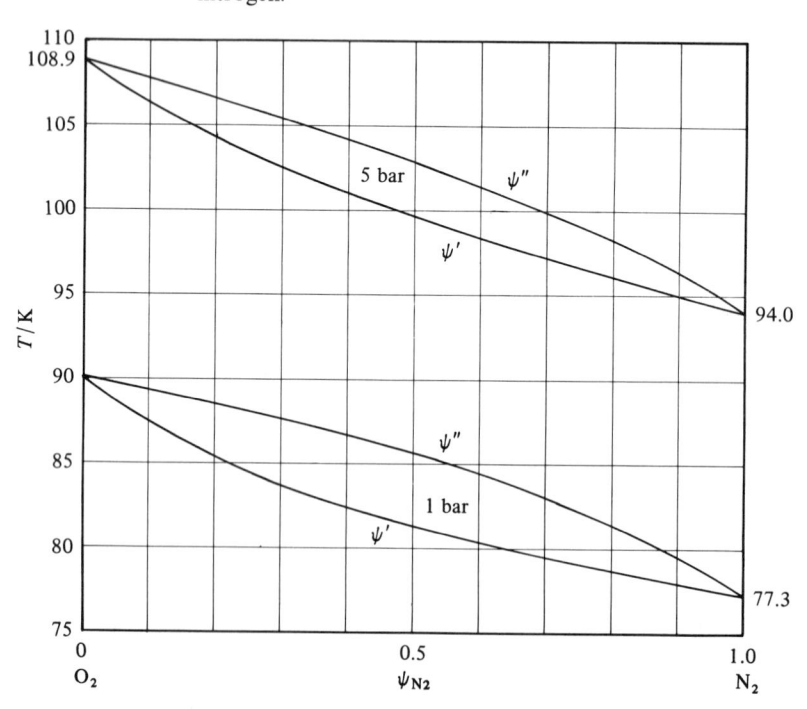

Fig. 8.6.2. Boiling point diagram for solutions of oxygen and nitrogen.

8.6 The separation of gases

Chapter 6 but in mole terms. The equations are (a) the total mole balance for all streams (b) the mole balance for one component which will be taken as nitrogen and (c) the steady-flow enthalpy equation. As a shorthand for these equations Keesom's stream notation will be used. Apart from this the treatment follows that of Bošnjaković (1937).

The stream algebra will first be applied to the question which was deferred in section 8.4, the working of the simple Linde–Hampson liquefier on air.

Using the nomenclature of Fig. 8.4.3, it is clear that apart from accidental heat gains the heat exchanger, valve and separator vessel form an adiabatic

Fig. 8.6.3. Molar enthalpy versus mole fraction of nitrogen for mixtures of oxygen and nitrogen. The enthalpies of the vapour must be regarded as approximate owing to lack of precise data. Datum: ideal gas at $T = 0$ K, $\bar{h} = 0$.

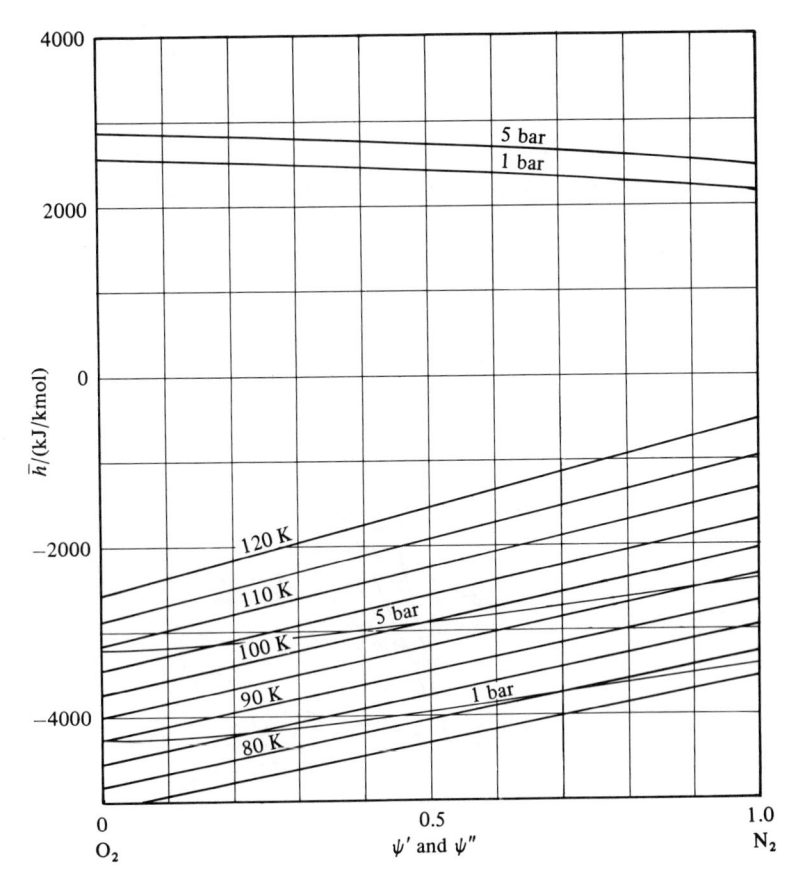

control surface. Consequently the points 1, 4 and 6 must lie on a straight line on the $\bar{h} - \psi$ diagram, as shown diagrammatically in Fig. 8.6.4. Point 6 lies on the line representing the molar enthalpy of mixtures at atmospheric pressure and temperature. Because the datum states used for the enthalpies are the ideal gas states at $T = 0 \, \text{K}$, and the molar heat capacities of oxygen and nitrogen are almost equal, this line is virtually horizontal. Point 1 represents the highly compressed air of normal composition. Point 4 lies on the bubble-point curve for liquid mixtures. The position of the line 4–1–6 is fixed by the facts that the vapour at 5 has to be in equilibrium with the liquid at 4, i.e. 4–5 is a two-phase isotherm, and that the composition at 5 has to be the same as that at 6. Points 4, 2 and 5 are also in a straight line, and this gives the state of the mixture before the valve at 2. Point 3 after the valve has the same composition and enthalpy as 2 but represents a two-phase mixture of 4 and 5.

By trial and error the points which satisfy the above construction can be found for a given $\bar{h}$ and T_6. For compression to 200 bar with complete recuperation, $T_6 = 300 \, \text{K}$, the mole fraction at 4 is found to be about $\psi' = 0.52$ with equilibrium vapour $\psi'' = 0.815$ at 5.

The simple Linde–Hampson liquefier thus already causes some enrichment of the oxygen in the liquid to about 0.48, and for some purposes this may be sufficient. This enrichment does not occur to the same extent in

Fig. 8.6.4. The states in the simple Linde–Hampson process for air shown on the $\bar{h} - \psi$ diagram.

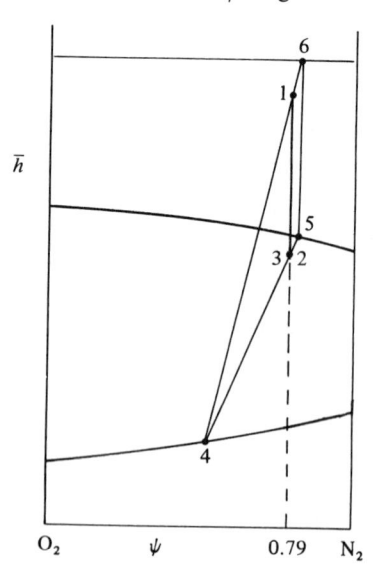

8.6 The separation of gases

the more efficient processes, however. An oxygen mole balance shows that any improvement which increases the fraction of liquid formed at the valve tends to shift the composition of the liquid towards that of normal air.

A simple way of enriching the liquid further is to allow some to evaporate, the vapour carrying away more nitrogen than oxygen, but this is rather wasteful. Suppose an amount of liquid n of mole fraction ψ' loses by evaporation an amount $-dn$ of mole fraction ψ''. The nitrogen balance gives:

$$d(n\psi') = \psi'' \, dn \tag{8.6.1}$$

and:

$$\frac{dn}{n} = \frac{d\psi'}{\psi'' - \psi'} \tag{8.6.2}$$

Integrating from an initial amount n_1 to a final (lesser) amount n_2:

$$\ln \frac{n_1}{n_2} = \int_1^2 \frac{d\psi'}{\psi'' - \psi'} \tag{8.6.3}$$

which can be evaluated graphically or otherwise when the relation between ψ' and ψ'' is known.

The enrichment of liquid air by evaporation into the atmosphere is not a serious method of obtaining oxygen, but the point is of some practical importance in connection with safety and fire risk, and stringent pre-

Fig. 8.6.5. Production of gaseous oxygen by a simple distillation process.

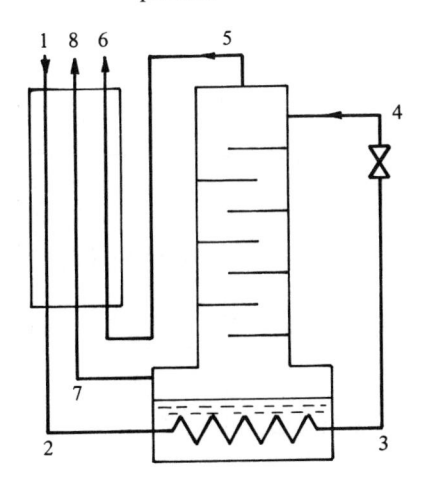

cautions have to be taken to prevent liquid air from coming into contact with combustible materials.

Nearly complete purification of the oxygen can be obtained in a simple still of the form shown in Fig. 8.6.5, which was introduced by Friedrich Linde in 1902 for making gaseous oxygen. High-pressure air at 1 enters the heat exchanger and is cooled to 2. It then enters the coil in the reboiler of the still and is cooled by the oxygen-rich liquid boiling there to 3. It passes through the expansion valve to 4 and is fed to the top of the column, being enriched on the way down by the rising vapour. Oxygen is taken off at 7 and returns through the heat exchanger. A nitrogen-rich gas is taken from the top of the column at 5 and passes back through the exchanger. The trays in the column are represented diagrammatically. In the first plants the columns were packed with glass balls, but bubble-caps and sieve trays were used later.

Considering the whole apparatus as adiabatic, the points representing the states 1 entering, and 6 and 8 leaving, must lie on a straight line on the $\bar{h} - \psi$ diagram, as shown in Fig. 8.6.6. Since streams 6 and 8 are at atmospheric pressure, for the same temperature they lie on a line of constant enthalpy which is almost horizontal when the datum is chosen as the ideal gas state at absolute zero, as explained earlier. Point 1 also has the same molar enthalpy so long as the apparatus is adiabatic, but in practice

Fig. 8.6.6. $\bar{h} - \psi$ diagram for the process of Fig. 8.6.5 with the states numbered to correspond.

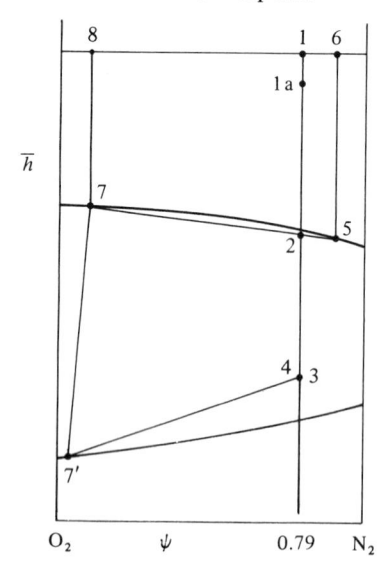

8.6 The separation of gases

the initial molar enthalpy has to be somewhat lower, at point 1a say, to provide the refrigeration to counter the unavoidable heat transfer to the column and exchangers. This means that the pressure at 1a has to be greater than atmospheric, to provide some Joule–Thomson cooling.

Considering now a control surface drawn around the column alone, stream 2 enters whilst streams 5 and 7 leave. For an adiabatic column these points lie on a straight line on the $\bar{h} - \psi$ diagram, as shown in Fig. 8.6.6. Point 2 is not of course in a two-phase region, because its pressure is higher than that for which the bubble and dew point lines are drawn. In the reboiler the air can be cooled to approach the temperature of the oxygen-rich liquid boiling there, shown as the isotherm at about 90 K from the liquid line at 7' to point 3. State 4 after throttling has the same enthalpy as 3 but is at a lower pressure. It lies in the two-phase region at a temperature of about 79 K and equilibrium mole fractions $\psi' = 0.775, \psi'' = 0.925$. The gases at the top of the column where this mixture is supplied cannot attain a stronger concentration than would be in equilibrium with this feed, and hence the nitrogen leaving the column still contains at least 7.5%, and in practice up to 9%, of oxygen. An important fraction of the oxygen in the air is thus wasted.

If the oxygen product is required as liquid, at state 7' say, then point 1a representing the air supply state must lie on the straight line joining 7' to 6, indicating a much higher compression of the air to provide the extra refrigeration. This is the requirement during start-up of the apparatus when enough liquid must be made to fill the reboiler and column plates. The pressure has to be high during this period, about 150 bar, but it can subsequently be reduced to about 5 bar if gaseous oxygen is to be made.

The loss of oxygen at the top of the column is caused by the fact that there is no liquid anywhere in the plant which is cold enough to condense the rich nitrogen at about 78 K and so provide a reflux of liquid strong enough in nitrogen to rectify the arising vapour. In the rectifier of the ammonia absorption machine this difficulty does not occur because the ammonia can be condensed by water at ambient temperature.

The first solution to the problem was reached by Georges Claude in 1902, and a second solution was found by Richard Linde and Rudolph Wucherer in 1908. Both methods embody the same idea, in different forms: that nitrogen can be condensed by boiling oxygen if the pressures are sufficiently different. For example, the boiling point of nitrogen at a pressure of 5 bar is 94 K whereas the boiling point of oxygen at 1 bar is 90 K. Consequently, oxygen boiling at 1 bar can condense nitrogen under a pressure of 5 bar.

In the Claude method the nitrogen-rich liquid stream is achieved by partial condensation of the incoming air, called 'liquéfaction avec retour en arrière' by Claude but now usually called 'dephlegmation'. Referring to the boiling point diagram Fig. 8.6.2, the air at a pressure of 5 bar would begin to deposit liquid when its temperature is lowered to 98.4 K, and this liquid would have a composition of $\psi' = 0.6$. The vapour remaining is enriched in nitrogen. If the vapour is allowed to continue up a vertical tube whilst the condensed liquid descends, it is possible to obtain about one half of the original air stream as nitrogen vapour of 98% purity.

The arrangement of the Claude column is shown in Fig. 8.6.7. Air at a pressure of about 5 bar, assuming oxygen gas is to be made, is cooled in the heat exchangers and admitted to the bottom of the column where it is dephlegmated in the vertical tubes cooled by oxygen boiling around them at a pressure of 1 bar. The oxygen-rich fraction falls back into the sump and is supplied as the feed, via the expansion valve, to the column at an appropriate point. The nitrogen-rich fraction passes out of the top of the tubes and is totally condensed in the tubes at the side. The liquid collects in the annular sump and is supplied via the expansion valve as a reflux stream to the top of the column. Oxygen gas can be drawn from the bottom of the column and nitrogen from the top.

Fig. 8.6.7. The Claude column, diagrammatic, incorporating dephlegmation.

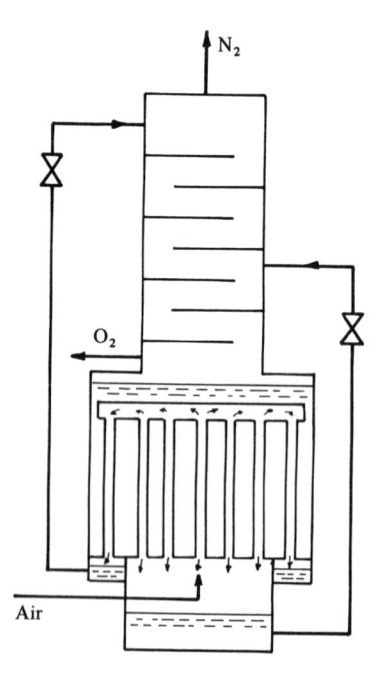

8.6 The separation of gases

The working of the Claude column depends not only on the partial condensation but also on some rectification occurring in the tubes. As a rectifier the arrangement is not very efficient because there is no hold-up of liquid as in a normal column. The Claude column is also very sensitive to operating conditions. It did not therefore come into such general use as the Linde column, described next. Dephlegmation as envisaged by Claude, however, has a role in other gas separation processes.

The double column of Linde and Wucherer is shown diagrammatically in Fig. 8.6.8. Compressed air already cooled in heat exchangers by the revert gases enters at 1 and is condensed at 2 by heat transfer to the oxygen-rich liquid boiling in the sump of the lower column. After throttling to the pressure in the lower column, about 5 bar, the condensed air provides the feed for the lower column at 3, which is introduced at an appropriate level. In its descent it is stripped of nitrogen by counterflow with rising vapour, to give the oxygen-rich liquid in the sump and an arising vapour enriched in nitrogen. This condenses at the top of the lower column in tubes surrounded by oxygen boiling at a pressure of about 1 bar and a temperature of 90 K. Part of the condensed nitrogen, which is known as the 'poor liquid', descends the lower column as a reflux stream, whilst part of it

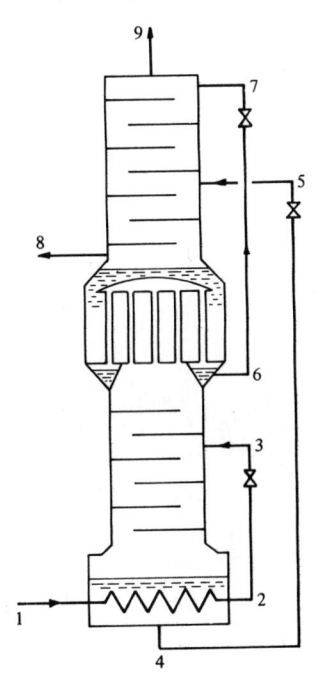

Fig. 8.6.8. The Linde double column, diagrammatic.

Cryogenic engineering 584

Fig. 8.6.9. The states in the Linde double column on the $\bar{h}-\psi$ diagram for the conditions of example 8.6.1. In the circled detail around point X, the straight line from S_U to 9 passing through X is omitted for clarity.

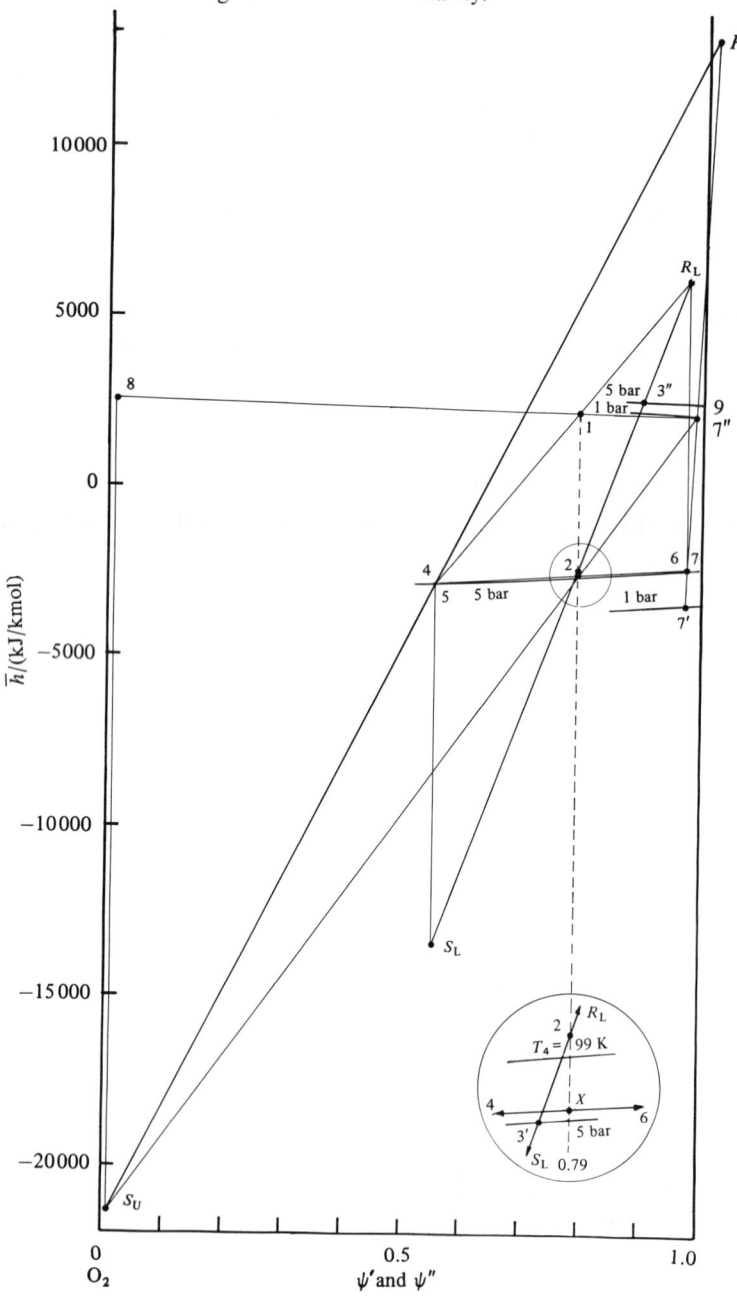

8.6 The separation of gases

is collected in the trough shown and is fed to the top of the upper column. After throttling to 7 it forms the reflux stream for the upper column. The rich liquid from the sump of the lower column is throttled to 5 and provides the feed for the upper column. Nitrogen gas of arbitrary purity is taken from the top of the upper column at 9 and oxygen gas is taken from the bottom of the upper column at 8.

The pressure in the upper column is nominally 1 bar, but because of the pressure drop through the heat exchangers it has to be somewhat higher than this, say 1.5 bar. The pressure in the lower column need ideally be only sufficiently above that in the upper column to allow the condensation of nitrogen at the higher pressure by oxygen boiling at the lower pressure. This difference is about 3.5 bar, but taking into account the necessary temperature difference for heat transfer the pressure in the upper column has to be about 5 bar.

The operation of the column will be discussed using Keesom's stream algebra. For simplicity, adiabatic operation will be assumed.

Beginning with the lower column and using the nomenclature for the streams shown in Fig. 8.6.8:

$$[1] = [4] + [6] + [\dot{Q}] \tag{8.6.4}$$

where $\dot{Q}$ is the rate of heat transfer from the nitrogen condensing at the top of the lower column to the oxygen boiling in the upper column. Grouping $\dot{Q}$ with stream 1:

$$[1 - q_1] = [4] + [6] \tag{8.6.5}$$

Consequently the points $(\bar{h}_1 - q_1), \psi_1$; $\bar{h}_4, \psi_4$; and $\bar{h}_6, \psi_6$ lie on a straight line on the $\bar{h} - \psi$ diagram, as shown in Fig. 8.6.9. The diagram is drawn approximately to scale for the conditions of example 8.6.1. The point $(\bar{h}_1 - q_1), \psi_1$ is denoted by X.

The compressed air at 1 is cooled to a temperature at 2 which is close to that of the boiling rich liquid at 4. The isotherm for liquid at $T = T_4$ is shown on the diagram with state 2 just a little warmer than this to allow for a terminal temperature difference. Point 3 after throttling at constant enthalpy and composition is the same as point 2 and the lowest position of the rectifier pole for the lower column R_L can be established by producing the two-phase isotherm through 3 to meet the ordinate at the composition for point 6. Similarly the highest position for the stripping pole of the lower column S_L is obtained by producing the two-phase isotherm through 3 back to meet the ordinate at the composition of the liquid at 4. The straight line S_L–3–R_L is then the operating line for the lower column assuming minimum reflux.

Associating $\dot{Q}$ with stream 6 in eq. (8.6.4) gives:

$$[1] = [4] + [6 + q_6] \tag{8.6.6}$$

which shows that the points 1, 4 and $(6 + q_6)$ must lie on a straight line. But $(6 + q_6)$ is also R_L the rectifying pole for the lower column. Thus points 1, 4 and R_L must be in line.

Turning now to the upper column at a lower pressure, points 4, 5 and 6, 7 are in the two-phase region with respect to this low pressure. The dryness fractions can be determined by the methods of section 6.13. The product states 8 and 9 are shown at the required purities on the dew point curve for the pressure in the upper column.

For the whole apparatus, assumed adiabatic:

$$[1] = [8] + [9] \tag{8.6.7}$$

Hence point 1 must lie on the straight line joining 8 to 9 at the composition $\psi_1 = 0.79$. In practice to make up for heat transfer to the apparatus point 1 would have to lie below this line.

Subtracting $\dot{Q}$ from each side of eq. (8.6.7) and associating it with stream 1 on one side and stream 8 on the other:

$$[1 - q_1] = [8 - q_8] + [9] \tag{8.6.8}$$

Hence the points $(1 - q_1)$, already located as X, $(8 - q_8)$ and 9 lie on a straight line.

Considering point 4, 5 as the feed state for the upper column, the highest position for the stripping pole S_U lies on the two-phase isotherm through 4, 5 produced to meet the ordinate at the oxygen product composition ψ_8. The ordinate of the stripping pole is $(h_8 - q_8)$ and, as shown above, the stripping pole lies on the straight line through X and 9.

The minimum position for the rectifier pole for the upper column also lies on the two-phase isotherm through 4, 5 produced. Its actual position has to be found by considering the balance for a control surface across streams 9 and 7 and a section of the upper column above the feed point. Let V and L denote the ascending vapour and descending liquid streams at this arbitrary section, then:

$$[V] + [7] = [9] + [L] \tag{8.6.9a}$$

or:

$$[L] - [V] = [7] - [9] \tag{8.6.9b}$$

Since 7 and 9 are fixed points the right-hand side of eq. (8.6.9b) is constant and the line joining L and V must intersect the line 7–9 at the rectifier pole

8.6 The separation of gases

for the upper column R_U. The minimum position of R_U thus lies at the intersection of 7–9 and the two-phase isotherm through 4, 5 produced.

The compositions are now fixed at all points and the molar flow rates can be determined. From eq. (8.6.7):

$$\frac{\dot{n}_8}{\dot{n}_1} = \frac{\psi_9 - 0.79}{\psi_9 - \psi_8} \tag{8.6.10}$$

and:

$$\dot{n}_9 = \dot{n}_1 - \dot{n}_8 \tag{8.6.11}$$

From eq. (8.6.4):

$$\frac{\dot{n}_4}{\dot{n}_1} = \frac{\psi_6 - 0.79}{\psi_6 - \psi_4} \tag{8.6.12}$$

and:

$$\dot{n}_6 = \dot{n}_1 - \dot{n}_4 \tag{8.6.13}$$

Example 8.6.1

Air enters a Linde double column at the rate 0.1 kg/s and emerges as gaseous nitrogen and oxygen each of 99% purity. The compressed air is condensed to liquid (point 2) at a temperature of 100 K. Determine the positions of the rectifying and stripping poles for the columns for minimum reflux, the heat transfer rate in the intermediate condenser/boiler and the mass flow rates of the streams at points 8, 9, 4 and 6. Assume adiabatic operation and equilibrium conditions at the tops of the columns. The pressure in the lower column is 5 bar and in the upper column 1 bar.

Assuming reversible isothermal compression of the air at a temperature of 298 K, determine the power taken and the exergetic efficiency of the plant with reference to this temperature and a pressure of 1 bar at which the air enters the compressor and at which the products leave.

For liquid air at 2, $T = 100$ K:

$$\bar{h}_2 = -2400 \text{ kJ/kmol}$$

On throttling to a pressure of 5 bar this splits into two phases, 3' and 3'' say. By trial construction on the $\bar{h} - \psi$ chart:

$$\psi'_3 = 0.785 \qquad \bar{h}'_3 = -2660 \text{ kJ/kmol}$$
$$\psi''_3 = 0.897 \qquad \bar{h}''_3 = 2520 \text{ kJ/kmol}$$

and the operating line for minimum reflux in the lower column must pass through these points.

The liquid at 6, throttled to 7, produces a two-phase mixture of which the liquid, 7' say, is in equilibrium with the product nitrogen at 9 at a pressure of 1 bar. From the $\psi'-\psi''$ diagram:

$$\psi'_7 = 0.975$$

Plotting this point on the $\bar{h}-\psi$ chart, the isotherm through it and $\psi_9 = 0.99$ cuts the bubble-point curve at the pressure 5 bar at point 6, which is the same as 7:

$$\psi_6 = 0.977 \qquad \bar{h}_6 = -2440 \text{ kJ/kmol}$$

Hence the rectifier pole R_L for the lower column lies at:

$$\bar{h} = \left[2520 + \frac{2520 + 2660}{0.897 - 0.785}(0.977 - 0.897)\right] \text{kJ/kmol}$$

$$= 6220 \text{ kJ/kmol}$$

$$\psi = 0.977$$

For the product gases at 8 and 9:

$$\bar{h}_8 = 2560 \text{ kJ/kmol} \qquad \psi_8 = 0.01$$
$$\bar{h}_9 = 2180 \text{ kJ/kmol} \qquad \psi_9 = 0.99$$

The state at 1 is given by:

$$\bar{h}_1 = \left[2180 + \frac{2560 - 2180}{0.99 - 0.01}(0.99 - 0.79)\right] \text{kJ/kmol}$$

$$= 2260 \text{ kJ/kmol}$$

$$\psi_1 = 0.79$$

Point 4 is on the bubble point curve at the pressure 5 bar, and also on the line joining the rectifying pole to point 1 produced. From construction on the $\bar{h}-\psi$ chart:

$$\bar{h}_4 = -2860 \text{ kJ/kmol} \qquad \psi_4 = 0.55$$

(The temperature of this liquid is 99 K, thus allowing an approach of 1 K between it and the liquefied air at 2.) The stripping pole for the lower column is at this composition and on the line from R_L, the rectifying pole for the lower column, through the feed state 2 produced. Hence:

$$\bar{h} = \left[6220 - \frac{6220 + 2400}{0.977 - 0.79}(0.977 - 0.55)\right] \text{kJ/kmol}$$

$$= -13\,460 \text{ kJ/kmol}$$

8.6 The separation of gases

is the ordinate of the stripping pole for the lower column S_L at $\psi = 0.55$.
The point X on the line 4–6 at $\psi = 0.79$ has the enthalpy:

$$\bar{h}_X = \left[-2440 - \frac{2860 - 2440}{0.977 - 0.55}(0.977 - 0.79) \right] \text{kJ/kmol}$$

$$= -2620 \text{ kJ/kmol}$$

$$\bar{q}_1 = \bar{h}_1 - \bar{h}_X = (2260 + 2620) \text{ kJ/kmol} = 4880 \text{ kJ/kmol}$$

The stripping pole for the upper column lies at $\psi = 0.01$ and on the line joining 9 and X produced, i.e. at:

$$\bar{h} = \left[2180 - \frac{2180 + 2620}{0.99 - 0.79}(0.99 - 0.01) \right] \text{kJ/kmol}$$

$$= -21\,340 \text{ kJ/kmol}$$

The rectifying pole for the upper column lies on the line through S_U and 4 produced, (condition 1), and on the line joining 7 and 9 produced (condition 2). The first condition gives:

$$\frac{\bar{h} + 2860}{\psi - 0.55} = \frac{-2860 + 21\,340}{0.55 - 0.01}$$

i.e.

$$\bar{h} = 34\,220\,\psi - 21\,680 \tag{1}$$

The second condition gives:

$$\frac{\bar{h} - 2180}{\psi - 0.99} = \frac{2180 + 2440}{0.99 - 0.977}$$

i.e.

$$\bar{h} = 355\,400\,\psi - 349\,600 \tag{2}$$

Solving (1) and (2) gives:

$$\bar{h} = 13\,260 \text{ kJ/kmol} \qquad \psi = 1.021$$

for the coordinates of the rectifying pole for the upper column R_U.
The molar mass of air, neglecting the argon, is:

$$[0.79(28) + 0.21(32)] \text{ kg/kmol} = 28.84 \text{ kg/kmol}$$

The molar flow rate of the air entering is then:

$$\dot{n}_1 = \frac{0.1}{28.84} \text{ kmol/s} = (3.47)10^{-3} \text{ kmol/s}$$

and the rate of heat transfer between the columns is:

$$(3.47)(4880)10^{-3} \text{ kW} = 16.9 \text{ kW}$$

The mole flow rate of the oxygen product is given by:

$$\frac{\dot{n}_8}{\dot{n}_1} = \frac{0.99 - 0.79}{0.99 - 0.01} = 0.204$$

and:

$$\dot{n}_8 = (0.204)(3.47)10^{-3} \text{ kmol/s} = (7.08)10^{-4} \text{ kmol/s}$$

The mass flow rate of the oxygen product is then:

$$7.08[0.01(28) + 0.99(32)]10^{-4} \text{ kg/s} = 0.0226 \text{ kg/s}$$

The mass flow rate of the product nitrogen is:

$$(0.1 - 0.0226) \text{ kg/s} = 0.0774 \text{ kg/s}$$

The mole flow rate at 4 is given by:

$$\frac{\dot{n}_4}{\dot{n}_1} = \frac{0.977 - 0.79}{0.977 - 0.55} = 0.438$$

and:

$$\dot{n}_4 = (0.438)(3.47)10^{-3} \text{ kmol/s} = (1.52)10^{-3} \text{ kmol/s}$$

The mass flow rate at 4 is then:

$$\dot{m}_4 = 1.52[0.55(28) + 0.45(32)]10^{-3} \text{ kg/s} = 0.0453 \text{ kg/s}$$

and the mass flow rate at 6 is:

$$\dot{m}_6 = (0.1 - 0.0453) \text{ kg/s} = 0.0547 \text{ kg/s}$$

The states of the mixture at the several points and the rectifying and stripping poles are set out in Fig. 8.6.9. The pressure required to liquefy the air at 100 K is approximately 6.9 bar. For reversible isothermal compression at a temperature of 298 K the power is:

$$\frac{(0.1)(8314)(298)}{28.84} \ln(6.9) \text{ W} = 16.6 \text{ kW}$$

Using the result of example 8.2.5, the rate of production of exergy is:

$$(0.1)(39.4) \text{ kW} = 3.94 \text{ kW}$$

The exergetic efficiency is therefore:

$$3.94/16.6 = 0.237$$

8.6 *The separation of gases* 591

Fig. 8.6.10. A modern version of the Heylandt process for making liquid oxygen and liquid nitrogen. *F1* – air filter; *C1* – air compressor; *D1* – moisture separator; *D2* – oil vapour trap; *J1* – Freon refrigerator; *J2* – molecular sieve adsorber; *E1* – Freon cooler; *E2* – precooler; *E3* – main exchanger; *E4* – liquid oxygen subcooler; *E5* – condenser-reboiler; *E6* – poor liquid subcooler; *P1* – liquid oxygen pump; *T1* – lower column; *T2* – upper column; *X1* – expansion engine. (By courtesy of Cryoplants Ltd.)

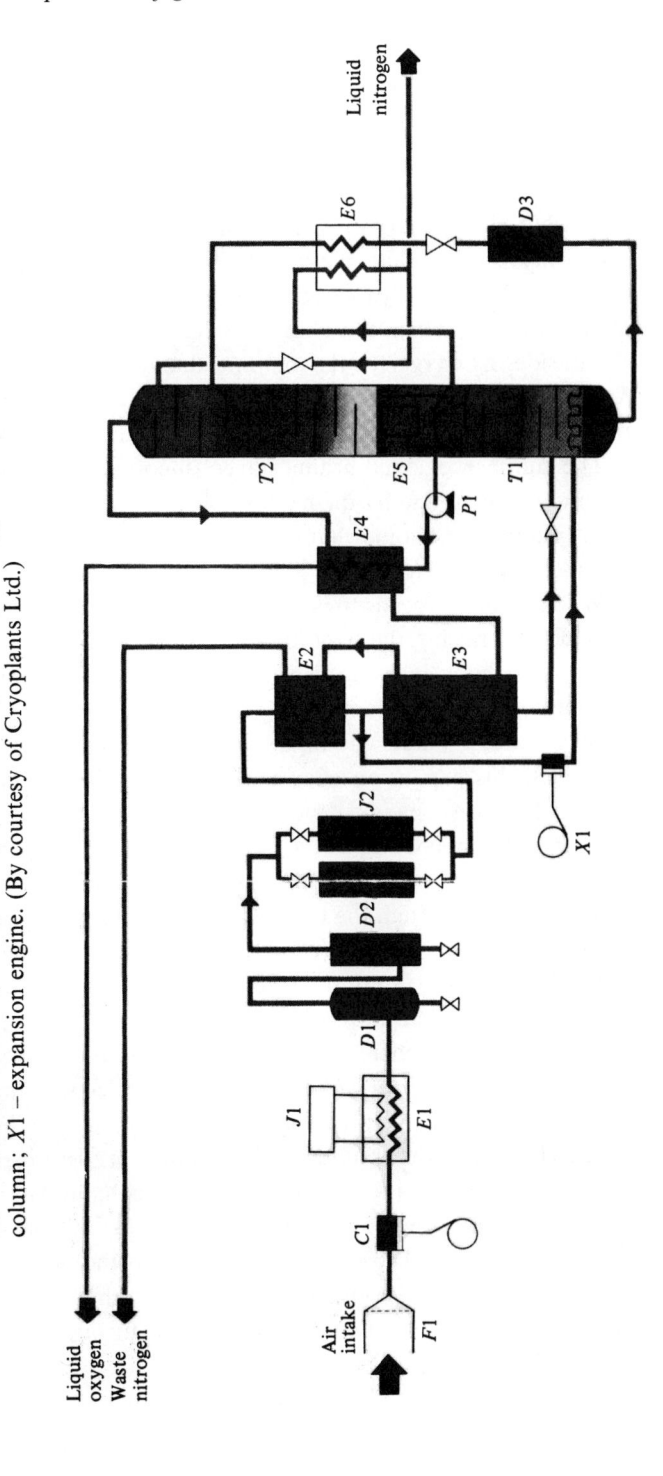

In practice this efficiency has to be reduced by a factor representing the efficiency of compression. Furthermore, because of heat transfer through the insulation to the column of an actual plant, and imperfect recuperation in the heat exchangers, the supply point 1 has to lie somewhat lower than the position calculated above. The rectifying pole R_L also has to be located somewhat higher than the position shown to ensure that the section lines are always steeper than the two-phase isotherms. Consequently the line from R_L through 1 produced intersects the bubble point curve at a composition ψ_4 which is higher than that found above, usually about 0.6.

Considering oxygen as the main product of the plant, the specific work for the (impure) gaseous oxygen produced is about 730 kJ/kg. Very large modern plants can in fact achieve values of about 1000 kJ/kg.

The number of equilibrium stages (theoretical plates) needed in the column can be found by the method of section 6.17. Practical calculations usually employ the somewhat simplified procedure of the McCabe–Thiele method, details of which are given in texts on chemical engineering.

When liquid oxygen is to be made, the extra refrigeration needed may be provided by raising the inlet air pressure so as to generate some Joule–Thomson cooling, by using an expansion machine, or by pre-cooling with a vapour compression plant; or by a combination of all three. One of the most successful plants for liquid oxygen was that of C.W.P. Heylandt employing his expansion engine, described earlier, in conjunction with Joule–Thomson expansion and a double column. A modern version of this process with precooling by a vapour compression plant is shown in Fig. 8.6.10. It is capable of making liquid oxygen at purities up to 99.6%.

Air is drawn through the filter F1 and compressed to about 175 bar by reciprocating compressors. The air is pre-cooled by the refrigerator J1 depositing moisture in D1 and traces of oil in the trap D2. Carbon dioxide and more water are removed by absorption in J2. The air is then cooled against revert nitrogen in E2 and split into two streams, one being cooled by further heat exchange in E3 before being expanded into the lower column, the other being expanded via the expansion engine into the lower column.

The working of the columns is substantially as described earlier, except that there is no coil in the reboiler of the lower column. The rich liquid at the bottom contains about 30% oxygen ($\psi = 0.7$) and passes through the adsorber D3 and subcooler E6 as the feed to the upper column. Liquid oxygen and nitrogen products are taken off as shown, and waste nitrogen from the top of the upper column passes back through the heat exchangers to cool the compressed air.

The coil in the reboiler of the lower column shown in Fig. 8.6.8 was

8.6 The separation of gases

necessary there because the enthalpy of the air at point 1 was not low enough to produce a substantial amount of liquid on direct throttling to the lower column pressure. In the Heylandt plant the refrigeration produced by the expansion engine renders the coil unnecessary.

The purpose of the adsorber D3 is to remove the traces of hydrocarbons which could accumulate in the plant with dangerous results. Although the mole fraction of these in the ingoing air might be as little as 10^{-6}, this amounts to 1 g/t for a hydrocarbon such as acetylene with a molar mass of 26, and a plant producing 100 t/day of oxygen would accumulate about half a kilogramme of acetylene per day.

For the production of gaseous oxygen in large amounts for metallurgical and chemical processes, the double column has been adopted employing regenerators instead of heat exchangers and incorporating an expansion turbine to provide the refrigeration. Regenerators, described in section 7.5 with reference to ideal gas cycles, were introduced into gas separation plants by Mathias Fränkl in the 1920s. Regenerators can be economically employed in plants for gaseous products because the pressure of the revert gas is low, so that very thick shells are not needed, and because they are self-cleaning. Ice and carbon dioxide deposited from the air being cooled are subsequently sublimed and carried away by the revert oxygen and nitrogen

Fig. 8.6.11. Diagram of the Linde – Fränkl process for making tonnage (gaseous) oxygen. 1 – Turbo-compressor; 2 – air – nitrogen regenerators; 3 – air – oxygen regenerators. 4 – carbon dioxide adsorbers; 5 – expansion turbine; and 6 – double column.

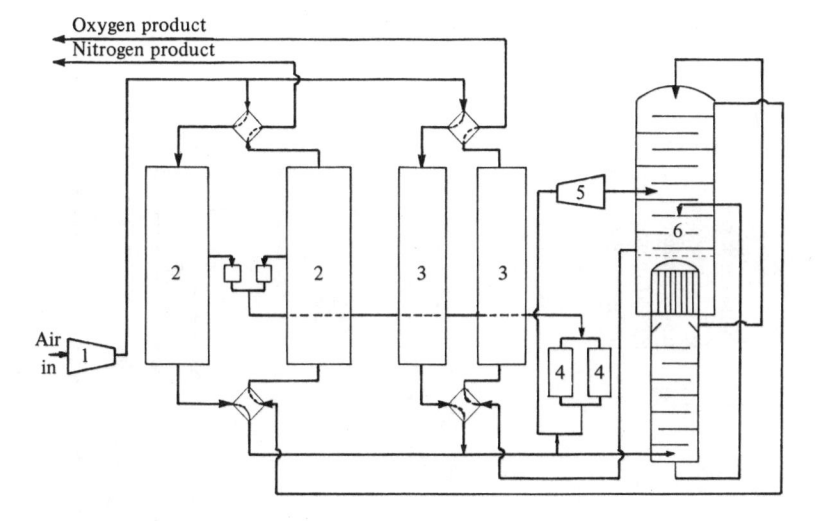

Fig. 8.6.12. A modern process for making tonnage oxygen incorporating reversing heat exchangers. C1 – Air turbo compressor; D1 – direct cooler; D2 – adsorber (rich liquid); D3 – adsorber (liquid oxygen); E1 – 2 – reversing heat exchangers; E3 – condenser; E4 – poor liquid-rich liquid subcooler; E5 – oxygen heater; E6 – nitrogen heater; F1 – air filter; P1 – direct cooler water pump; P2 – liquid oxygen pump; T1 – lower column; T2 – upper column; X1 – expansion turbine. (By courtesy of Cryoplants Ltd.)

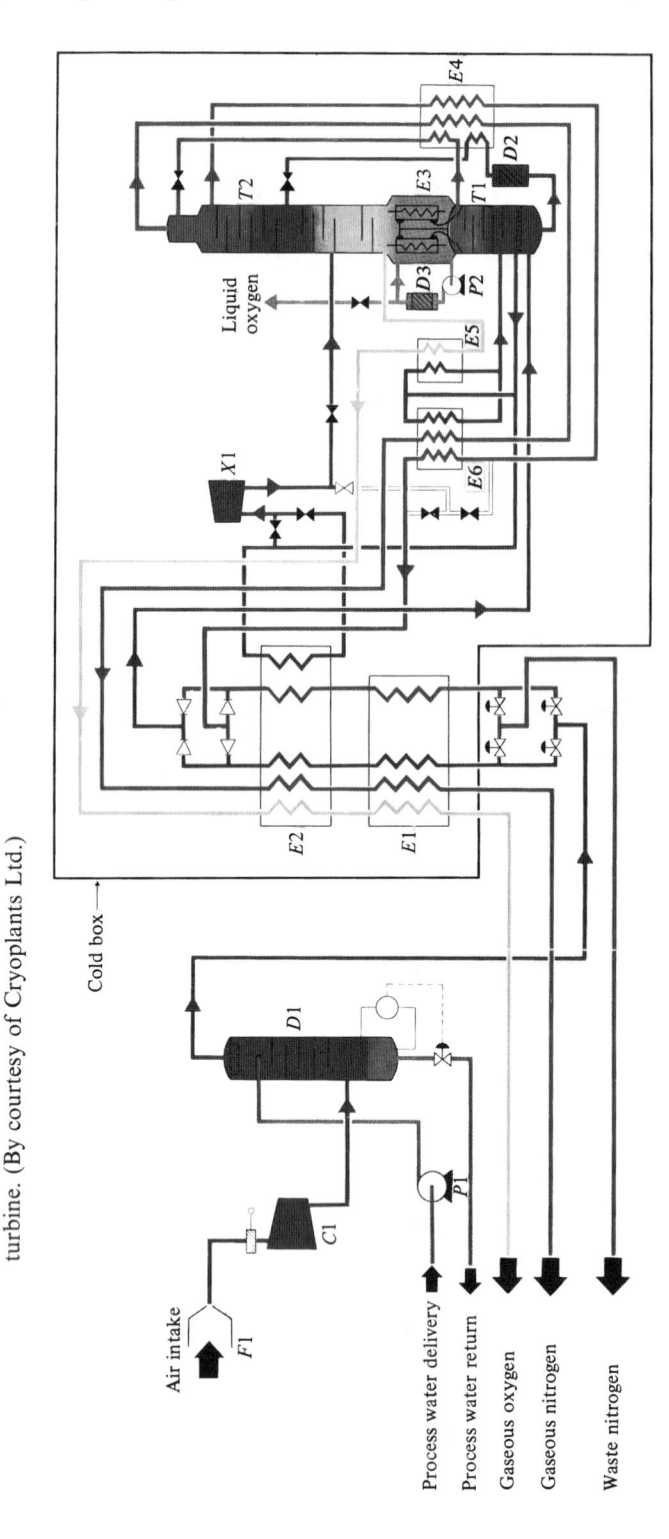

8.6 The separation of gases

streams. A simplified diagram of a Linde–Fränkl plant is shown in Fig. 8.6.11. The expansion turbine operates between the pressures in the lower and upper columns.

For a regenerator to be self-cleaning it has to work with a small temperature difference at the cold end and a revert mass flow rate which is at least equal to that of the forward stream. Because the specific heat capacity of compressed air at low temperature is higher than that of air at atmospheric pressure, if all of the forward stream passed through the regenerator it would be unbalanced. Some of the forward stream is diverted before it becomes very cold, as shown in Fig. 8.6.11 and passed straight to the turbine. Because this diverted fraction has not been cleaned of carbon dioxide, however, it has to go through the adsorbers shown.

It is clear that both the product gases will be contaminated with water and carbon dioxide in the plant of Fig. 8.6.11. This may not matter for some purposes, but there is now an increasing demand for uncontaminated oxygen. It is then necessary to provide separate oxygen coils embedded in the regenerator packing. If some of the nitrogen is also wanted free of water and carbon dioxide another pair of coils is needed. Furthermore, the whole of the moisture and carbon dioxide have to be carried away by a waste nitrogen stream which is much less than the air flow in the forward direction. In these circumstances regenerators are not so attractive, and the trend is now towards reversing heat exchangers in which the forward air stream and the revert waste nitrogen stream are interchanged at intervals. Gaseous products pass out through non-reversing passes and are not contaminated.

A modern system for producing tonnage oxygen and nitrogen gases with the possibility of extracting some liquid oxygen for storage against peak requirements is shown in Fig. 8.6.12.

Air is drawn in through the filter F1 and is compressed to about 6 bar by a turbo-compressor. In the reversing heat exchangers E1 and E2 the air is cooled to about 100 K against the revert products and waste nitrogen, with simultaneous deposition of moisture and carbon dioxide. These are subsequently removed when the waste nitrogen stream is interchanged with the air stream. Some cooling is also applied in the exchanger E2 by the stream from the bottom of the column which is warmed to about 170 K before being expanded in the turbine X1 to produce a cold gas stream for the upper column. The double column operates on the principles already described.

The cold components of cryogenic plants are usually grouped together in an insulated enclosure which is called the 'cold box', indicated by the boundary in Fig. 8.6.12.

Air as a ternary mixture

Air has so far been treated as a mixture of oxygen and nitrogen. The 0.93% of argon present in real air has a boiling point at atmospheric pressure of 87.3 K, i.e. between those of nitrogen and oxygen. It does not therefore necessarily pass through with the nitrogen as do the more volatile helium and neon, or remain largely in the oxygen as do the less volatile krypton and xenon, but tends to accumulate in some part of the plant and contribute to both streams. On a superficial view of the matter, if the argon ascends too high in the upper column it will be condensed at the nitrogen temperature, whereas if it descends too low it will be reboiled at the oxygen temperature. It is known, however, that the presence of argon has an important influence on the working of the upper column, which will not be treated here. Further details are given by Ruhemann (1949).

Because of the considerable industrial importance of argon it is often worth separating. A fraction containing about 10% of argon can be withdrawn from a point in the upper column where there is not much nitrogen present. This mixture can be rectified to give a product containing only about 2% of oxygen, which can then be purified further by chemical means.

8.7 Heat transfer equipment

It will now be clear that the processes described in this chapter depend largely for their success on effective heat transfer between fluid streams. In one of these, the simple Linde–Hampson process, it has been shown that a high efficiency of the heat exchanger is vital for the process to work at all; in others a high efficiency is essential for economic working. Since a high efficiency implies a large area for heat transfer, the heat transfer equipment accounts for a large part of the capital cost of a cryogenic plant.

Examples have been given of heat exchangers in which heat transfer takes place directly between two or more fluid streams, and of regenerators in which heat transfer takes place indirectly via a matrix which can store enthalpy. Direct heat exchangers are sometimes called recuperators to distinguish them from regenerators, but the name is not in common use and the term 'heat exchanger' without qualification will be taken to mean a direct heat exchanger.

Heat exchangers

Heat exchangers are usually classified into three main types as regards the flow arrangements: (a) counterflow, (b) parallel or co-current flow and (c) cross-flow. In pure counterflow and parallel flow exchangers

8.7 Heat transfer equipment

the temperatures of the fluids depend on only one dimension, say a length measured from one end, whereas in cross-flow two dimensions are needed to specify the position. In counterflow and parallel flow the temperatures can be visualised as profiles along the length of the exchanger, whilst in cross-flow one has to consider two temperature fields. The treatment of cross-flow exchangers is therefore much more difficult than that of counterflow and parallel flow ones.

For a close approach between the outlet temperature of one stream and the inlet temperature of another when both of the fluids are changing in temperature, only counterflow exchangers come into consideration. As explained earlier it is possible in these to bring the temperature of the fluid having the lower capacity rate, $\dot{m}c$, as near as desired to the inlet temperature of the other stream. This is impossible in parallel flow, when an intermediate temperature only can be attained. It is also impossible in cross-flow exchangers. When the number of passes is large, however, a sufficiently close approach can be obtained provided that the main directions of flow are opposed.

Fig. 8.7.1. (a) and (b) Types of Linde heat exchangers. (c) A paired-tube arrangement.

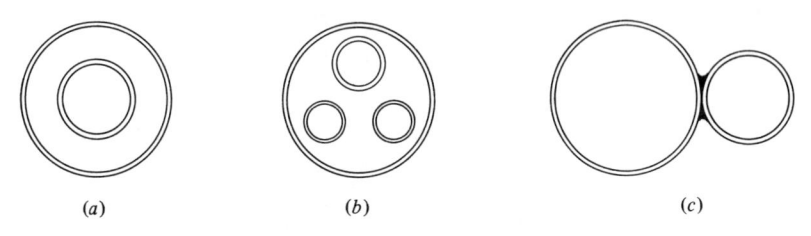

Fig. 8.7.2. A coil-in-shell heat exchanger. Heat transfer can take place between one shell-side fluid and three tube-side fluids. Each tube-side stream divides into a number of parallel paths.

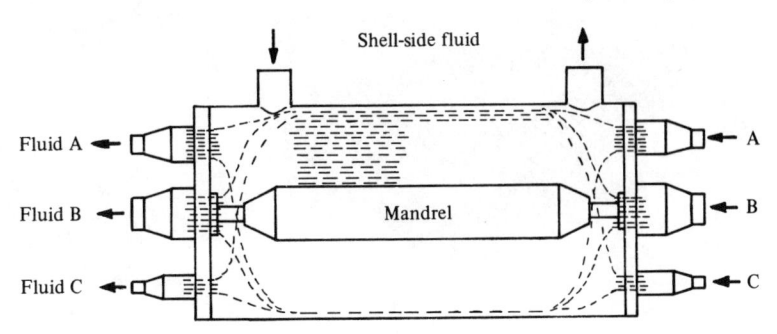

Cryogenic engineering 598

The traditional forms of heat exchanger for cryogenic plant have already been referred to in section 8.1 as the Linde and Hampson types. The Linde exchanger consists of one or more tubes carrying fluids inside the bore of a larger tube, as shown in Fig. 8.7.1(a) and (b). A variation is the paired-tube arrangement of (c) in which two tubes are soldered together.

Exchangers like these have to be used when all of the streams are at a high pressure because large shells are then impractical and expensive. When one of the streams is at a low pressure, as the revert gas in the simple Linde-Hampson process, a shell-and-coil type of exchanger is preferred. Hampson's original form of this was illustrated in Fig. 8.1.3, but it has been made in various forms since then. Several coils carrying different fluids can be incorporated in one shell, as shown in Fig. 8.7.2. Although the Hampson exchanger is strictly a cross-flow one of a rather complicated type, because of the large number of coils it approximates to the counterflow arrangement.

In recent years aluminium plate fin exchangers have been much used. They are assembled from modules consisting of separator sheets, corrugated fins and side bars, as shown in Fig. 8.7.3. Several modules are assembled in one block with appropriate headers, the whole being brazed together. A variety of flow arrangements can be obtained.

The coefficients of heat transfer in cryogenic exchangers are found using the correlations of dimensionless groups which are given in texts on heat

Fig. 8.7.3. Brazed aluminium fin exchangers. (a) Exploded view of a single element, (b) elements assembled to form a counterflow arrangement, (c) elements assembled to form a crossflow exchanger. (By courtesy of Marston Excelsior Ltd.)

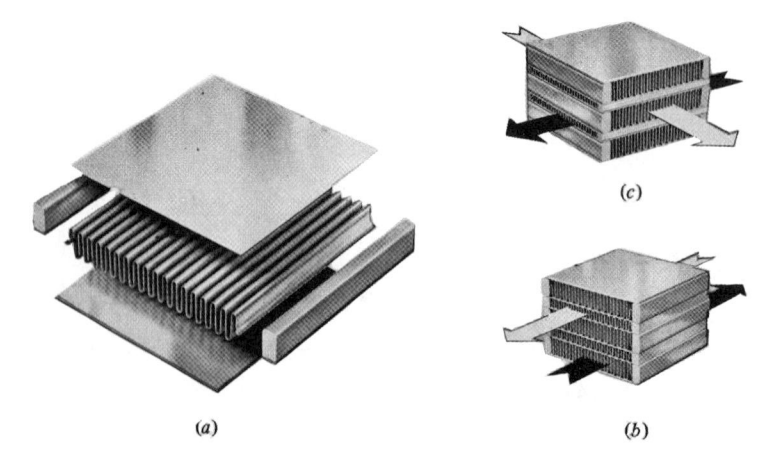

8.7 Heat transfer equipment

transfer. It must be remembered however that the range of temperature in a cryogenic heat exchanger is often much larger than that in orthodox refrigeration ones, and the physical properties may undergo a considerable change between the inlet and the outlet. The overall coefficient of heat transfer is not therefore usually constant. Furthermore the variation of the specific heat capacity of the fluids with temperature implies that the plots of temperature versus enthalpy are not straight lines. The usual logarithmic expression for the mean temperature difference does not hold in these circumstances.

Consider the counterflow heat exchanger shown diagrammatically in Fig. 8.7.4(a), with an element of heat transfer surface dA shown in Fig. 8.7.4(b). The fluid properties are denoted by subscripts a and b and the steady mass flow rates $\dot{m}_a$ and $\dot{m}_b$ are taken as positive in the directions shown. The heat rate d$\dot{Q}$ from fluid a to fluid b across an element of area, neglecting longitudinal conduction, is:

$$d\dot{Q} = K(T_a - T_b)dA \qquad (8.7.1)$$

where K is the overall coefficient of heat transfer referred to the area A, and T_a and T_b are the temperatures of the streams. From the point of view of heat transfer either absolute temperature T or conventional temperature t can be used since $dT = dt$, but in conformity with the rest of this chapter T will be preferred.

Fig. 8.7.4. (a) and (c) Diagram of counterflow exchanger with temperature profiles for the case when fluid b has the lower capacity rate; (b) an element of heat transfer surface separating the fluids.

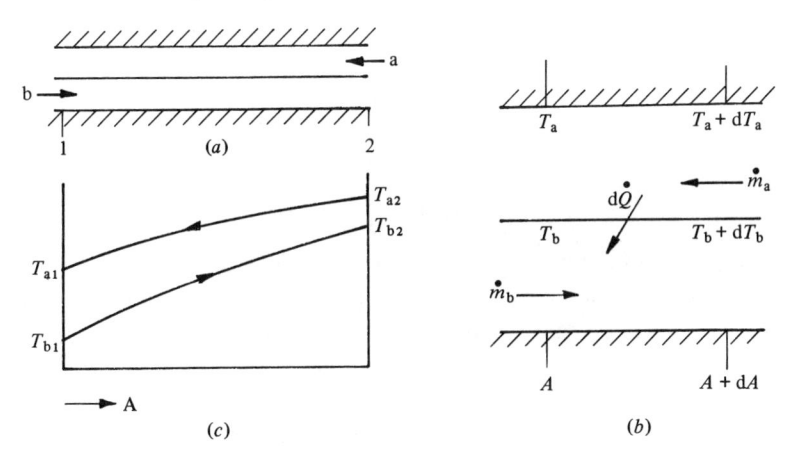

The elemental heat rates are also given by:

$$d\dot{Q} = \dot{m}_a c_a dT_a \tag{8.7.2}$$

and:

$$d\dot{Q} = \dot{m}_b c_b dT_b \tag{8.7.3}$$

where c is the specific heat capacity, taken at constant pressure for a gas. Noting that:

$$dT_a - dT_b = d(T_a - T_b) \tag{8.7.4}$$

eqs. (8.7.1), (8.7.2) and (8.7.3) may be combined to give:

$$\frac{d(T_a - T_b)}{T_a - T_b} = K\left(\frac{1}{\dot{m}_a c_a} - \frac{1}{\dot{m}_b c_b}\right) dA \tag{8.7.5}$$

Equation (8.7.5) holds whether K and the c are constant or not. When they are constant it can be integrated to give:

$$\ln \frac{T_{a2} - T_{b2}}{T_{a1} - T_{b1}} = KA\left(\frac{1}{\dot{m}_a c_a} - \frac{1}{\dot{m}_b c_b}\right) \tag{8.7.6}$$

where $A = A_2$ is the total area for heat transfer in the exchanger.

In eq. (8.7.6) subscripts 1 and 2 denote the ends of the exchanger, 1–2 being the direction of T increasing from T to $T + dT$, with $\dot{m}_b$ positive in this direction.

For the whole exchanger, for constant c_a and c_b:

$$\dot{Q} = \dot{m}_a c_a (T_{a2} - T_{a1}) \tag{8.7.7}$$

and:

$$\dot{Q} = \dot{m}_b c_b (T_{b2} - T_{b1}) \tag{8.7.8}$$

Using these to eliminate $\dot{m}_a c_a$ and $\dot{m}_b c_b$ from eq. (8.7.6) gives:

$$\dot{Q} = KA(\Delta T)_{\log} \tag{8.7.9}$$

where:

$$(\Delta T)_{\log} = \frac{(T_{a2} - T_{b2}) - (T_{a1} - T_{b1})}{\ln[(T_{a2} - T_{b2})/(T_{a1} - T_{b1})]} \tag{8.7.10}$$

is the logarithmic mean of the temperature differences at the ends of the exchanger. The sketched temperature profiles are shown in Fig. 8.7.4(c) for the case when stream b has the lower capacity rate, i.e. it undergoes a greater change of temperature than stream a.

Equation (8.7.9) with (8.7.10) is commonly used to determine the surface area required for a given heat rate $\dot{Q}$, but it must be appreciated that it can

8.7 Heat transfer equipment

cause considerable errors when K or, more especially, c_a and c_b vary with temperature. In an exchanger with a plot of temperature against enthalpy of the type shown in Fig. 8.5.5, for example, the logarithmic mean overestimates the true mean temperature difference in the exchanger because it is based on linear relations for the temperatures of both fluids against enthalpy. It thus underestimates the required surface area. If the curve 3–1 in Fig. 8.5.5 were convex upwards, as occurs towards the end of the heat exchanger in the Claude process, the use of the logarithmic mean overestimates the required area.

Variations of K, c_a and c_b can be taken into account in two ways. In one method eq. (8.7.1) with (8.7.2) or (8.7.3) is used as the basis of a step by step procedure. Thus from eq. (8.7.1) with (8.7.2):

$$dA = \frac{\dot{m}_a c_a \, dT_a}{K(T_a - T_b)} \qquad (8.7.11)$$

Taking steps δT_a say the corresponding steps δT_b are found and the temperature difference $(T_a - T_b)$ is calculated at the middle of the step. Using appropriate values of K, c_a and c_b the increment of area δA is found and the increments are summed for the whole exchanger. An example of the procedure for the more complicated case of three fluids is given later.

The other method, and the more usual one in practice, is to divide the exchanger into sections within which K and the c may be regarded as constant. The areas are found from eq. (8.7.6) and summed.

A useful concept for treating heat exchangers is the number of transfer units. It expresses in dimensionless form the heat transfer capability of an exchanger in relation to the job which it has to do in warming or cooling one of the fluids. For a two-fluid exchanger the dimensionless groups:

$$N_a = KA/\dot{m}_a c_a \qquad (8.7.12)$$

and:

$$N_b = KA/\dot{m}_b c_b \qquad (8.7.13)$$

can be formed. These may be called the number of transfer units for the a-stream and the b-stream, respectively. It will be seen later that the larger of these N, i.e. the one for the stream with the lower capacity rate, is of special importance, and when the expression 'number of transfer units' is employed without qualification it is taken to mean the larger of the two.

N_a and N_b can be considered as dimensionless areas which increase in the direction 1–2. In terms of these, eq. (8.7.5) may be written as:

$$\frac{d(T_a - T_b)}{T_a - T_b} = d(N_a - N_b) \qquad (8.7.14)$$

with the integrated form:

$$\ln \frac{T_{a2} - T_{b2}}{T_{a1} - T_{b1}} = N_a - N_b \tag{8.7.15}$$

Off-design conditions

The straightforward design problem for a heat exchanger treated above is to find the area A when the temperatures at the ends are known. In many circumstances one meets the converse problem of finding two unknown outlet temperatures when the area is given. This occurs when the exchanger is used at conditions different to those of the design or when a standard manufactured exchanger is being applied.

When K, c_a and c_b are constant the problem can be solved using eq. (8.7.6) above with the enthalpy balance from eqs. (8.7.7) and (8.7.8):

$$\dot{m}_a c_a (T_{a2} - T_{a1}) = \dot{m}_b c_b (T_{b2} - T_{b1}) \tag{8.7.16}$$

Supposing T_{a2} and T_{b1} to be known, the two simultaneous equations enable T_{a1} and T_{b2} to be found. The result of eliminating T_{a1} is:

$$\frac{T_{b2} - T_{b1}}{T_{a2} - T_{b1}} = \frac{1 - \exp(N_a - N_b)}{1 - (N_a/N_b)\exp(N_a - N_b)} \tag{8.7.17}$$

On eliminating T_{b2} the result is:

$$\frac{T_{a1} - T_{a2}}{T_{a2} - T_{b1}} = \frac{1 - \exp(N_b - N_a)}{1 - (N_b/N_a)\exp(N_b - N_a)} \tag{8.7.18}$$

Supposing that the b-stream has the lower capacity rate, then the ratio given by eq. (8.7.17) is also the efficiency of the heat exchanger, η_x, as defined by eq. (8.4.6), when constant specific heat capacities are assumed. For T_{a2} is the highest temperature which T_b can attain at the end 2. On the other hand if the a-stream has the lower capacity rate the efficiency is given by eq. (8.7.18). If it is agreed to denote by N, without subscript, the larger of the values N_a and N_b, the efficiency in all cases is given by:

$$\eta_x = \frac{1 - \exp[-N(1-r)]}{1 - r\exp[-N(1-r)]} \tag{8.7.19}$$

where r, less than unity, is the ratio of the capacity rates. When the capacity rates are equal, eq. (8.7.19) becomes on taking the limit as $r \to 1$:

$$\eta_x = \frac{N}{1 + N} \tag{8.7.20}$$

8.7 Heat transfer equipment

It should be noted that when the off-design condition consists of a moderate change of the inlet temperatures only, without changing the flow rates or the specific heat capacities, the efficiency of the exchanger remains constant provided that the change of temperature is not sufficient to alter the overall coefficient appreciably.

When K, c_a and c_b are not constants the prediction of the unknown outlet temperatures is not straightforward for a counterflow exchanger. The basic difficulty is that temperatures are specified at opposite ends of the exchanger and a step by step procedure cannot be used since no temperature difference is known. (The difficulty does not of course arise with parallel flow exchangers.)

For a counterflow exchanger the following approximative procedure may be used. Supposing T_{a2} and T_{b1} are known, eq. (8.7.17) with average values of K, c_a and c_b is used to find an approximate T_{b2}, and eq. (8.7.16) is then used to find an approximate T_{a1}. With the approximate $(T_{a1} - T_{b1})$ known at one end, the step by step procedure can be carried out. It is continued until the sum of the δA equals the given total area A. At this stage the temperature T_{a2} at the other end of the exchanger will not in general equal the actual given value. With this T_{a2} and the approximate T_{b2} so found, an efficiency is calculated. This efficiency is then applied to the specified T_{b1} and T_{a2} to determine more accurate T_{a1} and T_{b2}. The procedure can be repeated until convergence is obtained, but in many cases one application is enough.

Cross-flow exchangers

There are many possible variations on the basic cross-flow arrangement. For example one of the fluids may make several passes back and forth across the other stream. When the number of passes is large, so that the temperature change in one pass is relatively small, the configuration approaches the counterflow or parallel arrangement, depending on whether the main directions of flow are opposite or the same. In other arrangements the streams may be divided into several parallel paths, each of which makes several passes.

When one of the fluids in a cross-flow heat exchanger has a constant temperature, as for example a pure vapour condensing or a pure liquid boiling at constant pressure, the treatment given above for the counterflow exchanger can be applied. The mean temperature difference is given by eq. (8.7.10) with $T_{b1} = T_{b2}$ if the b-stream is the constant temperature one, or $T_{a1} = T_{a2}$ if the a-stream is at constant temperature. The efficiency of the

Cryogenic engineering 604

exchanger is given by eq. (8.7.19) with $r = 0$ and N calculated for the stream which changes in temperature.

In the more general case of both fluids changing in temperature, the counterflow treatment applies when there are a large number of passes, as in the Hampson exchanger. When this is not so, however, the problem is one of considerable difficulty which has been worked out for only a few cases. The treatment has to take into account the extent to which the streams are mixed at each cross-section. Useful collections of results for constant K and c are given by Jakob (1957) and by Kays & London (1964).

When these results are not applicable because K and c vary, a result can be computed by finite difference methods. Dusinberre (1951) has shown that useful results can be obtained from quite a coarse grid.

Three-fluid exchangers

Heat exchangers with several streams are very common in cryogenic plants. Examples were shown in Figs. 8.4.6 and 8.4.9 of exchangers with three streams. The following discussion is restricted to these, but the same principles apply to exchangers with more streams.

The exchanger is shown diagrammatically in Fig. 8.7.5(a). A stream of fluid a flows in counterflow with streams b and c. Heat transfer takes place from a to b and from a to c, but not from b to c. The situation would be realised physically in a Linde exchanger with the fluid a in the outer pipe and fluids b and c in the inner pipes. The mass flow rate $\dot{m}_a$ is taken as positive to the left, and $\dot{m}_b, \dot{m}_c$ positive to the right. Subscripts 1 and 2 denote the left- and right-hand ends of the exchanger, as before.

Fig. 8.7.5. (a) A three-fluid exchanger shown diagrammatically, and (b) elements of heat transfer area.

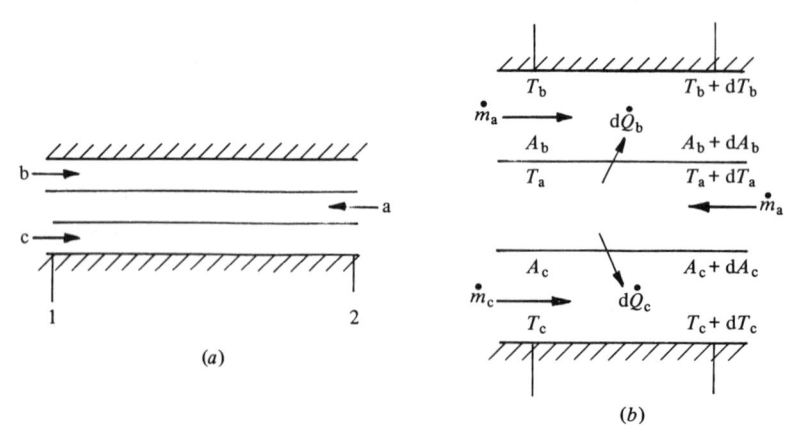

8.7 Heat transfer equipment

For the whole exchanger the enthalpy balance gives:

$$\dot{m}_a(h_{a2} - h_{a1}) = \dot{m}_b(h_{b2} - h_{b1}) + \dot{m}_c(h_{c2} - h_{c1}) \qquad (8.7.21)$$

or, on supposing the specific heat capacities constant:

$$\dot{m}_a c_a(T_{a2} - T_{a1}) = \dot{m}_b c_b(T_{b2} - T_{b1}) + \dot{m}_c c_c(T_{c2} - T_{c1}) \qquad (8.7.22)$$

The capacity rate of the combined b- and c-streams may be greater or less than that of the a-stream. If it be supposed that:

$$\dot{m}_b c_b + \dot{m}_c c_c < \dot{m}_a c_a \qquad (8.7.23)$$

it will then be possible in principle to raise the temperature of the b- and c-streams to the inlet temperature of the a-stream.

It will be noted that eq. (8.7.22) does not uniquely fix the exit temperatures of the b- and c-streams when their inlet temperatures and the inlet and outlet temperature of the a-stream are known. In fact the temperature changes of the b- and c-streams depend on how the heat transfer surface is apportioned between them and on how it is distributed through the exchanger. Supposing that very little surface were allocated to the c-stream say, then clearly the c-stream could not undergo a large change of temperature.

Taking as before an element of the exchanger, Fig. 8.7.5(b) the basic equations are:

$$\dot{m}_a c_a dT_a = \dot{m}_b c_b dT_b + \dot{m}_c c_c dT_c \qquad (8.7.24)$$

$$\dot{m}_b c_b dT_b = K_b(T_a - T_b)dA_b \qquad (8.7.25)$$

$$\dot{m}_c c_c dT_c = K_c(T_a - T_c)dA_c \qquad (8.7.26)$$

where K_b and K_c are the overall coefficients of heat transfer on the b and c sides with elemental areas dA_b and dA_c to correspond. Introducing the ratios:

$$\beta = \dot{m}_b c_b / \dot{m}_a c_a \qquad (8.7.27)$$

and:

$$\gamma = \dot{m}_c c_c / \dot{m}_a c_a \qquad (8.7.28)$$

and the numbers of transfer units, considered as variables:

$$N_b = K_b A_b / \dot{m}_b c_b \qquad (8.7.29)$$

and:

$$N_c = K_c A_c / \dot{m}_c c_c \qquad (8.7.30)$$

the equations become:

$$dT_a = \beta dT_b + \gamma dT_c \tag{8.7.31}$$

$$dT_b = (T_a - T_b)dN_b \tag{8.7.32}$$

$$dT_c = (T_a - T_c)dN_c \tag{8.7.33}$$

On dividing the last two:

$$\frac{dT_b}{dT_c} = \frac{(T_a - T_b)}{(T_a - T_c)} \frac{dN_b}{dN_c} \tag{8.7.34}$$

If the specific heat capacities and the overall coefficients remain constant, and the surface areas for the two streams are distributed uniformly throughout the exchanger, i.e. the ratio dA_b/dA_c remains constant, then the ratio dN_b/dN_c is also constant and equal to N_b/N_c for the whole exchanger. For an assumed value of this ratio a solution to the above set of simultaneous differential equations can be found. Its use, however, involves considerable numerical work and, when variable properties are to be taken into account, it is better to work numerically *ab initio*. To illustrate the method, however, the following treatment and the example will be restricted to constant K and c.

It will be supposed that the temperatures T_{a2}, T_{b1} and T_{c1} are given, and that the coefficients, mass flow rates and specific heat capacities are known and constant. The ratio N_b/N_c is known for the whole exchanger with the uniform distribution dN_b/dN_c mentioned above. The actual values of N_b and N_c to bring about a desired result, for example a given efficiency, are to be found. Assuming inequality (8.7.23) to hold, the efficiency is:

$$\eta_x = \frac{\dot{m}_b c_b(T_{b2} - T_{b1}) + \dot{m}_c c_c(T_{c2} - T_{c1})}{\dot{m}_b c_b(T_{a2} - T_{b1}) + \dot{m}_c c_c(T_{a2} - T_{c1})} \tag{8.7.35}$$

Using the overall enthalpy balance and β and γ this may be written:

$$\eta_x = \frac{T_{a2} - T_{a1}}{\beta(T_{a2} - T_{b1}) + \gamma(T_{a2} - T_{c1})} \tag{8.7.36}$$

For a given η_x this equation determines T_{a1}.

The numerical procedure begins at end 1 where the temperatures are now known. Taking steps of either T_b or T_c the increments of N_b or N_c are calculated.

Example 8.7.1

In a counterflow three-fluid exchanger like that of Fig. 8.7.5(a) the streams

8.7 Heat transfer equipment

are as follows:

a-stream $\dot{m}_a = 1 \text{ kg/s}$ $c_a = 1.7 \text{ kJ/kg K}$ $T_{a2} = 300 \text{ K}$
b-stream $\dot{m}_b = 0.7 \text{ kg/s}$ $c_b = 1.4 \text{ kJ/kg K}$ $T_{b1} = 120 \text{ K}$
c-stream $\dot{m}_c = 0.2 \text{ kg/s}$ $c_c = 1.0 \text{ kJ/kg K}$ $T_{c1} = 80 \text{ K}$

Assuming constant K and c, and that the number of transfer units for the two streams are equal and uniformly distributed, determine the number of transfer units required for an efficiency of 0.98.

$$\beta = \frac{(0.7)(1.4)}{(1.0)(1.7)} = 0.5765$$

$$\gamma = \frac{(0.2)(1.0)}{(1.0)(1.7)} = 0.1176$$

Since $(\beta + \gamma)$ is less than unity, inequality (8.7.23) holds. The efficiency is:

$$0.98 = \frac{300 - T_{a1}}{\beta(300 - 120) + \gamma(300 - 80)}$$

from which:

$$T_{a1} = 172.95 \text{ K} \quad \text{(say 173 K)}$$

The working will be given for the first step of $\delta T_c = 0.5$ K. Using the starting values of $T_{a1} = 173$ K, $T_{b1} = 120$ K, $T_{c1} = 80$ K, by eq. (8.7.34):

$$\delta T_b = \frac{173 - 120}{173 - 80} \, 0.5 \text{ K} = 0.285 \text{ K}$$

and the new T_b is 120.28 K.

By eq. (8.7.31):

$$\delta T_a = [(0.5765)(0.285) + (0.1176)(0.5)] \text{ K} = 0.223 \text{ K}$$

and the new T_a is 173.22 K.

The temperature difference between the a-stream and the c-stream at the middle of the step is:

$$[(173 + 0.11) - (80 + 0.25)] \text{ K} = 92.86 \text{ K}$$

and the incremental contribution to the number of transfer units is by eq. (8.7.33):

$$\delta N_c = 0.5/92.86 = 0.0054$$

Since $N_b = N_c$ this is also the increment of N_b. The process is repeated until the temperature T_a reaches the value 300 K. The sum of the increments then gives the total N_c.

This procedure is the simple Euler method in which the gradient is evaluated at the beginning of the step. In the more sophisticated Runge–Kutta methods an average gradient for the step is calculated, and this enables accurate results to be obtained with a larger step. The results of the calculation using the simplest of these, in which the gradient is evaluated at the middle of the step, are:

$$N_c = N_b = 9.00 \quad \text{(at the end 2)}$$
$$T_{b2} = 296.3 \text{ K}$$
$$T_{c2} = 296.3 \text{ K}$$

The temperatures in the exchanger plotted against the increasing sum of the transfer units are shown in Fig. 8.7.6(a). It will be seen that the temperatures of the b- and c-streams are eventually pulled into equality. With $N_b = N_c$ this follows from eq. (8.7.34), because if $(T_a - T_c)$ is greater than $(T_a - T_b)$ at the end 1, the temperature of c changes more rapidly than that of b until they become (asymptotically) equal.

When $N_c < N_b$, however, the temperature of c changes much more slowly towards that of b. The result of the calculation for this condition is shown in Fig. 8.7.6(b) for $N_b = 2N_c$, with a total at the end $N_b + N_c = 16.4$.

When $N_b < N_c$ the temperature of c catches up with that of b and overtakes it. The separation increases until $dT_b = dT_c$ when, according to eq. (8.7.34):

$$N_b(T_a - T_b) = N_c(T_a - T_c) \tag{8.7.37}$$

Fig. 8.7.6. Temperature profiles in a three-fluid exchanger for (a) the conditions of example 8.7.1 with $N_b = N_c$; (b) with $N_b = 2N_c$ and (c) with $2N_b = N_c$.

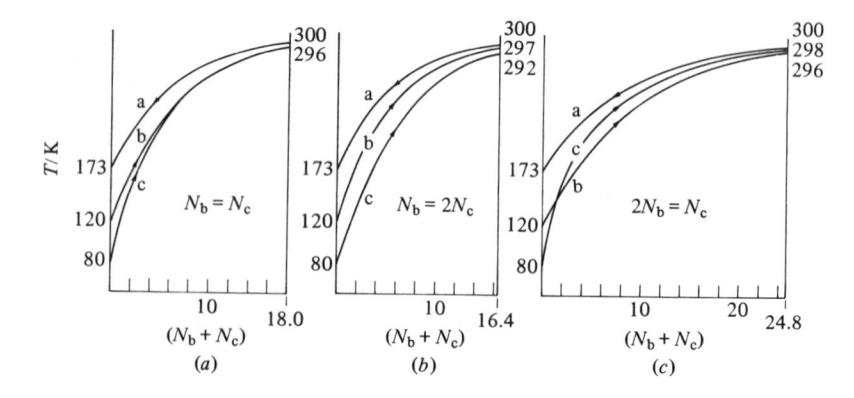

8.7 Heat transfer equipment

This is shown for $2N_b = N_c$ in Fig. 8.7.6(c) with a total at the end $N_b + N_c = 24.8$.

The cost of the exchanger is related to the sum of the areas $A_b + A_c$. In general this is different for the three cases shown, although they all have the same efficiency and the same heat rate. Supposing for illustration that K_b and K_c have the same value, the total area is proportional to:

$$N_b \dot{m}_b c_b + N_c \dot{m}_c c_c$$

giving in the three cases, with $\dot{m}_b c_b = 0.98 \text{ kW/K}$, $\dot{m}_c c_c = 0.2 \text{ kW/K}$:

(a) $[(1/2)0.98 + (1/2)0.2]\, 18.0 = 10.6$

(b) $[(2/3)0.98 + (1/3)0.2]\, 16.4 = 11.8$

(c) $[(1/3)0.98 + (2/3)0.2]\, 24.8 = 11.4$

When the fluid properties and the overall coefficients vary with temperature, the same calculation procedure applies, but it must be remembered that β and γ then depend on temperature, as do K_b and K_c. As the temperatures change, these values must be continuously corrected. The increments δN have to be converted into increments of area δA before summing them for the whole exchanger. The point of difference is that dN_b/dN_c is now not constant throughout.

Regenerators

Storage-type heat exchangers, or regenerators, are used in many kinds of engineering plant, but their principal use on a large scale in cryogenics is in gas separation plants, as described in section 8.6. The following description and treatment applies mainly to these.

The Fränkl regenerator has packing composed of coiled aluminium strip, shown in Fig. 8.7.7. Typically this is 0.2–0.5 mm thick, 20–50 mm

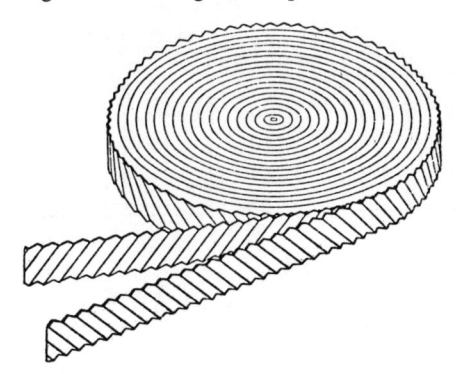

Fig. 8.7.7. Corrugated strip used in the Fränkl regenerator.

wide with corrugations 1.5–3 mm deep. The strip is coiled into flat rolls which are assembled, with suitable spacers to reduce longitudinal conduction, into a matrix containing some 200 or more layers. The warm and cold streams are blown alternatively in opposite directions for about 2–3 min.

The surface area which can be packed into a given volume is very large. Supposing that φ is the void fraction, the ratio of surface area to volume is:

$$\frac{A}{V} = \frac{2(1-\varphi)}{d} \qquad (8.7.38)$$

where d is the thickness of the strip. A typical void fraction is 0.8 giving $A/V = 1333 \text{ m}^2/\text{m}^3$ with strip 0.3 mm thick, but higher values are possible. This ratio is much higher than can be obtained with a direct heat exchanger. On the other hand two regenerators are needed, working alternatively.

Later designs of regenerator often employ a packing of smooth quartz pebbles graded 6–18 mm. A longer blow time of about 15 min is used with these.

Referring to Fig. 8.7.8, a flow of gas a at temperature T_{a2} enters a regenerator at end 2 and leaves at end 1 with the temperature T_{a1}. This continues for a time τ_a. The gas a is then shut off (and passed through another regenerator) whilst a flow of gas b enters at end 1 with temperature

Fig. 8.7.8. (a) Diagram showing nomenclature for regenerator treatment and (b) an element of the matrix.

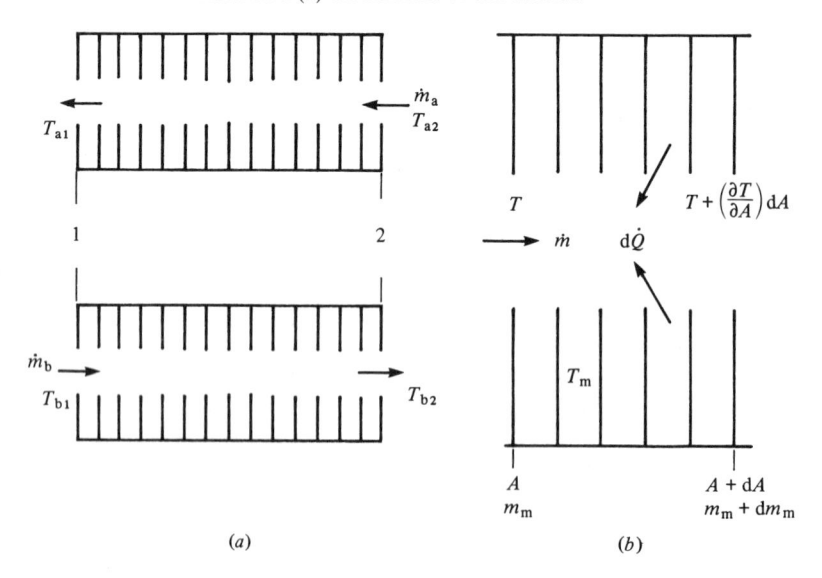

8.7 Heat transfer equipment

T_{b1} and leaves at end 2 with temperature T_{b2}. This continues for time τ_b, after which gas a is introduced again. Supposing the a-stream to be warmer than the b-stream, the matrix rises in temperature during time τ_a and falls in temperature during time τ_b. Because of this the temperatures of the gases leaving, T_{a1} and T_{b2}, are not constant in time.

If only two regenerators are in use it is clear that $\tau_a = \tau_b$. This is usually so in cryogenic plants. In some other applications of regenerators the times may be different.

An element of the matrix at temperature T_m with a gas passage is shown in Fig. 8.7.8(b), presenting area dA to the gas and having a mass dm_m and specific heat capacity c_m. Quantities referring to the gas, absolute temperature T, mass flow rate $\dot{m}$, specific heat capacity c and coefficient of heat transfer f will be written without subscripts on the understanding that the same equations will serve for the a- and b-streams when subscripts are introduced.

The assumptions are (i) no conduction in the matrix in the direction of flow, (ii) the temperature of the matrix at any cross-section is uniform and (iii) the change of internal energy of the gas in an elementary control volume is negligible compared with that of the matrix in the same time. With these assumptions, the heat transfer equation for the element is:

$$d\dot{Q} = f(T_m - T) \, dA \tag{8.7.39}$$

and the elementary enthalpy balances give:

$$d\dot{Q} = \dot{m}c \left(\frac{\partial T}{\partial A}\right) dA \tag{8.7.40}$$

and:

$$d\dot{Q} = -c_m \left(\frac{\partial T_m}{\partial \tau}\right) dm_m \tag{8.7.41}$$

The heat transfer coefficient f is a surface coefficient between the matrix and the gas, but it may include an allowance for the resistance of deposits on the surface. It may also be modified to take into account the fact that assumption (ii) above is not satisfied completely, and be specified in terms of a mean matrix temperature at any cross-section.

The boundary conditions to be imposed are that $T = T_{a2}$ at end 2 during the time τ_a and that $T = T_{b1}$ at end 1 during the time τ_b. The initial conditions consist of a statement of the temperature distribution in the matrix at a given time. The temperature distribution at any subsequent time can then be determined by solution of the equations. After a sufficient time has elapsed, however, which for a Fränkl regenerator may be several hours, the temperatures of the matrix and of the gases at any cross-section undergo regular cyclic variations. When this cyclic steady state has been reached an overall cyclic enthalpy balance can be established. For adiabatic operation and equal blow times $\tau_a = \tau_b$:

$$\dot{m}_a(h_{a2} - h_{a1,av}) = \dot{m}_b(h_{b2,av} - h_{b1}) \tag{8.7.42}$$

where subscript av indicates the time average of the specific enthalpy h. For constant specific heat capacities:

$$\dot{m}_a c_a(T_{a2} - T_{a1,av}) = \dot{m}_b c_b(T_{b2,av} - T_{b1}) \tag{8.7.43}$$

This equation has exactly the same form as eq. (8.7.16) for a heat exchanger, except that the average temperatures of the outflowing streams have to be used. It has the same implications for the temperature ranges of the streams. Suppose for example that the b-stream has the lower capacity rate, the range of the b-stream must be greater than the range of the a-stream. The regenerator efficiency can be defined as:

$$\eta_{reg} = \frac{T_{b2,av} - T_{b1}}{T_{a2} - T_{b1}} \tag{8.7.44}$$

exactly as for the counterflow exchanger.

Before considering the available solutions of eqs. (8.7.39)–(8.7.41), a simplified case will be presented which turns out to be relevant to Fränkl regenerators.

When the heat capacity of the gas which passes through during one blow is negligible in comparison with the heat capacity of the matrix, the temperature of the matrix will not change with time. Equation (8.7.40) then becomes an ordinary differential equation. For the flow of the a-stream:

$$d\dot{Q}_a = f_a(T_m - T_a)dA \tag{8.7.45}$$

and for the b-stream:

$$d\dot{Q}_b = f_b(T_m - T_b)dA \tag{8.7.46}$$

For a steady matrix temperature and equal blow times the heat rate from the a-stream, $-d\dot{Q}_a$, must equal the heat rate to the b-stream, $d\dot{Q}_b$, i.e.:

$$-d\dot{Q}_a = d\dot{Q}_b = d\dot{Q} \tag{8.7.47}$$

say. Using these in eqs. (8.7.45) and (8.7.46) solving for $(T_a - T_m)$ and $(T_m - T_b)$ and adding gives:

$$T_a - T_b = \frac{d\dot{Q}}{dA}\left(\frac{1}{f_a} + \frac{1}{f_b}\right) \tag{8.7.48}$$

8.7 Heat transfer equipment

or:
$$dQ̇ = K(T_a - T_b)dA \qquad (8.7.49)$$

where:
$$\frac{1}{K} = \frac{1}{f_a} + \frac{1}{f_b} \qquad (8.7.50)$$

The K so-defined is analogous to the overall coefficient of heat transfer in a heat exchanger with f_a and f_b the coefficients on the two sides of the heat transfer surface. In the regenerator however they act on the same surface but at different times.

The enthalpy equations (8.7.40) become for steady conditions
$$dQ̇ = ṁ_a c_a dT_a \qquad (8.7.51)$$
and:
$$dQ̇ = ṁ_b c_b dT_b \qquad (8.7.52)$$

Equations (8.7.49), (8.7.51) and (8.7.52) are identical to those for the counterflow heat exchanger, eqs. (8.7.1)–(8.7.3), and have the same solutions. In particular the efficiency is given by eq. (8.7.19) with the special form (8.7.20) when the capacity rates are equal.

When the heat capacity of the gas per blow is not negligible in comparison with the heat capacity of the matrix, solutions to eqs. (8.7.39)–(8.7.41) must be attempted either analytically or numerically. The results are usually expressed in dimensionless form with the dimensionless area:
$$\Lambda = fA/ṁc \qquad (8.7.53)$$
and the dimensionless time:
$$\Pi = fA\tau/m_m c_m \qquad (8.7.54)$$

with subscripts a and b as required. The dimensionless area Λ defined by eq. (8.7.53) is of exactly the same form as the number of transfer units in a heat exchanger, but it contains the matrix-to-gas coefficient instead of the overall coefficient K.

The first results for the regenerator efficiency under regular cycling conditions were obtained analytically by Hausen (1929) who treated the problem as one of forced vibrations of temperature. Numerical calculations were performed by Saunders & Smoleniec (1948) using a relaxation method, and by Johnson (1952). There is reasonable agreement between these results in the regions where they overlap and comparison can be made, but they are restricted to the condition of $ṁ_a c_a = ṁ_b c_b$, $\Lambda_a = \Lambda_b$ and $\Pi_a = \Pi_b$. The regenerator efficiencies are shown in Fig. 8.7.9, which is based mainly on Hausen's results, but using some of the others to establish the shape of the curves in places.

The values of η_{reg} at $\Pi = 0$ are simply those given by eq. (8.7.20). The condition of negligible heat capacity of the gas per blow compared with that of the matrix implies that Π/Λ is zero. Since Λ is finite, $\Pi = 0$. For equal capacity rates eq. (8.7.20) gives:

$$\eta_{\text{reg}}(\Pi = 0) = \frac{1}{1 + (1/\Lambda_a) + (1/\Lambda_b)} \tag{8.7.55}$$

on using eq. (8.7.50) for K in N. For $\Lambda_a = \Lambda_b = \Lambda$ say:

$$\eta_{\text{reg}}(\Pi = 0) = \frac{\Lambda}{2 + \Lambda} \tag{8.7.56}$$

This equation gives the intercepts on the axis $\Pi = 0$ in Fig. 8.7.9.

There is no exact analytical expression for η_{reg} when Π is not equal to zero, but an empirical expression of Corbitt (1948) gives a useful approximation

$$\frac{\eta_{\text{reg}}}{\eta_{\text{reg}}(\Pi = 0)} = 1 - \frac{\Pi^2}{9\Lambda^2} \tag{8.7.57}$$

Fig. 8.7.9. Dependence of regenerator efficiency on Π and Λ.

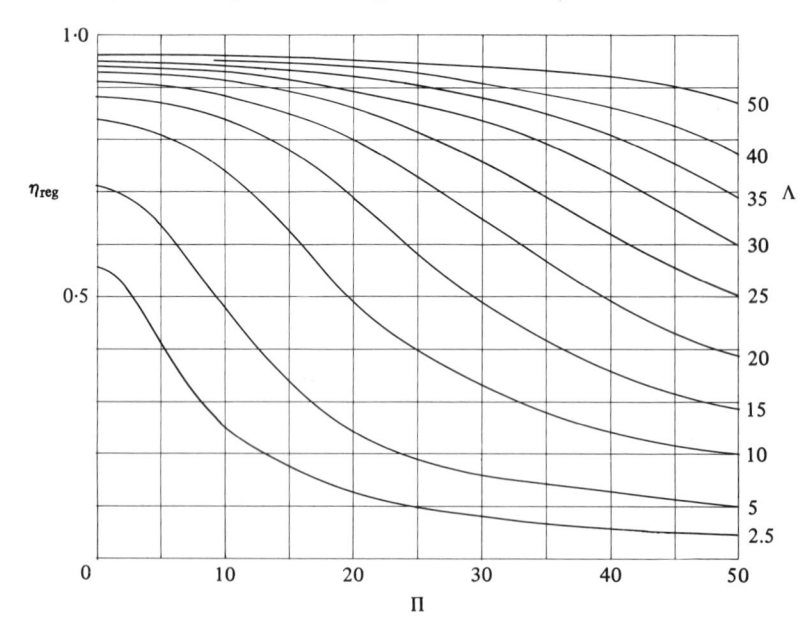

8.7 Heat transfer equipment

So far as regenerators in gas separation plants are concerned, only those having an efficiency over 0.96 or so come into consideration. This implies a dimensionless area of 50 or more. Since in this region the shape of the curves shows that Π has comparatively little effect on η_{reg}, eqs. (8.7.56) and (8.7.57) will be used to estimate the area required in a Fränkl regenerator.

Example 8.7.2

The packing in a Fränkl regenerator consists of aluminium strip, 0.4 mm thick, $\rho_m = 2780$ kg/m³, $c_m = 880$ J/kg K. The blow time is 3 min each way, and the streams have equal capacity rates, $\dot{m}_a = \dot{m}_b = 1.2$ kg/s, $c_a = c_b = 1050$ J/kg K. The forward (air) stream enters at 300 K and leaves at 98 K (average). The revert (nitrogen) stream enters at 94 K. The coefficient of heat transfer is 60 W/m² K in each direction. Determine the surface area required.

The equality of the capacity rates and of the coefficients of heat transfer imply that $\Lambda_a = \Lambda_b$, $\Pi_a = \Pi_b$. The required efficiency is:

$$\eta_{reg} = \frac{300 - 98}{300 - 94} = 0.9806$$

For the matrix:

$$\frac{A}{m_m} = \frac{2(1000)}{0.4(2780)} \text{ m}^2/\text{kg} = 1.799 \text{ m}^2/\text{kg}$$

and:

$$\Pi = \frac{60(1.799)(3)(60)}{880} = 22.1$$

By eqs. (8.7.56) and (8.7.57):

$$0.9806 = \frac{\Lambda}{2 + \Lambda}\left[1 - \frac{(22.1)^2}{9\Lambda^2}\right]$$

Solving for Λ:

$$\Lambda = 123$$

and the required area is:

$$A = \frac{1.2(1050)(123)}{60} \text{ m}^2 = 2580 \text{ m}^2$$

The condition of equal capacity rates in the forward and revert direction is approximately met in regenerators in gas separation plants. As mentioned in section 8.6 this does not necessarily imply equality of the mass flow rates because of the somewhat higher specific heat capacity of the compressed air at the low temperature end.

Few numerical results are available for unequal capacity rates. Provided that these are not too different, it would seem that no great error would be introduced for regenerators of high efficiency by using the exact expression for $\Pi = 0$, eq. (8.7.56), modified by Corbitt's factor, eq. (8.7.57). Otherwise each case has to be computed separately by methods such as those of Iliffe (1948), Willmott (1964) or Edwards, Evans & Probert (1971).

The processes inside regenerators are complicated by the deposition of carbon dioxide from the forward stream and its subsequent sublimation into the revert stream. The blow time in regenerators has to be a compromise between a low value of Π to obtain a high efficiency and a high value of Π to obtain a pure oxygen product and economic use of the compressed air. Each time the regenerators reverse, a volume of compressed air is let down uselessly to atmospheric pressure; at the same time it contaminates the oxygen product stream. It is usually said that regenerators are not suitable for oxygen purities over 98%, and this fact, together with the relative inflexibility of operation, has produced the trend towards reversing heat exchangers as mentioned in section 8.6.

8.8 Liquefied natural gas

Natural gas as supplied to the consumer is mainly methane, CH_4, with some ethane C_2H_6, propane C_3H_8, butane C_4H_{10}, nitrogen and carbon dioxide. At the well, however, it contains higher hydrocarbons ('heavies') which are extracted in a sweetening process. The proportions of the components depend on the origin of the gas, but North Sea type gas contains after purification about 0.93 mole fraction of methane, 0.04 of ethane and 0.02 of nitrogen. Some gases, particularly Dutch gas, contain up to 0.15 of nitrogen, at the wellhead. This nitrogen fraction has to be reduced because it lowers the boiling point from about 112 K to 91 K, which is inconvenient during storage. It also makes the gas incompatible with other gases which may be supplying the same region.

Liquefaction of natural gas is undertaken for two main reasons: to reduce its bulk for transport by sea and overland; and to provide a store of gas to balance supply and demand, a practice which is called 'peak shaving'.

Much of the natural gas in the world occurs where there is little demand. It may be pumped overland at pressures of about 50–60 bar to centres of population; but underwater pipelines are expensive and have not yet been laid over great distances at depth. By liquefying the gas its volume is

8.8 Liquefied natural gas

reduced to about 1/600, and this makes overseas transport in special ships practicable.

Peak shaving is required to meet the variation of demand, the maximum of which may reach 50 times the minimum. In a large pipeline system some storage can be obtained by increasing the line pressure, known as 'line stack', but the scope for this is limited. High-pressure storage vessels are expensive, so liquid storage is often preferred. A peak shaving plant may be a liquefaction plant taking gas from a pipeline, or it may be simply a storage facility receiving liquid gas by road tankers from a large plant. It is then called a 'satellite plant'.

Considering natural gas mainly as methane with a critical temperature of 191 K and a normal boiling point of 112 K, it is not of course possible to liquefy it by pressure at an ambient temperature of, say, 300 K. Reference to Fig. 8.8.1 shows that if it is liquefied at atmospheric pressure there is a large liquefaction step in the process indicating that a large amount of refrigeration has to be produced at the low temperature of 112 K. In practice, however, the gas comes from the well at considerable pressure, about 40 bar, and the cooling curve at this pressure in Fig. 8.8.1 shows that the liquefaction step is much reduced. Furthermore, the presence of higher boiling components reduces the step for natural gas as compared with pure methane; in North African gas it is almost absent. Cooling is thus required

Fig. 8.8.1. Cooling curves, temperature versus specific enthalpy, of methane at pressures of 1 bar and 40 bar.

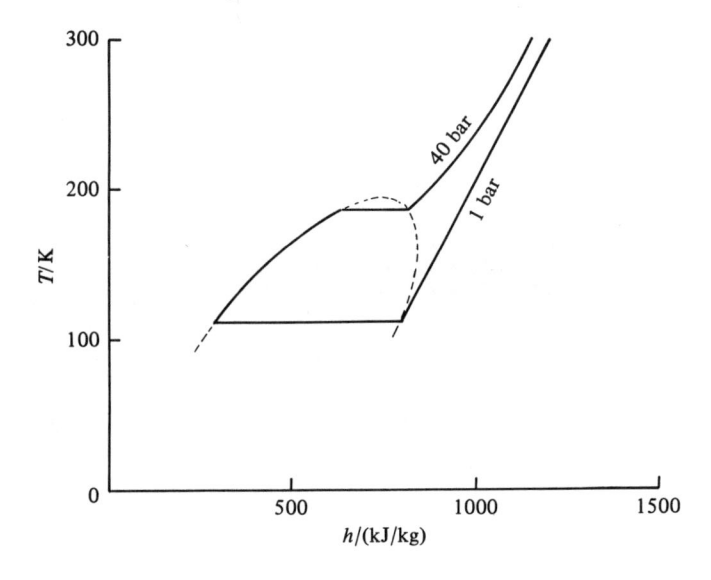

over a range of temperature with the régime of the refrigerating plant matching that of the natural gas stream. This suggests a liquefaction process of the Claude type using expansion machines. Such cycles have been used in fact, but mainly in small peak shaving plants in which only a portion of the gas, 10% or so, is to be liquefied, the remainder going into the mains. For large plants the vapour compression cascade process and its development, the auto-cascade process, have proved to be just as economical in power and cheaper in first cost.

The principles of the cascade process were outlined in section 5.11. As applied to natural gas liquefaction it is usual to employ propane, ethylene and methane as refrigerants. A simplified diagram of the process is shown in Fig. 8.8.2. Natural gas after purification, compression and cooling if necessary, is cooled in the propane evaporator to about 238 K by propane boiling at atmospheric pressure at 231 K. In the ethylene evaporator it is cooled to 176 K by ethylene boiling at 169 K, and finally cooled to about 122 K by methane boiling at 112 K. In the propane evaporator ethylene is condensed and the methane is desuperheated before it is condensed in the ethylene evaporator. The natural gas is finally flashed down to atmospheric pressure before going to storage. The flash gas and the boil-off from the storage tank may be passed back through the heat exchangers to recover their refrigerating effect and, depending on circumstances, reliquefied or burnt as fuel.

The refrigerants propane and methane can be extracted from the natural gas, but ethylene is not normally present and it has to be manufactured or bought in. At some stage in the process a knock-out pot is incorporated to remove the heavies.

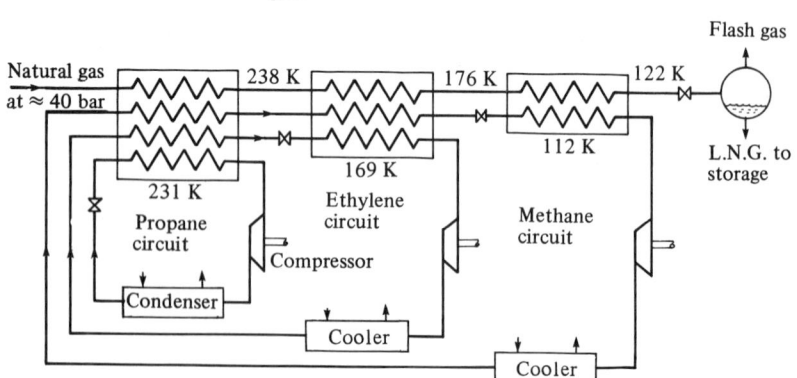

Fig. 8.8.2. Simplified diagram of a cascade system for liquefying natural gas.

8.8 Liquefied natural gas

A typical plant of this kind is the one at Arzew in Algeria, completed in 1964, which has three identical lines, each capable of liquefying gas at the rate of 10.5 kg/s, the compressors taking 11.65 MW in each propane circuit, 8.6 MW in each ethylene circuit and 3.5 MW in each methane circuit.

The actual flow sheet of these plants is more complicated than that shown in Fig. 8.8.2 because multi-stage compression is employed in each circuit. Three stages in the propane circuit, four stages in the ethylene circuit and three stages in the methane circuit are used at Arzew. This makes it possible to have eight levels of evaporating temperature for cooling the gas in stages and an approximate match between the refrigeration and the cooling curve can be obtained. To achieve this, however, a rather complicated and expensive plant is needed with ten compressors and numerous heat exchangers. These factors have led to the adoption of the mixed refrigerant auto-cascade system. Only one large compressor is required, which may be large enough to allow the use of an axial flow machine with an efficiency superior to that of centrifugal machines. The number of separate heat exchangers can be very much reduced, though the unit size is larger; and the refrigerants used are already present in the natural gas.

The idea of using refrigerant solutions to boil over a range of temperature at constant pressure was mentioned in section 6.23 in connection with the resorption–compression system of Osenbrück. Interest in mixed refrigerants returned in the 1950s when Haselden & Klimek (1958) reported tests on a small unit employing propane and normal-butane. Meanwhile, Ruhemann had proposed his auto-cascade system which was described in

Fig. 8.8.3. Simplified diagram of an auto-cascade system for liquefying natural gas.

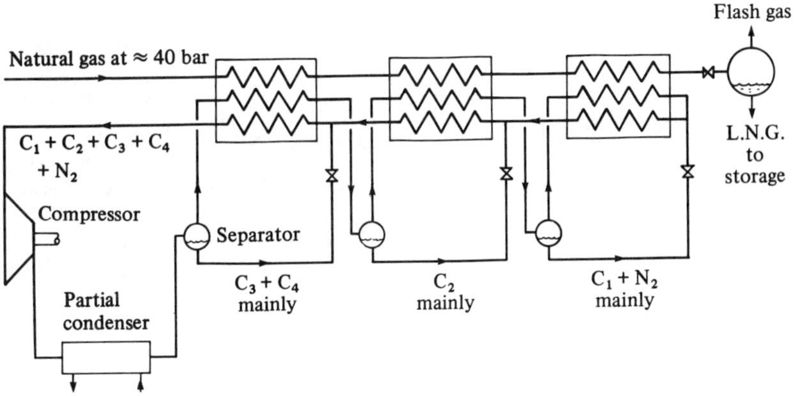

section 5.11. These ideas were combined in a pilot plant for liquefying natural gas by Kleemenko (1959), and since then several large ones have been built.

A simplified diagram of the process with three stages is shown in Fig. 8.8.3. As in the classical cascade system the natural gas stream is progressively cooled in successive evaporators before being flashed to atmospheric pressure; the difference is that in the auto-cascade the pressure is the same in all of the evaporators, the differences of temperature being brought about by differences in composition. The single compressor for the refrigerant cycle takes in a mixture of, typically, butane, propane, ethane and methane, usually denoted C_4, C_3, C_2 and C_1, respectively, and some nitrogen. The function of the nitrogen, which is more volatile than the methane, is to lower the evaporating temperature in the last stage, so that the natural gas can be cooled very nearly to its normal boiling point temperature. There will then be very little flash-gas produced when the natural gas is flashed down to atmospheric pressure. The low side pressure, which is uniform throughout the plant apart from pressure drops, is about 5–6 bar; the high side pressure, which is also the same throughout, is about 30–40 bar. The high-pressure gas delivered by the compressor is partially condensed by heat transfer to cooling water during which a less volatile fraction containing mainly butane and propane is liquefied. The liquid after separation from the uncondensed gas is throttled to the low pressure and provides the refrigeration for the first evaporator. This cools the natural gas stream and also partially condenses the fraction remaining as gas from the first partial condensation. This fraction after separation into liquid and gas goes on to the second evaporator, the liquid providing the refrigeration to cool the natural gas and to condense a further fraction of the high-pressure refrigerant stream.

Fig. 8.8.4. Arrangement of the streams in one of the evaporators in the auto-cascade system.

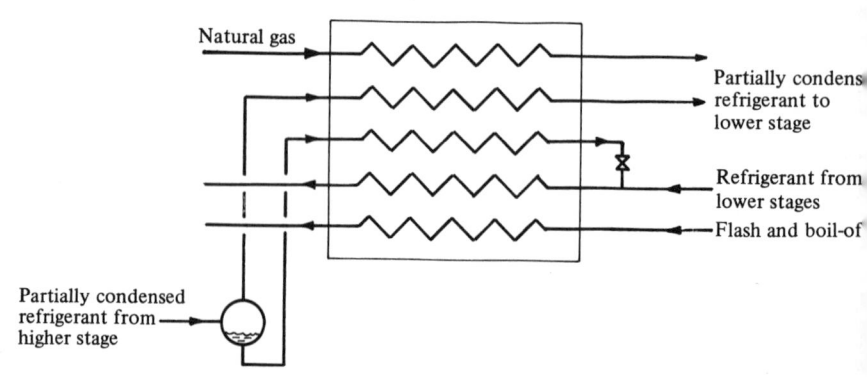

8.8 Liquefied natural gas

The evaporating temperature thus diminishes continuously towards the right of Fig. 8.8.3. Within each evaporator it diminishes towards the right because the mixture of refrigerant moving towards the left, in counterflow with the natural gas, rises in temperature as the more volatile components are boiled away. It will be seen that the vapour from each evaporator joins the two-phase mixture from the expansion valve which feeds the next higher evaporator.

It was shown in section 6.23 that when the refrigerant is a solution it is beneficial to use a part of the refrigerating effect to subcool the liquid going to the expansion valve. Adopting this practice the arrangement of a stage is then as shown in Fig. 8.8.4 which shows the liquid from the separator being subcooled in counterflow with the boiling refrigerant. If in addition the flash-gas and the boil-off from the tank are returned through the exchangers there are five streams in all, the top three in Fig. 8.8.4 being cooled and the bottom two providing refrigeration.

The calculation of mixed refrigerant cycles will not be treated here: in fact the design and optimisation of the process can scarcely be done without a computer. The outline of the procedure and flow sheets for several types of plant are given by Linnett & Smith (1969). The basic data required are those for the pure refrigerants discussed in section 3.18 together with the phase equilibrium data, i.e. the relations between the composition, pressure and temperature of the multi-component mixture, and a set of mixing rules for the properties specific volume or density in the equation of state. A more detailed description of mixed refrigerant processes is given by Thorogood (1972).

Storage

Liquefied natural gas (L.N.G.) is stored in larger amounts than any other cryo-fluid, tanks holding 20 000 t being more or less standard and several holding 30 000 t being in use. The tanks are constructed in or above the ground.

In-the-ground tanks have been made simply by excavating a hole and covering it with an insulated domed roof. Refrigerated probes are sunk around the periphery during excavation, as described by Miller & Brown (1968), but thereafter the ground is kept frozen by the L.N.G. Experience with frozen-earth tanks has been disappointing, and the above-ground tank is now mostly preferred.

Typically an above-ground tank, Fig. 8.8.5, consists of two shells, the inner one about 50 m diameter by 30 m high for a tank of 20 000 t, made of material such as aluminium, 9% nickel steel or stainless steel which is able to withstand the low temperature without embrittlement. The outer shell does not go much below ambient temperature and may be made of carbon steel. The insulation consists of a blanket of glass fibre next to the inner shell surrounded by a loose fill of expanded perlite. Perlite is a volcanic rock which on being heated to 820 °C expands owing to its trapped internal water to give light particles. The bulk density is about 48–64 kg/m^3 and the thermal conductivity is 0.036 W/m K at the mean temperature between the L.N.G. and the ambient atmosphere. The purpose of the glass-fibre blanket is to provide some resilience and to allow for thermal contraction of the inner shell, about 40 mm on a radius of 25 m for 9% nickel steel between ambient and L.N.G. temperature, which would leave voids and allow the perlite to settle. The insulation under the base has to be load bearing and is usually of foam (cellular) glass blocks covered with paving slabs. When it is

Fig. 8.8.5. Diagrammatic representation of the construction of a typical liquefied natural gas tank with a capacity of 20 000 t. Diameter of inner tank 50 m, internal height to deck 30 m. 1 – inner tank of 9% nickel steel; 2 – suspended deck; 3 – loose-fill expanded perlite insulation; 4 – anti-compaction blanket of glass fibre; 5 – seal; 6 – domed roof and roof framing; 7 – relief valve showing typical internal pipe and connection to the deck; 8 – pump handling system, including lifting gear and stowage for electric cables etc; 9 – outer tank of carbon steel; 10 – concrete bund wall capable of taking the contents of the tank in the event of rupture; 11 – in-tank discharge pumps, located in the stand pipe, which can be lifted out with the tank full; 12 – load-bearing base insulation; 13 – perlite concrete ringwall; 14 – base heaters. (With acknowledgement to Motherwell Bridge Engineering Ltd.)

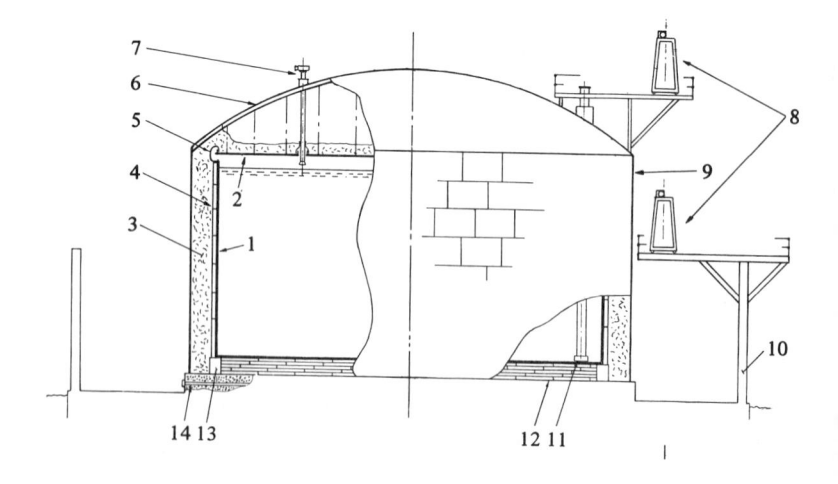

8.8 Liquefied natural gas

necessary to drive piles into the ground to support the tank it is possible to leave an air space below the tank so that freezing of the ground cannot occur. Otherwise a heated layer is provided as under cold stores.

The insulation of above-ground tanks is purged of air by dry nitrogen or by the natural gas itself in order to prevent the ingress of water vapour and subsequent freezing on the inner shell. The pressure in the interspace is maintained above atmospheric pressure. The upthrust of this pressure on the domed roof of the outer shell has to be taken by anchor straps to the ground concrete.

The rate of boil-off from a large tank caused by heat through the insulation is of the order 0.1% per day. In most cases this gas is reliquefied and returned to the tank. (A reliquefaction using a helium gas cycle was the subject of example 7.4.1.) The reasons for reliquefaction are that the L.N.G. is not pure methane but contains the less volatile ethane, etc., which would tend to be left in the tank and gradually change the composition of the liquid; and that tanks used for peak shaving, which are normally filled during the summer, would need replenishing before the winter.

Furthermore, the boil-off can exceed 0.1% on days when there is a rapid change of barometric pressure. Although the lower part of the shell is stressed to take the hydrostatic pressure, about 1 bar in a tank holding 20 000 t, the upper part of the shell and particularly the flat deck are stressed to take a maximum pressure difference of about 200 mbar only, with a normal operating value of about 55 mbar. Consequently a change of barometric pressure has to be compensated either by gasifying some of the liquid to bring the temperature and pressure down, or in the opposite direction by allowing nitrogen into the tank. The amount involved is shown in the following example.

Example 8.8.1

A tank containing 5000 t of methane is in equilibrium when the barometric pressure is 1030 mbar and the tank pressure is 50 mbar above this. Estimate the mass of methane which has to vaporise to maintain equilibrium when the barometric pressure falls to 1000 mbar, assuming that part of the fall can be accommodated by allowing the tank pressure to rise to 60 mbar above atmospheric pressure.

Initially the absolute pressure is 1080 mbar and finally it is 1060 mbar. The change of temperature corresponding to this small change of vapour pressure will be determined using the Clapeyron equation (eq. (3.18.32)):

$$\frac{dp_s}{dT} = \frac{h_{fg}}{T(v_g - v_f)}$$

Cryogenic engineering

For methane:

$T_n = 112 \text{ K}$

$h_{fg} = 510 \text{ kJ/kg}$

$v_g = 0.544 \text{ m}^3/\text{kg}$

Neglecting v_f the specific volume of the liquid:

$$\frac{dp_s}{dT} = \frac{510(10^3)}{112(0.544)} \text{ N/m}^2 \text{ K} = 84 \text{ mbar/K}$$

A change of vapour pressure of 20 mbar therefore requires a change of temperature of:

$$(20/84) \text{ K} = 0.238 \text{ K}$$

to maintain equilibrium.

The specific heat capacity of liquid methane is 3.5 kJ/kg K and the mass of vapour to be boiled off to reduce the tank temperature by 0.238 K is approximately:

$$\frac{5000(0.238)(3.5)}{510} \text{ t} = 8.17 \text{ t}$$

Excessive vaporisation can also take place owing to a phenomenon called 'roll-over' in which an unstable stratification of liquid reverts to a stable condition. It is accompanied by a very rapid rise in tank pressure which cannot normally be accommodated by the re-liquefaction plant and under these extreme conditions some of the gas has to be vented into the mains.

Marine Transport of L.N.G.

The first generation of L.N.G. tankers had capacities of the order 25 000–50 000 m³, but the new ones in service and under construction are of about 125 000 m³. Because of the low density of L.N.G., about 420 kg/m³, and the need for separate ballast water spaces, these are large ships comparable in size with crude oil carriers of about 145 000 t deadweight but with a much greater freeboard.

Two main methods of containing the L.N.G. are used:

(i) The independent tank system in which the insulated tank is self supporting and independent of the ship's structure except at support points. The tanks may be prismatic to fit the hull, or spherical, in which case they can be constructed completely on shore and lowered into position.

(ii) The integrated tank, also called the membrane tank, which receives support from the hull through the insulation.

8.8 Liquefied natural gas

Whichever type of tank is used, a basic requirement is that the containment arrangements must prevent any part of the ship's hull from being cooled below the temperature for which it is designed. With normal steel there is the possibility of low-temperature embrittlement at or below 0 °C, depending on the grade of steel. This absolutely demands that L.N.G. must not come into direct contact with the hull. Heating may have to be provided at heat bridges in the insulation.

The primary barrier which is in direct contact with L.N.G. has to be made of one of the limited number of materials suitable for use at 112 K. These are: 5083-0 aluminium alloy, 9% nickel steel, stainless steel (18–20% chromium, 8–10.5% nickel) or Invar (36% nickel steel). Invar, with its low thermal expansion of about $\frac{1}{4}$ that of stainless steel and $\frac{1}{5}$ that of aluminium between 112 K and 273 K has an advantage in this respect, but it is relatively expensive. It is used in one membrane design. The secondary barrier, which has to be capable of containing the L.N.G. in the event of failure of the primary barrier, is a second Invar membrane at the mid-depth of the insulation.

A stainless-steel membrane, 1.2 mm thick, is used in the design shown in Fig. 8.8.6. The membrane has 'waffle' corrugations to provide for the

Fig. 8.8.6. The Technigaz Mark I containment system for L.N.G., employing a stainless steel 'waffle' membrane, balsa wood sandwich panel insulation and a plywood secondary barrier. This has now been superseded by an improved Mark III version incorporating the same type of membrane. (With acknowledgement to Technigaz S.A.)

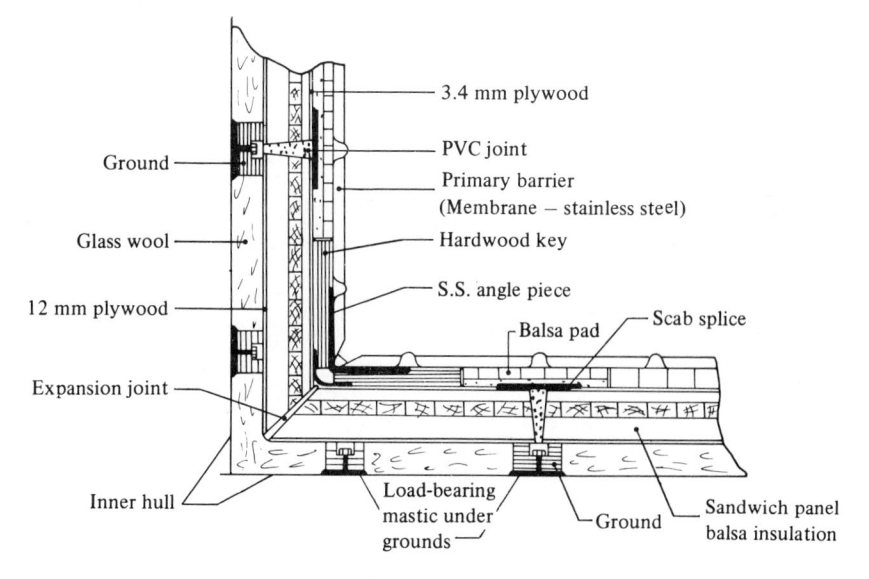

expansion and contraction which take place with temperature change, and also to provide for the ship's structural deflections. The secondary barrier in this design is of balsa panels faced with plywood.

A variety of insulating materials have been used. The design with the Invar membrane uses perlite-filled plywood boxes. The design shown in Fig. 8.8.6 uses balsa wood in the form of sandwich panels with glass fibre between the grounds which are bolted to the hull. Spaces surrounding tanks are purged with nitrogen which is monitored for hydrocarbons.

The boil-off during a voyage may be 2% of the cargo: it is normally burnt in the ship's boilers or engines.

Further details on the marine transport of L.N.G. are given by Wilson (1974), Ffooks (1979) and Lom (1974), who also treats the other aspects of L.N.G.

Appendix

A1 Properties of saturated ammonia liquid and vapour.

t_s °C	p bar	v_f m³/kg	v_g m³/kg	h_f kJ/kg	h_g kJ/kg	s_f kJ/kg K	s_g kJ/kg K
-60	.2190	.001401	4.703	-88.4	1354.7	-.3962	6.3755
-59	.2339	.001403	4.424	-84.0	1356.4	-.3756	6.3522
-58	.2495	.001406	4.164	-79.6	1358.2	-.3551	6.3292
-57	.2660	.001408	3.922	-75.2	1360.0	-.3347	6.3065
-56	.2834	.001410	3.697	-70.8	1361.7	-.3143	6.2840
-55	.3017	.001413	3.487	-66.4	1363.5	-.2941	6.2618
-54	.3210	.001415	3.290	-62.0	1365.2	-.2740	6.2399
-53	.3413	.001417	3.107	-57.6	1366.9	-.2539	6.2182
-52	.3627	.001420	2.936	-53.2	1368.7	-.2339	6.1967
-51	.3852	.001422	2.775	-48.7	1370.4	-.2140	6.1756
-50	.4088	.001424	2.625	-44.3	1372.1	-.1942	6.1546
-49	.4336	.001427	2.485	-39.9	1373.8	-.1744	6.1339
-48	.4596	.001429	2.353	-35.5	1375.5	-.1547	6.1134
-47	.4869	.001432	2.230	-31.1	1377.2	-.1351	6.0932
-46	.5155	.001434	2.114	-26.6	1378.8	-.1156	6.0732
-45	.5455	.001437	2.005	-22.2	1380.5	-.0962	6.0534
-44	.5769	.001439	1.903	-17.8	1382.2	-.0768	6.0338
-43	.6098	.001442	1.807	-13.3	1383.8	-.0575	6.0144
-42	.6442	.001444	1.717	-8.9	1385.5	-.0383	5.9952
-41	.6801	.001447	1.632	-4.4	1387.1	-.0191	5.9763
-40	.7177	.001449	1.552	.0	1388.7	.0000	5.9575
-39	.7570	.001452	1.477	4.5	1390.3	.0190	5.9389
-38	.7980	.001454	1.405	8.9	1391.9	.0380	5.9206
-37	.8408	.001457	1.339	13.4	1393.5	.0569	5.9024
-36	.8854	.001460	1.276	17.8	1395.1	.0757	5.8844
-35	.9320	.001462	1.216	22.3	1396.6	.0944	5.8666
-34	.9805	.001465	1.160	26.7	1398.2	.1131	5.8490
-33	1.031	.001468	1.105	31.2	1399.7	.1317	5.8315
-32	1.084	.001470	1.056	35.7	1401.3	.1503	5.8142
-31	1.138	.001473	1.009	40.2	1402.8	.1688	5.7971
-30	1.195	.001476	.9635	44.7	1404.3	.1872	5.7801
-29	1.255	.001478	.9209	49.1	1405.8	.2056	5.7634
-28	1.316	.001481	.8805	53.6	1407.3	.2239	5.7467
-27	1.380	.001484	.8422	58.1	1408.7	.2422	5.7303
-26	1.447	.001487	.8059	62.6	1410.2	.2604	5.7139
-25	1.516	.001489	.7715	67.1	1411.6	.2785	5.6978
-24	1.588	.001492	.7388	71.6	1413.1	.2966	5.6818
-23	1.662	.001495	.7078	76.1	1414.5	.3146	5.6659
-22	1.739	.001498	.6783	80.6	1415.9	.3326	5.6502
-21	1.819	.001501	.6503	85.2	1417.3	.3505	5.6346
-20	1.902	.001504	.6237	89.7	1418.7	.3683	5.6191
-19	1.988	.001507	.5984	94.2	1420.0	.3861	5.6038
-18	2.077	.001510	.5743	98.7	1421.4	.4038	5.5886
-17	2.169	.001513	.5514	103.3	1422.7	.4215	5.5736
-16	2.264	.001515	.5296	107.8	1424.0	.4391	5.5586
-15	2.363	.001518	.5088	112.3	1425.3	.4567	5.5438
-14	2.465	.001521	.4889	116.9	1426.6	.4742	5.5292
-13	2.571	.001525	.4701	121.4	1427.9	.4917	5.5146
-12	2.680	.001528	.4520	126.0	1429.2	.5091	5.5002
-11	2.792	.001531	.4349	130.6	1430.4	.5265	5.4859
-10	2.908	.001534	.4185	135.1	1431.6	.5438	5.4717
-9	3.029	.001537	.4028	139.7	1432.9	.5611	5.4576
-8	3.153	.001540	.3878	144.3	1434.1	.5783	5.4436
-7	3.280	.001543	.3735	148.8	1435.2	.5955	5.4297
-6	3.412	.001546	.3599	153.4	1436.4	.6126	5.4160
-5	3.549	.001550	.3468	158.0	1437.6	.6297	5.4023

A1 Properties of saturated ammonia liquid and vapour. (cont.)

t_s °C	p bar	v_f m³/kg	v_g m³/kg	h_f kJ/kg	h_g kJ/kg	s_f kJ/kg K	s_g kJ/kg K
−5	3.549	.001550	.3468	158.0	1437.6	.6297	5.4023
−4	3.689	.001553	.3343	162.6	1438.7	.6467	5.3888
−3	3.834	.001556	.3224	167.2	1439.8	.6637	5.3753
−2	3.983	.001559	.3109	171.8	1440.9	.6806	5.3620
−1	4.136	.001563	.3000	176.4	1442.0	.6975	5.3487
0	4.294	.001566	.2895	181.1	1443.1	.7143	5.3356
1	4.457	.001569	.2795	185.7	1444.2	.7312	5.3225
2	4.625	.001573	.2698	190.3	1445.2	.7479	5.3096
3	4.797	.001576	.2606	194.9	1446.3	.7646	5.2967
4	4.975	.001580	.2517	199.6	1447.3	.7813	5.2839
5	5.157	.001583	.2433	204.2	1448.3	.7980	5.2712
6	5.345	.001587	.2351	208.9	1449.2	.8146	5.2587
7	5.538	.001590	.2273	213.6	1450.2	.8311	5.2461
8	5.736	.001594	.2198	218.2	1451.1	.8477	5.2337
9	5.940	.001597	.2126	222.9	1452.1	.8641	5.2214
10	6.149	.001601	.2056	227.6	1453.0	.8806	5.2091
11	6.364	.001604	.1990	232.3	1453.9	.8970	5.1969
12	6.585	.001608	.1926	237.0	1454.8	.9134	5.1849
13	6.812	.001612	.1864	241.7	1455.6	.9297	5.1728
14	7.044	.001616	.1805	246.4	1456.5	.9460	5.1609
15	7.283	.001619	.1748	251.1	1457.3	.9623	5.1490
16	7.528	.001623	.1693	255.8	1458.1	.9785	5.1373
17	7.779	.001627	.1641	260.6	1458.9	.9947	5.1255
18	8.037	.001631	.1590	265.3	1459.7	1.0109	5.1139
19	8.301	.001635	.1541	270.0	1460.4	1.0270	5.1023
20	8.571	.001639	.1494	274.8	1461.2	1.0432	5.0908
21	8.849	.001643	.1448	279.6	1461.9	1.0592	5.0794
22	9.133	.001647	.1405	284.3	1462.6	1.0753	5.0680
23	9.424	.001651	.1363	289.1	1463.3	1.0913	5.0567
24	9.722	.001655	.1322	293.9	1464.0	1.1073	5.0455
25	10.03	.001659	.1283	298.7	1464.6	1.1232	5.0343
26	10.34	.001663	.1245	303.5	1465.2	1.1391	5.0232
27	10.66	.001667	.1208	308.3	1465.9	1.1550	5.0121
28	10.99	.001671	.1173	313.1	1466.4	1.1708	5.0011
29	11.32	.001676	.1139	318.0	1467.0	1.1867	4.9902
30	11.66	.001680	.1106	322.8	1467.6	1.2025	4.9793
31	12.02	.001684	.1075	327.7	1468.1	1.2182	4.9685
32	12.37	.001689	.1044	332.5	1468.6	1.2340	4.9577
33	12.74	.001693	.1014	337.4	1469.1	1.2497	4.9469
34	13.12	.001698	.0986	342.2	1469.6	1.2653	4.9362
35	13.50	.001702	.0958	347.1	1470.0	1.2810	4.9256
36	13.89	.001707	.0931	352.0	1470.4	1.2966	4.9149
37	14.29	.001711	.0905	356.9	1470.8	1.3122	4.9044
38	14.70	.001716	.0880	361.8	1471.2	1.3277	4.8938
39	15.12	.001721	.0856	366.7	1471.5	1.3433	4.8833
40	15.54	.001726	.0833	371.6	1471.9	1.3588	4.8728
41	15.98	.001731	.0810	376.6	1472.2	1.3742	4.8623
42	16.42	.001735	.0788	381.5	1472.4	1.3897	4.8519
43	16.88	.001740	.0767	386.5	1472.7	1.4052	4.8414
44	17.34	.001745	.0746	391.4	1472.9	1.4206	4.8310
45	17.81	.001750	.0726	396.4	1473.0	1.4360	4.8206
46	18.30	.001756	.0707	401.4	1473.2	1.4514	4.8102
47	18.79	.001761	.0688	406.4	1473.3	1.4668	4.7998
48	19.29	.001766	.0669	411.4	1473.3	1.4822	4.7893
49	19.80	.001771	.0652	416.5	1473.4	1.4977	4.7789
50	20.33	.001777	.0635	421.6	1473.4	1.5131	4.7684

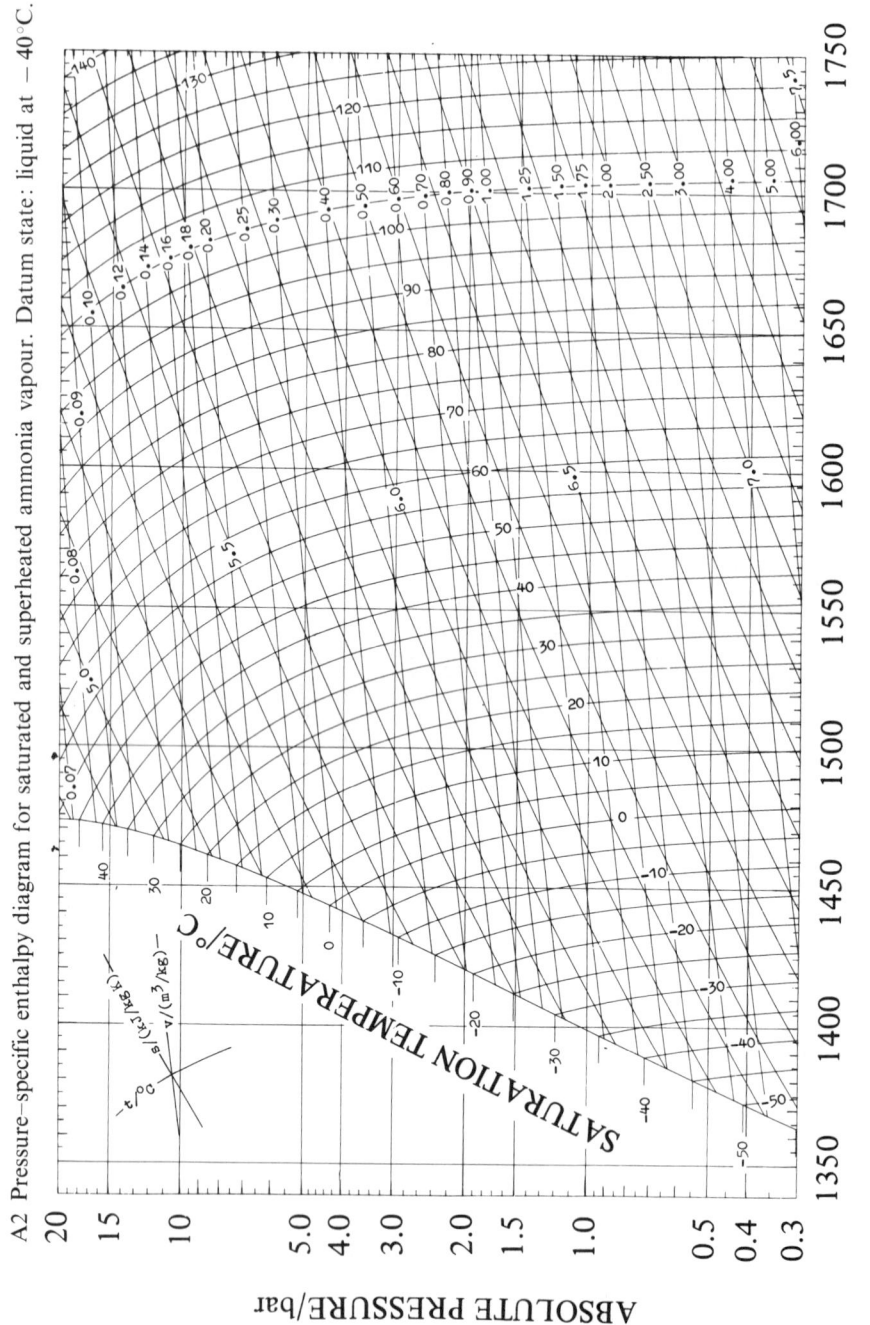

A2 Pressure–specific enthalpy diagram for saturated and superheated ammonia vapour. Datum state: liquid at −40 °C.

Appendix 631

A3 Pressure–specific enthalpy diagram for R12. Reproduced with permission of 'Freon®' Products Division of Du Pont.

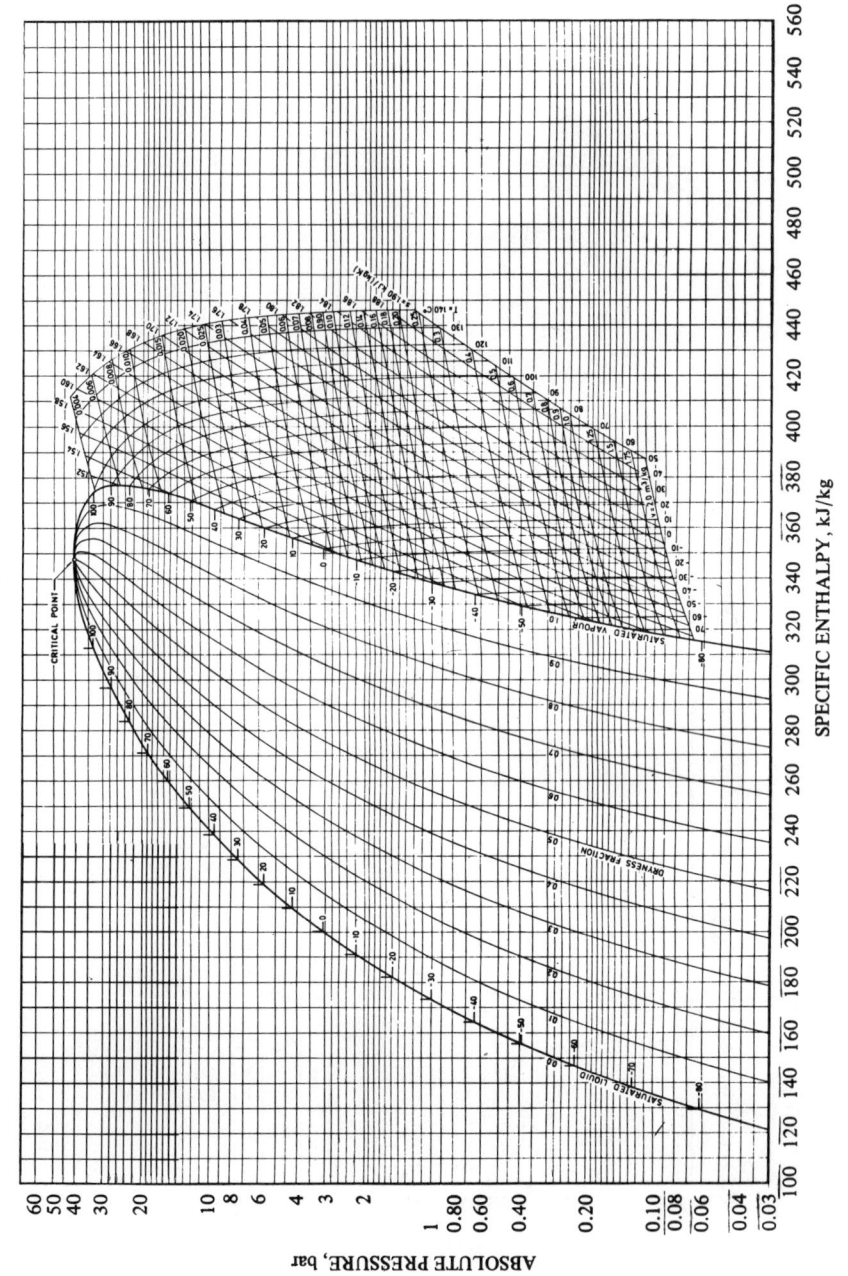

A4 Pressure–specific enthalpy diagram for R22. Reproduced with permission of 'Freon^R' Products Division of Du Pont.

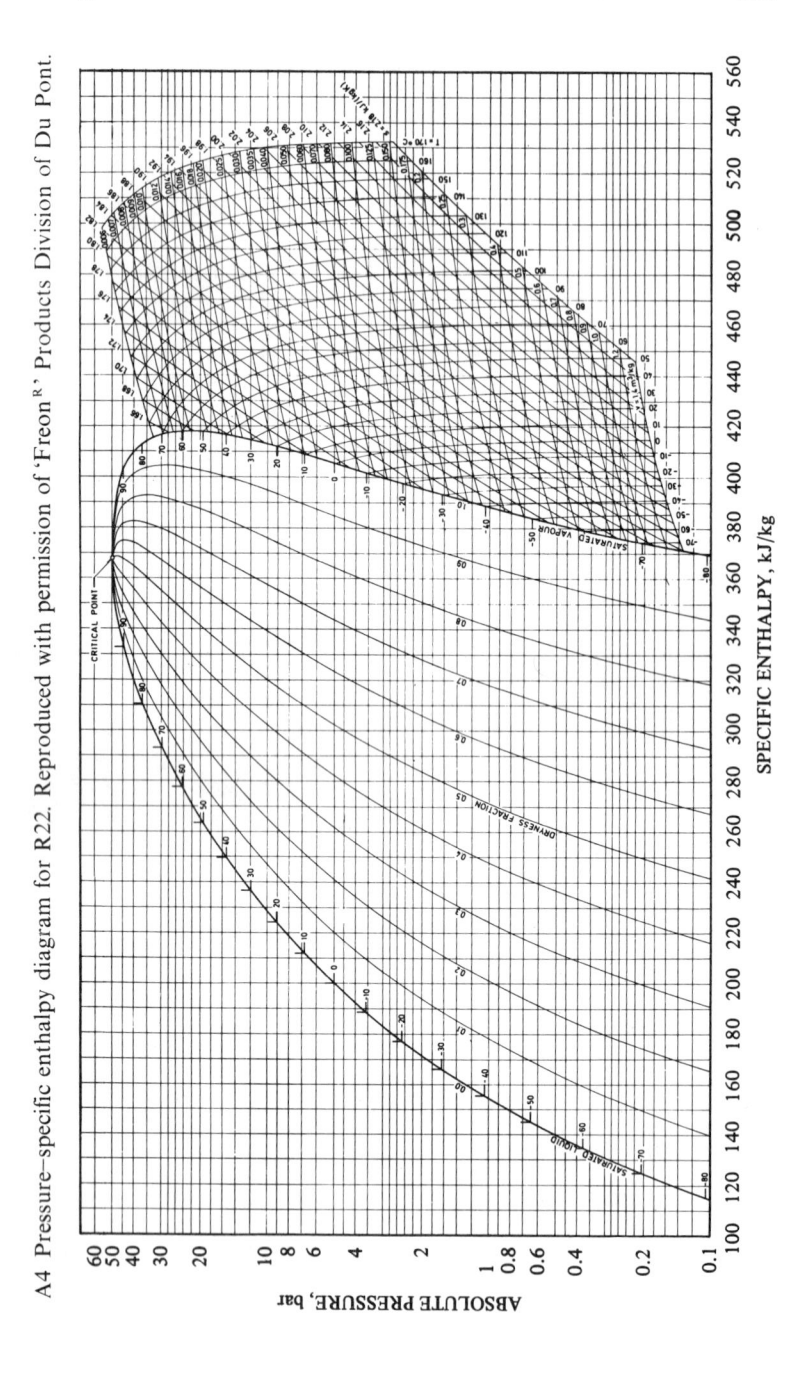

Appendix 633

A5 Pressure–specific enthalpy diagram for R502. Reproduced with permission of 'Freon®' Products Division of Du Pont.

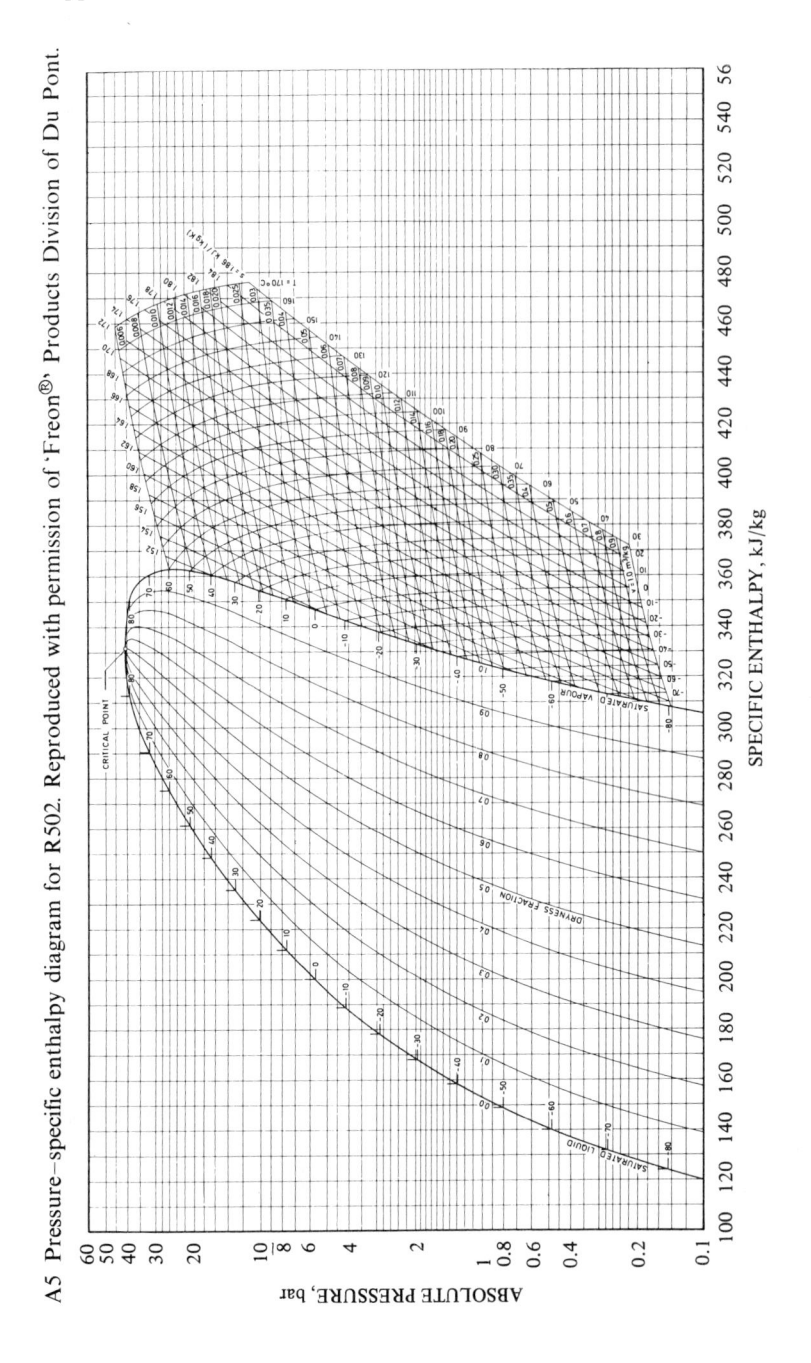

References

Chapter 1

Anderson, O. E., Jr, 1953. *Refrigeration in America: a History of a New Technology and its Impact.* Princeton University Press.

Beazley, E., 1977. Technology of beautiful simplicity. Iranian ice-houses. *Country Life*, **162**, 1229–31.

Bramwell, Sir F., 1882. *J. Roy. Soc. Arts*, **61**, 76–7. (See also contribution by Cowper in the discussion on Coleman's paper.)

Carré, F., 1860. *C.R. Acad. Sci.* **51**, 1023.

Coleman, J. J., 1882. Air refrigerating machinery and its applications. *Proc. Inst. Civ. Eng.* **68**, 146–215. (The date of the arrival of the 'Strathleven' is wrongly given as 1879.)

Cummings, R. O., 1949. *The American Ice Harvests: a Historical Study in Technology, 1800–1918.* University of California Press, Berkeley.

Forster, L. L., 1954. Steam jet refrigeration. *Proc. Inst. Refrig.* **50** (1953–4), 119–152.

Goldsmid, H. J., 1964. *Thermoelectric Refrigeration.* Heywood, London.

Harrison, James, 1856. British Patent 747. Producing cold by the evaporation of volatile liquids in vacuo, the condensation of their vapours by pressure, and the continued re-evaporation and re-condensing of the same materials.

Harrison, James, 1857. British Patent 2362. Improvements in apparatus for producing cold by the evaporation of volatile liquids in vacuo.

Hopkinson, J., 1882. Ice making and refrigeration. *J. Roy. Soc. Arts*, **31**, 19–27.

Ioffe, A. F., 1957. *Semiconductor Elements and Thermoelectric Cooling.* Infosearch, London.

Keenan, J., 1941. *Thermodynamics.* Wiley, New York.

Kirk, A. C., 1874. On the mechanical production of cold. *Proc. Inst. Civ. Eng.* **37**, (1873–4) part 1, 244–315.

Lightfoot, T. B., 1881. Machines for producing cold air. *Proc. Inst. Mech. Eng.* 105–32.

Lightfoot, T. B., 1886. Refrigerating and ice-making machinery and appliances. *Proc. Inst. Mech. Eng.* 201.

Linge, K., 1966. In *Handbuch der Kältetechnik*, ed. R. Plank, vol. 5, p. 81. Springer-Verlag, Berlin.

Lorenz, H., 1894. Beitrage zur Beurteilung von Kühlmaschinen. *Z.V.D.I.* **38**, 62.
Maiuri, G., 1932. The Maiuri cold multiplier. *Cold Storage and Produce Review* **35**, 270–1.
Mitford, N., 1966. *The Sun King*. H. Hamilton, London.
Nolcken, W. G., 1930. British Patent 359064. Improved methods and means for refrigeration.
Oldham, B. C., 1947. Evolution of machine and plant design. *Proc. Inst. Refrig.* **43**, (1946–7), 59–82.
Perkins, Jacob, 1834. British Patent 6662. Improvements in the apparatus and means for producing ice, and cooling fluids.
Thomson, W., 1852. On the economy of the heating and cooling of buildings by means of currents of air. *Proc. Phil. Soc. Glasgow*, **3**, 269.
Vahl, L., 1966. Dampfstrahl-Kältemaschinen. In *Handbuch der Kältetechnik*, ed. R. Plank. vol. 5, pp. 393–432. Springer-Verlag, Berlin.

Chapter 2

Clarke, R. J., Hodge, J. M., Hundy, G. F. & Zimmern, B., 1976. A new generation of screw compressors for refrigeration. *Proc. Inst. Refrig.* **72** (1975–76), 67–75.
Griffiths, E. A., 1936. Direct expansion air cooler for cold stores. *Engineering*, **141**, 331.
Hundy, G. F., 1978. Part load operation and testing of the single screw compressor with refrigerant vapours. In *Design and Operation of Industrial Compressors*, vol. 1, pp. 75–80. Institution of Mechanical Engineers.
Laing, P. D., 1969. The place of the screw compressor in refrigeration. In *An Engineering Review of Refrigeration Technology and Equipment*, pp. 91–105. Institution of Mechanical Engineers.
Laing, P. D. & Perry, E. J., 1964. The development of oil-injected screw compressors for refrigeration. *Proc. Inst. Refrig.* **60** (1963–4). 95–125.
Lorentzen, G., 1963. Conditions of cavitation in liquid pumps for refrigerant circulation. *Proc. 11th Int. Cong. Refrig. Munich*, vol. 1, pp. 497–511.
Lorentzen, G., 1976. The design of refrigerant recirculation systems. *Proc. Inst. Refrig.* **72** (1975–6), 58–66.
Lundberg, A., 1968. Die Eigenschaften des leistungsgeregelten Schraubenkompressors mit Öleinspritzung. *Kältetechnik–Klimatisierung*, **20**, 102–7.
Lundberg, A., 1975. A new concept of refrigeration screw compressors. *Proc. Inst. Refrig.* **71** (1974–5), 27–35.
Lundvik, B., 1967. Versatile applications of the screw compressor. *Proc. Inst. Refrig.* **63** (1966–7), 49–59.
Lysholm, A. J. R., 1943. New rotary compressor. *Proc. Inst. Mech. Eng.* **150**, 11–16, + plates; **151** (1944), 179–184.
Murdoch, A., 1957. The detection of refrigerant leaks. *Proc. Inst. Refrig.* **53** (1956–7), 155–69.
Nolcken, W. G., 1948. Halogen refrigerant leak detection equipment. *Modern Refrigeration*, **51**, 88–9, 100, 118–19, 148–9, 161, 174–5, 188.
Perry, E. J. & Laing, P. D., 1961. Positive displacement rotary compressors as applied to refrigeration. *Proc. Inst. Refrig.* **57** (1960–1), 25–46.
White, W. C., 1950. Positive-ion emission, a neglected phenomenon. *Proc. Inst. Rad. Eng.* **38**, 852–8.

Chapter 3

ASHRAE Handbook: Systems 1980. Chapter 32, Lubricants in refrigerant systems. American Society of Heating, Refrigerating and Air Conditioning Engineers Inc., New York.

Bambach, G., 1955. The behaviour of mineral oil–F12 mixtures in refrigerating machines. *Kältetechnik*, **7**, 187.

Barho, W., 1965. Die Molwärme der Fluor-Chlor Derivate des Methans im Zustand idealer Gase. *Kältetechnik*, **17**, 219–222.

Eiseman, B. J. Jr, 1952. Pressure–volume–temperature properties of the Freon compounds. *Ref. Eng.* **60**, 496–503.

Ewing, A., 1914. Report of the Refrigeration Research Committee. Institution of Mechanical Engineers.

Gatecliff, G. W. & Lady, E. R., 1971. Mathematical representation of the thermodynamic properties of Refrigerants 12 and 22 in the superheat region. *Proc. 13th Int. Cong. Refrig., Washington*, vol. 2, pp. 481–7.

Gosney, W. B., 1967. On the maximum coefficient of performance of a refrigerant. *Proc. 12th Int. Cong. Refrig., Madrid*, vol. 2, pp. 449–54.

Grindley, J. H., 1912. A contribution to the theory of refrigerating machines. *Proc. Inst. Mech. Eng.* no. 4, 1033–53.

Haar, L., 1968. Thermodynamic properties of ammonia as an ideal gas. *National Bureau of Standards Reference Data Series, NBS–19*.

Haywood, R. W., 1972. *Thermodynamic Tables in SI (Metric) Units*, 2nd edn. Cambridge University Press.

Inokuty, H., 1928. Zeichnerisches Verfahren zum Auffinden des günstigsten Kondesatordruckes in Kohlensäure Kältemaschinen. *Z. ges. Kälte-Industrie*, h 9.

Jaeger, H.-P. & Löffler, H. J., 1970. Thermodynamische Eigenschaften von Öl-Kältemittel-Gemischen. *Kältetechnik-Klimatisierung*, **22**, 246–56.

Linge, K., 1956. Der Einfluss des Ansaugezustandes auf die volumetrische und die spezifische Kälteleistung. *Kältetechnik*, **8**, 75–9.

Löffler, H. J., 1970. Zur Berechnung des Dampfdruckes von Öl-Kältemittel-Gemischen. *Kältetechnik-Klimatisierung*, **22**, 242–5.

Macintire, H. J., 1937. *Refrigeration Engineering*, Wiley, New York.

Mayhew, Y. R. & Rogers, G. F. C., 1980. *Thermodynamic and Transport Properties of Fluids*, 3rd edn. Blackwell, Oxford.

Schmidt, Ernst, *1969, Properties of Water and Steam in SI Units*. Springer-Verlag, Berlin.

Tsidzik, V. E., Barmin, V. P. & Veinberg, V. S., 1946. *Refrigerating Machines and Apparatus*, Moscow.

von Cube, H. L., Benke, K. & von Sambeck, J. A. A., 1958. Die Anwendbarkeit des i-ξ Diagrammes für Öl–Kältemittelgemische zur Berechnung von Kälteprozessen. *Kältetechnik*, **10**, 209–16.

Watson, J. T. R., 1975. *Thermophysical Properties of Refrigerant 12*. National Engineering Laboratory, H.M.S.O. Edinburgh.

Chapter 4

Reciprocating compressors

Bannister, F. K., 1959. Induction ramming of small high-speed air compressor. *Proc. Inst. Mech. Eng.* **173**, 375–97.

Bendixen, I., 1959. The volumetric efficiency of refrigerating compressors. *Proc. 10th Int. Congr. Refrig., Copenhagen*, vol. 2, pp. 190–8.

Brown, J. & Kennedy, W. K., 1971. Cyclical heat transfer in the cylinder of a refrigeration compressor. *Proc. 13th Int. Congr. Refrig., Washington*, vol. 2, pp. 545–9.

Brown, J. & Pearson, S. F., 1963. Piston leakage in reciprocating compressors. *J. Refrig.* **6**, 104–7.

Cheglikov, A. G., 1963. Effect of suction vapour superheat on the volumetric efficiency of a propane compressor. Abstract in *J. Refrig.* **6**, 112–14.

Costagliola, M., 1950. The theory of spring-loaded valves for reciprocating compressors. *J. Applied Mech.* **17**, 415–20.

Giffen, E. & Newley, E. F., 1940. Refrigerator performance: an investigation into volumetric efficiency. *Proc. Inst. Mech. Eng.* **143**, 227–36.

Gosney, W. B., 1953. An analysis of the factors affecting performance of small compressors. *Proc. Inst. Refrig.* **49** (1952–3), 185–216.

Gosney, W. B., 1972. Refrigerant 502: a study in power and performance. *Proc. Inst. Refrig.* **68** (1971–2), 58–66.

Kalitenko, F., 1960. The influence of the clearance between piston and cylinder on the performance of a compressor. (In Russian.) *Kholodilnaya Tekhnika*, no. 4, 25.

Löffler, G., 1941. Die rechnerische Erfassung des Ausnutzungsgrades (Liefergrades) von Ammoniak Kolbenverdichtern. *Z. ges. Kälte-Industrie*, **48**, 169–73.

Lorentzen, G., 1949. Leveringsgrad og virkningsgrad for kjølekompressorer, Fiskeridirektoratet, Bergen.

Lorentzen, G., 1951. The influence of speed on the volumetric efficiency of reciprocating compressors. *Proc. 8th Int. Congr. Refrig., London*, pp. 463–8.

Lorge, R. A., 1968. Technical aspects of refrigeration compressor performance. In *An Engineering Review of Refrigeration Technology and Equipment*, pp. 134–43, Institution of Mechanical Engineers.

MacLaren, J. F. T., Kerr, S. V. & Tramschek, A. B., 1975. Modelling of compressors and valves, *Proc. Inst. Refrig.* **71** (1974–5), 42–59.

Pearson, S. F., 1958. Losses in small refrigerating compressors. *Engineering*, **186**, 485–8.

Pierre, Bo., 1952. Eigenschaften moderner amerikaner Freonverdichter. *Kältetechnik*, **4**, 83–7.

Shipley, Th., 1910. Investigations as to the efficiency of ammonia compressors when running under dry and wet conditions. *Proc. 2nd Int. Congr. Refrig., Vienna*, p. 91.

Smith, E. C., 1935. The determination of mass flow in the vapour-compression refrigerator. *Proc. Inst. Mech. Eng.* **129**, 477–505.

Smith, H. J., 1961. Effect of piston/bore clearance in small refrigeration compressors. *Proc. Inst. Mech. Eng.* **175**, 991–1006.

Soumerai, H., 1965. Trends in compressor design, part 2. *Air Conditioning, Heating and Refrigeration News*, 22 November, 1965, 17.

Thomsen, E. G., 1951. Predicting maximum power in reciprocating compressors. *Refrig. Eng,* **59**, 269–71, 302–4.

Williams, T. J., 1959. Contribution to discussion on Bannister's paper, *Proc. Inst. Mech. Eng.* **173**, 392–3.

Wirth, G., 1933. Über den Einfluss der Wandtemperatur auf den Arbeitsprozess des Kältekolbenmaschinen. *Z. ges. Kälte-Industrie*, **40**, 138.

Turbo-compressors

Anderson, J. H., 1962. Refrigerants for centrifugal water-cooling systems. *J. Refrig.* **5**, 100–4.

Ferguson, T. B., 1963. *The Centrifugal Compressor Stage*. Butterworth, London.

Jassniker, K., 1962. Trends in the design of turbo-compressors for refrigeration plants. *Sulzer Tech. Rev.* no. 2, 17–24.

Schultz, J. M., 1962. The polytropic analysis of centrifugal compressors. *Trans. A.S.M.E.* **84**, series A (Journal of Engineering for Power, Jan. 1962) 69–82.

Stepanoff, A. J., 1955. Turboblowers. Wiley, New York.

Wiesner, F. J., 1960. Practical stage performance correlations for centrifugal compressors. *ASME* (*American Society of Mechanical Engineers*) Paper 60-Hyd-17 (1960).

Wiesner, F. J. & Caswell, H. E., 1959. How refrigerant properties affect impeller dimensions. *ASHRAE* (*American Society of Heating, Refrigerating and Air Conditioning Engineers. Inc.*) Journal **1**, no. 10 (Oct. 1959) 31–7, 104–6, 114.

Chapter 5

Baumann, K. & Blass, E., 1961. Beitrag zur Ermittlung des Optimalen Mitteldruckes bei zweistufigen Kaltdampf Verdichter-Kältemaschinen. *Kältetechnik*, **13**, 210–16.

Behringer, H., 1928. Berechnung des günstigsten Zwischendruckes bei Verbundkompression für NH_3-Kältemaschinen. *Z. ges. Kälte-Industrie*, **35**, 111–13.

Brier, H., 1914. Multiple effect compression as applied to CO_2 refrigerating machines. *Proceedings of the Cold Storage and Ice Association*, **11** (1913–14), 55–90. Plates 1–11.

Brier, H. & Brier, J. H., 1933. The possibilities of three-stage compression in one compound multiple effect compression cylinder for low temperatures, etc. *Proc. Brit. Assoc. Refrig.* **29**, no. 2, 5–26.

Cervenka, O. & Cvejn, M., 1967. Direct contact freeze-crystallisation as applied to the production of paraxylene. *Proc. 12th Int. Cong. Refrig. Madrid*, vol. **1**, pp. 425–31.

Chaikovsky, V. F. & Kuznetsov, A. P., 1963. The use of mixtures of refrigerants in compression refrigerating machines. *Khol. Tekh.* **1**, 9–11. (English abstract in *J. Refrig.* **6**, 66–7.)

Czaplinski, St., 1959. Über den optimalen Zwischendruck bei Kälteprozessen. *Allgemeine Wärmetechnik*, **9**, 93–6.

Dauser, A., 1941. Der Zwischendruck bei Verbundverdichtung. *Z. ges. Kälte-Industrie*, **48**, 77–80.

Din, F., 1956. (Editor) *Thermodynamic Functions of Gases*, vol. 1. Butterworths, London.

Gosney, W. B., 1967. Compound compression refrigeration cycles. *Proc. Inst. Refrig.* **63** (1966–7), 60–72.

Newitt, D. M., Pai, M. N., Kuloor, N. R. & Huggill, J. A. W., 1956. Carbon dioxide. In *Thermodynamic Functions of Gases*, vol. 1, ed. F. Din, pp. 102–34. Butterworths, London.

Serdakov, G. S., 1961. Determination of the optimum intermediate temperature in two-stage refrigerating machines. *Khol. Tekh.* **38**, no. 3, p. 25. (Abstract in *J. Refrig.* **5** (1962), 134.)

Soumerai, H., 1953. Simplified analysis of two-stage low-temperature refrigeration systems. *Refrig. Eng.* **61**, 746–50, 802.

Sparks, N. R., 1935. A thermodynamic analysis of the multiple effect compressor. *Refrig. Eng.* **30**, 149–51.

Stickney, A. B., 1932. The thermodynamics of CO_2 cycles. *Refrig. Eng.* **24**, 334–42.

Vahl, L., 1961. A novel process for the manufacture of dry ice in crystal-clear blocks. *Bull. Int. Inst. Refrig.*, Annexe 1961–3, 359.

Voorhees, G. T., 1905. Improvements relating to systems of fluid compression and to compressors thereof. British Patent 4448.

Vukalovich, M. P. & Altunin, V. V., 1968. *Thermophysical Properties of Carbon Dioxide.* Collet's (Publishers) Ltd., London.

Windhausen, F. Sr & Windhausen, F. Jr., 1901. *Improvements in Carbonic Anhydride Refrigerating Machines.* British Patent 9084.

Chapter 6

Altenkirch, E., 1913. Reversible Absorptionsmaschinen. *Z.f.d. gesamte Kälte-Industrie*, **20**, 1–9, 114–19, 150–161.

Altenkirch, E., 1914. Reversible Absorptionsmaschinen. *Z.f.d. gesamte Kälte-Industrie*, **21**, 7–14, 21–27.

Bäckstrom, Matts & Emblik, E., 1965. *Kältetechnik.* G. Braun, Karlsruhe.

Badylkes, I. C. & Danilov, R. L., 1966. *Absorption Refrigerating Machines* (in Russian). Moscow.

Bošnjaković, Fr., 1937. *Technische Thermodynamik*, vol. 2. 3rd edn. 1960. Steinkopff, Leipsig. Translated into English with supplementary material, but without the charts, by P. L. Blackshear. Holt, Rinehart & Winston, New York, 1965.

Dannies, J. H., 1951. *Die Absorptionskältemaschine.* Brücke-Verlag Kurt Schmersow. Hannover.

Jennings, R. H. & Shannon, F. P., 1938. The thermodynamics of absorption refrigeration. *Refrig. Eng.* **35**, 333–6.

Keesom, W. H., 1930. Etude sur la représentation graphique du processus de la rectification. *Communications from the Kamerlingh Onnes Laboratory of the University of Leiden*, supplement 72.

Kohloss, F. H. & Scott, G. L., 1951. Equilibrium properties of aqua-ammonia in chart form. *Refrig. Eng.* **58**, 970 and chart.

Lange, E. & Schwartz, E., 1928. Lösungs-und Verdünnungswärmen von Salzen von der äussersten Verdünnung bis zur Sättigung. *Z.f. Phys. Chem.*, **133**, 129–50.

Löwer, H., 1960. Thermodynamische und physikalische Eigenschaften der wässrigen Lithiumbromid Lösung. Dissertation, Technische Hochschule Karlsruhe.

Macriss, R. A., Eakin, B. E., Ellington, R. T. & Huebler, J., 1964. Physical and thermodynamic properties of ammonia water mixtures. Research Bulletin 34, Institute of Gas Technology, Chicago.

Merkel, Fr. & Bošnjakovic, Fr., 1929. *Diagramme und Tabellen zur Berechnung der Absorptionskältemaschinen.* Springer-Verlag, Berlin.

Mollier, H., 1909. Dampfdruck über wässrigen Ammoniaklösungen. Lösungswärme von Ammoniak und Wasser. *Mitteil. Über Forschungsarb. des VDI.* Heft 63/64, Berlin.

Niebergall. W., 1949. *Arbeitsstoffpaare für Absorptions-Kälteanlagen und Absorptions-Kühlschränke.* Verlag für Fachliteratur Rich. Markewitz, Mühlhausen.

Niebergall, W., 1959. *Sorptions-Kältemaschine.* In *Handbuch der Kältetechnik*, vol 7 ed. R. Plank. Springer, Berlin.

Pennington, W., 1955. How to find accurate vapor pressure of LiBr water solutions. *Refrig. Eng.* **63**, 57–61.

Perman, E. P., 1901. Vapour pressure of aqueous ammonia solution. Part I. *J. Chem. Soc.* **79**, 718–25.

Perman, E. P., 1903. Vapour pressure of aqueous ammonia solution. Part II. *J. Chem. Soc.* **83**, 1168–84.

Pierre, Bo., 1958. Jamviksdiagram och tabeller för ammoniak–vatten lösningar. *Kylteknisk Tidskrift*, no. 2, April 1958, 22–5, and Data Sheet 14.

Ruhemann, M., 1947. The ammonia absorption machine. *Trans. Inst. Chem. Eng.* **25**, 143–61.

Scatchard, G., Epstein, L. F., Warburton, J. & Cody, P. J., 1947. Thermodynamic properties – saturated liquid and vapor of ammonia–water mixtures. *Refrig. Eng.* **53**, 413–19, 446, 448, 450, 452.

Schulz, S., (1972). Die Berechnung und Optimierung von Absorptionskälte-maschinen-prozessen mit Hilfe von EDV-Anlagen. *Kältetechnik-Klimatisierung*, **24**, 181–8.

Tandberg, J. & Widell, N., 1937. Diagram för ammoniak–vatten lösningar. *Teknisk Tidskrift*, September 1937, 117–20.

Wucherer, J. 1932. Messungen von Druck, Temperatur und Zusammensetzung der flüssigen und dampf-förmigen Phase von Ammoniak–Wasser–Gemischen im Sättigungsgebiet. *Z.f.d. gesamte Kälte-Industrie*, **39**, 97–104, 136–40.

Zinner, K., 1934. Wärmetönung beim Mischen von Ammoniak und Wasser in Abhängigkeit von Zusammensetzung und Temperatur. *Z.f.d. gesamte Kälte-Industrie*, **41**, 21–9.

Chapter 7

Brown, P. M. T., 1971. LNG storage plant at Ambergate. *Cryogenics*, **11**, 82–4.

Fiedler, W. J., 1965. Air cycle refrigeration. In *Manual of Refrigeration Practice*. Technical Productions (London) Ltd.

Hilsch, R., 1946. Die Expansion von Gasen in Zentrifugalfeld als Kälteprozess. *Z.f. Naturforschung*, **1**, 208–14.

Kirk, A. C., 1873. On the mechanical production of cold. *Proc. Inst. Civ. Eng.* **37** (1873–4), part 1, 244–315.

Köhler, J. W. L. & Jonkers, C. O., 1954. I. Fundamentals of the gas refrigerating machine. II. Construction of a gas refrigerating machine. *Philips Tech. Rev.* **16**, no. 3 (Sept. 1954) 69–78. No. 4 (Oct. 1954) 105–15.

Martinovsky, V. S. & Dubinsky, M. G., 1964. Air turbo-refrigerating machines with additional cooling in the regenerative heat exchanger. *Kholodilnaya Tekhnika*, no. 6, 16–17. Abstract in *J. Refrig.* **8** (1965) 246–7

Martynov, A. V. & Brodyansky, V. M., 1964. Vortex tubes with external cooling. *Kholodilnaya Tekhnika*, no. 5, 46–51. Abstract in *J. Refrig.* **8** (1965), 354–5.

Parulekar, R. B., 1961. The short vortex tube. *J. Refrig.* **4**, 74–80.

Ranque, G. J., 1933. Expériences sur la détente giratoire avec production simultanée d'un échappement d'air chaud et d'un échappement d'air froid. *J. de Physique (et Le Radium)*, **7**, no. 4, 112.

Stirling, R. & J. 1827, B, P. 5456, *Air Engines*.

Walker, G., 1973. *Stirling-Cycle Machines*. Clarendon Press, Oxford.

References

Chapter 8

Appleton, A. D., 1977. Superconducting machines. *Proc. Inst. Refrig.* **73** (1976–7), 32–42.

Bošnjaković, Fr., 1937. *Technische Thermodynamic*, vol. 2. 3rd edn 1960. Steinkopff, Leipsig. Translated into English with supplementary material but without the charts, by P. L. Blackshear. Holt, Rinehart & Winston. New York. 1965.

Collins, S. C. & Cannaday, R. L., 1958. *Expansion Machines for Low Temperature Processes*. Oxford University Press.

Corbitt, R. W., 1948. Discussion on paper by Iliffe. (see Iliffe. 1948.)

Daglish, H. N., 1966. Applications of low temperatures in satellite communications. *Proc. Inst. Refrig.* **62** (1965–6), 25–31.

Darlington, M. E., 1969. The use of liquid nitrogen in the field of refrigeration. *Proc. Inst. Refrig.* **65** (1968–9), 59–71.

Din, F., 1960. The liquid–vapour equilibrium of the system nitrogen + oxygen at pressures up to 10 atm. *Trans. Faraday Soc.* **56**, 668–81.

Din, F. & Cockett, A. H., 1960. *Low Temperature Techniques*. Newnes, London.

Dusinberre, G. M., 1951. *General Discussion on Heat Transfer*, pp. 304–5. Institution of Mechanical Engineers.

Edwards, J. V., Evans, R. & Probert, S. D., 1971. Computation of transient temperatures in regenerators. *Int. J. Heat and Mass Transfer*, **14**, 1175–202.

Ffooks, R. C., 1979. *Natural Gas by Sea*. Gentry Books Ltd., London.

Foner, S. & Schwartz, B. B., 1974. *Superconducting Machines and Devices*. Plenum Press, New York.

Haselden, G. G., ed., 1971. *Cryogenic Fundamentals*. Academic Press, New York.

Haselden, G. G. & Klimek, L., 1958. An experimental study of mixed refrigerants for non-isothermal refrigeration. *Proc. Inst. Refrig.* **54** (1957–8), 129–54.

Hausen, H., 1929. Über der Theorie des Wärmeaustausches in Regeneratoren. *Z. Ang. Mathematik und Mechanik*. **9**, 173–200.

Hausen. H., 1957. Erzeugung sehr tiefer Temperaturen. *Handbuch der Kältetechnik*, vol. 8. Springer Verlag, Berlin.

Iliffe, C. E., 1948. Thermal analysis of the contra-flow regenerative heat exchanger. *Proc. Inst. Mech. Eng.* **159**, 363–72.

Jakob, M., 1957. *Heat Transfer*, vol. 2. Wiley, New York.

Johnson, J. E., 1952. *Regenerator Heat Exchangers for Gas Turbines*. A.R.C. Technical Report 2630, H.M.S.O.

Kays, W. M. & London, A. L., 1964. *Compact Heat Exchangers*, 2nd edn, McGraw Hill, New York.

Kleemenko, A. P., 1959. One flow cascade cycle. *Proc. 10th Int. Cong. Refrig.* Copenhagen, vol. 1, pp. 34–9.

Linnett, K. C. & Smith, K. C., 1969. The process design and optimisation of a mixed refrigerant cascade plant. *Proc. Int. Conf. on L.N.G.*, London, pp. 267–87. Institution of Mechanical Engineers.

Lom, W. L., 1974. *Liquefied Natural Gas*. Applied Science Publishers, London.

Mendelssohn, K., 1977. *The Quest for Absolute Zero*. 2nd edn. Taylor & Francis. London.

Miller, H. W. & Brown, T. P. G., 1968. Recent developments in ground freezing. *Proc. Inst. Refrig.* **64**, 20–9.

Rabes, M., 1930. Beitrag zur Theorie der Luftverflüssigung. *Z. ges. Kälte-Industrie*, **37**, 7–12, 26–9, 48–54.

Rose-Innes, A. C., 1964. *Low Temperature Techniques*, English Universities Press, London.

Ruhemann, M., 1949. *The Separation of Gases*. 2nd edn. Oxford University Press.

Saunders, O. A. & Smoleniec, S., 1948. Heat regenerators. *Proc. 7th Int. Cong. Applied Mechanics*, **3**, 91–5.

Thorogood, R. M., 1972. Mixed refrigerant processes for natural gas liquefaction. *Proc. Inst. Refrig.* **68**, 32–40.

Usyukin, I. P., 1965. *Plant and Machinery for the Separation of Air by Low temperature Methods*. Translation edited by M. Ruhemann, Pergamon Press, Oxford.

White, G. K., 1979. *Experimental Techniques in Low Temperature Physics*. 3rd edn. Oxford.

Willmott, A. J., 1964. Digital computer simulation of a thermal regenerator. *Int. J. Heat and Mass Transfer*, **7**, 1291–302.

Wilson, J. J., 1974. An introduction to the marine transportation of bulk LNG and the design of LNG carriers. *Cryogenics*, March 1974, 115–20.

Zemansky, M. W., 1968. *Heat and Thermodynamics*, 5th edn, McGraw-Hill. New York.

General Bibliography

Publications of the American Society of Heating, Refrigerating and Air-Conditioning Engineers Inc. Handbooks, one issued per year: 1977 *Fundamentals*: 1978 *Applications*: 1979 *Equipment*: 1980 *Systems. Transactions*, two volumes per year, containing the reports of the semi-annual meetings. Journal, *ASHRAE Journal*.

The Institute of Refrigeration, London. *Proceedings*, one volume per year.

Publications of the Institut International du Froid, Paris. Bulletin, six per year, containing abstracts of papers and occasional review articles. Annexes to the Bulletin, containing reports of meetings of Commissions, a variable number per year. Proceedings of the International Congresses, held every four years, published in the host country.

Periodicals: *Refrigeration and Air Conditioning*, London: The *International Journal of Refrigeration*, London: *La Revue Générale du Froid*, Paris: *La Revue Pratique du Froid*, Paris: *Kältetechnik-Klimatisierung*, Karlsruhe

Handbuch der Kältetechnik, 13 volumes, edited by R. Plank, Springer-Verlag, 1954–1969.

INDEX

The figures given in brackets after the names of chemical substances are the freezing or triple point temperature, the standard boiling point temperature and the critical temperature, in that order. A dash is inserted where the value is uncertain, unknown or irrelevant.

absolute zero of temperature
 defined 35
 unattainability of 517
absorbent, water as 14, 406–7
absorber 16, 406–8
 analysis of 415
 departure from equilibrium at 419, 447
accessible compressor 68–9
accumulator, see separator
acetylene, C_2H_2 (− 80.8/ − 84.0 sublimes/36 °C), compression of 79
acrolein (CH_2:CH. CHO) (− 86.9/52.5 °C/–) 123
Addison, Joseph 1
adiabatic demagnetisation 516, 517–18
adiabatic process
 defined 178
 see compression
adsorbent charged thermostatic expansion valve 111
aerosol cans 12
Ahrendts, J. 266
air (–/ − 194.3/ − 140.6 °C)
 examples on 318, 481, 615
 heat capacities and gas constant of 475
 as ideal gas 474
 inversion temperature of 537
air conditioning
 chilled water for 17, 20, 22, 106
 condenser pressure control for 95–6
 turbo-compressors for 85–6
air cycle refrigeration
 analysis of cycles 479–91
 at sub-atmospheric pressure 494–5
 development of 17
 open and closed cycles 18–19, 478
 principle of 17–18
 uses of 3
air, liquid 510–13
 composition of 574

example on separation of 587–90
 exergy of 528
 separation of oxygen from 572–3, 577–95
aircraft cooling 19, 495–7
algae, in cooling towers 93
Allen, Frank 19
Altenkirch, Edmund (1880–1953) 472, 640
Altunin, V. V. 389, 640
aluminium
 alloy for L. N. G. tanks 625
 in systems 125
 tube 134
American National Standard
 flared fittings 129–30
 pipe thread 131
 safety code, 166
American Society of Heating, Refrigerating and Air Conditioning Engineers, see ASHRAE
American Society of Refrigerating Engineers, defined Ton 28
amines, filming, for water treatment 93
ammonia, NH_3 (− 77.7/ − 33.3/132.5 °C) 9
 with centrifugal compressor 347
 coefficient of performance of 205, 207, 211
 examples on 185–9, 201–2, 217–18, 244–5, 247–8, 271, 278, 281, 285–6, 303–5, 306–7, 319–20, 323–4, 356–7, 359–60, 367–8, 371–2, 374
 flammability of 122–3
 high delivery temperature of 207, 212
 h–v diagram for 228
 isentropic index of 272
 incompatible with copper 125
 leaks of 123–4
 low miscibility with oil 126, 248
 p–h diagram for 630
 refrigerating effect of 203, 207

645

saturation properties of 628-9
as secondary refrigerant 23
staged expansion of 363
toxicity of 121-2
work of compression of 204, 207
for mixtures of ammonia with water see aqua-ammonia
amount of substance, defined 402
analyser 429
Anderson, J. H. (1726-96), his college 8
Andrews, Thomas (1813-85) 509-10
anode, sacrificial 88
Appleton, A. D. 524, 642
approach of temperatures, *see* temperature approach
aqua-ammonia
 examples on 422-4, 436-7, 444, 445-7, 451-2, 454, 466
 freezing points of 421
 introduced 14-16
 liquid, enthalpy of 433
 vapour, enthalpy of 437
 vapour from 424
 vapour pressure of 421
'Arcton', trade name 11
argon ($-189.3/-185.9/-122.3\,°C$)
 in atmosphere 574
 example on 533
 exergy of liquid 528
 in lamps and for welding 572
 separation of 533, 596
Arzew, L. N. G. plant at 619
ASHRAE (American Society of Heating, Refrigerating and Air Conditioning Engineers), publications cited 248, 439, 470, 637, 643
atmosphere
 composition of 574
 ozone in 12
atmospheric condenser 90-1
atmospheric pressure
 as reference pressure 114
 variations of 143
auto-cascade, *see* cascade systems
automatic control
 introduction of 13
 of refrigerating capacity 157-66
 see also expansion valves, valves.
'automatic' expansion valve 113-14
axial flow compressor, *see* compressors, axial flow
azeotrope 12

back pressure valves 147-9, 243
'backing-up', in condensers 88, 97, 121
Bäckström, Matts 438, 640
Badylkes, I. C. 470, 640
Baehr, H. D. 266
baffle plates 105
balance pipes
 with float control 117
 for sight glasses 140-1
'balance loading' 254
balanced seal 61-2
Bambach, G. 250, 637
Bannister, F. K. 294, 637
Barho, W. 257, 637
barium chloride, $BaCl_2$
 in eutectic mixture 25
Barmin, V. P. 238, 637
Beattie-Bridgman equation of state 258
Beazley, E. 1, 635
Behringer, H. 369, 639
Beilstein test 124
Bell, Henry (1848-1931) 18
bellows
 in control valves and thermostats 108, 110, 113, 148, 158
 in shaft seals 59, 61-2
Bendixen, I. 289, 638
Benedict-Webb-Rubin equation of state 258-9
Benke, K. 250, 637
bimetallic strip 157-9 *passim*
bismuth telluride 21
Black, Joseph (1728-99) 4
Blackford, J. 572
Blass, E. 370, 639
'blow-by', *see* leakage, internal
blow-time, in regenerators 610-11
Boileau, Nicholas 1
boiler, of absorption system 16, 408-9, 410
 analysis of 416-17, 445-7, 452, 459
 in Platen-Munters system 464
boiling point
 elevation of 403
 of halogenated refrigerants 11
 of mixtures, *see* bubble point
 relation to critical temperature 392
 relation to vapour pressure 4
boil-off, from L. N. G. tanks 623-4, 626
 treatment of 491-2, 623
bolometer 523
booster compressor 352
 ejector as 20

Index

rotary machine as 72
'bootstrap' system 496–7
Bošnjaković, Fr. 438, 440, 527, 577, 640, 642
Bourne, J. 7
Bramwell, Sir F. J. (1818–1903) 7, 635
brazed fin heat exchangers 598
Brier, H. 375, 388, 639
Brier, J. H. 388, 639
Briggs thread 131
brine 22–3
  ammonia leaks into 124
Brin's process 572
British Cryogenics Council 166, 525
British Standards Institution 166
Brodyansky, V. M. 506, 641
brominated refrigerants, designation of 11
  see also Refrigerant 13B1
Brown, J. 291, 293, 638
Brown, P. M. T. 491, 641
Brown, T. P. G. 621, 642
bubble cap 461–2
bubble chamber 523
bubble point 425
bubble point curves
  absorption cycle on 430–2
  on h–ξ diagram 433–6, 440–4
  to illustrate boiling and distillation 426–9
  of oxygen–nitrogen mixtures 576, 577
  rectification on 449–50, 458–60
  on t–ξ diagram 425–6
bubble pump 17, 410, 464
built-in pressure and volume ratios, see pressure ratio, volume ratio
'Bundy' tube, trade name 134
bursting disc 167
butane, C_4H_{10} ($-138.5/-0.5/152\,°C$)
  in auto-cascade system 620
  in natural gas 616
by-pass
  of hot gas 83, 250–4
  in pump circulation 102

Cailletet, Louis (1832–1913) 477, 509, 510
calcium chloride, $CaCl_2$
  in eutectic mixture 25
  solution as secondary refrigerant 22–3
cancer, possibility of 12
Cannaday, R. L. 572, 642
'canned' pumps 103

capacity rates
  ratio of 602; in three-fluid exchangers 605
  significance of lower capacity rate 546–7, 554, 597, 601
  of streams in absorption system 432
  of streams in compression system 224
capacity reduction
  automatic regulation of 160–6
  by 'balance loading' 254
  by hot-gas by-pass 250–4
  by partial duty ports 285–6
  of reciprocating compressors 64–5
  of screw compressors 75–7, 79–80
  by suction throttling 241–3
  of turbo-compressors 84, 348–50
  by variation of clearance volume 284–5
capillary fittings 131
'capillary tube' restrictor
  mode of operation 120–1
  replaces expansion valve 66
carbon dioxide CO_2 ($-56.6/-78.5$ sublimes/$31.0\,°C$)
  in atmosphere 574
  behaviour with superheat of 230, 234, 396
  coefficient of performance 211, 303
  cycle quantities of 207, 395
  dual-effect compression of 375, 377–9
  examples on 219, 377–9, 388–91, 527–8
  as expendable refrigerant 26–7
  introduced 9–10
  inversion temperature of 537
  low toxicity and non-flammability of 121–2
  operation near critical point 235–9
  p–h diagram 385
  phase diagram of 26
  solid, manufacture of 385–91: exergy of 527–8
  staged expansion of 363
  vapour pressure of 4
carbon tetrachloride, CCl_4 ($-23.0/76.5/283.2\,°C$) 11
carcinogen, R22 as possible 122
Carnot, Sadi (1796–1832)
  coefficient of performance 35, 37
  cycle for ideal gas 497–9
  cycle for liquid and vapour 195:
    comparison with vapour compression cycle 195–202

Index

maximum efficiency of engine 33
Carré, Ferdinand (1824–1900) 14–16, 635
Carrier, W. H. (1876–1950) 81
cascade systems
   auto-cascade 398–400, 619–21
   discussed 391–8
   explained 352
   for L. N. G. 618–19
Caswell, H. E. 335, 338, 342, 346, 639
cavitation 103
Celsius scale of temperature 33–4
centrifugal compressors, *see* compressors, centrifugal
Cervenka, O. 394, 639
Chaikovsky 400, 639
characteristic functions 266
check valves, *see* non-return valves
Cheglikov, A. G. 291, 638
chilled water
   by absorption system 410
   as secondary refrigerant 22
   by vapour jet system 19–20
chlorine ($-101/-34.6/144\,°C$)
   compression of 79
chloroform, $CHCl_3$ ($-63.5/61.7/263\,°C$) 11
'circuits', in evaporators 100, 111
circulation factor
   in absorption system 407, 445
   in pumped circulation evaporator 103
Clapeyron's equation 262
   applied to L. N. G. tank 623–4
Clarke, R. J. 80, 636
Claude, Georges (1870–1960) 513
Claude process
   of air liquefaction 513, 558–70
   of air separation 581–3
Clausius' statement of the Second Law 5
clearance, in compressors
   effect on mean effective pressure 279–84
   effect on volumetric efficiency 277–80
   liquid in 291
   variation of 284–5
   volume and fraction 54, 276
clearance pocket 284
closed superheat 111
Clouet, J. F. (1751–1801) 5
Cockett, H. A. 524, 642
codes, for construction and safety 166
Cody, P. J. 439, 641
coefficient of performance
   analysis of 302–3
   approximate expression for 37
   Carnot value of 35
   defined 30
   of heat pump 45
   of ideal gas cycles 479–82, 484–5, 488–90
   of vapour compression refrigerator
      178–80: dependence on evaporating and condensing temperatures 205–6; dependence on refrigerant properties 207, 210–11; dependence on suction state 231–5
   variation near critical point 237–9
cold multiplier 47–9
Coleman, J. J. (1838–88) 18, 635
Collins, S. C.
   cited 572, 642
   expander 570, 572
   helium liquefier 516, 569–70
commissioning of plant 51
compound compressor 352, *see* staged compression
compression
   adiabatic 178, 315–16
   adiabatic and reversible 178–80, 268–76, 315–16
   heat transfer during 179, 288–92, 360
   index of 287
   isothermal and reversible 179, 550
   on p–h diagram 182
   in steady flow 313–21
   wet and dry 57, 199, 289
   see also compressors, staged compression
compression fittings 134
compressors
   function of 51, 172
   main types 51–2
compressors, axial flow
   in air cycles 495
   limited use of 51
   in L. N. G. plants 619
compressors, centrifugal
   description of 81–6
   need for staging 354
   theory of 324–50
compressors, reciprocating
   double acting 54–5
   hermetic 13, 66–8
   ideal 267–76: with clearance 276–86
   limitation of piston speed 305–7
   at low absolute pressure 392–3

648

Index

principle of 53-4
single acting 56-65
semi-hermetic (accessible) 68-9
V-block 63-5
see also capacity reduction, efficiency, volumetric, *and* power, compressor
compressors, reciprocating, valves for
  lifting of 64-5
  poppet 55
  pressure drops through 57, 293
  reed 57-8
  ring plate 57, 58, 64
  velocity through 294
compressors, rotary (vane) 70-3
compressors, screw
  built-in volume and pressure ratio of 307-13
  described 73-80
  with dual-effect compression 381-5
  may not need staging 354-5
  mean effective pressure of 312
  swept volume of 307
concentration ratio, in cooling towers 93
concrete, cooling of 20
condensers
  air-cooled 94-5
  atmospheric 90
  control of pressure in 95-8
  evaporative 92
  function of 7, 172-3
  water cooled, shell-and-tube 86-90:
    shell-and-coil 89
  water treatment for 92-3
condensing pressure
  in 'capillary tube' systems 120-1
  of refrigerants 207, 395
condensing unit 89, 94-5
contactor, electrical 157
cooling towers
  described 91
  range and approach in 467
  water treatment for 92-3
copper
  incompatibility with ammonia 125
  not brittle at low temperatures 131-2 521
  plating 125
  tubes 88, 128-9
Corbitt, R. W. 614, 616, 642
Coriolis component of acceleration 329
corrosion

  of condenser tubes 88
  inhibitor for brine 23
  in recirculated water 93
Costagliola, M. 294, 638
Cox plot 405
CPRV 150, 243
crankcase heater 65-6
crankcase pressure regulating valve, *see* CPRV
critical point
  Andrews' experiments on 509-10
  dependence of coefficient of performance on 210-11
  operation near 235-9
  on $p-h$ diagram 181, 385
critical solution temperature 126
cross-ambient failure, *see* reversal
cross-over connections 55, 56, 70
cryo-pumping 522
cryo-quenching 521
Cullen, William (1710-90) 3
Cummings, R. O. 2, 635
cut-outs, electrical 166
Cvejn, M. 394, 639
cyclic compounds, designation of 11
cyclical condensation 290-1
Czaplinski, St. 369, 639

Daglish, H. N. 523, 642
Danilov, R. L. 470, 640
Dannies, J. H. 470, 640
Darlington, M. E. 521, 642
datum state
  of components in solution 413-14
  of exergy 525
  reduction to 265
  of thermodynamic properties 183, 439
Daunt, J. G. 517
Dauser, A. 370, 639
DBP, dibutylphthalate, $C_6H_4(CO_2C_4H_9)_2$ (-/340 °C/-) 470
'dead state' 525
Debye, Peter (1884-1966) 516
decade of temperature scale 517
defrosting 168-71
degasser 471-2
dehydrators 142-3
delay angle 295
delivery temperature
  dependence on evaporating and condensing temperatures 206

Index

dependence on refrigerant properties 207, 212
increased by suction throttling 242–3; by hot gas by-pass 251–3; by leakage 292–3
may require staged compression 354
practical limit 212
for reduction of delivery temperature, see desuperheating
dense air machine 19
density, of liquid
of brines 22–3
equation for 260
dephlegmation 582
see also partial condensation
desuperheating
of vapour from a stage: in condenser 215; by direct heat transfer 355–7; by liquid injection 357–61; by mixing 365, 366; combined with liquid cooling 366; in cascade systems 397–8
of vapour to a stage 252–3
deuterium ($-254.4/-249.5/-234.8\,°C$) separation of 573
dew point 425
Dewar, Sir James (1842–1923) 513–14
differential, of thermostat 157–9
differential enthalpy of solution 412–13
diffuser
in centrifugal compressor 81–2, 328, 344–5: moveable vanes in 84–95, 350
in vapour jet system 19–20
diffusion absorption system 464
dilution, heat of 412–13
dilution refrigerator 518–21
dimethyl ether, $(CH_3)_2O$ ($-138.5/-23.6/126.9\,°C$) 9
Din, F. 524, 575, 576, 639, 642
direct action
of pilot valves 119
of solenoid valves 147
direct expansion, defined 22
distillation
described 424–9
of liquid air 579–96
rectification 448–53
stripping and rectification 455–62
distillation columns 461–2, 580
distributor 99–100, 111
domestic refrigerators
introduced 13

Dotto, L. 12
double acting compressor 54
double bellows thermostatic expansion valve 110
Douglas–Conroy pump 103
drainage
to high-side float 116
of liquid from condenser 88
see also backing-up
driers, see dehydrators
dry charged thermostatic expansion valve 110
dry expansion evaporators 98–100
oil return from 137
'dry ice' 26
dryness fraction (quality)
leaving expansion valve 174
of two-component mixture 426, 441, 443–4
dual-effect compression 375–81
combined with staged expansion 382–5
dual pressure process 551–3
Dubinsky, M. G. 494, 641
Dühring plot
for aqua-ammonia 421–2
auto-cascade on 398–9
explained 405–6
for water and lithium bromide 404
'dummy load' 253
duplex arrangement, of compressors 306
Du Pont 259, see also 'Freon'
Dusinberre, G. M. 604, 642
duty, refrigeration 29

Eakin, B. E. 439, 640
earthing, of motors 156
Edwards, J. V. 616, 642
efficiency, built-in 310–11
efficiency, exergetic
defined 527
of Linde–Hampson process 549, 552
of solid carbon dioxide manufacture 528
efficiency, of heat exchanger
defined 486, 546
expression for 602
efficiency, isentropic 37
defined 180, 315–16
discussed 180–1
of expansion process 567
peak, in centrifugal compressors 346: implications of 347–8
relation to small stage and polytropic

Index

efficiency 317–18
  in terms of stagnation values 324
  of vortex tube 505
efficiency, isothermal 490, 550
efficiency, plate 462
efficiency polytropic
  application of 316–20, 330–1
  defined 315
  discussed 316
  in expansion process 492
  peak 334, 346
  in terms of stagnation values 324
efficiency, rectifier 451, 452
efficiency, regenerator 612, 614
efficiency, relative 37, 211
efficiency, small stage 314–15
efficiency, of temperature drop 505–6
efficiency, volumetric
  apparent (indicated), defined 287
  of ideal compressor 277–8
  introduced 52
  real 288–98
  of rotary (vane) compressors 71
  of screw compressors 308
  zero value of 279, 353
Eiseman, B. J. Jr. 211, 637
ejector (thermo-compressor)
  for purging air 410
  in vapour jet system 19
electromagnets, superconducting windings for 524
Ellington, R. T. 439, 640
Emblik E. 438, 640
embrittlement, low temperature 132, 625
enthalpy
  approximate values for water and vapour 414
  of evaporation 4
  general expression for 256
  of ideal gases 475
  interpolation of 191
  of melting of eutectic ice 25
  of oxygen and nitrogen solutions 577
  of solution and mixtures 411–13
enthalpy–composition diagrams
  of aqua-ammonia 433–4, 437
  discussion of 440–3
  distillation on 449–51, 455–6, 458–60
  introduced 412
  of oxygen and nitrogen solutions 577

separation of air on 578, 580, 584
  of water and lithium bromide solution 413
enthalpy–entropy diagram 231–3
enthalpy–volume diagram 228–9
entropy
  at absolute zero 517
  defined 41
  general expressions for 255
  of ideal gas 531
  of ideal gas mixture 531
environment temperature 45–6
  and pressure 525
EPRV, see back pressure valves
Epstein, L. F. 439, 641
equations of state
  of ideal gas 474
  Redlich–Kwong, virial forms 529
  van der Waals, Beattie–Bridgman, Benedict–Webb–Rubin, Martin–Hou 258
equilibrium
  of helium I and II 515
  of helium-3 and -4 518
  of liquid and vapour 3–4
  of phases of carbon dioxide 26
  of solution and solid 24, 421
  of solution and vapour 425–6
equilibrium, departure from
  at absorber exit 419, 447–8
  in compression 179
  in expansion valve 120
  at top of stripper 429
equilibrium stage 459–62
erosion, in tubes 88
ethane, C_2H_6 ($-183.3/-88.6/32.3$ °C)
  in cascade system 395–6, 620
  coefficient of performance of 211
  derivatives of 11
  exergy of liquid 528
  in natural gas 616
  vapour pressure of 4
ethanol (ethyl alcohol), C_2H_5OH ($-117.3/78.5/243.1$ °C) 23
ether (sulphuric, di-ethyl ether) $(C_2H_5)_2O$ ($-116.2/34.5/193.8$ °C) 3, 4, 8–9
ethyl chloride, C_2H_5Cl ($-136.4/12.3/187.2$ °C) 11
ethylene, C_2H_4 ($-169.1/-103.7/9.5$ °C)
  in cascade system 618–19
  exergy of liquid 528

Index

sources of 573
vapour pressure of 4
ethylene glycol (1–2 ethanediol)
  $(CH_2OH)_2$ ($-11.5/198\,°C/-$) 23
Euler head 321
Euler method 608
eutectic
  behaviour 23–5
  mixtures 25
  plates 25
  points, of aqua-ammonia 421
  temperature 22
Evans, Oliver (1755–1819) 5
Evans, R. 616, 642
evaporating temperatures, multiple
  using one compressor 149
  with staged compression 374–5
evaporation
  into a gas 463–5
  principle of refrigeration by 3–5
evaporative condenser 90, 92
evaporator feed regulation, see expansion valves
evaporator pressure regulating valve, see back pressure valve
evaporators
  dry-expansion 98–100
  oil return from 137–8
  with recirculation by gravity 100–1: by pump 101–4; by jet 104
  shell-and-tube 104–5
  spray type of 105, 411
  types of 98
  for water chilling 106
Ewing, Sir J. A. (1855–1935) 232, 233, 637
exergy
  defined and discussed 525–30
  loss of in heat exchanger, and in expansion valve 548–9
  of mixed gases 530–5
expansion
  adiabatic: in cylinder 475; in steady flow 476; from a vessel 477
  of air 17–19
  of ideal gases 475
  isentropic 558; efficiency of 567
expansion machines 491–2, 513, 570–2, 591, 593, 594
expansion valves
  'automatic' 113–14
  capacity of 119–20

controlled by high-side level 114–16
controlled by low-side level 116–17
function of 7, 106, 173
hand regulated 106–7
pilot operation of 117–19
replaced by U-tube 409
superheat controlled (thermostatic) 108–13
types of control 107
expendable refrigerants 26, 521
external régime 39–44
externally equalised thermostatic expansion valve 111

face area, defined 94
Faraday, Michael (1791–1867) 509
Ferguson, T. B. 324, 329, 639
Ffooks, R. C. 626, 642
Fiedler, W. J. 497, 641
finned length 94
fins
  for condensers 94
  spacing of, in evaporators 168
flammability of refrigerants 122–3
flanges, pipe 132–4
flare fittings 129–31
flash chamber, see separator
flash gas
  in Claude process 564
  in expansion valve 173
  from L. N. G. 618, 620–1
  reduced by staged expansion 361–2
flashing, in pumps 103
floating control with neutral zone 163
flooded evaporators 98, 100–4
  oil return from 137–8
flow coefficient, in centrifugal compressors 226, 333–5, 346
fluorinated refrigerants
  in atmosphere 12
  delivery temperature of 212
  heat capacities in ideal gas state 257
  introduction and designation of 10–11
  low condensing coefficient of 86
  low sonic velocity in 328
  see also Refrigerant 11, 12, etc.
fluorine ($-219.6/-188.1/-129°C$), liquid 514, 522, 525
foaming, of oil/refrigerant mixture 65, 127
  remedy 65–6
Foner, S. 524, 642

Index

Forster, L. L. 20, 635
Frank–Caro process 572
Fränkl regenerator 593, 609–10, 615
  see also Linde–Fränkl process
free energy, defined 266
free enthalpy, defined 266
freezing
  damage by 106
  protection against 150
'Freon', trade name
  introduced 10
  properties used in examples xiii
  thermodynamic tables for 183
frozen foods
  freezing of, by carbon dioxide 27
  refrigeration of vehicles for 25
fuses 156

gadolinium sulphate, in adiabatic
  demagnetisation 516
gas, ideal
  approach to ideal gas state 182
  cycles employing ideal gas
    477–495, 497–502
  defined, internal energy and enthalpy of
    474
  heat capacities of 475
  isentropic index of 271
  polytropic compression of 317–18
  reduction of temperature of 475–6
  reversible adiabatic expansion of 475
gas, non-condensable
  in absorption systems 16, 463–5
  estimation of 247–8
  in high-side float chamber 116
  purging of 139–40
gas constant
  for air 475
  defined 257, 474
  for Refrigerant 12 264
  related to heat capacities 257, 475
  used 190, 209, 248, 264
Gatecliff, G. W. 266, 637
generator
  of absorption system 16, 429
  analysis of 445, 447, 457–61
  for power production 473
Geppert, H. 16
Giauque, F. 516
Gibbs–Dalton law 531
Gibbs function 266

Giffard, Paul (1837–97) 17
Giffen, E. 289, 638
Goddard, R. H. (1822–1945) 522
Goldsmid, H. J. 21, 635
Gorrie, John (1803–1855) 17, 510
Gosney, W. B. 233, 289, 303, 637, 638
gravity circulation
  in absorption systems 464
  in evaporators 100–1
Griffiths, E. A. 104, 636
Grindley, J. H. 224, 637
Guldberg number 392

Haar, L. 257, 637
Haber process 572
Halford, P. 572
Hall, J. and E. Ltd. 10
halogenated refrigerants
  designation of 10–11
  see also fluorinated refrigerants
Hampson, William (1854–1926)
  air liquefier 510–13
  heat exchanger 513, 598
  see also Linde – Hampson process
hand expansion valve 106–7
  by-passing float valve 101, 102
hardness of water 92–3
Harrison, James (1815?–1893) 8–9
Haselden, G. G. 524, 572, 619, 642
Haslam, Sir A. S. (1844–1927) 18
Hausen, H. 552, 613, 614, 642
Haywood, R. W. 184, 637
head, of pumps and compressors 320–1
  see also hydrostatic and static head
head coefficients
  application of 335–41, 347
  defined 332
  dependence on flow coefficient 333–5
Health and Safety Executive 122
heat capacities
  of air 475
  in ideal gas state 257–8
  related to Joule–Thomson effect 540
heat exchangers
  in absorption system 16, 409, 416, 420,
    432, 462
  brazed fin 598
  in cascade systems 394, 397–8
  efficiency of, defined 486, 546
  Hampson heat exchanger 513, 598
  in ideal gas cycles 486–95

Index 654

with Joule–Thomson effect 541–2
Linde heat exchanger 511, 513, 598
in Linde–Hampson process 542–4: effect of efficiency on yield fraction 545–6; temperatures in 547; balanced and unbalanced 548, 549; loss of exergy in 548–9; pressure drop in 549–50; temperatures in, with pre-cooling 554–5
parallel, counter- and cross-flow 596–7
pinch points in 567
theory of 599–604
theory of three-fluid heat exchanger 604–9
in vapour compression system 113, 135–6, 221–4
heat pump 44–7
heat ratio
defined 31, 45
maximum values of 38, 45: general expression for 48
heat transfer, in compressors 288–92
amount of, in water-cooled machines 360
heat transfers, incidental 240–5
Heer, C. V. 517
helium (–/ – 268.93/ – 267.95 °C)
in atmosphere 574
dilution refrigerator using 518–21
example on 491–2
exergy of liquid 528
discovery of 514
liquefaction of 516, 569, 555–6
maximum inversion temperature of 537
miscibility of He-3 and -4 518
in natural gas 516
phase diagram for He-I and -II 515
Helmholtz function 266
hermetic construction
of absorption systems 410, 462–4
of pumps 103
of reciprocating compressors 13, 66–8
of turbo-compressors 82
Hesketh, Everard (1884–1942) 10
Heylandt, Paul (1884–1947)
process of 555, 566: comparison with dual pressure process 568–9; modern version of 591–3
high-side float valve 114–16
Hilsch, R. 502–7 *passim* 641
Hodge, J. M. 80, 636

Hopkinson, J. 14, 635
hot gas
by-pass 83, 250–4
defrosting by 169
injection valve for 150–1
Huebler, J. 439, 640
Huggill, J. A. W. 378, 639
Hundy, G. F. 80, 636
hunting, of expansion valves 95, 104
hydrates
of ammonia 421–2
of lithium bromide 403
of sodium and calcium chlorides 24
hydrocarbons, as refrigerants 122, 126
see also butane, ethane, ethylene, methane, propane
hydrochloric acid, test for ammonia 123
hydrogen (– 259.14/ – 252.76/ – 239.9°C)
in bubble chamber 523
example on 556–7
exergy of liquid 528
for leak detection 125
liquefaction of 555–8
maximum inversion temperature of 514, 537
ortho- and para- 557–8
uses of 522, 573
hydrometer, for brine density 22–3
hydrostatic equilibrium, in float control 117
hydrostatic head, circulation by 17, 464
see also static head

ice
in evaporator tubes 106
in expansion valve 127
use of and trade in 1–2
ice-cream, freezing of 104
ideal gas, *see* gas, ideal
Iliffe, C. E. 616, 642
impeller, of compressor 81–2, 83
blade angles of 327
inlet blade angle for minimum Mach number 343–4
number of blades in 329
slip factor of 328–9
torque taken by 326
velocity diagrams for 325, 342, 349
index of compression
actual, defined 287
isentropic, defined 270; for ammonia,

Index

R12, R22 and R502 272-4
  polytropic 317-18
indicator diagrams
  of dual-effect compressor 376
  of ideal compressor 268: with clearance 276
  of real compressor 286
  of screw compressor 309
inert gases, see noble gases
Inokuty, H. 238, 637
Institute of Refrigeration
  proceedings 643
  safety code 166
Institut International du Froid,
  publications 643
integral enthalpy of solution 412
intermediate pressure
  choice of 369-70
  determined by system balance 370-4
  in dual effect compression 382-4
internal energy
  general expression for 256
  of ideal gas 475
internal régime 39-44
  related to external regime 200
internally equalised thermostatic expansion valve 111
International Exhibition of 1862 9
interpolation, in tables 189-91
Invar, for L. N. G. tanks 625
inversion
  described 514, 537
  locus of 537-40
  maximum inversion temperature 537
Ioffe, A. F. 21, 635
irreversibility
  in absorption systems 468-9
  in compression 183
  due to superheat 200
  in expansion valve 198, 200
  reduction by staged expansion 361-2
  in steady flow compression 313-16
isomers, as refrigerants, designation of 10

Jaeger, H. -P. 248, 637
Jakob, M. 604, 642
Jassniker, K. 344, 639
Jennings, R. H. 439, 640
Johnson, J. E. 613, 642
Jonkers, C. O. 502, 641
Joule,, J. P. (1818-89) 536
Joule-Thomson effect

  applied in refrigerator 540-2
  coefficient defined 536
  in Linde-Hampson process 542-4
  mentioned 237, 475, 510, 558
  treated 535-40
  used in most liquefiers 559-60

Kalitenko, F. 293, 638
Kältemaschinen-Regeln 288
Kays, W. M. 604, 642
Keenan, J. 38, 527, 635
Keesom, W. H. (1876-1956) 455, 640
  Keesom's stream algebra 455-7; applied to distillation 457-61; to Linde-Hampson process 577-8; to Linde double column 584-9
Kelvin, Lord, see Thomson, William
kelvin, defined 33
Kennedy, A. B. W. 73
Kennedy, W. K. 291, 638
Kerr, S. V. 294, 295, 301, 638
king valve 135, 144
Kirk, A. C. (1830-92) 7, 17, 502, 635, 641
Kleemenko, A. P. 620, 642
Klimek, L. 619, 642
knock-out pot, see separator
Kohler, J. W. L. 502, 641
Kohloss, F. H. 439, 640
Krigar 73
krypton ($-156.6/-151.7/-63.1$ °C)
  in atmosphere 574
  in discharge tubes 572
  liquefied 514
Kuloor, N. R. 378, 639
Kurti, Nicholas 517
Kuznetsov, A. P. 400, 639

Lady, E. R. 266, 637
Laing, P. D. 79, 636
lambda point 515
Lange, E. 413, 640
lantern ring 54
lead
  as superconductor 515
  as thermal switch 518
leakage, external
  of air inwards 9
  from compressors 57-8
  detection of 123-5
leakage, internal
  between piston and cylinder 293
  effects of 292-3

Index

in screw compressors 308
through valves, static and dynamic (blow-by) 293–4
Leblanc, Maurice (1857–1923) 20
Le Fevre, E. J. 178
Leiden laboratory 510
Leidenfrost phenomenon 521
Lenz, Heinrich (1804–65) 20, 21
Leslie, Sir John (1766–1832) 13
level indication 140–2
Liefergrad 288
Lightfoot, Thomas B. (1849–1921) 17, 18, 635
limit charged thermostatic expansion valve 110–11
Linde, Carl von (1842–1934)
　air liquefier 510–11
　heat exchanger 597
　introduced ammonia 9
Linde, Friedrich 580–1
Linde, Richard 581
　double column of 583–90, 592
Linde–Fränkl process 593, 595
Linde–Hampson process 513
　separation in 577–9
　studied 540–58
'line stack' 617
Linge, K. 37, 229, 635, 637
Linnett, K. C. 621, 642
Linz–Donawitz process 573
liquid injection valve 151, 361
litharge, as thread sealant 131
lithium bromide, LiBr
　in absorption system 17, 406–20
　enthalpy of solution of 413–14
　examples on 404–5, 406, 417–20
　hydrates of 403
　with methanol 470
　vapour pressure of solutions of 403–6
Lloyd's Register of Shipping 166
L. N. G. (liquefied natural gas), see natural gas
load, refrigerating, see refrigerating load
Lockyer, Sir. J. N. 514
Löffler, G. 296, 298, 638
Löffler, H. J. 248, 250, 637
logarithmic mean
　of absolute temperatures 43
　of temperature differences 227, 600
Lom, W. L. 626, 642
London, A. L. 604, 642

London, Heinz (1907–70) 519
Lorentzen, G. 103, 289, 293, 636, 638
Lorenz, Hans (1865–1940) 41, 42–4, 375, 473, 636
Lorge, R. A. 283, 638
Louis XIV 1
Löwer, H. 413, 640
Lowe, T. S. C. (1832–1913) 9
low-fin tubing 87
low-side float valve 101, 102, 116–17
lubrication
　of reciprocating compressors 54, 57, 65–6
　of screw compressors 78
　of turbo-compressors 84
Lundberg, A. 79, 636
Lundvik, B. 79, 636
Lurgi process 573
Lysholm, A. J. R. (1893–1973) 73, 636

McCabe–Thiele method, mentioned 592
Mach number
　in impeller 342–5
　in wind tunnels 522
Mach number, rotational 308
　defined 334
　little effect on performance 342
Macintire, H. J. 237, 637
MacLaren, J. F. T. 294, 295, 301, 638
Macriss, R. A. 439, 440, 640
magnesium, in hermetic systems 125
magnesium chloride, $MgCl_2$, solution of as secondary refrigerant 22
magnetic valves, see solenoid valves
Maiuri, G. (1877–1940) 47, 472, 636
manganese, in hermetic systems 125
Marcet, Alexander 10
marine refrigeration 10, 18–19, 624–6
Martin–Hou equation of state 258–9
Martinovsky, V. S. 494, 641
Martynov, A. V. 506, 641
masers 523
mass fraction
　defined 138, 402, 403
　on linear scale 401
mass spectrograph
　for leak detection 125
matching, of internal and external régimes
　approach by multiple evaporating temperatures 44
　approach by use of mixtures 471, 473

Index

effect of mismatch 40–4
mismatch with vapour compression system 200
Mayhew, Y. R. 184, 637
mean effective pressure (m.e.p.)
 defined 269
 of dual effect compressor 380
 of ideal compressor 269–70: with clearance 280–1
 of pumping and friction 301–2
 of real compressor 300–2
 of screw compressor 312
Mendelssohn, K. 524, 642
mercury
 not used with ammonia 144
 superconductivity of 515
mercury switch 159
meridional component of velocity 326
Merkel, Fr. 438, 440, 640
methane, CH_4 ($-182.2/-161.5/-82.1\,°C$)
 in cascade system 618–20
 derivatives of 11
 examples on 529, 623–4
 exergy of liquid 528
 in natural gas 616–17
methanol (methyl alcohol), CH_3OH ($-97.1/65.0/240\,°C$)
 with lithium bromide 470
 as secondary refrigerant 23
methyl chloride, CH_3Cl ($-97.7/-24.2/143.1\,°C$) 11, 13
 coefficient of performance of 211
 flammability of 123
 incompatible with aluminium 125
Midgley, Thomas (1889–1944) 10
Miller, H. W. 621, 642
miscibility
 of helium-3 and -4 518
 of oil and refrigerant 126–7
 of water and refrigerant 127
Mitford, N. 1, 636
mixed refrigerants
 in auto-cascade 399, 619–21
 mentioned 41
 in resorption system 471–3
mixtures of gases
 in condenser 139–40, 247–8
 enthalpy of 435–8
 see also separation of gases
molar mass xii

of oils 249
mole fraction
 defined 249, 402
molecular sieve, *see* synthetic zeolites
Mollier, Hilde 438, 640
Mollier, Richard (1863–1935) 181
Monge, Gaspard (1746–1818) 5
MOP (maximum operating pressure) 110
motive steam 19–20
motors, electric 151–7
 methods of limiting power of 110
 winding temperatures in 136
multiple effect compression, *see* dual effect compression
multi-stage compression, *see* staged compression
Munters, Carl Georg 17, 462–6 *passim*
Murdoch, A. 125, 636

National Board of Underwriters, toxicity scale of 121
National Bureau of Standards, ammonia tables xii–xiii, 266, 439
Natterer, Johannes (1821–1901) 509
natural gas
 composition of 616
 liquefaction of 616–21
 liquid: storage of 621–4; transport of 624–6
N. E. L. (National Engineering Laboratory), R12 tables 259, 262–5
neon ($-248.6/-246.06/-228.75\,°C$)
 in atmosphere 574
 in discharge tubes 572
 exergy of liquid 528
 liquefied 514
Neoprene tubing 134
Nessler's solution 124–5
net positive suction head 103
neutral zone 163
Newitt, D. M. 378, 388, 389, 639
Newley, E. F. 289, 638
nickel steel 621, 622, 625
Niebergall, W. 438, 470, 640
niobium and compounds 523–4
nitrogen ($-210.0/-195.79/-146.8\,°C$)
 in atmosphere 574
 in cascade system 620
 examples on 532–3, 534–5, 545–7, 548–9, 550–2, 563–5, 567–8, 587–90, 615

Index

liquefaction of 545
liquid, as expendable refrigerant 27
521: exergy of 528; in solution with oxygen 575–7
maximum inversion temperature of 537
in natural gas 616
properties of 544
nitrous oxide, N_2O ($-102/-88.3/36.1\,°C$) 211, 395–6
noble gases, discovery of 514
see argon, helium, krypton, xenon
Nolcken, W. G. (1890–1973) 47, 124, 636
non-condensable gas, see gas, non-condensable
non-return valves (NRV) 150
applications of 72, 96, 149, 168, 170
N. P. T. (National Pipe Thread) 131
nuclear refrigeration 517
nylon tubing 134

'Odessa' system 494–5
'off-cycle' 157
oil
behaviour of, in plants 65–6, 126–7
effects of 248–50: on volumetric efficiency 291; on leakage 293
in screw compressors 78, 308
separation of 136–8
see also lubrication
oil still 137–8
Oldham, B. C. 9, 636
Olszewski, K. S. (1846–1915) 510
'once-through' evaporator 98–100
'on-cycle' 157
Onnes, Heike Kamerlingh (1853–1926) 510, 515, 516
on–off control 157, 160
open superheat, of thermostatic expansion valve 111
Osenbrück system 472–3, 619
overload devices 155–6
oxygen ($-218.8/-182.97/-118.38\,°C$)
in atmosphere 574
enrichment of, in Linde–Hampson process 577–9
examples on 532–3, 534–5, 587–90
from liquid air 579–92
liquid: exergy of 528; modern process for 592–3; in solution with nitrogen 575–7
'tonnage' oxygen 573, 593–6
uses of 522, 573
ozone, in atmosphere 12, 574

packed column 462
'packless' valves 146
Pai, M. U. 378, 639
paramagnetic salts, in adiabatic demagnetisation 516, 517–18
parametric amplifier 523
Parsons, Sir Charles A. (1854–1931) 19
partial condensation
in auto-cascade 399–400, 619–20
in Claude process for oxygen 581–2
in rectifier 429, 464
partial duty ports 56, 285–6
partial pressure
of gas in condenser 247–8
of gas in a mixture 531
of refrigerant in solution 248
Parulekar, R. B. 507, 641
passes
in condenser 86
in heat exchangers 603
pass-plates
in condenser 87
in water chillers 101, 105
peak shaving 617
Pearson, S. F. 293, 638
Peltier, J. (1785–1845) 3, 20
Pennington, W. 404, 405, 641
Perkins, Jacob (1766–1849) 5–6, 6–7, 636
perlite, expanded 622, 626
Perman, E. P. 439, 641
Perry, E. J. 79, 636
Philips air liquefier 17, 501–2
pH value, of cooling water 93
Pictet, Raoul (1846–1929) 9, 510
Pierre, Bo 296, 439, 638, 641
pilot operation
of expansion valves 117–19
of solenoid valves 147
'pinch point' 567
piping 128–135
piston speed, mean 54
defined 306
limitation and effect on compressor design 306–7
Plank, R. 635, 636, 640, 643
Platen, Balzar von 17, 462–6 *passim*
polyamide tubing 134
polyphosphates, glassy 93

Index

Ponchon–Savarit method 460, 461
porous plug experiment 536
positive displacement compressor 51
positive ion emission 124
power, compressor
  of centrifugal compressors 326, 328, 329–30, 338–41
  of ideal compressors 269–70, 280
  limitation of 110, 282: by limit charged expansion valve 110; by CPRV 150; by cutting out cylinders 282
  of real compressors 300–2
  remarks on calculation of 275
  in terms of isentropic efficiency 180
power, pump 416
pre-cooling
  in gas liquefaction 553–6
  in multi-stage systems, *see* staged expansion
pre-rotation vanes, *see* pre-whirl vanes
pressostat 157–8, 160
pressure drop
  in absorption systems 410, 419, 448
  after expansion valve and in evaporator 244–7; requires externally equalised valve 111
  in Claude process 568
  in Linde – Hampson process 549–50
  in suction and delivery pipes 240–5, 298
pressure–enthalpy diagram
  for ammonia 630
  for carbon dioxide 385
  introduced 181
  for R12 631
  for R22 632
  for R502 633
pressure gauges 143–4
pressure ratio
  built-in: of rotary (vane) compressors 72; of screw compressors 73–5, 308–9, 381
  dependence on refrigerant 207, 213, 395
  effect on volumetric efficiency 278–80
pressure snubber 143
pressure–temperature diagram, inversion locus on 537
see also vapour pressure
pressure–volume diagram, vapour compression cycle on 173
pre-whirl vanes 84, 349–50

priming, in evaporators 98
Probert, S. D. 616, 642
propane, C_3H_8 ($-187.7/-42.1/96.8°C$)
  behaviour with superheat 230, 234
  in natural gas 616
  as refrigerant 207, 211, 618–19
  shape of T–s diagram 199
proportional band, of thermostatic expansion valve 111
propylene glycol (1.2-propanediol) $CH_3CHOHCH_2OH$ ($-/189°C/-$)
  in solution as secondary refrigerant 23
PTC (positive temperature coefficient thermistor) 154–5
PTFE (polytetrafluoroethylene), as thread sealant 131
pulses
  damping of, in pressure gauges 143
  in suction and delivery pipes 294–5
pump circulation, of refrigerant 101–4
pump-down 144
  condensing pressure control with pump down cycle 97–8
pumps
  bubble 17, 410, 464
  change of state across 432
  power of 416
  for refrigerants 103–4
  for solution 16, 409
purging
  of non-condensable gas 139–40: from absorption system 410
  of oil from evaporators 137–8
  of water from cooling towers 92–3
  of water from evaporators 454–5

quality, *see* dryness fraction
quantities, physical, equations and symbols for xi–xii

Rabes, M. 569, 643
'ram air' 496
ramming effects, in suction and delivery pipes 294–5
Ramsay, Sir William (1852–1916) 514
range
  of temperatures in heat exchanger 546, 548
  of thermostat 157
  of water temperature in cooling tower 467

Index

Ranque, George J. (1898–1973) 502, 503, 641
Ranque–Hilsch tube, *see* vortex tube
Rant, Zoran 527
Raoult's law
  applied to oil–refrigerant solutions 248–9
  deviations from 404–5, 422–3;
    significance of 470
  not exact 403–4
  predicts vapour composition 423–4
  stated 403
Rateau, Auguste (1863–1930) 81
Rayleigh, Lord (J. W. Strutt) (1842–1919) 514
real gases 535–40
reboiler 429
receiver, liquid
  drainage into 88, 96–8
  function of 135
recirculation evaporators 98, 100–4
  oil return from 137–8
recovery, temperature and factor 477
rectification, *see* distillation
recuperators 596
Redlich–Kwong equation of state 259
reed valves 57, 58, 60
reflux
  controlled by superheat 453
  explained 429
reflux condenser 429
  heat transfer in 449–52
  as part of main condenser 452
refrigerants
  behaviour with oil and water 126–7
  for cascade systems 394–6
  comparison of 207, 395: of behaviour with superheat 230, 234, 396
  cost of 127
  designation of 10–11
  detection of 123–5
  flammability of 122–3
  hydrocarbons as 122, 126
  stability and behaviour with materials 125
  storage and handling of 128
  toxicity of 121–2
  see also under name of refrigerant
Refrigerant 11, CCl_3F ($-111/23.8/198.0\,°C$)
  with centrifugal compressor 347
  cycle quantities of 207

miscible with oil 137
purging air from 140
as secondary refrigerant 23
with staged expansion 363
uses of 12
Refrigerant 12, CCl_2F_2 ($-158/-29.8/112.0\,°C$)
  behaviour with superheat of 230, 234
  with centrifugal compressor 347
  coefficient of performance of 211
  cycle quantities for 207
  examples on 225–7, 242–3, 249, 251–2, 262–5, 331–2, 336–8, 345
  introduced 10
  isentropic index of 272
  miscible with oil 126, 137
  N. E. L. tables for 259
  p–h diagram for 631
  solubility of water in 142
  in solution with oil 248, 250
  staged expansion of 363
  uses of 11–12
  vapour pressure of 4
Refrigerant 13, $CClF_3$ ($-181/-81.4/28.9\,°C$)
  in auto-cascade 398–400
  in cascade system 395, 396
  example on 394
  with oil 248
  in single stage 393
  uses of 12
  vapour pressure of 4
Refrigerant 13B1, $CBrF_3$ ($-168/-57.7/67.0\,°C$) 11
  in cascade system 395–6
  with centrifugal compressor 83, 347
  with oil 248
Refrigerant 14, CF_4 ($-184/-128.0/-45.7\,°C$) 11, 12
Refrigerant 21, $CHCl_2F$ ($-135/8.9/178.5\,°C$) 11, 211
Refrigerant 22, $CHClF_2$ ($-160/-40.8/96.0\,°C$)
  in absorption system 470
  in auto-cascade 398–400
  behaviour with superheat 230, 234
  with centrifugal compressor 347
  coefficient of performance of 211
  cycle quantities for 207
  examples on 338–41, 353
  isentropic index of 273

Index

low temperature limits of 392
with oil 248
p–h diagram for 632
partially miscible with oil 126
solubility of water in 142
staged expansion of 363
uses of 12
vapour pressure of 4
Refrigerant 23, CHF_3 ($-155/-82.0/25.9\,°C$) 11
Refrigerant 113, CCl_2F—$CClF_2$ ($-35/47.6/214.1\,°C$) 11, 199, 211
Refrigerant 114, $CClF_2$—$CClF_2$ ($-94/3.8/145.7\,°C$) 11, 199, 200, 211
Refrigerant 115, C_2ClF_5 ($-106/-38.7/80.0\,°C$) 11, 199, 200, 211
Refrigerant 142b, CH_3—$CClF_2$ ($-131/-9.8/137.1\,°C$) 11
Refrigerant 143a, CH_3—CF_3 ($-111/-47.6/73.1\,°C$) 11
Refrigerant C318, C_4F_8 cyclic ($-41.4/-5.8/115.3\,°C$) 11
Refrigerant 502, 0.63 $CHClF_2$ + 0.37 C_2ClF_5 ($-/-45.4/179.9\,°C$) azeotrope 12, 422
behaviour with superheat 230, 234
coefficient of performance of 211
cycle quantities for 207
example on 283–4
isentropic index of 274
low delivery temperature of 212
p–h diagram for 633
shape of T–s diagram for 199–200
solubility of water in 142
staged expansion of 363
vapour pressure of 4
refrigerants, secondary, *see* secondary refrigerants
refrigerating capacity
defined, units of 28
of centrifugal compressors 349–50
of dilution refrigerator 521
expressions for 177, 178
near critical point 237
of real compressors 298–300
see also capacity reduction
refrigerating effect, specific
defined 29, 177
dependence on evaporating and condensing temperatures 202–3
dependence on refrigerant 206–8

in dilution refrigerator 520
of expendable refrigerants 27
improved by staged expansion 363–6
in Linde–Hampson process 544–5
reduced near critical point 235–6
refrigerating effect, volumic
defined 178
dependence on evaporating and condensing temperatures 203–4
dependence on refrigerant 207, 208–9
dependence on suction state 228–31
refrigerating load 29
simulated 254
refrigeration duty 29
regenerative cooling
combined with Joule–Thomson effect 541–2
in Linde–Hampson process 542–4
suggested 510
regenerators
in air cycles 495
Fränkl regenerator 593, 609–10
in oxygen plants 593–5
in Stirling machines 499, 501
theory of 610–16
reheat factor 316
relay
for motor starting 154–5, 157, 159–60
reversing 147
relief valves 167
in defrosting 168–9
on surge drum 102
residual volume 190
resorption system 471–3
return flow compressor 57
heat transfer in 289
Reuleaux, F. 73
reversal (cross ambient failure) 110–11
reverse action
of pilot valves 119
of solenoid valves 147
reversed cycle defrosting 170
reversibility
in compression 179
defined 37
in expansion 198
revert gas 387, 542
temperature of 546–7, 554, 560, 564
waste of refrigeration by 545–6
Révillion 73
Reynolds number, rotational 334, 342

Richard I 1
ring plate valves 56, 57, 58
Rogers, G. F. C. 184, 637
rolling diaphragm seal 502
Rose-Innes, A. C. 524, 643
'roll-over' 624
Root's blower 52
  efficiency of 311
roto-dynamic compressor 51
rounding-off, caution in 184
Royal Society, The xi
Ruhemann, M. 398, 399, 455, 552, 596, 619, 641, 643
'run-around system' 253–4
Runge–Kutta method 608
Russell, P. N. 9

S. A. E. (Society of Automotive Engineers), flare fittings 129–31
safety
  with cryo-products 525
  in handling refrigerants 128
  in refrigerating plants 166–7
safety head, in compressors 54, 56, 57, 63, 64
Saladin 1
satellite plant 617
saturated vapour 4
  see also vapour pressure
Saunders, O. A. 613, 643
Scatchard, G. 439, 641
Schiff, H. 12
Schmidt, Ernst 266, 637
Schultz, J. M. 320, 639
Schulz, S. 439, 641
Schwartz, B. B. 524, 642
Schwartz, E. 413, 640
Scott, G. L. 439, 640
screwed joints 134
seals, see shaft seals
secondary refrigerants 22–5
semi-conductors 21
semi-hermetic compressors 68–9
separation of gases
  applications of 572–3
  minimum work for 530–5
  principles of 573–96
separator
  for liquid in recirculation evaporators (surge drum or accumulator) 100–2, 104

for liquid in suction pipe (knockout pot) 116, 138
for oil 136–7
with staged expansion (flash chamber) 361–3, 382, 386
sequential control 161
Serdakov, G. S. 370, 639
service valves 69, 70
serviceable compressor 69
shaft seals
  for reciprocating compressors 59–62
  for turbo-compressors 82
Shannon, F. P. 439, 640
shell-and-coil condensers 89
shell-and-tube
  condensers: horizontal 86–8; vertical 89–90
  evaporators 104–6: oil return from 137–8
Shipley, Th. 289, 638
shock
  in diffuser 82, 344
  in impeller 342
shut-off (stop) valves 144–7
Siebe, Daniel Edward (1827–66) 8, 9
Siemens, Sir William (1823–83) 510
sight glasses 140–2
silica gel 127, 143
Simon, Sir Francis (1893–1956) 477, 517
single acting compressor 56–7
single-phasing, protection against 156
skin condenser 95
slip factor, in impeller 329
Smith, E. C. 289, 638
Smith, H. J. 293, 638
Smith, K. C. 621, 642
Smoleniec, S. 613, 643
snowmaker 386–7
sodium chloride, NaCl
  in eutectic mixture 24, 25
  in solution as secondary refrigerant 22
sodium dichromate, $Na_2Cr_2O_7$, as inhibitor 23
sodium sulphate, Na_2SO_4, in eutectic mixture 25
solar heat
  for absorption system 470
  upgrading of 47
solenoid (magnetic) valves
  for capacity reduction 161–4
  for defrosting 168–9

Index

described 147
  with two evaporators 149
  as expansion valves 117
solution, enthalpy of
  ammonia in water 432–3
  defined 411–12
  lithium bromide in water 413
solutions
  in absorption systems 14–16, 406–7
  enthalpy of 411–14
  freezing of 24–5
  mixing of 401–2, 455–6
  of oil in refrigerant 126–7, 137–8, 248–50
  as secondary refrigerants 22–5
  of water in refrigerant 125, 127, 142–3, 422
  vapour pressure of 403–6, 422–3
Solvay, Ernest (1838–1922) 510
Soumerai, H. 306, 370, 638, 639
space simulation 522–3
sparging of coolers 168, 171
Sparks, N. R. 377, 640
specific quantities, defined xi
sprayed refrigerant evaporator 105
staged compression
  in cascade systems 398, 619
  with centrifugal compressors 84, 348
  with desuperheating 355–61
  reasons for 352–5
  with staged expansion 361–9
staged expansion
  benefits from 361–4
  with desuperheating 366–9
  with dual effect compression 375, 382–5
  in liquefaction of gases 551–3
  procooling and subcooling 364–6
stagnation
  enthalpy, pressure and temperature 322–4: for ideal gas 476
  temperature, meeting aircraft 496
  temperature, in vortex tube 504
stainless steel
  for L. N. G. tanks 625
  for refrigerant piping 134–5
starter (starting relay), for motors 154–5, 157, 159–60
star-delta 156–7
starting, unloaded
  of reciprocating compressors, 65
  of rotary (vane) compressors 72

static condenser 95
static head
  in boiler of absorption system 419
  in evaporators, 105, 246–7, 411
static quantities, distinguished from stagnation (total) quantities 323
steady flow energy equation, quoted 175
steady flow momentum equation 326
steam jet refrigeration, *see* vapour jet refrigeration
steel tube 131–5
Stepanoff, A. J. 345, 639
Stickney, A. B. 388, 640
Stirling cycle machines 17, 497–502
Stirling, R. and J. 500, 641
Stokes, Sir F.W.S. (1860–1927) 375
stop valves 144–7
storage
  of L. N. G. 621–4
  of refrigerant 128
straight charged thermostatic expansion valve 109–10
strainers 142
'Strathleven' 18
stripping, *see* distillation
subcooling
  at condenser outlet 247
  enthalpy of liquid with 217–19
  leaving absorber 448
  with mixed refrigerants 472–3, 621
  of solution 426, 442
  with staged expansion 364–6
  in vapour compression cycle 213–16
sublimation 26
suffocation 12, 128, 525
sugar, cane, in eutectic mixture 25
sulphur, test for ammonia 123
sulphur dioxide, SO_2
  ($-75.5/-10.0/157.2\,°C$)
  non-flammable 122
  as refrigerant 9, 13, 211
  toxicity of 121
  vapour pressure of 4
sulphuric acid, H_2SO_4, absorbent for water 13–14
superconductivity 515, 523–4
superfluidity 515
superheat
  acquired in heat exchanger 221–7
  control of expansion valves by 108–13
  defined 184

Index

deviation from Carnot cycle caused by 198–9
effect of on volumetric efficiency 288–92
effect on volumic refrigerating effect 228–231
effect on coefficient of performance 231–5
leaving generator 453
of suction vapour 219–21: useful and useless 221, 298
see also desuperheating
superheated vapour, from solutions 249, 414, 426, 435
surge drum (accumulator), see separator
surging, in compressors 350
swept volume
 defined 52
 rate 52, 268
 of twin-screw compressor 308
synthetic zeolite ('molecular sieve') 127, 143

Tandberg, J. 438, 641
tandem arrangement, of compressors 306
Tantam, D. H. 572
Tellier, Charles (1828–1913) 9
temperature, absolute thermodynamic
 defined 33
 logarithmic mean 43
temperature approach
 in condensers 215
 in cooling towers 467
 in evaporators 112–13
 in heat exchangers generally 489, 546, 597
 in subcooling coil 366
temperature difference, logarithmic mean 227, 600
'temperature elevator' 46–7
temperature–entropy diagram
 exergy displayed on 530
 inversion shown on 538
 near absolute zero 516–17
 symmetry of 200
temperature, evaporating
 effect of oil on 249
 related to pressure 4, 173
 variable: by diffusion into gas 464–5; by use of solution 471–2, 620–1
temperature nip, see pinch point

TEV, see thermostatic expansion valve
theoretical plate 459–62
thermal valve (heat switch) 518
thermistor, as motor starter 155–6
thermo-compressor, see ejector
thermodynamic properties
 computation of 254–6
 tables and charts of 183–4
thermodynamics
 applied to refrigeration 31–9
 Second Law of 5
 Third Law of 517, 531
thermo-electric refrigeration 2, 3, 20–1
thermostat
 mode of operation of 157–60
 used in automatic control 160–4
thermostatic expansion valve
 described 108–13
 as liquid injection valve 361
 for liquid subcooler 365
Thomsen, E. G. 283, 638
Thomson, William (Lord Kelvin) (1824–1907) 44, 536, 636
Thorogood, R. M. 621, 643
throttling
 deliberate 83, 241–3
 exergy loss in 548–9
 in expansion valve 176
 loss due to 198, 200: reduction by staged expansion, 361–5
 of a real gas 535–8
 recovery of temperature in 477
throttling range (proportional band), of thermostatic expansion valve 111
timing gears, in screw compressors 78, 79
tip speed
 in centrifugal compressors 83–4, 326–7, 347
 in screw compressors 78
TLV–STEL 122
TLV–TWA 122
Ton of Refrigeration, defined by American Society of Refrigerating Engineers 28
'tonnage' oxygen 573, 593–6
torque
 in centrifugal compressors 326, 328
 of electric motors 151–3, 282
 maximum, in reciprocating compressors 283–4
total quantities, see stagnation
toxicity, of refrigerants 121–2

Index

Tramschek, A. B. 294, 295, 301, 638
transfer passages, in turbo-compressors 84
transfer units, number of
  in three-fluid exchangers 605–9
  in two-fluid exchangers 601–2
Trevithick, Richard (1771–1833) 17
trichlorethylene, C_2HCl_3
  ($-73/87.2/271°C$) as secondary refrigerant 23
triple-effect compression 388
triple point
  of carbon dioxide 26, 385
  defined 33
Tripler 510
Tsidzik, V. E. 238, 637
tube plates, in condenser 86–7
Tudor, F. (1783–1864) 2
turbines
  for aircraft cooling 496–7
  comparison of vortex tube with 507
  for gas liquefaction 570–2, 593, 594
  for helium 492
turbo-compressors, see compressors, axial flow and compressors, centrifugal
Twaddell scale of density 23
Twining, A. C. (1801–84) 9

ultra violet radiation
  at earth's surface 12
  for leak detection 124–5
uniflow compressor
  described 57
  heat transfer in 288–9
Usyukin, I. P. 572, 643

vacuum cooling 20
vacuum pump 279
  rotary compressor as 72
Vahl, L. 20, 388, 636, 640
valves
  back pressure 147–50, 243
  crankcase pressure regulating 150, 243
  hot gas injection 151, 253
  liquid injection 151, 253, 361
  non-return 150
  service 69
  shut-off 144–7
  solenoid 117, 147, 149, 161–4, 168–9
  water regulating 96, 97–8
  see also expansion valves and compressors, reciprocating, valves for

van der Waals equation of state 258
van Marum, Martin (1750–1837) 5
van Troostwijk, A. P. (1752–1837) 5
vaned and vaneless diffusers 82
vapour jet refrigeration 3, 19–20
vapour pressure
  of aqua-ammonia solution 421–3
  equations for 190, 259–60
  explained 3–4
  interpolation of 190
  of lithium bromide in water solution 404–5
  of oil–refrigerant mixtures 248–9
  plots of 405–6
  of solutions generally 403–4
Veinberg, V. S. 238, 637
velocity
  in impellers 325, 342, 349
  through valve ports 294
  of vapour in suction pipes 137, 241
  of water in condenser tubes, 86, 88
Vergne, de la, ammonia joint 133
Versailles, court of 1
virial equations of state 259
volume ratio, built-in
  of rotary (vane) compressors 72
  of screw compressors 73–5, 308–9
volume, specific
  of aqua-ammonia liquid 447
  interpolation of 190
  of liquid, equation for 260
  of lithium bromide in water solution 416
  of wet mixture 174
volumetric efficiency, see efficiency, volumetric
volumic quantities, defined 178
volute (scroll) 81
von Braun, Wernher 522
von Cube, H. L. 250, 637
von Sambeck, J. A. A. 250, 637
Voorhees, G. T. (1869–1937) 375, 640
vortex tube 502–7
'Votator', trade name 104
VSA (vertical single acting) compressor 56–63
Vukalovich, M. P. 389, 640

Walker, G. 502, 641
Warburton, J. 439, 641
water
  as absorbent, see aqua-ammonia

enthalpy of 414
in evaporators 453–4
as impurity 125, 127, 142–3, 422
as refrigerant 14, 17, 19–20, 406–11
as secondary refrigerant 22
water chillers 101, 106
water, cooling 88, 91–3
for absorption systems 431, 466–7
leaks into 123–4
regulating valves for 97–8
treatment of 93
Watson, J. T. R. 259, 262, 637
welding
oxygen and argon for 572
of pipework 134
Wenham Lake Ice Co. 2
wet bulb thermometer 14, 465
whirl component of velocity 326
White, G. K. 524, 643
White, W. C. 124, 636
Whitehead compressor 511
Widell, N. 438, 641
Wiesner, F. J. 334, 335, 338, 342, 346, 639
Williams, T. J. 294, 295, 638
Willmott, A. J. 616, 643
Wilson, J. J. 626, 643
Windhausen, Franz Sr. (1829–1904) 9, 14, 361, 640
Windhausen, F. Jr. 361, 640
'wiredrawing' 301
Wirth, G. 289, 290, 638
work coefficient

defined 326
dependence on blade angle 327
dependence on flow coefficient 327, 333–5
expressions for 326, 329, 332
work of isentropic compression, specific and volumic
defined 180
dependence on evaporating and condensing temperatures 204
dependence on refrigerant properties 207, 209–10, 395
as integrals 269
in terms of isentropic index 270, 275
Wroblewski, Sigmund von (1845–1888) 510
Wucherer, J. 421, 422, 424, 438, 439, 641
Wucherer, R. 581, 583

xenon ($-140.2/-109.1/16.6\,°C$)
in atmosphere 574
liquefied 514

yield fraction
in Claude process 564
in Linde–Hampson process 545
Young, Meldrum and Binny 9

Zemansky, M. W. 524, 643
Zimmern, B. 79, 80, 636
zinc bromide, $ZnBr_2$, with lithium bromide and methanol 470